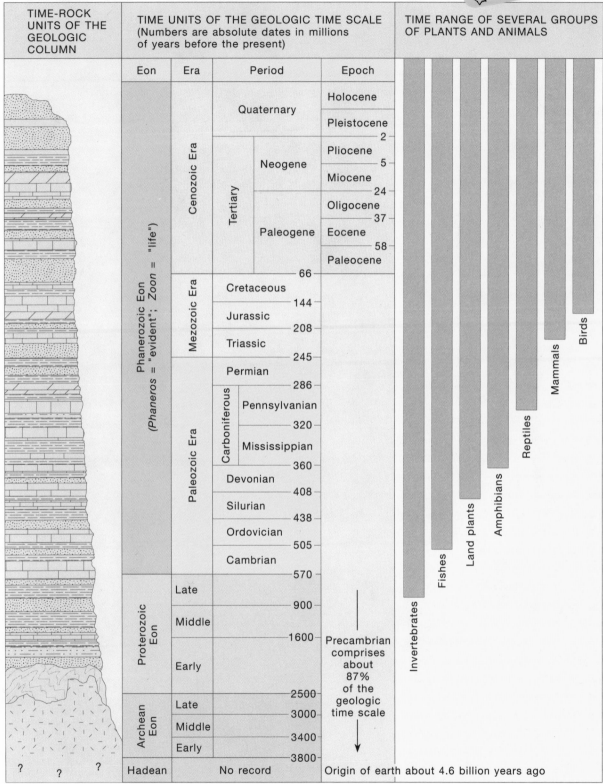

| TIME-ROCK UNITS OF THE GEOLOGIC COLUMN | TIME UNITS OF THE GEOLOGIC TIME SCALE (Numbers are absolute dates in millions of years before the present) | | | | TIME RANGE OF SEVERAL GROUPS OF PLANTS AND ANIMALS |
|---|---|---|---|---|---|
| | Eon | Era | Period | Epoch | |

Geologic time scale. *(from A. R. Palmer, The decade of North American geology, geologic time scale, Geology 11:503–504, 1983.)*

# THROUGH
# TIME

# THE EARTH

# THROUGH

# TIME

*FOURTH EDITION*

*Harold L. Levin*

**Professor of Geology**
**Washington University**
**St. Louis**

**SAUNDERS COLLEGE PUBLISHING**
**A Harcourt Brace Jovanovich College Publisher**
**Fort Worth   Philadelphia   San Diego   New York   Orlando   Austin   San Antonio**
**Toronto   Montreal   London   Sydney   Tokyo**

Text Typeface: Berkeley Oldstyle
Compositor: York Graphic Services
Acquisitions Editor: John Vondeling
Developmental Editor: Christine Connelly
Managing Editor: Carol Field
Project Editor: Anne Gibby
Copy Editor: Kathy Walker
Manager of Art and Design: Carol Bleistine
Art and Design Coordinator: Doris Bruey
Text and Cover Designer: Lawrence R. Didona
Text Artwork: Tasa Graphic Arts, Inc.
Director of EDP: Tim Frelick
Production Manager: Bob Butler
Marketing Manager: Marjorie Waldron

Cover Credit: *Center:* Painting of Pleurocoelus by Eleanor M. Kish. Reproduced with permission of the Canadian Museum of Nature, Ottawa, Canada. *Background:* Photograph of boundary clay (detail), Berwing Canyon by Harry Foster. Reproduced with permission of the Canadian Museum of Nature, Ottawa, Canada.
Title Page Photo Credit: Sea stacks composed of Quaternary sediments, Crescent City, California, at sunset. (Photo by Stephen D. Levin)
Page vi Photo Credit: Pinnacle of Jurassic Wingate Formation underlain by the Chinle Formation (Triassic), central Utah. (Courtesy of Lehi Hintze)

Printed in the United States of America

Library of Congress Catalog Card Number: 91–052946

THE EARTH THROUGH TIME, fourth edition
ISBN 0-03-053689-8

891 032 987654321

*This book is dedicated to the memory
of our parents.*

*They set us on the right path.*

*Marshall and Dorothy
Harry and Ann*

# Preface

While reflecting on the progress of science in his time, Charles Lyell concluded that never before had geology revealed "so many novel and unexpected truths and overcome so many preconceived opinions." The great nineteenth century geologist had in mind the changing view of the earth from a static planet unaltered since its origin to a "theatre of reiterated change." Lyell believed that he lived during a time of revolution in the geologic sciences. Today, we are experiencing another such revolution, and Lyell's remarks about the progress of geology are even more appropriate now than they were when he wrote his monumental *Principles of Geology* nearly 160 years ago. As a result of bold new discoveries, geologists are now able to recognize the deeper and more fundamental causes for the many changes the earth has undergone. They have synthesized older interpretations of geologic events into a grand scenario of a dynamic planet in which ocean floors are in constant movement, continents break apart and reassemble, and great mountain systems rise from quiet seas. With some amazement, they have discovered pieces of Africa firmly anchored within the state of Georgia and splinters of China and Russia in the mountains of western North America. It is an exciting time to study the history of the earth.

*The Earth Through Time* is a college-level textbook for a course in historical geology. It is broad in scope so as to serve both the beginning geology major and the undergraduate seeking to learn about the earth while completing a college science course requirement. Few would disagree that the components of a course in the history of the earth and its life will enrich any student's liberal education.

## Approach

The geologic history of the earth, like the planet's brief cultural history, includes information about the nature and time of occurrence of important events. Cultural history is largely dependent upon written documents, whereas geologic history must be inferred from the careful study of fossils and rocks. The historical geologist sees the results of past events and then searches for their causes. An important objective of this book is to provide the student with an understanding of how this is accomplished. The approach is first to provide the facts and then to examine the meaning of those facts. For example, the characteristics of rocks formed during a particular interval of time are described, not as self-important entities, but rather to demonstrate their value in revealing conditions at particular times and places. Although interesting in themselves, fossils are described so as to indicate their value in understanding evolution and interpreting past environments.

## Organization

For the student to fully understand how geologists learn about the history of the earth, a considerable amount of background information is needed. The first chapter of the text provides an introduction to earth materials and an overview of plate tectonics. Readers who have already taken a course in physical geology may wish to use the section dealing with rocks and minerals as a review. In subsequent chapters, the text explores the historical significance of sedimentary rocks and fossils, the methods used by geologists in ascertaining the age of strata, and the operation of plate tectonics and sea floor spreading. After a thorough consideration of the essential background material in the first six chapters, the text proceeds to a traditional chronologic sequence of topics dealing with the events and life of the Archean and Proterozoic eons and the eras of the Phanerozoic. The final chapter, "Meteorites, Moons, and Planets," is written as an independent unit that can be inserted into a topic sequence of the instructor's choosing. This chapter reminds the reader that the earth is but one of nine planets and shares many aspects of its history with neighboring bodies in space.

## Changes in the Fourth Edition

Preparation of a new edition of a textbook provides the opportunity to refashion explanations, find better illustrations, and provide information about new discoveries. Readers will find that all of the drawings and most of the photographs are in color. We have utilized the exceptional talents of Dennis Tasa in the art program. Also new in this edition are interviews with four geologists (Preston Cloud, Tanya Atwater, Samuel Bowring, and B. Clark Burchfiel), who have made significant contributions to our understanding of the earth's history. Students are often keenly interested in the people that have been influential in advancing science. For those undergraduates thinking about a career in geology, the interviews provide a glimpse of the intellectual rewards and challenges afforded by a career in geology.

Throughout the text, the reader will find discussions of interesting topics in new special boxes entitled **Commentaries**. Their purpose is to reinforce ideas, present applications of some of the concepts discussed in the text, illustrate methods used in geologic research, and, most of all, to spark student interest.

In this edition, the treatment of Precambrian history has been extensively revised and a new chapter added so that students can examine the Archean (Chapter 7) and Proterozoic (Chapter 8) separately. Discussion of such topics as Precambrian life, the Burgess Shale fauna, the origin of mountain ranges,

important figures in the history of geology, plate tectonics, methods of stratigraphy, evolution in populations, and dinosaur paleoecology have all been augmented. The description of rocks and minerals in the third edition's Chapter 2 has been shortened and included in Chapter 1 ("The Science of Historical Geology"), and the discussion of planets and meteorites has been moved to the final chapter (Chapter 17) to be included in the course at the instructor's discretion.

## Learning Aids

Many pedagogic aids have been included to help the student in mastering the diverse information of a historical geology course. New terms are set in **boldface** for emphasis. At the end of each chapter, there is a **Summary**, a set of **Questions for Review**, **Terms to Remember**, and **Supplemental Readings and References**. The boxes entitled **Commentaries** and **Interviews** provide additional insight and relevance. Color is used throughout the text, not as a cosmetic, but as an aid in understanding diagrams and photographs. A visit to an outcrop is the best way to teach geology, but when confined to the classroom, the student may find that a color photograph can be the next best thing.

At the back of the text is an illustrated **Glossary of Terms** containing over 400 entries, all defined in conformity with the *Dictionary of Geological Terms* published by the American Geological Institute. When the need arises to check the age of a rock unit mentioned in the text or to find its equivalents in other locations, the reader can refer to the **Formation Correlation Charts for Representative Sections** in the appendices. Also included in the appendices are a **Classification of Living Things**, a map depicting the **Physiographic Provinces of the United States**, a **Periodic Table of Chemical Elements**, and **Convenient Conversion Factors**.

## Ancillaries

*The Earth Through Time* is accompanied by an extensive set of support materials. To enhance the students' understanding of the text, a **Study Guide** has been prepared by Vicki Harder of Texas A&M University. The Study Guide includes chapter overviews, learning objectives, detailed chapter outlines and study questions. An **Instructor's Manual with Test Bank** written by Michael S. Smith of Montana State University provides learning objectives, teaching aids, answers to the review questions in the text, and a variety of test questions for each chapter including true/false, multiple choice, and essay questions. For use in the classroom or laboratory, a set of 100 **overhead transparencies** in color are provided, as well as a set of 100 **35-mm slides** of text illustrations and photographs.

## Acknowledgments

Many of the modifications in this new edition were suggested by our reviewers, as well as colleagues who have used the previous edition in their historical geology courses. I extend my thanks to all of these earth scientists, including:

William Berry
University of California, Berkeley

Roger Cuffey
University of Pennsylvania

James H. Darrell
Georgia Southern University

Larry E. Davis
Washington State University

George F. Engelmann
University of Nebraska at Omaha

Vicki Harder
Texas A&M University

John Howe
Bowling Green State University

Gary D. Johnson
University of South Dakota

Roger Kaesler
University of Kansas

Ralph L. Langenheim, Jr.
University of Illinois

David N. Lumsden
Memphis State University

Donald Marchand
Old Dominion University

Eldridge Moores
University of California, Davis

Peter Neilsen
Keene State College

Cathryn Newton
Syracuse University

John Pope
Miami University

Jennifer Smith Prouty
Corpus Christi State University

Hi Il Yi
University of Delaware

I am grateful to professors Sam A. Bowring, the late Preston Cloud, Burrell C. Burchfiel, and Tanya M. Atwater for the interviews they graciously provided.

The support and prudent advice I received from the staff and editors of Saunders College Publishing were indispensable in the preparation of this edition. In particular, the book has benefited from the guidance of John J. Vondeling. Developmental Editor Christine Connelly, Project Editor Anne Gibby, Art Director Carol Bleistine, and Copy Editor Kathy Walker have accomplished a multitude of tasks with professional commitment to quality, precision, and efficiency.

My thanks conclude with a special note of gratitude to my wife Kay, who has cheerfully endured my preoccupation with the writing of textbooks for over fifteen years.

# Contents Overview

# Contents

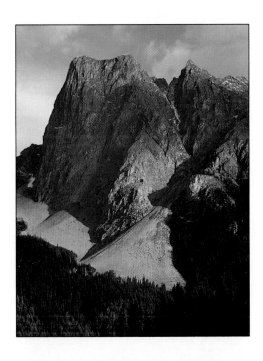

## 10  Late Paleozoic Events    337

## 11  Life of the Paleozoic    373

## 12  The Mesozoic Era    423

## 13  The Mesozoic Biosphere    465

# THROUGH
# TIME

Photograph of the earth taken by the crew of Apollo 17 in 1972. The orange-red deserts of Africa and Saudi Arabia stand in contrast to white clouds and the deep blue of the ocean. Snow and ice covered Antarctica is clearly visible at the bottom. *(Courtesy of NASA)*.

# The Science of
# Historical Geology

## Changing Views of the Third Planet

We live on the third planet from the sun (Fig. 1–1). Our planet was formed 4.6 billion years ago, and since that time it has circled around the sun like a small spacecraft observing a rather average star. The sun, in turn, carries the earth and its companion planets around the center of our galaxy, while the great galactic spiral that contains the solar system moves silently through the universe.

Scarcely 100,000 years ago, primates called humans evolved on earth. Unlike earlier animals, these creatures with oversized brains and nimble fingers asked questions about themselves and their surroundings. Their questioning has continued to the present day. How was the earth formed? Why do earthquakes occur? What lies beneath the lands we live on and beneath the floor of the ocean? Even ancient people sought answers for these questions. In frail wooden ships they probed the limits of the known world, fearing that they might tumble from its edge. Their descendants came to know the planet as an imperfect sphere, and they began an examination of every obscure recess of its surface. In harsher regions, exploration proceeded slowly. It has been only within the last 100 years that humans have penetrated the deep interior of Antarctica. Today, except for a few areas of great cold or dense forest, the continents are well charted. New frontiers for exploration now lie beneath the oceans and outward into space.

A major step in the advance of science in the oceanic frontier occurred in the late 1800s with the scientific expeditions of British oceanographic vessels. Prominent among these was the naval ship *H.M.S. Challenger* (Fig. 1–2). Although limited by the relatively primitive technology of their time, the *Challenger* scientists obtained an extraordinary amount of information about water depths, movement of ocean currents, marine life, and the nature of sediments on the ocean floor. The advancement of technology since that time has allowed that base of knowledge to be enormously expanded. Criss-crossing the seas, ships equipped with precision depth recorders that employed the principle of echo sounding (Fig. 1–3) plotted continuous topographic profiles of the sea floors. Related methods using stronger energy sources provided images of layers of rock and sediment beneath the ocean floor. Such cross-sections of the earth's crust are called seismic profiles (Fig. 1–4). A panoramic view of undersea geology emerged that was more magnificent and complex than anyone had imagined (Fig. 1–5). The magnetic characteristics of the ocean floors were also examined, and the scientific community was startled to learn that the earth's magnetic polarity had occasionally reversed itself during the past.

1

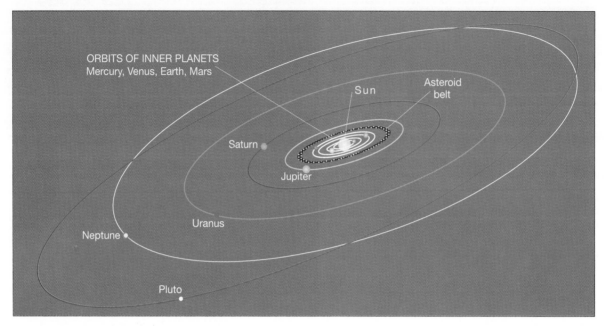

**Figure 1–1** Orbits of the nine planets in the solar system.

To further study the effects of these phenomena on earth history, a unique deep-sea drilling ship, the *Glomar Challenger,* was constructed and quickly put into operation. With this research vessel, geologists were able to penetrate deeply into the sea floor and bring on deck the sediment and rock needed to decipher the history of the ocean basins. Samples from beneath the sea were combined with interpretations of seismic profiles and magnetic data, stimulating the development of exciting new hypotheses of drifting crustal segments, splitting continents, and continuously changing global geography. The success of the *Glomar Challenger* led to the outfitting of a new drilling ship, the *JOIDES Resolution* (Fig. 1–6), which began operation in 1985.

Exploration of space has also contributed to our new view of the earth. Intricately engineered spacecraft circled the earth and set down on the moon. Members of our inquisitive species reached down and held lunar rocks in their gloved hands. Analysis of those rocks has yielded important clues to the early history of the earth. Satellites and spacecraft continue to circle the earth, transmitting thousands of images of our planet to scientists below. These images have given us an unprecedented synoptic view of a dynamic, vibrant planet, and on a day-to-day basis, they have provided information useful in finding mineral deposits, assessing worldwide agricultural productivity, and ascertaining global changes in environmental conditions. Perhaps most important, however, these data have made all of us aware of the stunning beauty and fragility of our planetary home.

**Figure 1–2** *H.M.S. Challenger (From the* Report of the Scientific Results of the Exploring Voyage of H.M.S. Challenger during the years 1873–1876, *Narrative, Part II, 1885.)*

*(text continued on page 6)*

**Figure 1–3**  The principle of echo sounding. A transmitter sends a sound wave, which is reflected back to the surface by the ocean bottom and is picked up by a receiver. By knowing the total time involved and the speed of sound in the ocean (1500 meters per second), water depth can be determined. *(From J. M. McCormick and J. V. Thiruvathukal.* Elements of Oceanography, *2nd ed. Philadelphia, Saunders College Publishing, 1981.)*

**Figure 1–4**  Continuous seismic profiles showing abyssal hills and abyssal plains along the edge of the Mid-Atlantic Ridge. *(From Hayes, D. E., and Pimm, A. C.,* Initial Reports of Deep Sea Drilling Project, *1972. National Science Foundation publication, vol. 14, pp. 341–376.)*

**Figure 1–5** The physical features of the ocean floors and the continents. *(Copyright © Marie Tharp.)*

**Figure 1–6** The *JOIDES Resolution,* successor to the *Glomar Challenger.* This oceanographic research vessel is designed for taking drill cores from the bottom of the floor of the ocean. The vessel is a floating oceanographic research center, with a seven-story laboratory stack that occupies 12,000 square feet. On-board facilities include laboratories for sedimentology, paleontology, geochemistry, and geophysics. The ship can suspend as much as 9000 meters of drill pipe to obtain core samples. *(Courtesy of Ocean Drilling Program, National Science Foundation.)*

## An Eclectic Science

Geologists concern themselves with an exceptional variety of scientific tasks and must employ knowledge from diverse fields. Some examine the composition and texture of meteorites and moon rocks. With magnifiers and computers, others scrutinize photographs of planets to understand the features that characterize their surfaces. Still other geologists are busily unraveling the structure of mountain ranges or attempting to understand the behavior of glaciers, underground water, volcanoes, and streams. Large numbers of geologists search for the fossil fuels and ores we require. They worry, as do you and I, about the fate of humans in a world of diminishing resources. To help in their work, geologists draw upon the knowledge of astronomy, physics, chemistry, mathematics, and biology. For example, petroleum geologists must understand the physics of moving fluids, the chemistry of hydrocarbons, and the biol-ogy of the invertebrate fossils they use to trace subsurface rock layers. Because geology incorporates the information of so many other scientific disciplines, it can be termed an **eclectic science.** *Eclectic* is a word useful in describing a body of selected information drawn from a variety of sources. All sciences are eclectic to some degree, and geology is decidedly so.

For convenience of study, the body of knowledge called geology can be divided into **physical geology** and **historical geology.** The origin, classification, and composition of earth materials, as well as the varied processes that occur on the surface and in the deep interior of the earth, are the usual subjects of physical geology. Historical geology addresses itself to the earth's evolution, changes in the distribution of lands and seas, growth and destruction of mountains, succession of animals and plants through time, and chronologic changes in other planets in our solar system. Clearly, the division between physical and historical geology is somewhat arbitrary, and it is necessary to understand the physical aspects of our planet if we are to know its complete history.

## The Task of the Historical Geologist

Simply stated, the task of historical geologists is to examine planetary materials and structures and to discover how they came into existence. Geologists have in the world about them the tangible *result* of past events, and they must work backward in time to discover the causes of those events. In their work, they use the usual procedures of science—namely, the collection of observations, the formulation of hypotheses to explain those observations, and the validation of the hypotheses by means of further observations. Thus, observed facts serve both as the basis for building hypotheses and as the ultimate check on their accuracy.

As illustrations of the way in which geologic investigations may be conducted, we can examine two actual research projects. One of these provided evidence to confirm astronomic calculations that the number of days in the year have been decreasing throughout geologic time. Another provided evidence that 6 million years ago the Mediterranean Sea was a deep, dry basin.

**Figure 1–7** Growth banding displayed by a specimen of the extinct coral *Heliophyllum halli*. The finer lines may represent daily growth increments. There are approximately 200 growth lines per centimeter. Together with the annual bands, the growth increments can be used to estimate the number of days in a year at the time the animal lived. *(Photograph courtesy of G. R. Clark, II.)*

0    1 cm

## Growth Lines and Time

In the early 1960s, Professor John Wells of Cornell University was musing over the fact that although paleontology (the study of fossils and ancient life forms, from the Greek *palaios,* meaning ancient) provided a way to determine *relative* geologic age, fossils could not be used in determining absolute geologic time. His thoughts led him to consider the astronomic basis for time and to search for some detectable effect of movement in the sun–earth–moon system on the fossil remains of organisms. On theoretical grounds, geophysicists and astronomers had determined that the present 24-hour period of the earth's rotation on its axis has not been constant throughout geologic time. Because of tidal friction, the earth has been slowing down at a rate of about 2 seconds per 100,000 years. Thus, the length of the day has been increasing, and the number of days in a year has been steadily decreasing. On the basis of work completed by earlier researchers, Wells knew that the fine growth lines on the exoskeletons of coral organisms might represent daily growth increments (Fig. 1–7). Thus, a coral might secrete one thin ridge of calcium carbonate each day. In addition to the fine daily growth lines, there are coarser monthly bands (presumably related to monthly breeding cycles during which carbonate secretion was inhibited) and still broader annual bands in cor-

als living in areas of seasonal fluctuations. Wells counted the growth lines on several species of living corals and found that the count "hovers around 360 in the space of a year's growth." Proceeding next to fossil corals of successively older geologic age, Wells counted correspondingly larger numbers of growth lines in the yearly segment of the exoskeleton. For example, on corals that had been determined to be about 370 million years old, Wells found 398 daily growth lines in a yearly increment. This evidence suggesting that there were nearly 400 days in a year 370 million years ago correlated astonishingly well with the calculations provided by the astronomers. Historical geologists could now state with considerable confidence that long ago when the first land vertebrates began to appear, the days were shorter, but there were more of them in a year. Furthermore, once the relationship between number of days in the year and absolute geologic age had been plotted on a graph (Fig. 1–8), then it might be possible to derive the age of a stratum by counting the growth lines on the fossil corals collected from that stratum. Of course, the method would be valid only if the slowdown of the earth's rotation has been uniform.

As in all forms of geochronology, the Wells method has its limitations. Only exceptionally well-preserved fossils are suitable, and large numbers must be subjected to growth band counts in order to obtain statistically valid results. Many specimens

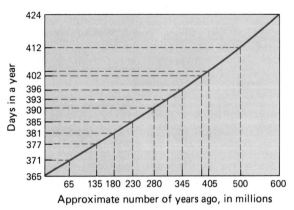

**Figure 1–8** The changing length of the day through geologic time. *(From J. W. Wells,* Nature *197:948–950, 1963.)*

show variations caused by factors in the environment unrelated to ordinary daily changes. Nevertheless, the method has provided an interesting correlation between astronomic and biologic phenomena.

## A Study of Mediterranean History

A second research effort that exemplifies the way scientists work is embodied in the ongoing work of Kenneth J. Hsu of the Swiss Federal Institute of Technology and of William B. F. Ryan of the Lamont-Doherty Geological Observatory. These two geologists were chief scientists aboard the 1970 *Glomar Challenger* cruise to investigate the floor of the Mediterranean Sea.

Long before the cruise, paleontologists had discovered evidence that the invertebrate fauna of the Mediterranean had changed abruptly about 6 million years ago. The older organisms of the region had nearly suffered annihilation, although a few species had managed to migrate into the Atlantic, and a still smaller group of hardy species had survived. Somewhat later, the refugees returned, bringing new species with them. What was the reason for the migrations and extinctions?

A second question concerned the origin of an enormous buried gorge that lay beneath—and nearly parallel to—the present course of the Rhone River, which flows from Switzerland to France. The buried canyon is nearly 1000 meters deep. What had happened in the geologic past to give a stream sufficient erosive powers to cut so deeply?

The third task for the *Challenger*'s scientists was to try to determine the origin of the domelike masses buried deep beneath the Mediterranean sea floor. These structures had been detected years earlier by echo-sounding instruments, but they had never been penetrated in the course of drilling. Were they salt domes such as are common along the United States Gulf Coast, and if so, why should there have been so much solid crystalline salt beneath the floor of the Mediterranean?

Finally, there was the presence—again detected on seismic profiles—of a hard layer of sedimentary rock 100 meters or so below the present sea floor. This hard layer was a strong acoustic reflector (layer of sediment or rock capable of reflecting sound waves) and was easily recognized nearly everywhere in the Mediterranean. Most deep-sea deposits are soft, poorly reflective oozes. What was the composition and origin of the hard layer?

With such questions as these clearly before them, the scientists aboard the *Glomar Challenger* proceeded to the Mediterranean to search for the answers. On August 23, 1970, they recovered a sample from the surface of the hard layer. The sample consisted of pebbles of hardened sediment that had once been soft, deep-sea mud, as well as granules of gypsum and fragments of volcanic rock. Not a single pebble was found that might have indicated that the pebbles came from the nearby continent. In the days following, samples of solid gypsum were repeatedly brought on deck, as drilling operations penetrated the hard layer at its predicted depth. Clearly, the strong acoustic reflector was a layer of gypsum. Furthermore, the gypsum was found to possess peculiarities of composition and structure that suggested it had formed on desert flats. Sediment above and below the gypsum layer contained tiny marine fossils, indicating open-ocean conditions. As they drilled into the central and deepest part of the Mediterranean basin, the scientists took solid, shiny, crystalline salt from the core barrel. Interbedded with the salt were thin layers of what appeared to be wind-blown silt.

The time had come to formulate a hypothesis. The investigators theorized that about 20 million years ago, the Mediterranean was a broad seaway linked to the Atlantic by two narrow straits. Crustal movements closed the straits, and the landlocked Mediterranean began to evaporate. Increasing salinity caused by the evaporation resulted in the extermination of scores of invertebrate species. Only a few organisms especially tolerant of very salty conditions remained. As evaporation continued, the remaining brine became so dense that the calcium sulfate of the hard layer was precipitated. In the central deeper part of the basin, the last of the brine evaporated to precipitate more soluble sodium chloride. Later, under the weight of overlying sediments, this salt flowed plastically upward to form salt domes. Before this happened, however, the Mediterranean was a vast "Death Valley" 3000 meters deep. Then, about $5\frac{1}{2}$ million years ago came the deluge. As a result of crustal adjustments and faulting, the Strait of Gibraltar opened, and water cascaded spectacularly back into the Mediterranean. Turbulent waters tore into the hardened salt flats, broke them up, and ground them into the pebbles observed in the first sample taken by the *Challenger*. As the basin was refilled, normal marine organisms returned. Soon layers of oceanic ooze began to accumulate above the old hard layer.

The salt and gypsum, the faunal changes, and the unusual gravel provided abundant evidence that the Mediterranean was once a desert. But how did this knowledge relate to the buried gorge beneath the Rhone Valley? The answer was now clear. If the Mediterranean had been emptied, the surrounding lands would have stood high above the floor of the basin. The gradients of streams would have markedly increased, and their swiftly flowing waters would have incised rapidly, eroding deep canyons. Such canyons should also have been cut by other streams that once flowed into the Mediterranean. As news of the *Glomar Challenger*'s findings spread, geologists who had worked in the region alerted the investigators to the presence of gorges in North Africa. A Russian geologist recalled a 240-meter chasm that had been encountered at the time the foundations had been prepared for the Aswan Dam in Egypt. Oil company geologists in Libya sent word of many similar gorges that they had detected from seismograph records. A final piece of the puzzle seemed to fall into place.

In the pages ahead, the methods of historical geology will be further demonstrated. Many of these methods enable geologists to place past events in their proper chronologic order and to discern changes in the earth's past climate, atmosphere, and geography.

# The Founders of Historical Geology

Today, geologists are aided in their studies by a variety of sophisticated instruments. Yet, the basic style of thinking used by a geologist working with computers, mass spectrometers, and electron microprobes has not changed much from that of seventeenth century naturalists armed only with compass and hammer. Also, the progress of science is cumulative, and we build our new theories by using observations and postulates of those who preceded us. Four centuries ago Leonardo da Vinci (1452–1519) recognized the true significance of fossils and understood the impermanence of mountains. Robert Hooke (1635–1703), who was also intrigued by fossils, perceived that "it would not be impossible to raise a chronology out of them." Around the year 1669, three very important concepts were clearly enunciated by Nicolaus Steno.

## Nicolaus Steno

Nicolaus Steno (1638–1687) was a Danish physician who settled in Florence, Italy. There he became physician to the Grand Duke of Tuscany. Since the duke was a generous employer, Steno had ample time to tramp across the countryside, visit quarries, and examine strata. His investigations of sedimentary rocks led him to formulate such basic principles of historical geology as superposition, original horizontality, and original lateral continuity.

The **principle of superposition** states that in any sequence of undisturbed strata, the oldest layer is at the bottom and successively higher layers are successively younger. It is a rather obvious axiom. Yet Steno, on the basis of his observations of strata in northern Italy, was the first to explain the concept formally. The fact that it is self-evident does not diminish the principle's importance in deciphering earth history. Furthermore, the superpositional relationship of strata is not always apparent in regions where layers have been steeply tilted or overturned (Fig. 1–9). In such instances, the geologist must examine the strata for clues useful in recognizing their uppermost layer. The way fossils lie in the rock and the evidence of mudcracks and ripple marks are par-

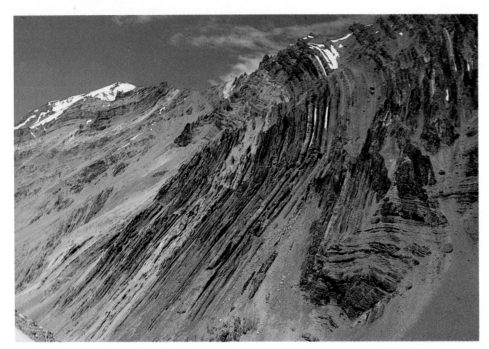

**Figure 1–9** Steeply dipping strata grandly exposed in the Himalayan Mountains. It is often difficult to recognize the original tops of beds in strongly deformed sequences such as this. *(Courtesy D. Bhattacharyya.)*

ticularly useful clues when one is trying to determine which way was up.

The observation that strata are often tilted led Steno to his **principle of original horizontality**. He reasoned that most sedimentary particles settle from fluids under the influence of gravity. The sediment then must have been deposited in layers that were nearly horizontal and parallel to the surface on which they were accumulating (Fig. 1–10). Hence, steeply inclined strata indicate an episode of crustal disturbance after the time of deposition (Fig. 1–11).

The **principle of original lateral continuity** was the third of Steno's stratigraphic axioms. It pertains to the fact that, as originally deposited, strata extend in all directions until they terminate by thinning at the margin of the basin, end abruptly against some former barrier to deposition, or grade laterally into a different kind of sediment. This observation is significant in that whenever one observes the exposed cross-section of strata in a cliff or valley wall, one should recognize that the strata continue laterally for a distance that can be determined by field work and drilling. When geologists stand on a sandstone ledge at one side of a canyon, it is the principle of original lateral continuity that leads them to seek out the same ledge of sandstone on the far canyon wall and then to realize that the two exposures were once continuous.

Today, we recognize that Steno's principles are basic to the geologic specialty known as **stratigraphy**, which is the study of layered sedimentary rocks, including their texture, composition, arrangement, and correlation from place to place. Because stratigraphy enables geologic events as recorded in rocks to be placed in their correct sequence, it is the key to the history of the earth.

## Interpreters of the Geologic Succession

The stratigraphic principles formulated by Steno in the seventeenth century were rediscovered several decades later by other European scientists. Among the most prominent of these early geologists were John Strachey (1671–1743), Giovanni Arduino (1714–1795), Johann G. Lehmann (1719–1767), Georg Füchsel (1722–1776), and Peter Simon Pallas (1741–1811). John Strachey is best remembered for

**Figure 1–10** Superpositional sequence of undisturbed, nearly horizontal Paleozoic strata, Grand Canyon National Park, Arizona. Photographed from the Kaibab Trail. *(Courtesy D. Bhattacharyya.)*

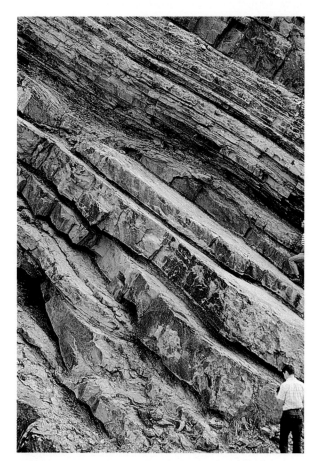

**Figure 1–11** Steeply dipping shale and sandstone strata, Ouachita Mountains of Oklahoma.

his use of the principles of superposition and original lateral continuity in deciphering the stratigraphic succession of coal-bearing formations in Somerset and Northumberland, England. He clearly illustrated the sequence of formations encountered at the surface and in mines and described the manner in which horizontal strata rested upon the eroded edges of inclined older formations. Years later this type of stratigraphic relationship would be termed *unconformable*.

Whereas John Strachey was particularly interested in the local stratigraphic succession, other naturalists developed a broader, more global view of the geologic succession. In Italy, Giovanni Arduino classified mountains according to the most abundant type of rock that composed them. He defined Primary mountains as constructed of crystalline rocks of the kinds later to be named igneous and metamorphic. Arduino recognized that rocks of the Primary group were likely to be the oldest in a mountain system and were usually exposed along the central axis of the ranges. Secondary mountains were constructed of layered, well-consolidated, fossiliferous rocks. Such rocks were later to be named sedimentary. Arduino's Tertiary designation was reserved for mountains composed of unconsolidated gravel, sand, and clay, as well as lava beds.

Classifications similar to that of Arduino also appeared in the works of the German scientists Lehmann and Füchsel. These men were not rocking-chair theorists, for both devoted long hours to making careful observations in the field. Füchsel worked chiefly in the mountains of Thuringia, whereas his contemporary Lehmann tramped on the rocks of the Harz and Erz Gebirge. They prepared excellent summaries of the stratigraphic succession in these mountains and further developed a remarkably perceptive understanding of some of the events involved in the making of mountain ranges.

This insight into the history of mountains was improved as a result of the work of an indefatigable field geologist named Peter Simon Pallas. Under the patronage of Catherine II of Russia, Pallas traveled across the whole of Asia and made careful studies of the Ural and Altai Mountains. He recognized the threefold division of mountains formulated by his predecessors. In addition, Pallas was able to construct a general geologic history of the Urals, and he provided a lucid description of how the rock assemblages change as one travels from the center to the flanks of mountain systems.

## Abraham Gottlob Werner

One of the most influential geologists working in Europe near the close of the eighteenth century was Professor Abraham Gottlob Werner (1750–1817). Werner's eloquent and enthusiastic lectures at the Freiberg Mining Academy in Saxony transformed that school into an international center for geologic studies. Werner was a competent mineralogist, and many geologists of his day used his scheme for the identification of minerals and ores. Werner is not, however, remembered as much for his contributions to mineralogy as he is for his interpretation of the geologic history of the earth. The cornerstone of that interpretation was his insistence that all rocks of the earth's crust were deposited or precipitated from a great ocean that once enveloped the entire planet. Today we know that some rocks (the group called sedimentary) are indeed often of marine origin. Others, however, are decidedly not formed in water. Because they believed in neptunism, which posited an aqueous origin for all rocks, Werner and his followers became known as neptunists (after Neptune, the Roman god of the sea).

Werner envisioned his universal ocean in the earliest stage of earth history as being hot, seething, and saturated with all the dissolved substances needed to form the rocks of his oldest division. He called these *Primitive Rocks,* or *Urgebirge.* Most of these rocks would later come to be known as igneous and metamorphic.

The second stage in the Wernerian interpretation of earth history witnessed a subsidence and cooling of the primitive ocean. It came to resemble the ocean of today. Werner told his students that this change was marked by the deposition of fossil-bearing, well-consolidated, stratified rocks that lay above the Urgebirge. These he designated *Transition Rocks* and suggested that they were deposited when the earth had passed from an uninhabitable to an inhabitable condition. The fossils proved the planet had become suitable for life. Today, we would refer to these rocks as part of Europe's Paleozoic sequence of strata.

Above the Transition Rocks, Werner noted the occurrence of sandstones, shales, coal beds, very fossiliferous limestones, and occasional layers of a black rock later determined to be basalt. These basal layers were actually old lava flows. For all of these rocks lying above the Transition Rocks, Werner employed Johann Lehmann's term *Flötzgebirge.* A final term, *Alluvium,* was used for the unconsolidated sand, gravel, and clay that rested on the Flötzgebirge.

Although initially received with great interest and enthusiasm, Werner's ideas were soon to draw criticism. His theory failed to explain what had become of the immense volume of water that once covered the earth to so great a depth that all continents were submerged. An even greater problem was his insistence that basaltic lava layers such as those in the Flötzgebirge were deposited in precisely the same manner as the enclosing limestones and shales. With visible, indisputable field evidence, geologists such as J. F. D'Aubisson de Voisins in France clearly demonstrated the volcanic origin of these basaltic layers. Geologists with this opposing view came to be known as plutonists (after Pluto, the Roman god of the Underworld). According to the plutonists, fire rather than water was the key to the origin of igneous rocks. James Hutton of Scotland was a prominent plutonist who clearly stated that rocks like basalt and granite "formed in the bowels of the earth of melted matter poured into rents and openings of the strata."

## James Hutton

James Hutton (1726–1797), an Edinburgh physician and geologist, is remembered not only as a staunch opponent of neptunism, but also for his penetrating comprehension of how geologic processes alter the earth's surface. For Hutton, the earth was a dynamic, ever-changing place in which new rocks, lands, and mountains arise continuously as a balance against their destruction by erosion and weathering. He took a cyclic view of our planet, as opposed to Werner's more static concept of an earth that had changed very little from its beginning down to the present time. In addition, Hutton believed that "the past history of our globe must be explained by what can be seen to be happening now." This simple yet powerful idea was later to be named **uniformitarianism** by William Whewell. Charles Lyell (1797–1875) became the principal advocate and interpreter of uniformitarianism. We will speak of this great man again in the pages ahead.

Perhaps because it is so general a concept, uniformitarianism has been reinterpreted and altered in a variety of ways by scientists and theologians from Hutton's generation down to our own. Some of the ideas about what uniformitarianism now implies would seem strange to Hutton himself. If the term uniformitarianism is to be used in geology (or any science), one must clearly understand what is uniform. The answer is that the physical and chemical laws that govern nature are uniform. Hence, the history of the earth may be deciphered in terms of present observations on the assumption that natural laws are invariant with time. These so-called natural laws are merely the accumulation of all our observational and experimental knowledge. They permit us to predict the conditions under which water becomes ice, the behavior of a volcanic gas when it is expelled at the earth's surface, or the effect of gravity on a grain of sand settling to the ocean floor. Uniform natural laws govern such geologic processes as weathering, erosion, transport of sediment by streams, movement of glaciers, and the movement of water into wells.

Hutton's use of what later was termed uniformitarianism was simple and logical. By observing geologic processes in operation around him, he was able to infer the origin of particular features he discovered in rocks. When he witnessed ripple marks being produced by wave action along a coast, he was able to state that an ancient rock bearing similar markings was once a sandy deposit of some equally ancient shore. And if that rock now lay far inland from a coast, he recognized the existence of a sea where Scottish sheep now grazed.

Hutton's method of interpreting rock exposures by observing present-day processes was given the catchy phrase "the present is the key to the past" by Sir Archibald Geike (1835–1924). The methodology implied in the phrase works very well for solving many geologic problems, but it must be remembered that the geologic past was sometimes quite unlike the present. For example, before the earth had evolved an atmosphere like that existing today, different chemical reactions would have been prevalent during weathering of rocks. Life originated in the time of that primordial atmosphere under conditions that have no present-day counterpart. As a process in altering the earth's surface, meteorite bombardment was once far more important than it has been for the last three billion years or so. Many times in the geologic past continents have stood higher above the oceans, and this higher elevation resulted in higher rates of erosion and harsher climatic conditions compared with intervening periods when the lands were low and partially covered with inland seas. Similarly, at one time or another in the geologic past volcanism was more frequent than at present.

Nevertheless, ancient volcanoes disgorged gases and deposited lava and ash just as present-day volcanoes do. Modern glaciers are more limited in area than those of the recent geologic past, yet they form erosional and depositional features that resemble those of their more ancient counterparts. All of this suggests that present events do indeed give us clues to the past, but we must be constantly aware that in the past, the rates of change and intensity of processes often varied from those to which we are accustomed today, and that some events of long ago simply do not have a modern analogue.

The eighteenth-century concept of uniformitarianism was not the only contribution James Hutton made to geology. In his *Theory of the Earth,* published in 1785, he brought together many of the formerly separate thoughts of the naturalists who preceded him. He showed that rocks recorded events that had occurred over immense periods of time and that the earth had experienced many episodes of upheaval separated by quieter times of denudation and sedimentation. In his own words, there had been a "succession of former worlds." Hutton saw a world of cycles in which water sculptured the surface of the earth and carried the erosional detritus from the land into the sea. The sediment of the sea was compacted into stratified rocks, and then by the

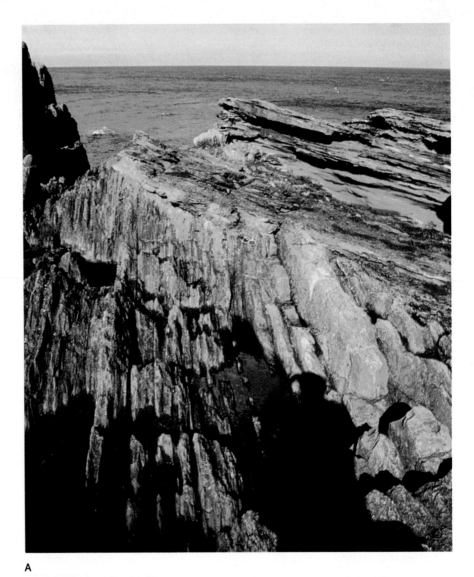

A

**Figure 1–12A** Angular unconformity at Siccar Point, eastern Scotland. It was here that James Hutton first realized the historical significance of an unconformity. The drawings on page 15 indicate the sequence of events documented in this famous exposure. *(Photograph courtesy of E. A. Hay.)*

action of enormous forces, the layers were cast up to form new lands. In this endless process, Hutton found "no vestige of a beginning, no prospect of an end." In this phrase, we see Hutton's intoxicating view of the immensity of geologic time. No longer could geologists compress earth history into the short span suggested by the Book of Genesis.

At Scotland's Isle of Arran and at Jedburgh, Scotland, Hutton came across an exposure of rock where steeply inclined older strata had been beveled by erosion and covered by flat-lying younger layers (Fig. 1–12). It was clear to Hutton that the older sequence was not only tilted but also partly removed by erosion before the younger rocks were deposited.

Uplift, tilting, erosion

5

Deposition of younger strata (Devonian)

4

Erosion to produce surface of unconformity

3

Deformation of strata in mountain-building event

2

Deposition of older strata (Silurian)

1

Younger strata (Devonian)

Older strata (Silurian)

**Figure 1–12B**   Sequence of events indicated by the Siccar Point exposure.

The erosional surface meant that there was a gap or hiatus in the rock record. In 1805, Robert Jameson named this relationship an **unconformity**. More specifically, Hutton's Jedburgh exposure was an angular unconformity because the lower beds were tilted at an angle to the upper. This and other unconformities provided Hutton with evidence for periods of denudation in his "succession of worlds." Although he did not use the word unconformity, he was the first to understand and explain the significance of this concept.

During most of his career, Hutton's published reports attracted only modest attention, and a good part of that attention came from opponents who preferred to follow the views of Abraham Gottlob Werner. To remedy this situation, British scientists who appreciated the value of Hutton's ideas convinced his friend John Playfair, a professor of mathematics and natural philosophy, to publish a summary of and commentary on *Theory of the Earth*. The work by Playfair was published in 1802 under the title *Illustrations of the Huttonian Theory of the Earth*. Whereas Hutton's writing was often complex and difficult to follow, Playfair's text was lucid, easy to read, unburdened by lengthy quotations from foreign sources, and altogether delightfully persuasive.

Indeed, subsequent geologists of the nineteenth century based much of their understanding of Hutton's ideas not from their reading of Hutton's original publications, but from the intelligent and convincing phrases of John Playfair.

Hutton died five years before the publication of *Illustrations*. Throughout his life he had been a man absorbed in the investigation of the earth. He was seen frequently in the field, scrutinizing every rock exposure he happened upon, and soon became so familiar with certain strata that he was able to recognize them at different localities. What he was unable to do well, however, was determine if dissimilar-looking strata were roughly equivalent in age. He had not discovered how to correlate beds that did not have a similar composition and texture (lithology). This problem was soon to be resolved by William Smith (1769–1839).

## William Smith

William Smith was an English surveyor and engineer who devoted 24 years to the task of tracing out the strata of England and representing them on a map. Small wonder that he acquired the nickname "Strata Smith." He was employed to locate routes of canals, to design drainage for marshes, and to restore springs. In the course of this work, he independently came to understand the principles of stratigraphy, for they were of immediate use to him. By knowing that different types of stratified rocks occur in a definite sequence and that they can be identified by their lithology, the soils they form, and the fossils they contain, the was able to predict the kinds and thicknesses of rock that would have to be excavated in future engineering projects. His use of fossils was particularly significant. Prior to Smith's time, collectors rarely noted the precise beds from which fossils were taken. Smith, on the other hand, carefully recorded the occurrence of fossils and quickly became aware that certain rock units could be identified by the particular assemblages of fossils they contained. He used this knowledge first to trace strata over relatively short distances and then to extend over great distances his "correlations" to strata of the same age but of different lithology. Ultimately, this knowledge led to the **principle of biologic succession**, which stipulates that the life of each age in the earth's long history was unique for particular periods, that the fossil remains of life permit geologists to recognize

contemporaneous deposits around the world, and that fossils could be used to assemble the scattered fragments of the record into a chronologic sequence.

Smith did not know why each unit of rock had a particular fauna. This was 60 years before the publication of Darwin's *Origin of Species*. Today, we recognize that different kinds of animals and plants succeed one another in time because life has evolved continuously. Because of this continuous change, or evolution, only rocks formed during the same age could contain similar assemblages of fossils.

News of Smith's success as a surveyor spread widely, and he was called to all parts of England for consultation. On his many trips, he kept careful records of the types of rocks he saw and the fossils they contained. Armed with his notes and observations, in 1815 he prepared a geologic map of England and Wales that has remained substantially correct even today. His work clearly demonstrated the validity of the principle of biologic succession.

## Cuvier and Brongniart

The use of fossils for the correlation and recognition of formations was not exclusively William Smith's discovery. At the same time that Smith was making his observations in England, two scientists across the English Channel in France were diligently advancing the study of fossils. They were Baron Georges Léopold Cuvier (1769–1832) and his close associate Alexander Brongniart (1770–1847). Cuvier was an expert in comparative anatomy, and with this knowledge he became the most respected vertebrate paleontologist of his day. Brongniart was a naturalist who worked not only on fossil vertebrates, but on plants and minerals as well. Together these men established the foundations of vertebrate paleontology. They validated Smith's findings that fossils display a definite succession of types within a sequence of strata and that this succession remains more or less constant wherever found. They also noticed that certain large groupings of strata were often separated by unconformities. As one would pass from one group of strata across the unconformity into the overlying group, a dramatic change in the kinds of animals preserved as fossils was apparent. From this observation, the two French scientists erroneously concluded that the history of life was marked by frightful catastrophes involving sudden violent flooding of the continents and abrupt crustal upheavals of stupendous magnitude. The last of these catastrophic episodes was considered to be the Noachian deluge. Cuvier and Brongniart believed that each catastrophe resulted in the total extinction of life and was then followed by the appearance of new animals and plants. Cuvier did not speculate on how each of the many new species originated. Many geologists of the time, including the eminent Charles Lyell, held that geologic history was a uniform and gradual progression and could not accept Cuvier's concept of catastrophism. Thus began a catastrophism versus uniformitarianism controversy that rivaled the earlier neptunist–plutonist debates in scope and passion.

Eventually catastrophism was found to be an untenable hypothesis. As stratigraphic correlations were extended to other regions, it became apparent that the seemingly abrupt changes in fossil faunas were often caused by missing strata or other imperfections in the geologic record. On close study, some of the apparent breaks in the fossil record were truly not sudden, and the precursors of each animal group could be found as fossils in underlying beds. Many catastrophists, including Cuvier, were unable to explain how each new group of animals suddenly appeared on earth. Perhaps the most difficult for the catastrophists to defend was their insistence that the earth was only several thousand years old and that all the catastrophes and regenerations of life occurred in so incredibly short a span of time. They failed to grasp Hutton's view of the great age of the earth.

### Charles Lyell

In the early nineteenth century, a geologist wrote a book that would both amplify the ideas of Hutton and present under one title the most important geologic concepts of the day. His name was Charles Lyell (Fig. 1–13), and he authored the classic *Principles of Geology*. The first volume of this work was printed in 1830. It grew to four volumes and became immensely important in the Great Britain of Queen Victoria. In these volumes one can recognize Lyell's skill in explaining and synthesizing the geological findings of contemporary geologists. As his friend Andrew C. Ramsey remarked, "We collect the data, and Lyell teaches us the meaning of them." Lyell's *Principles* became the indispensable handbook of every English geologist. In it are amplified many of the principles expressed earlier by Hutton regarding the recognition of the relative ages of rock bodies.

**Figure 1–13**  Sir Charles Lyell. *(Courtesy of Geological Society of London.)*

For example, Lyell discusses the general principle that a geologic feature that cuts across or penetrates another body of rock must be younger than the rock mass penetrated. In other words, the feature that is cut is older than the feature that crosses it. This generalization, called the **principle of cross-cutting relationships,** applies not only to rocks but also to geologic structures like faults and unconformities. Thus, in Figure 1–14, fault *b* is younger than stratigraphic sequence *d*; the intrusion of igneous rock *c* is younger than the fault, since it cuts across it; and by superposition, rock sequence *e* is youngest of all.

Another generalization to be found in Lyell's *Principles* relates to **inclusions.** Lyell logically discerned that fragments within larger rock masses are older than the rock masses in which they are enclosed. Thus, whenever two rock masses are in contact, the one containing pieces of the other will be the younger of the two.

In Figure 1–15A, the pebbles of granite (a coarse-grained igneous rock) within the sandstone tells us that the granite is older and that the eroded granite fragments were incorporated into the sandstone. In Figure 1–15B, the granite was intruded as a

**Figure 1–14** An example of how the sequence of geologic events can be determined from cross-cutting relationships and superposition. From first to last, the sequence indicated in the cross-section is deposition of *d*, faulting to produce fault *b*, intrusion of igneous rock mass *c*, and erosion followed by deposition of *e*. Strata labeled *d* are oldest, and strata labeled *e* are youngest.

melt into the sandstone. Because there are sandstone inclusions in the granite, the granite must be the younger of the two units.

## Charles Darwin

As noted earlier, William Smith and some of his contemporaries were able to recognize that strata were often characterized by particular fossils and that there was a general progression toward more

modern-looking assemblages of shells in higher, younger strata. It was Charles Darwin (1809–1882) who provided a general theory that would account for the changes seen in the fossil record.

As a young man, Darwin had acquired an impressive knowledge of both biology and geology. That knowledge was the basis for his securing an unpaid position as a naturalist aboard the *H.M.S. Beagle,* bound for a five-year mapping expedition around the world. On his return from the voyage in

A    B

**Figure 1–15** (A) Granite inclusions in sandstone indicate that granite is the older unit.
(B) Inclusions of sandstone in granite indicate that sandstone is the older unit.

1836, Darwin (Fig. 1–16) had assembled volumes of notes in support of his theory of evolution of organisms by natural selection. His theory was based on a logical system of observations and conclusions. He observed that all living things tend to increase their numbers at prodigious rates. Yet, in spite of their reproductive potential to do so, no one group of organisms has been able to overwhelm the earth's surface. In fact, the actual size of any given population remains fairly constant over long periods of time. Because of this, Darwin concluded that not all the individuals produced in any generation can survive. In addition, Darwin recognized that individuals of the same kind differ from one another in various morphologic and physiologic features. From this and the previous observation, he concluded that those individuals with the most favorable variations would have the best chance of surviving and transmitting their favorable traits to the next generation. Darwin had no knowledge of genetics, and therefore did not know the cause of the variation that was so important to his theory. Gregor Mendel's 1865 report of experiments in heredity had escaped his attention. In the decades following Darwin's death, geneticists clearly established that the variability essential to Darwin's natural selection is derived from new gene combinations that occur during reproduction and from genetic mutation.

Possibly because he was reluctant to face the controversy that his theory would provoke, Darwin did not publish his findings on his return to England. He did, however, confide in Charles Lyell and the great botanist Joseph Hooker, and these friends urged him to quickly publish before someone else anticipated his discoveries. Yet, Darwin continued to procrastinate. Then, in 1858, a comparatively unknown young naturalist named Alfred Russel Wallace sent Darwin a manuscript for review that contained the basic concepts of natural selection. Wallace had conceived of natural selection while on a biological expedition to Indonesia. The idea came to him while suffering with malaria shortly after he read an essay on overpopulation by Thomas Malthus.

Understandably disturbed by Wallace's letter, Darwin sought the help of his friends Lyell and Hooker. Recognizing the importance of giving Darwin the credit he deserved for his discovery and the long years of assembling supporting evidence, the two scientists arranged for a presentation of Darwin's

**Figure 1–16** Charles Darwin as a young man. This portrait was made shortly after Darwin returned to England from his voyage around the world on the *H.M.S. Beagle.* Observations made during this voyage helped him formulate the concept of evolution by natural selection.

work and Wallace's paper before the Linnaean Society. Thus, the theory of natural selection was credited simultaneously to both scientists. Darwin now worked at top speed to complete his famous *Origin of Species.* The book was published in 1859. In that volume, Darwin hoped to accomplish two things. The first was to convince the world that evolution had occurred. Organisms had evolved or changed throughout geologic time. The second was to propose a mechanism for evolution. That mechanism was natural selection. Darwin's success in achieving his objectives can be measured by the fact that, within a decade, organic evolution had become the guiding principle in all paleontologic and biologic research. His book changed the way people viewed the world, and for this reason it has been described as one of the greatest books of all time.

Darwin died at his home in Down, England, in 1882. By that time, geologists everywhere were using their knowledge of evolution, the biologic succession, superposition, cross-cutting relationships, and inclusions to decipher earth history.

# Minerals

The history of the earth has been deduced not only from rocks and fossils, but from minerals as well. Every mineral was formed within a specific range of physical and chemical conditions. They therefore provide tangible evidence of past events. Some minerals develop exclusively in ocean water and provide the geologist with evidence of the incursion of seas across former areas of dry land. Others form under conditions of excessive aridity and can be used to locate warm tropical belts of long ago. The magnetic properties of certain minerals provide clues to the shifting of ancient continents. Many minerals form within a narrow range of physical conditions and can therefore be used to diagnose the pressures and temperatures involved in the formation of crustal rocks and mountains. Yet others contain radioactive elements that permit us to determine the ages of various parts of the earth's crust.

## *Common Rock-Forming Minerals*

A **mineral** can be defined as a naturally occurring element or compound formed by inorganic processes that has a definite chemical composition or range of compositions as well as distinctive properties that reflect its internal atomic structure. Minerals can usually be identified by such physical properties as color, hardness, density, crystal form, and cleavage. **Cleavage** refers to the tendency of some minerals to break smoothly along certain planes of weakness.

More than 3000 minerals have already been discovered and described, but most of these are rarely encountered. For our present purposes, it is important only to consider those minerals that compose the bulk of common rocks. Among these, the most important are the silicate minerals.

## Silicate Minerals

About 75 per cent by weight of the earth's crust is composed of the two elements oxygen and silicon (Table 1–1). These elements usually occur in combination with such abundant metals as aluminum, iron, calcium, sodium, potassium, and magnesium to form a group of minerals called the silicates. A single family of silicates, the feldspars, comprises about one-half of the material of the crust, and a single mineral species called quartz represents a sizable portion of the remainder.

**Quartz**   The principal silicate minerals and some of their properties are listed in Table 1–2. The mineral **quartz** ($SiO_2$) is one of the most familiar and important of all the silicate minerals. It is common in many different families of rocks. In general appearance,

### Table 1–1   Abundances of Chemical Elements in the Earth's Crust*

| Element and Symbol | Percentage by Weight | Percentage by Number of Atoms | Percentage by Volume |
|---|---|---|---|
| Oxygen (O) | 46.6 | 62.6 | 93.8 |
| Silicon (Si) | 27.7 | 21.2 | 0.9 |
| Aluminum (Al) | 8.1 | 6.5 | 0.5 |
| Iron (Fe) | 5.0 | 1.9 | 0.4 |
| Calcium (Ca) | 3.6 | 1.9 | 1.0 |
| Sodium (Na) | 2.8 | 2.6 | 1.3 |
| Potassium (K) | 2.6 | 1.4 | 1.8 |
| Magnesium (Mg) | 2.1 | 1.9 | 0.3 |
| All other elements | 1.5 | — | — |
| | 100.0 | 100.0† | 100.0† |

\* Based on B. Mason 1966. *Principles of Geochemistry.* New York, John Wiley. Note the high percentage of oxygen in the earth's crust.
† Includes only the first eight elements.

COMMENTARY COMMENTARY COMMENTARY COMMENTARY COMMENTARY
MMENTARY COMMENTARY COMMENTARY COMMENTARY COMMEN
COMMENTARY COMMENTARY COMMENTARY COMMENTARY
MMENTARY COMMENTARY COMMENTARY COMMENTARY COMMEN

## COMMENTARY

### The Quartz We Wear

Many varieties of the mineral quartz are so attractive when cut and polished that they are highly valued as gems. Single crystals may be cut into faceted geometric shapes that greatly enhance their luster and color. Quartz minerals may also be cut and polished into rounded, convex shapes called cabochons. Among the quartz family gems that are cut from single crystals are amethyst (purple quartz), citrine (yellow quartz), rose quartz, smoky quartz, and rock crystal (colorless quartz). The purple of amethyst is attributed to impurities of colloidal iron. Rose quartz gets its color from impurities of molybdenum, and often has a cloudy appearance because of myriads of tiny cracks and inclusions. Yellow and brown colors are attributed to defects in the crystal structure of the quartz rather than impurities.

In addition to quartz gems cut from single crystals, many varieties used in jewelry are composed of an aggregate of microcrystalline crystals. The individual crystals are far too small to see except at high magnification. Chalcedony is a common type of microcrystalline quartz that often exhibits a waxy luster. Chalcedony that displays banding is called banded agate. Moss agate is unbanded, and contains green or black moss-like inclusions. Carnelian is red microcrystalline quartz, and chrysophase is an attractive green variety. Tiger's-eye quartz is known for its distinctive play of colors that seem to change as the stone is turned in the light. This characteristic, termed chatoyancy, results from light that is reflected off the mineral's fibrous structure. The fibers are relics of asbestos fibers that have been replaced by silica. Jasper is an opaque variety of microcrystalline quartz whose rich red, yellow, or brown colors are the result of a high content of iron oxide.

**A**

**B**

**C**

**D**

(A) Amethyst; (B) Tiger's-eye; (C) Agate; (D) Moss agate (chalcedony).

**Table 1–2   Common Rock-Forming Silicate Minerals**

| Silicate Mineral | Composition | Physical Properties |
|---|---|---|
| Quartz | Silicon dioxide (silica, $SiO_2$) | Hardness of 7 (on scale of 1 to 10)*; will not cleave (fractures unevenly); specific gravity: 2.65 |
| Potassium feldspar group | Aluminosilicates of potassium | Hardness of 6.0–6.5; cleaves well in two directions; pink or white; specific gravity: 2.5–2.6 |
| Plagioclase feldspar group | Aluminosilicates of sodium and calcium | Hardness of 6.0–6.5; cleaves well in two directions; white or gray; may show striations on cleavage planes; specific gravity: 2.6–2.7 |
| Muscovite mica | Aluminosilicates of potassium with water | Hardness of 2–3; cleaves perfectly in one direction, yielding flexible thin plates; colorless; transparent in thin sheets; specific gravity: 2.8–3.0 |
| Biotite mica | Aluminosilicates of magnesium, iron, potassium, with water | Hardness of 2.5–3.0; cleaves perfectly in one direction, yielding flexible, thin plates; black to dark brown; specific gravity: 2.7–3.2 |
| Pyroxene group | Silicates of aluminum, calcium, magnesium, and iron | Hardness of 5–6; cleaves in two directions at 87° and 93°; black to dark green; specific gravity: 3.1–3.5 |
| Amphibole group | Silicates of aluminum, calcium, magnesium, and iron | Hardness of 5–6; cleaves in two directions at 56° and 124°; black to dark green; specific gravity: 3.0–3.3 |
| Olivine | Silicates of magnesium and iron | Hardness of 6.5–7.0; light green; transparent to translucent; specific gravity: 3.2–3.6 |
| Garnet group | Aluminosilicates of iron, calcium, magnesium, and manganese | Hardness of 6.5–7.5; uneven fracture; red, brown, or yellow; specific gravity: 3.5–4.3 |

*(Biotite mica, Pyroxene group, Amphibole group, and Olivine are bracketed together as "Ferromagnesian minerals.")*

*The scale of hardness used by geologists was formulated in 1822 by Frederich Mohs. Beginning with diamond as the hardest mineral, he arranged the following table:

| | |
|---|---|
| 10 Diamond | 5 Apatite |
| 9 Corundum | 4 Fluorite |
| 8 Topaz | 3 Calcite |
| 7 Quartz | 2 Gypsum |
| 6 Feldspar | 1 Talc |

quartz is a glassy, colorless, gray, or white mineral. Quartz is relatively hard (see Mohs Scale in the footnote of Table 1–2) and will easily scratch glass. When quartz grows in an open cavity, it may form the hexagonal crystals. (Fig. 1–17) that are prized by mineral collectors. More frequently, the crystal faces cannot be discerned because the orderly addition of atoms during crystal growth was interrupted by contact with other growing crystals and results in an aggregate that exhibits crystalline texture. Such common minerals as chert, flint, jasper, and agate are sedimentary varieties of quartz. *Chert* (Fig. 1–18) is a dense, hard, white mineral composed of submicrocrystalline quartz that forms as a result of the precipitation of silicon dioxide by either biologic or chemical means. *Flint* is the popular name for a dark gray or black variety of chert in which the dark color results from inclusions of organic matter. *Jasper* is an opaque variety of quartz and is red, brown, green, or yellow. *Agate* is a variety of quartz that exhibits distinct banding (Fig. 1–19).

The origin of chert is a complex problem made even more difficult by the fact that different varieties are formed by somewhat different processes. Some cherts are replacements of earlier carbonate rocks. Others appear to have formed as a result of the solution and reprecipitation of silica from the siliceous skeletal remains of organisms. Small amounts of chert may also be precipitated directly from concentrated aqueous solutions.

**Figure 1–17**  A large crystal of quartz. This specimen is 14 cm (about 5.5 inches) tall.

**Figure 1–18**  Chert. This variety (called novaculite) has a grainy texture that makes it useful as a grinding stone.

**Figure 1–19**  Agate (banded outer portion) with crystalline quartz in the center.

**The Feldspars**  The **feldspars** are the most abundant constituents of rocks, composing about 60 per cent of the total weight of the earth's crust. There are two major families of feldspars: the potassium feldspar group (Fig. 1–20), which comprises the potassium aluminosilicates, and the plagioclase group, which comprises the aluminosilicates of sodium and calcium (Fig. 1–21). Members of the plagioclase group exhibit a wide range in composition—from a calcium-rich end member called *anorthite* ($CaAl_2Si_2O_8$) to a sodium-rich end member called *albite* ($NaAl_2Si_3O_8$). Between these two extremes, plagioclase minerals containing both sodium and calcium occur. The substitution of sodium for calcium, however, is not random, but rather is governed by the temperature and composition of the parent material. Thus, by examining the feldspar content of a once molten rock, it is possible to infer the physical and chemical conditions under which it originated. Feldspars are nearly as hard as quartz and

range in color from white or pink to bluish gray. They have good cleavage in two directions, and the resulting flat, often rectangular surfaces are useful in identification. Those plagioclase feldspars with abundant sodium tend, along with potassium feldspars, to occur in silica-rich rocks such as granite. Calcium-rich plagioclases occur in such rocks as the Hawaiian lavas (called basalts).

**Mica**  Mica is a silicate mineral easily recognized by its perfect and conspicuous cleavage along one directional plane. The two chief varieties are the

A

**Figure 1–20** (A) Crystals of potassium feldspar. (B) A cleavage fragment of potassium feldspar.

B

colorless or pale-colored **muscovite** mica, which is a hydrous potassium aluminum silicate, and the dark-colored **biotite** mica (Fig. 1–22), which also con-

tains magnesium and iron. Both muscovite and biotite are common rock-forming silicates.

Identification of large specimens of mica is rarely a problem because of its perfect planar cleavage and the way cleavage flakes snap back into place when they are bent and suddenly released. The micas are common constituents of igneous and metamorphic

**Figure 1–21** The variety of plagioclase feldspar known as labradorite. The mineral often displays beautiful blue and gold reflections, as well as fine striations on cleavage planes. *(Courtesy of Wards National Science Establishment, Inc., Rochester, New York.)*

**Figure 1–22** Biotite mica.

**Figure 1–23** Hornblende. The black scaley flakes on the surface are biotite mica. Specimen is 6 cm long.

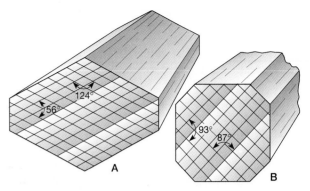

**Figure 1–24** (A) In crystals of the ferromagnesian mineral hornblende, planes of cleavage intersect at 56° and 124°. (B) In crystals of augite, planes of cleavage intersect at 87° and 93°.

rocks, in which they can be recognized by their shiny surfaces and the ease with which they can be plucked loose with a fingernail or penknife.

**Hornblende**    Hornblende (Fig. 1–23) is a vitreous, black or very dark green mineral. It is the most common member of a larger family of minerals called **amphiboles**, which have generally similar properties. Because of its content of iron and magnesium, hornblende (along with biotite, augite, and olivine) is designated a *ferromagnesian* or *mafic* mineral. Crystals of hornblende tend to be long and narrow. Two good cleavage planes are developed parallel to the long axis and intersect each other at angles of 56° and 124° (Fig. 1–24A).

**Augite**    Just as hornblende is only one member of a family of minerals called amphiboles, **augite** is an important member of the **pyroxene** family, in which

many other mineral species also occur. Like hornblende, it is a ferromagnesian mineral and thus dark colored. An augite crystal (Fig. 1–24B) is typically rather stumpy in shape, with good cleavages developed along two planes that are nearly at right angles. Thus, the cross-section of a crystal appears nearly square rather than rhombic, as in hornblende.

**Olivine**    As you might guess from its name, this glassy looking iron and magnesium silicate often has an olive green color. It is a common mineral in dark mafic rocks. When not dispersed throughout such rocks, **olivine** may occur as masses of tiny intergrown crystals (Fig. 1–25). Olivine is also an important constituent of stony meteorites. Larger crystals

**Figure 1–25** Olivine.

are cut and polished into attractive gemstones called *peridots*.

**Clay Minerals**    Clay minerals are silicates of hydrogen and aluminum with additions of magnesium, iron, and potassium. Their basic structure is similar to that of mica, but because individual flakes are extremely small, their micalike form can only be seen with the magnification provided by an electron microscope (Fig. 1–26). Unlike many of the silicate minerals already described, clay minerals form as a result of weathering of other aluminum silicate minerals such as feldspars. As a group, clays are known by their amorphous form, softness, low density, and ability to become plastic when wet. To identify individual species of clay minerals, it is necessary to employ X-ray diffraction techniques.

## Nonsilicate Minerals

Approximately 8 per cent of the earth's crust is composed of nonsilicate minerals. These include a host of **carbonates**, sulfides, sulfates, chlorides, and oxides. Among these groups, the carbonates such as **calcite** and **dolomite** are the most important. Calcite ($CaCO_3$), which is the main constituent of limestone and marble, forms in many ways. It is secreted as skeletal material by certain invertebrate animals, precipitated directly from sea water, or formed as

**Figure 1–27**    Calcite, showing the characteristic rhombohedral shape of cleavage pieces. *(Courtesy of Wards Natural Science Establishment, Inc., Rochester, N.Y.)*

dripstone in caverns. Calcite is easily recognized by its rhombohedron-shaped cleaved fragments (Fig. 1–27) and by the fact that an application of dilute hydrochloric acid on its surface will cause effervescence.

The term *dolomite* is used both for the carbonate mineral having the chemical formula $CaMg(CO_3)_2$ and for the rock composed largely of that mineral. (The alternate term *dolostone* is also occasionally used to designate dolomite rock.) In the mineral dolomite, calcium and magnesium occur in approximately equal proportions. The mineral is usually white and is slightly harder than calcite. It will not effervesce when dilute hydrochloric acid is applied to its surface, as does calcite.

**Aragonite** ($CaCO_3$) is another carbonate mineral that occurs more rarely than either calcite or dolomite. It develops in a different crystal form and has poor cleavage.

Other common nonsilicate minerals include common rock salt, or **halite** (NaCl), and **gypsum**. Halite is easily recognized by its salty taste and the fact that it crystallizes and cleaves to form cubes (Fig. 1–28). Gypsum is a soft, hydrous calcium sulfate ($CaSO_4 \cdot 2\,H_2O$). The variety of gypsum called satinspar has a fine, fibrous structure, whereas the variety known as selenite (Fig. 1–29) will split into thin plates. The finely crystalline, massive variety known as alabaster is widely used in carvings and sculpture because of its uniform texture and softness.

**Figure 1–26**    Electron micrograph of the clay mineral kaolinite. The flaky, stack-of-cards character of the clay crystals is a manifestation of their silicate sheet structure. Magnified about 2000 times. *(Courtesy of Kevex Corporation.)*

**Figure 1–28**  Halite. *(Courtesy of Wards Natural Science Establishment, Inc., Rochester, N.Y.)*

Halite and the various gypsum minerals are sometimes referred to as **evaporites** because they are often precipitated from bodies of water that have been subjected to intense evaporation.

If you were to make a list of the elements that compose the minerals described thus far, the list would be a surprisingly short one. Only eight elements make up the bulk of these minerals, and only the same eight are abundant in the earth's crust as well. As was shown in Table 1–1, these abundant elements are oxygen, silicon, aluminum, iron, calcium, sodium, potassium, and magnesium.

# Rocks

## *Rock Conversions*

We have noted that minerals can provide information about the environment in which they formed. It is therefore possible to know the origin of the rocks containing those minerals. Geologists are agreed on a fundamental division of rocks into three great families according to difference of origin. **Igneous rocks** are those that have cooled from a molten state. **Sedimentary rocks** consist of materials derived from other rocks and deposited primarily by water, but also by ice and wind. **Metamorphic rocks** are any that have been changed from previously existing rocks by the action of heat, pressure, and associated chemical activity. The changes may include a recrystallization of the previous minerals or growth of entirely new minerals.

It is important to remember that the rocks of any of the three major rock families are not immutable. The earth's crust is dynamic and ever changing. Any sedimentary or metamorphic rock may be partially

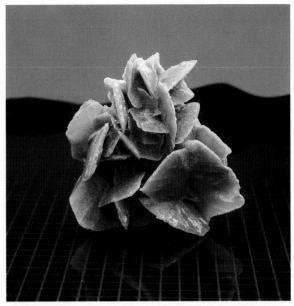

**Figure 1–29**  A cluster of selenite gypsum crystals. *(Courtesy of Wards Natural Science Establishment, Inc., Rochester, N.Y.)*

melted to produce igneous rocks, and any previously existing rock of any category can be compressed and altered during mountain building to produce metamorphic rocks. The weathered and eroded residue of any family of rock can be observed today being transported to the sea for deposition and conversion into sedimentary rocks. These changes can be incorporated into a schematic diagram that is designated the *rock cycle* (Fig. 1–30).

Although rocks are classified into groups that have had similar origin, the identification of rocks is not based on origin, but on description. For identification it is necessary to know general appearance as well as mineral composition and physical properties. **Texture** (size, shape, and arrangement of constituent minerals) and **mineral composition** are essential for identification. Inferences regarding the origin of the rock are based on geologic observations and experimentation.

## *Igneous Rocks*

Igneous rocks constitute over 90 per cent by volume of the earth's crust, although their great abundance may go unnoticed because they are extensively covered by sedimentary rocks. Much of our mountain scenery is sculptured in igneous rocks formed long

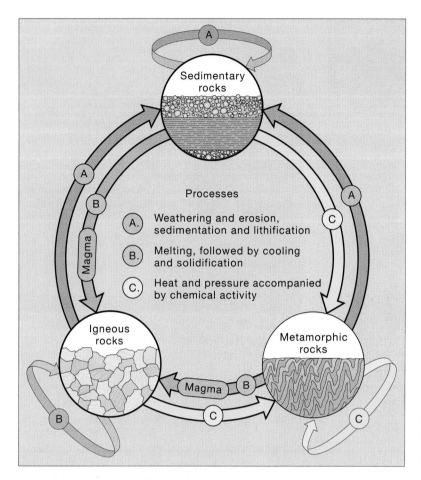

**Figure 1–30**   Geologic processes act continuously on earth to change one type of rock into another.

**Figure 1–31**   Aerial view of erupting Mount St. Helens volcano in southwestern Washington State, May 18, 1980. The plume of steam and ash rose over 20,000 meters above the crater rim. *(Photograph courtesy of U.S. Department of the Interior, Geological Survey.)*

**Figure 1–32**   Lava flow entering ocean. Kilauea Volcano, 1970. *(Photograph courtesy of D. A. Swanson, U.S. Geological Survey.)*

ago. Recent volcanic activity (Fig. 1–31) provides an often spectacular reminder of the processes that produce igneous rocks (the word igneous comes from the Latin *ignis*, meaning fire) (Fig. 1–32). It is an appropriate name for rocks that develop from cooling masses of molten material derived from exceptionally hot parts of the earth's interior. **Magma** is the term used to describe this mixture of molten silicates and gases while it is still beneath the surface. If it should penetrate to the surface, it becomes **lava** (Fig. 1–33).

## Cooling History of Igneous Rocks

Igneous rocks that formed from magma that had penetrated into other rocks and solidified before reaching the surface are termed **intrusive igneous rocks.** Very large masses of such rocks are sometimes called plutons. Their exposure at the earth's surface results from crustal uplift and erosional removal of overlying rocks. In contrast, **extrusive igneous**

**Figure 1–33**   Front of an advancing lava flow over an older flow, Kilauea Volcano, Hawaii, February 1990. The advancing flow is breaking into a jagged, rough kind of lava termed *aa*. The underlying older flow exhibits the ropy texture of *pahoehoe* lava. *(Courtesy of J. Plaut.)*

A

B

C

**Figure 1–34**   Differences in the texture of igneous rocks. (A) A coarse-grained intrusive rock called gabbro. (B) A fine-grained igneous rock known as andesite. (C) A porphyry with large phenocrysts of potassium feldspar. It is used here in the ornamental posts outside St. Paul's Cathedral in London.

**rocks** form from melts that have reached the earth's surface. This group includes rocks formed from lava erupted from volcanoes or lava that has welled out of fissures.

The grain size of igneous rocks is an index to their history of cooling (Fig. 1–34). Magmas lose heat slowly and retain water. This tends to inhibit formation of crystal nuclei. Thus, there is time and space for the growth of larger crystals around fewer nuclei. In typical intrusive rocks such as granite, dio-

rite, and gabbro (Table 1–3), the intergrown crystals are large enough to be seen readily without magnification. In contrast, extrusive igneous rocks have a finer texture in which crystals are too small to be seen with the unaided eye. The structure of such rocks reflects sudden chilling of molten silicates as they were ejected at the surface of the earth. When a lava is extruded, there is not enough time for the growth of large crystals. Furthermore, water from the melt is quickly lost to the atmosphere, and without

### Table 1–3    Common Igneous Rocks

| Coarsely Crystalline Intrusive Igneous Rocks | Approximate Extrusive Equivalents | Common Silicate Mineral Components | Average Specific Gravity |
|---|---|---|---|
| Granite | Rhyolite | Quartz<br>Potash feldspar<br>Sodium plagioclase<br>(Minor biotite, amphibole,<br>     magnetite) | 2.7 |
| Diorite | Andesite | Sodium-calcium plagioclase<br>Amphibole, biotite<br>(Minor pyroxene) | 2.8 |
| Gabbro | Basalt | Calcium plagioclase<br>Pyroxene<br>Olivine<br>(Minor amphibole, ilmenite) | 3.0 |

water, myriads of tiny crystals form rather than fewer crystals that can grow to a larger size.

One such extrusive rock is **basalt**, composed of ferromagnesian minerals and tiny rectangular grains of plagioclase feldspars. Obsidian is an extrusive rock that cooled so rapidly that there was insufficient time for crystallization; the melt therefore froze into a glass (Fig. 1–35).

If coarsely crystalline igneous rocks indicate slow cooling and finely crystalline ones indicate rapid cooling, then what would be the cooling history of a rock with large crystals immersed in a very fine grained matrix? Such rocks are said to have **porphyritic texture.** The large crystals (phenocrysts) were formed slowly at depth and were then swept upward and incorporated in the lava as it hardened at the surface.

### Composition of Igneous Rocks

The minerals in an igneous rock reflect the proportions of such elements as silicon, oxygen, aluminum, calcium, iron, magnesium, sodium, and potassium present in the melt from which the rock was formed. These eight elements combine in various ways to form feldspars, ferromagnesian minerals, micas, and quartz minerals, which are the constituents of igneous rocks. If the magma is particularly rich in silica, for example, it is likely that quartz will be present in the rock formed by solidification (crystallization) of the melt. In general, intrusive igneous rocks tend to be richer in silica than are extrusive rocks. This ob-

servation is related to the fact that greater amounts of silica in a magma increase its viscosity (resistance to flow) so that the thick liquid cannot readily work its way to the surface.

**Figure 1–35**  The volcanic glass obsidian.

**Figure 1–36** The coarse-grained intrusive rock granite. The larger, light-colored grains are potassium feldspar, the dark grains are mainly hornblende and biotite, and the grayish grains are quartz.

**Granite** (Fig. 1–36) is a silica-rich, relatively light-colored intrusive rock composed primarily of potassium feldspar, quartz (at least 25 per cent), sodium plagioclase, hornblende, and mica. It is derived from magmas so rich in silica that after all chemical linkages with metallic atoms are satisfied, enough silicon and oxygen still remain to form quartz grains. These form an interlocking network with the other minerals to create the crystalline texture of granite.

Another abundant quartz-bearing intrusive rock is called **granodiorite**. Plagioclase is the dominant feldspar in granodiorite. Quartz-bearing igneous rocks such as granite and granodiorite are loosely termed granitic rocks.

Basalt (Fig. 1–37) is a fine-grained extrusive rock derived from a low-silica melt that is rich in iron and magnesium. This composition makes extensive oxygen sharing between silica tetrahedra unnecessary, and the silica tetrahedra tend not to link up into the chain structures that make a melt viscous. Thus, basaltic lavas are relatively thin and have been observed to flow with the approximate consistency of motor oil, a fact accounting for the frequency with which it is found at the earth's surface. Since the silica percentage is low, quartz grains are rarely found in basalt.

There is one final important relationship in the chemistry of magmas and lavas. The silica-rich melts also tend to include ample quantities of potassium and sodium and therefore yield crystals of potassium

A

B

**Figure 1–37** (A) Hand specimen of basalt. (B) Thin section of basalt viewed with polarized light. Tabular crystals are plagioclase and brightly colored crystals are iron-magnesian minerals. Width of field is 2 mm.

feldspar, sodium plagioclase, and mica, along with the ubiquitous quartz. Calcium, magnesium, and iron are rather minor constituents of silica-rich magmas, but they increase in abundance in rocks deficient in silica. The low-silica rocks utilize these elements in the formation of calcium plagioclases and ferromagnesian minerals. Because of the ferromagnesians and gray plagioclase minerals, low-silica rocks tend to be dark gray, black, or green.

## Volcanic Activity

Although deciphering the history of a granitic mass is certainly intellectually stimulating, it is unlikely to evoke the feelings of awe and excitement one experiences when viewing volcanic activity. Volcanoes are basically vents in the earth's surface through which hot gases and molten rock flow from the earth's interior. The extrusions may be quiet or explosive. Quiet eruptions are exemplified by the Hawaiian volcanoes and are frequently characterized by truly enormous outpourings of low-viscosity lava (Fig. 1–38). The lava spreads widely to form the gentle slopes of a shield volcano. Explosive eruptions are caused by the sudden release of molten rock driven upward by large pockets of compressed gases. Such explosive eruptions can literally blow the volcanic cone to bits. The catastrophic eruption of Krakatoa in 1883 was heard 5000 kilometers away and was responsible for the death of 36,000 humans. Fortunately, most eruptions are not so violent.

There are, of course, all stages of volcanic activity between quiet and violently explosive. Perhaps in response to changes in the composition of the melt, some volcanoes have even been known to shift from one type of activity to another. Volcanic activity also includes successive outpourings of lava from great fissures so as to form lava plateaus that extend over thousands of square kilometers. Such fiery floods long ago produced the Columbia and Snake River Plateau as well as the Deccan Plateau of India.

By far, the most abundant kind of volcanic rock is basalt. It underlies the ocean basins, has been built into mid-oceanic ridges, and has accumulated sufficiently in such places as Hawaii and Iceland to have produced substantial land area.

How and at what depth did this great volume of basalt originate? To answer this question, it is neces-

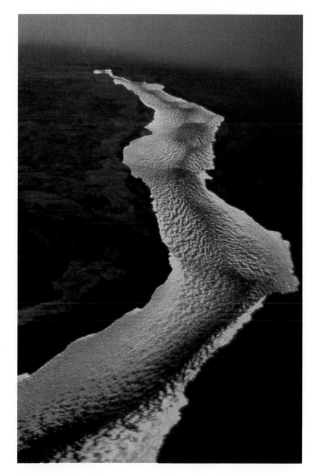

**Figure 1–38**   River of lava formed during the 1984 dual eruption of the Mauna Loa and Kilauea volcanoes, Hawaii. *(Courtesy of I. Duncan.)*

sary to refer briefly to a model of the earth's interior that has been formulated by the study of earthquake waves. The model depicts the earth's basaltic crust as a thin zone averaging about 6 km thick and overlying the mantle of denser olivine- and pyroxene-rich rocks. The boundary between the crust and the mantle is recognized by an abrupt change in the velocity of earthquake waves as they travel downward into the earth. For many years geologists believed that basaltic lavas originated from the lower part of the basaltic crust. However, several recent lines of evidence suggest that the basaltic lavas may have come from molten pockets of upper mantle material. For example, present-day volcanic activity is closely as-

COMMENTARY COMMENTARY COMMENTARY COMMENTARY
COMMENTARY COMMENTARY COMMENTARY COMMENTARY COMMENT
COMMENTARY COMMENTARY COMMENTARY COMMENTARY
MENTARY COMMENTARY COMMENTARY COMMENTARY COMMENT

## COMMENTARY

COMMENTARY COMMENTARY COMMENTARY COMMEN

## *Mount Pelée*

Ten kilometers north of the city of St. Pierre on the island of Martinique in the Caribbean Sea is a volcano known as Mount Pelée. The eruption of Mount Pelée in May of 1902 probably qualifies as one of the great natural catastrophes in human history. Eruptions of the volcano began in late March and were characterized by loud explosions and ash showers. The rim of the crater broke on May 5, and water that had been contained within the crater rushed down into the surrounding low coastal areas, incorporating mud and debris and destroying everything in its path. However, the most destructive part of the eruption was yet to come. On May 8th, following a series of deafening explosions, a great purplish cloud rose from the top of the mountain and rolled rapidly down the slopes, spreading fanlike as it engulfed the entire town of St. Pierre. Within two minutes all but two inhabitants of the town were killed by the searing heat, gases, and suffocating ash. One of the survivors was a murderer imprisoned in the town's dungeon, and the other a shoemaker whose escape simply defies explanation. In the wake of the hot gases, the entire town was left in flames (see accompanying figure).

A cloud of incandescent dust, ash, and gases such as the one that flowed down the slope of Mount Pelée is known as a **nuée ardente** (fiery cloud). A nuée ardente may take a high toll in lives whenever it passes through a heavily populated area. The extreme mobility of the cloud is caused by expanding hot gases, which contain cinders and ash. The solid particles give the cloud the density required to keep it close to the slope. Cooler air, trapped within the front of the hot moving mass, rapidly expands and increases the turbulence. It is this storm of turbulence that provides the destructive force of nuée ardente.

The eruption of Mount Pelée did not stop after the destruction of the city. A second nuée ardente swept through the desolated town on May 20, and yet another eruption on August 30 killed 2000 people in the small village of Morne Rouge.

The smoldering ruins of St. Pierre as it appeared on May 14, 1902, six days following the May 8 eruption of Mount Pelee. *(Courtesy of the Institute of Geological Sciences, London.)*

sociated with deep earthquakes that occur within the mantle far beneath the crust. It is quite likely that fractures produced by these earthquakes could serve as passages for the escape of molten material to the surface. A detailed study of earthquake shocks from particular volcanic eruptions in Hawaii indicates that the erupting lavas were derived from pockets of molten material within the upper mantle at depths of about 100 km. A weak plastic zone (called the "low-velocity zone") in the mantle appears to represent the level at which the lavas originated. The mechanism by which they developed is called partial melting. **Partial melting** is that general process by which a rock subjected to high temperature and pressure is partly melted and the liquid component is moved to another location. At the new location, the separated liquid may solidify into rocks having different composition from the parent mass. The word *partial* in the expression *partial melting* refers to the fact that some minerals melt at lower temperatures than others, and so for a time the material being melted resembles a hot slush composed of liquid and still solid crystals. The molten fraction is usually less dense than the solids from which it was derived and thus tends to separate from the parent mass and work its way toward the surface. In this way, melts of basaltic composition separated from denser rocks of the upper mantle and eventually made their way to the surface to form volcanoes.

Many complex and interrelated factors control where in the mantle partial melting may occur or even if it will occur at all. Generally, heat in excess of 1500°C is required, but the precise temperature for melting is also influenced by pressure and the water content of the rock. As pressure increases, the temperature at which particular minerals melt also rises. Thus, a rock that would melt at 1000°C near the surface will not melt in deeper zones of higher pressure until it reaches far greater temperatures. Water has an effect opposite to that of pressure, for its presence will allow a rock to start melting at lower temperatures and shallower depths than it would have otherwise. Laboratory experiments indicate that the melting of "dry" mantle rock can occur at depths of about 350 km, but that the presence of only a little water can cause partial melting and yield basaltic liquid from depths as shallow as 100 km.

Not all lavas found at the earth's surface are basaltic. Volcanoes of the more explosive type that are located at the edge of continents around the Pacific and in the Mediterranean extrude a lava called **andesite.** Andesite contains more silica than basalt does, and its lava is more viscous. This greater resistance to flow contributes to the gas containment that precedes explosive volcanic activity. Andesites are considered to be intermediate in silica content between the rocks of the continental crust and those of the oceanic crust of the earth.

Andesitic rocks may originate in more than one way. Some emplacements result from originally basaltic magmas in which minerals such as olivine and pyroxene form early and settle out, thus leaving the remaining melt relatively richer in silica. This process is called **fractional crystallization**. Other andesites may result from the melting of basaltic ocean crust and siliceous marine sediments as they descend into hot zones of the mantle. The water-rich and silica-enriched melts of andesitic composition might then rise buoyantly and erupt along volcanic island arcs. We will examine this interesting idea more fully in Chapter 6, which deals with plate tectonics.

## *Sedimentary Rocks*

### Introduction

Sedimentary rocks are simply rocks composed of consolidated sediment—particles that are the product of weathering and erosion of any previously existing rock or soil. The components of sedimentary rocks may range from large boulders to the molecules dissolved in water. Sediment is deposited through such agents as wind, water, ice, or mineral-secreting organisms. The loose sediment is converted to coherent solid rock by any of several processes: precipitation of a cementing material around individual grains, compaction, or crystallization. These processes constitute **lithification**.

The most obvious feature of sedimentary rocks is their occurrence in beds or layers called **strata.** Stratification is commonly the result of changes in the conditions of deposition that cause materials of a somewhat different nature to be deposited for a period of time. For example, the velocity of a stream might decrease, causing particles to settle out that might otherwise have stayed in suspension. In another situation, the kind of materials brought into a given depositional site by streams might change, and there would then be a corresponding change in composition of the accumulating layers.

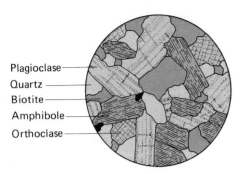

Plagioclase
Quartz
Biotite
Amphibole
Orthoclase

**Figure 1–39**  Sketch of a thin section of granodiorite as viewed through a petrographic microscope. Diameter is about 3 mm.

Sandstone, shale, and carbonate rocks (such as limestone) constitute the most abundant sedimentary rocks. **Sandstones** are composed of grains of quartz, feldspar, and other particles that are cemented or otherwise consolidated. **Shale** consists largely of very fine particles of quartz and abundant clay. The carbonates are rocks formed when the carbon dioxide contained in water combines with oxides of calcium and magnesium.

### Derivation of Sedimentary Materials

Sedimentary rocks must have originally come from the decomposition of older rocks. Commonly, the older rocks are igneous; indeed, these were once the only rocks on earth. It is therefore instructive to review the manner in which the common components of sedimentary rocks might be derived from an abundant kind of igneous rock, such as granodiorite (Fig. 1–39).

Consider, first, the quartz in the granodiorite. Quartz will persist almost unchanged during weathering. It is one of the most chemically stable of all the common silicate materials. As the parent rock is gradually decomposed, quartz grains tend to be washed out and carried away to be deposited as sand that will one day become sandstone (Fig. 1–40).

The feldspars decay far more readily than quartz. They are primarily aluminum silicates of potassium, sodium, and calcium. In the weathering process, the last three elements are largely dissolved and carried away by solutions (although some may remain in soils within clay minerals). Ultimately, they reach the sea, where they may stay in solution, or they are deposited as layers of limestone. If large quantities of

lake or sea water are evaporated, evaporites such as halite or gypsum may be formed. Of course, not all the feldspars and micas in the granodiorite necessarily decay. Some may persist as detrital grains that become incorporated into sandstones and other sediments.

Biotite is another mineral in the granodiorite source rock. Decomposition of biotite, which is a potassium, magnesium, and iron aluminosilicate, yields soluble potassium and magnesium carbonates, small amounts of soluble silica, and iron oxides. The iron oxides serve to color many sedimentary rocks in shades of brown and red.

### Variety Among Sedimentary Rocks

Sedimentary rocks are classified according to their composition and texture. The term *texture* refers to the size and shape of the individual grains and to

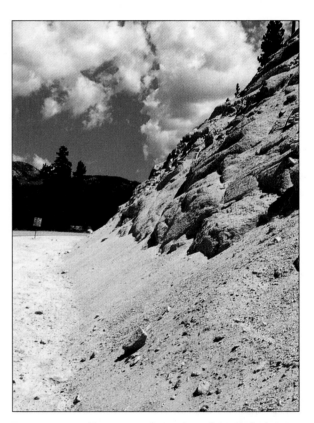

**Figure 1–40**  Exposure of weathered granodiorite showing accumulation of grains of quartz and partially decomposed feldspar resulting from the disintegration of the parent rock.

their arrangement in the rock. A rock that has a **clastic texture** is composed of grains and broken fragments (clasts) of preexisting minerals, rocks, and fossils. Other sedimentary rocks are composed of a network of intergrown crystals and therefore have a crystalline texture.

**Clastic Rocks**    The fragments of preexisting materials that compose clastic sedimentary rocks range in size from huge boulders to microscopic particles. Particle size is particularly useful in classifying these rocks, which include conglomerates, sandstones, siltstones, and shales. **Conglomerate** (Fig. 1–41) is composed of water-worn, rounded particles larger than 2 mm in diameter. The term **breccia** is reserved for clastic rocks composed of fragments that are angular (Fig. 1–42) but similar in size to those of conglomerates. In **sandstones**, grains range between $\frac{1}{16}$ mm and 2 mm in diameter (see Table 2–1). The varieties of sandstone are then subdivided mainly according to composition. **Siltstones** are finer than sandstones ($\frac{1}{16}$ mm to $\frac{1}{256}$ mm), and **shales** are composed of particles finer than $\frac{1}{256}$ mm. Shales may contain abundant clay minerals, which are flaky minerals that align parallel to bedding planes. As a result, shales characteristically split into thin slabs parallel to bedding planes. This property is termed **fissility**. Rocks lacking fissility but composed of clay-sized particles are called claystones or mudstones.

**Carbonate Rocks**    The minerals that compose carbonate rocks are calcite, $CaCO_3$ (which crystallizes

**Figure 1–42**    A polished slab of breccia. Note the angularity of the larger clasts. Area shown is approximately 8 cm across.

in the hexagonal system); aragonite, $CaCO_3$ (which crystallizes in the orthorhombic crystal system), and dolomite, $CaMg(CO_3)_2$ (which, like calcite, forms hexagonal crystals). Calcite is the predominant mineral in limestone, and dolomite predominates in the rock dolomite. Aragonite is relatively unstable and is usually converted to calcite. Carbonate rocks usually also contain variable amounts of impurities including iron oxide, clay, and clastic particles that were swept into the environment of deposition by currents.

*Limestones.*    The most abundant limestones are of marine origin and have formed as a result of precipitation of calcite or aragonite by organisms and the incorporation of skeletons of those organisms into sedimentary deposits. Inorganic precipitation of carbonate minerals may also form deposits of limestone. The importance of this process is questionable, however, because the precipitation is nearly always closely associated with photosynthetic and respiratory activities of organisms or with the release of tiny particles of aragonite upon the decay of green algae. Strictly speaking, it appears very few marine limestones are the result of direct chemical precipitation.

After the calcium carbonate has accumulated, it becomes recrystallized or otherwise consolidated into indurated rock that may be variously colored—from white, through tints of brown, to gray. Limestones tend to be well stratified (Fig. 1–43), frequently contain nodules and inclusions of chert, and

**Figure 1–41**    Conglomerate. The pebbles in this particular sample are chert in a carbonate matrix. The largest pebbles are about 2.5 cm in diameter.

**Figure 1–43**   Roadcuts exposing limestone strata are familiar to travelers on interstate highways throughout the Mississippi valley region.

are often highly fossiliferous (containing fossils). The rock may range in texture from coarsely granular to very fine grained and aphanitic.

In general, limestones consist of one of a combination of textural components such as micrite, carbonate clasts, oolites, or carbonate spar. Micrite is exceptionally fine grained carbonate mud. Carbonate clasts are sand- or gravel-sized pieces of carbonate. The most common clasts are either bioclasts (skeletal fragments of marine invertebrates) or oöids (Fig. 1–44), which are spherical grains formed by the precipitation of carbonate around a nucleus. Sparry carbonate is a clear, crystalline carbonate that is normally deposited between the clasts as a cement or has developed by replacement of calcite. These textural categories of limestone as seen through the microscope are shown in Figure 1–45. They permit classification of particular samples as micritic limestone, clastic limestone, oölitic limestone, or sparry (crystalline) limestone.

There are many varieties of limestone. Chalk is a soft, porous variety that is composed largely of extremely minute calcareous skeletal elements called *coccoliths* (Fig. 1–46). Coccoliths are secreted by marine golden brown algae. Lithographic limestone is a dense, micritic limestone once widely used as an etching surface in printing illustrations. Some limestones consist almost entirely of skeletal remains of reef corals and other frequently skeletonized marine invertebrates.

**Figure 1–44**   Oöids. Diameter of field is 2 cm.

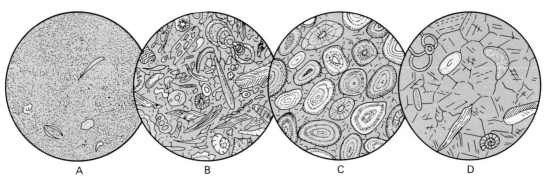

**Figure 1–45** Textures of limestones as seen in thin section under a microscope. (A) Aphanitic limestone or micrite. (B) Bioclastic limestone with fine-grained sparry calcite as cement. (C) Oölitic limestone. (D) Sparry or crystalline limestone.

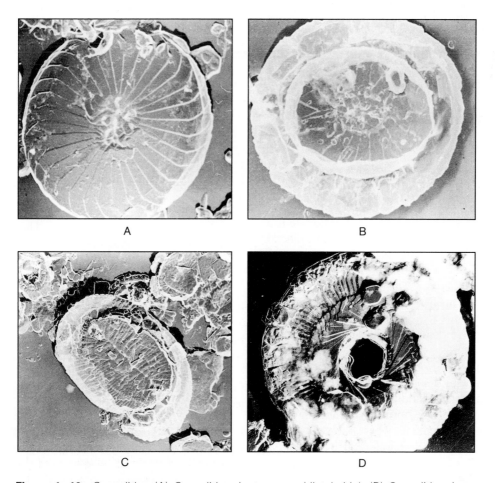

**Figure 1–46** Coccoliths. (A) *Coccolithus leptoporous* (distal side). (B) *Coccolithus leptoporous* (proximal side). (C) *Helicosphaera carteri.* (D) *Cyclococcolithus* sp. For scale, D is 22 micrometers in diameter.

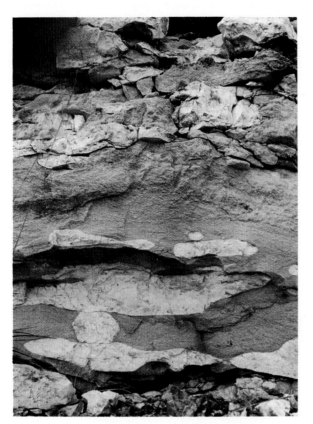

**Figure 1–47** White nodular chert in tan limestone. Fern Glen Formation west of St. Louis, Missouri.

*Dolomite.* As we have already noted, dolomite is a rock composed of the calcium–magnesium mineral of the same name. Like limestone, it occurs in extensive stratified sequences, and is not easily distinguishable from limestone. The usual field test for distinguishing dolomite from limestone is to apply cold, dilute hydrochloric acid. Unlike limestone, which bubbles readily, dolomite will effervesce only slightly, if at all.

The origin of dolomites is somewhat problematic. The mineral dolomite is not secreted by organisms during shell building. Direct precipitation from sea water does not normally occur today, except in a few environments where the sediment is steeped in abnormally saline water. Such an origin is not considered adequate to explain the thick sequences of dolomitic rock commonly found in the geologic record. The most widely believed theory for the origin of dolomites is that they result from partial replacement of calcium by magnesium in the original calcareous sediment. However, it is not known for

how long or at what time in the history of the rock this dolomitization occurs.

**Other Sedimentary Rocks** *Chert.* We have previously mentioned a form of microcrystalline quartz called chert ($SiO_2$), noting its occurrence as nodules in limestones (Fig. 1–47). The origin of these nodules is still being debated among petrologists, although the majority believe that they form as replacements of carbonate sediment by silica-rich sea water trapped in the sediment. Some cherts occur in areal-extensive layers and thus qualify as monomineralic rocks. These so-called bedded cherts are thought to have formed from the accumulation of the siliceous remains of diatoms and radiolaria and from subsequent reorganization of the silica into a microcrystalline quartz. Silica from the dissolution of volcanic ash is believed to enhance the process; indeed, many bedded cherts are found in association with ash beds and submarine lava flows.

*Evaporites.* As indicated by their name, evaporites are chemically precipitated rocks that are formed as a result of evaporation of saline water bodies. Only about 3 per cent of all sedimentary rocks consist of evaporites. Evaporite sequences of strata are composed chiefly of such minerals as gypsum, anhydrite, halite, and associated calcite and dolomite. Extensive ancient deposits of evaporites are currently being commercially worked in Michigan, Kansas, Texas, New Mexico, Germany, and Israel. The conditions required for precipitation of thick sequences of evaporites include warm, relatively arid climates and a physiographic situation that would provide periodic additions of sea water to the evaporating marine basin. In the Gulf coastal region of the United States, as a result of the pressure of overlying rocks, deeply buried deposits of salt have flowed plastically upward to form underground domes of salt. In the process, the salt has arched overlying strata, thereby producing structures in which petroleum could collect (Fig. 1–48).

*Coal.* Coal is a carbonaceous rock resulting from the accumulation of plant matter in a swampy environment combined with alteration of that plant tissue by both biochemical and physical processes until it is converted to a consolidated carbon-rich material. The biochemical and physical changes may produce a series of products ranging from peat and lignite to bituminous and anthracite coal. For coal to

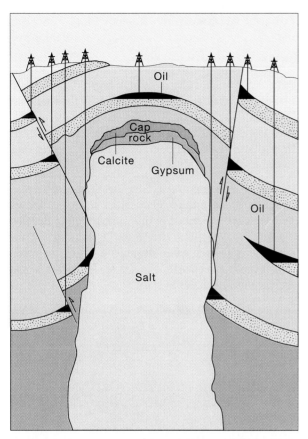

**Figure 1–48** Salt dome, illustrating possibilities for oil entrapment in domelike structures (top center), by faults, and by pinchout of oil-bearing strata.

form, plant tissue must be accumulated under water or be quickly buried because vegetable matter, if left exposed to air, is readily oxidized to water and carbon dioxide. With underwater accumulation or quick burial of plant material, a major part of the carbon can be retained.

## *Metamorphic Rocks*

### All Things Change

Sir Charles Lyell recognized that igneous or sedimentary rocks, if subjected to high temperature, pressure, and the chemical action of solutions and gases, can be altered to quite different kinds of rocks. Lyell employed the term *metamorphism* (from the Latin *metamorphosis*, meaning change of form) to describe this process. It is still used today to describe alterations in rocks brought about by physical or chemical changes in the environment that are intermediate between those that result in igneous rocks and those that produce sedimentary rocks. Any previously existing rock may be converted to a metamorphic rock, and the changes primarily involve recrystallization of minerals in the rock while it remains in the solid state. In the process of recrystallization, the textural characteristics of the parent rock may be changed while at the same time new minerals develop that are stable under the new conditions of pressure and temperature. New elements need not be introduced; instead, those that are already present are incorporated into different and often denser minerals. Variations in heat and pressure may result in different kinds of metamorphic rocks, even from the same parent material.

Most metamorphic rocks exhibit a layering called **foliation**, which results from the parallel alignment of mineral grains. Whether this foliation is very fine or coarse depends upon the size and shapes of the constituent minerals. There are a few nonfoliated metamorphic rocks as well, for many marbles and quartzites do not exhibit foliation.

### Metamorphism

Alterations of rock immediately adjacent to igneous intrusions constitute **contact metamorphism** (Fig. 1–49). The changes that occur in the intruded rock are largely the result of high temperatures and the emanation of chemically active vapors that accompany igneous intrusions. Such factors as the size of the magmatic body, its composition and fluidity, and the nature of the intruded rock also influence the kind and degree of contact metamorphism. Important ore deposits are commonly situated in the metamorphosed rock surrounding the intrusives. Examples of such deposits include magnetite and copper ores in metamorphic zones around granite intrusives in the Urals, central Asia, the Appalachian Mountains, Utah, and New Mexico.

**Regional** or **dynamothermal metamorphism** is a type of rock alteration that is areally extensive and occurs under the conditions of great confining pressures and heat accompanying deep burial and mountain building. In a subsequent chapter, we will discuss how rocks deposited in crustal troughs adjacent to continents may be compressed into mountain systems and thus be regionally metamorphosed. Metamorphic index minerals known to form under spe-

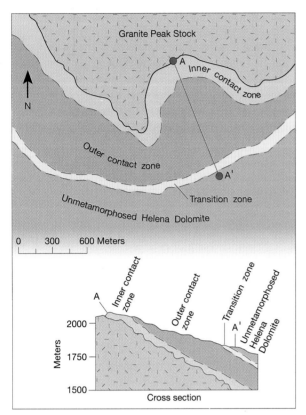

**Figure 1–49** The Granite Peak aureole. An aureole is the zone of contact metamorphism surrounding an igneous intrusion. This metamorphic aureole is developed around a granitic stock that has intruded into dolomite country rock. The intensity of metamorphism diminished outward from the granite margin, and particular assemblages of metamorphic minerals occur within each contact metamorphic zone. *(Simplified from W. G. Melson. American Mineralogist 51:404, 1966, Fig. 1.)*

cific temperature and pressure conditions are used to decipher the history of growth of these ancient mountainous regions, even when only the roots of the ranges remain.

## Kinds of Metamorphic Rocks

As any rock can be metamorphosed in a number of different ways, there are hundreds of different kinds of metamorphic rocks. However, for our purposes, we need only consider several that occur extensively at the earth's surface. It is convenient to divide metamorphic rocks into two groups based on the presence or absence of foliation.

**Foliated Metamorphic Rocks**   *Slate.*   In slate, the foliation is microscopic and caused by the parallel alignment of minute flakes of silicates with sheet structure (Fig. 1–50). The planes of foliation are quite smooth, and the rock may be split along these planes of "slaty cleavage." The planes of foliation may lie at any angle to the bedding in the parent rock, and this characteristic helps to differentiate slate from dense shale. Slate is derived from the regional metamorphism of shale.

*Phyllite.*   The texture in phyllite is also very fine, although some grains of mica, chlorite, garnet, or quartz may be visible. Phyllite surfaces often develop a wrinkled aspect and are more lustrous than slate. Phyllite represents an intermediate degree of metamorphism between slate and schist. The parent rocks are commonly shale or slate.

*Schist.*   The platy or needle-like minerals in schist are sufficiently large to be visible to the unaided eye;

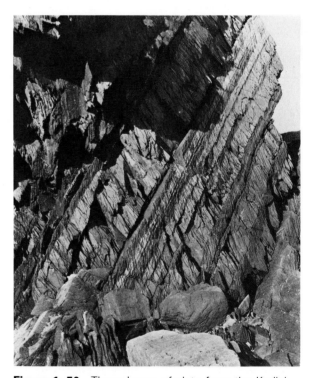

**Figure 1–50** These layers of slate from the Kodiak Peninsula of Alaska exhibit well-developed slaty cleavage. Note that the cleavage planes are vertical and are not parallel to bedding, as in the fissility of shale. *(Courtesy of U.S. Geological Survey.)*

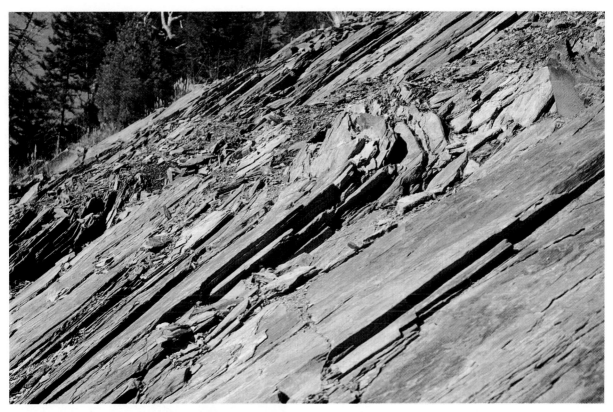

**Figure 1–51** Steeply inclined mica schists splitting apart along foliation planes. Great Smoky Mountain National Park, North Carolina-Tennessee.

**Figure 1–52** Coarse (gneissic) foliation developed in a quartz–feldspar–biotite gneiss.

the minerals tend to be segregated into distinct layers. Schists are named according to the most conspicuous mineral present. Thus, there are mica schists (Fig. 1–51), amphibole schists, chlorite schists, and many others. Shales are the usual parent rocks for schists, although some are derived from fine-grained volcanic rocks.

*Gneiss.* This is a coarse-grained, evenly granular rock. Foliation (Fig. 1–52) results from segregation of minerals into bands rich in quartz, feldspar, biotite, or amphibole. Foliation is coarse and appears less distinct than in schist. High-silica igneous rocks and sandstones are the usual parent rocks for gneisses.

**Nonfoliated Metamorphic Rocks** *Marble.* A fine to coarsely crystalline rock, marble (Fig. 1–53) is composed of calcite or dolomite and therefore is relatively soft. (It can be scratched with steel.) Marble is derived from limestone or dolostone.

**Figure 1–53**  Marble. On close examination one can discern the lustrous cleavage surfaces of the calcite crystals. Width of specimen is about 9.0 cm.

*Quartzite.*  A fine-grained, often sugary-textured rock, quartzite (Fig. 1–54) is composed of intergrown quartz and therefore is very hard. The rock will break through constituent grains; it may be any color. Quartzite is derived from quartz sandstone.

*Greenstone.*  A dark green rock, greenstone has a texture so fine that mineral components, except for scattered larger crystals, cannot be seen without magnification. It is derived by the low-grade metamorphism of low-silica volcanic rocks such as basalt.

## Historical Significance of Metamorphic Rocks

We have noted that the conditions for metamorphism are developed in regions that have been subjected to intense compressional deformation. Such regions of the earth's crust either now have, or once had, great mountain ranges. Thus, where large tracts of low-lying metamorphic terrane are exposed at the earth's surface, geologists conclude that crustal uplift and long periods of erosion leveled the mountains. Metamorphic rock exposures at many localities across the eastern half of Canada represent the truncated foundations of ancient mountain systems.

From studies of the mineralogic composition of metamorphic rocks, it is often possible for geologists to reconstruct the conditions under which the rocks were altered and then to make inferences about the directions of compressional forces, pressures, temperatures, and the nature of parent rocks. Investiga-

tors are aided in these studies by the knowledge that specific metamorphic minerals form and are stable within finite limits of temperature and pressure. Maps of **metamorphic facies**, or zones of rocks that formed under specific conditions, can be constructed. Commonly, such maps delineate broad bands of metamorphic rocks, each of which formed under sequentially more intense conditions of pressure and temperature. Imagine a terrane that was once underlain by a thick sequence of calcareous shales, was subjected to compression so as to produce mountains, and then experienced loss of those mountains by erosion. One might then begin a traverse across this eroded surface on unmetamorphosed shales that were not involved in the mountain building (Fig. 1–55). These shales would contain only unaltered sedimentary minerals. Progressing farther toward the area of most intense metamorphism, one might see that the shales had given way to slates bearing the green metamorphic mineral chlorite. Still farther along the traverse, schists containing such intermediate-grade metamorphic minerals as biotite and garnet would appear. Finally, one might come upon coarsely foliated schists containing kyanite, staurolite, and sillimanite—minerals that develop under high temperature and pressure.

Metamorphic minerals do not always appear in the orderly fashion indicated in the preceding example. Depending on the nature and depth of metamor-

**Figure 1–54**  Hand specimen of an attractive green quartzite.

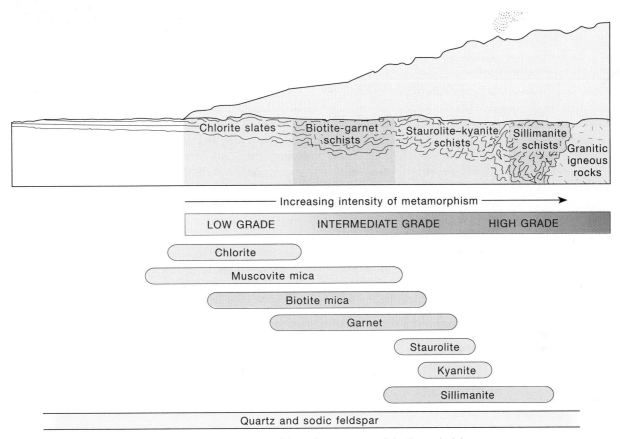

**Figure 1–55** Changes in the mineralogic composition of a terrane originally underlain by shales following regional metamorphism.

phism, temperature may increase at a faster or slower rate than does pressure, resulting in the growth of different index minerals. It is also possible for a previously metamorphosed terrane to experience a second, less severe episode of metamorphism. In such cases, a lowering of metamorphic grade might result. Yet another factor that influences the kinds of minerals produced is the mineralogic composition of the parent rock and the amount of water present during metamorphism.

Metamorphic rocks may contain a wealth of historical information. A marble containing flakes and veins of chlorite that is dated as 1 billion years old tells the geologist many things. It records the existence of an ancient, somewhat clayey limestone that experienced a relatively low level of metamorphism. Because limestone was the parent rock, the geologist may infer that conditions in the earth's atmosphere and ocean a billion years ago were similar to those that permit carbonate deposition today.

## SUMMARY

Simply stated, geology is the study of all naturally occurring processes, phenomena, and materials of the past and present earth. Historical geology is that branch of the science concerned particularly with

decoding the rock record of the earth's long history. In the last few decades, advances in technology have added immensely to the store of geologic knowledge. Interpretation of the new data requires an exceptional understanding not only of geology but also of physics, chemistry, mathematics, and biology. However, the historical inferences that are drawn from the data are frequently derived from the fundamental geologic and paleontologic concepts introduced by Nicolaus Steno (superposition), James Hutton (uniformitarianism), Charles Lyell (cross-cutting relationships, inclusion), William Smith (correlation), Charles Darwin (organic evolution), and others. These scientists formulated the principles by which geologists are able to determine the relative age of a rock outcrop, its history of deposition and deformation, and its spatial and chronologic relationship to strata in other regions of the earth.

Rocks are the materials of which the earth is composed. The rocks are themselves composed of minerals, and minerals in turn are constructed of chemical elements. Rocks and minerals are the materials from which geologists make interpretations of ancient environments and past geologic happenings. Silicate minerals are the most important of the rock-forming minerals. Quartz, mica, feldspar, hornblende, augite, and olivine are silicates that were initially crystallized from molten material. Clay minerals, calcite, dolomite, and certain varieties of quartz (such as chert and flint) are formed by processes of weathering and precipitation at temperatures that prevail at the earth's surface.

Rocks are aggregates of minerals. Igneous rocks are those that have cooled from a molten silicate material. Consolidated materials derived from other rocks by weathering, erosion, transport, and deposition constitute sedimentary rocks. Metamorphic rocks are those altered from older rocks that have been subjected to heat, pressure, and chemically active solutions.

## QUESTIONS FOR REVIEW

1. Summarize the evidence indicating that the Mediterranean Sea was once a deep desert basin.

2. Describe the general procedures followed by geologists and other scientists as they attempt to solve particular problems.

3. Why is the concept of uniformitarianism useful in understanding the history of the earth?

4. How did William Smith demonstrate the equivalency of strata from one location to another in England?

5. What biologic mechanism did Charles Darwin propose that would account for the biologic succession of fossils observed by William Smith?

6. What criteria might a geologist seek in distinguishing the top from the bottom of strata that had been forced into vertical alignment by mountain-building forces?

7. If a granitic intrusion cuts into or across sedimentary strata, which is older, the granite or the strata?

8. What are the three major categories of rocks? How does a rock differ from a mineral?

9. Why are silicate minerals so important in geology?

10. What type of igneous rock best approximates the composition of the oceanic crust? The continental crust?

11. What differences in texture might exist between an igneous rock that cooled slowly and one that cooled rapidly?

12. What inferences can be drawn from the color of igneous rocks?

13. What are the differences between the rocks limestone and dolomite? How might limestone be altered to dolomite?

14. Of the mineral groups discussed in this chapter, which are particularly characteristic of sedimentary rocks?

15. What is the origin of foliation in metamorphic rocks? List the foliated metamorphic rocks in order of increasing coarseness.

16. What historical interpretation can be made about regions in which extensive tracts of metamorphic and igneous rocks are exposed at the earth's surface?

17. Formulate a definition of stratification. What causes stratification in a sedimentary rock?

## TERMS TO REMEMBER

amphibole
andesite
aragonite
augite
basalt
breccia
calcite
carbonate
chert
cleavage
conglomerate
contact metamorphism
dolomite

evaporite
extrusive igneous rock
feldspar
fissility
foliation
fractional crystallization
gneiss
granite
granodiorite
greenstone
gypsum
halite
hornblende

igneous rock
intrusive igneous rock
lava
limestone
magma
marble
metamorphic rock
mica
mineral
olivine
partial melting
phyllite
porphyritic texture

pyroxene
quartz
quartzite
regional metamorphism
sandstone
schist
sedimentary rock
shale
siltstone
slate
stratigraphy
unconformity
uniformitarianism

## SUPPLEMENTAL READINGS AND REFERENCES

Adams, F. D. 1938. *The Birth and Development of the Geological Sciences.* Baltimore: Williams & Wilkins Co. (Reprinted by Dover Press, 1954.)

Blatt, H., Middleton, G., and Murray, R. 1972. *Origin of Sedimentary Rocks.* Englewood Cliffs, N. J.: Prentice-Hall, Inc.

Dietrich, R. V., and Skinner, B. J. 1979. *Rocks and Minerals.* New York: John Wiley.

Eiseley, L. C. 1959. Charles Lyell, *Sci. Am.* 201(8):98–106. (Offprint No. 846. San Francisco: W. H. Freeman Co.)

Eiseley, L. 1961. *Darwin's Century.* Garden City, N. Y.: Doubleday & Co.

Ernst, W. G. 1969. *Earth Materials.* Englewood Cliffs, N. J.: Prentice-Hall, Inc.

Geikie, A. 1905. *The Founders of Geology* (2nd ed.). New York: Macmillan. (Reprinted by Dover Press, 1954.)

Hallam, A. 1983. *Great Geologic Controversies.* Oxford, England: Oxford Univ. Press.

Hsu, K. J. 1972. When the Mediterranean dried up. *Sci. Am.* 227(6):26–36.

Mather, K. F., and Mason, S. L. 1939. *A Source Book in Geology, 1400–1900.* Boston: Harvard Univ. Press.

Pough, F. H. 1960. *A Field Guide to Rocks and Minerals.* Cambridge, Mass.: Houghton Mifflin Co.

Zoltai, T., and Stout, J. H. 1984. *Mineralogy: Concepts and Principles.* Minneapolis: Burgess.

INTERVIEW INTERVIEW INTERVIEW INTERVIEW INTERVIEW
INTERVIEW INTERVIEW INTERVIEW INTERVIEW INTER
INTERVIEW INTERVIEW INTERVIEW INTERVIEW
VIEW INTE INTERVIEW INTERVIEW INTER
INTERVIEW INTERVIEW INTERVIEW INTERVIEW
VIEW INTERVIEW INTERVIEW INTERVIEW INTE

# PRESTON CLOUD

## INTERVIEW

Preston Cloud was born in 1912 in West Upton, Massachusetts. He completed his undergraduate work at George Washington University (1938) and his graduate work at Yale University (1940). Cloud taught at the Missouri School of Mines, Harvard University, University of Minnesota, University of California–Los Angeles, and University of California–Santa Barbara. He also worked extensively for the U.S. Geological Society on such projects as geologic mapping and field studies of calcareous rocks, studies of Pacific tropical reefs and islands, marine geology of the Bahamas, petroleum reconnaissance of northeastern Spain, and studies of the ocean's continental shelves and coastal zones.

Cloud's best known research was on the interactions between early biospheric, atmospheric, and crustal evolution in which he focused on the origin and early evolution of life, and the influence of life on the development of the atmosphere. By incorporating studies of invertebrate paleontology, marine ecology and paleoecology, and marine sedimentation and geochemistry, Cloud developed a modern integrated approach to the study of life processes in geology called biogeology. He was the first to advance and defend the now-accepted view that metazoa evolved rapidly and from a variety of different ancestors beginning in late Proterozoic time. He shared in demonstrating the existence of a long premetazoan record of microbial life, and showed how that life influenced surface processes on the earth during the Precambrian.

Cloud served on the National Academy of Science Committee on Science and Public Policy, the council of the National Academy of Science and its executive committee, the governing board of the National Research Council, and the National Research Council Assembly of Mathematical and Physical Sciences. During his long and active career, he chaired various national committees concerned with resources and environment and was also chairman of the National Academy of Science Section of Geology. He was elected to the National Academy of Sciences, the American Academy of Arts and Sciences, and the American Philosophical Society. Among his many honors are five medals, including the Paleontological Society Medal and the Penrose Medal of the Geological Society of America. Preston Cloud died on January 16, 1991. He will be remembered for his extraordinary geologic versatility and his concern for the earth and its inhabitants.

48

INTERVIEW INTERVIEW INTERVIEW INTERVIEW INTERVIEW
VIEW INTERVIEW INTERVIEW INTERVIEW INTERVIEW INTER
INTERVIEW INTERVIEW INTERVIEW INTERVIEW INTERVIEW
VIEW INTERVIEW INTERVIEW INTERVIEW INTERVIEW INTER
INTERVIEW INTERVIEW INTERVIEW INTERVIEW INTERVIEW
VIEW INTERVIEW INTERVIEW INTERVIEW INTER

**What originally got you interested in geology?**

From boyhood I have loved the out-of-doors, all aspects of natural history, and history itself. In the university I was torn between biology, chemistry, and geology. Geology offered the broadest prospects. To understand the history of our planet appealed to me as an appropriate goal.

**Tell us about your research specialties and more recent interests.**

In geology I have been more oriented toward problem-solving than toward subject matter or methods. I followed my heart into problems that appealed to me at the time and that seemed amenable to solution given the necessary working knowledge of related subject matter and methods. I went from invertebrate paleontology and stratigraphy into sedimentology and marine geology and finally into deep earth history, employing a strategy I refer to as biogeology. In biogeology, all relevant variables are brought to bear on the study of life processes in geology and their connections with atmospheric, hydrospheric, and lithospheric evolution.

My most recent research interests have focused on the conditions under which life began on the primitive earth and the role played by early life in the evolution of the atmosphere. I have been continually engaged in the study of Archean and Proterozoic geologic history.

Having long believed that the future belongs to posterity and that the passing of the baton is best facilitated by a readable account of existing knowledge, I committed my last active decade to the preparation of a summary of current outlooks on the whole of earth history. The result of my

efforts in this regard is to be found in the chapters of *Oasis in Space,* published in 1988 by W. W. Norton.

**How has the study of the history of the earth changed over the past several decades?**

The science of earth history since mid-century has broadened, deepened, become more quantitative,

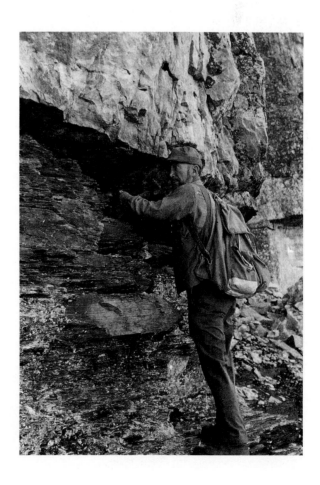

and incorporated far-reaching new applications and variables—plate tectonics, meteoritics, paleomagnetics, paleoclimatology, and paleoceanology. Earth scientists have become increasingly involved in the study of our neighboring planets and in the problems relating to the environment, conservation, and the management of natural resources.

**What advice would you offer an undergraduate who is thinking of becoming a geology major? A professional geologist?**

Before deciding whether or not to major in geology, it is advisable to take an introductory course or two. Should you then choose a different major, the geology learned will still enrich the joys of travel and sharpen your perceptions of current international and environmental issues. Should you instead decide on a career in geology, plan to go on to graduate school, preferably in a region different enough from that of your undergraduate work that you will experience new geology and encounter new outlooks. And be sure to get as much chemistry, physics, and biology as time allows, as well as math up through differential equations. In case you choose a career in research, a good rule of operation is to be very careful

about your facts but bold with the hypotheses you draw from them or relate them to. For without the first, you have no foundation, and without the second, even a mountain of evidence alone asks no questions, introduces no targets, and fails to advance understanding. Indeed, the value of a hypothesis rests not so much on its ultimate validity as on the degree to which it stimulates reflection and research among others, that is, its testability. It is no disgrace to be mistaken, only to refuse revision when mistakes are found.

**What good is an understanding of earth science to someone who does not plan to be a geologist?**

Understanding how our planet works and how it has worked to create the human habitat and the resources that sustain us is a basic element of a liberal arts education and of human culture. It enlightens our understanding of international and environmental issues and enhances enjoyment of travel on and above earth's ever variable surface. Its scope is far reaching. All landscapes become interesting. Building and decorative stones tell stories of their age and formation. Mysterious natural events take their places in the natural order

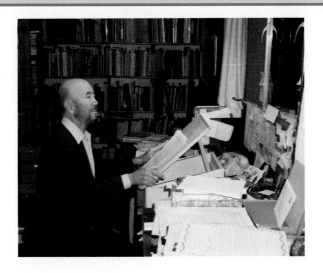

of things when we learn about their geologic origins.

**Has your work changed any of the traditional views of the earth sciences? If so, how?**

Let the words of an active successor speak to the influence of my work in the field (Knoll, A. H., *American Journal of Science,* v.190A, 1990). He writes that "Cloud's enduring contribution has been to establish the very way in which we think about early evolution. He has insisted that traditional paleontology be married to geology and geochemistry so that early biological history can be understood in the context of the developing planetary surface." The main product of that linkage is a now repeatedly tested and widely accepted model of an initially oxygen-deficient earth that underwent a dramatic rise in free atmospheric oxygen beginning about 2 billion years ago and reached present levels only in Paleozoic time. This new paradigm explains many previously puzzling geologic observations, including the stratigraphic distribution of red beds, detrital uraninite deposits, and banded iron formations. It denies the previously widespread belief in a long pre-Phanerozoic metazoan fauna, while concurrently favoring a prevalence of microbial life as far back as the oldest sedimentary rocks.

*Author's Note.* In his book, *Oasis in Space,* Professor Cloud closes with an eloquently expressed proposal for students everywhere. "If you ask, as thinking young people sometimes do, 'what is the purpose of life?' I would propose a goal within reach of all: to live it with as much grace and integrity as possible, to enjoy and improve it while you have it, and to leave the world no worse for having been there."

Dolomite of the Ordovician Joachim Formation forms a resistant ledge over massive quartz sandstone of the St. Peter Formation. The cave was excavated for the extraction of the sandstone used in making glass.

# The Historical Significance of Sedimentary Rocks

## 2

> In the high mountains, I have seen shells. They are sometimes embedded in rocks. The rocks must have been earthy materials in days of old, and the shells must have lived in water. The low places are now elevated high, and the soft material turned into hard stone.
>
> Chu-Hsi, A.D. 1200

## The Sedimentologic Archives

Ever since the earth has had an atmosphere and hydrosphere, sediments have been accumulating on its surface. The sediments, now formed into sedimentary rocks, contain features that tell us about the environment in which they were deposited. By interpreting these bits of evidence in successively higher strata, one can decipher the geologic history of a part of the earth.

### The Tectonic Setting

Many factors determine the kind of sedimentary rock that will be formed in a particular area. Among these are the method of transport of sedimentary materials; the physical, chemical, and biological processes operating in the place of deposition; the climate under which processes of weathering and erosion take place; and the changes that occur to sediment as it is being converted to solid rock. On a grander scale, the characteristics of an entire assemblage of rocks is also influenced by the tectonics of the region in which deposition takes place. **Tectonics** may be defined as the study of the deformation or structural behavior of a large area of the earth's crust over a long period of time. For example, a region may be tectonically stable, subsiding, rising gently, or more actively rising to produce mountains and plateaus.

Where a source area has recently been compressed and uplifted, an abundance of coarse sediment (forming sandstones and conglomerates) derived from the rugged upland source area will be supplied to the basin. In the geologic past, such a tectonic setting has resulted in the accumulation of great "clastic wedges" of sediment that thinned and became finer away from the former mountainous source area. In other tectonic settings of the past, the source area has been stable and topographically more subdued, so that finer particles and dissolved solids became the most abundant components being carried by streams.

The tectonic setting influences not only the size of clastic particles being carried to sites of deposition, but also the thickness of the accumulating deposit. For example, if a former marine basin of deposition had been provided with an ample supply of sediment and was experiencing tectonic subsidence (sinking), enormous thicknesses of sediments might accumulate. Over a century ago, James Hall (1811–1898), an eminent American geologist and paleontologist, recognized that the thick accumulations of shallow water sedimentary rocks in the Appalachian region required that crustal subsidence had accompanied deposition. His reasoning was quite straightforward. It was easy to visualize filling a basin that was 40,000 ft deep with 40,000 ft of sediment. However, where fossils indicated a basin never more than several hundred feet deep, the only way to get tens of thousands of feet of sediment into it would be to have subsidence occurring simultaneously with sedimentation.

In a marine basin of deposition that is stable or subsiding very slowly, the surface on which sedimentation is occurring is likely to remain within the zone of wave activity for a long time. Wave action and currents will wear, sort, and distribute the sediment into broad, blanketlike layers. If the supply of sediment is small, this type of sedimentation will continue indefinitely. Should the supply of sediment become too great for currents and waves to transport, however, the surface of sedimentation would rise above sea level and deltas would form.

It is also important to consider the tectonic framework of entire continents as well as of particular areas. The principal tectonic elements of a continent are *cratons* and *orogenic belts* (Fig. 2–1). Cratons have two components: large areas of exposed ancient crystalline rocks called *shields*, as well as surrounding regions called *platforms* in which these ancient rocks are covered by flat-lying or gently warped layers of sedimentary rocks. Cratons have been undisturbed by mountain-building events since Precambrian time (about 570 million years ago). They comprise the stable interiors of continents.

Orogenic belts are elongate regions that border the craton and have been deformed by compressional forces since Precambrian time. Today, young orogenic belts are recognized by their high frequency of earthquakes and volcanic eruptions, whereas older belts are marked by severely deformed strata, crustal displacements, metamorphic terranes, and huge bodies of intrusive igneous rocks.

The history of the earth provides many examples of marginal orogenic belts that were once sites for the accumulation of great thicknesses of sediment. Such elongate tracts of sedimentation are found today along the continental shelves and rises. Following long episodes of deposition, the sediments along these tracts may be deformed as a result of an encounter with an oncoming tectonic plate, and a mountain range may form where there was once only a sedimentary basin. These events and their causes will be examined in Chapter 6.

The tectonic setting of deposition largely determines the nature of sedimentary deposits. Conversely, the kind of tectonic setting often can be inferred from a rock's textural and structural features and from its color, composition, and fossils. The tectonic and historical significance of some of these characteristics of sedimentary rocks will be frequently examined in the pages to follow.

## Environments of Deposition

An *environment of deposition* refers to all the physical, chemical, biological, and geographic conditions under which sediments are deposited. Each environment of deposition is characterized by geographic and climatic conditions that modify or determine the properties of sediment that is deposited within it. Thus, the type of sediment becomes the key to the environment. Some sediments, such as chemical precipitates in water bodies, are solely the products of their environment of deposition. Their component minerals formed and were deposited at the same place. Other sediments consist of materials formed elsewhere and transported to the site of deposition. Geologists are keenly interested in the sediment of today's depositional environments because the features they find in the deposits of these modern areas can also be seen in ancient sedimentary rocks. Comparing present-day sedimentary deposits to old sedimentary rocks permits one to reconstruct conditions in various parts of the earth as they were hundreds of millions of years ago.

The three major environments of deposition are designated *marine*, *continental*, and *transitional*. Each of these contains a number of subenvironments. For example, the marine environment can be divided into shallow, deep, and continental slope. The shallow marine realm corresponds to the oceanic topographic regions called continental shelves. **Continental shelves** (Fig. 2–2) are nearly flat, smooth surfaces that fringe the continents in widths that range from only a few kilometers to about 300 km and depths that range from low tide to about 200 meters. In a geologic sense, these shallow areas are not part of the oceanic crust but resemble the continents in their structure and composition. They are, in fact, the submerged edges of the continents and have a readily apparent continuity with the coastal plains. The outer boundaries of the shelves are defined by a marked increase in slope to greater depths. The smoothness of parts of the continental shelves seems to have been produced in part by the action of waves and currents during the last Ice Age. At that time, sea level was periodically lowered by as much as 140 meters as a result of water being locked

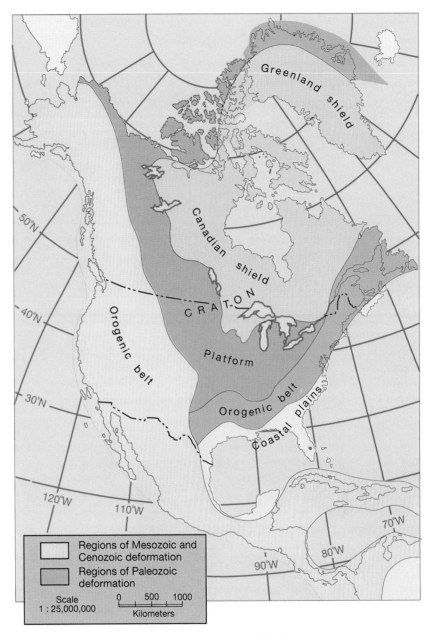

**Figure 2–1** The craton and orogenic belts of North America.

in glacial ice. Waves and currents sweeping across the shelves shifted sediment into low places and generally leveled the surface.

For geologists specializing in the study of sedimentary rocks, the continental shelves hold great interest. All of the sediment eroded from the continents and carried to the sea in streams must ultimately cross the shelves or be deposited on them.

Many factors influence the kind of sediment deposited on the shelves, including the nature of the source rock on adjacent land masses, the elevation of source areas, the distance from shore, and the presence of carbonate-secreting organisms. Because of wave turbulence and currents, sediment deposited in the shallow marine environment tends to be coarser than material laid down in the deeper parts of the

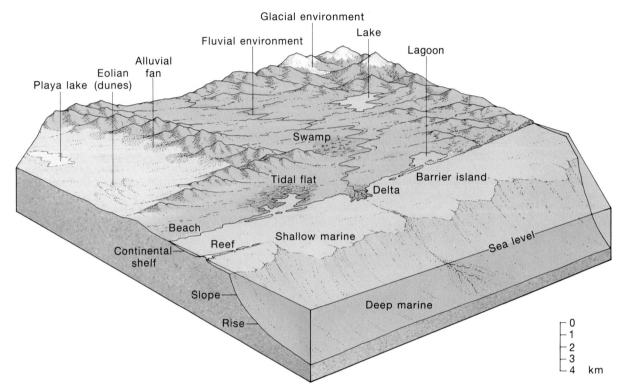

**Figure 2–2**    Diagram illustrating some marine and continental environments of deposition.

**Figure 2–3**    Marine life flourishes in many areas of the continental shelves, for light is required for the growth of marine plants and plants are the basic components of the food chain that supports marine animals. *(Courtesy of L. E. Davis.)*

ocean. Sand, silt, and clay are common. Where there are few continent-derived sediments and the seas are relatively warm, lime muds of biochemical origin may be the predominant sediment. Coral reefs are also characteristic of warm, shallow seas. They remind us of the enormous biological importance of the shallow marine environment. Over most of this realm, sunlight penetrates all the way to the sea floor. Algae and other forms of plant life proliferate. Here one finds the "pastures of the sea" on which, directly or indirectly, a multitude of swimming and bottom-dwelling animals are dependent (Fig. 2–3).

Those areas of the ocean floor that extend from the seaward edge of the continental shelves down to the ocean depths are named the **continental slopes.** Physiographic diagrams of the ocean floor are usually drawn with a large amount of vertical exaggeration, so that the continental slopes appear as steep escarpments. Actually, the inclination of the surface is only 3° to 6°. From the sharply defined upper boundary of the continental slope, the surface of the ocean floor drops to depths of 1400 to 3200 meters.

COMMENTARY *COMMENTARY* COMMENTARY *COMMENTARY*
*MMENTARY* COMMENTARY *COMMENTARY* COMMENTARY *COMMEN*
COMMENTARY *COMMENTARY* COMMENTARY *COMMENTARY*
*MENTARY* COMMENTARY *COMMENTARY* COMMENTARY *COMMEN*

## COMMENTARY

# A Tale of Two Deltas

Deltas are accumulations of sediment formed by the entrance of a stream into quiet water such as a lake or the sea. The term was proposed about 25 centuries ago by Herodotus who noted that the Nile Delta had the general shape of the Greek letter Δ. Not all deltas, however, have this shape, for every delta responds differently to depositional and erosional processes that act in opposition to one another. All deltas are characterized by a restless interplay between the rate of sediment supply brought in by streams and the rate at which that sediment is removed by marine processes.

The Mississippi Delta is termed a bird-foot delta because its many divergent streams (distributaries) extend seaward to form an areal pattern of lowland that resembles the outstretched toes of a bird. In this great delta, the sediment supply vastly exceeds the reworking capabilities of ocean waves and currents. The delta, therefore, builds itself seaward or **prograges** at the mouths of distributaries.

In contrast to the Mississippi Delta where the amount of sediment brought in by streams greatly exceeds the ability of the marine environment to rework and transport that sediment, the Niger Delta on the west coast of Africa provides an example of a delta in which marine destructional processes are, to a far greater degree, in balance with the supply of sediment. The action of waves and longshore currents along the delta front is capable of reworking newly deposited sand, silt, and clay almost as fast as it is supplied. Over time, however, sedimentation has somewhat exceeded removal so that there has been concentric growth along the entire front of the delta. If the reverse were the case, there would be no delta at all.

The sediments of deltas are rich in organic debris, and include many alternate and intersecting bodies of permeable sands and impermeable clays. Because of this, the rocks of ancient deltas have yielded huge volumes of petroleum and natural gas. Subsidence followed by marine incursions and then rebuilding of deltas has resulted in many-leveled sequences of oil reservoirs, such as those in many of the oil fields of the Texas and Louisiana Gulf Coast.

**A**    **B**

The Mississippi (A) and Niger (B) deltas.

At these depths the inclination of the ocean floor becomes gentler. The less pronounced slopes comprise the **continental rises** (see Fig. 2–2).

Sediment deposited on slopes and rises are mostly fine sand, silt, and clay. These materials are often transported to sites of deposition by **turbidity currents**. Water in a turbidity current is denser than surrounding water because it is laden with suspended sediment. It therefore flows down the slope of the ocean floor beneath the surrounding clear water. Upon reaching more level areas, the current slows and drops its load of suspended particles. Often the deposits, called turbidites, form submarine fans at the base of the continental slope (Fig. 2–4). In addition to turbidites, slope and rise deposits include fine clay that has settled out of the water column and large masses of material that have slid or slumped down the slope under the influence of gravity.

In the deep marine environment far from the continents, only very fine clay, volcanic ash, and the calcareous or siliceous remains of microscopic organisms settle to the ocean floor. The exceptions are sporadic occurrences of coarser sediments that are carried down continental slopes into the deeper parts of the ocean by turbidity currents. Coarse sediments may also be dropped into deep water as they are released from melting icebergs.

The shoreline of a continent is the transitional zone between marine and nonmarine environments. Here one finds the familiar shoreline accumulations of sand or gravel we call beaches. Mud-covered **tidal flats** that are alternately inundated and drained of water by tides are also found here, as are deltas. **Deltas** form when streams enter bodies of standing water, undergo an abrupt loss of velocity, and drop their load of sediment. Provided shoreline currents and waves do not remove the deposits as quickly as they are supplied, the delta will grow seaward. In general, progressively finer sediment will be deposited in progressively deeper or quieter water as the current provided by the stream diminishes. At the same time, the stream channel is extended over former deposits, periodically chokes in its own debris, and breaks out to form new branches of the delta.

In addition to deltas, the transitional zone includes such features as **barrier islands**, which are built parallel to shorelines by wave and current action, and **lagoons**, which are often found between the mainland and a barrier island. Swamps are also frequent features of low-lying areas adjacent to the sea.

Continental environments of deposition include river **floodplains**, alluvial fans, lakes, glaciers, and eolian (wind) environments. The silt, sand, and clay

**Figure 2–4**   Deep-sea fan built of land-derived sediment emerging from the lower part of a submarine canyon. Such fans occur in association with such large rivers as the Amazon, Congo, Ganges, and Indus. (Vertical exaggeration 200:1.)

**Figure 2–5**  Coalescing alluvial fans covering part of the floodplain, Gulf of Suez area, Egypt. The stream is unable to transport the huge amount of debris supplied to it and is dry during part of the year. *(Courtesy D. Bhattacharyya.)*

found along the banks, bars, and floodplains of streams are familiar to most of us. In general, stream deposits develop as elongate lenses that are oriented downstream and that grade abruptly from sediment of one particle size to another. In another setting, stream-transported materials may accumulate quickly when a rapidly flowing river emerges from a mountainous area onto a flat plain. The result of the abrupt deposition is an **alluvial fan** (Fig. 2–5). Somewhat quieter deposition occurs in lakes, which are ideal traps for sediment. Silt and clay are common lake sediments, although a variety of sediment is possible, depending on water depth, climate, and the character of the surrounding land areas. The **playa lakes** of arid regions are shallow temporary lakes. Periodically they become dry as a result of evaporation. Perhaps the most dramatic environments of deposition are found in areas having active glaciers. Glaciers have the ability to transport and deposit huge volumes and large fragments of rock detritus. Deposits are characteristically unsorted

mixtures of boulders, gravel, sand, and clay (Fig. 2–6). Where such materials have been reworked by glacial meltwater, however, they become less chaotic and resemble stream deposits.

Whereas glaciers can move materials of great size, wind is much more selective in the particle size it can transport. Environments where wind is an important agent of sediment transport and deposition are called **eolian environments** (Fig. 2–7). They are characterized by an abundance of sand and silt, little plant cover, and strong winds. These characteristics typify many desert regions.

## Color of Sedimentary Rocks

We have seen that color in igneous rocks can be used to indicate the approximate amount of ferromagnesian minerals present. Color in sedimentary rock can also provide useful clues to identification. For exam-

A             B

**Figure 2–6** Glacial deposits. (A) In the foreground are glacial gravels being washed and sorted by meltwater. At the center left is a ridge of glacial debris called a *moraine.* The glacier in the background is Athabasca Glacier in Jasper National Park, Canada. The glacier's toe (terminus) can be seen in the center of the photograph. (B) Unsorted glacial debris called *till* in a moraine of a valley glacier, Kenai Peninsula, Alaska.

ple, varieties of chert can be identified as flint if they are gray or black, or as jasper if they are red. Color is also useful in providing clues to the environment of deposition of sedimentary rocks. Of the sedimentary coloring agents, carbon and the oxides and hydroxides of iron are clearly the most important.

## Black Coloration

Black and dark gray coloration in sedimentary rocks—especially shales—usually results from the presence of compounds containing organic carbon and iron. The occurrence of an amount of carbon sufficient to

**Figure 2–7** Sand dune in the Gran Desierto region of the Sonoran Desert, Mexico. The wind responsible for this dune blew from left to right. Note wind ripples on the windward side. *(Courtesy of Jeff J. Plaut.)*

result in black coloration implies an abundance of organisms in or near the depositional areas as well as environmental circumstances that kept the remains of those organisms from being completely destroyed by oxidation or bacterial action. These circumstances are present in many marine, lake, and estuarine environments today. In a typical situation, the remains of organisms that lived in or near the depositional basin settle to the bottom and accumulate. In the quiet bottom environment, dissolved oxygen needed by aerobic bacteria to attack and break down organic matter may be lacking. There also may be insufficient oxygen for scavenging bottom dwellers that might feed on the debris. Thus, organic decay is limited to the slow and incomplete activity of anaerobic bacteria; consequently, incompletely decomposed material rich in black carbon tends to accumulate. In such an environment, iron combines with sulfur to form finely divided iron sulfide (pyrite, $FeS_2$), which further contributes to the blackish coloration. Such environments of deposition are likely to yield toxic solutions of hydrogen sulfide ($H_2S$). The lethal solutions rise to poison other organisms and thus contribute to the process of accumulation. Black sediments do not always form in restricted basins. They may develop in relatively open areas, provided the rate of accumulation of organic matter exceeds the ability of the environment to cause its decomposition.

## Red Coloration

Hues of brown, red, and green often occur in sedimentary rocks as a result of their iron oxide content. Few, if any, sedimentary rocks are free of iron, and less than 0.1 per cent of this metal can color a sediment a deep red. The iron pigments are not only ubiquitous in sediments, but are also difficult to remove in most natural solutions.

The iron present in sediment often occurs as either ferrous iron compounds or ferric iron compounds. Ferrous iron oxide (FeO) frequently occurs in oxygen-deficient environments. It is unstable and may slowly oxidize to form ferric iron oxide ($Fe_2O_3$). When oxygen is is short supply, ferric iron may be similarly reduced to ferrous iron. Ferric minerals such as **hematite** tend to color the rock red, brown, or purple, whereas the ferrous compounds impart hues of gray and green. Hydrous ferric oxide (**limonite**) is often yellow.

## Red Beds

Strata colored in shades of red, brown, or purple by ferric iron are designated **red beds** by geologists. Oxidizing conditions required for the development of ferric compounds are more typical of nonmarine than marine environments; most red beds are floodplain, alluvial fan, or deltaic deposits. Some, however, are originally reddish sediment carried into the open sea. Electron microscope studies of red beds forming today in Baja California indicate that the red coloration developed long after the sediment was deposited. After burial, the decay of clastic ferromagnesian minerals released iron that was oxidized by the oxygen in underground water circulating through the pore spaces. Thus, red coloration may be imparted in the subsurface and may be independent of climate. The paleoenvironmental interpretations one can draw from red beds should be based to a large degree on the associated rocks and sedimentary structures. Red beds interspersed with evaporite layers indicate warm and arid conditions.

Although red beds are more likely to represent nonmarine than marine deposition, red beds are occasionally interbedded with fossiliferous marine limestones. In such cases, the color may be inherited from red soils of nearby continental areas. Lands located in warm, humid climates often develop such red soils. When the soil particles arrive at the marine depositional site, they will retain their red coloration if there is insufficient organic matter present to reduce the ferric iron to the ferrous state. Otherwise, they will be converted to the gray or green colors of ferrous compounds.

In summary, sedimentary rocks of red coloration may be a product of the source materials, may have developed after burial as a result of a lengthy period of subsurface alteration, or may be the result of subaerial oxidation. Geologists are suspicious of the last possibility because most modern desert sediments are not red unless composed of materials from nearby outcrops of older red beds.

## Historical Significance of Texture

The size, shape, and arrangement of mineral and rock grains in a rock constitute its **texture**. In addition to the larger grains themselves, the textural appearance of a rock is influenced by the materials that hold the particles together. **Matrix** is one category of bonding material that consists of finer clastic particles (often clay) that were deposited at the same time as the larger grains and that fill the crevices between them. **Cement,** on the other hand, is a chemical precipitate that crystallizes in the voids between grains following deposition. *Silica* ($SiO_2$) and *calcium carbonate* ($CaCO_3$) are common natural cements.

Texture can provide many clues to the history of a particular rock formation. In carbonate rocks, extremely fine grained textures such as developed in lithographic limestones probably indicate deposition in quiet water. Fine carbonate muds, which are the source sediment of such rocks, are not likely to settle to the bottom in turbulent water. Whole, unbroken fossil shells confirm the quiet water interpretation. Limestones containing the worn and broken fragments of fossil shells are likely to be the products of reworking by wave action. They are turbulent water deposits.

### Size and Sorting of Clastic Grains

Geologists universally use a scale of particle sizes known as the **Wentworth scale** to categorize clastic sediments (Table 2–1). After disaggregation of a rock in the laboratory, the particles can be passed through a series of successively finer sieves, and the weight percentage of each size range in the rock can be determined. It is obvious that a stronger current of water (or wind) is required to move a large particle than a small one. Therefore, the size distribution of grains tells the geologist something about the turbulence and velocity of currents. It can also be an indicator of the mode and extent of transportation. If sand, silt, and clay are supplied by streams to a coastline, the turbulent nearshore waters will winnow out the finer particles, so that gradations from sandy nearshore deposits to offshore silty and clayey deposits frequently result (Fig. 2–8). Sandstones formed from such nearshore sands may retain considerable porosity and provide void space for petroleum accumulations. For this reason, one often finds petroleum geologists assiduously making maps showing the grain size of deeply buried ancient formations to determine areas of coarser and more permeable clastic rock.

One aspect of a clastic rock's texture that involves grain size is sorting. **Sorting** is an expression of the range of particle sizes deviating from the aver-

**Table 2–1  Size Range of Sedimentary Particles**

| Wentworth scale (in millimeters) | Fractional equivalents (in millimeters) | Particle name |
|---|---|---|
| | | Boulders |
| 256 — | | |
| 128 | | Cobbles |
| — 64 — | | |
| 32 | | |
| 16 | | Pebbles |
| 8 | | |
| — 4 — | | |
| | | Granules |
| — 2 — | | |
| | | Very coarse sand |
| 1.0 | | |
| | | Coarse sand |
| 0.5 | 1/2 | |
| | | Medium sand |
| 0.25 | 1/4 | |
| | | Fine sand |
| 0.125 | 1/8 | Very fine sand |
| — 0.0625 — | 1/16 — | |
| 0.0313 | 1/32 | |
| 0.0156 | 1/64 | |
| | | Silt |
| 0.0078 | 1/128 | |
| — 0.0039 — | 1/256 — | |
| | | Clay |

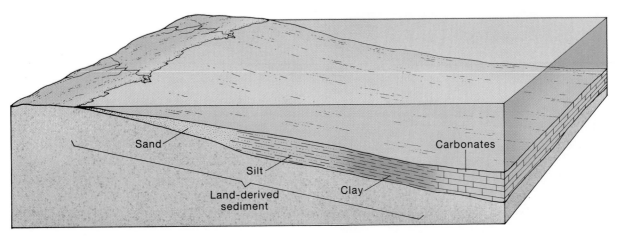

**Figure 2–8** Idealized gradation of coarser nearshore sediments to finer offshore deposits.

age size. Rocks composed of particles that are all about the same average size are said to be *well sorted* (Fig. 2–9) and those that include grains with a wide range of sizes are termed *poorly sorted*. Sorting often provides clues to conditions of transportation and deposition. Wind, for example, winnows the dust particles from sand, producing grains that are all of about the same size. Wind also sorts the particles that it carries in suspension. Only rarely is the velocity of winds sufficient to carry grains larger than 0.2 mm. While carrying grains of that size, winds sweep finer particles into the higher regions of the atmosphere. When the wind subsides, well-sorted

silt-sized particles drop and accumulate. In general, windblown deposits are better sorted than deposits formed in an area of wave action, and wave-washed sediments are better sorted than stream deposits. It must be kept in mind, however, that if a source sediment is already well sorted, the resulting deposit will be similarly well sorted. Poor sorting occurs when sediment is rapidly deposited without being selectively separated into sizes by currents (see Fig. 2–6B). Poorly sorted conglomerates and sandstones are deposited at the foot of mountains, where stream velocity is suddenly checked. Another example of a poorly sorted conglomerate is **tillite**, a rock depos-

A                         B

**Figure 2–9** Sorting of grains in sandstones as seen under the microscope may range from good sorting (A) to poor sorting (B). (A) Quartz (light tan grains) sandstone with carbonate (pink) cement. (B) A sandstone known as graywacke composed of poorly sorted angular grains of quartz (light tan), feldspar (green), and rock fragments (red). The graywacke lacks cement; spaces between grains are filled with a matrix of clay and silt. Width of fields is 1.5 mm.

**Figure 2–10** Well-rounded grains of quartz viewed under the microscope. From the St. Peter Formation near Pacific, Missouri. Width of field is 1.85 mm.

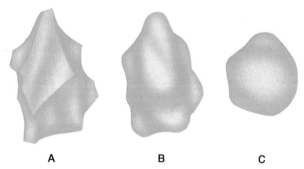

**Figure 2–11** Shape of sediment particles. (A) An angular particle (all edges sharp). (B) A rounded grain that has little sphericity. (C) A well-rounded, highly spherical grain. *Roundness* refers to the smoothing of edges and corners, whereas *sphericity* measures the degree of approach of a particle to a sphere.

ited by glacial ice containing all particle sizes in a heterogeneous mixture.

## Shape of Clastic Grains

The **shape** of particles in a clastic sedimentary rock can also be useful in determining its history. Shape can be described in terms of **rounding** of particle edges and **sphericity** (how closely the grain approaches the shape of a sphere) (Figs. 2–10 and 2–11). A particle becomes rounded by having sharp corners and edges removed by impact with other particles. The relatively heavy impacts between pebbles and granules being transported by water cause rapid rounding. Lighter impacts occur between sand grains in water transport; the water provides a cushioning effect. The result is far slower rounding. In conjunction with other evidence, the roundness of a particle can be used to infer the history of abrasion. It is a reflection of the distance the particle has traveled, the transporting medium, and the rigor of transport. It can also be used as evidence for recycling of older sediments.

## Arrangement of Clastic Grains

The third element in our definition of texture is the arrangement of the grains in the clastic rock. Geolo-

gists examine the rock to ascertain whether the grains are the same size and if they are clustered into zones or heterogeneously mixed. These observations may help to determine whether the sediment had been winnowed and sorted by currents or had been dumped rapidly. Such factors as the medium of transport, surface of deposition, and direction and velocity of currents control grain orientation. Geologists are particularly interested in studying grain orientation as a means of determining the current direction of the transporting medium. In general, sand grains deposited in moving water tend to acquire a preferred orientation in which the long axes of most of the elongate grains are aligned parallel to the direction of flow. Such particles also tend to be slightly tilted in the upstream direction.

To determine whether preferred orientation exists in a rock unit, geologists must collect many carefully oriented samples (Fig. 2–12) and subject the samples to a statistical analysis of the orientation of constituent grains. Many times, because of weakness or variability of currents, there may be little or no preferred orientation. Also, orientation in wind-blown sands is almost always less well developed than in water-carried sands. Several grain orientation studies of modern beach sands show that grain orientation in this environment is mainly controlled by backwash; in general, grains become aligned parallel to the shoreline. In steam-deposited sands, grain orientation is usually parallel to the elongation of the sand body. In glacial sediment, both elongate sand

**A**

**B**

**Figure 2–12**  Grain orientation study. One method of studying grain orientation is to prepare an oriented thin section of a rock whose field orientation has been recorded. Grain orientations are then measured under a microscope equipped with a rotating stage. The angle of the long axis of each elongated grain from the north line is determined. From many individual measurements, a mean orientation—in this example, about N 45° E or S 45° W—is determined and its statistical significance evaluated. A thin section cut perpendicular to bedding might reveal the tilt of the grains and might then be used to determine that the transportation medium flowed northeastward rather than southwestward.

grains and pebbles show a longitudinal orientation parallel to the direction of ice movement. Grain orientation analyses are useful in determining sediment distributional patterns in the geologic past and have occasionally provided important clues to the subsurface location and trend of petroleum-bearing sandstone strata.

# Inferences from Sedimentary Structures

**Sedimentary structures** are those larger features of sediments that are formed during or shortly after deposition and before lithification. Because particular sedimentary structures result from specific depositional processes, the structures are extremely useful to geologists interested in reconstructing ancient environments. For example, **mud cracks** indicate

**Figure 2–13**  Modern and ancient mudcracks. (A) This modern mudcrack formed in soft clay around the margins of an evaporating pond. (B) Mudcracks and wave ripples (caused by wind blowing over a shallow lake) in mudstones of the Oneonta Formation of Devonian age near Unadilla, New York. Divisions on the scale are 1.0 cm. *(Modern mudcracks courtesy of L. E. Davis; ancient mudcracks courtesy of W. D. Sevon.)*

drying after deposition. These conditions are common on valley flats and in tidal zones. Mud cracks (Fig. 2–13) develop by shrinkage of mud or clay on drying and are most abundant in the subaerial environment. **Cross-bedding** is an arrangement of beds

**Figure 2–14**   Two types of cross-bedding. The upper block shows planar laminations, as seen in beach deposits and dunes; the lower block represents trough laminations, as often formed in river channels. The line at the base of each set of laminations represents a surface of erosion that truncates older sets below.

**Figure 2–15**   Cross-bedding (cross-stratification) of the tabular-planar type in the Mountain Lakes Formation of Proterozoic age, Northwest Territories, Canada. *(Courtesy of G. Ross.)*

**Figure 2–16**  Cross-bedding in a delta. The succession of inclined foreset beds are deposited over bottomset beds that were laid down earlier. Topset beds are deposited by the stream above the foreset beds.

or laminations in which one set of layers is inclined relative to the others (Figs. 2–14 and 2–15). The cross-bedding units can be formed by the advance of a delta (Fig. 2–16) or a dune (Fig. 2–17). A depositional environment dominated by currents is inferred from cross-bedding. The currents may be wind or water. In either medium, the direction of the inclination of the sloping beds is a useful indicator of the direction taken by the current. By plotting these directions on maps, geologists have been able to determine the pattern of prevailing winds at various times in the geologic past (Fig. 2–18).

**Graded bedding** consists of repeated beds, each of which has the coarsest grains at the base and suc-

**Figure 2–17**  Sand being transported by wind action. In this dune at Great Dunes, Colorado, sand is being swept up the windward side (the gentler slope) and deposited on the leeward slope to form the steeper slip face of the dune.

Late Triassic

EXPLANATION                                      N

(A) Wingate Sandstone

(B) Tongues of Wingate Sandstone in
    upper part of Chinle Formation

(C) Dinosaur Canyon Sandstone Member
    of Moenave Formation

Approximate limit of deposition
Gray band on side of nondeposition
or erosion

↘ Generalized direction of wind

**Figure 2–18**   Wind directions determined from cross-stratified Late Triassic sandstones of the Colorado Plateau. The dune beds indicate, on the average, paleowinds blowing from the northwest to the southeast during the Triassic Period. *(After F. G. Poole, U.S. Geological Survey Prof. Paper 450-D, 1962, pp. 147–151.)*

cessively finer grains nearer the top (Fig. 2–19). Although graded bedding may form simply as the result of faster settling of coarser, heavier grains in a sedimentary mix, it appears to be particularly characteristic of deposition by the turbidity currents discussed earlier. Turbidity currents are often triggered by submarine earthquakes and landslides that occur along steeply sloping regions of the sea floor. The forward part of the turbidity current contains coarser debris than does the tail. As a result, the sediment deposited at a given place on the sea bottom grades from coarse to fine as the "head" and then the "tail" of the current passes over it. The presence of graded beds may indicate former turbidity currents. Geologists believe that turbidity currents frequently characterize unstable, tectonically active environments.

**Ripple marks** are commonly seen sedimentary features that developed along the surfaces of bedding planes (Fig. 2–20). *Symmetric ripple marks* are

fine

coarse

**Figure 2–19**   Graded bedding.

A                                          B

**Figure 2–20**   (A) Ripple marks formed in sand along a modern beach. (B) Ripple marks on a bedding surface of the Tensleep Formation of Pennsylvanian age, Wyoming. *(Courtesy of L. E. Davis.)*

formed by the oscillatory motion of water beneath waves. *Asymmetric ripple marks* are formed by air or water currents and are useful in indicating the direction of movement of currents (Fig. 2–21). For example, ripple marks form at right angles to current directions; the steeper side of the asymmetric variety faces the direction in which the medium is flowing. Although there are instances of ripple marks developed at great depths on the sea floor, these features occur more frequently in shallow water areas and in streams.

## The Meaning of Sandstones

Among the clastic group of sedimentary rocks, sandstones have been studied in great detail and provide

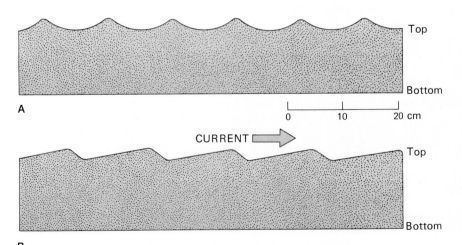

**Figure 2–21**   Profiles of ripple marks. (A) Symmetric ripples. (B) Asymmetric ripples.

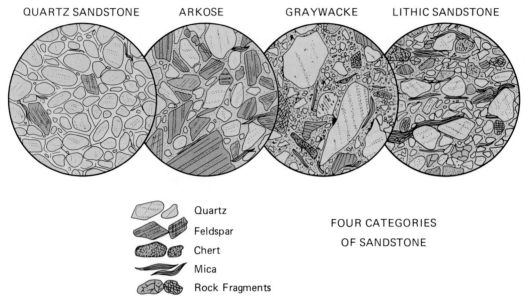

QUARTZ SANDSTONE   ARKOSE   GRAYWACKE   LITHIC SANDSTONE

Quartz
Feldspar
Chert
Mica
Rock Fragments

FOUR CATEGORIES
OF SANDSTONE

**Figure 2–22** Four categories of sandstone as seen in thin section under the microscope. Diameter of field is about 4 mm.

an extraordinary amount of information about conditions in and near the site of deposition. In particular, the mineral composition of sandstone grains can be used to identify source areas and to interpret what may have occurred prior to deposition. Often, by closely studying the grains, one can ascertain whether the source material was metamorphic, igneous, or sedimentary. The mineral content also provides a rough estimate of the amount of transport and erosion experienced by the sand grains. Rigorous weathering and long transport tend to reduce the less stable feldspars and ferromagnesian minerals to clay and iron compounds and tend to cause rounding and sorting of the remaining quartz grains. Hence, one can assume that a sandstone rich in these less durable and angular components underwent relatively little transport and other forms of geologic duress. Such sediments are termed *immature* and are most frequently deposited close to their source areas. On the other hand, quartz can be used as an indicator of a sandstone's maturity; the higher the percentage of quartz, the greater the maturity.

In addition to providing an indication of a rock's maturity, composition is an important factor in the classification of sandstones into quartz sandstone, arkose, graywacke, and lithic sandstone (sometimes termed subgraywacke) (Fig. 2–22).

**Quartz sandstones** (Fig. 2–23) are clastic rocks characterized by a dominance of quartz with little or no feldspar, mica, or fine clastic matrix. The quartz grains are well sorted and well rounded. They are most commonly held together by such cements as calcite and silica. Chemical cements such as these tend to be more characteristic of "clean" sandstones like quartz sandstones and are not as prevalent in

**Figure 2–23** Quartz sandstone. The color in this rock is caused by minor amounts of iron oxide.

**Figure 2–24** Idealized geologic conditions under which quartz sandstone may be deposited. There is little tectonic movement in this environment. Water depth is shallow, and the basin subsides very slowly.

"dirtier" rocks containing clay. The presence of a dense, clayey matrix seems to retard the formation of chemical cement, perhaps because fine material fills pore openings where crystallization might occur.

Calcite cement may develop between the grains as a uniform finely crystalline filling or large crystals may form, and each may incorporate hundreds of quartz grains. Silica cement in quartz sandstone commonly develops as overgrowths on the original grain surfaces.

Quartz sandstones reflect deposition in stable, quiet, shallow-water environments such as the shallow seas that inundated large parts of low-lying continental regions in the geologic past or some parts of our modern continental shelves (Fig. 2–24). These sandstones, as well as clastic limestones, exhibit sedimentary features, such as cross-bedding and ripple marks, which permit one to infer shallow-water deposition.

Sandstones containing 25 per cent or more of feldspar (derived from erosion of a granitic source area) are called **arkoses** (Fig. 2–25). Quartz is the most abundant mineral, and the angular to subangular grains are bonded together by calcareous cement, clay minerals, or iron oxide. The presence of abundant feldspars and iron imparts a pinkish-gray color to many arkoses. In general, arkoses are coarse, moderately well-sorted sandstones. They may de-

**Figure 2–25** Thin section of an arkose viewed through a petrographic microscope. The clear grains are mostly quartz, whereas the grains that show stripes or a plaid pattern are feldspars. The matrix consists of kaolinite clay and fine particles of mica, quartz, and feldspar. *(Courtesy of the U.S. Geological Survey; photo by J. D. Vine.)*

**Figure 2–26** Geologic environment in which arkose may be deposited.

velop as basal sandstones derived from the erosion of a granitic coastal area experiencing an advance of the sea, or they may accumulate in fault troughs or low areas adjacent to granitic mountains (Fig. 2–26).

**Graywackes** (from the German term *wacken,* meaning waste or barren) are immature sandstones consisting of significant quantities of dark, very fine grained material (Fig. 2–27). Normally, this fine matrix consists of clay, chlorite, micas, and silt. There is little or no cement, and the sand-sized grains are not in close contact because they are separated by the finer matrix particles. The matrix constitutes approximately 30 per cent of the rock, the remaining coarser grains consisting of quartz, feldspar, and rock particles. Graywacke has a dirty, "poured in" appearance. The poor sorting, angularity of grains (Fig. 2–28), and heterogeneous composition of graywackes indicate an unstable source and depositional area in which debris resulting from accelerated erosion of highlands is carried rapidly to subsiding basins. Graded bedding (Fig. 2–19), interspersed layers of volcanic rocks, and cherts (which may indirectly derive their silica from volcanic ash) further attest to dynamic conditions in the area of deposition. The inferred tectonic setting is unstable, with deposition occurring offshore from an actively rising mountainous region (Fig. 2–29). Graywackes and associated shales and cherts may contain fossils of deep-water organisms, indicating deposition at great depth. Such shallow-water sedimentary structures as

**Figure 2–27** Graywacke.

**Figure 2–28** Thin section of a poorly sorted sandstone (graywacke) observed under petrographic microscope with crossed polarizers. Note wide range of grain sizes. Width of field is 9.0 mm.

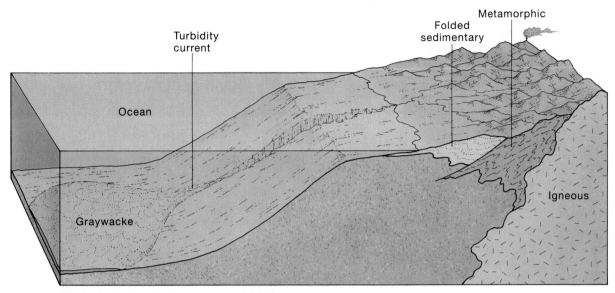

**Figure 2–29** Tectonic setting in which graywacke is deposited. Frequently graywackes are transported by masses of water highly charged with suspended sediment. Because of the suspended matter, the mass is denser than surrounding water and moves along the sloping sea floor or down submarine canyons as a turbidity current. The graywacke sediment characteristically accumulates in deep-sea fans at the base of the continental slope.

cross-bedding and ripple marks are rarely found. To the experienced petrologist, graywackes clearly indicate dynamic, unstable conditions.

Quartz sandstone, arkose, and graywacke are rather distinct kinds of sandstones. A rock that has a more transitional composition and texture is termed a **lithic sandstone** (subgraywacke). In lithic sandstones (see Fig. 2–22), feldspars are relatively scarce, whereas quartz, muscovite, chert, and rock frag-

ments are abundant. There is a fine-grained detrital matrix that does not exceed 15 per cent, and the remaining voids are filled with mineral cement or clay. In lithic sandstone, the quartz grains are better rounded and more abundant, the sorting is better, and the quantity of matrix is lower than in graywacke.

The characteristic environment for lithic sandstones are deltaic coastal plains (Fig. 2–30), where

**Figure 2–30** Deltaic environment in which lithic sandstones may be deposited.

they may be deposited in nearshore marine environments or swamps and marshes. Coal beds and micaceous shales are frequently associated with lithic sandstones.

## The Meaning of Carbonate Sedimentary Rocks

Limestones are the most abundant of carbonate sedimentary rocks. Although limestone lake deposits do occur, most limestones originated in the seas. Nearly always, the formation of these marine limestones appears to have been either directly or indirectly associated with biologic processes. In some limestones, the importance of biology is obvious, for the bulk of the rock is composed of readily visible shells of molluscs and skeletal remains of corals and other marine organisms. In other limestones, skeletal remains are not present, but nevertheless the calcium carbonate that forms the bulk of the deposit was precipitated from sea water because of the life processes of organisms living in that water. For example, the relatively warm, clear ocean waters of tropical regions are usually slightly supersaturated with calcium carbonate.

**Figure 2–31** Carbonate mud accumulating on the sea floor in the shallow warm waters of the Bahama Banks. *(Courtesy L. Walters.)*

In this condition, only a slight increase in temperature, loss of dissolved carbon dioxide, or influx of supersaturated water containing $CaCO_3$ "seeds" can bring about the precipitation of tiny crystals of calcium carbonate. Organisms do not appreciably affect temperature, but through photosynthesis, myriads of microscopic marine plants remove carbon dioxide from the water and thus may trigger the precipitation of calcium carbonate. Bacterial decay may also enhance precipitation of calcium carbonate by generating ammonia and thereby changing the alkalinity of sea water. In either case, the precipitate can be considered an inorganic product of organic processes. Usually, the mineral is precipitated as an orthorhombic crystal form of calcium carbonate called **aragonite.** Aragonite is unstable and changes to calcite, which is hexagonal in crystal form.

Carbonate sedimentation today is most rapid in shallow, clear-water tropical marine areas such as the Bahama Banks east of Florida (Fig. 2–31). The lime muds that are prevalent in such environments may develop in more than one way. Some are derived from the death and decay of calcareous algae that grow on the sea floor. Additional fine carbonate particles consist of the microscopic shells of unicellular marine organisms. Oöids are also abundant. As noted in the previous chapter, oöids are tiny spheres composed of calcium carbonate that are formed when particles roll back and forth on the sea floor and acquire concentric rings of carbonate (see Fig. 1–44). Some of the sediment results from the precipitation of tiny crystals of lime from sea water that has been chemically altered by the biologic processes of marine plants. Coarser particles result from the abrasion of the shells of invertebrates or consist of fecal pellets produced by burrowing organisms.

The requirement of warm, clear, shallow seas for the accumulation of modern carbonates seems to apply equally well to ancient deposits. Major sequences of ancient limestones are relatively free of clay and frequently contain an abundance of fossils thought to be representative of shallow, warm seas. Ancient carbonate rocks have developed in a variety of tectonic settings. Thick sections of limestones and dolomites have formed in subsiding basins in West Texas, Alberta, and Michigan. In such areas, optimum conditions for carbonate sedimentation resulted in a rate of accumulation that approximately equaled subsidence. Thick deposits of limestones are also known to have occurred along the cratonic margins of continents. Thinner but very extensive blankets of limestones have accumulated on the stable platform regions of the continents when those regions were subjected to marine inundations.

One type of carbonate sedimentary rock that continues to perplex geologists is the magnesium–calcium carbonate rock dolomite. You will recall from Chapter 1 that dolomite is a rock composed primarily of the mineral dolomite, $CaMg(CO_3)_2$. Dolomite is believed to form when magnesium that has been concentrated in sea water replaces a portion of the calcium carbonate in previously deposited calcium carbonate sediment. Supporting this interpretation are fossil shells in dolomite strata that were originally composed of calcite, but that subsequently have been changed to dolomite. Today, dolomite formation occurs in only a few areas, usually where evaporation of sea water is sufficiently intense to concentrate magnesium. Yet during parts of the Precambrian, Paleozoic, and Mesozoic eras, dolomites were extensively developed. To account for these ancient deposits, it would appear that extensive areas of evaporative conditions would be required. Recently, geologist David Lumsden discovered a correlation between ancient periods of high dolomite formation and episodes of high sea level. He suggests that during times of eustatic (worldwide) rise in sea level, broad low-lying tracts of the continents were inundated by shallow seas. Where climatic conditions were favorable, these shallow seas provided the ideal evaporative environments needed to enrich sea water with magnesium. As the magnesium replaced part of the calcium in the calcium carbonate sediment that blanketed the sea floor, calcite was converted to dolomite. When the sea level subsequently fell, the magnesium enrichment process halted and dolomite formation ceased.

A

B

**Figure 2–32** (A) A hand specimen of shale. This variety of shale contains waxy hydrocarbons and is called oil shale. (B) A thin section of shale cut perpendicular to bedding. The brown color results from the organic matter that is mixed with the clay and other fine particles.

## The Meaning of Shales

**Shale** (Fig. 2–32) is a general term for a very fine textured, fissile (capable of being split into thin layers) rock composed of clay, mud, and silt. In general, the environmental significance of shales parallels that of the sandstones with which they are associated.

Frequently, the silt-size particles in shales are similar in composition and shape to the associated sandstone beds. These silty components can frequently be extracted for study by disaggregating the shale in water and repeatedly pouring off the muddy liquid, retaining the larger particles as a residue.

In shales associated with quartz sandstones the silt fraction often consists predominantly of rounded quartz grains. Such quartz shales result from the reworking of older residual clays by transgressing shallow seas. Their association with thin, widespread limestones and quartz sandstones provides evidence for their deposition under stable tectonic conditions.

Feldspathic shales contain at least 10 per cent feldspar in the silt size and tend to be rich in the clay mineral kaolinite. Feldspathic shales are frequent associates of arkoses and are presumed to have formed in a similar environment. Such shales are representative of the finer sediment winnowed from coarser detritus and deposited in quieter locations.

Chloritic shales are usually associated with graywackes. As implied by their name, flakes of chlorite are common among the silt-sized components. The less flaky particles tend to be angular. Fissility in these rocks is less well developed than in other shales. The clay and silt particles in chloritic shales are generally derived from mountainous, unstable source areas nearby.

A shale type that is the approximate equivalent of a lithic sandstone is designated a micaceous shale. Mica flakes, quartz, and feldspar are all among its silt-sized components. Micaceous shales are depos-

ited under conditions somewhat less stable than the environment for quartz shales. They are particularly characteristic of ancient deltaic deposits.

The clay minerals that occur in shales are complex hydrous aluminosilicates with constituent atoms arranged in silicate sheet structures. Kaolinites, smectites, and illites are the three major groups of clay minerals. Kaolinites are the purest and seem to have a preferred occurrence in terrestrial environments. Smectite may contain magnesium, calcium, or sodium or any combination of these three, whereas potassium is an essential constituent of illites. Illites are the predominant clay mineral in more ancient shales. Unfortunately, positive identification of clay minerals in hand specimens or under the microscope is not possible and can be accomplished only by means of X-ray, thermal, or chemical analyses.

# The Sedimentary Rock Record

## Rock Units

William Smith, the British surveyor mentioned in Chapter 1, demonstrated that distinctive bodies of strata could be traced over appreciable distances and therefore could be mapped. He produced an exceptionally fine geologic map of England and Wales in 1815. Smith's map, the first ever made of such high quality and accuracy, was accompanied by a comprehensive table of the rock units encountered in the area. Each of these units was given a particular name, such as the "Clunch Clay," the "Great Oölyte," or the "Cornbrash Limestone." Thus originated the concept of a fundamental unit in geology that was lithologically distinctive, that had recognizable contacts with other units both above and below, and that could be traced across the countryside from exposure to exposure (or in the subsurface from well to well). Smith referred to such a unit as a stratum, but today it is universally known as a **formation** (Fig. 2–33). Furthermore, formations and groupings or subdivisions of formations all constitute rock units. **Rock units** are formally defined as bodies of rock identified by their distinctive lithologic and structural features *without regard to time boundaries*. Such features as texture (grain size, clastic or crystalline, etc.), color, composition, thickness, type of bedding, general na-

**Figure 2–33** Formations. The diagram shows three formations. In practice, these formations would be formally named, often after a geographic location near which they are well exposed. For example, the three formations shown here might be designated the Cedar City Limestone, Big Springs Sandstone, and Plattsburgh Shale.

**Figure 2–34** Diagram showing how the original deposits of a formation may vary in age from place to place.

ture of organic remains, and the appearance of the unit in surface exposures (or in the subsurface on well logs that indicate rock properties) are all used to define a rock unit and recognize it in the field.

As mentioned above, an important element in the definition of a formation is that it may or may not be the same age everywhere it is encountered. The nearshore sands deposited by a sea slowly advancing (transgressing) across a low coastal plain may deposit a single blanket of sand (perhaps later to be called the Oriskany Sandstone); however, that layer will be older where the sea began its advance and younger where the advance ended (Fig. 2–34).

### Naming Rock Units

Formations are given two names: first, a geographic name that refers to a locality where the formation is well exposed or where it was first described, and second, a rock name if the formation is primarily of one lithologic type. For example, the St. Louis Limestone is a carbonate formation named after exposures at St. Louis. When formations have several lithologic types within them, the locality name is simply followed by the name Formation.

There are other rock units in addition to formations. Distinctive smaller units within formations may be split out as **members,** and formations may be combined into larger units called **groups** because of related lithologic attributes (or their position between distinct stratigraphic breaks). For example, in Grand Canyon National Park, the Whitmore Wash, Thunder Springs, Mooney Falls, and Horseshoe Mesa rock units are *members* of the massive Redwall Limestone. In another part of the canyon wall, one finds the Tapeats Sandstone, Bright Angel Shale, and Muav Limestone combined to form a larger mappa-

ble rock unit known as the Tonto Group (Table 2–2).

### Time-Stratigraphic Units

Historically, geologists attempting to reconstruct the biologic and physical environment of a particular region for a particular time interval quickly recognized that rock units (formations, groups, members) were not adequate. As we have noted, a given rock unit might be of different ages at different localities. Thus, it was necessary to formulate the term **time-stratigraphic unit** (also known as a *chrono-stratigraphic unit*) for sedimentary layers deposited during a particular time interval. The fundamental time-stratigraphic unit is the system. A **system** corresponds to a unit of pure geologic time called a **period** (Table 2–3). Both terms have the same "first name." Thus, it is correct, for example, to refer to the *Devonian Period* (Table 2–4) as a certain length of time in geologic history, whereas the *Devonian System* refers to the rocks that were deposited during the Devonian Period. Geologic periods can be divided into **epochs** and further into **ages.** Similarly, a system is divisible into **series,** which in turn may include yet smaller units termed **stages.**

Time units of higher rank than the period are the era and eon. An **era** is a division of geologic time that includes two or more periods. The *Cenozoic Era,* for example, includes the *Tertiary* and *Quaternary periods* (see Table 2–4). An **eon** is the largest formal unit of geologic time. Presently, three eons are recognized: the *Archean, Proterozoic,* and *Phanerozoic.* (The time-stratigraphic terms for era and eon are erathem and eonothem, respectively, but these terms are rarely used.)

**Table 2–2 Rock Units of the Paleozoic Section in Grand Canyon National Park, Arizona**

| System | Group | Formation | Member |
|---|---|---|---|
| Permian | | Kaibab | |
| Permian | | Toroweap | |
| Permian | Hermit | Coconino | |
| Permian | Hermit | Hermit Shale | |
| Pennsylvanian | | Supai | |
| Mississippian | | Redwall Limestone | Horseshoe Mesa |
| Mississippian | | Redwall Limestone | Mooney Falls |
| Mississippian | | Redwall Limestone | Thunder Springs |
| Mississippian | | Redwall Limestone | Whitmore Wash |
| Devonian | | Temple Butte | |
| Cambrian | Tonto | Muav Limestone | |
| Cambrian | Tonto | Bright Angel Shale | |
| Cambrian | Tonto | Tapeats Sandstone | |

The areal extent of a time-stratigraphic unit can be determined by examination of surface exposures and records of rocks encountered in deep drill holes. Upper and lower boundaries are intangible time planes inferred largely from fossil evidence. Thus, time-stratigraphic units are interpretive and less objective than are rock units.

The determination of both rock and time-stratigraphic boundaries and their correlation is an important component of a branch of geology known as stratigraphy. **Stratigraphy** deals with the study of stratified rocks, their interpretation, and their mutual relationships, description, and identification. Because stratified rocks cover approximately three-fourths of the earth's total land area and because these strata contain some of our most readily interpreted clues to past events, stratigraphy is the essential core of geologic history.

**Table 2–3 Hierarchy of Time and Time-Stratigraphic Terms**

| Time Divisions | Equivalent Time-Stratigraphic Divisions |
|---|---|
| **Era** | Erathem (rarely used) |
| **Period** | **System** |
| **Epoch** | **Series** |
| **Age** | **Stage** |
| Chron | **Zone** (Chronozone) |

Table 2–4    The Geologic Time Scale

**Time Units of the Geologic Time Scale (Numbers are absolute dates in millions of years before present)**

| Eons | Eras | Periods | Epochs | |
|------|------|---------|--------|---|
| | | | Recent or Holocene | |
| | | Quaternary | Pleistocene | |
| | | | | 2.0 |
| | | | Pliocene | |
| | | | | 5 |
| | | | Miocene | |
| | Cenozoic | | | 24 |
| | | Tertiary | Oligocene | |
| | | | | 37 |
| | | | Eocene | |
| | | | | 58 |
| | | | Paleocene | |
| | | | | 66 |
| | | Cretaceous | | |
| | | | | 144 |
| | Mesozoic | Jurassic | | |
| | | | | 208 |
| | | Triassic | | |
| | | | | 245 |
| | | Permian | | |
| | | | | 286 |
| | | Pennsylvanian (Upper Carboniferous) | | |
| | | | | 320 |
| | | Mississippian (Lower Carboniferous) | | |
| | Paleozoic | | | 360 |
| | | Devonian | | |
| | | | | 408 |
| | | Silurian | | |
| | | | | 438 |
| | | Ordovician | | |
| | | | | 505 |
| | | Cambrian | | |
| | | | | 570 |

Eon — Phanerozoic (Cenozoic, Mesozoic, Paleozoic)

| Eons | Eras | |
|------|------|---|
| Proterozoic | Late | 900 |
| | Middle | 1600 |
| | Early | 2500 |
| Archean | Late | 3000 |
| | Middle | 3400 |
| | Early | 3800 |
| Hadean (no record) | | |

## Facies

There is yet another term used by geologists in reference to those characteristics of rocks that relate to their depositional environments. The term is **facies**. A sedimentary facies is the general appearance or aspect of a rock from which its environment of deposition can be inferred. For example, a body of rock might consist of a clastic limestone along one of its lateral margins and lithographic limestone elsewhere. Geologists might then delineate a "clastic limestone facies" and interpret it as a nearshore part of the rock body, whereas they might interpret the micritic limestone facies as a former offshore deposit. In this case, the distinguishing characteristics are lithologic (rather than biologic); therefore, the facies can be further designated as a **lithofacies**. In other cases, the rock unit may be lithologically uniform, but the fossil assemblages differ and permit recognition of different **biofacies** that reflect differences in the environment. A limestone unit, for example, might contain abundant fossils of shallow-water reef corals along its thinning edge and elsewhere be characterized by remains of deep-water sea urchins and snails. There would thus be two biofacies developed— one reflecting deeper water than the other. They could, therefore, be designated the "coral" and the "echinoid–gastropod biofacies."

These examples illustrate that facies are clearly the products (sediment, shells of organisms) of particular environments of deposition. Today, as we travel across swamp, floodplain, and sea, we traverse different environments of deposition (Fig. 2–35). Each of these environments of deposition changes laterally into the adjacent environment of deposition, and each provides its present-day facies that likewise change to adjacent synchronous facies. Geologists record these facies changes on lithofacies and biofacies maps. Because the basis for these maps are time–rock units, they provide a view of different facies of essentially the same age. If one were able to make such maps of successively different times, it would become apparent that ancient facies have shifted their localities as the seas advanced or retreated or as environmental conditions changed.

Consider for a moment an arm of the sea slowly transgressing (advancing over) the land. The sediment deposited on the sea floor may ideally consist

**Figure 2–35** Sedimentary facies (lithofacies) developed in the sea adjacent to a land area. The upper surface of the diagram shows present-day facies, whereas the front face shows the shifting of facies through time. Notice that bottom-dwelling organisms also differ in environments having different bottom sediment and water depth.

of a nearshore sand facies, an offshore mud facies, and a far offshore carbonate facies. As the shoreline advances inland, the boundaries of these facies also shift in the same direction, thereby developing an **onlap** sequence (Fig. 2–36) in which coarser sediments are covered by finer ones. Should the sea subsequently begin a withdrawal (regression), the facies boundaries will again move in the same direction as the shoreline, creating as they do so an **offlap** sequence of beds (Fig. 2–37). In offlap situations, coarser nearshore sediment tends to lie above finer sediments. Also, because offlap units are deposited during marine regressions, recently deposited sediment is exposed to erosion, and part of the sedimentary sequence is lost. Study of sequential vertical changes in lithology, such as those represented by offlap and overlap relationships, is one method by which geologists recognize ancient advances and retreats of the seas and chart the positions of shorelines.

If the pattern of sediments spread seaward from shorelines always graded from nearshore sands to shales and carbonates, as depicted in Figure 2–36, predicting the locations of particular facies in ancient rocks would be comparatively easy. In reality, however, the task is usually more complex. For example, nearshore sandy facies are not present at all along some coastlines. This may be because little sand is being brought to the coast by streams or because vigorous wave and current action has carried the sand grains away, or possibly because sand grains are trapped in submarine canyons farther up the coast.

The nature of sedimentation along a coastline is also controlled by the direction of longshore currents, the location of the mouths of major streams that dump their sedimentary load into the sea, the amount of sediment supplied, the presence of barriers to dispersal of sediment, and whether the agents bringing the sediments to the sea are running water, wind, or glacial ice. Any of these factors complicate the study of facies, but they also provide fascinating problems for the geologist to solve.

## Changes in Sea Level

Whenever a change in sea level occurs that is worldwide and independent of continental tracts, the change is termed **eustatic.** Ice accumulating on the continents during an ice age will cause lowering of sea level because much of the water making up the ice ultimately came from the ocean. Conversely, during warmer episodes, water from melting ice flows back into the ocean, causing a rise in sea level and the landward advance of shorelines along many coasts around the world. In addition to eustatic changes associated with ice ages, upwarping of the floor of an ocean basin or the development of extensive mid-oceanic ridges can result in a eustatic change in sea level. Mid-ocean ridges are submarine mountain ranges extending for thousands of kilometers across the floors of the major ocean basins. They are composed of basaltic lavas derived from the mantle through a split or rift that widens to admit the basalt. As will be described in Chapter 6, the newly

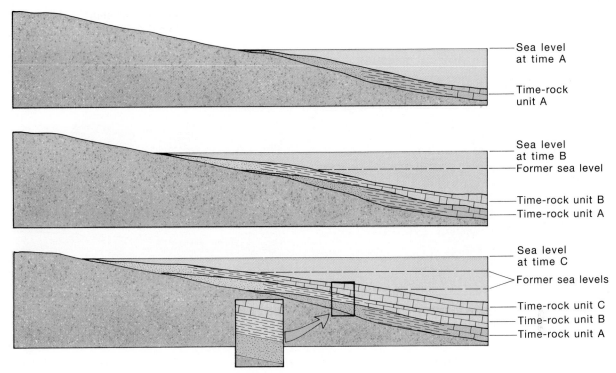

**Figure 2–36** Sedimentation during a transgression produces an onlap relationship in which finer offshore lithofacies overlie coarser nearshore facies (see inset), nearshore facies are progressively displaced away from a marine point of reference, and older beds are protected from erosion by younger beds.

formed basaltic ocean floor moves laterally by a process called sea floor spreading, and at a distance may plunge back into the mantle. The basaltic rocks formed along the mid-ocean ridges are hot and thermally expanded. As a result, they take up space that otherwise would be occupied by water and thus cause a eustatic rise in sea level. When the rate of extrusion of the basalts is rapid, there is likely to be a significant worldwide rise in sea level. Subsequent slower rates of extrusion and spreading would conse-

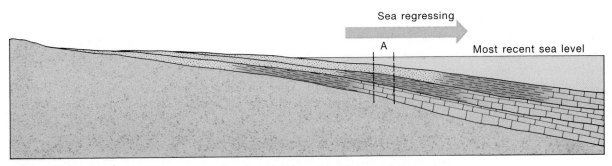

**Figure 2–37** Sedimentation during a regression produces an offlap relationship in which coarser nearshore lithofacies overlie finer offshore lithofacies, as shown in A. The sandy nearshore facies is progressively displaced toward the marine point of reference. Older beds are subjected to erosion as the regression of the sea proceeds.

**Figure 2–38**  A rise or fall in sea level will affect a far greater area along a low coastline than along coastlines composed of highlands that rise steeply adjacent to the sea.

quently result in lowering of sea level. These changes profoundly influence the geologic history of continental areas, for they determine when they are inundated by inland seas, when those seas regress, when there is deposition, and when there is erosion. The alternate advance and retreat of seas associated with events along mid-oceanic ridges permit geologists to recognize distinct packages or sequences of strata having erosional boundaries that reflect global cycles of sea level fluctuations having a frequency on the order of 10 to 80 million years. The fluctuations are sometimes called *Vail cycles* after P. R. Vail who demonstrated their importance in regional stratigraphic studies.

Changes in the elevation of land areas bordering the ocean can cause effects similar to those resulting from eustatic changes. For example, a coastal area may experience either tectonic uplift or subsidence. The former is likely to cause a retreat of the sea from the rising land area, whereas subsidence might allow the sea to slowly advance across subsiding tracts of land.

Precisely how far the sea will advance or retreat during a change in sea level is determined by the amount of change in sea level and the topography of the land. A low-lying, gently sloping terrain would have a much greater area inundated by a small increase in sea level than would a steeply sloping mountainous tract (Fig. 2–38).

Ultimately, the amount of inundation or regression along a coast must necessarily be related to the interaction between tectonic movements on the continents and eustatic sea level changes. If the land area along a continental margin rises at the same time and amount as a eustatic rise in sea level, the tectonic change will cancel out the eustatic one. The effect on shoreline displacement will be minimal. In contrast, if the land adjacent to the ocean is subsiding while sea level rises, a greater marine advance can be expected.

Anyone examining a map depicting one of the world's greatest deltas would quickly recognize that tectonic and eustatic changes are not the only causes of a shift in shorelines. The rapid accumulation of sediment along a coast or at the mouths of rivers will cause land area to be built seaward in the form of a delta. Such seaward migration of the shoreline resulting from nearshore deposition of sediment brought to the sea by rivers is termed **progradation**.

In the geologic past, there have been repeated advances of the sea into low-lying regions of the interiors of continents. Many appear to be directly related to the Vail cycles just discussed. At times these marine transgressions covered as much as two-thirds of North America. The resulting inland seas are termed **epeiric**, meaning "a sea over a continent." In these epeiric seas were deposited the sedimentary rock record of much of the Paleozoic and Mesozoic eras. The advance and retreat of the epeiric seas were characteristically rather irregular, often interrupted by partial regressions, and ultimately followed by gradual withdrawal back into the major ocean basins. The average rates of advance and retreat, however, were in no sense catastrophic, for they rarely exceeded a few inches per century. A slow advance indeed, but ample to cause extensive inundation of the continents over the tens of millions of years encompassed by a geologic period.

## Correlation

When examining an isolated exposure or rock in a road cut or the bank of a stream, a geologist is aware that the rock may continue laterally beneath the cover of soil and loose sediment and that the same stratum or rock body, or its equivalent, is likely to be found at other localities. The determination of the equivalence of bodies of rock in different localities is called **correlation**. The rock bodies may be equivalent in their lithology (composition, texture, color, and so on), in their age, in the fossils they contain, or in a combination of these attributes. Because there is more than one meaning for the term, geologists are careful to indicate the kind of correlation used in solving a particular geologic problem. In some cases it is only necessary to trace the occurrence of a lithologically distinctive unit, and the age of that unit is not critical. Other problems can be solved only through the correlation of rocks that are of the same age. Such correlations involve time–rock units and are of the utmost importance in geology. They are the basis for the geologic time scale and are essential in working out the geologic history of any region.

The correlation of strata from one locality to another may be accomplished in several ways. If the strata are well exposed at the earth's surface, as in arid regions, where soil and plant cover is thin, then it may be possible to trace distinctive rock units for many kilometers across the countryside by actually walking along the exposed strata. In using this straightforward method of correlation, the geologist can sketch the contacts between units directly on topographic maps or aerial photographs. The nota-

tions can then be used in the construction of geologic maps. It is also possible to construct a map of the contacts between correlative units in the field by using appropriate surveying instruments.

In areas where bedrock is covered by dense vegetation and a thick layer of soil, geologists must rely upon intermittent exposures found here and there along the sides of valleys, in stream beds, and in road cuts. Correlations are more difficult to make in these areas, but can be facilitated by recognizing the similarity in position of the bed one is trying to correlate with other units in the total sequence of strata. A formation may have changed somewhat in appearance between two localities, but if it always lies above or below a distinctive stratum of consistent appearance, then the correlation of the problematic formation is confirmed (Fig. 2–39).

A simple illustration of how correlations are used to build a composite picture of the rock record is provided in Figure 2–40. A geologist working along the sea cliffs at location 1 recognizes a dense oölitic limestone (formation F) at the lip of the cliff. The limestone is underlain by formations E and D. Months later the geologist continues the survey in the canyon at location 2. Because of its distinctive character, the geologist recognizes the oölitic limestone in the canyon as the same formation seen earlier along the coast and makes this correlation. The formation below F in the canyon is somewhat more clayey than that at locality 1, but is inferred to be the same because it occurs right under the oölitic limestone. Working upward toward location 3, the geologist maps the sequence of formations from G to K. Questions still remain, however. What lies below the lowest formation thus far found? Perhaps years later an oil well, such as that at location 3, might provide the answer. Drilling reveals that formations C, B, and A lie beneath D. Petroleum geologists monitoring the drilling of the well would add to the correlations by matching all the formations penetrated by the drill to those found earlier in outcrop. In this way, piece by piece, a network of correlations across an entire region is built up.

For correlations of time–rock units, one cannot depend upon similarities in lithology to establish equivalence. Rocks of similar appearance have been formed repeatedly over the long span of geologic time. Thus, there is the danger of correlating two apparently similar units that were deposited at quite

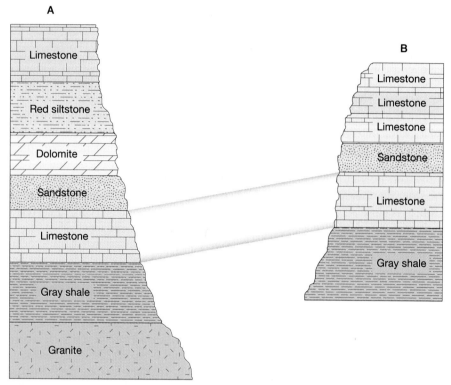

**Figure 2–39** If the lithology of a rock is not sufficiently distinctive to permit its correlation from one locality to another, its position in relation to distinctive rock units above and below may aid in correlation. In the example shown here, the limestone unit at locality A can be correlated to the lowest of the four limestone units at locality B because of its position between the gray shale and the sandstone units.

**Figure 2–40** An understanding of the sequence of formations in an area usually begins with examination of surface rocks and correlation between isolated exposures. Study of samples from deep wells permits the geologist to expand the known sequence of formations and to verify the areal extent and thickness of both surface and subsurface formations.

different times. Fortunately, the use of fossils in correlation may help to prevent mismatching. Methods of correlation based on fossils (biostratigraphic correlation) are fully described in Chapter 3. They are based on the fact that animals and plants have undergone change through geologic time, and therefore the fossil remains of life are recognizably different in rocks of different ages. Conversely, rocks of the same age but from widely separated regions can be expected to contain similar assemblages of fossils.

Unfortunately, there are complications to these generalizations. For two strata to have similar fossils, they would have to have been deposited in rather similar environments. A sandstone formed on a river floodplain would have quite different fossils than one formed at the same time in a nearshore marine environment. How might one go about establishing that the floodplain deposit could be correlated to the marine deposit? In some cases this might be done by physically tracing out the beds along a cliff or valley side. Occasionally, one is able to find fossils that actually do occur in both deposits. Pollen grains, for example, could have been wafted by the wind into both environments. Possibly, both deposits occur directly above a distinctive, firmly correlated stratum such as a layer of volcanic ash. Ash beds are particularly good time markers because they are deposited over a wide area during a relatively brief interval of time. Such key beds are exceptionally useful in establishing the correlation of overlying strata. Finally, the geologist may be able to obtain the actual age of the strata using radioactive methods, and these values can then be used to establish the correlation.

## Depicting the Past

### Geologic Columns and Cross-Sections

To aid in the synthesis and interpretation of field observations of sedimentary and other rocks, geologists prepare a variety of maps, graphs, and charts designed to show relationships of rock bodies to one another, their thickness, the manner in which they have been disturbed, and their general composition. The most important graphic devices for communicating such information are columnar sections, cross-sections, and geologic maps.

Columnar sections show the vertical sequence of strata (layers of rock) at any given locality. After geologists have compiled columnar sections at several locations within a region, they may next prepare a *composite geologic column* (Fig. 2–41) that combines the data from all of the various individual sections. Individual columnar sections are also used in the construction of cross-sections. As implied by their name, *cross-sections* show the vertical dimension of a slice through the earth's crust. Some cross-sections, namely the stratigraphic type, emphasize the age or lithologic equivalence of strata. The vertical measurements for such stratigraphic cross-sections are made from a horizontal line representing the surface of a definite bed or stratum. That surface is called the *datum*. In nature, the surface used as datum may be inclined or folded, and so stratigraphic sections do not validly show the tilt or position of beds relative to sea level. Stratigraphic cross-sections are most effective in showing the way beds correlate and vary in thickness from exposure to exposure or well to well. To show the way beds are folded, faulted, or tilted, a *structural cross-section* can be prepared (Fig. 2–42). In drawing the structural cross-section, the datum is a level line parallel to sea level, and the tops and bottoms of rock units are plotted according to their true elevations. If the vertical and horizontal scales are similar, the attitude of the beds will be correctly depicted. Many times, however, it is useful to have a larger vertical then horizontal scale to emphasize geologic features.

### Geologic Maps

Geologic maps show the distribution of rocks of different kinds and ages that lie directly beneath the loose rock and soil covering most areas of the surface. Assume for a moment that all of this loose material and the vegetation growing on it were miraculously removed from your home state, so that bedrock would be exposed everywhere. Imagine, further, that the surfaces of the formations now exposed were each painted a different color and photographed vertically from an airplane. Such a photograph would constitute a simple geologic map. In actual practice, a geologic map is prepared by locating contact lines between formations in the field and then plotting these contacts on a base map (Fig. 2–43A, B). Symbols are added to the colored areas to

Composite Geologic Column
Salt River basin

**Figure 2–41**   Example of a composite geologic column.

Alluvium & soil
Glacial deposits & loess
Pleasanton Grp.

Marmaton Grp.

Cherokee Grp.

Cheltenham Fm.
Basal Atokan

Burlington-Keokuk Fm.

Chouteau Fm.

Hannibal Fm.

| | Limestone |
| | Cherty limestone |
| | Shale |
| | Clay |
| | Sandstone |
| | Coal |

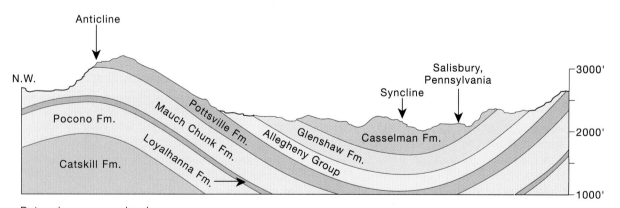

Datum is mean sea level
No vertical exaggeration

**Figure 2–42**   Geologic structural cross-section across Paleozoic rocks in the Appalachian Mountains, southeastern Pennsylvania, extending northwestward from the town of Salisbury.

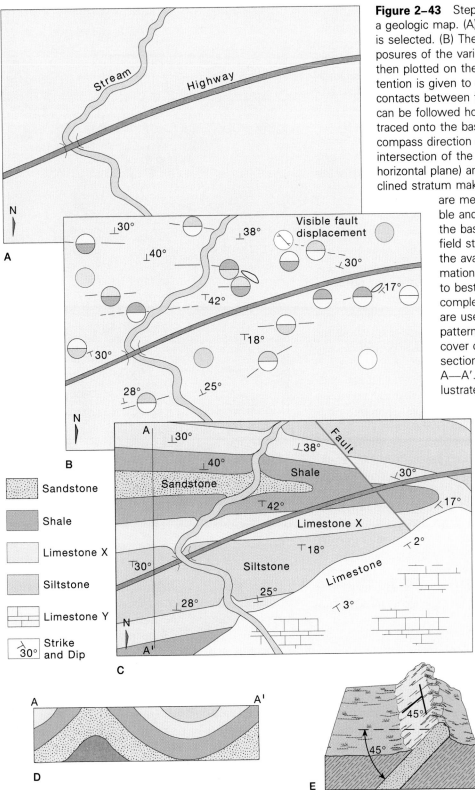

**Figure 2–43** Steps in the preparation of a geologic map. (A) A suitable base map is selected. (B) The locations of rock exposures of the various formations are then plotted on the base map. Special attention is given to exposures that include contacts between formations; where they can be followed horizontally, they are traced onto the base map also. *Strike* (the compass direction of a line formed by the intersection of the surface of a bed and a horizontal plane) and *dip* (the angle an inclined stratum makes with the horizontal) are measured wherever possible and added to the data on the base map. After careful field study and synthesis of all the available information, formation boundaries are drawn to best fit the data. (C) On the completed map, color patterns are used to show the areal pattern of rocks beneath the cover of soil. (D) A cross-section is shown along line A—A'. (E) A block diagram illustrates strike and dip.

indicate formations and lithologic regions, mineral deposits, and structures such as folds and faults. Once the geologic map is completed, a geologist can tell a good deal about the geologic history of an area. The formations depicted represent sequential "pages" in the geologic record. From the simple geologic map shown in Figure 2–43C, the geologist is able to deduce that there was an ancient period of compressional folding, that the folds were subsequently faulted, and that an advance of the sea resulted in deposition of younger sedimentary layers above the more ancient folded strata.

## Paleogeographic Maps

A map showing the geography of a region or area at some specific time in the geologic past is termed a **paleogeographic map**. Such maps are really interpretations based on all available paleontologic and geologic data. The majority of such maps show the distribution of ancient lands and seas (Fig. 2–44). Paleogeographic maps are, at best, of limited accuracy, because since seas advance and retreat endlessly through time, the line drawn at the sea's edge may represent an average of several shoreline positions. They are nevertheless useful for showing general geographic conditions within regions or continents. To prepare a paleogeographic map, one would plot all occurrences of rocks of a given time interval on a map and enclose the area of occurrence in boundary lines (Fig. 2–45A and B). Areas of nonoccurrence may be places of no deposition or places where deposits once existed but were subsequently eroded away. With the help of fossils, the nature of the sediments—that is, whether marine or nonmarine—is determined and plotted on the map. The final step is to complete the paleogeographic reconstruction (Fig. 2–45C).

## Isopachous Maps

**Isopachous maps** are prepared by geologists to illustrate changes in the thickness of a formation or time-stratigraphic unit. The lines on an isopachous map (Fig. 2–46) connect points at which the unit is of the same thickness. On a base map, the geologists plot the thickness of units as they are revealed by drilling or in measured surface sections. Isopachous

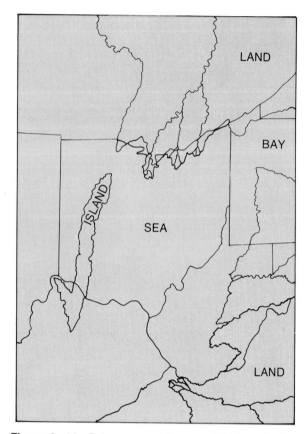

**Figure 2–44**   Paleogeographic map of Ohio and adjoining states during an early part of the Mississippian Period. The data for this study were obtained from outcrops and over 40,000 well records. *(After J. F. Pepper, W. J. de Witt, and D. F. Demarest,* U.S. Geological Survey Prof. Paper *259, 1954.)*

lines are then drawn to conform to the data points. Ordinarily, the upper surface of the unit being mapped is used as the horizontal plane or datum from which thickness measurements are made. An isopachous map may be very useful in determining the size and shape of a depositional basin, the position of shorelines, and areas of uplift. Figure 2–47 is an isopachous map of Upper Ordovician formations in Pennsylvania and adjoining states. The map indicates a semicircular center of subsidence in southern New York and Pennsylvania in which over 2000 ft of sediment accumulated. The isopachous pattern further indicates a highland source area to the southeast.

A

B

C: Continental sediments containing fossils of fresh water clams and land plants

Mss: Marine sandstone with fossils of marine invertebrates

Msh: Marine shale containing abundant fossils of marine microorganisms

C

**Figure 2–45**   Stages in the construction of a paleogeographic map. (A) Area of occurrence of a particular time–rock unit. (B) Plot of rock types within the time–rock unit. (C) Paleogeographic reconstruction.

**Figure 2–46** Diagram illustrating the construction of a simple isopachous map in an area of undeformed strata.

## Lithofacies Maps

Maps constructed to show areal variations in facies can provide additional details and validity to paleogeographic interpretations. Such graphic representations are called **lithofacies maps.** Figure 2–48 is a hypothetical base map of an area subjected to exploratory drilling by oil companies. The logs for the wells are shown below their locations. Geologists first correlate the formations. Then, assuming that the unconformity represents one time plane and the ash bed another, they define the time–rock unit as "X." Paleontologic study of the rocks between the time planes confirms the validity of the time–rock unit. Geologists may now prepare the lithofacies map. Time–rock unit X is missing at well number 11; this may be the result of its not being deposited there or, having been deposited, of its being eroded away. It is logical that the sandy facies was deposited adjacent to a north–south trending shoreline.

Figure 2–49 is a lithofacies map of rocks deposited over 400 million years ago in the eastern United States. From this map, one can infer the existence of a highland area that existed at that time along our eastern seaboard and that supplied the coarse clastics. Detrital sediments from the source area become fine, and the section thins as one proceeds westward from the source highlands. Finally, as far west as Indiana, the map indicates that only carbonate precipitates were laid down. The conglomerates were probably the deposits of great alluvial fans built out from the ancient mountain system.

The lithofacies maps just described provide a qualitative interpretation of areal changes in rock bodies. Quantitative lithofacies maps can also be constructed and are frequently used in the study of subsurface formations that are known primarily from

**Figure 2–47** Isopach map of Upper Ordovician formations in Pennsylvania and adjoining states. *(After M. Kay,* Geol. Soc. Am. Mem. *No. 48, 1951.)*

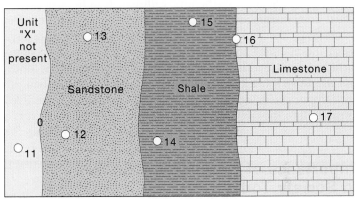

**Figure 2–48** Diagram illustrating the preparation of a lithofacies map from a subsurface time–rock unit. Well locations are indicated by small circles. Correlation of rock units between wells is indicated by dashed lines. Because of the few control points, the exact position of lithofacies boundaries on this map is somewhat arbitrary.

**Figure 2–49** Lithofacies map of Lower Silurian rocks in the eastern United States. *(After T. W. Amsden,* Bull. Am. Assoc. Petrol. Geol. *39:60–74, 1955.)*

well records. By means of contour lines, such maps show the areal distribution of some measurable characteristic of the unit being mapped. For example, contours may be drawn on the percentage of one lithologic component (such as clay) compared to the total unit or on the ratio of one rock type (such as sandstone) to the others within the unit. An isopachous map is ordinarily the base map for any of the quantitative maps, since one must know the total stratigraphic thickness of the unit with which individual components are compared.

Perhaps the most important fact to remember about the mapping techniques used in deciphering earth history is that they are limitless in variety. New and inventive schemes for mapping the past appear regularly in geologic journals as geologists continue to probe into the details of ancient geologic events.

# Importance of Sedimentary Rocks to Society Today

From caveman to astronaut, humans have learned to use sedimentary rocks to improve their lot. Men and women of antiquity fashioned tools from chert and flint, and they later learned the art of manufacturing useful containers from clay. Their successors developed the means of extracting copper and iron from the rocks and from these metals made more sophisticated implements. In addition to the metals they sometimes contain, sedimentary formations include coal seams and contain the oil, gas, and groundwater important to our present way of life.

## Mineral Fuels

The mineral fuels—petroleum, gas, and coal—are our most important sedimentary resources. They are essential for heat and power and metal refining; they are also the source of many chemicals useful in the manufacturing of plastics and fertilizers. **Coal** is a brownish-black to black combustible rock that occurs in beds from a few inches to many feet in thickness and is interbedded with shale, sandstone, and other sedimentary rocks. Extensive coal-bearing sequences characterize the Pennsylvanian and Cretaceous systems. Pennsylvanian sequences of strata in the Appalachians include over 100 individual beds of coal, each composed of the compressed and altered remains of land plants. Coal is abundant in the United States and will become increasingly more important as our reserves of petroleum are consumed.

Commercial accumulations of oil and gas require rather special geologic conditions. These conditions are found almost exclusively in sedimentary rocks. First, there must be a **source rock** for the petroleum. Ordinarily, this is a series of beds rich in the organic remains of unicellular organisms that accumulated along with the particles of sediment. Second, there must be a permeable **reservoir rock** such as sandstone or porous limestone to provide passage and storage for the gas and oil. The reservoir rock is covered by an impermeable **cap rock** to prevent the upward escape of hydrocarbons. Finally, there must be an oil-trapping structure (Fig. 2–50), so that the hydrocarbons can be concentrated in one place.

Petroleum geologists seek out these structures in many ways. Long before drilling programs are begun, geologists conduct a variety of exploratory surveys, the most useful of which are *seismic reflection surveys* (Fig. 2–51). In this technique, seismic waves are sent into the earth either by the detonation of dynamite in shallow drill holes or by the thumping action of hydraulic vibrators. The latter method is more frequently used today because it is safer and less disturbing to the environment. The depth to the stratum that reflects these waves is determined from the time required for the waves to travel down to a reflecting stratum and back to the surface, where they are detected by portable seismometers. Repeating the process along a predetermined traverse permits one to detect potential oil-bearing geologic structures deep beneath the surface.

Similar seismic exploration can also be accomplished for rock layers beneath the ocean floor by a process called *sub-bottom profiling*. In this process a survey vessel employs devices that emit strong signals produced by air guns or electric sparks. The signals are transmitted to the sea floor, penetrate into it, and are reflected back from the surfaces of sedimentary layers as deep as 10 km below the ocean floor. Hydrophones trailing behind the ship pick up the returned energy as the vessel steams along, and the moving graph paper of the recorder provides an

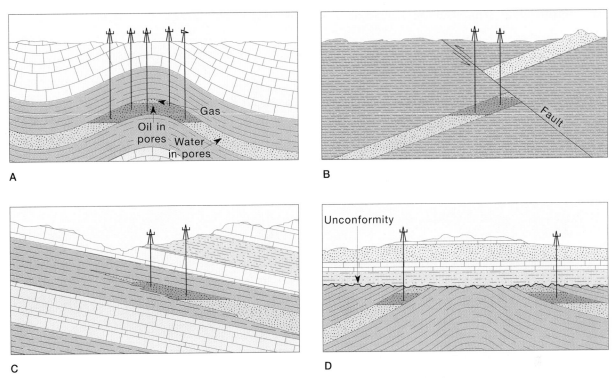

**Figure 2–50** Idealized cross-sections depicting geologic conditions that may lead to the accumulation of petroleum and natural gas. (A) Anticlinal trap. (B) Fault trap. (C) Stratigraphic trap. (D) Entrapment beneath an unconformity.

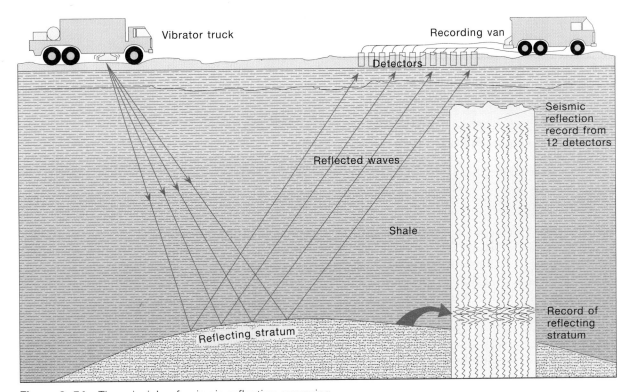

**Figure 2–51** The principle of seismic reflection surveying.

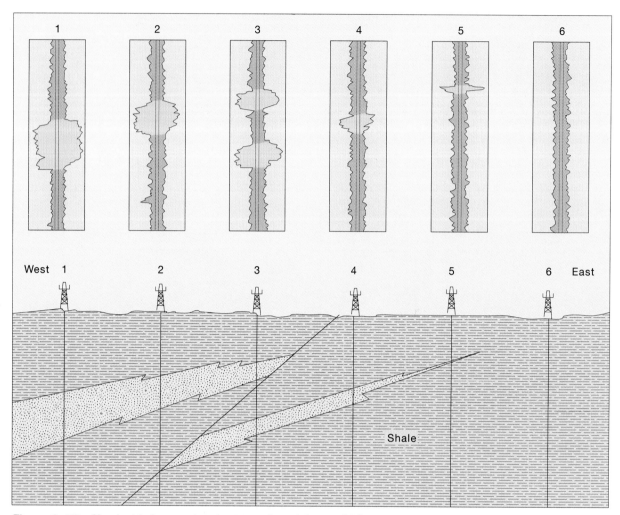

**Figure 2–52** Six electric logs and an interpretation of the geologic structure penetrated by wells from which the logs were prepared. The prominent "kicks" in the logs record a petroliferous sandstone. From a study of the electric logs, geologists developed an interpretation of a petroliferous sandstone that dips westward, pinches out toward the east, and is disrupted by a reverse fault.

uninterrupted cross-section of what lies deep beneath the bottom of the sea.

Later, as exploratory wells are drilled in an offshore or land area, the lithologic, paleontologic, and geophysical properties of the rocks penetrated are carefully recorded on well logs. Among the instrument-based logs used by geologists, none is more useful than electric logs. They are produced by devices wired to a cylindrical tool called a **sonde**. As the sonde is raised out of the drill hole, it measures the natural electrical properties of the formations with their contained fluids as well as their resistance to an induced flow of electricity. Rocks of different composition and fluid content (oil, fresh water, salt water) provide distinct patterns of lines on the electric log. The correlation of these patterns from well to well often reveals an accurate picture of the configuration of subsurface rock layers (Figure 2–52), and it is indispensible to modern exploration for oil and the development of oil fields.

## Construction Materials and Ores

Other sedimentary materials useful to humans include clay for use in ceramics and bricks, limestone for building stones and cement, sand and gravel for concrete and glass, and evaporites for use in the chemical industry. Some sedimentary strata are rich in iron oxide. Those near Birmingham, Alabama, have been mined and smelted for several decades. Because minerals such as gold, chromite, diamond, cassiterite (tin oxide), and magnetite are heavier than quartz and other common silicates, they tend to accumulate in quiet areas of streams or to be concentrated by waves as **placer deposits**. Finally, there are sedimentary ores that form in place by the deep chemical decay of a parent rock. The most important metal concentrated in this manner is aluminum.

Bauxite, an almost pure hydrous aluminum oxide, is the ore mineral of aluminum. Bauxites developed in once humid, tropical regions as a result of the weathering of source rocks rich in aluminum silicates.

## Limited Supply of Mineral Resources

Any commentary on mineral resources would be incomplete without noting that ore bodies and oil fields are limited in extent and are the result of unusual associations of geologic conditions. Furthermore, it is not theory but fact that they are exhaustible and irreplaceable. Their effect upon the standards of living and society is so enormous that every citizen should become involved in political decisions relating to the management of our natural resources.

## SUMMARY

Sedimentary rocks represent the material record of environments that once existed on the earth. For this reason, they are of great importance to the science of historical geology. All sedimentary rocks are formed by the accumulation and consolidation of the products of weathering derived from older rock masses, as well as by chemical precipitation and the accumulation of organic debris. Because of their mode of formation, the composition of sedimentary rocks provides information about source areas. The materials of sedimentary rocks are often transported by wind, water, or ice, or they are carried in solution to be precipitated in a particular environment of deposition. The transporting media imparts characteristics of texture or composition that can be used to reconstruct the depositional history and tectonic setting. Fossils in sedimentary rocks are splendid environmental indicators. They tell us if strata are marine or nonmarine, if the water was deep or shallow, or if the climate was cold or warm.

Color in a rock provides clues to the chemistry of the depositional medium. In the case of clastic rocks, the size, shape, and arrangement of grains can provide data about the energy of the transporting agent, the distance the grains had traveled, erosional recycling, and the degree to which movements of the earth's crust had disturbed a basin of deposition. Similar kinds of information are elucidated by such primary sedimentary structures as graded bedding, cross-bedding, and current ripple marks. Sandstones are particularly useful in paleoenvironmental studies. Graywackes, arkoses, quartz sandstones, and lithic sandstones accumulate in particular paleogeographic and tectonic situations. It is the geologist's task to discover the details of those situations by examining the rocks.

Geologists usually divide successions of sedimentary rocks into rock units that are sufficiently distinctive in color, texture, or composition to be recognized easily and mapped. Such rock units are called formations and are not necessarily of the same age throughout their areal extent. A time–stratigraphic unit differs from a rock unit in that it is an assemblage of strata deposited within a particular interval of time. The Cambrian System, for example, is a time–stratigraphic unit including all the rocks deposited in the Cambrian Period.

Within any given time–stratigraphic unit, one may find rocks that vary in composition, texture, organic content, or other features from adjacent rocks. These rock bodies of distinctive appearance or aspect are called facies. Facies reflect deposition in a particular environmental setting. For example, along a coastline one may find nearshore sand facies that change seaward to shale facies and carbonate facies. As shorelines shift landward or seaward, facies shift accordingly, maintaining their association with a particular local set of environmental conditions.

Shoreline migrations may be the result of such factors as worldwide changes in sea level (eustatic changes), tectonic movements of continental borderlands, or progradation—the seaward advance of the coastline resulting from rapid deposition of sediment brought to the sea by rivers. Progradation is particularly evident in deltas.

The study of facies is of great importance in the development of reconstructions of conditions on earth long ago. Geologists employ various graphic methods to record variations in facies and other attributes of sedimentary rocks. These methods include the preparation of lithofacies, biofacies, geologic, and isopach maps. If examined in chronologic sequence, such maps are useful not only in reconstructing ancient geography, but also in providing a picture of the changing patterns of lands and seas to which parts of the earth have been subjected.

Sedimentary rocks are the source of a great many materials necessary to modern civilization. Some of these materials, such as clay and building stones, are rather commonplace. Others, such as placer deposits of diamonds, chromite, and gold, are more limited in occurrence. Aluminum, a metal derived from a residual mineral named bauxite, is another extensively exploited sedimentary resource. However, the fossil fuels—coal and petroleum—are the most important of all sedimentary materials today. Indeed, oil, coal, and gas power our industrialized world. These resources are present on the earth in finite quantities and must be carefully managed.

## QUESTIONS FOR REVIEW

1. What is meant by the expression tectonic setting? What kind of past tectonic activity can be inferred for a region underlain by a great thickness (over 10,000 meters) of sediment that had been deposited in a marine basin no deeper than 200 meters?

2. Why are sandstones and siltstones of nonmarine desert environments rarely black or dark gray in color?

3. Under what circumstances might a reddish siltstone containing fossils of marine invertebrates have originated?

4. How does matrix in a rock differ from cement? What kinds of cements can bind a sedimentary rock together? Which cement is most durable?

5. What properties of a sedimentary rock can be used to determine its pre-existing parent rock, its mode of transportation, and its depositional environment?

6. What sort of history might you infer for a sandstone composed of very angular grains, poorly sorted and indurated with a 30 per cent matrix of mud?

7. What is the difference between a rock unit and a time-stratigraphic unit? How is each defined and recognized?

8. Distinguish between an immature and a mature sandstone.

9. Compare the probable tectonic environments in which a graywacke and a quartz sandstone might be deposited.

10. Why, in general, are fine-grained sediments deposited farther from a shoreline than coarse-grained sediments?

11. A geologist studying a sequence of strata discovers that limestone is overlain by shale, which in turn is overlain by sandstone. What might this signify with regard to the advance or retreat of a shoreline with time?

12. What features or properties of sedimentary rocks are useful in determining the direction of currents of the depositing medium? What features indicate relatively shallow-water deposition?

13. What are turbidity currents? How might turbidity current deposits be recognized?

14. What sort of paleogeographic information can be derived from lithofacies maps? From isopach maps?

15. What are the essential natural requirements for the occurrence of an oil pool?

## TERMS TO REMEMBER

age (geologic)
aragonite
arkose
biofacies
cap rock
cement
continental rise
continental shelf
continental slope
cross-bedding (cross-
  stratification)
eolian environment
eon
epeiric sea
  (epicontinental sea)

epoch
era
facies
formation
geologic map
graded bedding
graywacke
illite
isopachous map
kaolinite
lithic sandstone
lithofacies
matrix
offlap
onlap

paleogeographic map
period (of geologic
  time)
placer deposit
progradation
quartz sandstone
red beds
reservoir rock (for
  petroleum)
ripple marks
rock unit
rounding (of
  sedimentary particles)
sedimentary structures
series

smectite
sonde
sorting
source rock (for
  petroleum)
sphericity
stage
stratigraphy
system
tectonics
time-stratigraphic unit
turbidity current

## SUPPLEMENTAL READINGS AND REFERENCES

Davis, R. A. 1978. *Coastal Sedimentary Environments.* New York: Springer–Verlag.

Dunbar, C. O., and Rogers, J. 1957. *Principles of Stratigraphy.* New York: John Wiley.

Gallowy, W. E., and Hobday, D. K. 1983. *Terrigenous Clastic Depositional Systems.* New York: Springer–Verlag.

Hsu, K. J. 1989. *Physical Principles of Sedimentology: A Readable Textbook for Beginners and Experts.* New York: Springer–Verlag.

Miall, A. D. 1984. *Principles of Sedimentary Basin Analysis.* New York: Springer–Verlag.

Prothero, D. R. 1990. *Interpreting the Stratigraphic Record.* San Francisco: W. H. Freeman.

Scholle, P. A., and Spearing, D. 1982. *Sandstone Depositional Environments.* Tulsa: American Association of Petroleum Geologists.

Scholle, P. A., Bobout, D. G., and Moore, C. H. 1983. *Carbonate Depositional Environments.* Tulsa: American Association of Petroleum Geologists.

Tucker, M. E. 1981. *Sedimentary Petrology.* New York: John Wiley.

The trilobite *Isotelus brachycephalus* from strata of Ordovician age near Cincinnati, Ohio. The specimen is 23 cm in length. *(Courtesy of Wards Natural Science Estab-*

# Life Through Time

*The search for a fossil may be considered at least as rational as the pursuit of a hare.*

William Smith (1769–1839)

## The Nature of the Fossil Record

The earth is about 4.6 billion years old. Life has been present on it for about 3.5 billion years. The science that seeks to understand all aspects of the succession of plants and animals over that great expanse of geologic time is called **paleontology.** It is a science based on the study of **fossils,** the remains or traces of ancient life. As paleontologists study these relics of former living organisms, they endeavor to learn how these organisms lived and how they grew. What was the organism's true fleshed-out appearance? How did it interact with other organisms and its physical environment? What were its ancestors like, and what changes occurred in its descendants? Why did these changes occur?

### Preservation

When one considers the many ways by which organisms are completely destroyed after death, it is remarkable that fossils are as common as they are. Attack by scavengers and bacteria, chemical decay, and destruction by erosion and other geologic agencies make the odds against preservation very high. However, the chances of escaping complete destruction are vastly improved if the organism happens to have a mineralized skeleton and dies in a place where it can be quickly buried by sediment. Both of these conditions are often found on the ocean floors, where shelled invertebrates flourish and are covered by the continuous rain of sedimentary particles. Although most fossils are found in marine sedimentary rocks, they also are found in terrestrial deposits left by streams and lakes. On occasion, animals and plants have been preserved after becoming immersed in tar or quicksand, trapped in ice or lava flows, or engulfed by rapid falls of volcanic ash.

The term *fossil* often implies petrifaction, literally a transformation into stone. After the death of an organism, the soft tissue is ordinarily consumed by scavengers and bacteria. The empty shell may be left behind, and if it is sufficiently durable and resistant to dissolution, it may remain basically unchanged for a long period of time. Indeed, unaltered shells of marine invertebrates are known from deposits over 100 million years old. In many marine creatures, however, the skeleton is composed of a mineral variety of calcium carbonate called aragonite. Although aragonite has the same composition as the more familiar mineral known as calcite, it has a different crystal form, is relatively unstable, and in time changes to the more stable calcite.

Many other processes may alter the shell of a clam or snail and enhance its chances for preservation. Water containing dissolved silica, calcium carbonate, or iron may circulate through the enclosing sediment and be deposited in cavities once occupied by veins, canals, nerves, or tissues. In such cases, the

**Figure 3–1** Petrified wood from the Petrified National Forest in Arizona. This is an example of petrification by replacement; the original wood has been replaced by silica. The upper specimen is a sawed and polished log, about 46 cm in diameter. The lower photograph depicts the surface texture of a similarly replaced log. These fossils occur in the Chinle Formation of Triassic age.

**Figure 3–2** This fossil seed fern from rocks of Pennsylvanian age has been preserved by carbonization. The frond is approximately 27 cm in length.

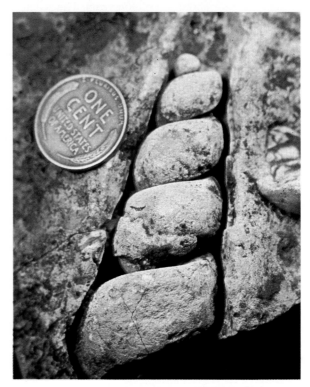

**Figure 3–3** An internal mold (or steinkern) formed by filling of the spiral cavity of an ancient marine snail.

original composition of the bone or shell remains, but the fossil is made harder and more durable. This addition of a chemically precipitated substance into pore spaces is termed **permineralization.**

Petrifaction also may involve a simultaneous exchange of the original substance of part of a dead plant or animal with mineral matter of a different composition. The process is termed **replacement** because solutions have dissolved the original material and replaced it with an equal volume of a new substance. Replacement can be a marvelously precise process, so that even fine growth lines on shells and delicate structures in petrified wood may be accurately preserved (Fig. 3–1).

Another type of fossilization, known as **carbonization,** occurs when soft tissues are preserved as thin films of carbon. Leaves and tissue of soft-bodied organisms such as jellyfish or worms may accumulate, become buried and compressed, and lose their volatile constituents. The carbon often remains behind as a blackened silhouette (Fig. 3–2).

Fossils may also take the form of molds, imprints, and casts. Any organic structure may leave an impression of itself if it is pressed into a soft material and if that material is capable of retaining the imprint. Commonly, among shell-bearing invertebrates, the shell is dissolved after burial and lithification, leaving a vacant **mold** bearing surface features of the original shell that are opposite to those on the shell itself (ridges are represented by grooves, knobs by depressions, and vice versa). If external features (growth lines, ornamentation) of the fossil are visible, the mold is an *external mold*. Conversely, the *internal mold* shows features of the inside of a shell, such as muscle scars or supports for internal organs. Many invertebrate shells enclose a hollow space that may be either left empty or filled with sediment (Fig. 3–3). The filling is called a *steinkern* (stone core). Finally, molds may be subsequently filled, forming **casts** that faithfully show the original form of the shell (Fig. 3–4).

Although it is certainly true that the possession of hard parts enhances the prospects of preservation, organisms having soft tissues and organs are also occasionally preserved. Insects have been found preserved in the hardened resins of conifers and certain other trees (Fig. 3–5). X-ray examination of thin

**Figure 3–4** A cast (above) and mold (below) of a trilobite formed in a nodule of calcareous shale. The trilobite (*Calliops*) is 4 cm long.

**Figure 3–5** Worker ant (*Sphecomylma freyi*) preserved in Cretaceous amber. *(Photograph courtesy of Professor Frank M. Carpenter.)*

**Figure 3–6** In the summer of 1977, the carcass of this baby mammoth was dug from frozen soil (permafrost) in northeastern Siberia. The mammoth stood about 104 cm tall at the shoulders, was covered with reddish hair, and was judged to be only several months old at the time of death. Dating by the radiocarbon method indicates death occurred 44,000 years ago. *(Photograph courtesy of Klavdija Novikova, Biologopoczvennyj Institut, Vladivostok, U.S.S.R.)*

slabs of rock sometimes reveals the ghostly outlines of tentacles, digestive tracts, and visual organs of a variety of marine creatures. Soft parts, including skin, hair, and viscera of ice age mammoths have been preserved in frozen soil (Fig. 3–6) or in the oozing tar of oil seeps. A striking example of preservation of soft parts was discovered in 1984 at Lindow Moss, England, when an excavating machine uncovered the body of a 2000-year-old human. Named Lindow Man (Fig. 3–7), the Briton had been ritually slaughtered (possibly by Druids), stripped, garroted, and then bled. The body was placed in a pool in the bog where the skin was preserved and the body flattened by the weight of accumulating peat. Only the upper part of the body was recovered, the lower part having been destroyed by the machine before the remains were detected.

Evidence of ancient life does not consist solely of petrifactions, molds, and casts. Sometimes the paleontologist is able to obtain clues to an animal's appearance and how it lived by examining tracks, trails (Fig. 3–8), burrows, and borings. Such markings are called **trace fossils**, and the study of trace fossils is termed **ichnology**. The tracks of an ancient vertebrate animal may indicate whether the animal that made them was bipedal (walked on two legs) or

**Figure 3–7** Preserved torso, arms, and head of the 2000-year-old Lindow Man. This example of preservation of soft tissue was found in a peat bog in 1984 at Lindow Moss, England. The lower half of the body was destroyed by the excavating machine. *(British Museum.)*

**Figure 3–8**  Trails made by millipedes about 250 million years ago during the Permian Period. These trace fossils were discovered in the Coconino Sandstone, Grand Canyon National Park. Millipedes are elongated segmented animals with a large number of paired legs. Along with insects and crustaceans, they are members of the Phylum Arthropoda. *(Courtesy of U.S. Geological Survey, photograph by E. D. McKee.)*

quadrupedal (walked on four legs), digitigrade (walked on toes) or plantigrade (walked on the flat of the foot), whether it had an elongate or a short body, if it was lightly built or ponderous, and sometimes whether it was aquatic (with webbed toes) or possibly a flesh-eating predator (with sharp claws).

Trace fossils of invertebrate animals are more frequently found than the tracks of vertebrates, and they are also useful indicators of the habits of ancient creatures. One can sometimes infer if the trace-making invertebrate was crawling, resting, grazing,

feeding, or simply living within a relatively permanent dwelling. For example, crawling traces, as might be expected, are linear and show directed movement (Fig. 3–9A). Shallow depressions that reflect the shape of the animal may be resting traces (Fig. 3–9B). Simple or **U**-shaped structures approximately perpendicular to bedding are often dwelling traces (Fig. 3–9C). Grazing traces (Fig. 3–9D) occur along bedding planes and are characterized by a systematic meandering or concentric and parallel patterns that represent the animal's effort to cover the

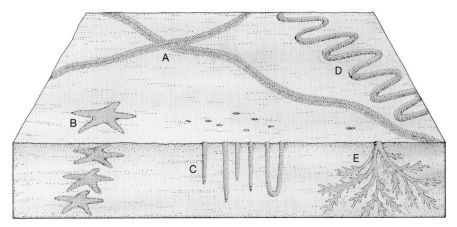

**Figure 3–9**  Traces that reflect animal behavior: (A) crawling traces, (B) resting traces, (C) dwelling traces, (D) grazing traces, and (E) feeding traces.

area containing food in an efficient manner. The three-dimensional counterparts of grazing traces are called feeding traces. Feeding traces are made by animals that consume sediment for the organic nutrients within it. The traces consist of systems of branched or unbranched burrows, as shown in Figure 3–9E.

## The Incomplete Record of Life

The fossil record of life is incomplete. If it were a complete and total record, it would include information on all past forms of life for every increment of time and for every place on earth. Clearly this is unattainable. Only a limited number of animals and plants have been preserved. Others that were preserved have never been exposed to our view by erosion or drilling. Still others have simply not yet been discovered.

The record is rather more complete for marine life having hard external skeletons and less complete for land life lacking bone or shell. Where burial is rapid and dead organisms are protected from scavengers and agents of decay, the probability of preservation improves. As we have noted, the ocean floor and low-lying land areas where deposition predominates are often favorable places for fossil preservation. Many dinosaur remains, for example, have been discovered in the sands and clays deposited by streams that flowed across the low plains that lay east of the Rocky Mountains during the Cretaceous Period. There were certainly animals and plants living in nearby highland areas as well, but in such places erosion rather than deposition predominated and living things were less likely to escape destruction.

In spite of the many factors that prevent fossilization, the fossil record is remarkably comprehensive. On the basis of this impressive record, paleontologists have been able to piece together a history of past life that is balanced and largely accurate.

# The Rank and Order of Life

## The Linnaean System of Classification

Because of the large number of living and fossil animals and plants, random naming would be confusing and inefficient. Realizing this, the Swedish naturalist Carl von Linné, also known as Carolus Linnaeus, (1707–1778) established a carefully conceived system for naming animals and plants. The **Linnaean system** uses morphologic structure as the basis of classification and employs binomial nomenclature at the species level. In this scheme, the first, or **generic**, name was used to indicate a general group of creatures that were visibly related, such as all doglike forms. The second, or trivial, name denoted a definite and restricted group—a species. *Canis lupus,* for example, designates the wolf among all canids. Linné went on to recognize larger divisions such as classes and orders. His groupings were based on traits that seemed most basic or natural; the modern nomenclature system, however, is based on an attempt to be consistent with evolutionary relationships.

## Concepts Involved in Classification

### The Species

The **species** is the fundamental category in the modern classification of living things. A species consists of a population or group of populations composed of individuals that are actually or potentially able to interbreed. The ability to interbreed proves that individuals of a population are closely related genetically. Species exist during a relatively short period of geologic time, are derived from ancestral populations (Fig. 3–10), and in turn are the population from which new species arise.

In deciding on the validity of giving a species name to a group of seemingly related and similar organisms, biologists have the advantage of being able to determine whether or not the organisms can interbreed. This, of course, is not possible for an extinct population known only from fossils. Thus, the designation of a paleontologic species is unavoidably subjective. The paleontologist must determine the range of variation within the group of organisms being studied and make a logical inference about their inclusion within a designated species. In doing this, one compares the range of variation among individuals of the proposed fossil species with the range of variation within an analogous living species.

### Taxonomy

The species is the basic unit in biologic classification, or **taxonomy.** In this system, the various categories

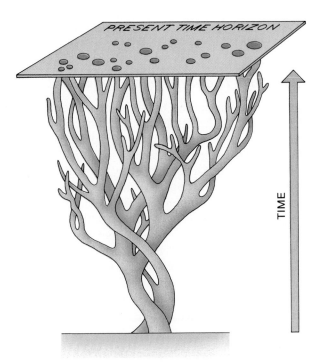

PRESENT TIME HORIZON

TIME

**Figure 3–10** Diagram of a family tree of organisms. Living species appear as mostly isolated groups. Many of these, however, can be traced backward to branch junctions that represent common points of ancestry. Extinctions are represented by branches which do not reach the present time horizon. The tree is drawn according to the phyletic gradualistic model of evolution. For an example of the punctualistic model, see Figure 3–14.

of living things are arranged in a hierarchy that expresses levels of kinship. For example, a **genus** (pl. *genera*) is a group of species that have close ancestral relationships; a **family** is a group of related genera; an **order** is a group of related families; a **class** is a group of related orders; a **phylum** (pl. *phyla*) is a group of related classes; and a **kingdom** is a large group of related phyla. To use an example familiar to all, individual humans are members of the Kingdom Animalia, Phylum Chordata, Class Mammalia, Order Primates, Family Hominidae, Genus *Homo,* and species *Homo sapiens.*

Biologic classification is more than a mere system for cataloging various organisms. Because it is based on organic structures, it also depicts the broad outlines of ancestral evolutionary relationships. In a sense, the classification is a depiction of the tree of life. (For reference, a simplified classification of the major groups of frequently fossilized living and extinct organisms is provided in Appendix A).

**Rules of Nomenclature.**   Because of the incredible variety of living and once-living organisms, scientists at the turn of the century recognized the need for a set of rules governing scientific nomenclature. To meet this need, the international congresses of zoologists and biologists established Commissions on Nomenclature and these commissions drafted the *International Code of Zoological and Botanical Nomenclature.* The code states that no two genera within a Kingdom may have the same name. It stipulates that scientific names must be either Latin or Latinized and be printed in italics. Further, the name of the genus must be a single word (nominative singular) and must begin with a capital letter. The species name begins with a small letter and must agree grammatically with the generic name. To prevent duplication of names, only the generic or specific name that is first published is considered valid, provided it is suitably described in an acceptable publication.

## Organic Evolution

The Roman poet Ovid (43 B.C.E.–A.D. 17) once wrote that "There is nothing constant in the universe, all ebb and flow, and every shape that's born bears in its womb the seeds of change." The words are remarkably relevant when one considers the way life has changed through time as revealed by the fossil record. At times that record indicates that change may have been startlingly sudden, whereas at other times evidence shows a more gradual change. In either case, older forms have changed or evolved into newer forms so as to better cope with changes in their environment. This is not to say that the early, relatively simpler forms of life are gone, for many persist along with their often more complex contemporaries.

The concept that animals and plants have changed through time is not a new one. Anaximander (611–547 B.C.) alluded to the idea 25 centuries ago. This famous Greek taught that life arose from mud warmed in the sun; that plants came first, then animals, and finally humans. Unfortunately, during the Middle Ages, complete faith in the theologic doctrine of special creation of "species" effectively stifled imaginative thinking about evolution. It was not until the eighteenth century in Europe that intellectuals such as Jean Lamarck and Georges Buffon began to challenge the concept of special creation. Buffon was convinced that evolution had occurred, but offered no explanation of its cause. He believed, however, that the environment was somehow involved in the process that resulted in evolution. Buffon also set down a careful definition of species, noting that species were separate entities that were not able to interbreed successfully with other dissimilar species. The first general theory of evolution to excite considerable public attention was developed by the celebrated naturalist Jean Baptiste de Lamarck in the period between 1801 and 1815. Lamarck's writings were followed a half century later by those of Charles Darwin and Alfred R. Wallace.

### Lamarck's Theory of Evolution

The Lamarckian theory of evolution stipulated correctly that all species, including humans, are descended from other species. His theory, however, was based on the mistaken assumption that new structures in an organism appear because of needs or "inner want" of the organisms and that structures once acquired in this way are somehow inherited by later generations. In similar but reverse fashion, little-used structures would disappear in succeeding generations. For example, Lamarck believed that snakes evolved from lizards that had a strong preference for crawling. Because of this inner need to have long, thin bodies, certain lizards developed such bodies, and because legs became less and less useful in crawling, these structures gradually disappeared. Lamarck's ideas were challenged almost immediately. There was no way to prove by experimentation that such a thing as "inner want" existed. More important, Lamarck's belief that characteristics acquired during the life span of an individual could then be inherited was tested and shown to be invalid. We know that some organisms can "adapt" to environmental conditions in a narrowly limited way within a life span. For example, a fair-skinned woman on the beach can be tanned by exposure to the sun, but no amount of sunbathing will result in her first child being born with a suntan. There is no way that somatic (or body) cells can pass characteristics over to reproductive cells and thereby on to the next generation. Thus, the Lamarckian concept of

evolution based on use or disuse of organs was discredited.

### Darwin's Theory of Natural Selection

Charles Darwin praised Lamarck for his courage in perceiving that the descendents of former creatures have undergone biologic change and are different from their ancestors. As described in Chapter 1, Darwin and his younger contemporary Alfred R. Wallace jointly proposed natural selection as an important mechanism for evolution. Darwin, with his years of careful data collection, and Wallace with his sudden insight, had discerned that competition for food, shelter, living space, and sexual partners among species with individual variations and surplus reproductive capacity will inevitably result in elimination of the less well fitted and survival of those that are better fitted to their environments.

Darwin's *The Origin of Species by Means of Natural Selection* was published in 1859. It appeared at a time when at least part of the European intellectual atmosphere was more liberal and less satisfied with the theologic doctrine that every species had been independently created. Although the ideas of Darwin and Wallace continued to disturb the religious feelings of some, they nevertheless increasingly acquired adherents among nineteenth and twentieth century scientists. Darwin did not consider his views impious. He saw a "grandeur in this view of life with its several powers, having been originally breathed by the Creator into a few forms or into one; and that . . . from so simple a beginning endless forms most beautiful and most wonderful have been and are being evolved."

### Mendelian Principles of Inheritance

Every theory has its strong and weak points. In Darwin's theory, the weakness lay in an inability to explain the *cause* of variability in a way that could be experimentally verified. The cause of at least a part of that variability was discovered in the course of elegant experiments on garden peas conducted by a Moravian monk named J. Gregor Mendel (1822–1884). Mendel discovered the basic principles of inheritance. His findings, printed in 1865 in an obscure journal, were unknown to Darwin and unheeded by the scientific community until 1900, when the article was rediscovered. Mendel described the mechanism by which traits are transmitted from adults to offspring. In his experiments with garden peas, he demonstrated that heredity in plants is determined by "character determiners" that divide in the pollen and ovules and are recombined in specific ways during fertilization. Mendel called these hereditary regulators "factors." They have since come to be known as *genes*.

As currently understood, genes are chemical units or segments of a nucleic acid—specifically **deoxyribonucleic acid (DNA)**. As suggested by careful chemical and X-ray studies, the DNA molecule is conceived as two parallel strands twisted somewhat like the handrails of a spiral staircase (Fig. 3–11). The twisted strands are made up of phosphate and sugar compounds and are linked with cross-members composed of specific nitrogenous bases.

The importance of DNA is evident when we realize that it indirectly controls the production of proteins, the essential components of many basic structures and organs. Even the activities of organisms are regulated by specific catalytic proteins called enzymes. Without DNA and its products, there would be no life as we know it. Its ability to precisely replicate itself is the basis for heredity, and organic evolution ultimately depends upon this remarkable molecule.

The part of the DNA molecule that is active in the transmission of hereditary traits is called a **gene**. In nearly all organisms, genes are linked together to form larger units termed **chromosomes**, the central axis of which consists of a very long DNA molecule comprising hundreds of genes.

## Reproduction and Cell Division

Reproduction of an organism may be sexual or asexual or may involve an alternation of sexual and asexual methods. All reproductive methods involve the division of cells. In sexual reproduction there is a union of reproductive or sex cells from separate individuals, whereas in asexual reproduction cells do not unite. The more usual methods of asexual reproduction are binary fission, budding, and spore production. **Binary fission** is found in single-celled organisms such as *Amoeba* that divide to form genetically identical daughter organisms. **Budding** occurs in some unicellular as well as multicellular organisms. In this method, the parent organism simply sprouts a bulge or appendage that may either remain attached

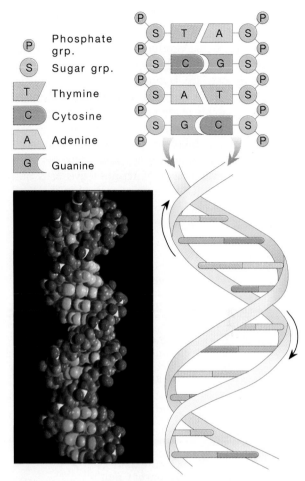

**Figure 3–11**   Representations of portions of the deoxyribonucleic acid (DNA) molecule. At the lower left is a computer reconstruction by R. J. Feldmann. The drawing depicts the twisted, double-stranded helix, with side rails composed of alternate sugar (deoxyribose) and phosphate molecules. Each rung of the twisted ladder is composed of one pair of nitrogenous bases. Of these, thymine links to adenine, and cytosine to guanine.

to the parent or separate and grow as an isolated individual. The third principal method involving asexual reproduction is the formation of tiny reproductive cells called **spores** by a parent organism. The spores are formed by division of special spore parent cells and are shed by the parent organism (called the *sporophyte*) when they mature. When a spore settles onto moist soil, the floor of a water body, or some other suitable surface, it germinates and grows into a tiny plant (called a *gametophyte*) that in turn produces male and female sex cells. The union of these cells produces a new plant resembling the original sporophyte.

Organisms that produce asexually have the advantage of increasing their numbers rapidly whenever conditions are favorable. Some disadvantage exists, however, in that they are not able to develop as much variability among their offspring as do sexually reproducing organisms. To understand the reasons for that variability, we need to know some simple facts about chromosomes and their behavior during the development of reproductive cells and during fertilization. We note initially that the kind and number of chromosomes are constant for each species and differ between different species. Except in bacteria and blue-green algae, the chromosomes are located in the nucleus of the cells and occur in duplicate pairs. Thus, each chromosome has a homologous mate. In humans, for example, there are 46 chromosomes, or 23 pairs. Cells with paired homologous chromosomes are designated **diploid** cells. In all living things, new cells are being produced constantly to replace worn-out or injured cells and to permit growth. In asexual organisms, and in all the **somatic** or body cells of sexual organisms, the process of cell division that produces new diploid cells with exact replicas of the chromosomal components of the parent cells is called **mitosis** (Fig. 3–12).

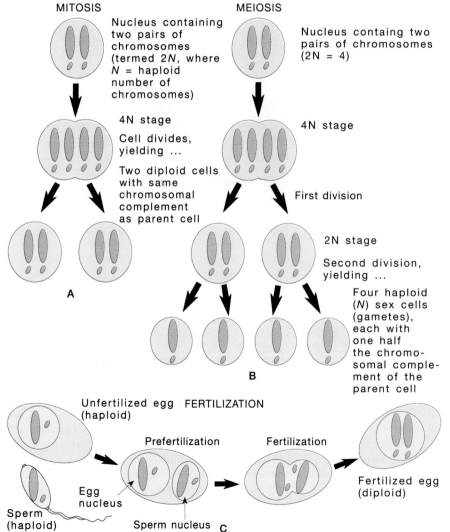

**Figure 3–12** Greatly simplified summary comparison of the major features of (A) mitosis, (B) meiosis, and (C) fertilization.

In most organisms with sexual reproduction, a second type of division, called **meiosis**, takes place when **gametes** (egg cells or sperm cells) are formed. Meiosis may occur in unicellular organisms, but in multicellular forms it takes place only in the reproductive organs (testes or ovaries). Meiosis consists of two cell divisions. In the initial division, the chromosomal pairs are divided in random fashion so that each of the potential new cells receives one member of each original pair. These daughter cells are termed **haploid** because they do not have paired chromosomes. The haploid cells then reproduce themselves exactly as in mitosis to form a total of four haploid cells, which are the gametes, or reproductive cells. When two gametes meet during sexual reproduction, the sperm enters the egg to form a single cell, which can now be called the **fertilized egg**. Because two gametes have been combined into one cell, there is now a full complement of chromosomes and genes representing a mix from both parents. The fertilized egg now begins a process of growth by mitotic cell division that eventually leads to a complete organism.

The organism that develops from the union of haploid cells will already have some variability relative to its parents. There is, however, yet another aspect of meiosis that is important in producing variation in offspring. During the initial chromosome division while the chromosomes are still paired, they may break at corresponding places and exchange their severed segments in a process called **crossing over**. The result is an additional mixing of genes.

## Mutations

Some of the variation we see among individuals of the same species results from the mixing of genes that occurs during reproduction. However, if genes were never altered, then the number of variations they could produce would be limited. For organisms to evolve a truly new variation, a process is required that will change the genes themselves. That process is known as **mutation**. Mutations can be caused by ultraviolet light, cosmic and gamma rays, chemicals, and certain drugs. They may also occur spontaneously without a specific causative agent. Mutations may occur in any cell, but their evolutionary impact is greater when they occur in the sex cells, for then succeeding generations will be affected.

To understand how mutations occur, it is useful to re-examine the configuration of DNA (see Fig. 3–11). The steps on the twisted DNA ladder are composed of two kinds of nitrogenous bases: *purines* and *pyrimidines*. Each step is made up of one purine joined to one pyrimidine. The purine called *adenine* is normally coupled with the pyrimidine called *thymine*. Similarly, the purine *guanine* joins with the pyrimidine *cytosine*. Thus, adenine–thymine and guanine–cytosine form *base pair* steps of the ladder. Genes owe their specific characteristics to a particular order of these base-pair steps. If the order is disrupted, a mutation may result. For example, during cell division, when the twisted strands separate and then proceed to attract bases to rebuild the DNA molecule, there may be a mishap that results in the positioning of a guanine–cytosine base pair in what should have been the location of an adenine–thymine base pair. At that location, the gene is altered and may result in an inheritable change.

Clearly, organic evolution may involve change from at least three different sources: mutations, gene recombinations, and natural selection. Mutations are the ultimate source of new and different genetic material. Recombination (mixing) spreads the new material through the population and mixes the new with the old. Natural selection sorts out the multitude of varying traits, preserving those that by chance are best fitted to a particular environment.

## Evolution in Populations

Organisms do not live as totally isolated individuals. Each is part of a **population**, a term used to describe a group of individuals that live close enough together so that each individual has an equal chance to mate with all members of the opposite sex within the group. Populations whose members can interbreed when they come into contact are grouped into that larger biologic entity called species. A species may be composed of a small geographically limited population, such as the sea snail *Retusa obtusa* which is found only in the waters west of Great Britain, or the species may include populations with wide geographic distributions, such as the brine shrimp *Artemia salina* which occurs in saline lakes on all continents except Antarctica and Australia. Because it contains the total genetic composition of an entire group of interbreeding individuals, a population can be considered the basic unit of evolution. Each of the individuals within the population has its own particular combination of genes, and the sum of all these combinations constitutes the *gene pool* of the population. The gene pool is divided up in each generation and partitioned out to the new offspring. Inevitably, in each successive generation, new mutations and genetic combinations occur that manifest themselves in the offspring. As natural selection comes into play, some of these inherited traits will be passed on to the next generation in greater numbers, others in lesser numbers, so that ultimately the gene pool is altered. Thus, a source of evolutionary change becomes the impact of natural selection upon the gene pool.

## Speciation and Adaptive Radiation

The entire course of evolution depends upon the origin of new species. The process by which this occurs is called *speciation*. We have seen that a species is a collection of populations within which there is a free flow of genes. From this definition, it is implicit that there is no gene exchange between two different species. Barriers between different species keep the gene pools separate. These barriers may be reproductive barriers that prevent fertilization, mating, or development of offspring, or they may be geographic barriers, such as islands that separate and isolate animals. Wherever reproductive or geographic barriers exist, various segments of the population become isolated for many generations. During this period, they are likely to accumulate enough genetic differences so that interbreeding between the segments is no longer possible. Once this has occurred, the isolated segments have become different species.

After a new species has appeared, its success will depend upon the advantages of its particular attributes relative to the environment. Species may encounter severe competition for food, shelter, living space, and other vital needs, and this may shorten their life span. If, however, a species is able to cope well with its environment, it may increase and persist. The average duration for most fossil invertebrate species has been 5 to 10 million years, a short time from a geologic point of view.

Once a new population has become established, pioneering segments of the population located around the fringes of the habitats may, like their parent species, undergo additional speciation. With innumerable successive speciations, diverse organisms characterized by diverse living strategies emerge. This branching of a population so as to produce descendants adapted to particular environments and living strategies is termed **adaptive radiation.**

The honey creepers of Hawaii provide an illustration of adaptive radiation. These birds, comprising many different species, are believed to have descended from a common ancestor. The most striking differences between the species occur in the sizes and shapes of the beaks, which are adaptively related to the kind of food the birds are dependent upon (Fig. 3–13). Some have stout beaks for crushing hard seeds, other have beaks well adapted for seeking out insects in cracks and crevices or for sucking nectar. All of these variations in beak morphology are examples of **adaptations**, a word that means the acquisition of heritable characteristics that are advantageous to an individual and a population.

On a higher taxonomic level, one finds adaptive radiations among classes, orders, and families of organisms. All the orders of mammals, for example, had a common origin in an ancestral species that lived during the early Mesozoic. By adapting to different ways of life, the descendent forms came to diverge more and more from the ancestral stock, thereby providing today's rich diversity of mammalian plant eaters, flesh eaters, insect eaters, walkers, climbers, swimmers, and flyers.

## Gradual or Punctuated Evolution

A question that is presently being widely discussed by paleontologists is whether evolution proceeds gradually by means of an infinite number of subtle steps, as Charles Darwin proposed, or if there are sudden sporadic advances during which most of the change in a lineage occurs over a relatively short span of time. Gradual progressive change is referred to as **phyletic gradualism.** Proponents of phyletic gradualism believe that a given lineage of organisms changes from each ancestral to descendent group by slow degrees, and where striking evolutionary breaks occur, they are not true breaks at all, but merely what appear to be sudden changes because of imperfections in the geologic record. Darwin would insist that the transitional forms actually once existed, but were either not preserved, were destroyed, or have not yet been discovered.

Evidence for phyletic gradualism has been found in several groups of invertebrates, including lineages

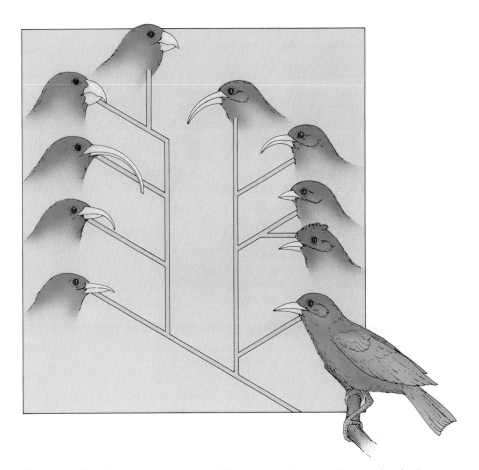

**Figure 3–13** The honey creepers of Hawaii provide a good example of adaptive radiation. When the ancestor of today's honey creepers first reached Hawaii, few birds were present. Succeeding generations diversified to occupy various ecologic niches. Their diversity is most apparent in the way their beaks have become adapted to different diets. Some have curved bills to extract nectar from tubular flowers, whereas others have short, sturdy beaks for cracking open seeds, or pointed bills for seeking out insects in tree bark. *Himatione sanguinea,* at the lower right, has a relatively unspecialized beak and appears to be similar to the original honey creeper ancestor.

of well-preserved trilobites from the Ordovician of Wales, as well as among the marine planktonic organisms known as foraminifers. In the latter study, the fossil foraminifers were recovered from complete and uninterrupted well-dated sections of deep sea cores. The fossils indicated a gradualistic line of descent that produced at least four species of the genus *Globorotalia* over a span of about 8 million years.

Paleontologists who oppose phyletic gradualism argue that the fossil record for the past 3 billion years contains many examples of new groups appearing suddenly in places where there is no evidence of any imperfections in the geologic record. Indeed, they find that the slow and stately advance of evolution can be documented only rarely and that evolutionary progress is more often sporadic than slow and uniform. Although this concept was suggested more than a half century ago, it has recently received new and strong support from paleontologists such as Stephen J. Gould and Niles Eldridge in their research on Pleistocene snails and Devonian trilobites. Gould and Eldridge have suggested the term **punctuated**

**equilibrium** for evolution that consists of fitful sudden advances that punctuate long episodes of little evolutionary progress.

According to this concept, the punctuation or sudden morphologic change that interrupts equilibrium occurs at the periphery of the geographic area occupied by the population. These segments of the population are referred to as **peripheral isolates**. Gene flow is rapid among the individuals of peripheral isolates, and changes in morphology or physiology leading to speciation may occur within a short span of time. Species do not usually originate in the places where their parental stock exists, but rather in boundary zones where variations can be tested against new environmental situations. Should the parent species suffer extinction or severely decline, the new species may move back into the parental domain or they may expand into new territory. With adaptations that are a significant advantage, they are likely to enter a period of rapid evolution. This stage is usually followed by a period of moderate evolutionary rates and stability that lasts until the population begins to decline. Subsequently, yet another new group may begin its evolutionary expansion.

The final chapter on the question of punctuated evolution versus phyletic gradualism has not been written. At present, the proponents of punctuated evolution appear to be more numerous than those of phyletic gradualism. Like most controversies in science, however, the answer need not lie totally in one camp, and it is evident that instances of phyletic gradualism can also be recognized in the fossil record of certain groups of plants and animals.

## Patterns of Evolution

The retrospective view of life through time provided by the fossil record affords an opportunity to detect overall patterns of evolution. The most familiar way of depicting these patterns is by a diagram called a **phylogenetic tree**, in which various taxa are traced from outermost branches (representing lower taxonomic divisions such as species, genera, or families) through a succession of junctures downward (and backward in time) to the major trunks of the higher taxa. On such a diagram, time is plotted on the vertical axis and biologic change on the horizontal. For the punctuated equilibrium model of evolution (Fig. 3–14A), the phylogenetic tree has short hori-

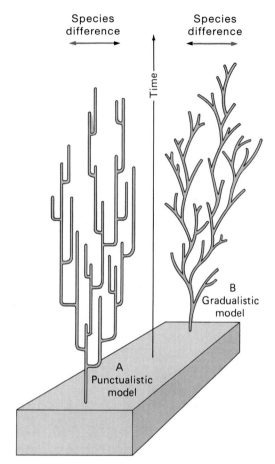

**Figure 3–14** Phylogenetic trees depicting (A) the punctualistic model and (B) the gradualistic model of evolution. Morphologic change occurs in sideward directions. Time is depicted by the vertical direction. The short horizontal side branches of the punctualistic model depict sudden change, whereas the inclined branches of the gradualistic model suggest slow and uniform change through time.

zontal branches (reflecting sudden morphologic change) that quickly become vertical (to reflect subsequent stability through time). In contrast, the phyletic gradualism model (Fig. 3–14B) shows gently inclined branches that indicate slow and gradual change through time. Branching occurs by gradual deviation of forms.

The phylogenetic tree serves as a model for depicting particular styles of evolution. As is evident from Figure 3–14B, a branch representing a single lineage may simply divide to form two or more line-

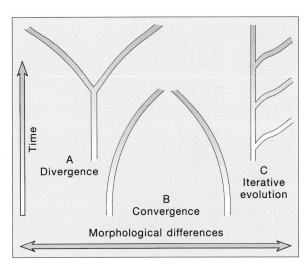

**Figure 3–15** Three common evolutionary patterns as indicated by morphologic change among different lineages. (A) The *divergence* of two lineages. (B) Two lineages evolve toward morphologic similarity as an example of *convergence.* Presumably, this convergence results from selective pressures to adapt to similar environmental conditions. (C) In *iterative evolution,* morphologically similar forms evolve repeatedly from a root stock.

ages. This kind of evolutionary pattern is called **divergence** (Fig. 3–15A). **Convergence** (Fig. 3–15B) occurs when two or more unrelated lineages acquire similar morphologic traits, usually because of adaptations to similar modes of life and environments. A familiar example of convergence is seen in the evolution of superficially similar structures—wings—in birds, bats, and flying reptiles of the Mesozoic known as pterosaurs. **Iterative evolution** refers to the evolution of similar morphology more than once from the parental lineage (Fig. 3–15C). An example is provided by the coiled chambered mollusks known as cephalopods. Three or more times during the history of this class of invertebrates, irregularly coiled forms evolved from the uniformly coiled parental lineage.

## Evidence for Evolution

### Paleontologic Clues

The proponents of the concept of evolution had a difficult time convincing some of their nineteenth century contemporaries of the validity of Darwin's theory. One reason was that evolution is an almost imperceptibly slow process. The human life span is too short to witness evolutionary changes across generations of plants or animals. Fortunately, we can overcome this difficulty by examining the remains of organisms left in rocks of successively younger age. If life has evolved, the fossils preserved in consecutive

formations should exhibit those changes. Indeed, many examples are known of sequential morphologic changes among related creatures during successive intervals of geologic time. The most famous example is provided by fossil horses recovered from formations of the Cenozoic Age (Fig. 3–16). The oldest horses thus far discovered as fossils were about the size of fox terriers. Unlike modern horses, they had four toes on their front feet and three on the rear. As one examines the many branching lineages of the horse "family tree" in successively higher (hence, younger) formations, one is able to see the results of evolutionary change. The animal shows an increase in size, a reduction of side toes with emphasis on the middle toe, an increase in the height and complexity of teeth, and a deepening and lengthening of the skull.

The change in horse dentition through time provides an interesting example of the important link between environment and organic evolution. During the time that the horse family was evolving, there is paleontologic evidence that grasslands were becoming increasingly widespread in North America and Eurasia. Grass is a rather harsh food for an animal that must feed upon plants. The blades contain silica, and because grass grows closely to the ground, it is usually coated with abrasive dust. Early members of the horse family lived in forested areas and fed upon the less abrasive leaves of trees and shrubs. They were browsers and had the low-crowned teeth of leaf eaters. Later members of the horse family were af-

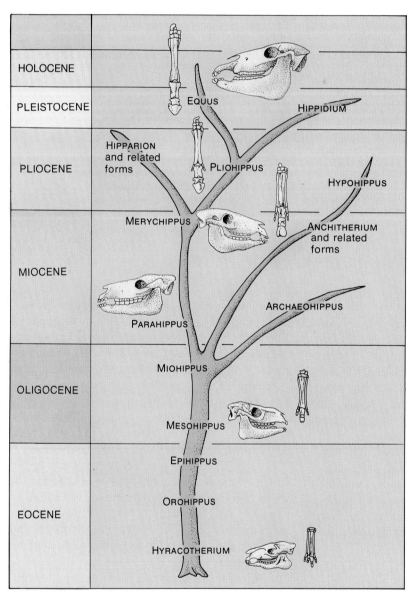

| | |
|---|---|
| HOLOCENE | |
| PLEISTOCENE | |
| PLIOCENE | |
| MIOCENE | |
| OLIGOCENE | |
| EOCENE | |

EQUUS

HIPPIDIUM

HIPPARION and related forms

PLIOHIPPUS

HYPOHIPPUS

MERYCHIPPUS

ANCHITHERIUM and related forms

ARCHAEOHIPPUS

PARAHIPPUS

MIOHIPPUS

MESOHIPPUS

EPIHIPPUS

OROHIPPUS

HYRACOTHERIUM

**Figure 3–16** Simplified phylogenetic tree (gradualistic model) showing the evolutionary relationships of major genera of horses. *(Data from G. G. Simpson,* Horses. *New York: American Museum of Natural History, 1962.)*

fected by selection processes that favored variants better able to cope with the problem of tooth wear from eating grasses. Over many generations, they had evolved the high-crowned teeth with complex patterns of enamel that characterize grazing animals (Fig. 3–17). These horses were likely to have been well nourished and hence lived longer lives and were able to produce more progeny. Their evolution must also have affected the evolution of the predators that pursued them and may even have contributed to the evolution of species of grasses that were resistant to damage by grazing. Evolution is an intricate process in which every animal and plant interacts with its neighbors and with the physical environment.

## COMMENTARY

## *Earbones Through the Ages*

Humans and other terrestrial mammals hear because receptor cells in the inner ear generate electrical impulses in response to vibrations caused by sound waves. These waves travel through the external ear canal to the tympanic membrane or ear drum (see accompanying figure). The tympanic membrane vibrates in response, and the vibrations are transmitted through a chain of three small bones located in the middle ear, to the sound organ or cochlea of the inner ear. These ear bones, called auditory ossicles, provide one of evolution's best examples of anatomical structures that served a particular function in ancestral animals but are changed in function and form in the descendent animals.

The first of the three auditory ossicles is the **stapes**. One can trace its transformation backward in time to the Silurian when jaws evolved from cartilagenous or bony supports for gills. The bone that was to become the stapes was originally an upper element of one of these gill supports. In the evolution of fish with true jaws, the gill support just posterior to those that evolved into jaws became a supporting prop between the upper jaw and braincase. It is called the hyomandibular. With the advent of amphibians during the Devonian, the old fish hyomandibular was transformed into the stapes, where it functioned as a transmitter of vibrations from the tympanic membrane to the inner ear. The stapes persisted as the only auditory ossicle throughout the evolutionary history of both amphibians and reptiles. With the advent of mammals, however, two bones in reptiles that had served as the joint between the upper and lower jaws were changed in function to transmit vibrations. They became the **incus** and **malleus**. Thus, over the past 400 million years or so, the bones that were to evolve into our present hearing apparatus served such diverse functions as supports for the gills, props for the braincase, and articulation for the jaws.

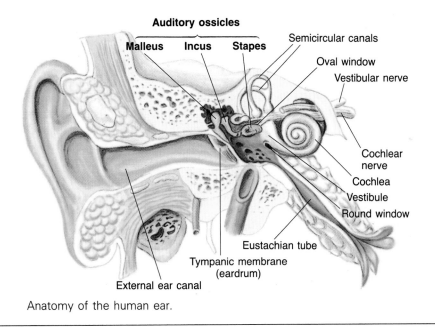

Anatomy of the human ear.

**Figure 3–17** Development of high-crowned grinding molars in horses. Enamel shown in black. From left to right, *Hyracotherium, Merychippus,* and *Equus. (From H. L. Levin,* Life Through Time. *Dubuque, IA: William C. Brown Co., 1975.)*

Although fossil remains of horses provide a fine illustration of paleontologic evidence for evolution, hundreds of other examples representing every major group of animals and plants would serve as well. The marine invertebrates known as cephalopods provide exceptionally fine examples of progressive evolutionary changes (Fig. 3–18). Because of this, they are also one of the fossil groups most useful in correlation.

## Biologic Clues

Supplemental to the paleontologic clues favoring evolution are several persuasive arguments that are more directly biologic in nature. For example, in studies of the comparative morphology of organisms, it is not at all unusual to find body parts that are evidently of similar origin, structure, and development, even though they may be adapted for different functions in various species. For example, in seed plants, leaves are found as petals, tendrils, and thorns. In four-limbed vertebrates, the bones of the limbs may vary in size and shape, but they are fundamentally similar and in similar relative positions in birds, horses, whales, and humans (Fig. 3–19). Such basically similar structures in superficially dissimilar organisms are referred to as **homologous**. The differences in homologous structures are the result of vari-

**Figure 3–18** An example of progressive evolutionary change in a group of Permian ammonoid cephalopods. According to this interpretation, two evolutionary lineages originated from *Paraceltites elegans,* one terminating in *P. rectangularis* and a second producing *Cibolites waageni.* The latter was the ancestral stock for three additional lineages. The curved lines beneath each drawing are tracings of the suture lines; the terms along the left border are the names of Permian stages. *(From C. Spinosa, W. M. Furnish, and B. F. Glenister,* J. Paleontol. *49(2):239–283, 1975.)*

Xenodiscus carbonarius

Prototoceras tropitum

Araxoceras latissimum

Xenodiscus wanneri

Kingoceras kingi

Eoaraxoceras ruzhencevi

Paraceltites rectangularis

Cibolites waageni

Paraceltites hoeferi

Cibolites africanus

Paraceltites elegans

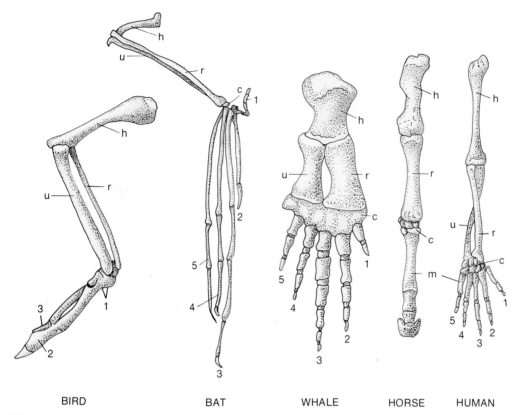

BIRD                 BAT                 WHALE           HORSE        HUMAN

**Figure 3–19**   Skeleton of right forelimb of several vertebrates to show similarity of structure. Key: *c,* carpals; *h,* humerus; *m,* metacarpals; *r,* radius; *u,* ulna; *1–5,* digits.

ations and adaptations to particular environmental conditions, but the similarities indicate genetic relationship and common ancestry.

As another line of evidence, biologists point to the existence in animals and plants of useless and usually reduced structures that in other related species are well developed and functional. These **vestigial organs** are the "vestiges" of body parts that were utilized in earlier ancestral forms. Genetic material inherited from those ancestors still produces the vestigial organs, but the importance of these structures to the well-being of the organism has diminished with changes in environment and habits. Humans have over 100 of these vestigial structures, including the appendix, the ear muscles, and the coccyx (tail vertebrae). The vestigial pelvic bones in such different animals as the boa constrictor and whale (Fig. 3–20) clearly suggest that they were evolved from four-legged animals.

Biologists have provided further evidence of evolution through comparative studies of the embryos of vertebrate animals. In their early stages of development, embryos of fish, birds, and mammals are strikingly similar. It would seem that all these animals received basic sets of genes from remote common ancestors. These genes control embryologic development for a time. Later in the developmental process, other genes begin to assume control and to cause each different species to develop in its own unique way.

Many plants and animals resemble one another not only in structure and embryologic development, but in biochemistry as well. Present knowledge indicates that all organisms produce nucleic acids—especially DNA—and that all use a compound called **adenosine diphosphate (ADP)** in life processes that involve energy storage and transfer. Digestive enzymes and hormone secretions are similar in many

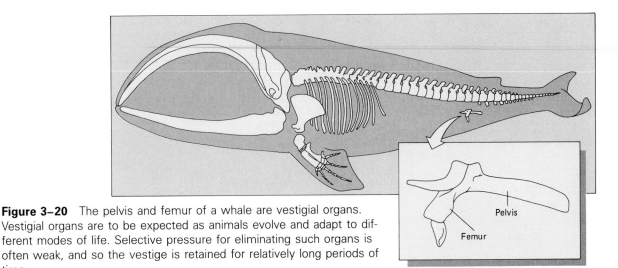

**Figure 3–20**   The pelvis and femur of a whale are vestigial organs. Vestigial organs are to be expected as animals evolve and adapt to different modes of life. Selective pressure for eliminating such organs is often weak, and so the vestige is retained for relatively long periods of time.

forms of life. Proteins extracted from corresponding tissues of closely related animals show definite similarities. The antigenic reactions of blood from the various groups of humans have been found to be practically identical or rather similar to such reactions in the blood of anthropoid apes.

# Fossils and Stratigraphy

## Establishing Age Equivalence of Strata with Fossils

One of William Smith's major contributions to geology was his recognition that individual strata contain definite assemblages of fossils. Because of the change in life through time, superposition, and the observation that once species have become extinct they do not reappear in later ages, fossils can be used to recognize the approximate age of a unit and its place in the stratigraphic column. (Such a method for judging the age of a unit would not be possible with inorganic characteristics of strata because they frequently recur in various parts of the geologic column.) Further, because organic evolution occurs all around the world, rocks formed during the same age in identical environments but diverse localities often contain similar fauna and flora if there has been an opportunity for genetic exchange. This fact permits geologists to match chronologically or correlate strata from place to place. It provides a means for establishing the age equivalence of strata in widely separated parts of the globe.

## The Geologic Range

Before fossils could be used as indicators of age equivalence, it was first necessary to determine the relative ages of the major units of rock on the basis of superposition. Geologists began by working out the superpositional sequences locally, then they added sections from other localities around the world, fitting the segments together into a composite geologic column (Fig. 3–21). The next step was to determine the fossil assemblage from each time–rock unit and to identify the various genera and species. This work began well over a century ago and is still very much in progress. Gradually, it became possible to recognize the oldest (or first appearance of) particular species as well as their youngest occurrence (last to appear) stratigraphically. The interval between the first and last appearance of a species constitutes the **geologic range**. Clearly, the geologic range of any ancient organism is not known *a priori,* but is determined only by recording its occurrence in numerous stratigraphic sequences from hundreds of locations. Figure 3–22 illustrates just one such study that records the geologic ranges of cephalopods (a class of mollusks having chambered conchs) from a Cretaceous formation in Antarctica. Today, ranges are well known for some species and groups of species, but

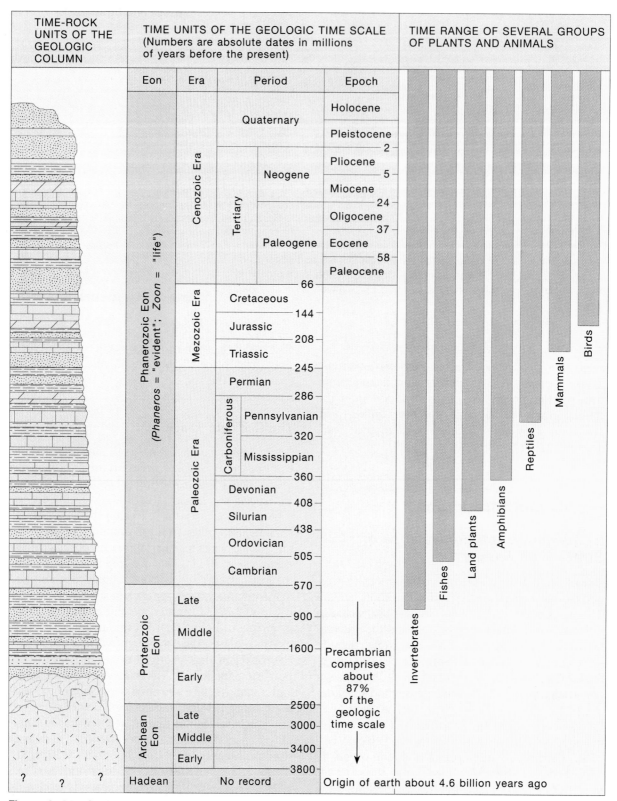

**Figure 3–21** Geologic time scale. *(From A. R. Palmer, The decade of North American geology, geologic time scale, Geology 11:503–504, 1983.)*

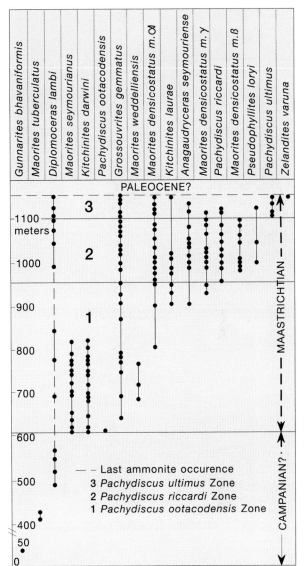

**Figure 3–22** Example of a range chart showing the ranges of Late Cretaceous ammonite cephalopods (chambered mollusks) from the Lopez de Bertodano Formation, Seymour Island, Antarctic Peninsula. The Campanian and Maastrichtian are stages of the Upper Cretaceous. Zonal boundaries are based on the first appearances of the species after which each zone is named. *(From Macellari, E. E., 1986, J. Paleontol. Mem. 18, Part 2.)*

relatively poorly known for others. Fortunately, there are now enough data so that isolated sections of strata in geologically unexplored areas can be located in the composite geologic column by use of their contained fossils.

## Identification of Time-Stratigraphic Units

The method for identifying time-stratigraphic units is illustrated in Figure 3–23. Geologists working in Region 1 come upon three time–rock systems of

strata designated *O, D,* and *M.* Perhaps years later in Region 2, they again find units *O* and *D*, but in addition, they recognize an older unit—namely, *C*, below and hence older than *O*. Finally, while working in Region 3, they find a "new" unit, *S*, sandwiched between units *O* and *D*. The section, complete insofar as can be known from the available evidence, consists of five units that decrease in age as one progresses upward from *C* to *M*. The geologists may then plot the ranges of fossil species found in *C* and *M* alongside the geologic column. If they should

**Figure 3–23** Use of geologic ranges of fossils to identify time–rock units.

next find themselves in an unexplored region, they might experience difficulty in attempting to locate the position of this rock sequence in the standard column, especially if the lithologic traits of the rocks had changed. However, on discovering a bed containing species **A**, they are at least able to say that the rock sequence might be *C, O,* or *S.* Should they later find fossil species **B** in association with **A**, they might then state that the outcrop in the unexplored region correlates in time with unit *S* in Region 3. In this way, the strata around the world are incorporated into an accounting of global stratigraphy.

## Paleontologic Correlation

Of course, the preceding illustration is greatly simplified, and future discoveries may extend known fossil ranges. Geologists must always keep in mind that the chance inability to find key fossils might lead to erroneous interpretations. Fortunately, the chances of making such mistakes are diminished by the practice of using entire assemblages of fossils. Two or three million years from now, geologists might have difficulty in firmly establishing on fossil evidence that the North American opossum, the Australian wallaby, and the African aardvark lived during the same episode of geologic time. However, if they found fossils of *Homo sapiens* with each of these animals, it would indicate their contemporaneity. In this example, *Homo sapiens* can be considered the **cosmopolitan species,** for it is not restricted to any single geographic location within the terrestrial environment. The aardvark and wallaby are said to be **endemic species,** in that they are confined to a particular area.

In the case of fossilized marine animals, cosmopolitan species have been especially useful in establishing the contemporaneity of strata, whereas endemic species are generally good indicators of the environment in which strata were deposited. Endemic species may slowly migrate from one locality to another. For example, the peculiar screw-shaped marine fossil bryozoan known as *Archimedes* (Fig. 3–24) was endemic to central North America during the Mississippian Period, but migrated steadily westward, finally reaching Nevada by Pennsylvanian time and Russia by the Permian Period.

One must assume a tentative attitude in making correlations based on fossils. The validity of a correlation increases each time the sequence of faunal changes is found again at different locations around the world. Geologists must also be aware that all changes are not because of evolution, but instead may indicate fauna migrations or shifts in flora that accompanied ancient environmental changes. In this regard, the sudden disappearance of a fossil need not mean that it became extinct, but rather that it moved elsewhere. One might also note that the earliest appearance of a fossil in rocks of a given region might

A

B

**Figure 3–24** Two specimens of the guide fossil *Archimedes*. Specimen (A) shows only the screwlike axis of this bryozoan animal. In specimen (B) below, one can see the fragile, lacy skeleton of the colony that is attached to the sharp, helical edge. Specimens are 8 to 10 cm in height.

mean it had evolved there; however, it might also signify only that an older species had come into the new locality.

In addition to being on the alert for faunal changes that are related to shifting of ancient environments (as opposed to changes resulting from evolution), paleontologists must also be concerned about the possibility of **reworked fossils**. At various times in the geologic past, weathering and erosion have freed fossils from their host rock. Much as would be the case with any clastic particle, these fossils might then be reworked into younger beds, and the younger strata might then be mistakenly assigned to an older geologic time. Some fossils are particularly resistant to erosion and chemical decay and therefore are more susceptible to reworking. Among such resistant fossils are conodonts (see Fig. 3–50), tiny toothlike structures of phosphatic composition and uncertain origin, which are abundant in some Paleozoic strata.

**Guide Fossils.**   Many fossils are rare and restricted to a few localities. Others are abundant, widely dispersed, and derived from organisms that lived during a relatively short span of geologic time. Such short-lived but widespread fossils are called **guide fossils** because they are especially useful in identifying time–rock units and correlating them from one area to another. A guide fossil with a short geologic range is clearly more useful than one with a long range. A fossil species that lived during the total duration of a geologic era would not be of much use in identifying the rocks of one of the subdivisions of lesser duration within that era. It is apparent that rate of evolution is an important factor in the development of guide fossils. Simply stated, the **rate of evolution** is a measure of how much biologic change has occurred with respect to time. Figure 3–25 provides an example of a change in the rate of evolution among horned dinosaurs. The data indicate that, although increase in size was a persistent evolutionary trend, the rate of that change diminished with time.

Obviously, different animals have evolved at different rates. Human beings and their ancestors, for example, are considered to have had a relatively rapid rate of evolution. In general, lineages with rapid rates of evolution provide greater numbers of guide fossils. Usually, rates of evolution can be related to changes in the environment or to the inher-

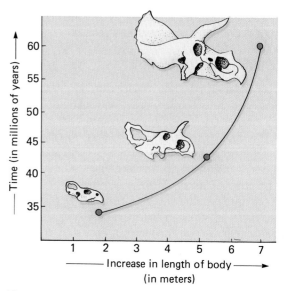

**Figure 3–25**   A study of rate of evolutionary change. In this lineage of horned dinosaurs, the rate of increase in size was intially rapid and subsequently much slower. *(Adapted from E. H. Colbert,* Evolution, *2:145–163, 1948.)*

ent reproductive or genetic characteristics of the evolving populations.

Although guide fossils are of great convenience to geologists, correlations and interpretations based on assemblages of fossils are often more useful and less susceptible to error or uncertainties caused by undiscovered, reworked, or missing individual species. Also, by using the **overlapping geologic ranges** of particular members of the assemblage, it is often possible to recognize the deposits of smaller increments of time. The advantage of overlapping ranges can be illustrated with the help of Figure 3–26. Here, rather than using actual ranges of species, larger animal categories are used. It is apparent from this chart that a rock containing both stromatoporoids and goniatites can only be considered Devonian, whereas the occurrence of members of only one of the fossil groups would not provide as narrow a limit to the age of the rock.

**Zones.**   Geologists frequently use the term **biozone** when describing a body of rock that is identified strictly on the basis of its contained fossils. Paleontologic biozones can vary in thickness or lithology and can be either local or global in lateral extent. Without formally naming them, we have already described

**Figure 3–26**  Advantage of using overlapping ranges of fossil taxa to recognize smaller increments of geologic time.

the two major kinds of biostratigraphic zones: the range zone and the assemblage zone. A **range zone** is simply the rock body representing the total geologic life span of a distinct group of organisms. For example, in Figure 3–27 the *Assilina* range zone is marked by the first (lowest) occurrence of that genus at point A and its extinction at point B. Geologists may also designate **assemblage zones** selected on the basis of several coexisting taxa and named after an easily recognized and usually common member of the assemblage. In certain areas, however, even though the guide fossil for the assemblage zone may not be present, the other members permit recognition of the zone. A variation of the assemblage zone is the **concurrent range zone.** It is recognized by the overlapping ranges of two or more *taxa* (sing. *taxon*). (The

**Figure 3–27** Geologic ranges of three genera of fo-raminifers within the family Nummulitidae. The interval between *A* and *B* is the total range of Assilina. The interval between *X* and *Y* could be designated the Assilina–Heterostegina concurrent range zone.

term **taxon** refers to a group of organisms that constitute a particular taxonomic category, such as species, genus, and family.) For example, the interval between X and Y in Figure 3–27 might be designated the *Assilina–Heterostegina* concurrent range zone.

Biozones are of fundamental importance in stratigraphy. They are the basic unit for all biostratigraphic classification and correlation. They are also the basis for time–stratigraphic terms because zones are aggregated into stages, stages into series, and series into systems. The boundaries of these units are usually zonal boundaries.

# Fossils as Clues to Ancient Environments

## Paleoecology

Although inert and mute, fossils are vestiges of once lively animals and plants that nourished themselves, grew, reproduced, and interacted in countless ways with other organisms and their physical environment. The study of the interaction of ancient organisms with their environment is called **paleoecology**. Paleoecologists attempt to discover precisely where and how ancient creatures lived and what their habits and morphology reveal about the geography and climate of long ago. This scientific detective work is accomplished in various ways. One can, for example, compare species known only from fossils with living counterparts. The assumption is made that both living and fossil forms had approximately the same needs, habits, and tolerances. One can also examine the anatomy of the fossil and attempt to identify structures that were likely to have developed in response to particular biologic and physical conditions in the environment. As noted earlier, such modifications for living in a certain way and performing particular functions are called **adaptations**. For example, the broad, spiny valve of the brachiopod *Marginifera ornata* (Fig. 3–28) was an adaptation that served to support the animal on a sea bed composed of soft mud. In another example, reduction of the skeleton to mere spines in the Silurian trilobite *Deiphon* (Fig. 3–29) was probably an adaptation to provide buoyancy in an animal that fed near the surface of the sea. Coiling in living and fossil cephalopods appears to be an adaptation designed to bring the center of buoyancy above the center of gravity

**Figure 3–28**  A fossil of the spinose brachiopod *Marginifera ornata.* The shell has been replaced by silica. The enclosing limestone has been dissolved so as to provide excellently preserved fossils with delicate spines intact. The animal rested on its ventral valve (shown here), and its spines provided anchorage. *(Photograph courtesy of Richard E. Grant, U.S. Geological Survey.)*

and thereby permit these creatures to keep on an "even keel" while swimming. A list of additional examples of adaptations would be immense, for every creature is a result of hundreds of adaptations. Not all adaptations are morphologic. Biochemical and physiologic adaptations occur as well, although these are more difficult to recognize in fossils.

The language of paleoecology is derived from **ecology**, the study of the present relationships between organisms and their environments. In ecologic studies, one tends to concentrate on the **ecosystem**, which is any selected part of the physical environment together with the animals and plants in it. An ecosystem maybe as large as the earth or as small as a garden pond. Paleoecologists are particularly interested in the ocean ecosystem because of the richness of the fossil record of marine life. The physical aspects of the ocean ecosystem include the water itself, its dissolved gases (especially carbon dioxide and oxygen), salts (phosphates, nitrates, chlorides, and carbonates of sodium, potassium, and calcium), various organic compounds, turbidity, pressure, light penetration, and temperature.

The biologic components of the ocean ecosystem are usually classified according to so-called **trophic** or feeding levels (Fig. 3–30). For example, **producer organisms** such as green plants manufacture compounds from simple inorganic substances by means of photosynthesis. These organisms in the sea are mostly very small (less than 0.1 mm in diameter) and include diatoms and other forms of algae. Producer organisms are eaten by **consumer organisms**

such as mollusks, crustaceans, and fish. The primary consumers commonly feed directly on unicellular plants and thus may be further designated as **herbiv-**

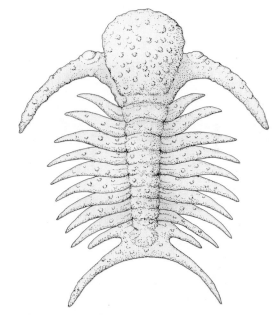

**Figure 3–29**  *Deiphon* was a Silurian trilobite whose extreme spinosity suggests it was a swimmer and floater rather than a bottom dweller like many of its trilobite cousins. Spinosity increases surface area without adding weight, thus enhancing buoyancy. It is also possible that the swollen anterior structure (glabella) may have been filled with a liquid of low specific gravity, thus providing additional buoyancy. The length of the specimen is 27 mm.

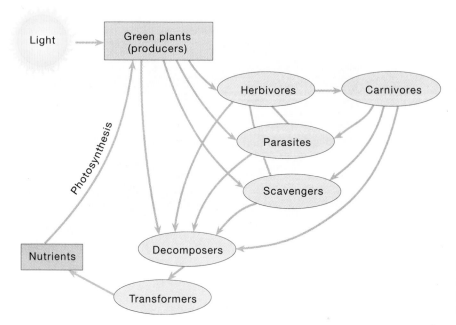

**Figure 3–30** The movement of materials through an ecosystem. Components within ovals are consumers.

**ores**. The secondary consumers that eat the herbivores are called **carnivores**. Tertiary and even quaternary consumers feed on carnivores from lower trophic levels. The ecosystem also contains decomposers and transformers, including bacteria and fungi, which are able to break down the organic compounds in dead organisms and waste matter and produce simpler materials that can be used by the producers. Thus, in an ecosystem, the basic chemical components of life are continuously being recycled. The marine ecosystem also contains parasites, which feed on other organisms without necessarily killing them, and scavengers, which derive their nourishment from dead organisms.

Within any given ecosystem, one finds the specific environments or habitats where certain organisms live. In the ocean, the habitats range from the cold, dark realm of the abyss to the warm, illuminated areas of tropical coral reefs. And within every habitat there are ecologic niches in which particular organisms make their living (Fig. 3–31). Each contains all the biologic, chemical, and physical conditions that permit the organism occupying the niche to survive. Many niches exist in the coral reef habitat. Some are occupied by tiny, tentacled coral animals that build the framework of the reef and feed as carnivores on even smaller crustaceans. Certain marine snails occupy a niche along the surface of algal mats, where with snail-like sluggishness they graze on films of algae. Behind the reef in areas of soft mud, plump lugworms consume the soupy sediment for its content of organic nutrients. Each creature has its characteristic ecologic niche.

To facilitate the study of the ocean ecosystem, ecologists have developed a simple classification of

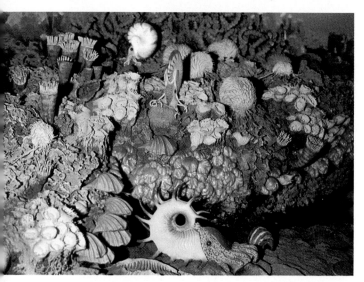

**Figure 3–31** A Permian patch reef. In this diorama one can recognize strings of bead sponges (top) and compact sponges (lower left), horn corals (upper left), cephalopods (center and bottom center), and many aggregates of brachiopods. The leaf-shaped shells to the left of the spiny cephalopod are remains of oysterlike brachiopods (*Leptodus*). *(Courtesy of the National Museum of Natural History.)*

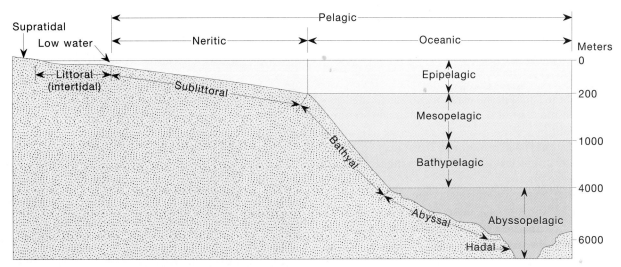

**Figure 3–32**  Classification of marine environments. *(After J. W. Hedgpeth (ed.), Treatise on Marine Ecology and Paleoecology, Geol. Soc. Am. Mem. 67(1):18, 1957.)*

marine environments. It begins with a twofold division of the entire ocean into pelagic and benthic realms. The **pelagic realm** consists of the water mass lying above the ocean floor. It can be divided into a **neritic zone**, which overlies the continental shelves, and an **oceanic zone**, which extends seaward from the shelves (Fig. 3–32).

Within the pelagic realm one finds myriads of small animals and plants that float, drift, or feebly swim. These are **plankton**. **Phytoplankton** consist of plants and plantlike protists and include algae such as diatoms (see Fig. 13–6) and coccolithophorids (see Fig. 1–46). Planktonic animals constitute **zooplankton** and include protists such as radiolaria (see Fig. 13–19), foraminifers (Fig. 3–33), certain tiny mollusks, small crustaceans, and the motile larva of many different families of invertebrates that live the adult stage of their life cycles on the sea floor.

The pelagic realm is also the home of **nekton**, or true swimming animals. Nekton are able to travel where they choose under their own power, and this is clearly advantageous. A swimming creature can search for its food and does not have to depend upon food particles carried in chance currents. It can use its mobility to escape predators and can move to more favorable areas when conditions become difficult. The nekton is a diverse group that includes invertebrates such as shrimp, cephalopods, and certain extinct trilobites, as well as vertebrates such as fish, whales, and marine turtles.

The second great division of the ocean ecosystem is the bottom or **benthic** realm. It begins with a narrow zone above high tide called the **supralittoral zone**. Several kinds of marine plants and ani-

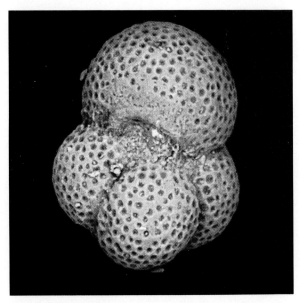

**Figure 3–33**  Electron micrograph of the planktonic foraminifer *Globigerinoides*. The height of the specimen is about 0.09 mm.

mals have adapted themselves to this harsh environment where ocean spray provides vital moisture, but where drying poses a constant danger. Seaward of the supralittoral zone is the area between high and low tide. This is the **littoral zone.** Like those creatures of the supralittoral area, organisms in the littoral zone must be able to tolerate alternate wet and dry conditions. Some avoid the drying by burrowing into wet sand, whereas others have adapted themselves in various ways to retain body moisture when exposed to air.

Benthic animals and plants are most abundant seaward of the littoral zone in the continuously submerged **sublittoral zone.** This zone extends from low tide levels down to the edge of the continental shelf (about 200 meters deep). Depending on the clarity of the water, light may penetrate to the sea floor in the sublittoral zone, although the base of light penetration is usually slightly less than 200 meters. Various kinds of algae thrive here, as well as abundant protozoans, sponges, corals, worms, mollusks, crustaceans, and sea urchins. These animals are never subjected to desiccation. Their adaptations are mainly associated with food gathering and protection from predators. Some of these benthic animals live on top of the sediment that carpets the sea floor and are called **epifaunal.** Others, termed **infaunal,** burrow into the soft sediment for food and

**Figure 3–35** The stalked animals with radiating arms depicted in this diorama are crinoids. They are *sessile,* meaning that they remain attached to the substrate and do not move about as do *vagile* animals. Crinoids were especially abundant during the Mississippian, as suggested in this view of part of a Mississippian shallow sea floor. *(Courtesy of the National Museum of Natural History.)*

protection (Fig. 3–34). Those bottom dwellers capable of locomotion are called **vagile** in contrast to immobile creatures, which are **sessile** (Fig. 3–35).

Beyond the continental shelves, the benthic environment is subjected to low temperatures, little or no light, and high pressures. Without light, plants are unable to live at these depths. One encounters this **bathyal** environment from the edge of the shelf to a depth of about 4000 meters. Still deeper levels constitute the **abyssal** environment. The term **hadal** is reserved for the extreme depths found in oceanic trenches. As might be expected, animals are less abundant in the abyssal and hadal environments. Most of these deep water creatures are scavengers that depend upon the slow fall of food from higher levels and predators that feed on the scavengers.

## Use of Fossils in Reconstructing Ancient Geography

The geographic distribution of present-day animals and plants is closely controlled by environmental limitations. Any given species has a definite range of conditions under which it can live and breed, and it

**Figure 3–34** Burrows made by infaunal organisms (probably worms) are preserved in the Northview Formation (Mississippian) of Missouri.

**Figure 3–36**  Major land and sea regions in North and South America during the Carboniferous Period. *(Adapted from C. A. Ross, J. Paleontol. 41(6):1341–1354, 1967.)*

Appalachian orogen

Cordilleran orogen

Andean orogen

Regions of marine deposition

Regions of erosion

Regions of continental deposition

Carboniferous fusulinid collecting localities

Enlarged sketch of a fusulinid with section cut out to show internal structure

is generally not found outside that range. Ancient organisms, of course, had similar restrictions on where they could survive. If we note the locations of fossil species of the same age on a map and correctly infer the environment in which they lived, we can produce a paleogeographic map for that particular time interval. One might begin by plotting on a simple base map the locations of fossils of marine organisms that lived at a particular time. This provides an idea of which areas were occupied by seas and might even suggest locations for ancient coastlines. Figure 3–36 shows major land and sea regions during the middle Carboniferous Period. The locations of marine protozoans called *fusulinids* are plotted as open

circles. Notice that the lofty ranges of the Rocky Mountains were nonexistent, their present locations occupied by a great north–south seaway.

Having obtained a fair idea of the marine regions, one might next give attention to the locations of fossils diagnostic of land areas. The fossilized bones of land animals such as dinosaurs or mastodons would suggest a terrestrial environment, as would their preserved footprints. Fossil remains of land plants, including fossilized seeds and pollen, provide excellent evidence of terrestrial paleoenvironments.

By an analysis of the fossils and the nature of the enclosing sediment, it is often possible to recognize deeper or shallower parts of the marine realm or to discern particular land environments such as ancient floodplains, prairies, deserts, and lakes. River deposits may yield the remains of freshwater clams and fossil leaves. A mingling of the fossils of land organisms and sea organisms might be the result of a stream entering the sea and perhaps building a delta in the process.

The migration and dispersal patterns of land animals, as indicated by the fossils they left behind, is one important indicator of former land connections as well as mountainous or oceanic barriers that once existed between continents. Today, for example, the Bering Strait between North America and Asia prevents migration of animals between the two continents. The approximately 80 km between the two shorelines (Fig. 3–37), however, is covered by less

than 50 meters of water; this might lead one to wonder whether Asia was once connected to North America. The fossil record shows that a land bridge did connect these two continents on several occasions during the Cenozoic Era. The earliest Cenozoic strata on both continents have a fossil fauna uncontaminated by foreign species. Somewhat higher and younger rocks contain fossil remains of animals that heretofore had been found only on the opposite continent. These remains mark a time of land connection. The last such connection may have existed only 14,000 to 15,000 years ago, when Stone Age hunters used the route to enter North America.

Another familiar example of how fossils aid in paleogeographic reconstructions is found in South America. Careful analyses of fossil remains indicate that, in the early Cenozoic Era, South America was isolated from North America; as a result, a uniquely South American fauna of mammals evolved over a period of 30 to 40 million years. The establishment of a land connection between the two Americas is recognized in strata only a few million years old (late Pliocene) by the appearance of a mixture of species formerly restricted to either North or South America. (The migrations were decidedly detrimental to South American species.)

In addition to deciphering the positions of shorelines or locations of land bridges, paleontologists can also provide data that help to locate the equator, parallels of latitude, and pole positions of long ago. It has been observed that in the higher latitudes of the globe, one is likely to find large numbers of individuals but that these are members of relatively few different species. In contrast, equatorial regions tend to develop a large number of species, but with comparatively fewer individuals within each species. Stated differently, the variety or **species diversity** for most higher categories of plants and animals increases from the poles toward the equator. This is probably related to the fact that relatively few species can adapt to the rigors of polar climates. Conversely, there is a stable input of solar energy at the equator, less duress caused by seasons, and a more stable food supply. These warmer areas place less stress on organisms and provide opportunity for continuous, uninterrupted evolution. Of course, in particular areas, generalizations about species diversity can be upset by special circumstances of altitude, upwelling, or other vagaries of air and water currents.

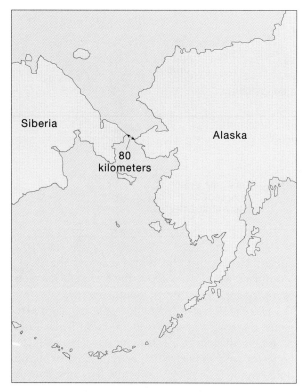

**Figure 3–37** The Bering Strait between North America and Asia. The area between the two continents would become land if sea level were lowered by only 45 meters. Such a lowering of sea level occurred at various times during the Cenozoic Era and permitted migrations across the land bridge.

**Figure 3–38** A view looking down the steep frontal face of a modern reef. Admiralty Islands, Papua, New Guinea. *(Courtesy of L. E. Davis.)*

Another way to locate former equatorial regions (and therefore also the polar regions that lie 90° of latitude to either side) is by plotting the locations of fossil coral reefs of a particular age on a world map. Nearly all living coral reefs (Fig. 3–38) lie within 30° of the equator. It is not an unreasonable assumption that the ancient reef corals had similar geographic preferences.

## Use of Fossils in the Interpretation of Ancient Climatic Conditions

Among the many climatic factors that limit the distribution of organisms, *climate* (especially the temperature component of climate) is of great importance. There are many ways by which paleoecologists gain information on ancient climates. An analysis of fossil spore and pollen grains (Fig. 3–39) often provides outstanding evidence of past climatic conditions. Living organisms with known tolerances can often be directly compared with fossil relatives. As noted above, corals thrive in regions where water temperatures rarely fall below 18°C, and it is likely their ancient counterparts were similarly constrained. However, even when a close living analogue is not found, plants and animals may exhibit morphologic traits useful in determining paleoclimatology. Plants with aerial roots, lack of annular rings, and large wood cell structure indicate tropical or subtropical climates lacking in strongly contrasting seasons. Marine mollusks (such as clams, oysters, and snails) with well-developed spinosity and thicker shells tend to occur in warmer regions of the oceans. In the case of par-

A

B

**Figure 3–39** Pollen grains. (A) Pollen grain of a fir tree. Note the inflated bladders that make the pollen grain more buoyant. The specimen is about 125 microns in its longest dimension. (B) Pollen grains of ragweed. These grains are about 25 microns in diameter.

ticular species of marine protozoans, variation in the average size of individuals or in the direction of coiling can provide clues to cooler or warmer conditions.

An example of environmentally induced changes in shell coiling directions is provided by the planktonic (floating) foraminifer *Globorotalia truncatulinoides* (Fig. 3–40). In the Pacific Ocean, left-coiling tests (shells) of these foraminifers have dominated in periods of relatively cold climate, whereas right-coiling tests predominated during warmer episodes. Another foraminifer, *Neoglobquadrina pachyderma,* exhibits similar changes in coiling directions, not only in the Pacific Ocean but in the Atlantic as well. Such reversals in coiling may occur quickly and over broad geographic areas. For this reason, they are exceptionally useful in correlation studies.

Aside from morphologic changes within species, entire assemblages of foraminifers are widely used in paleoecologic studies. Today, as in the past, benthic (bottom-dwelling) species have inhabited all major marine environments. As a result of the accumulated data on living foraminifers, inferences about salinity and water depths can be made. Fossil foraminifers have frequently provided the means for recognition of ancient estuaries, coastal lagoons, and nearshore or deep oceanic deposits.

Sometimes, when the overall morphology of a fossil does not provide clues to temperature or climate, the compositions of the skeleton can be used. Magnesium, for example, can substitute for calcium in the calcium carbonate of shelled invertebrates. For particular groups of living marine invertebrates, it

**Figure 3–40** Shell coiling in the foraminifer *Globorotalia truncatulinoides.* Sketch of both sides of a single left-coiling specimen. (Diameter is about 0.9 mm.) *(From F. L. Parker,* Micropaleontology *8(2):219–254, 1962.)*

has been found that those living in warmer waters have higher magnesium values than do those residing in colder areas. This knowledge has been used to interpret the climate at the times fossil forms were living.

In another kind of biogeochemical study of shell matter, investigators have found that certain organisms build their skeleton of calcium carbonate in both of its two mineral forms: *aragonite* and *calcite.* Moreover, in colder realms, particular species have higher calcite-to-aragonite ratios, whereas their fellow species living in warmer waters have lower calcite-to-aragonite ratios. Secretion of calcium carbonate in the form of aragonite appears to be "easier" at higher temperatures.

There are, however, problems when geochemical means are used to deduce temperature. Aragonite is an unstable mineral that reverts to the more stable calcite. It is nearly absent in Paleozoic rocks. High-magnesium calcite is also unstable. Magnesium may be leached out of the sediment after burial, causing the conversion of high-magnesium calcite to its more stable low-magnesium counterpart.

Oxygen isotope analyses offer yet another way for learning the temperature of ancient seas. When water evaporates from the ocean, there is a fractionation of the oxygen-16 and oxygen-18 in the water. (Oxygen-16, the usual form of oxygen, has an atomic mass of 16, whereas oxygen-18 has an atomic mass of 18.) Oxygen-16, being lighter, is preferentially removed, while the heavier oxygen-18 tends to remain behind. When the evaporated moisture is returned to the earth's surface as rain or snow, water containing the heavier isotope precipitates first, often near the coastlines, and flows quickly back into the ocean. Inland, the precipitation from the remaining water vapor is depleted in oxygen-18 relative to its initial quantity. If the interior of a continent is cold and contains growing glaciers, the glacial ice will lock up the lighter isotope, preventing its return to the ocean and thereby increasing the proportion of the heavier isotope in sea water. As this occurs, the calcium carbonate shells of marine invertebrates such as foraminifers will also be enriched in oxygen-18 and thereby reflect episodes of continental glaciation.

Even in the absence of ice ages, the oxygen isotope method may be useful as a paleotemperature indicator. As invertebrates extract oxygen from sea water to build their shells, a temperature-dependent

A

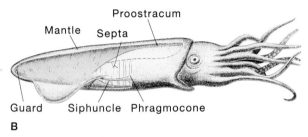

B

**Figure 3–41**  (A) The usual belemnite fossil consists only of the solid part of the internal skeleton called a guard. The guard here is 9.6 cm long. (B) Belemnites are extinct relatives of squid and cuttlefish. Shown here is an interpretation of the appearance of a belemnite as it would appear if living.

fractionation of oxygen-18 occurs between the water and the secreted calcium carbonate in the shell matter. Provided there has been no alteration of the shell, it is possible to find the temperature of the water at the time of shell formation. In a famous early study, the oxygen isotope ratio in the calcium carbonate skeleton of a Jurassic belemnite indicated that the average temperature in which the animal lived was 17.6°C, plus or minus 6°C of seasonal variation. Subsequent studies on both Jurassic and Cretaceous belemnites (Fig. 3–41) provided confirmation of the inferred positions of the poles during the Mesozoic and indicated that tropical and semitropical conditions were far more widespread during late Mesozoic time than they have been in subsequent geologic time.

## An Overview of the History of Life

As yet, no fossil evidence has been found of life on earth during the planet's first billion years. The old-

est direct indications of ancient life are remains of bacteria and primitive algae discovered in rocks over 3.5 billion years old. These early fossils are usually considered evidence that the long evolutionary march had begun. However, they also stand at the end of another long and remarkable period during which living things presumably evolved from nonliving chemical compounds. We have no direct geologic evidence to tell us how and when the transition from nonliving to living occurred. What we do have are reasonable hypotheses supported by careful experimentation. Some of these hypotheses will be examined in Chapters 7 and 8, which deal with the Precambrian eons. It will suffice for this brief overview to note that life came into existence early in that great 4-billion-year span of time prior to the beginning of the Cambrian Period.

For most of Precambrian time, living things left only occasional traces. Here and there paleontologists have been rewarded with finds of bacteria and filamentous or unicellular blue-green algae. Occasionally, filamentous organisms formed extensive

**Figure 3–42** The circular pods depicted in this mural of the margins of a Precambrian sea are calcareous algal structures called stromatolites. *(Courtesy of the National Museum of Natural History.)*

algal mats (Fig. 3–42), and these may have produced important quantities of oxygen by photosynthesis. However, all life was at or below the unicellular level until about 1 billion years ago, when the world's first multicellular organisms left their trails and burrows in rocks that would later be called the *Torrowangee Group* of Australia. In rocks deposited about 0.7 billion years ago, fossil metazoans recognizable as worms, coelenterates, and arthropods have been found—albeit rarely—in scattered spots around the world, including the United States, Australia (Fig. 3–43), Canada, England, Russia, China, and South Africa. Thus, near the end of the Proterozoic, the

**Figure 3–43** *Mawsonites,* a fossil found in the Pound Quartzite of Australia. The formation is late Proterozoic in age. *Mawsonites* is considered to be the mouth end of a jellyfish. *(Courtesy of B. N. Runnegar.)*

stage was set for a wide range of Paleozoic plants and animals.

## The Evolutionary History of Animals

Some general observations are possible after a brief scanning of the history of life following the Precambrian. One notes that the principal groups of invertebrates appear early in the Paleozoic (Fig. 3–44). Less advanced members of each phylum characterized the earlier geologic periods, whereas more advanced members came along later. Also, most of the principal phyla are still represented by animals today. In our review of fossil creatures, we are not startled by sudden appearances of bizarre or exotic animals and plants. Further, we are able to recognize periods of environmental adversity that caused extinctions. Such episodes were usually followed by much longer intervals of recovery and more or less orderly evolution. One final observation is that there has been a persistent gain in the overall diversity of life down through the ages.

After the Precambrian, life became so varied that it is difficult to describe it both briefly and adequately. We can, however, make some generalizations. Such marine invertebrate animals as trilobites, brachiopods, nautiloids, crinoids, horn corals, honeycomb corals, and twiglike bryozoans were abundant during the Paleozoic Era (Fig. 3–45). The Paleozoic was also the era when fishes, amphibians, and reptiles appeared and left behind a fascinating record of the conquest of the lands. Remains of more

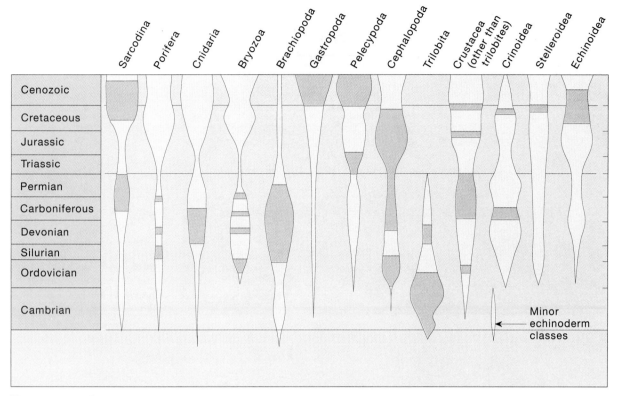

**Figure 3–44**   Geologic ranges and relative abundances of frequently fossilized categories of invertebrate animals. Width of range bands indicates relative abundance. Colored areas indicate where fossils of a particular category are widely used in zoning and correlation.

**Figure 3–45**   A large, straight-shelled cephalopod dominates this scene of an Ordovician sea floor. Also visible are corals, bryozoa, and crinoids. *(Courtesy of the National Museum of Natural History.)*

**Figure 3–46** Jurassic carnivorous dinosaurs in combat. *(Courtesy of the American Museum of Natural History.)*

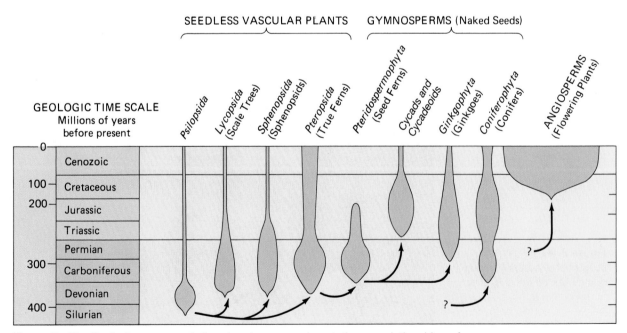

**Figure 3–47** Geologic ranges, relative abundances, and evolutionary relationships of vascular land plants.

**Figure 3–48**  Foraminifers obtained from the Monterey Formation of Miocene age, California.

modern corals, diverse pelecypods, sea urchins, and ammonoids characterize the marine strata of the Mesozoic Era. However, the Mesozoic is best known as the "age of reptiles," when dinosaurs and their kin dominated the continents (Fig. 3–46). No less important than these big beasts, however, were the rat-sized primitive mammals and earliest birds that skittered about perhaps unnoticed by the "thunder beasts."

The mollusks were particularly well represented in the marine invertebrate faunas of the Mesozoic, and they have continued in importance into the Cenozoic. However, no ammonoid cephalopods have occurred in this most recent era. Rather, rocks of the Cenozoic are recognized by distinctive families of protozoans (foraminifers) and a host of modern-looking snails, clams, sea urchins, barnacles, and encrusting bryozoa. Because the Cenozoic saw the expansion of warm-blooded creatures such as ourselves, it is appropriately termed the "age of mammals."

### The Evolutionary History of Plants

Long before the first animal appeared on earth, there were plants. Plants also experienced a complex and marvelous evolutionary history. That history began with the origin of the primitive unicellular aquatic algae of the Precambrian. Except for the blue-green algae and bacteria noted previously, the fossil record of this earliest but important part of plant evolution is extremely poor. It improves during the second stage in plant history with the origin in the Silurian Period of seedless (spore-bearing) vascular land plants (Fig. 3–47). These plants proliferated in the coal forests of the late Paleozoic Era. A third episode in plant evolution is marked by the advent of non-flowering pollen- and seed-producing plants. Although this was also a late Paleozoic event, the seed plants expanded and diversified during the Mesozoic Era. Finally, in the last period of Mesozoic, plants having both seeds and flowers evolved. Grasses were added to the flowering plant floras during the Cenozoic, and the lands in the age of mammals took on a decidedly modern appearance.

### Fossils and the Search for Mineral Resources

One might be surprised to learn that the majority of paleontologists in the world are not based in museums and universities, but rather are industrial scientists using their knowledge of fossils in the search for oil and gas. In their investigations, they generally utilize only very small fossils because these are not likely to be broken by drilling tools. During the drilling process, samples of the rock being penetrated are returned to the surface in the drilling mud or in rock cores. The rock is then broken down, and the microfossils are extracted for study. By far, the most useful microfossils are the protozoans known as *foraminifers* (Fig. 3–48). Other microfossils that have been

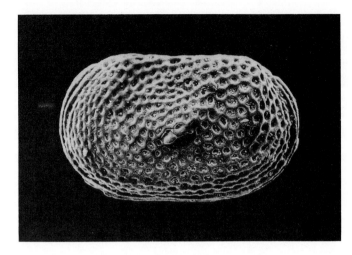

**Figure 3–49** Electron micrograph of an ostracode recovered from Pennsylvanian shales during the drilling of an oil well. The specimen is 0.32 mm long.

used frequently in petroleum exploration include tiny, toothlike structures called *conodonts,* small crustaceans known as *ostracodes* (Fig. 3–49), *pollen grains* and *spores,* and the calcareous remains of the golden brown marine calcareous algae named *coccolithophorids* (see Fig. 1–46). Micropaleontologists, specialists in the study of microscopic fossils, prepare subsurface logs that depict the depth from which each species of fossil was obtained. From the accumulated information of many such paleontologic logs, one is able to identify the geologic period of the formations through which the drill penetrated. Particular time–rock units can be recognized by their unique index fossils. Knowledge of the location of such key fossils relative to known oil-bearing strata is useful when wells are drilled in unproved territory. Indeed, the microfossils permit correlation from well to well and from area to area across entire petroliferous (oil-containing) provinces. The vertical successions of fossil assemblages in well samples often reflect changes in the depth of water in which the sediment was deposited and can therefore be used to infer tectonic conditions within the petroleum province. Along a single stratigraphic zone, knowledge of the depth of water preferences for microfossils may help in locating the porous reef or strand line sediments that are occasionally saturated with petroleum.

Fossils may also play a role in the exploration of resources other than oil and gas. Mining geologists use fossils to date strata above and below mineralized beds and to help determine directions of movements along faults that offset the ore-bearing strata. With the aid of fossils and careful field work, they are often able to pinpoint the geologic time in which valuable metals were emplaced and then to use that knowledge to seek out ores in unexplored areas. Mapping of ancient algal reefs has shown how these structures may sometimes control the emplacement of metal sulfides.

Sometimes the study of fossils provides unexpected information that is useful in mineral exploration. As an example, geologists working in the Appalachian Mountains became intrigued by the fact that a group of fossils called conodonts exhibited a range of colors from pale yellow to black. Conodonts (Fig. 3–50) are microscopic hard parts of organisms that lived about 560 to 200 million years ago. They are composed of apatite—the same mineral of which bone is made. The geologists suspected that the color changes were related to deep burial and attendant high temperatures in the sedimentary rocks containing the conodonts. By plotting the conodont color differences on maps, it became apparent that the colors changed systematically in the direction of rocks that had been most deeply buried. In the laboratory, heating of conodonts provides the same range of colors, from pale yellow (low temperature) to black (high temperature). Thus, conodonts provide indices of the depth of burial. Conodont colors are also an aid in exploration, for it has been shown that sedimentary rocks bearing black conodonts are less likely to yield commercial quantities of oil and gas. This information is now used routinely by oil companies in their search for petroleum.

**Figure 3–50**   Conodonts exhibiting varying degrees of color alteration resulting from deep burial and attendant increases in temperature. *(Courtesy of U.S. Geological Survey.)*

## Speculations About Life on Other Planets

While reflecting upon the long and complex history of life on earth, one might drift on to questions of whether or not other planets had—or now have—life as we know it. Is organic evolution a process that is unique to this planet? What properties of the earth have made it a suitable place for the origin and evolution of life? Could these conditions exist elsewhere in the universe?

To begin with, the earth is large enough that its gravitational attraction is sufficient to retain an atmosphere. The temperature of most of the earth's surface is low enough to provide an abundance of water in its liquid form. In addition, that temperature is suitable for chemical reactions required for life processes and is itself a consequence of the size of our sun and its distance from the earth. Our sun-star also has a life span sufficiently long to permit time for the emergence and evolution of life. Finally, the earth has always had all of the chemical elements required for life processes.

### Life in Our Solar System

We may first consider the possibilities for extraterrestrial life here in our own solar system. All evidence to date indicates that because of either size or distance from the sun, conditions are currently too harsh on neighboring planets to permit the evolution of *higher* forms of life. The Viking mission to Mars sampled the Martian surface for indications of life, but found nothing to suggest the existence of present or past organisms on the red planet.

With regard to our moon, scientists had predicted that no evidence of life would be found. The moon lacks such important attributes as an atmosphere and water. The predictions were correct. The moon is sterile. Satellites of other planets in the solar system, however, appear to have a greater potential for harboring life. One of these is Europa, a satellite of Jupiter. Data from the Voyager flybys of Europa suggest that the satellite has a liquid water ocean covered by a thin, fractured crust of ice. It is only conjecture, but perhaps these waters harbor simple forms of life. By analogy, here on earth certain species of algae and diatoms are known to be living in Antarctica in water that is overlain by impressive thicknesses of ice. They are able to photosynthesize even in the feeble light that has penetrated the overlying ice. Perhaps here and there beneath the icy surface of Europa, somewhat analagous conditions and organisms exist.

## Life in the Universe

Thus, we see that the earth is indeed biologically special when compared with other planetary bodies in our solar system. In contrast, however, it may not be so peculiar at all in the vast realms of the universe. Astronomers estimate that there are about 150 billion stars in our galaxy, and the number of galaxies now appears to be nearly limitless. Of those 150 billion stars in our galaxy, a not unreasonable estimate stipulates that at least 1 billion have planets with size and temperature conditions similar to the earth's. These are potentially habitable planets. For the entire known universe (of which our galaxy is but a small part), reputable scientists have estimated that there are as many as 100,000,000,000,000,000,000 planetary systems similar to our solar system. Such calculations indicate that it is probable that life does exist out there somewhere. Indeed, the universe may be rich in suitable habitats for life—but what kind of life? If we assume such life was formed from the same universal store of atoms and under physical conditions not too dissimilar to those that existed on earth, then we might very well recognize it as a living thing. But it is highly unlikely that duplicates of humans, cows, or butterflies exist on other planets. There are many variables in the evolutionary interactions of genetics, environment, and time involved in the making of a particular species. To say that the very same mutations, genetic recombinations, and environmental conditions producing a sparrow could occur in precisely the same sequential steps in time on a distant planet seems most improbable.

## SUMMARY

Fossils are the remains or traces of life of the geologic past. The processes that are important in fossilization are varied and include the precipitation of chemical substances into pore spaces (permineralization), molecular exchange for substances that were once part of the organisms with inorganic substances (replacement), or compression of the animal and plant into a thin film of carbonized remains (carbonization). Tracks, trails, molds, and casts are additional varieties of fossils. Paleontologists, the specialists that study fossils, are aware that the fossil record is not complete for all forms of life that have existed on earth. It is more complete for organisms that had hard parts such as teeth, bone, or shell and for organisms that lived on the floors of shallow seas, where the deposition of sediment is relatively continuous and rapid.

It has been repeatedly demonstrated that particular lineages of fossil organisms exhibit change with the passage of time. This observation is part of the evidence for Darwin's conclusion that life has developed from simple to more complex forms (evolution). Because of this change through time, rock layers from different periods can be recognized and correlated on the basis of the kinds of fossils they contain.

Our modern concept of evolution combines natural selection with hereditary mechanisms for change by means of recombinations of hereditary materials and mutations. The hereditary materials are chromosomes and genes. Chromosomes are located in the nuclei of cells, carry the genes that are composed of DNA, and are responsible for the total characteristics of an organism. A process called meiosis segregates the members of homologous pairs of chromosomes and halves the total number for each gamete. During fertilization, there is a random union of two gametes from parents of unlike sex that brings together assortments of chromosomes (and genes) from two parents, resulting in the production of individuals with different gene combinations. Changes in genes called mutations may occur. Changes may occur also by rearrangement of chromosomes. These changes and redistributions alter the assortment of genes and, hence, also the characteristics that are passed on to succeeding generations.

Adaptation is the process by which organisms become fitted to their environments. Commonly, adaptations involve a combination of characteristics, including not only structure, but physiology and behavior as well. Different vertebrates, for example, display adaptive modifications of limbs for either swimming or running, or of teeth for various kinds of food. The evolution and spread of a single lineage of organisms into several ecologic niches often results in different forms, each of which is distinctly adapted to a particular habitat. This process of diversification from a central stock is called adaptive radiation.

Within any large adaptive radiation, one finds many examples of an evolutionary pattern called divergence, in which a parent stock splits into two or more distinct lineages. Other evolutionary patterns include convergence, in which two morphologically different groups evolve similarities in morphology as they become adapted to similar environments, and iterative evolution, wherein organisms of similar morphology evolve more than once from the parental lineage.

Charles Darwin viewed evolution as a gradual stately change in lineages through time. This interpretation, called phyletic gradualism, persists today and is very probably valid for many groups of organisms. In addition, however, the present generation of paleontologists are finding considerable evidence that evolution entails sudden advances that punctuate periods of relative stability. The process is termed punctuated equilibrium and is thought to occur when new species arise abruptly from isolated populations. The subsequent expansion of new species from these isolated groups results in the sudden faunal or floral changes often seen in the fossil record.

An important goal for paleontologists is to understand the complex relationships between ancient plants and animals and their habitats. This area of paleontology is termed paleoecology. Paleoecology can, in turn, provide information about the distribution of ancient lands and seas, ancient climates, depth of water, natural barriers to migration, and former locations of continents.

Because fossils are so useful in inferring environments of deposition, as well as in correlating and determining relative geologic age of strata, they are widely employed by commercial paleontologists engaged in the search for fossil fuels. It is likely that this aspect of paleontology will continue in importance as the search for economically essential minerals intensifies. Paleontologists will continue their work in the future not only in such areas of exploration, but possibly also in the detection of once-living things in planets other than the earth. Statistics suggest the strong probability of present or past life somewhere in the universe.

## QUESTIONS FOR REVIEW

1. What is a fossil? When are fossils most valuable as guide fossils?

2. Account for the observation that there are more fossils of marine invertebrates than of any other group of organisms.

3. How has paleontology contributed to the concept of organic evolution?

4. How may fossils aid in deciphering ancient climates? Give an example.

5. Which would probably become better guide fossils: organisms with slow rates of dispersal or organisms with rapid rates of dispersal? Give reasons for your answer.

6. Fossil *A* occurs in rocks of the Cambrian and Ordovician periods. Fossil *B* occurs in rocks that range from Early Ordovician through Permian in age. Fossil *C* occurs in Mississippian through Permian rocks.

    a. What is the maximum possible range of age for a rock containing only fossil *B*?

    b. What is the maximum possible range for a rock containing both fossil *A* and fossil *B*?

    c. Which is the better guide fossil, *A* or *C*? Support your answer.

7. A time–rock unit contains a different fossil assemblage at one locality than it does at another one located 50 miles away. Suggest a possible cause for the dissimilarity.

8. Does the rate of evolution of a particular group of related organisms affect their usefulness as guide fossils? Substantiate your answer.

9. Explain why our knowledge of evolution suggests that although there may be life outside the earth, there will probably not be human beings precisely like ourselves.

10. What are the major stages in the evolutionary history of terrestrial plants?

11. What were the contributions of Darwin and Mendel to our modern concepts of organic evolution?

12. In fossilization by replacement, what are the usual, or most common, replacing substances?

13. Why is the determination of a fossil species more subjective than the determination of a still-living (not extinct) species?

14. Using fossils for age correlation is dependent upon *a priori* knowledge of their time ranges. How is this information obtained? What might cause the information to be in error?

15. What is meant by the term *adaptation*? Contrast the modern concept of adaptation with that of Jean Larmack.

16. What are peripheral isolates? Why are peripheral isolates likely to evolve into new species?

17. Distinguish between phyletic gradualism and punctuated equilibrium. What evidence might you present if you were required to defend the concept of punctuated equilibrium?

18. Provide an example of an adaptive radiation, from either the present world or the geologic past.

## TERMS TO REMEMBER

abyssal
adaptive radiation
bathyal environment
benthic
biozone
cast
chromosome
convergence
deoxyribonucleic acid (DNA)
diploid cell
divergence

ecology
evolution
family (taxonomic)
fossil
gene
genus
geologic range
guide fossil
haploid cell
homologous structure
ichnology
iterative evolution

littoral zone
meiosis
mitosis
mold (fossil)
mutation
natural selection
nekton
neritic zone
order (taxonomic)
paleontology
pelagic
peripheral isolate

permineralization
phyletic gradualism
phylogenetic tree
phylum
plankton
population
punctuated equilibrium
replacement
species
species diversity
vestigial organ

## SUPPLEMENTAL READINGS AND REFERENCES

Boardman, R. S., Cheetham, A. H., and Rowell, A. J. (eds.) 1987. *Fossil Invertebrates*. Palo Alto, CA: Blackwell Scientific Publications.

Carroll, R. L. 1988. *Vertebrate Paleontology and Evolution*. Palo Alto, CA: Blackwell Scientific Publications.

Clarkson, E. N. K. 1979. *Invertebrate Paleontology and Evolution*. London: Geo. Allen & Unwin, Ltd.

Cowen, R. 1990. *History of Life*. Palo Alto, CA: Blackwell Scientific Publications.

Darwin, C. 1859. *The Origin of Species* (1963 ed.). Introduction by H. L. Carson. New York: Washington Square Press.

Delevoryas, T. 1962. *Morphology and Evolution of Fossil Plants*. New York: Holt, Rinehart, and Winston.

Dodd, R. J., and Stanton, R. J., Jr. 1981. *Paleoecology, Concepts and Applications*. New York: John Wiley.

Eldredge, N., and Gould S. J. 1972. Punctuated equilibria: An alternative to phyletic gradualism. In *Models in Paleobiology*. Schopf, T. J. M. (ed). San Francisco: W. H. Freeman.

Gould, S. J. 1982. *The Panda's Thumb*. New York: W. W. Norton & Co.

Skinner, B. J. (ed). 1981. *Paleontology and Paleoenvironments*. Readings from *American Scientist*. Los Altos, CA: Wm. Kauffman, Inc.

Stearn, C. W., and Carroll, R. L. 1989. *Paleontology: The Record of Life*. New York: John Wiley.

Stewart, W. N. 1983. *Paleobotany and the Evolution of Plants*. Cambridge: Cambridge Univ. Press.

# Time and
# Geology

*Each grain of sand, each minute crystal in the rocks about us is a tiny clock, ticking off the years since it was formed. It is not always easy to read them, and we need complex instruments to do it, but they are true clocks or chronometers. The story they tell numbers the pages of earth history.*

Patrick M. Hurley, *How Old is the Earth?*
Doubleday Anchor Books, 1959

## Introduction

We humans have a fascination with the concept of time. Geology instructors are particularly aware of this fascination, for they are regularly asked the age of various rock and mineral specimens brought to the university by students, inquisitive artisans, and returning vacationers. If informed that the samples are tens or even hundreds of millions of years old, the collectors are often pleased, but they are also perplexed. "How can this person know the age of this specimen by just looking at it?" they think. If they insist on knowing the answer to that question, they may next receive a short discourse on the subject of geologic time. It is explained that the rock exposures from which the specimens were obtained have long ago been organized into a standard chronologic sequence based largely on superposition, evolution as indicated by fossils, and actual rock ages in years obtained from the study of radioactive elements in the rock. The geologist's initial estimate of age is based on experience. He or she may have spent a few hours kneeling at those same collecting localities and had a background of information to draw upon. Thus, at least sometimes, a geologist can recognize particular rocks as being of a certain age. The science that permits accomplishment of this feat is called **geochronology**. It is a science that began over four

centuries ago when Nicholas Steno described how the position of strata in a superpositional sequence could be used to show the relative geologic age of the layers. As described in Chapter 1, this simple but important idea was expanded and refined much later by William Smith and some of his contemporaries. These practical geologists showed how it was also possible to correlate strata. Outcrop by outcrop the rock sequences with their contained fossils were pieced together, one above the other, until a standard geologic time scale based on relative ages had been constructed.

## The Standard Geologic Time Scale

The early geologists had no way of knowing how many time units would be represented in the completed geologic time scale, nor could they know which fossils would be useful in correlation or which new strata might be discovered at a future time in some distant corner of the globe. Consequently, the time scale grew piecemeal, in an unsystematic manner. Units were named as they were discovered and studied. Sometimes the name for a unit was borrowed from local geography, from a mountain range in which rocks of a particular age were well exposed, or from an ancient tribe of Welshmen; sometimes the name was suggested by the kind of rocks that predominated.

## Divisions in the Geologic Time Scale

Geologists have proposed the term **eon** for the largest divisions of the geologic time scale. In chronologic succession, the first three eons of geologic time are the **Hadean, Archean,** and **Proterozoic** (see Fig. 3–21). The beginning of the Archean corresponds approximately to the ages of the oldest known rocks on earth. Although not universally used, the term Hadean refers to that period of time for which we have no rock record, which began with the origin of the planet 4.6 million years ago. The Proterozoic Eon refers to the time interval from 2500 to 570 million years ago.

The rocks of the Archean and Proterozoic are informally referred to simply as Precambrian. The antiquity of Precambrian rocks was recognized in the mid-1700s by Johann Lehman, a professor of mineralogy in Berlin, who referred to them as the "primary series." One frequently finds this term in the writing of French and Italian geologists who were contemporaries of Lehman. In 1833, the term appeared again when Sir Charles Lyell used it in his formulation of a surprisingly modern geologic time scale. Lyell and his predecessors recognized these "primary" rocks by their crystalline character and took their uppermost boundary to be an unconformity that separated them from the overlying—and therefore younger—fossiliferous strata.

All of the remainder of geologic time is included in the second eon, the **Phanerozoic Eon.** As a result of careful study of superposition accompanied by correlations based on the abundant fossil record of the Phanerozoic, geologists have divided it into three major subdivisions, termed **eras.** The oldest is the **Paleozoic Era,** which we now know lasted about 370 million years. Following the Paleozoic is the **Mesozoic Era,** which continued about 170 million years. The **Cenozoic Era,** in which we are now living, began about 66 million years ago.

As shown in Figure 3–21, the eras are divided into shorter time units called **periods;** periods can in turn be divided into **epochs.** In Chapter 2 we noted that eras, periods, epochs, and divisions of epochs, called **ages,** all represent intangible increments of time. The rocks laid down during a specific time

(geochronologic) interval are called time-stratigraphic units. For example, the rocks comprising a geologic system were deposited or emplaced during a geologic period. As indicated in Chapter 2, one may properly speak of climatic changes during the "Cambrian Period" as indicated by fossils found in rocks of the "Cambrian System." Each of the geologic systems is recognized by its distinctive fauna and flora of fossils. The fossils are different in stage of evolution from other fossils in both older and younger systems. **Series** is the time–stratigraphic term used for rocks deposited during an epoch, whereas **stage** represents the tangible rock record of an age (see Table 2–3).

## Recognition of Time Units

Units of geologic time bear the same names as the time–rock units to which they correspond. Thus, we may speak of the "Jurassic System" or the "Jurassic Period" according to whether we are referring to the rocks themselves or to the time during which they accumulated.

Time terms have come into use as a matter of convenience. Their definition is necessarily dependent upon the existence of tangible time–rock units. The steps leading to the recognition of time–rock units began with the use of superposition in establishing age relationships. Local sections of strata were used by early geologists to recognize beds of successively different age and, thereby, to record successive evolutionary changes in fauna and flora. (The order and nature of these evolutionary changes could be determined because higher layers are successively younger.) Once the faunal and floral succession was deciphered, fossils provided an additional tool for establishing the order of events. They could also be used for correlation, so that strata at one locality could be related to the strata of various other localities. No single place on earth contains a complete sequence of strata from all geologic ages. Hence, correlation to standard sections of many widely distributed local sections was necessary in constructing the geologic time scale (Fig. 4–1). Clearly, the time scale was not conceived as a coherent whole, but rather evolved piece by piece as a result of the individual studies of many geologists. Indeed, for some units at the series and stage level, the process continues even today. The fact that the time scale developed in

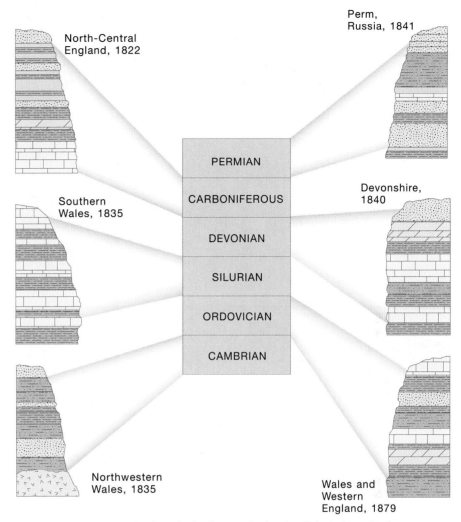

**Figure 4–1**  The standard geologic time scale for the Paleozoic and other eras developed without benefit of a grand plan. Instead, it developed by the compilation of "type sections" for each of the systems.

piecemeal fashion is apparent when one reviews its growth and development.

## The Cambrian System

The rocks of the **Cambrian System** take their name from *Cambria,* the Latin name for Wales. Exposures of strata in Wales (Fig. 4–2) provide a standard section with which rocks elsewhere in Europe and on other continents can be correlated. The standard section in Wales is named Cambrian *by definition.* All

other sections deposited during the same time as the rocks in Wales are recognized as Cambrian *by comparison* to the standard section.

Adam Sedgwick, a highly regarded professor of geology at Cambridge (Fig. 4–3) named the Cambrian in the 1830s for outcrops of poorly fossiliferous graywackes and dark siltstones and sandstones. The area in north Wales that Sedgwick studied was noted for its complexity, yet he was able to unravel its geologic history on the basis of spatial relationships and lithology.

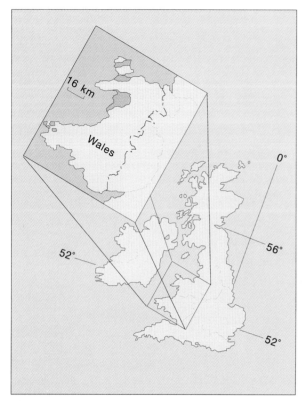

**Figure 4–2** Outcrop of Cambrian strata in north-western Wales where these rocks were named.

**Figure 4–3** Adam Sedgwick, one of the foremost geologists of the nineteenth century. A professor of geology at Cambridge University, Sedgwick is best remembered for deciphering the highly deformed system of rocks in northwestern Wales that he defined as the Cambrian System. He also founded the geologic museum at Cambridge which bears his name. *(Courtesy of the Cambridge Museum, Cambridge, England.)*

## The Silurian and Ordovician Systems

At about the same time that Sedgwick was laboring with outcrops that were to become the Cambrian System, his former student, Sir Roderick Impey Murchison, had begun studies of fossiliferous strata outcropping in the hills of south Wales. Murchison named these rocks the **Silurian System** (Fig. 4–4), taking the name from early inhabitants of western England and Wales known as the *Silures*. In 1835, Murchison and Sedgwick jointly presented a paper, *On the Silurian and Cambrian Systems, Exhibiting the Order in Which the Older Sedimentary Strata Succeed Each Other in England and Wales*. With this publication, the two geologists initiated the development of the early Paleozoic time scale. In the years that followed, a controversy arose between the two men that was to sever their friendship. Because Sedgwick had not described fossils distinctive of the Cambrian, the unit could not be recognized in other countries. Murchison argued, therefore, that the Cambrian was not a valid system. During the 1850s, he maintained that all fossiliferous strata above the "Primary Series" (the old name for Precambrian) and below the Old Red Sandstone (of Devonian age) belonged within the Silurian System. Sedgwick, of course, disagreed, but his opinion that the Cambrian was a valid system did not receive wide support until fossils were described from the upper part of the sequence. The fossils proved to be similar to faunas in Europe and North America. Hence, the Cambrian did meet the test of recognition outside England. Using these fossils as a basis for reinterpretation, the English geologist Charles Lapworth proposed combining the upper part of Sedgwick's Cambrian and the lower part of Murchison's Silurian into a new system. In 1879, he named the system **Ordovician** after the *Ordovices*, an early Celtic tribe (Fig. 4–5). The first three systems of the Paleozoic were now established (Fig. 4–6).

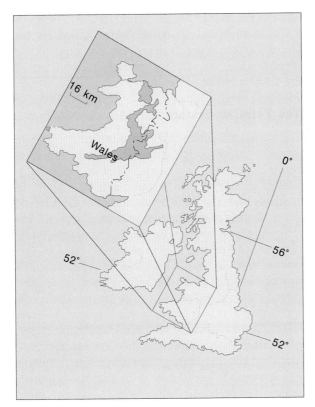

**Figure 4–4**  Outcrop of strata of the Silurian System in the classic region of Wales and western England where Sir Roderick Impey Murchison named them.

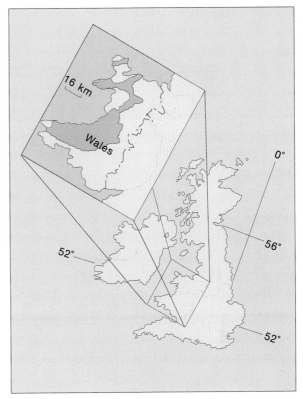

**Figure 4–5**  Outcrop pattern of strata of the Ordovician System in the type region of Wales.

## The Devonian System

The **Devonian System** was proposed for outcrops near Devonshire, England, (Fig. 4–7) by Sedgwick and Murchison in 1839 (prior to the years of their

bitter debate). They based their proposal on the fact that the rocks in question lay beneath the previously recognized Carboniferous System and contained a fauna that was different from that of the underlying Silurian and overlying Carboniferous. In their interpretation of the distinctive nature of the fauna, they

**Figure 4–6**  Generalized geologic cross-section for the Silurian type region. Uncomformities separate the Ordovician from the Cambrian and Silurian systems. Silurian strata are inclined toward the east, with more resistant rocks forming escarpments that face toward the west.

**Figure 4–7**   Outcrops of strata of the Devonian System in Wales and southwestern England where these rocks were named.

were aided by the studies of William Lonsdale, a retired army officer who had become a self-taught specialist on fossil corals. Further evidence that the new unit was a valid one came when Murchison and Sedgwick were able to recognize it in the Rhineland region of Europe. The Devonian rocks of Devonshire were also found to be equivalent to the widely known *Old Red Sandstone* of Scotland and Wales.

## The Carboniferous System

The term **Carboniferous System** was coined in 1822 by the English geologists William Conybeare and William Phillips to designate strata that included beds of coal in north-central England. Subsequently, it became convenient in Europe and Britain to divide the system into a *Lower Carboniferous* and *Upper Carboniferous*—the latter containing most of the workable coal seams. Two systems in North America, the **Mississippian** and **Pennsylvanian**, are broadly equivalent to these subdivisions. The American geologist Alexander Winchell formally proposed the name Mississippian in 1870 for the Lower Carboniferous strata that are extensively exposed in the upper

Mississippi River drainage region. In 1891, Henry S. Williams provided the name Pennsylvanian for the Upper Carboniferous System. The Mississippian and Pennsylvanian are now recognized as subsystems of the Carboniferous.

## The Permian System

The **Permian System** takes its name from a Russian province called Perm (from the nearby town of Perm) located on the western side of the Ural Mountains (Fig. 4–8). In 1840 and 1841, Murchison, in company with the French paleontologist Edouard de Verneuil and several Russian geologists, traveled extensively across western Russia. To his delight, Murchison found he was able to recognize Silurian, Devonian, and Carboniferous rocks by the fossils they contained. As a result he became even more convinced that groups of fossil organisms succeed one another in a definite and determinable order (this finding is now labeled the *principle of biologic succession*). Murchison established the new Permian System for rocks that overlay the Carboniferous System and contained fossils similar to those in German strata (the Zechstein beds), which had the same stratigraphic position as the Magnesian Limestone in England. Field studies had previously shown that the Magnesian Limestone rested upon Carboniferous strata. Thus Murchison was able to include the Magnesian Limestone within the Permian by correlation. The fossils of the new system appeared distinctly intermediate between those of the Carboniferous

**Figure 4–8**   Location of the basin near Perm (Molotov), Russia, where Murchison established the Permian System.

below and the Triassic above. In a letter to the Society of Naturalists of Moscow dated October 8, 1841, Murchison stated, "The Carboniferous System is surmounted, to the east of the Volga, by a vast series of marls, schists, limestones, sandstones, and conglomerates to which I propose to give the name 'Permian System.'" Murchison's establishment of the Permian system provides a fine example of the logic employed by early geologists in putting together the pieces of the standard time scale.

## The Triassic System

The influence of British geologists in providing names for the systems of the Paleozoic is by now obvious. However, their presence is not as evident in the development of Mesozoic nomenclature (Fig. 4–9). The **Triassic**, for example, was applied in 1834 by a German geologist named Frederich von Alberti. The term refers to a threefold division of rocks of this age in Germany. However, because the German strata in the type area are poorly fossiliferous, the standard of reference has been shifted to richly fossiliferous marine strata in the Alps.

## The Jurassic System

Another German scientist, Alexander von Humboldt, proposed the term **Jurassic** for strata of the Jura Mountains between France and Switzerland. However, in 1795, when he used the term, the concept of systems had not been developed. As a result, the Jurassic was redefined as a valid geologic system in 1839 by Leopold von Buch.

## The Cretaceous System

During the same year that Conybeare and Phillips were defining the Carboniferous, a Belgian geologist named Omalius d'Halloy proposed the term **Cretaceous** (from the Latin *Creta,* meaning "chalk") for rock outcrops in France, Belgium, and Holland. Although chalk beds are prevalent in some Cretaceous exposures, the system is actually recognized on the basis of fossils. Indeed, some thick sections of Cretaceous rocks contain no chalks whatsoever.

## The Tertiary System

The name **Tertiary** leads us back to the time when geology was just beginning as a science. Giovanni Arduino suggested a classification with four major

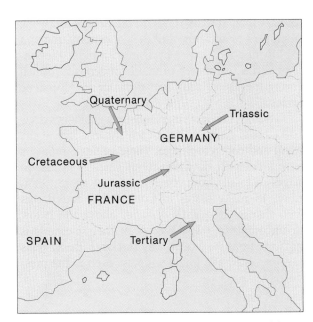

**Figure 4–9** Type areas for the systems of the Mesozoic and Cenozoic.

divisions: Primary, Secondary, Tertiary, and Quaternary. The Tertiary was derived from his 1759 description of unconsolidated *montes tertiarii* sediments at the foot of the Italian Alps. Later, the Tertiary was more precisely defined, and standard sections for series of the Tertiary were established in France. The **Eocene, Miocene**, and **Pliocene**, for example, were proposed by Charles Lyell in 1832 on the basis of the proportions of species of living marine invertebrates in the fossil fauna. By definition, 3 per cent of the fossil fauna of the Eocene still live, whereas the Miocene contained 17 per cent, and the Pliocene contained 50 to 67 per cent. The term **Oligocene** was proposed by August von Beyrich in 1854, and the term **Paleocene** was proposed 20 years later by Wilhelm Schimper. Other system names are also used in place of the Tertiary. Many geologists now use the terms **Paleogene System** (for the Paleocene, Eocene, and Oligocene) and **Neogene System** (for the Miocene and Pliocene).

## The Quaternary System

In 1829, the French geologist Jules Desnoyers proposed the term **Quaternary** for certain sediments and volcanics exposed in northern France. Although these deposits contained few fossils, Desnoyers was convinced on the basis of field studies that they were younger than Tertiary rocks. In the decade following Desnoyer's establishment of the Quaternary, the unit was divided by Charles Lyell into an older **Pleistocene Series**, composed primarily of deposits formed during the glacial ages, and the younger **Holocene Series**.

This brief review describing how geologists drew up a table of geologic time clearly shows a lack of any grand and coherent design. These geologic pioneers were influenced by conspicuous changes in assemblages of fossils from one sequence of strata to another. In many places in Europe they found that such changes frequently occurred above and below an unconformity. The success of their methods is apparent from the fact that, by and large, the systems have persisted and found wide use even to the present day.

## Quantitative Geologic Time

### Early Attempts at Absolute Geochronology

After having constructed a geologic time scale on the basis of relative age, it is understandable that geologists would seek some way to assign actual ages in millions of years to the various periods and epochs. From the time of Hutton, leaders in the scientific community were convinced that the earth was indeed very old, and certainly it was much older than the approximately 6000 years estimated by biblical scholars from calculations involving the ages of post-Adamite generations. But how old was the earth? And how might one quantify the geologic time scale?

To geologists of the 1800s it was apparent that to determine the absolute age of the earth or of particular rock bodies, they would have to concentrate on natural processes that continue in a single recognizable way and that also leave some sort of tangible record in the rocks. Evolution is one such process, and Charles Lyell recognized this. By comparing the amount of evolution exhibited by marine mollusks in the various series of the Tertiary System with the amount that had occurred since the beginning of the Pleistocene Ice Age, Lyell estimated that 80 million years had elapsed since the beginning of the Cenozoic. He came astonishingly close to the mark. However, for older sequences, estimates based on rates of evolution were difficult, not only because of missing parts in the fossil record, but also because rates of evolution for many taxa were not well understood.

In another attempt, geologists reasoned that if rates of deposition could be determined for sedimentary rocks, they might be able to estimate the time required for deposition of a given thickness of strata. Similar reasoning suggested that one could estimate total elapsed geologic time by dividing the average thickness of sediment transported annually to the oceans into the total thickness of sedimentary rock that had ever been deposited in the past. Unfortunately, such estimates did not adequately account for past differences in rates of sedimentation or losses to the total stratigraphic section during episodes of erosion. Also, some very ancient sediments were no longer recognizable, having been converted to igneous and metamorphic rocks in the course of mountain building. As a result of these uncertainties, estimates of the earth's total age based on sedimentation rates ranged from as little as a million to over a billion years.

Some types of sediment do permit one to know the rate of deposition quite accurately. For example, shales deposited in lakes often exhibit summer and winter layers that differ in color and texture and can be counted much like growth rings in trees. The regularly repeated layers are called **varves.** Each summer–winter couplet represents one year. Thus, one can count the successive couplets in a section of varved sediment and determine the time that elapsed during the deposition of the sediment in that section. Counting varves has been particularly useful in providing a chronology for the retreat of large continental glaciers during the Pleistocene Ice Age. Marginal lakes characteristically formed at the front of these glaciers as they receded. In spring and summer, the ice thawed, and the marginal lakes received the sediment which formed the somewhat coarser and thicker layer of the varve couplet. During winter, the lakes froze over. Fine sediment still in suspension ultimately settled to form the finer winter layer. Each year the process was repeated. As the ice sheet receded over many centuries, the marginal lakes followed. So also did the sections of varved sediments (Fig. 4–10) which were superimposed in repetitive layers rather like shingles on a roof, allowing one to determine the duration of a glacial recession.

In yet another geochronologic scheme, investigators attempted to determine the total age of the oceans. They speculated that the oceans' basins had been filled very shortly after the origin of the planet

**Figure 4–10** Varved clays from glacial Lake Barlow, Ojibway, Ontario. The section from the overhanging ledge to the contorted beds near the base is 48 cm thick. The contorted beds at the bottom resulted from aqueous slumping and flow. *(Courtesy of the Geological Survey of Canada.)*

and thus would be only slightly younger in age than the earth itself. The best known of the calculations for the age of the oceans were made by the distinguished Irish geologist John Joly. From information provided by gauges placed at the mouths of streams, Joly was able to estimate the annual increment of salt to the oceans. Then, knowing the salinity of ocean water and the approximate volume of water, he calculated the amount of salt already held in solution in the oceans. An estimate of the age of the oceans was derived from the following formula:

$$\text{Age of ocean} = \frac{\text{Total salt in ocean (in grams)}}{\text{Rate of salt added (grams per year)}}$$

Beginning with essentially nonsaline oceans, it would have taken about 90 million years for the oceans to reach their present salinity, according to Joly. The figure, however, was off the mark by a factor of 50, largely because there was no way to account accurately for recycled salt and salt incorporated into clay minerals deposited on the sea floors. Vast quantities of salt once in the sea had become extensive evaporite deposits on land; some of the salt being carried back to the sea had been dissolved, not from primary rocks, but from eroding marine strata on the continents. Even though in error, Joly's calculations clearly supported those geologists who insisted on an age for the earth far in excess of a few million years. The belief in the earth's immense antiquity was also supported by Darwin, Huxley, and other evolutionary biologists who saw the need for time in the hundreds of millions of years to accomplish the organic evolution apparent in the fossil record.

The opinion of the geologists and biologists that the earth was immensely old was soon to be challenged by the physicists. Spearheading this attack was Lord Kelvin, considered by many to be the outstanding physicist of the nineteenth century. Kelvin calculated the age of the earth on the assumption that it had cooled from a molten state and that the rate of cooling followed ordinary laws of heat conduction and radiation. Kelvin estimated the number of years it would have taken the earth to cool from a hot mass to its present condition. His assertions regarding the age of the earth varied over two decades of debate, but in his later years he confidently believed that 24 to 40 million years was a reasonable age for the earth. The biologists and geologists found Kelvin's estimates difficult to accept. But how could they do battle against his elegant mathematics when they were themselves armed only with inaccurate dating schemes and geologic intuition? For those geologists unwilling to capitulate, however, new discoveries showed their beliefs to be correct and Kelvin's to be unavoidably wrong.

A more correct answer to the question "how old is the earth?" was provided only after the discovery of radioactivity, a phenomenon unknown to Kelvin during his active years. With the detection of natural radioactivity by Henri Becquerel in 1896, followed by the isolation of radium by Marie and Pierre Curie two years later, the world became aware that the earth had its own built-in source of heat. It was not inexorably cooling at a steady and predictable rate, as Kelvin had suggested.

## Radiometric Methods for Dating Rocks

### Atoms and Isotopes

For those who have not had an introduction to the structure of atoms, a brief review will be useful in understanding how atoms are combined to make minerals, and how rocks can be dated radiometrically.

**Atoms** are the smallest particles of an element that can enter into a chemical reaction. They are also the smallest chemically indivisible particles of an element. An individual atom consists of an extremely minute but heavy nucleus surrounded by rapidly moving **electrons.** The electrons are relatively farther apart than are the planets surrounding our sun; consequently, the atom consists primarily of empty space. However, the electrons move so rapidly around the nucleus that they effectively fill the space within their orbits, giving size to the atom and repelling other atoms that may approach.

In the nucleus of the atom are closely compacted particles called **protons,** which carry a unit charge of positive electricity equal to the unit charge of negative electricity carried by the electron. Associated with the protons in the nucleus are electrically neutral particles having the same mass as protons. These are called **neutrons.** Modern atomic physics has made us aware of still other particles in the nucleus. For our understanding of the atom, however, knowledge of protons and neutrons is sufficient. The number of protons in the nucleus of an atom establishes

its number of positive charges and is called its **atomic number.** Each chemical element is composed of atoms having a particular atomic number. Thus, every element has a different number of protons in its nucleus. In naturally occurring elements, the number may range from 1 proton in hydrogen to 92 in uranium (Table 4–1).

The **mass** of an atom is approximately equal to the sum of the masses of its protons and neutrons. (The mass of electrons is so small that it need not be considered.) Carbon-12 is used as the standard for comparison of mass. By setting the atomic mass of carbon at 12, the atomic mass of hydrogen, which is the lightest of the elements, is just a bit greater than 1 (1.008, to be precise). The nearest whole number to the mass of a nucleus is called its **mass number.** Some atoms of the same substance have different mass numbers. Such variants are called **isotopes.** Isotopes are two or more varieties of the same element that have the same atomic number and chemical properties but differ in mass numbers because they have a varying number of neutrons in the nucleus. By convention, atomic mass is noted as a superscript preceding the chemical symbol of an element, and the atomic number is placed beneath it as

**Table 4–1  Number of Protons and Neutrons and Atomic Mass of Some Geologically Important Elements**

| Element and Symbol | Atomic Number (Number of Protons in Nucleus) | Number of Neutrons in Nucleus | Atomic Mass |
|---|---|---|---|
| Hydrogen (H) | 1 | 0 | 1 |
| Helium (He) | 2 | 2 | 4 |
| Carbon-12 (C)* | 6 | 6 | 12 |
| Carbon-14 (C) | 6 | 8 | 14 |
| Oxygen (O) | 8 | 8 | 16 |
| Sodium (Na) | 11 | 12 | 23 |
| Magnesium (Mg) | 12 | 13 | 25 |
| Aluminum (Al) | 13 | 14 | 27 |
| Silicon (Si) | 14 | 14 | 28 |
| Chlorine-35 (Cl)* | 17 | 18 | 35 |
| Chlorine-37 (Cl) | 17 | 20 | 37 |
| Potassium (K) | 19 | 20 | 39 |
| Calcium (Ca) | 20 | 20 | 40 |
| Iron (Fe) | 26 | 30 | 56 |
| Barium (Ba) | 56 | 82 | 138 |
| Lead-208 (Pb)* | 82 | 126 | 208 |
| Lead-206 (Pb) | 82 | 124 | 206 |
| Radium (Ra) | 88 | 138 | 226 |
| Uranium-238 (U) | 92 | 146 | 238 |

*When two isotopes of an element are given, the most abundant is starred.

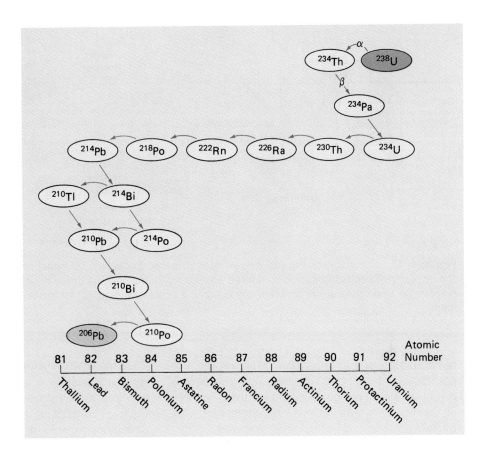

**Figure 4–11** Radioactive decay series of uranium-238 ($^{238}$U) to lead-206 ($^{206}$Pb).

a subscript. Thus, $^{40}_{20}$Ca is translated as the element calcium having an atomic number of 20 and an atomic mass of 40 (see Table 4–1).

## Radioactivity

The **radioactivity** discovered by Henri Becquerel was derived from elements such as uranium and thorium, which are unstable and break down or decay to form other elements or other isotopes of the same element. Any individual uranium atom, for example, will eventually decay to lead if given a sufficient length of time. To understand what is meant by "decay," let us consider what happens to a radioactive element such as uranium-238 ($^{238}$U). Uranium-238 has an atomic mass of 238. The "238" represents the sum of the weights of the atom's protons and neutrons (each proton and neutron having a mass of 1). Uranium has an atomic number (number of protons) of 92. Such atoms with specific atomic number and weight are sometimes termed **nuclides.** Sooner or later (and entirely spontaneously) the uranium-238 atom will fire off a particle from the nucleus

called an **alpha particle**. Alpha particles are positively charged ions of helium. They have an atomic weight of 4 and an atomic number of 2. Thus, when the alpha particle is emitted, the new atom will now have an atomic weight of 234 and an atomic number of 90. The new atom, which is formed from another by radioactive decay, is called a **daughter element.** From the decay of the parent nuclide, uranium-238, the daughter nuclide, thorium-234, is obtained (Fig. 4–11). A shorthand equation for this change is written:

$$^{238}_{92}U \rightarrow \,^{234}_{90}Th + \,^{4}_{2}He$$

This change is not, however, the end of the process, for the nucleus of thorium-234 ($^{234}$Th) is not stable. It eventually emits a **beta particle** (an electron discharged from the nucleus when a neutron splits into a proton and an electron). There is now an extra proton in the nucleus, but no loss of atomic weight because electrons are essentially weightless. Thus, from $^{234}_{90}$Th the daughter element $^{234}_{91}$Pa (protactinium) is formed. In this case, the atomic number

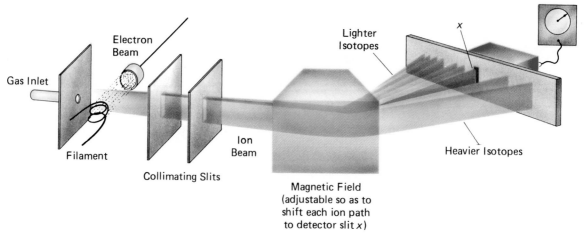

**Figure 4–12** Schematic drawing of a mass spectrometer. In this type of spectrometer, the intensity of each ion beam is measured electrically (rather than recorded photographically) to permit determination of the isotopic abundances required for radiometric dating.

has been increased by 1. In other instances, the beta particle may be captured by the nucleus, where it combines with a proton to form a neutron. The loss of the proton would decrease the atomic number by 1.

A third kind of emission in the radioactive decay process is called **gamma radiation**. It consists of a form of invisible electromagnetic waves having even shorter wavelengths than X-rays.

The rate of decay of radioactive isotopes is uniform and is not affected by changes in pressure, temperature, or the chemical environment. Therefore, once a quantity of radioactive nuclides has been incorporated into a growing mineral crystal, that quantity will begin to decay at a steady rate with a definite percentage of the radiogenic atoms undergoing decay in each increment of time. Each radioactive element has a particular mode of decay and a unique decay rate. As time passes, the quantity of the original or parent nuclide diminishes, and the number of the newly formed or daughter atoms increases, thereby indicating how much time has elapsed since the clock began its timekeeping. The beginning, or "time zero," for any mineral containing radioactive nuclides would be the moment when the radioactive parent atoms became part of a mineral from which daughter elements could not escape. The retention of daughter elements is essential, for they must be counted to determine the original quantity of the parent nuclide.

The determination of the ratio of parent to daughter nuclides is usually accomplished with the use of a *mass spectrometer,* an analytic instrument capable of separating and measuring the proportions of minute particles according to their mass differences. In the mass spectrometer, samples of elements are vaporized in an evacuated chamber where they are bombarded by a stream of electrons. This bombardment knocks electrons off the atoms, leaving them positively charged. A stream of these positively charged ions is deflected as it passes between plates that bear opposite charges of electricity. The degree of deflection depends on the charge-to-mass ratio. In general, the heavier the ion, the less it will be deflected (Fig. 4–12).

Of the three major families of rocks, the igneous clan is by far the best for radiometric dating. Fresh samples of igneous rocks are less likely to have experienced loss of daughter products, which must be accounted for in the age determination. Igneous rocks can provide a valid date for the time that a silicate melt containing radioactive elements solidified.

In contrast to igneous rocks, the minerals of sediments can be weathered and leached of radioactive components, and age determinations are far more prone to error. In addition, the age of a detrital grain in a sedimentary rock does not give an age of the sedimentary rock, but only of the parent rock that was eroded much earlier.

Dates obtained from metamorphic rocks may also require special care in interpretation. The age of a particular mineral may record the time the rock first formed or any one of a number of subsequent metamorphic recrystallizations.

Once an age has been determined for a particular rock unit, it is sometimes possible to use that data to approximate the age of adjacent rock masses. A shale lying below a lava flow that is 110 million years old and above another flow dated at 180 million years old must be between 110 and 180 million years of age (Fig. 4–13). Similarly, the age of a shale deposited on the erosional surface of a 490-million-year-old granite mass and covered by a 450-million-year-old lava flow must be between 450 and 490 million years old. The fossils in that shale might then be used to assign the shale to a particular geologic system or series. Then by correlation, the quantitative age obtained at the initial location (Fig. 4–14, Section A) could be extended to correlative formations at other locations (Fig. 4–14, Section B).

## Half-Life

One cannot predict with certainty the moment of disintegration for any individual radioactive atom in a mineral. We do know that it would take an infinitely long time for all of the atoms in a quantity of radioactive elements to be entirely transformed to stable daughter products. Experimenters have also shown that there are more disintegrations per increment of time in the early stages than in later stages (Fig. 4–15), and one can statistically forecast what percentage of a large population of atoms will decay in a certain amount of time.

Because of these features of radioactivity, it is convenient to consider the time needed for half of the original quantity of atoms to decay. This span of time is termed the **half-life**. Thus, at the end of the time constituting one half-life, $\frac{1}{2}$ of the original quantity of radioactive element still has not undergone decay. After another half-life, $\frac{1}{2}$ of what was left remains or $\frac{1}{4}$ of the original quantity. After a third half-life, only $\frac{1}{8}$ would remain, and so on.

Every radioactive nuclide has its own unique half-life. Uranium-235, for example, has a half-life of 704 million years. Thus, if a sample contains 50 per cent of the original amount of uranium-235 and 50 per cent of its daughter product, lead-207, then that sample is 704 million years old. If the analyses indicate 25 per cent of uranium-235 and 75 per cent of

A

B

**Figure 4–13** Igneous rocks that have provided absolute radiogenic ages can often be used to date sedimentary layers. (A) The shale is bracketed by two lava flows. (B) The shale lies above the older flow and is intruded by a younger igneous body. (Note: m.y. = million years.)

lead-207, two half-lives would have elapsed, and the sample would be 1408 million years old (Fig. 4–16).

## The Principal Geologic Timekeepers

At one time, there were many more radioactive nuclides present on earth than there are now. Many of these had short half-lives and have long since decayed to undetectable quantities. Fortunately for those interested in dating the earth's most ancient rocks, there remain a few long-lived radioactive nuclides. The most useful of these are uranium-238, uranium-235, rubidium-87, and potassium-40 (Table 4–2). There are also a few short-lived radioactive elements that are used for dating more recent events. Carbon-14 is an example of such a short-lived isotope. There are also short-lived nuclides that represent segments of a uranium or thorium decay series.

**Uranium–Lead Methods** Dating methods involving lead require the presence of radioactive nuclides of uranium or thorium that were incorporated into rocks when they originated. To determine the age of a sample of mineral or rock, one must know

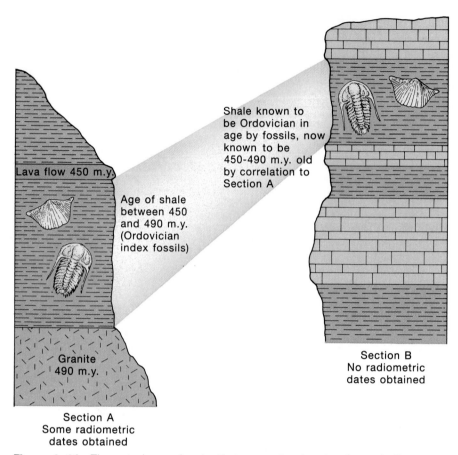

Shale known to
be Ordovician in
age by fossils, now
known to be
450-490 m.y. old
by correlation to
Section A

Lava flow 450 m.y.

Age of shale
between 450
and 490 m.y.
(Ordovician
index fossils)

Granite
490 m.y.

Section B
No radiometric
dates obtained

Section A
Some radiometric
dates obtained

**Figure 4–14** The actual age of rocks that cannot be dated radiometrically can sometimes be ascertained by correlation.

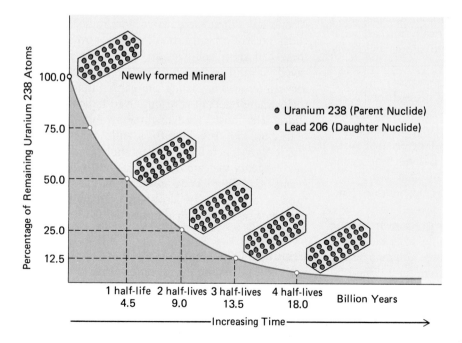

Newly formed Mineral

● Uranium 238 (Parent Nuclide)
● Lead 206 (Daughter Nuclide)

Percentage of Remaining Uranium 238 Atoms

100.0

75.0

50.0

25.0

12.5

1 half-life
4.5

2 half-lives
9.0

3 half-lives
13.5

4 half-lives
18.0

Billion Years

◀— Increasing Time —▶

**Figure 4–15** Rate of radioactive decay of uranium-238 to lead-206. During each half-life, one-half of the remaining amount of the radioactive element decays to its daughter element. In this simplified diagram, only the parent and daughter nuclides are shown, and the assumption is made that there was no contamination by daughter nuclides at the time the mineral formed.

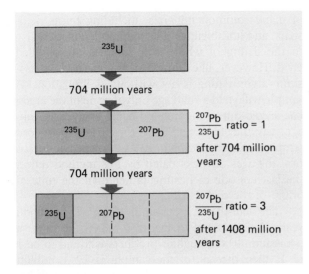

**Figure 4–16**   Radioactive decay of uranium-235 to lead-207.

recognize original lead and make the needed corrections.

As we have seen, different isotopes decay at different rates. Geochronologists take advantage of this fact by simultaneously analyzing two or three isotope pairs as a means to cross-check ages and detect errors. For example, if the $^{235}U/^{207}Pb$ radiometric ages and the $^{238}U/^{206}Pb$ ages from the same sample agree, then one can confidently assume the age determination is valid.

Radiometric ages that depend on uranium–lead ratios may also be checked against ages derived from lead-207 to lead-206. Because the half-life of uranium-235 is much less than the half-life of uranium-238, the ratio of lead-207 (produced by the decay of uranium-235) to lead-206 will change regularly with age and can be used as a radioactive timekeeper (Fig. 4–17). This is called a lead–lead age, as opposed to a uranium–lead age.

**The Potassium–Argon Method**   Potassium and argon are another radioactive pair widely used for dating rocks. By means of electron capture (causing a proton to be transformed into a neutron), about 11 per cent of the potassium-40 in a mineral decays to argon-40, which may then be retained within the parent mineral. The remaining potassium-40 decays to calcium-40 (by emission of a beta particle). The decay of potassium-40 to calcium-40 is not useful for obtaining radiometric ages because radiogenic calcium cannot be distinguished from original calcium in a rock. Thus, geochronologists concentrate their efforts on the 11 per cent of potassium-40 atoms that

the original number of parent nuclides as well as the number remaining at the present time. The original number of parent atoms should be equal to the sum of the present number of parent atoms and daughter atoms. The assumptions are made that the system has remained closed, so that neither parent nor daughter atoms have ever been added or removed from the sample except by decay, and that no daughter atoms were present in the system when it formed. The presence, for example, of original lead in the mineral would cause the radiometric age to exceed the true age. Fortunately, geochemists are able to

**Table 4–2   Some of the More Useful Nuclides for Radioactive Dating**

| Parent Nuclide* | Half-Life (years)[†] | Daughter Nuclide | Materials Dated |
|---|---|---|---|
| Carbon-14 | 5730 | Nitrogen-14 | Organic materials |
| Uranium-235 | 704 million ($7.04 \times 10^8$) | Lead-207 (and helium) | Zircon, uraninite, pitch-blende |
| Potassium-40 | 1251 million ($1.25 \times 10^9$) | Argon-40 (and calcium-40)[‡] | Muscovite, biotite, hornblende, volcanic rock, glauconite, K-feldspar |
| Uranium-238 | 4468 million ($4.47 \times 10^9$) | Lead-206 (and helium) | Zircon, uraninite, pitchblende |
| Rubidium-87 | 48,800 million ($4.88 \times 10^{10}$) | Strontium-87 | K-micas, K-feldspars, biotite, metamorphic rock, glauconite |

*A *nuclide* is a convenient term for any particular atom (recognized by its particular combination of neutrons and protons).
[†]Half-life data from Steiger, R. H., and Jäger, E., 1977. Subcommission on geochronology: Convention on the use of decay constants in geo- and cosmochronology. *Earth and Planetary Science Letters* 36:359–362.
[‡]Although potassium-40 decays to argon-40 and calcium-40, only argon is used in the dating method because most minerals contain considerable calcium-40, even before decay has begun.

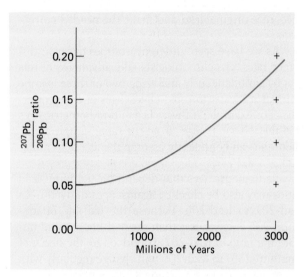

**Figure 4–17** Graph showing how the ratio of lead-207 to lead-206 can be used as a measure of age.

of many common minerals, including micas, feldspars, and hornblendes. The method lends itself also to the dating of whole-rock samples of solidified lavas. However, like all radiometric methods, potassium–argon dating is not without its limitations. A sample will yield a valid age only if none of the argon has leaked out of the mineral being analyzed. Leakage may indeed occur if the rock has experienced temperatures above about 125°C. In specific localities, the ages of rocks dated by this method reflect the last episode of heating rather than the time of origin of the rock itself.

The half-life of potassium-40 is 1251 million years. As illustrated in Figure 4–18, if the ratio of potassium-40 to daughter products is found to be 1 to 1, then the age of the sample is 1251 million years. If the ratio is 1 to 3, then yet another half-life has elapsed, and the rock would have a radiogenic age of two half-lives, or 2502 million years.

Potassium–argon dating is widely used in deciphering various types of geologic problems. Geologists are now using the method in studies relating to sea floor movements. For many years, scientists have been curious about the alignment of the major Hawaiian Islands and the adjacent seamounts. With the advent of the theory of sea floor spreading, scientists developed the concept that these volcanic islands

decay to argon. One advantage of using argon is that it is inert—that is, it does not combine chemically with other elements. Argon-40 found in a mineral is very likely to have originated there following the decay of adjacent potassium-40 atoms in the mineral. Also, potassium-40 is an abundant constituent

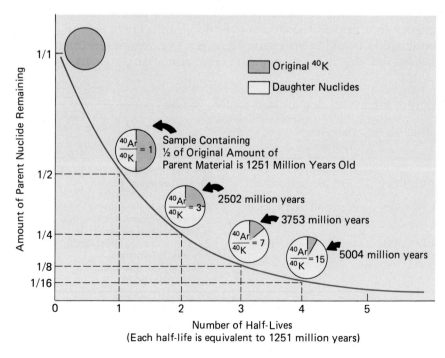

**Figure 4–18** Decay curve for potassium-40.

were built over a relatively fixed "hot spot" deep in the upper mantle. Conduits from the hot spot brought lavas up to the sea floor, where eruptions periodically occurred. Volcanoes that developed over the hot spot were then conveyed along by sea floor movement, and new volcanoes were produced over the vacated position. Geologists reasoned that if this process had taken place in the Hawaiian Islands, then potassium–argon radiometric ages should change in sequence along the island-seamount chain. The dates do indeed support the theory, and they even suggest that the direction of movement changed from a more northerly trend to a northwesterly trend about 40 million years ago (See Figure 6–46).

**The Rubidium–Strontium Method**   The dating method based on the disintegration by beta decay of rubidium-87 to strontium-87 can sometimes be used as a check on potassium–argon dates because rubidium and potassium are often found in the same minerals. The rubidium–strontium scheme has a further advantage in that the strontium daughter nuclide is not diffused by relatively mild heating events, as is the case with argon.

In the rubidium–strontium method, a number of samples are collected from the rock body to be dated. With the aid of the mass spectrometer, the amounts of radioactive rubidium-87, its daughter product strontium-87, and strontium-86 are calculated for each sample. Strontium-86 is an isotope not derived from radioactive decay. A graph is then prepared in which the $^{87}$Rb/$^{86}$Sr ratio in each sample is plotted against the $^{87}$Sr/$^{86}$Sr ratio (Figure 4–19). From the points on the graph, a straight line that is termed an **isochron** is constructed. The slope of the isochron results from the fact that, with the passage of time, there is continuous decay of rubidium-87, which causes the rubidium-87/strontium-86 ratio to decrease. Conversely, the strontium-87/strontium-86 ratio increases as strontium-87 is produced by the decay of rubidium-87. The older the rocks being investigated, the more the original isotope ratios will have been changed and the greater will be the inclination of the isochron. The slope of the isochron permits a computation of the age of the rock.

The rubidium–strontium and potassium–argon methods need not always depend upon the collection of discrete mineral grains containing the required isotopes. Sometimes the rock under investigation is so finely crystalline and the critical minerals

so tiny and dispersed that it is difficult or impossible to obtain a suitable collection of minerals. In such instances, large samples of the entire rock may be used for age determination. This method is called **whole-rock analysis.** It is useful not only for fine-grained rocks, but also for rocks in which the yield of useful isotopes from mineral separates is too low for analysis. Whole-rock analysis has also been useful in determining the age of rocks that have been so severely metamorphosed that the potassium-argon or rubidium–strontium radiometric clocks of individual minerals have been reset. In such cases, the age obtained from the minerals would be that of the episode of metamorphism, not the total age of the rock itself. The required isotopes and their decay products, however, may have merely moved to nearby locations within the same rock body, and therefore analyses of large chunks of the whole rock may provide valid radiometric age determinations.

**The Carbon-14 Method**   Techniques for age determination based on content of radiocarbon were first devised by W. F. Libby and his associates at the University of Chicago in 1947. The method is an

**Figure 4–19**   Whole-rock rubidium–strontium isochron for a set of samples of a Precambrian granite body exposed near Sudbury, Ontario. *(Modified from T. E. Krogh et al.* Carnegie Institute Washington Year Book *66:530, 1968.)*

COMMENTARY

## Dating Marine Sediments With Thorium and Proactinium

Recent decades of exploration of the ocean floor have prompted the development of dating methods that are especially useful for oceanic sediments too old to be dated with carbon 14. These new methods use isotopes that are produced in the intermediary stages of the uranium decay series. The geochronologists who developed these techniques recognized that most of the uranium brought to the oceans by streams remains in solution. While in solution it decays, eventually producing thorium 230. The thorium isotope is precipitated and becomes a component of ocean floor sediment. Thorium 230 itself decays with a half-life of 75,000 years. Because lower levels of the sediment have been undergoing decay longer than layers near the surface of the ocean floor, there is a decrease in quantity of thorium 230 at greater depths in a cored column of sediment.

The thorium 230 concentration of each measured interval is compared to the quantity in the surface layer, and a time scale relating age to quantity of remaining parent isotope is then formulated.

Deep ocean sediment can also be dated by means of a procedure based on the ratio of thorium to proactinium 231. The decay of uranium 235 produces proactinium 231, which like thorium 230 precipitates rapidly on the sea floor. Thorium 230 has a half-life of 75,000 years, whereas the half-life of proactinium is only 34,300 years. Both isotopes are precipitated in the same proportions, but because of their different rates of decay, the ratio of the two parent isotopes changes regularly with the passage of time. Thus, the greater differences in the quantity of undecayed parent isotopes, the older the sediment.

indispensable aid to the archeologic research and is useful in deciphering very recent events in geologic history. Because of the short half-life of carbon-14—a mere 5730 years—organic substances older than about 50,000 years contain very little carbon-14. New techniques, however, allow geologists to extend the method's usefulness back to almost 100,000 years.

Unlike uranium-238, carbon-14 is created continuously in the earth's upper atmosphere. The story of its origin begins with cosmic rays, which are extremely high energy particles (mostly protons) that bombard the earth continuously. Such particles strike atoms in the upper atmosphere and split their nuclei into small particles, among which are neutrons. Carbon-14 is formed when a neutron strikes an atom of nitrogen-14. As a result of the collision, the nitrogen atom emits a proton, captures a neutron, and becomes carbon-14 (Fig. 4–20). Radioactive carbon is being created by this process at the rate of about two atoms per second for every square centimeter of the earth's surface. The newly created carbon-14 combines quickly with oxygen to form

$CO_2$, which is then distributed by wind and water currents around the globe. It soon finds its way into photosynthetic plants because they utilize carbon dioxide from the atmosphere to build tissues. Plants containing carbon-14 are ingested by animals, and the isotope becomes a part of their tissue as well.

Eventually, carbon-14 decays back to nitrogen-14 by the emission of a beta particle. A plant removing $CO_2$ from the atmosphere should receive a share of carbon-14 proportional to that in the atmosphere. A state of equilibrium is reached in which the gain in newly produced carbon-14 is balanced by the decay loss. The rate of production of carbon-14 has varied somewhat over the past several thousand years (Fig. 4–21). As a result, corrections in age calculations must be made. Such corrections are derived from analyses of standards such as wood samples, whose exact age is known from tree ring counts.

The age of some ancient bits of organic material is not determined from the ratio of parent to daughter nuclides, as is done with previously discussed dating schemes. Rather, the age is estimated from the ratio of carbon-14 to all other carbon in the sample.

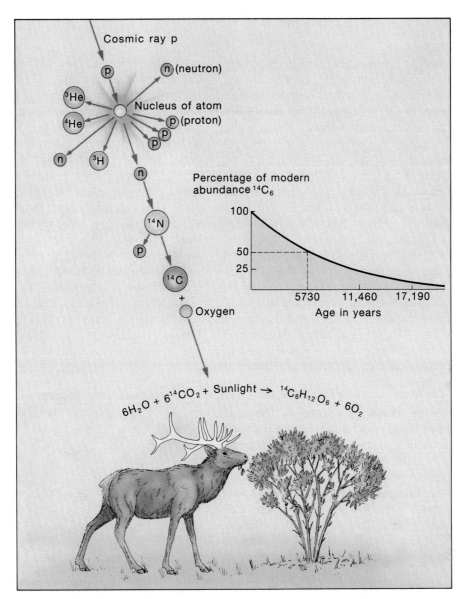

**Figure 4–20** Carbon-14 is formed from nitrogen in the atmosphere. It combines with oxygen to form radioactive carbon dioxide and is then incorporated into all living things.

**Figure 4–21** Deviation of carbon-14 ages to true ages from the present back to about 5000 B.C. Data are obtained from analysis of bristle cone pines from the western United States. Calculations of carbon-14 deviations are based on half-life of 5730 years. *(Adapted from Ralph, E. K., Michael, H. N., and Han, M. C., Radiocarbon dates and reality, MASCA Newsletter 9:1, 1973.)*

After an animal or plant dies, there can be no further replacement of carbon from atmospheric $CO_2$, and the amount of carbon-14 already present in the once living organism begins to diminish in accordance with the rate of carbon-14 decay. Thus, if the carbon-14 fraction of the total carbon in a piece of pine tree buried in volcanic ash were found to be about 25 percent of the quantity in living pines, then the age of the wood (and the volcanic activity) would be two half-lives of 5730 years each or 11,460 years.

The carbon-14 technique had considerable value to geologists studying the most recent events of the Pleistocene Ice Age. Prior to the development of the method, the age of sediments deposited by the last advance of continental glaciers was surmised to be about 25,000 years. Radiocarbon dates of a layer of peat beneath the glacial sediments provided an age of only 11,400 years. The method has also been found useful in studies of groundwater migration and in dating the geologically recent uppermost layer of sediment on the sea floors.

**Nuclear Fission Track Timekeepers** Nuclear **particle fission tracks** were discovered in the early 1960's when scientists using the electron microscope were able to examine the areas around presumed locations of radioactive particles that were embedded in mica. Closer examination showed that the tracks were really small tunnels—like bullet holes—that were produced when high-energy particles of the nucleus of uranium were fired off in the course of spontaneous fission (spontaneous fragmentation of an atom into two or more lighter atoms and nuclear particles). The particles speed through the orderly rows of atoms in the crystal, tearing away electrons from atoms located along the path of trajectory and rendering them positively charged. Their mutual repulsion produces the track (Fig. 4–22). The tracks are only a few atoms in width and are impossible to see without an electron microscope. Therefore, the sample is immersed for a short period of time in a suitable solution (acid or alkali), which rushes up into the tubes, enlarging the track tunnel so that it can be seen with an ordinary microscope.

The natural rate of track production by uranium atoms is very slow and occurs at a constant rate. For this reason, the tracks can be used to determine the number of years that have elapsed since the uranium-bearing mineral solidified. One first determines the number of uranium atoms that have al-

ready disintegrated. This number is obtained with the aid of a microscope by counting the etched tracks. Next, one must find the original number of uranium atoms. This quantity can be determined by bombarding the sample with neutrons in a reactor and thereby causing the remaining uranium to undergo fission. A second count of tracks reveals the original quantity of uranium. Finally, one must know the spontaneous fission decay rate for uranium-238. This information is determined by counting the tracks in a piece of uranium-bearing synthetic glass of known date of manufacture.

Fission track dating is of particular interest to geochronologists because it can be used to date specimens only a few centuries old, as well as to date rocks billions of years in age. The method helps to date the period between 50,000 and 1 million years ago—a period for which neither carbon-14 nor po-

**Figure 4–22** Fission tracks. These tracks were produced by plutonium-244 in a melilite crystal that was extracted from a meteorite. The small crystals are spinel inclusions. Melilite is a calcium–magnesium aluminosilicate. *(Photograph courtesy of Frank Podosek, Department of Earth and Planetary Sciences, Washington University.)*

tassium–argon methods are suitable. As with all radiometric techniques, however, there can be problems. If rocks have been subjected to high temperatures, tracks may heal and fade away.

## The Age of the Earth

Anyone interested in the total age of the earth must decide what event constitutes its "birth." Most geologists assume "year 1" commenced as soon as the earth had collected most of its present mass and had developed a solid crust. Unfortunately, rocks that date from those earliest years have not been found on earth. They have long since been altered and converted to other rocks by various geologic processes. The oldest materials known are grains of the mineral zircon taken from a sandstone in western Australia. The zircon grains are 4.1 to 4.2 billion years old. The zircons were probably eroded from nearby granitic rocks and deposited, along with quartz and other detrital grains, by rivers. Other very old rocks on earth include 3.7-billion-year-old granites of south-western Greenland, metamorphic rocks of about the same age from Minnesota, and 3.96-billion-year-old rocks from the Northwest territories of Canada (north of Yellowknife, Canada).

Meteorites, which many consider to be remnants of a shattered planet or asteroid that originally formed at about the same time as the earth, have provided uranium–lead and rubidium–strontium ages of about 4.6 billion years. From such data, and from estimates of how long it would take to produce the quantities of various lead isotopes now found on the earth, geochronologists feel that the 4.6-billion-year age for the earth can be accepted with confidence. Evidence substantiating this conclusion comes from returned moon rocks. The ages of these rocks range from 3.3 to about 4.6 billion years. The older age determinations are derived from rocks collected on the lunar highland, which may represent the original lunar crust. Certainly, the moons and planets of our solar system originated as a result of the same cosmic processes and at about the same time.

## SUMMARY

In no other science does time play as significant a part as in geology. Time provides the frame of reference necessary to the interpretation of events, processes, and inhabitants of the earth. In the earlier stages of its development, geology was dependent upon relative dating of events. James Hutton helped scientists visualize the enormous periods needed to accomplish the events indicated in sequences of strata, and the geologists who followed him pieced together the many local stratigraphic sections, using fossils and superposition. A scale of relative geologic time gradually emerged. Initial attempts to decide what the rock succession meant in terms of years were made by estimating the amount of salt in the ocean, the average rate of deposition of sediment, and the rate of cooling of the earth. However, these early schemes did little more than suggest that the planet was at least tens of millions of years old and that the traditional concept of a 6000-year-old earth did not agree with what could be observed geologically.

An adequate means of measuring geologic time was achieved only after the discovery of radioactivity at about the turn of the twentieth century. Scientists found that the rate of decay by radioactivity of certain elements is constant and can be measured and that the proportion of parent and daughter elements can be used to reveal how long they had been present in a rock. Over the years, continuing efforts by investigators as well as improvements in instrumentation (particularly of the mass spectrometer) have provided many thousands of age determinations. Frequently, these radiometric dates have shed light on difficult geologic problems, provided a way to determine rates of movement of crustal rocks, and permitted geologists to date mountain building or to determine the time of volcanic eruptions. In a few highly important regions, radiometric dates have been related to particular fossiliferous strata and have thereby helped to quantify the geologic time scale and to permit estimation of rates of organic evolution. Radiometric dating has also changed the way humans view their place in the totality of time.

The radiometric transformations most widely used in determining absolute ages are uranium-238 to lead-206, uranium-235 to lead-207, thorium-232 to lead-208, potassium-40 to argon-40, rubidium-87 to strontium-87, and carbon-14 to nitrogen-14. Methods involving uranium–lead ratios are of importance in dating the earth's oldest rocks. The short-lived carbon-14 isotope that is created by cosmic ray bombardment of the atmosphere provides a means to date the most recent events in earth history. For rocks of intermediate age, schemes involving potassium–argon ratios, those utilizing intermediate elements in decay series, or those employing fission fragment tracks are most useful. A figure of 4.6 billion years for the earth's total age is now supported by ages based on meteorites and on lead ratios from terrestrial samples.

Improvements in radiometric geochronology are being made daily and will provide further calibration of the standard geologic time scale in the future. Some of the time boundaries in the scale, such as that between the Cretaceous and Tertiary systems, are already well validated. Others, such as the boundary between the Paleozoic and Mesozoic, require additional refinement. Additional efforts to further incorporate radiometric ages into sections of sedimentary rocks are among the continuing tasks of historical geologists. The usual methods for determining the age of strata involve the dating of intrusions that penetrate these sediments or the dating of interbedded volcanic layers. Less frequently, strata can be dated by means of radiogenic isotopes incorporated within sedimentary minerals that formed in place at the time of sedimentation. At present, the best radiometric age estimates indicate that Paleozoic sedimentation began about 570 million years ago, the Mesozoic Era began about 245 million years ago, and the Cenozoic commenced about 66 million years ago.

## QUESTIONS FOR REVIEW

1. By what methods did geologists attempt to determine the age of the earth before the discovery of radioactivity? Why were these methods inadequate?

2. What types of radiation accompany the decay of radioactive isotopes?

3. In making an age determination based upon the uranium–lead method, why should an investigator select an unweathered sample for analysis?

4. How do the isotopes carbon-12 and carbon-14 differ from one another in regard to the following: (a) number of protons, (b) number of electrons, and (c) number of neutrons? What is the origin of carbon-14?

5. Define *half-life*. Why is this term used instead of the expression *whole-life* in radiometric dating? What are the half-lives of uranium-238, potassium-40, and carbon-14?

6. How do fission tracks originate? What geologic conditions might destroy these tracks?

7. Pebbles of basalt within a conglomerate yield a radiometric age of 300 million years. What can be said about the age of the conglomerate? Several miles away, the same conglomerate stratum is bisected by a 200-million-year-old dike. What now can be said about the age of the conglomerate in this location?

8. Has the amount of uranium in the earth increased, decreased, or remained about the same over the past 4.5 billion years? What can be stated about changes in the amount of lead?

9. State the estimated age of a sample of mummified skin from a prehistoric human that contained 12.5 per cent of an original quantity of carbon-14.

10. How are dating methods involving decay of radioactive elements unlike methods for determining elapsed time by the funneling of sand through an hour glass?

11. State the effect on the radiometric age of a zircon crystal being dated by the potassium–argon method if a small amount of argon-40 escaped from the crystal.

12. What is the advantage of having both uranium and thorium present in a mineral being used for a radiometric age determination?

13. What are *varves,* and how might they be used to determine the time span of existence of an ancient lake?

14. Minerals suitable for radiometric age determinations are usually components of igneous rocks. How, then, can absolute ages be obtained for sedimentary formations?

15. Discuss the contributions to the development of the geologic time scale made by Charles Lapworth, Frederick von Alberti, and Charles Lyell.

16. What observations were used by Sir Roderick Murchison in the establishment of the Permian System?

17. What is the essential difference between a time–stratigraphic unit and a time unit? Give an example of each.

18. What characteristics of the rocks of northern Wales caused Adam Sedgwick difficulty in establishing them as a valid geologic system?

## TERMS TO REMEMBER

| | | | |
|---|---|---|---|
| age (geologic) | Eocene | Mississippian | Pleistocene |
| alpha particle | eon | Neogene | Pliocene |
| Archean Eon | epoch | neutron | Proterozoic Eon |
| atom | fission track | nuclide | proton |
| atomic number | gamma radiation | Oligocene | Quaternary |
| beta particle | geochronology | Ordovician | series |
| Cambrian | half-life | Paleocene | Silurian |
| Carboniferous | Holocene | Paleogene | stage |
| Cenozoic | isotope | Paleozoic | Triassic |
| Cretaceous | Jurassic | Pennsylvanian | varves |
| daughter element | mass number | period (geologic) | whole-rock radiometric |
| Devonian | Mesozoic | Permian | analysis |
| electron | Miocene | Phanerozoic | |

## SUPPLEMENTAL READINGS AND REFERENCES

Berry, W. B. N. 1968. *Growth of a Prehistoric Time Scale.* San Francisco: W. H. Freeman Co.

Eicher, D. L. 1976. *Geologic Time* (2nd ed.). Englewood Cliffs, N.J.: Prentice-Hall, Inc.

Faul, H. 1966. *Ages of Rocks, Planets, and Stars.* New York: McGraw-Hill.

Faure, G. 1977. *Principles of Isotope Geology.* New York: John Wiley.

Fleischer, R. L., Price, P. B., and Walker, R. M. 1965. Tracks of charged particles in solids. *Science* 149(3682):383–393.

Hamilton, E. I. 1965. *Applied Geochronology.* New York: Academic Press, Inc.

Palmer, A. R. 1983. The decade of North American geology, geologic time scale. *Geology* 11:503–504.

Schaffer, O. A., and Zahringer, J. (eds.). 1966. *Potassium–Argon Dating.* New York: Springer-Verlag.

Speakman, C. 1982. *Adam Sedgwick, Geologist and Dalesman.* England: Broad Oak Press Ltd.

York, D., and Farquhar, R. M. 1972. *The Earth's Age and Geochronology.* Oxford, England: Pergamon Press.

The meeting of atmosphere, hydrosphere, and lithosphere. Oregon coast, just north of Brookings, Oregon.

# Development of the Earth's Major Features

*What was the earth like in the very beginning? How does one get from a conglomeratic ball of stardust to a planet with concentric structure—inner and outer core, silicate mantle, crust, oceans, and layered atmosphere?*

Preston Cloud, 1988,
*Oasis in Space*.
W. W. Norton & Co.

## The Earth's Place in Space

The earth is one of nine planets that revolve around a rather average star, our sun. The planets, in order of increasing distance from the sun, are Mercury, Venus, Earth, Mars, Jupiter, Saturn, Uranus, Neptune, and Pluto. Pluto is considered by many astronomers to be a large satellite that escaped from Neptune. A belt of asteroids orbits the sun in the region between the paths of Mars and Jupiter. This grouping of planets and asteroids around the sun constitutes our solar system (Fig. 5–1). Certainly, ours is not the only such system in the universe. It is true that the suspected planets of other systems are too distant to be detected directly, but we think that they are out there because of wobbling motions in distant stars. The wobbling is believed to be caused by the gravitational pull of orbiting, but not visible, planets.

Our solar system is a small part of a much larger aggregate of stars, planets, dust, and gases called a **galaxy.** Galaxies are also numerous in the universe, and their constituents are arranged into several different general forms: tightly packed elliptical galaxies, irregular galaxies, and the more familiar discoidal spiral galaxies with their glowing central bulge and great curving arms.

The galaxy in which our solar system is located is called the **Milky Way galaxy** because as we look toward its dense central bulge, we see a great milky haze of light in the heavens. The Milky Way is a spiral type galaxy (Fig. 5–2). It rotates slowly in space, completing one rotation about every 240 million years. Our sun is located about two-thirds of the distance (or 26,000 light years) outward from the center of the galaxy to its edge.

## The Sun

For many reasons, the sun is the most important member of our solar system. It is a modest star (in comparison to others in our galaxy) and is composed of hydrogen (about 70 per cent) and helium (about 27 per cent). The remaining 3 per cent consists of heavier elements that exist as gases in the hot interior, where temperatures may exceed 20 million °C.

Although an enormous amount of solar energy is intercepted by the earth, our planet is not roasted, but is able to maintain a range of temperatures roughly between −50 and +60°C. The maintenance of this vital temperature range is made possible by three factors. First, because of rotation, the earth receives energy from the sun on one hemisphere, whereas it returns heat to space over its entire surface. Second, some of the incoming radiation is reflected off the atmosphere and clouds and is directed

**Figure 5–1**    The solar system. *(From Turk, J., and Turk, A. P. Physical Science. Philadelphia, W. B. Saunders and Co., 1977.)*

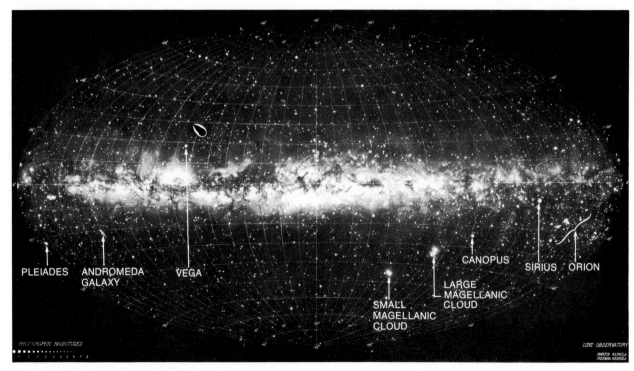

**Figure 5–2**    A drawing of the Milky Way, made under the supervision of Knut Lundmark at the Lund Observatory in Sweden. Several thousand stars plus the Milky Way are shown in this panorama, which is in coordinates such that the Milky Way falls along the equator. *(Courtesy of Lund Observatory, Sweden; from Pasachoff, J. M. Contemporary Astronomy, 3rd ed. Philadelphia, Saunders College Publishing, 1985.)*

**Figure 5–3**  Examples of two fusion reactions (n = neutron).

back into space without ever reaching ground level. Finally, a part of the intercepted radiation is absorbed by the atmosphere and radiated back into space without warming the earth's surface.

The energy that maintains the sun as a great glowing sphere of gases is derived from a continuous thermonuclear reaction called **fusion**. In the fusion process, hydrogen is changed to helium and excess mass is converted to energy (Fig. 5–3). Each second, the sun transmits an amount of energy equivalent to that which would result from the burning of 25 billion pounds of coal. This energy from the sun is the ultimate force behind the many geologic processes that sculpture the earth's surface. For example, the sun's rays aid in the evaporation of surface waters, which in turn results in clouds that provide the precipitation required for erosion. Along with the earth's rotation, the sun's radiation results in winds and ocean currents. Some scientists believe that protracted variations in the heat received from the sun may trigger episodes of continental glaciation or may reduce lush forests to barren wastelands. In company with the moon, the sun helps move the tides. Even primitive people knew the sun's importance and recognized it as the fountainhead and sustainer of life.

## The Birth of the Solar System

Any account of the origin of the solar system is obliged to conform to certain basic constraints within the system. These include dynamic constraints relating to the movements of planets, compositional constraints, and age constraints.

### Dynamic Constraints

The planets of our solar system all revolve around the sun in the same direction and lie approximately in the plane of the sun's rotation. Thus, the solar system is like a disk in shape. The direction taken by the planets as they revolve around the sun is counterclockwise (called **prograde**) when viewed from a hypothetical point in space above the sun's north pole. The direction of rotation of the planets on their axes is also counterclockwise, with the exception of Venus and Uranus. With few exceptions, satellites mimic the movements of the planets they orbit.

### Compositional Constraints

As indicated by their differing densities (see Table 5–1), as well as by earth, moon, and meteorite samples, the planets differ in composition. Mercury,

**Table 5–1  Size and Density of the Planets**

| Planet | Diameter (km) | Mean Density (g/cm³) |
|---|---|---|
| Mercury | 4,878 | 5.43 |
| Venus | 12,104 | 5.24 |
| Earth | 12,756 | 5.52 |
| Mars | 6,794 | 3.9 |
| Jupiter | 142,796 | 1.34 |
| Saturn | 120,000 | 0.69 |
| Uranus | 52,400 | 1.27 |
| Neptune | 50,450 | 1.58 |
| Pluto | 2,400 | 1.7 |

Venus, Earth, and Mars have mean densities of 5.4, 5.2, and 3.9, respectively. These four planets constitute the inner or **terrestrial planets.** They are small, dense, rocky, and rich in metals. Jupiter, Saturn, Uranus, and Neptune have lesser densities of 1.3, 0.7, 1.3, and 1.6, respectively. They are termed the outer or **Jovian planets.** Like the sun, the compositions of Jupiter and Saturn are dominated by hydrogen and helium.

## Age Constraints

As indicated in Chapter 4, there are rocks on the surface of the earth that have been present for at least 4.2 billion years, and samples from the moon that are 4.6 billion years old. Meteorites that have not been melted after they were originally formed (melting would reset their radioactive clocks) yield ages of about 4.6 million years. This age agrees with theoretical calculations of the age of the sun.

## The Solar Nebula Hypothesis

When all the dynamic, compositional, and age constraints are taken into account, the favored hypothesis is one that concludes that the solar system was born about 4.6 billion years ago from a rotating cloud of hot gases called the **solar nebula.** Immanuel Kant first suggested the idea in 1755. About 40 years later, Pierre LaPlace fully developed Kant's idea, and in recent years it has been improved with the rich flow of information emanating from the space program.

The solar nebula hypothesis suggests that the solar system began as a cold, rarefied cloud of dust particles and gases. The cloud gradually contracted and flattened as it took on a counterclockwise rotation. About 90 per cent of its mass remained concentrated in the thicker central part of the disklike cloud. While rotation and contraction were occurring, the phenomenon of turbulence (not considered in the earlier nebular hypothesis) began to affect the cloud system. Any departure from the smooth flow of a gas or fluid constitutes turbulence. In the dust cloud it consisted of chaotic swirling and churning movements that were superimposed on the grander primary motion of the entire cloud. Each turbulent eddy was affected by the movement of adjacent eddies, and some served as collecting sites for matter brought to their location by neighboring swirls.

When the disk had shrunk to a size somewhat larger than the present solar system, its denser condition permitted condensation of the concentrated knots of matter. Smaller particles merged to build chunks up to meters in diameter, and the dense swarms swept up finer particles within their orbital paths, thereby growing in size. This process of accumulation of bits of matter around an initial mass is called **accretion.** Because the particles available for accretion were cold and essentially similar throughout the period of accretion, the process is often further described as **cold homogeneous accretion.** The large globules formed from swarms of accreting bodies, dust, and gases are called **protoplanets.** They were enormously larger than present-day planets. Each rotated somewhat like a miniature dust cloud, and each eventually swept away most of the debris in its orbital path and was able to revolve around the central mass without collision with other protoplanets (Fig. 5–4). Their formation from the original dark, cold cloud required an estimated 10 million years.

While the protoplanets were in the process of condensation and accretion, material that had been pulled into the central region of the nebula condensed, shrank, and was heated to several million degrees by gravitational compression. The sun was born. Later, when pressures and temperatures within the core of the sun became sufficiently high, thermonuclear reactions began, pouring out additional energy derived from matter by the nuclear fusion of hydrogen atoms into atoms of helium. (Today, the sun converts about 596 million metric tons of hydrogen into 532 million metric tons of helium each second. The difference represents matter that has been converted to energy.)

As the sun began its thermal history, its radiation ionized surrounding gases in the nebula. These ionized gases in turn interacted with the sun's magnetic lines of force, causing the magnetic field to adhere to the surrounding nebula and drag it around during rotation. At the same time, the nebula caused a drag on the magnetic field and slowed down the sun's rotation. In this way, the old problem of why the sun rotates slowly was explained.

The stream of radiation from the sun sometimes called **solar wind** drove enormous quantities of the lighter elements and frozen gases outward into space. This solar force is what causes a comet's tail to show wavy streaming (Fig. 5–5) or to be bent away from the sun. As would be expected, the planets

**Figure 5–4** An illustration of the solar nebula hypothesis for the origin of the solar system. (A) The solar nebula takes form. (B) It begins to rotate, flatten, and contract. (C) The nebula is now distinctly discoidal and has a central concentration of matter that will become the sun. (D) Thermonuclear reactions begin in the primordial sun as material in the thinner area of the disk condenses and accretes to form protoplanets. (E) The late stage in the evolution of the solar system in which planets are fully formed, and materials not incorporated in planets are largely swept away by solar wind and radiation. *(From Abell, G. O., Morrison, D., and Wolff, S. C., 1988.* Realm of the Universe. *Philadelphia, Saunders College Publishing.)*

away by the solar wind. The Jovian planets were the least affected and retained their considerable volumes of hydrogen and helium.

## Differentiation of the Earth

### Cold Accretion Model

If the cold, homogeneous model for accretion is correct, then the earth would have consisted initially of an unsorted conglomeration of metals and silicates from its center to its surface. The earth, however, is not now homogeneous throughout. Neither are the other terrestrial planets of the solar system. They, along with the earth, have experienced **differentiation,** a process whereby a planet becomes internally zoned or layered.

To accomplish differentiation of an initially homogeneous planet, the planet must have been heated and become at least partially melted. Under such conditions, much of the iron and nickel in the planet would percolate downward to form the dense core. The remaining iron and other metals would combine with silicon and oxygen and separate out to form the less dense mantle that surrounds the core. Still lighter components would separate out of the mantle to form an uppermost crustal layer. Geologists do not insist on complete melting for this differentiation to have occurred. Partial melting on a vast scale would be a likely occurrence, as well as solid diffusion of ions mobilized by high temperatures and pressures. Indeed, it would be difficult to account for the present abundance of volatile elements on earth if the planet had melted completely, for under such

nearest the sun lost enormous quantities of lighter matter and thus have smaller masses but greater densities than the outer planets. They are composed primarily of rock and metal that could not be moved

**Figure 5–5** Halley's Comet on March 12, 1986, showing the dust and gas tails. The color image was composed from blue, green, and red plates, so the motion of the comet makes each star appear three different colors. (© 1986 Royal Observatory Edinburgh, photography by David Malin from U.K. Schmidt plates.)

circumstances these materials would have been driven off. Even for partial melting and solid diffusion to have occurred, however, an enormous amount of heat would have been required.

The heat needed for differentiation may have been derived from several different processes. Some of it may have been accretionary heat derived from bombardment by particles and meteoric materials when the planet was still actively growing. Much of the heat generated by the great infall of impacting bodies would have been preserved because subsequent showers of material provided blankets of insulating debris. Also, as more and more material accumulated at the surface of the planet, the weight of overburden on underlying zones would have provided heat by gravitational compression. Compression also served to reduce the earth's size.

Another extremely important source of heat for differentiation would result from the decay of radioactive isotopes incorporated within the materials of the planet during accretion. Much of this radioactively generated heat would have accumulated deep within the earth, where the poor thermal conductivity of enclosing rocks provided insulation. There was more heat derived from radioactivity in the early

days of the earth's development than today because many of the isotopes having relatively short half-lives had not yet disintegrated.

Geophysicists estimate that, about 1 billion years after the solid earth formed, there would have been sufficient heat from all these sources to melt iron at depths of 400 to 800 km. One can envision myriads of heavy, molten droplets of iron and nickel gravitating toward the planetary center, displacing lighter materials that had been there (Fig. 5–6). As the mantle began to form, it, too, experienced differentiation that eventually led to the development of the crust and the exhalation of gases that formed an atmosphere and hydrosphere.

## Hot Accretion Model

The idea of an initially cold birth for our planet has recently been reexamined as a result of new information emanating from the space program. A different model for the early accumulation of the earth's materials has emerged. In contrast to cold homogeneous accretion, the newer concept stipulates that the earth and neighboring inner planets accumulated material that was very hot. Furthermore, they were not ini-

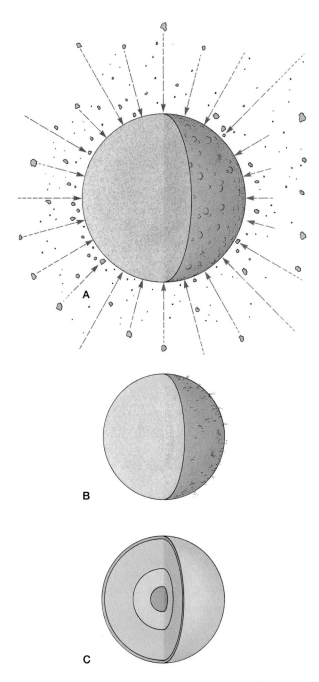

**Figure 5–6** Conceptual diagrams of stages in the earth's early history. (A) Representation of the growth of the planet by the aggregation of particles and meteorites that bombarded its surface. At this time, the earth was composed of a homogeneous mixture of materials. (B) The earth has lost volume because of gravitational compression. Temperatures in the interior have reached a level where differentiation has begun. Iron (red drops) sink toward the interior to form the core, whereas lighter silicates move upward. (C) The result of the differentiation of the planet is evident by the formation of core, mantle, and crust.

tially homogeneous solid bodies that later differentiated as a result of subsequent heating. On the contrary, it is thought that the internal zonation of planets developed *during*, not after, accretionary accumulation by a process termed **hot heterogeneous accretion** (Fig. 5–7).

The hot heterogeneous model for accretion begins with the solar nebula at a time when its gases were still very hot, possibly reaching temperatures of over 1000°C. It is possible that as it began to cool, heavy, iron-rich compounds condensed, precipitated, and accumulated in masses that would eventually become the iron-rich cores of the terrestrial planets. Next, with continued cooling, lower density silicate minerals began to condense and were swept up by the denser masses formed earlier. The silicates would have accumulated around the core as a thick covering that eventually became the mantle. During the accretionary process, the earth was hot and probably at least partially molten. In this state, chemical differentiation was able to proceed throughout the period of accumulation. Thus, differentiation was not a distinct, subsequent event. It is likely, however, that later heating by meteoric bombardment and radioactive decay could have further propelled the differentiation already established.

Currently, it is not possible to judge which model of differentiation is more valid. It is possible that components of both theories were in operation. Most geologists appear to favor cold homogeneous accretion followed by melting as an explanation for the earth's internal zonation.

## Messages from Inner Space

As described in the preceding section, the earth began to take form when particles of various sizes orbiting the sun aggregated to form a body sufficiently large to be termed a planet. The planet earth continued to grow over a span of 50 to 90 million years as a result of the ongoing shower of particles and meteorites that impacted upon its surface. Eventually, however, most of the material in or near the earth's orbital path had accreted onto the planetary surface, and the earth's active growth stage ended. By this time the earth's interior was either partially zoned as the result of hot heterogeneous accretion, or was in the process of differentiation by melting and gravity separation of iron-rich compounds from

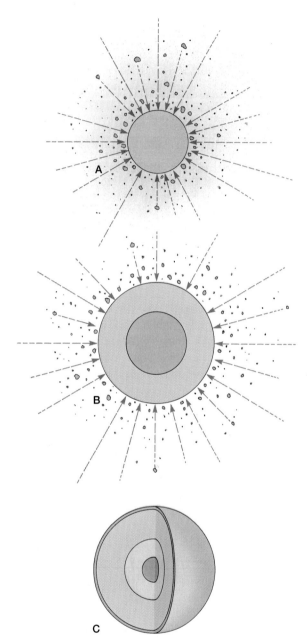

**Figure 5–7** Origin of the earth's core according to the hot heterogeneous model of accretion. (A) Primarily iron and nickel condense, collect, and form a core. (B) Silicates envelop the earlier formed core and form a mantle. (C) The mantle differentiates and provides the materials for the crust.

less dense silicates. The result was the density layering that now characterizes the earth's deep interior.

Much of what we know about the location, thickness, and properties of these layers has been indirectly determined from the study of earthquake waves (seismic waves). Such waves are generated when rocks are suddenly disturbed, as when they break or rupture. Vibrations spread out in all directions from the source of the disturbance. They move outward in waves that travel at different speeds through materials that differ in chemical composition or physical properties. These waves can be designated primary, secondary, or surface waves.

**Primary waves,** or P-waves, are the speediest of the three kinds of waves and therefore the first to arrive at a seismograph station after there has been an earthquake (Fig. 5–8). They travel through the upper crust of the earth at speeds of 4 to 5 km per second. Near the base of the crust they speed along at 6 or 7 km per second. In these primary waves, pulses of energy are transmitted in such a way that the movement of rock particles is parallel to the direction of propagation of the rock itself. Thus, a given particle of rock set in motion during an earthquake is driven into its neighbor and bounces back. The neighbor strikes the next particle and rebounds, and subsequent particles continue the motion. The result is a series of alternate compressions and expansions that speed away from the source of shock. Thus, P-waves are similar to sound waves in that they are *longitudinal* and travel by compression and rarefaction (Fig. 5–9). It is an accordian-like "push–pull" movement that can be transmitted through solids, liquids, and gases. Of course, the speed of P-wave transmission will differ in materials of different density and elastic properties. They tend to die out with increasing distance from the earthquake source and will echo or reflect off rock masses of differing physical properties.

**Secondary waves,** which are also termed S-waves or transverse waves, travel 1 to 2 km per second slower than do P-waves. The movement of rock particles in secondary waves is at right angles to the direction of propagation of the energy (Fig. 5–10). A demonstration of this type of wave is easily managed by tying a length of rope to a hook and then shaking the free end. A series of undulations will develop in the rope and move toward the hook—that is, in the direction of propagation. Any given particle or point along the rope, however, will move up and down in a direction perpendicular to the direction of propagation. It is because of their more complex motion that S-waves travel more slowly than P-waves. They are the second group of oscillations to appear on the seismogram (see Fig. 5–8).

**Figure 5-8**  Record of a magnitude 6 earthquake that occurred in Turkey on March 28, 1964, and was recorded by a vertical component seismograph located in northwestern Canada. Time increases from left to right. Small tick marks represent 1-minute intervals. P means primary waves, S indicates secondary waves. Notice that the first S-waves arrive about 10 minutes after the arrival of the first P-waves. *(Courtesy of the Earth Physics Branch, Department of Energy, Mines, and Resources, Canada.)*

Unlike P-waves, secondary waves will not pass through liquids or gases.

Both P- and S-waves are sometimes also termed **body waves** because they are able to penetrate deep into the interior or body of our planet. Body waves travel faster in rocks of greater elasticity, and their speeds therefore increase steadily as they move downward into more elastic zones of the earth's interior and then decrease as they begin to make their ascent toward the earth's surface. The change in velocity that occurs as body waves invade rocks of different elasticity results in a bending or refraction of the wave. The many small refractions cause the body

**Figure 5-9**  Movement of primary waves. In 1, the compression has moved the particles closer together at A. In 2, the zone of compression has moved to $A_1$. In 3, the zone of compression has moved to $A_2$ and a second compressional zone (B) has moved in from the left.

**Figure 5-10**  Analogy of propagation of S-waves by displacement of a rope. In rocks, as in this rope, particle movement is at right angles to the direction of propagation of the wave. A, B, and C show the displacement of the crest from left to right at successive increments of time.

## Recording Earthquakes

It is easy to record the passage of an automobile or train as we stand in one place and observe their movement. We have objects on the ground to use as reference in relation to which the motion can be readily seen. However, during an earthquake, we move in unison with the earth and there is no stationary frame of reference. To solve this problem, the **seismograph** was developed. Seismographs are the preeminent tools for examining the interior of the earth. They make use of the property of *inertia,* or the tendency of a heavy object to "stay put" as the earth moves around it. To accomplish this feat, the heavy body is suspended from springs and flexible wires which serve to absorb earth movements. The heavy body is able to lag behind, maintaining its position as the earth moves around it.

In order to keep the heavy body or weight as independent of earth motions as possible, there are several different designs of seismographs. For recording vertical movements, the seismograph contains a weight suspended by springs from an overhead support (see accompanying figure). Horizontal movements are recorded by seismographs containing a horizontal pendulum having a weight on a rigid arm supported by a wire. The pendulum is freely attached to a socket joint from the supporting column. In both devices, the weight remains motionless while the support vibrates with the earth. The difference in motion can be recorded in various ways. One method employs a narrow beam of light emanating from the weight. The beam of light traces a continuous line on photographic paper wrapped around a revolving drum that is fixed to the supporting framework of the seismograph. The drum turns as a clock so that the arrival times and durations of vibrations can be determined.

The accompanying figure illustrates the general principle of the seismograph. Actual seismographs located in earthquake monitoring stations are far more complex. The ray of light used to record vibrations is reflected off a series of mirrors that serve to amplify even very small motions of the ground. Other devices in the seismograph dampen the natural swinging motion of the pendulum and filter out unrelated background motions. To analyze all movements of earthquakes fully, at least three seismographs must be in simultaneous operation.

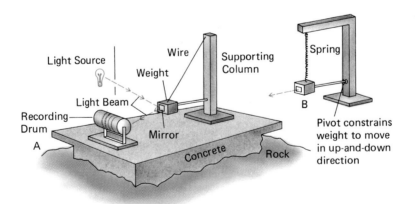

Principle of seismographs recording horizontal (A) and vertical (B) earth motion. A spot of light on the boom moves across photographic paper on the recording drum as the boom oscillates. (Springs are omitted for clarity.)

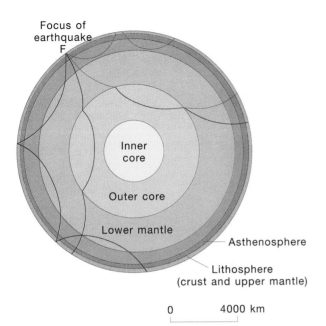

**Figure 5–11** Cross-section of the earth showing paths of some earthquake waves. P-waves (black) that penetrate to the core are sharply refracted, as shown in the path from F to 1. S-waves (red) end at the core, although they may be converted to P-waves, traverse the core, and emerge in the mantle again as P- and S-waves. Both P- and S-waves may also be reflected back into the earth again at the surface. *(From U.S. Geological Survey publication,* The Interior of the Earth.*)*

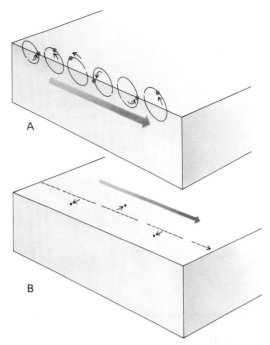

**Figure 5–12** (A) Elliptical particle motion in Rayleigh surface waves (ground surface shown without displacement). (B) Particle motion in Love wave is in horizontal plane and perpendicular to the direction of wave propagation.

waves to assume a curved travel path through the earth (Fig. 5–11).

Not only are body waves subjected to refraction, but they may also be partially reflected off the surface of a dense rock layer in much the same way as light is reflected off a polished surface. Many factors influence the behavior of body waves. An increase in the temperature of rocks through which body waves are traveling will cause a decrease in velocity, whereas an increase in confining pressure will cause a corresponding increase in wave velocity. In a fluid where no rigidity exists, S-waves cannot propagate and P-waves are markedly slowed.

**Surface waves** are large-motion waves that travel through the outer crust of the earth. Their pattern of movement resembles that of waves caused when a pebble is tossed into the center of a pond. They develop whenever P- or S-waves disturb the surface of the earth as they emerge from the interior.

There are actually several different types of motion in surface waves. *Rayleigh surface waves*, for example, have an elliptical motion that is opposite in direction to that of propagation, whereas *Love surface waves* vibrate horizontally and perpendicular to wave propagation (Fig. 5–12). Surface waves are the last to arrive at a seismograph station. They are usually the primary cause of the destruction that can result from earthquakes affecting densely populated areas (Fig. 5–13). This destruction results because surface waves are channeled through the thin outer region of the earth, and their energy is less rapidly dissipated into the large volumes of rock traversed by body waves. Indeed, surface waves may circle the earth several times before friction causes them to fade. As is the case with ordinary water waves, the motion of surface waves diminishes with depth. It has been demonstrated that the total depth to which surface wave motion can exist is about equal to the distance between two surface wave crests. Thus, surface waves that are 2000 meters from crest to crest will shake a section of crust about 2000 meters thick.

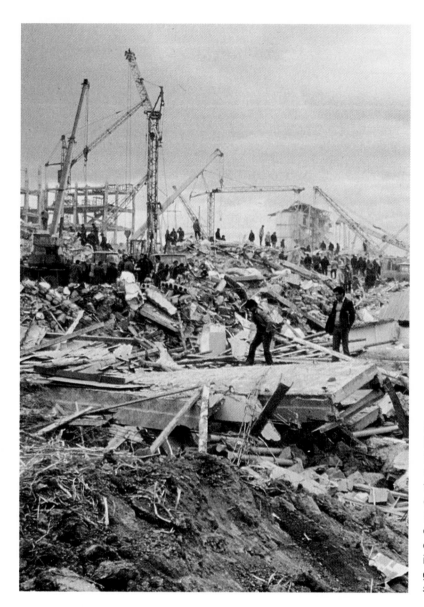

**Figure 5–13** Ruins of apartment buildings in the city of Leninakan following the December 1988 earthquake in Soviet Armenia near the Soviet–Turkish border. The earthquake, registering 6.9 on the Richter scale, caused widespread devastation and tens of thousands of fatalities in the villages and cities of this Caucasus mountain region. *(Wide World Photo, with permission.)*

## The Divisions of Inner Space

Most of what we know about the earth's deep interior is derived from the interpretation of countless seismograms documenting recent and past earthquakes around the globe. From such studies, geophysicists have found evidence of the gradual change in rock properties with depth as well as the relatively abrupt boundaries between major internal zones. Boundaries where seismic waves experience an abrupt change in velocity or direction are called **discontinuities**. Two widely known breaks of this kind are named, after their discoverers, the Mohorovičić and Gutenberg discontinuities.

The discontinuity discovered by **Mohorovičić** (pronounced Mō-hō-rō-vĭtch'-ĭtz) was based on his observation that seismograph stations located about 150 km from an earthquake received earthquake waves sooner than those nearer to the focus. Mohorovičić reasoned that below a depth of about

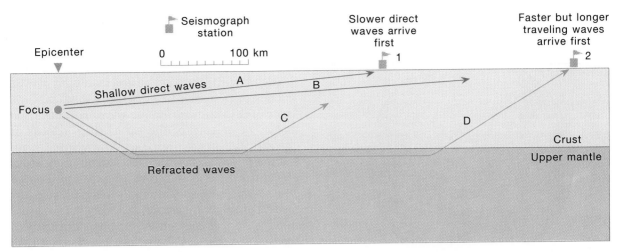

**Figure 5–14** Mohorovičić's conclusion about the location of the base of the earth's crust was based on this interpretation of the travel paths of early- and late-arriving seismic body waves. (*Focus* is the true center of an earthquake and the point at which the disturbance originates. *Epicenter* is a point on the earth's surface above the focus.)

30 km there must be a zone having physical properties that permit earthquake waves to travel faster. That layer is the upper mantle.

In Figure 5–14, the shallow direct waves (A), although traveling at only about 6 km per second, are the first to arrive at the closer seismograph station. At the farther station, however, the deeper but faster traveling wave (D) has caught up and moved ahead of the shallow direct wave and is therefore the first to arrive at seismograph station 2. The situation is analogous to what many of us do when we take a longer freeway route to our destination than a more direct route through city streets because the greater speed on the highway allows us to arrive at the destination sooner.

The Mohorovičić discontinuity (or Moho, for short) lies at about 30 or 40 km below the surface of the continents and at lesser depths beneath the ocean floors (Fig. 5–15). It marks the boundary between the crust and the mantle.

The **Gutenberg discontinuity** is located nearly halfway to the center of the earth at a depth of 2900 km. Its location is marked by an abrupt decrease in P-wave velocities and the disappearance of

**Figure 5–15** Generalized cross-section showing location of the Mohorovičić discontinuity.

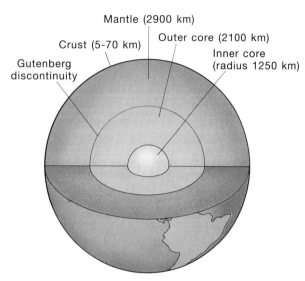

Mantle (2900 km)

Crust (5-70 km)  Outer core (2100 km)

Gutenberg discontinuity

Inner core (radius 1250 km)

**Figure 5–16** The interior of the earth. The crust appears as a line at this scale.

S-waves. The Gutenberg discontinuity marks the outer boundary of the earth's core (Fig. 5–16).

## The Core

### Inferences from Body Waves

As just noted, the precise boundary of the core was determined by the study of earthquake waves. Seismology has also provided a means for discerning subdivisions of the core and deciphering some of its physical properties. For example, at a depth of 2900 km (the core boundary), the S-waves meet an impenetrable barrier, while at the same time P-wave velocity is drastically reduced from about 13.6 km/second to 8.1 km/second. Earlier, we noted that S-waves are unable to travel through fluids. (Fluids cannot sustain shear.) Thus, if they were to enter a fluidlike region of the earth's interior, they would be absorbed there and would not be able to continue. Geophysicists believe this is what happens to S-waves as they enter the outer part of the core. As a result, the secondary waves generated on one side of the earth fail to appear at seismograph stations on the opposite side, and this observation is the principal evidence for an outer core that behaves as a fluid. The outer core barrier to S-waves results in an *S-wave shadow zone* on the side of the earth opposite an

earthquake. Within the **shadow zone**, which begins 105° from the earthquake focus, S-waves do not appear.

Unlike S-waves, primary waves are able to pass through liquids. They are, however, abruptly slowed and sharply refracted as they enter a fluid medium. Therefore, as primary seismic waves encounter the molten outer core of the earth, their velocity is checked and they are refracted downward. The result is a *P-wave shadow zone* that extends from about 105° to 143° from the focus (Fig. 5–17). Beyond 143°, P-waves are so tardy in their arrival as to further validate the inference that they have passed through a liquid medium. At the upper boundary of the core, P-waves are also reflected back toward the earth's surface. Such P-wave echos are clearly observed on seismograms.

The radius of the core is about 3500 km. The inner core is solid and has a radius of about 1220 km, which makes this inner core slightly larger than the moon. A transition zone approximately 500 km thick surrounds the inner core. Most geologists believe that the inner core has the same composition as the outer core and that it can exist only as a solid because of the enormous pressure at the center of the earth.

Evidence for the existence of a solid inner core is derived from the study of hundreds of seismograms produced over several years. These studies showed that weak late-arriving primary waves were somehow penetrating to stations that were within the P-wave shadow zone. Geophysicists recognized that this penetration could be explained by assuming the inner core behaved seismically as if it were solid.

### Composition of the Core

The earth has an overall density of 5.5 g/cm³, yet the average density of rocks at the surface is less than 3.0 g/cm³. This difference indicates that materials of high density must exist in the deep interior of the planet to achieve the 5.5 g/cm³ overall density. Calculations indicate that the rocks of the mantle have a density of about 4.5 g/cm³ and that the average density of the core is about 10.7 g/cm³. Under the extreme pressure conditions that exist in the region of the core, iron mixed with nickel would very likely have the required high density. In fact, laboratory experiments suggest that a highly pressurized iron–

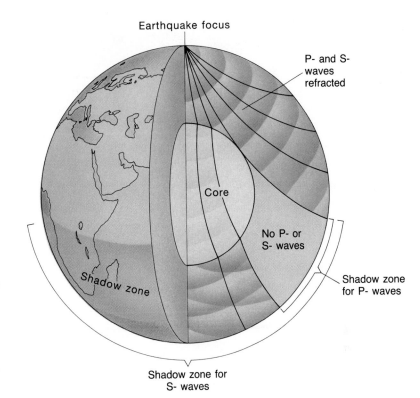

**Figure 5–17** Refraction of seismic body waves. The S-wave shadow zone is believed to be the result of absorption of S-waves in the liquid outer core. When the P-wave enters the liquid outer core, it slows and is bent downward, giving rise to the P-wave shadow zone within which neither P- nor S-waves are received.

nickel alloy might be too dense and that minor amounts of such elements such as silicon, sulfur, or oxygen may also be present to "lighten" the core material.

Support for the theory that the core is composed of iron (85 per cent) with lesser amounts of nickel has come from the study of meteorites. A large number of these samples of solar system materials are iron meteorites (Fig. 5–18) that consist of metallic iron alloyed with a small percentage of nickel. Some geologists believe that iron meteorites may be fragments from the core of a shattered planet. Their abundance in our solar system suggests that the existence of an iron–nickel core for the earth is plausible.

There is another kind of evidence for the earth's having a metallic core. Anyone who understands the functioning of an ordinary compass is aware that the earth has a magnetic field. The planet itself behaves as if there were a great bar magnet embedded within it at a small angle to its rotational axis. Geophysicists believe that the earth's magnetic field may, in some way, be associated with electric currents. This inter-

pretation is favored by the discovery, over 160 years ago, that a magnetic field is produced by an electric current flowing through a wire. The silicate rocks of the lithosphere and mantle are not good conductors, however, and therefore unlikely materials for the development of electromagnetism. In contrast, iron is an excellent conductor. If scientists are correct in the inference made from density considerations and study of meteorites that the core is mostly iron, then the magnetic field lends credence to that inference.

## *The Mantle*

### Materials of the Mantle

As was the case with the earth's core, our understanding of the composition and structure of the **mantle** is based on indirect evidence. As inferred from seismic data, the mantle's average density is about 4.5 g/cm$^3$, and it is believed to have a stony, rather than metallic, composition. Oxygen and silicon probably predominate and are accompanied by iron and magnesium as the most abundant metallic

A

**Figure 5–18** (A) Collecting an iron meteorite on the Antarctic ice. (B) Cut and polished iron meteorite found in Mohave County, Arizona, showing the interesting pattern of interlocking crystals, called the Widmanstatten pattern. *(A, courtesy of NASA; B, from the Washington University collection.)*

B

ions. The iron- and magnesium-rich rock **peridotite** (Fig. 5–19) approximates fairly well the kind of material inferred for the mantle. A peridotitic rock not only would be appropriate for the mantle's density, but also is similar in composition to stony meteorites as well as rocks that are thought to have reached the earth's surface from the upper part of the mantle itself. Such suspected mantle rocks are indeed rare.

They are rich in olivine and pyroxenes and contain small amounts of certain minerals, including diamonds, that can form only under pressures greater than those characteristic of the crust.

## Layers of the Mantle

The mantle is not merely a thick, homogeneous layer surrounding the core but is itself composed of sev-

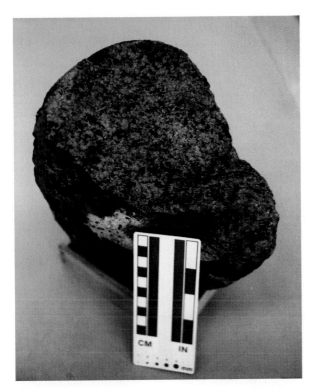

**Figure 5–19**  Peridotite. The green minerals are olivine and spinel, and the black grains are pyroxene. The specimen was collected at Kilbourne Hole maar, Dona Ana County, New Mexico. A *maar* is a crater formed during a violent volcanic explosion. The explosion brought chunks of mantle rock to the earth's surface.

**Figure 5–20**  Generalized graph of average P-wave velocities versus depth in the outer 70 km of the earth, showing the location of the low-velocity zone.

eral concentric layers that can be detected by studying earthquake data. In Figure 5–20, note that within the mantle there are three zones of increase in wave velocity. These sudden increases cannot be explained simply as the result of pressure increases with depth. Such increases would be too gradual to cause seismic wave velocity to increase so abruptly. There must, therefore, be some change in the physical nature of the material. One of the zones is encountered at about 400 km and is taken to mark the base of the **upper mantle**. Beneath the upper mantle is the **transition zone,** which extends downward to about 650 km. The **lower mantle** lies beneath the transition zone and above the core.

The composition of the lower mantle is difficult to infer. A plausible guess that is supported by seismic evidence is that it consists mostly of silicates and oxides of magnesium and iron. In the high-temperature and high-pressure environment of the lower mantle, iron, magnesium, silicon, and oxygen atoms are rearranged into denser and more compact crystals. For example, near the earth's surface, olivine and pyroxene are relatively stable minerals. However, in the high-temperature and high-pressure environment below a depth of about 400 km, olivine is likely to be converted to a form whose atoms are more closely packed (Fig. 5–21). Similar dense crystals have been formed from olivine in high-pressure laboratory experiments. Rocks composed of such dense materials would be capable of causing the increase in seismic velocities that characterize the mantle.

The upper mantle is of particular importance because its evolution and internal movements affect the geology of the crust. The most remarkable fea-

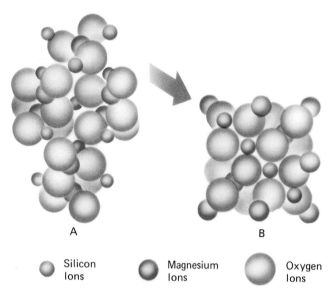

A             B

Silicon Ions    Magnesium Ions    Oxygen Ions

**Figure 5–21** High- and low-pressure forms of the mineral olivine. (A) At shallow depths in the mantle, olivine is stable in this form, which is the form of olivine found at the earth's surface. (B) When the pressure reaches a critical value corresponding to a depth of about 400 km, the molecule collapses into a more dense form in which oxygen ions are more closely packed.

ture of the upper mantle is the **low-velocity zone.** As suggested by its name, this is a region in which there is a decrease in the speed of S- and P-waves (see Fig. 5–20). The low-velocity zone occupies an upper region of the larger **asthenosphere.** The rock here is near or at its melting point, and magmas are generated.

Geophysicists believe the seismic waves are slowed in the low-velocity zone, not because of a decrease in density, but rather because they enter a region that is in the state of a crystalline–liquid mixture. In such a "hot slush," perhaps 1 to 10 per cent of the material would consist of pockets and droplets of molten silicates. Such material would be capable of considerable motion and flow and would serve as a slippery mobile layer on which overlying lithosphere plates could move.

## *The Crust*

The **crust** of the earth is seismically defined as all of the solid earth above the Mohorovičić discontinuity. It is the thin, rocky veneer that constitutes the continents and the floors of the oceans. The crust is not a homogeneous shell in which low places were filled with water to make oceans and higher places make continents. Rather, there are two distinct kinds of crust, which, because of their distinctive compositions and physical properties, determine the very existence of separate continents and ocean basins.

### The Oceanic Crust

Beneath the varied topography of the ocean floors lies an **oceanic crust** that is approximately 5 to 12 km thick and has an average density of about 3.0 g/cm$^3$. Three layers of this oceanic crust can be recognized. On the upper surface is a thin layer of unconsolidated sediment that rests on the irregular surface of the igneous basement layer. The second layer consists of basalts that had been extruded underwater (Fig. 5–22). The nature of the deepest layer of oceanic crust is not clear. Many suspect that it is metamorphosed basaltic mantle material that has become somewhat less dense by chemically combining with sea water. In rocks of the oceanic crust, we find a concentration of such common elements as iron, magnesium, and calcium. They are included in the plagioclase feldspars, amphiboles, and pyroxenes of basalts. The upper mantle is the ultimate source for the lavas that formed the oceanic crust.

### The Continental Crust

**Properties of the Continental Crust** At the boundaries of the ocean basins, the Mohorovičić discontinuity plunges sharply beneath the thicker **continental crust** (see Fig. 5–15). Depth to the Moho beneath the continents averages about 35 km, although it may be considerably deeper or shallower in particular regions. The continental crust is not only thicker than its oceanic counterpart, but also less

**Figure 5–22**   Pillow structures formed by lavas being extruded on the floor of the Pacific Ocean along the Juan de Fuca ridge, a plate tectonic spreading center west of Oregon. Photo covers 5 meters. *(Courtesy of W. R. Normark and the U.S. Geological Survey.)*

dense, averaging about 2.7 g/cm$^3$. As a result, continents "float" higher on the denser mantle than the adjacent oceanic crustal segments. Somewhat like great stony icebergs, the roots of continents extend downward into the mantle.

The concept of light crustal rocks "floating" on denser mantle rocks is called *isostasy*. Were it not for isostasy, mountain ranges would gradually subside, for there are no rocks having sufficient strength to bear the heavy load of mountain ranges. Thus, mountains are not supported by the strength of the crust, but rather are in a state of flotational equilibrium with denser underlying rocks.

Although the continental crust is referred to as being "granitic," it is really composed of a variety of rocks that approximate granite in composition. Igneous continental rocks are richer in silicon and potassium and poorer in iron, magnesium, and calcium than igneous oceanic rocks. Also, extensive regions of the continents are blanketed by sedimentary rocks.

**Origin of the Continental Crust**   One of the major questions relating to the history of the earth is how the lighter granitic masses that constitute the continents developed. Many geologists believe the development of the crust involved a series of events that began about 4 billion years ago with upwellings of lava derived from the partially molten upper mantle. This initial crust of lava was then subjected to repeated episodes of remelting, during which time lighter components were separated out and distributed near the earth's surface. Wherever uplands existed, such as along volcanic island arcs, the solidified lavas were subjected to erosion and the ordinary processes of oxidation, carbonation, and hydration that accompany chemical weathering. The products of this weathering and erosion were the earth's earliest sediments, which were then altered by rising hot gases and silica-rich solutions from below. Recycling and melting of this well-cooked and less dense mixture of earth materials led ultimately to the rocks of granitic character that formed the nuclei of continents. The new continents might then have provided a source for additional sediment that would have collected along continental margins. Subsequently, these sediments also might have been metamorphosed and melted during orogenic events and then welded or "accreted" onto the initial continental nuclei.

## Evolution of the Atmosphere and Hydrosphere

### The Primitive Atmosphere

We have described how the earth formed by aggregation of small and large bodies that were themselves once part of a nebula of gas and dust. The particles and meteorites incorporated into the growing planet contained significant amounts of gases as well. These

**volatiles** (substances easily driven off by heating) either were scattered uniformly throughout the accreting planet by the process of homogeneous accretion described earlier, or they were contained mostly in the impacting bodies that arrived during the late stages of accretion. Calculations indicate that meteorites could have brought sufficient quantities of gases to the earth to account for our present inventory of volatiles. Volatiles are especially abundant in the meteorites called *carbonaceous chondrites* (Fig. 5–23). These meteorites take their name from their content of organic compounds and from spherical bodies called chondrites which are conspicuously present. In addition to the nitrogen, hydrogen, and carbon present in organic compounds, some of the ferromagnesium minerals in carbonaceous chondrites contain water. Water and the gaseous elements of the organic compounds would have been readily released by heat associated with accretion or by subsequent melting and volcanism accompanying differentiation.

The process by which water vapor and other gases are released from the rocks that held them and then vented to the surface is termed **outgassing**. Calculations indicate that most of the outgassing of

**Figure 5–23** Small carbonaceous chondrite. This meteorite fell near Murray, Kentucky, in 1950. Scale divisions are in millimeters. *(Washington University collection.)*

water occurred within the first 1 billion years of earth history. As evidence of this, the presence of an early ocean is clearly indicated by marine sedimentary rocks dating from as long ago as 3.8 billion years.

The precise chemical nature of the gases released during the early phases of outgassing would have been dependent upon the temperatures of the rocks from which the volatiles were derived, as well as the state of oxidation of the rocks through which the gases passed. Geochemists believe the outgassed volatiles produced a primitive atmosphere rich in carbon dioxide and water vapor, with lesser amounts of carbon monoxide, hydrogen, and hydrogen chloride. Actually, the volatiles released by modern volcanoes provide an approximation of the composition of the early atmosphere, with the exception that the primitive exhalations probably contained a higher percentage of hydrogen and possibly traces of ammonia and methane as well. The gases from today's volcanoes differ because they are released from relatively young rocks that have been recirculated near the earth's surface by plate tectonic processes. Uncombined oxygen was not discharged into the early atmosphere, and if small amounts were released from the interior, these amounts would not have escaped immediate recombination with easily oxidized metals such as iron.

One might argue that the gases of our atmosphere were not degassed from the interior, but were merely late accumulations of light volatiles left over from the original nebula. The earth's relative scarcity of inert gases such as argon, neon, and krypton provides strong evidence for opposing this view. These chemically stable gases are too heavy to escape the earth's gravity. If they were present in gases collected near the end of the accretionary period, they would be unable to escape and would still be here. The inert gases, however, are about a million times less abundant in our present atmosphere than in the sun and other stars. This indicates that our atmosphere is not a residue of gases left over from the solar nebula, but was derived from the planet's interior by degassing.

Once at the earth's surface, the volatiles that had been vented were subjected to a variety of changes (Fig. 5–24). Water vapor was removed by condensation, and seas began to take form. The outgassing of carbon dioxide and other gases resulted in rainfall and ocean water that were considerably more acidic than today. When in contact with rocks, the acidic

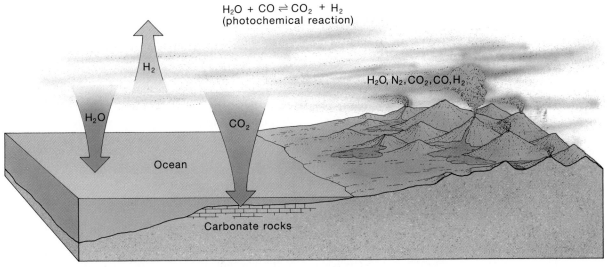

$$H_2O + CO \rightleftharpoons CO_2 + H_2$$
(photochemical reaction)

**Figure 5–24**   The relative amounts of gases in the primordial atmosphere were different from the abundances vented to the exterior of the earth during differentiation. Nitrogen tended to be retained in the atmosphere, whereas much of the water vapor was lost to the oceans by condensation. Some $CO_2$ was combined with calcium and magnesium derived from weathering and extracted to form carbonate rocks, and hydrogen, being light, was lost to space.

waters caused rapid chemical weathering and thereby brought calcium, magnesium, and other ions into solution. Much later, when the seas were less acidic and oxygen more prevalent in the atmosphere, these ions would be joined with carbon dioxide to form limestones, the calcareous structures of algae, and the shells of myriads of marine organisms.

## An Oxygen-Rich Atmosphere

We have seen that the early atmosphere of the earth was reducing and nonoxygenic in character. It was followed by an atmosphere rich in oxygen, an atmosphere that was and still is dependent upon life for its maintenance. We are reminded of the significance of this atmosphere with every breath we take, yet the earlier nonoxygenic atmosphere was also of vital importance, for in it the earliest forms of life evolved.

The change from an oxygen-poor to an oxygen-rich atmosphere occurred as a result of two processes. The first was dissociation of water molecules into hydrogen and oxygen. Termed **photochemical dissociation**, this process occurs in the upper atmosphere when water molecules are split by high-energy beams of ultraviolet light from the sun. Photochemical dissociation, however, does not produce

free oxygen at a rate sufficient to balance loss of the gas by dissipation into space. Another far more important oxygen-generating mechanism came with the advent of life on earth and, more specifically, with life that had evolved the remarkable capability of separating carbon dioxide into carbon and free oxygen. We now know this process as **photosynthesis**. It has made the earth a truly unique planet in our solar system.

## Geologic Clues to the Nature of Ancient Atmospheres

The most ancient rocks providing evidence of the atmospheric changes described previously are about 3.5 billion years old. Although they are metamorphic rocks, many of them still contain depositional features indicating that they were originally water-laid sediments. Mineral grains within these old rocks were formed or modified by weathering, and weathering cannot occur without an atmosphere. Further, it is highly probable that the atmosphere was not oxygenic, for there is an obvious absence of oxidized iron such as hematite. Rather, the sulfide of iron called pyrite is the usual iron mineral found in these rocks. A carbon dioxide rich atmosphere is a rather acidic one, in that carbon dioxide and water combine to form carbonic acid. In such an environment, alkaline rocks such as dolomite and limestone do not develop, and this may account for the absence of carbonate rocks from this early stage of earth history. Chert, a rock more tolerant of such conditions, is frequently found associated with these ancient rocks.

The advent of an oxygen-rich atmosphere is reflected by the state of oxidation of iron in the rocks overlying (and hence younger than) those described in the preceding paragraph. Beginning in rocks that are about 3 billion years old, one observes evidence of periodic abundance of oxygen. Many formations exhibit an alternation of rusty-red and gray bands. The former are colored by ferric iron oxide ($Fe_2O_3$) and indicate abundant free oxygen in the environment (Fig. 5–25). The grayish layers contain oxygen-deficient iron compounds and suggest a lesser availability of oxygen. The alternation of color bands might have been caused by a fluctuating and perhaps season-related supply of oxygen. In the iron-mining district north of Lake Superior, the ores above the banded iron deposits are uniformly oxidized. These rocks are about 1.8 billion years old and are evidence of the persistent presence of an oxygenic atmosphere by that time. In other parts of the world, one finds extensive outcrops of dolomite and limestone of about the same age. If carbon dioxide were still abundant in the atmosphere, these carbonate rocks would not have been deposited.

## The Ocean

If the water in the atmosphere was outgassed from the interior, then the bodies of water that accumulated on the surface of the earth were derived from the atmosphere after it had cooled sufficiently to condense and allow rain to fall. Gradually, the great depressions in the primordial crust began to fill, and oceans began to take form. One cannot help but wonder if the enormous volume of oceanic water could have all come from the interior. Calculations provide a very strong "yes" answer to this question. As noted earlier, vast amounts of water were locked within hydrous minerals that were ubiquitous within the primordial earth. These waters, "sweated" from the earth's interior and precipitated onto the uplands, immediately began to dissolve soluble minerals and carry the solutes to the sea. The initial high acidity of the oceans enhanced and speeded the process so that the oceans quickly acquired the saltiness that is their most obvious compositional characteristic. They have maintained a relatively consistent composition by precipitating surplus solutes at about the same rate as they are supplied. Of course, because of its high solubility, sodium remains in sea water longer than do other common elements. However, the fossil record of marine organisms suggests that even this element has not varied appreciably in sea water for at least the past 600 million years.

Having outgassed its present quantity of water early in its history, the earth has been recycling that water ever since. Water is continuously recirculated by evaporation and precipitation—processes powered by the sun and gravity. Some of the water in the oceans is temporarily lost by being incorporated into

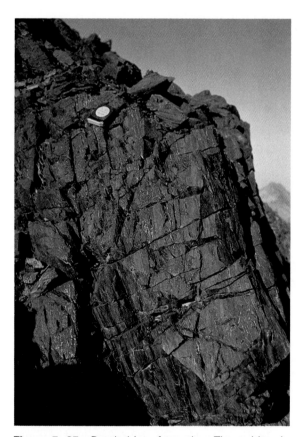

**Figure 5–25**  Banded iron formation. The red bands are hematite and are interbedded with chert. Wadi Kareim, Egypt. *(Courtesy of D. Bhattacharyya.)*

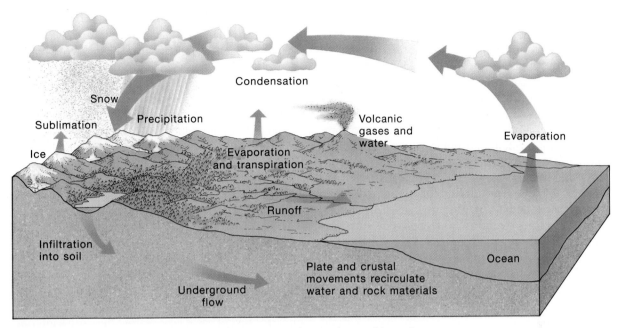

**Figure 5–26**  The constant circulation of water at and near the earth's surface.

hydrous clay minerals that settle to the ocean floors. However, even this water has not permanently vanished, for the sediments may be moved to orogenic belts and melted into magmas that return the water to the surface in the course of volcanic eruptions (Fig. 5–26).

## SUMMARY

The planet earth is one of a group of nine planets revolving around a modest-sized star, our sun. The planets and sun form our solar system, which, together with myriads of other stars, planets, dust, and gas, make up the Milky Way galaxy. Our sun provides energy to the earth, which drives geologic processes and which has ultimately determined the character of terrestrial life. The sun is composed mostly of hydrogen and helium. The helium in the core results from the fusion of hydrogen nuclei with the conversion of a part of the hydrogen into solar energy.

Theories for the origin of the solar system must conform to the known facts of the system's distribution of mass as well as the orbital and rotational characteristics of the planets. Currently, the most widely held theory is that the planets originated by conden-sation and accretion from a rotating nebula of gases and dust particles.

Knowledge of the earth's deep interior is derived from the study of earthquake waves. Among the various kinds of earthquake or seismic waves are primary, secondary, and surface waves. Primary and secondary waves (also called body waves) pass deep within the earth and therefore are the most instructive. Study of abrupt changes in the velocities of seismic waves at different depths provides the basis for a threefold division of the earth into a central core; a thick, overlying mantle; and a thin, enveloping crust. Sudden changes in earthquake wave velocities are termed discontinuities.

The core of the earth, as indicated by the Gutenberg discontinuity, begins at a depth of 2900 km. The core is divided into an inner and outer core. It is likely that the outer core is molten and that the entire core is composed of iron with small amounts of nickel and possibly either silicon or sulfur. In the hot heterogeneous model for the accretion of planetary materials, the core is thought to have formed early on when the nebula was hot by initial condensation of mostly iron and nickel, followed by condensation of less dense silicates around the earlier formed core. In the more favored cold homogeneous model, the earth was formed by accretion of essentially cold particles. In this model, the core originated after the planet had accumulated most of its matter as a result of an episode of heating during which heavier constituents of the protoplanet were drawn by gravity to the center and lighter components rose to the surface.

Constituting about 80 per cent of the earth's volume, the mantle is composed of iron and magnesium silicates such as olivine and pyroxene. A hot, plastic zone in the mantle is recognized by lower seismic wave velocities. The presence of this low-velocity layer, which is part of the asthenosphere, is an important element in the modern theory of plate tectonics. The asthenosphere may function as a weak plastic layer upon which horizontal motions in more rigid surface layers can occur.

The seismic boundary that separates the mantle from the overlying crust is the Mohorovičić discontinuity. It lies far deeper under the continents than under the ocean basins. Thus, the continental crust is thicker than the oceanic crust. There are compositional and density differences as well. The continental crust has an overall granitic composition and is less dense than the oceanic crust, which is composed of basaltic rocks. It is likely that earliest patches of continental crust were products of weathering and erosion of preexisting basaltic crust, as well as of conversion of sediment to rocks of granitic character during episodes of metamorphism and melting.

The earth's primitive atmosphere was rich in carbon dioxide and nitrogen, with lesser but significant amounts of water vapor, carbon monoxide, and hydrogen. This atmosphere and the hydrosphere were formed of gases vented from the earth's interior. The present atmosphere of the earth was initiated by additions of large amounts of uncombined oxygen, most of which originated as a byproduct of photosynthesis.

## QUESTIONS FOR REVIEW

1. What are the names of the planets in our solar system? In what galaxy is our solar system located?

2. What is the source of the sun's heat? Given the amount of solar radiation intercepted by the earth, why is the earth's surface not much hotter than it is?

3. What is the difference between the theories of cold homogeneous accretion and hot heterogeneous accretion?

4. What major internal zones of the earth would be penetrated if one were able to drill a well from the North Pole to the center of the earth?

5. What are the three main kinds of earthquake waves? Describe their characteristics.

6. What does the presence of an S-wave shadow zone indicate about the interior of the earth?

7. What is the Mohorovičić discontinuity? The Gutenberg discontinuity? Where is each located?

8. How does continental crust differ from oceanic crust in thickness, composition, and density?

9. What is the inferred composition of the mantle? Of the core? On what evidence are these inferred compositions based?

10. What is the origin of most of the free oxygen in the earth's atmosphere? What geologic evidence attests to the presence of oxygen in ancient rocks?

11. What geologic interpretation might one make about the source of ultrabasic rocks such as peridotite that fill the pipe or conduit of an old volcano?

12. If we consider melting to cause gravity separation and internal zonation of the earth, what are some of the likely sources for the heat required to cause melting?

## TERMS TO REMEMBER

asthenosphere
body wave (seismic)
cold homogeneous
    accretion
continental crust
core
crust
differentiation
    (planetary)
discontinuity (seismic)

Gutenberg discontinuity
hot heterogeneous
    accretion
Jovian planets
low-velocity zone
mantle
Mohorovičić
    discontinuity
oceanic crust
outgassing

peridotite
photochemical
    dissociation
photosynthesis
primary seismic wave
    (P-wave)
prograde planetary
    revolution
protoplanets
secondary seismic wave

shadow zone
solar nebula
solar wind
surface seismic wave
    (S-wave)
terrestrial planets
transition zone (in
    mantle)
volatiles

## SUPPLEMENTAL READINGS AND REFERENCES

Bolt, B. A. 1982. *Inside the Earth*. San Francisco: W. H. Freeman and Co.

Cloud, P. E. The biosphere. *Sci. Am.* 249(3):176–189.

Cocks, L. R. M., ed. 1981. *The Evolving Earth*. Cambridge, England: Cambridge Univ. Press.

Holland, H. D. 1984. *Chemical Evolution of the Atmosphere and Oceans*. Princeton, N.J.: Princeton Univ. Press.

Kaufmann, W. J. 1979. *Planets and Moons*. New York: W. H. Freeman and Co.

Moorbath, S. 1978. The oldest rocks and the growth of continents. *Sci. Am.* 236:92–104.

Ozima, M. 1981. *The Earth: Its Birth and Growth*. New York: Cambridge Univ. Press

Pollack, J. B. 1981. Atmospheres of the terrestrial planets. *In* Beatty, J. K., O'Leary, B., and Chaikin, A. (eds.). *The New Solar System*. Cambridge, Mass.: Sky Publishing Corp.: London: Cambridge Univ. Press.

Taylor, S. R., and McLennan, S. M. 1985. *The Continental Crust: Its Composition and Evolution*. Oxford: Blackwell Scientific Publications.

INTERVIEW INTERVIEW INTERVIEW INTERVIEW INTERVIEW
VIEW INTERVIEW INTERVIEW INTERVIEW INTE
INTERVIEW INTERVIEW INTERVIEW INTERVIEW
INTERVIEW INTERVIEW INTERVIEW INTE
INTERVIEW INTERVIEW INTERVIEW INTERVIEW INTE
INTERVIEW INTERVIEW INTERVIEW INTERVIEW INTE

# TANYA ATWATER

## INTERVIEW

*A native Californian, Tanya Atwater was born on August 27, 1942, in Los Angeles. As an undergraduate, she attended Massachusetts Institute of Technology for three years before transferring to the University of California, Berkeley, where she was awarded a B.A. degree in 1965. She completed her Ph.D. at Scripps Institution of Oceanography in 1972. Since July 1980, she has been a Professor of marine geophysics at the University of California, Santa Barbara.*

*Earlier in her career, Professor Atwater served as a Research Associate in paleomagnetism at Stanford University and as a member of the faculty of both Scripps Institution of Oceanography and Massachusetts Institute of Technology. In 1973 she studied in the Soviet Union as a USA–USSR Exchange Scientist.*

*In 1980, Professor Atwater was awarded an American Association for the Advancement of Science Newcomb Cleveland Prize. She is a fellow of both the American Geophysical Union and the Geological Society of America, and she has served on several national scientific planning committees. Her many publications are often at the forefront of geologic thought and are the reason for the high regard with which she is held by earth scientists in North America and abroad.*

**What events in your pre-college years initiated your interest in science?**

When I was growing up, I wanted to be an artist. I took all the art classes I could in school, on weekends, and during summers. What attracted me most to art was a natural, inborn love for spatial relationships. On family trips I always hogged the maps and was the navigator, taking great pleasure in translating between the paper map and the passing countryside. My father was an engineer and my mother a botanist, so I grew up in a household where science and technical subjects were natural dinner-time fare. Thus, it was easy to move over to science when I decided to change. I have to admit, however, that to this day, many of my science research projects look suspiciously like art projects, with an especially strong emphasis on maps.

**As an undergraduate at MIT, what influenced your decision to major in geology?**

During my undergraduate years, I tried a number of different majors—physics, chemistry, mechanical engineering, electrical engineering—and I did well enough in them, but I didn't "catch fire." In my junior year, I took some geology courses, pretty much by accident, and attended the Indiana University geology summer field camp in Montana, just for fun. I was in heaven. Also, it was a revelation to me that people could get paid for hiking around in the mountains and coaxing out their secrets.

By the end of the field camp I was hooked on geology, but I was nervous about abandoning my high-tech, pure science ambitions for this relatively grubby, descriptive, obscure field. I dropped out of school for half a year to travel and think about this, but my fondness of geology only grew. I decided to switch schools to the University of California at Berkeley. I had already completed a lot of math and physics, so I chose geophysics, but I crammed in all the extra geology courses that would fit into my schedule. As it turned out, this combination of strengths in math–physics–geophysics and classical geology was somewhat unusual, and it put me in a special position to make discoveries and connections that others could not make.

## What advice do you offer undergraduates who are thinking of becoming geology majors?

I often tell students (and their nervous parents) about my shifty undergraduate career: my several major changes, drop-out times, school changes. I believe that the primary job of undergraduate students is to try out many subjects, give each their best effort, find out what suits them and their particular set of talents, and find out what they enjoy doing. Finding a good career and doing things that one enjoys and does well is a prime ingredient for a happy, productive life. And I often encourage students to consider unusual double majors or double emphases, for example, geology and communication studies, geophysics and economics, geochemistry and political science. There are many extremely important and exciting opportunities for people with cross-disciplinary skills.

## After completing your undergraduate work, what events led to your research interests in plate tectonics and sea floor spreading?

After my undergraduate studies, I spent a year working at the Institute of Geophysics and Seismology at the University of Chile. It gave me a chance to gain some practical experience in my new field and to live, work, and travel in a different culture. I also took part in the summer internship program at Woods Hole Oceanographic Institute and discovered that one could combine geophysics with the adventure of sailing the seas. Thus, when I was ready to go back, I applied to oceanographic graduate schools and was accepted at Scripps Institution of Oceanography in La Jolla, California.

When I arrived at Scripps, the "Deep Tow" group was about to go to sea to make a very detailed survey of the Gorda Rise, off northern California. They needed a student to come along and work up the survey data. It would be the first close-up look at a sea floor spreading center! I signed on at once and never looked back. By this stroke of luck, I became embroiled for life in the effort to understand the process of sea floor spreading and the details of the formation and aging of the world's oceanic crusts. I have never lost my fascination with this subject and have continued to explore it with new technologies as they came along: mapping with various sophisticated sonars, deep sea cameras, and deep-diving submersibles. I have visited and studied the deep ocean floor, 2.5 to 3.5 km deep, on twelve different dives in the small submersible *Alvin*. It is amazingly different down there. It's a little like going to Mars! With each new improvement in our surveying techniques, we learn a new level of refinement; we answer many of our previous questions and formulate a new, deeper set.

**Tell us more about your most recent investigations.**

In the last decade I have been working in a group led by Richard Hey that has been documenting the phenomenon of propagating rifts. We had known for a long time that spreading centers were able to change their trends in response to changes in plate motion direction, but we didn't understand how they accomplished these reorientations. It seems that a rift segment begins to crack and extend itself in the new direction, steadily propagating past other segments and establishing the new trend. Now that we have documented the existence of these features and learned to recognize the patterns they produce, we are finding their traces in many ancient sea floor records all over the globe.

My other primary research interest has concerned the history of plate tectonic interactions in western North America. It was clear from the beginning of the plate tectonic revolution that the San Andreas fault system in California was a major strike-slip plate boundary between the Pacific and North American plates. However, when we examined the magnetic anomalies over the adjacent Pacific ocean basin, they implied a very different plate configuration for past times. They showed that the ocean floor here was formed by sea floor spreading between the Pacific and North American plates but had since been completely subducted along much of its length, allowing the other two to come into contact only rather recently and only then to form the San Andreas fault. The geologic record in California told a similar story: a long history of subduction and a relatively young origin for the San Andreas system. My combined background in marine geophysics and land geology put me in the perfect position to explore these relationships. I could see that the oceanic and continental histories should be intricately interrelated here, and I could not resist thinking, reading, and talking about it and experimenting with it. When the dates of the magnetic anomalies were finally established first-hand by the Deep Sea Drilling Project in 1969, the final piece fell in place. I went into a frenzy of writing

and drafting. The result was a paper about the origin and evolution of the San Andreas fault system.

**In what way have some of your views about plate tectonics changed over the years?**

Much of the definitive data for the plate tectonic revolution came from the oceanic realm. This was natural since the ocean floors have relatively simple histories, being formed by sea floor spreading, traveling a while, then being destroyed at subduction zones. The continents, on the other hand, are much more permanent features, so that the record of each new tectonic disturbance is overprinted upon older records, and the resulting histories, structures, and compositions of the continental crusts are complex and heterogeneous. Over the decades, it has become more and more clear that simple rigid plates cannot be used to describe most events that can be observed in the continental geologic record. The plate motions can give us general ideas—for example, Africa and North America collided in the late Paleozoic, causing continental shortening in the southern Appalachians, then split apart again in the Triassic, forming rifts—but such generalizations do not tap the quantitative power of plate tectonics. For these reasons, the origin and evolution of the San Andreas fault system is particularly interesting. It is one of the few continental tectonic problems that can be approached in a quantitative way using the step-by-step predictions of rigid plate motions.

**How do you feel about being a woman in a male-dominated field of science?**

I am the third of four children, the second of three girls. When we were growing up, our parents encouraged us to do our very best in anything we wanted to do. When I got out into the world, I was astonished to discover that women weren't supposed to be able to do all sorts of things that I had been doing all my life and that my mother had done all her life before me. So, of course, I didn't believe it! In those days, there were a myriad of restrictions barring women from many activities and especially from field situations.

Women were not allowed on many of the ships in the oceanographic fleet. (Women on ships were considered unlucky.) The early support from my family gave me great strength to stand firm and smile and insist, knowing that I was right, that the rules and restrictions were unjust. I had important work to do and I needed to do it from a ship. Several antidiscrimination laws and court cases in the 1950s and 1960s also played a crucial role. I never threatened anyone with a lawsuit, but I know that the possibility was mentioned from time to time by my male colleagues to prod various recalcitrant administrators. (Administrators hate lawsuits.) It has been a great pleasure to me through the 1970s and 1980s to watch the numbers and diversity of women grow in the physical sciences. Of course, there are still vestiges of prejudice. The intensity and isolation inherent in research efforts and especially in field work accentuate all interpersonal interactions, good and bad. For the most part, I find that the women and the men are happier with an integrated approach, one that capitalizes on the talents and strengths of a diverse working group.

**Do you believe an understanding of the geosciences is important for students that do not intend to major in science?**

One of the things that is becoming more and more clear to me is that everyone in modern society needs to have a fundamental understanding of science and technology. Ours is a technological age and one in which the results of our technology are fundamentally altering our planet. The users of this technology must understand the consequences of that use. Geology, in particular, offers a unique long-term view that all world citizens need to incorporate into their thinking. The earth was here long before humans evolved and will continue, indifferent to whether we thrive as a species. Many geologic processes such as earthquakes, floods, landslides, beach-cliff erosion, and volcanic eruptions have been occurring for billions of years and will continue, no matter what we do. We now have enough understanding of many of these phenomena to take them into account when planning our structures and our activities. We

know how to minimize the impacts of many natural disasters, but we will not benefit from that knowledge unless our citizens understand what is involved.

Another aspect that geologists need to communicate and citizens need to learn concerns the very real limitations of our earth's resources and environment. It particular, we are presently using up the world's supplies of fossil fuels, especially gas and oil, at an alarming rate and with reckless abandon, and few of our leaders are even thinking about it, much less curtailing our wastefulness. The limitations of the environment—the land, water, and air—to absorb pollutants are only beginning to claim our attention, and many aspects of this problem require geologic understanding.

Beyond these practical matters, there is a more fundamental reason that people should learn about the natural world. It is fun and interesting and empowering to learn the reasons that things are the way they are. The shapes, colors, and textures of the countryside tell their own stories to the viewer who is geologically aware. The configurations and locations of continents, oceans, mountains, plains, swamps, and deserts are all understandable and somewhat predictable. A basic knowledge of geology can enhance every cross country trip and can deepen every person's appreciation of their planet and of their backyard.

The central igneous and metamorphic core of the Himalayan Mountains. The Himalayas are the highest and youngest of the earth's major mountain belts. They were formed by the collision of the Indian tectonic plate with the Eurasian plate. *(Courtesy of D. Bhattacharyya.)*

# Dynamic Crust

*What was solid earth has become the sea, and solid ground has issued from the bosom of the waters.*

Ovid, *Metamorphoses*

## A Changeable Planet

Christopher Columbus was a visionary. It is easy to imagine him standing straddle-legged on the deck of the Santa Maria speculating about the unseen mysteries that lay beyond the horizon. Although an imaginative man, Columbus would not have guessed that the ocean floor far beneath him once did not exist and that the waters over which he sailed had filled the widening chasm formed where a continent had been torn apart.

Great ocean basins, continents, mountains, and plains appear to us to be everlasting. The human life span is far too short to permit a view of lands heaving and moving about. People are allowed time only to witness an occasional earthquake or volcanic eruption. Fortunately, the myopic view of the earth can be corrected with the immense perspectives provided by historical geology. Along the banks of streams and in the rocks exposed at the sides of our highways, it is possible to examine the cumulative effects of scores of geologic events. One may also observe the unrelenting degradation of the lands by water and ice, and it becomes apparent that, if continents were unchangeable, plateaus and mountains would have long ago been reduced to low-lying plains. That highlands remain at all attests to the fact that the degradational forces of weathering and erosion must be balanced by internal forces that uplift and rebuild.

The occurrence of upheavals is further demonstrated by layers of fossiliferous marine sediments that are now poised in cloud-flanked mountain peaks. Crustal uplift is implicit in exposures of originally deep-seated rocks, such as granite. These granitic terranes are exposed to view only after removal of miles of overlying rocks beneath which the granite had congealed. This removal could not have been accomplished without a comparable amount of uplift. Elsewhere, one may observe broadly meandering streams with a pattern of sinuosity normally found in low-lying regions of the continents. They are now incising deep canyons, for the low-lying lands over which they once wandered have been elevated high above sea level. Such uplifts, along with the many other movements of the earth's crust, break and deform rocks and produce geologic structures such as faults, folds, and unconformities. A brief discussion of these geologic features will provide the necessary background for discussion of our principal topic, plate tectonics.

## Geologic Structures

A **fault** is a break in the earth's crustal rocks along which there has been movement (Figs. 6–1 and 6–2). According to the relative movement of the rocks

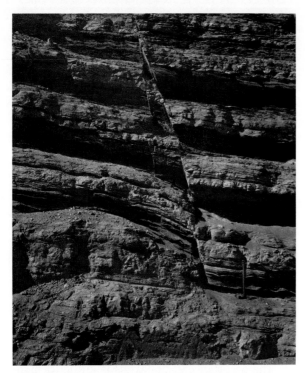

**Figure 6–1**    Close-up view of the edge of a fault plane that cuts through beds of sandstone and silt-stone near Dry Gulch, Idaho. Beds on the right have slipped downward relative to beds on the left. This is a normal fault. There is a small splinter fault visible just to the left of the larger fault. *(Courtesy of U.S. Geological Survey.)*

on either side of the plane of breakage, faults can be classified as **normal, reverse,** or **lateral** (Fig. 6–3). Lateral faults can be further designated as *right lateral* or *left lateral* if one looks across the fault zone to see if the opposite block has moved to the right or left. The San Andreas fault in California is a strike-slip fault of the right-lateral type (Fig. 6–4). In normal faults, the mass of rock that lies above the shear plane, called the **hanging wall,** appears to move downward relative to the opposite side or foot wall. Such faults occur where rocks are subjected to tensional forces—forces that tend to stretch the crust. On the other hand, reverse faults exhibit a hanging wall that has moved up relative to the **foot wall.** If we imagine ourselves holding two blocks cut from wood like those of Figure 6–3, we find that they must be shoved together, or compressed, in order to cause reverse as well as thrust faulting. Thus, regions of the earth's crust containing numerous reverse faults (and folded strata) are likely to have been compressed at some time in the geologic past. Reverse faults in which the shear zone is inclined only a few degrees from horizontal are termed **thrust faults.** Although examples of all kinds of faults may be found in any mountain belt, compressional structures are decidedly the most prevalent in the world's great mountain ranges.

No less important than faults as evidence for the earth's instability are the bends in rock strata that are

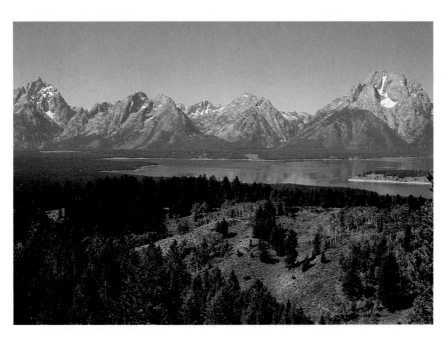

**Figure 6–2**    A mountain range formed by faulting—the Grand Tetons of Wyoming viewed from Signal Mountain. The Grand Tetons were formed by normal faulting. The western block was uplifted to create the mountains, whereas the east block (represented by the foreground) dropped to form Jackson Hole with its picturesque Jenny Lake. Grand Teton at the extreme left rises over 2000 meters above the valley floor of the downfaulted block.

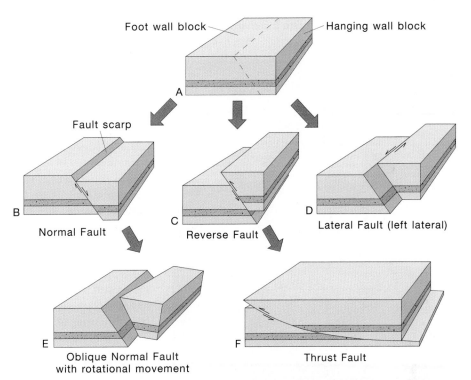

**Figure 6–3** Types of faults. (A) The unfaulted block with the position of the potential fault shown by dashed line. In nature, movements along faults may vary in direction, as shown in (E). A thrust fault (F) is a type of reverse fault that is inclined at a low angle from the horizontal.

Foot wall block      Hanging wall block

A

Fault scarp

B

Normal Fault

C

Reverse Fault

D

Lateral Fault (left lateral)

E

Oblique Normal Fault
with rotational movement

F

Thrust Fault

**Figure 6–4** Air photo mosaic in which the trace of the right-lateral strike-slip San Andreas fault is clearly visible. Notice the offset streams (arrows). San Luis Obispo County, California. *(Courtesy of U.S. Geological Survey.)*

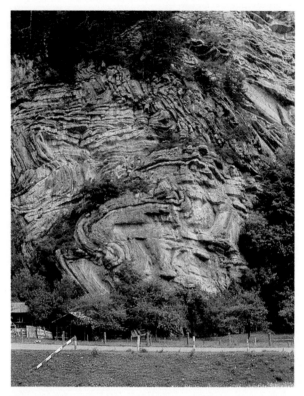

**Figure 6–5**   Severely folded strata exposed on the north side of the Rhone River valley west of Sierra, Switzerland.

termed **folds** (Fig. 6–5). Sediments are, of course, originally laid down in approximately level layers. As noted earlier, the perceptive naturalist Nicolaus Steno referred to this observation as the "law of original horizontality." Steno correctly inferred that if strata were originally horizontal, then folds, flexures, or inclinations of those beds were direct evidence of crustal movement. The principal categories of folds are **anticlines** (Fig. 6–6), **synclines** (Fig. 6–7), **domes, basins,** and **monoclines** (Fig. 6–8).

Although most folds (and particularly those found in mountain belts) are formed by compression, some may develop under a variety of other circumstances. For example, some relatively small folds may result from differential compaction as loose sediment is converted to sedimentary rock, from slumping of mounds of sediment at the time of its accumulation, from an upward protuberance of rock masses from below, from draping over uplifted fault blocks, and from crumpling as great blankets of strata slide over an older rock surface.

## Unconformities

Unconformities are geologic features distinctly different from faults and folds, but they may also provide evidence of crustal movements. An **unconformity** is

**Figure 6–6**   Anticline exposed in the north bank of the Sun River, Teton County, Montana. *(Courtesy of R. R. Mudge and U.S. Geological Survey.)*

**Figure 6–7** Geologic structures often have no relationship to surface topography. Here a downfold in beds (syncline) lies beneath a high ridge. Sideling Hill road cut, 6 miles west of Hancock, Maryland, along U.S. 48, Ridge and Valley Province. *(Courtesy of K. A. Schwartz.)*

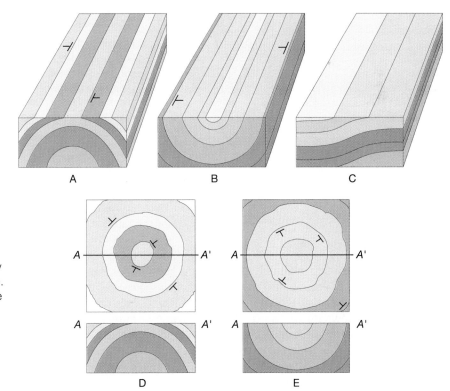

**Figure 6–8** Types of folds. (A) Anticline. (B) Syncline. (C) Monocline. (D) Map view and cross-section of a dome. Notice older strata are in the center of the outcrop pattern. (E) Map view and cross-section of a basin. Younger rocks occur in the central area of the outcrop pattern.

**A**

**B**

**C**

**Figure 6–9** Three types of unconformities. (A) Angular unconformity. (B) Nonconformity. (C) Disconformity.

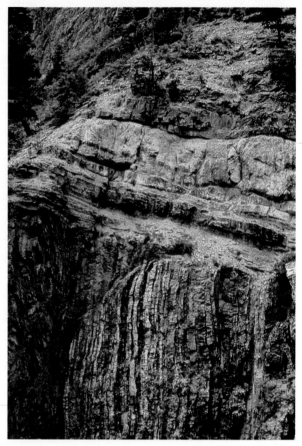

**Figure 6–10** An angular unconformity separates vertical beds of the Precambrian (Proterozoic) Uncompahgre Sandstone from overlying, nearly horizontal Devonian strata of the Elbert Formation at Box Canyon Falls, Ouray, Colorado. *(Photograph by Thomas E. Williams, with permission.)*

a buried erosion surface with older rocks lying beneath the surface of erosion and markedly younger rocks resting above the surface. Because erosion has removed an unknown thickness from the underlying strata and because no accumulation of sediment and fossils has occurred for a period of time, the unconformity represents a gap or break of significant duration in the geologic record of the region in which it is found. Indeed, the lost record may encompass tens or even hundreds of millions of years.

The three major kinds of unconformities illustrated in Figure 6–9 differ with regard to the orientation of the rocks beneath the erosional surface. Of the types shown, the **angular unconformity** (Fig. 6–10) provides the most readily apparent evidence for crustal deformation. As noted in Chapter 1, James Hutton described a now famous angular unconformity at Siccar Point on the Scottish coast of the North Sea (see Fig. 1–12).

Examples of unconformities are abundant on every continent. Some do not reflect the degree of deformation apparent in the strata at Siccar Point,

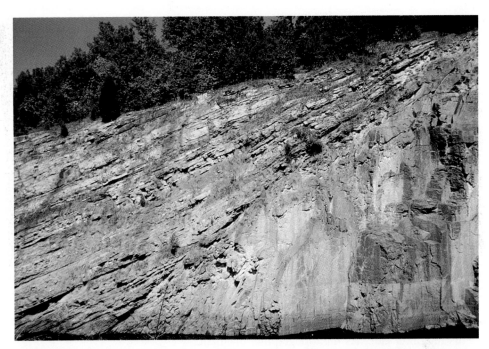

**Figure 6–11**  The erosional surface of this nonconformity is inclined at about 45° and separates Precambrian rhyolitic rock from overlying Upper Cambrian Bonneterre Dolomite. Taum Sauk Mountain, southeastern Missouri. *(Courtesy of D. Bhattacharyya.)*

but rather document the simple withdrawal of the sea for a period of time, followed by another marine transgression. The result may be a **disconformity**, in which parallel strata are separated by an erosional surface. The withdrawal and advance of the sea may be caused by fluctuations in the volume of ocean water, but more commonly it is the result of crustal uplift and subsidence. **Nonconformities** are surfaces where stratified rocks rest on older intrusive igneous or metamorphic rocks (Fig. 6–11). In many nonconformities, crystalline rocks were emplaced deep within the roots of ancient mountain ranges that subsequently experienced repeated episodes of erosion and uplift. Eventually, the igneous and metamorphic core of the mountains lay exposed and provided the surface upon which the younger strata were deposited.

# Plate Tectonics

Perhaps more than other scientists, geologists are accustomed to viewing the earth in its entirety. It is their task to assemble the multitude of observations about the origin of mountains, the growth of continents, and the history of ocean basins into a coherent view of the whole earth. Exciting and revolutionary discoveries made over the past three decades have provided that kind of integrated understanding of our planet. The new view of the earth's dynamic geology, called **plate tectonics**, is based on the movements of plates of lithosphere driven by convection in the underlying mantle.

As with many scientific breakthroughs, the plate tectonics theory was not fully conceptualized until a foundation of relevant data had been assembled. In the years following World War II, research related to naval operations produced submarine detection devices that also proved useful in measuring magnetic properties of rocks. In the mid-1940s to late 1950s, the need to monitor atomic explosions resulted in the establishment of a worldwide network of seismometers. This network provided precise information about the global pattern of earthquakes. The magnetic field over large portions of the sea floor was soon to be charted by the use of newly developed and delicate *magnetometers*. Other technologic advances ultimately permitted scientists to examine rock that had been carefully dated by radiometric methods and then to determine the nature of the earth's magnetic field at the time those rocks had formed. Geologically recent reversals of the magnetic field were soon detected, correlated, and accurately dated. A massive federally funded program to map the bottom of the oceans was launched, and depth information from improved echo depth-sounding devices poured into data-collecting rooms to be translated into maps and charts.

A new picture of the ocean floor began to emerge (see Figure 1–5). It was at once awesome, alien, and majestic. Great chasms, flat-topped, submerged mountains, boundless abyssal plains, and interminable volcanic ranges appeared on the new maps and begged an explanation. How did the volcanic mid-oceanic ridges and deep sea trenches originate? Why were both so prone to earthquake activity? Why was the mid-Atlantic ridge so nicely centered and parallel to the coastlines of the continents on either side? As the topographic, magnetic, and geochronologic data accumulated, the relationship of these questions became apparent. An old theory called **continental drift** was reexamined, and the new, more encompassing theory of plate tectonics was formulated. It was an idea whose time had come.

## Fragmented Continents

It requires only a brief examination of the world map to notice the remarkable parallelism of the continental shorelines on either side of the Atlantic Ocean. If the continents were pieces of a jigsaw puzzle, it would seem easy to fit the great "nose" of Brazil into the re-entrant of the African coastline. Similarly, Greenland might be inserted between North America and northwestern Europe. It is not surprising, therefore, that earlier generations of mapgazers also noticed the fit and formulated theories involving the breakup of an ancient supercontinent.

In 1858 there appeared a work titled *La Création et ses Mystères Devoilés*. Its author, A. Snider, postulated that before the time of Noah and the biblical flood, there existed a great region of dry land. This antique land developed great cracks encrusted with volcanoes, and during the Great Deluge, a portion separated at a north–south trending crack and drifted westward. Thus, North America came into existence.

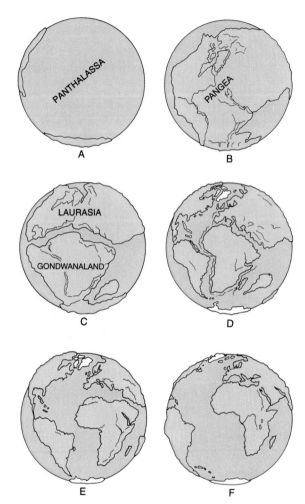

**Figure 6–12**   (A) and (B) Alfred Wegener's view of the earth as it existed 200 million years ago. At this time there was one ocean (Panthalassa) and one continent (Pangea). (C) By about 20 million years later, the supercontinent had begun to split into a northern Laurasia and southern Gondwanaland. (D) Dismemberment continued, and (E) looked this way about 65 million years ago. (F) Further widening of the Atlantic and northward migration of India brings the earth to its present state.

Near the close of the nineteenth century, the Austrian scientist Eduard Suess became particularly intrigued by the many geologic similarities shared by India, Africa, and South America. He formulated a more complete theory of a supercontinent that drifted apart following fragmentation. Suess called that great land **Gondwanaland** after Gondwana, a geologic province in east-central India.

The next serious effort to convince the scientific community of the validity of these ideas was made in the early decades of the twentieth century by the energetic German geophysicist Alfred Wegener. His book, *Die Entstehung der Kontinente und Ozeane* ("The Origin of the Continents and Oceans"), is considered a milestone in the historical development of the concept of continental drift.

Wegener's hypothesis was quite straightforward. Building on the earlier notions of Eduard Suess, he argued again for the existence in the past of a supercontinent that he dubbed **Pangea**. That portion of Pangea that was to separate and form North America and Eurasia came to be known as **Laurasia**, whereas the southern portion retained the earlier designation of Gondwanaland. According to Wegener's perception, Pangea was surrounded by a universal ocean named **Panthalassa**, which opened to receive the shifting continents when they began to split apart about 200 million years ago (Fig. 6–12). The fragments of Pangea drifted along like great stony rafts on the denser material below. In Wegener's view, the bulldozing forward edge of the slab might be expected to crumple and produce mountain ranges like the Andes.

The assiduous investigations of Wegener were not to go unchallenged. Criticism was leveled chiefly against his notion that the continents were able to slide along through an unyielding oceanic crust. The eminent and scientifically formidable Sir Harold Jeffries calculated that the ocean floor was far too rigid to allow for the passage of continents, no matter what the imagined driving mechanism. It is now known that Jeffries was correct in asserting that continents cannot—and do not—plow through oceanic crust. Shortly, it will be shown that continents do move, but they move only as passive passengers on large rafts of lithosphere that glide over a comparatively soft and plastic upper layer of the earth's mantle. Nevertheless, much of the evidence Wegener and others had assembled can be used to substantiate both the old and the newer concepts. We should have a look at this evidence.

## Evidence for Continental Drift

The most convincing evidence for continental drift remains the geologic fit of the continents. Indeed, the correspondence is far to good to be fortuitous, even when one considers the expected modifications of shorelines resulting from erosion, deformation, or intrusions following the breakup of Pangea about 200 million years ago. A still closer match results when one fits the continents together to include the continental shelves, which are really only submerged portions of the continents. Such a computerized and error-tested fitting of continents was carried out by Sir Edward Bullard, J. E. Everett, and A. G. Smith of the University of Cambridge (Fig. 6–13). Remarkably, this work showed that over most of the boundary the average mismatch was no more than a degree—a snug fit indeed.

Another line of evidence favoring the drift theory involves sedimentologic criteria indicating similarity

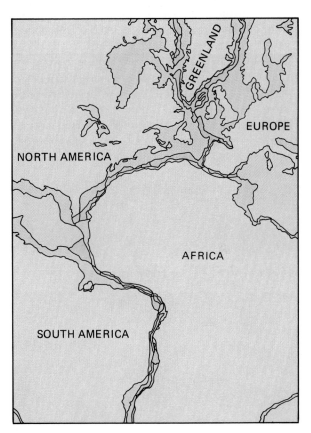

**Figure 6–13** Fit of the continents as determined by Sir Edward Bullard, J. E. Everett, and A. G. Smith. The fit was made along the continental slope (green color) at the 500-fathom contour line. Overlaps and gaps (shown in orange) are probably the result of deformations and sedimentation after rifting. *(Adapted from Bullard, E. C., et al.* Philos. Trans. R. Soc. Lond. *258:41, 1965.)*

of climatic conditions for widely separated parts of the world that were once closely adjacent to one another. For example, in such widely separated places as South America, southern Africa, India, Antarctica, Australia, and Tasmania, one finds glacially grooved rock surfaces and deposits of glacial rubble developed in the course of the late Paleozoic continental glaciation. The deposits of poorly sorted clay, sand, cobbles, and boulders are called **tillites.** They, along with the grooves and scratches on rock surfaces that were apparently beneath the moving ice, attest to a great ice age that affected Gondwanaland at a time when it was as yet unfragmented and lying at or near the south polar region (Fig. 6–14). Furthermore, if the directions of the grooves on the bedrock are plotted on a map, they indicate the center of ice accumulation and the directions in which it moved. Unless the southern continents are reassembled into Gondwanaland, this center would be located in the ocean, and great ice sheets do not develop centers of accumulation in the ocean. Hence, the existence of Gondwanaland seems plausible. In a few instances, oddly foreign boulders in the tillites of one continent are found in the deposits of another continent now located thousands of miles across the oceans. Petrologists are able to trace their parent outcrops to the distant land masses.

There are additional clues to paleoclimatology that can be used to test the concept of moving land masses. Trees that have grown in tropical regions of the globe characteristically lack the annual rings resulting from seasonal variations in growth. Exceptionally thick coal seams containing fossil logs from such trees imply a tropical paleoclimate. The locations of such coal deposits should approximate an equatorial zone relative to the ancient pole position for that age. It is evident that such coal seams now being exploited in northern latitudes must have been moved to those locations from the equatorial zones along which the source vegetation accumulated.

Because of the decrease in solubility of calcium carbonate with rising temperatures, thick deposits of marine limestone also imply relatively warm climatic conditions. Arid conditions, such as those existing today on either side of the equatorial rain belt, can be recognized in ancient rocks by desert sandstones and evaporites. (Evaporites are chemical precipitates such as salt and gypsum that characteristically form when a body of water containing dissolved solids is evaporated.) Today such deposits are ordinarily formed in warm, arid regions located about 30° north or south of the equator. If one believes continents have always been where they are now, then it is difficult to explain the great Permian salt deposits now found in northern Europe, the Urals, and the southwestern United States (Fig. 6–15). The evaporites had probably been precipitated in warmer latitudes before Laurasia migrated northward.

A similar kind of evidence can be obtained by examining the locations of Permian reef deposits. Modern reef corals are restricted to a band around the earth that is within 30° of the equator. Ancient reef deposits are now found far to the north of the latitudes at which they had originated.

At least some of the paleontologic support for continental drift was well known to Suess and Wegener. In the Gondwana strata overlying the tillities or glaciated surfaces, there can often be found nonmarine sedimentary rocks and coal beds containing a distinctive assemblage of fossil plants. Named after a prominent member of the assemblage, the plants are referred to as the ***Glossopteris*** flora (Fig. 6–16). Paleobotanists who have supported the idea of shifting continents have argued that it would be virtually impossible for this complex temperate flora to have developed in identical ways on the southern continents as they are separated today. The seeds of *Glossopteris* were far too heavy to have been blown over such great distances of ocean by the wind.

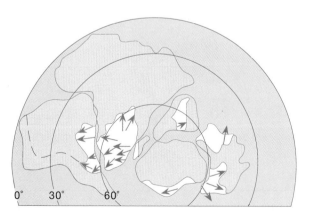

**Figure 6–14**   Reconstruction of Gondwanaland near the beginning of Permian time, showing the distribution of glacial deposits (shown in white). Arrows show direction of ice movement as determined from glacial scratches on bedrock. *(Modified from Hamilton, W., and Kinsley, D. Geol. Soc. Am. Bull. 78:783–799, 1967.)*

**Figure 6-15** Location of prominent Permian evaporite deposits.

**Figure 6-16** Fossil *Glossopteris* leaf associated with coal deposits and derived from glossopterid forests of Permian age. This fossil was found on Polarstar Peak, Ellsworth Land, Antarctica. *(Courtesy of U.S. Geological Survey, J. M. Schopf, and C. J. Craddock.)*

**211**

**Figure 6–17** *Mesosaurus.*

Silurian and Devonian fishes, for example, are comparable in such now distant locations as Great Britain, Germany, Spitzbergen, eastern North America, and Quebec. These similarities, not only in fishes but also in amphibians, persist into Carboniferous rocks. In Permian and Mesozoic age rocks, one again finds striking similarities in the reptilian faunas of Europe and North America. The fossil evidence implies not only the former existence of continuous land connections, but also a uniformity of environmental conditions between the elements of Laurasia that were once at approximately similar latitudinal zones.

When one turns to Cenozoic mammalian faunas, however, one finds the situation to be quite different. Distinctive faunal elements are evident on separate continents and are especially evident in Australia, South America, and Africa. Apparently, as the continents became separated from each other by ocean barriers, genetic isolation resulted in morphologic divergence. The modern world's enormous biologic diversity is at least partially a result of evolutionary processes operating on more or less isolated continents. The faunas and floras during the periods prior to the breakup of Pangea were less diverse.

The character, sequence, age, and distribution of rock units have also been examined for insights into concepts of drift. One might presume that locations close to one another on the hypothetic supercontinent would have environmental similarities that would result in resemblances in the kinds of rocks deposited. As indicated by the correlation chart of southern continents (Table 6–1), there is such a similarity in the geologic sections of now widely separated land masses. The sections begin with glacial deposits such as the Dwyka Tillite of Africa and the Itararé Tillites of South America. These cobbly layers are overlain by nonmarine sandstones and shales containing the *Glossopteris* flora and layers of coal. This overlaying may indicate a warming trend from cold glacial to more temperate climates. However, one must be careful with such interpretations, for our most recent continental glaciations occurred under a cool, temperate climatic regime. The next higher group of strata shows evidence of deposition under terrestrial conditions with an abundance of alluvial, eolian, and stream deposits that contain fossils of an interesting group of mammal-like reptiles. The uppermost beds of the sequences often include basalts and other volcanic rocks that are of similar age and composition in India and Africa,

Another element of paleontologic corroboration for the concept of moving land masses is provided by the distribution in the Southern Hemisphere of fossils of a small aquatic reptile named **Mesosaurus** (Fig. 6–17). An interpretation, based on its skeletal remains, of this animal's habits, as well as the nature of the sediment in which its fossils are found, strongly suggests that it once inhabited lakes and estuaries and was not an inhabitant of the open ocean. The discovery of fossil remains of *Mesosaurus* both in Africa and in South America lends credence to the notion that these continents were once attached. Perhaps for a time, as they were just beginning to separate, the location of the present coastlines became dotted with lakes and protected bodies of water that harbored *Mesosaurus*. This seems a more plausible explanation than one requiring the creature to navigate the South Atlantic or to wend its way across formidable latitudinal climatic barriers by following the shorelines northward and around the perimeter of the Atlantic.

Before fragmentation of a supercontinent such as Laurasia, one would expect to find numerous similar plants and animals living at corresponding latitudes on either side of the line of future separation. Such similarities do occur in the fossil record of continents now separated by extensive oceanic tracts. Faunas of

**Table 6–1    Gondwanaland Correlations**

| System | Southern Brazil | South Africa | Peninsular India |
|--------|-----------------|--------------|------------------|
| Cretaceous | Basalt | Marine sediments | Volcanics<br><br>Marine sediments |
| Jurassic | (Jurassic rocks not present) | Volcanics | Sandstone and shale<br><br>Volcanics<br><br>Sandstone and shale |
| Triassic | Sandstone and shale<br><br>with reptiles | Sandstone and shale<br><br>with reptiles | Sandstone and shale<br><br>with reptiles |
| Permian | Shale and sandstones<br>with<br>coal and *Glossopteris*<br><br>Shale with<br>*Mesosaurus* | Shale and sandstones<br>with<br>coal and *Glossopteris*<br><br>Shale with<br>*Mesosaurus* | Shale and sandstones<br>with<br>coal and *Glossopteris*<br><br><br>Shale |
| Carboniferous | Sandstone, shale, and<br>coal with *Glossopteris*<br><br>Tillite | Sandstone, shale, and<br>coal with *Glossopteris*<br><br>Tillite | Tillite |

but are slightly younger in South America. The absence of the reptile-bearing Triassic strata and the Jurassic volcanics in Australia is believed to be the result of Australia separating from Africa at an early date—probably early in the Permian—and long before the separation of Africa and South America. Australia must have retained just enough of a connection with Africa to permit the entry of dinosaurs and marsupials.

Yet another way to test the notion of a supercontinent is to see whether geologic structures, such as the trends of folds and faults, match up when now distant continents are hypothetically juxtaposed. Such correlative trends do exit. Folds and faults are often difficult to date, although if successful, one can establish the contemporaneity of a fault lineage or fold axis now on separated continents. One can also examine the outcrop patterns of Precambrian basement rocks and discern correlative boundaries be-

tween now widely separated continents. A folded geosynclinal sequence of Precambrian strata in central Gabon in Africa, for example, can be traced into the Bahia Province of Brazil. Also, isotopically dated Precambrian rocks of west Africa can be correlated with rocks of similar age in northeastern Brazil.

## The Testimony of Paleomagnetism

Although Alfred Wegener was unable to provide a satisfactory explanation for what breaks continents apart and moves them across the face of the earth, he had assembled a convincing body of evidence that such events occurred. Discoveries in the late 1950s and 1960s would provide further substantiating evidence. The new information came from the study of magnetism imparted to ancient rocks at the time of their formation and preserved down to the present time. To understand this **paleomagnetism**, as it is called, it is necessary to digress for a moment and consider the general nature of the earth's present magnetic field.

## The Earth's Present Magnetism

It is common knowledge that the earth has a magnetic field. It is this field that causes the alignment of a compass needle. The origin of the magnetic field is still a question that has not been fully resolved, but many geophysicists believe it is generated as the rotation of the earth causes slow movements in the liquid outer core. The magnetic lines of force resemble those that would be formed if there were an imaginary bar magnet extending through the earth's interior. The long axis of the magnet would be the conceptual equivalent of the earth's magnetic axis, and the ends would correspond to the north and south geomagnetic poles (Fig. 6–18). Although today the geomagnetic poles are located about 11° of latitude from the rotational axis, they slowly shift position. When averaged over several thousand years, the geomagnetic poles and the geographic poles do coincide. If we assume that this relationship has always held true, then by calculating ancient magnetic pole positions from paleomagnetism in rocks, we have coincidentally located the earth's geographic poles. It should be kept in mind, however, that such interpretations are based on the supposition that the rotational and magnetic poles have always been relatively close together. This seems a reasonable assumption based on the modern condition, as well as on paleontologic studies that have shown inferred ancient climatic zones in plausible locations relative to an-

cient pole positions. Another assumption is that the earth has always been dipolar. Paleomagnetic studies from around the world thus far support this supposition.

## Remanent Magnetism

The magnetic information frozen into rocks may originate in several ways. Imagine for a moment the outpouring of lava from a volcano. As the lava begins to cool, magnetic iron oxide minerals form and align their polarity with the earth's magnetic field. That alignment is then retained in the rock as its crystallization is completed. In a simple analogy, the magnetic orientations of the minerals responded as if they were tiny compass needles immersed in a viscous liquid. Because they are aligned parallel to the magnetic lines of force surrounding the earth, they not only point toward the poles (magnetic *declination*), but also become increasingly more inclined from the horizontal as the poles are approached (Fig. 6–19). This *inclination,* when detected in paleomagnetic analysis, can be used to determine the latitude at which an igneous body containing magnetic minerals cooled and solidified.

Magnetism frozen into ancient rocks is called **remanent magnetism.** The type described previously, in which igneous rocks cool past the Curie temperature (also called the *Curie point*) of its magnetic minerals, is further classified as *thermoremanent magnetism.* The **Curie temperature** is simply that temperature above which a substance is no longer magnetic. A few minerals, the most important of which is magnetite ($Fe_3O_4$), have the property of acquiring remanent magnetism. Magnetite is widespread in varying amounts in virtually all rocks. The manner in which remanent magnetism is acquired is complicated but can be explained in a general way. Remanent magnetism in a mineral is ultimately due to the fact that some atoms and ions (charged atoms) have so-called magnetic moments; this means that they behave like tiny magnets. The magnetic moment of a single atom or ion is produced by the spin of its electrons. When magnetite takes on its remanent magnetism, the iron ions align themselves within the crystal lattice so that their magnetic moments are parallel. Now most igneous rocks crystallize at temperatures in excess of 900°C. At these extreme temperatures, the atoms in the minerals have a large amount of energy and are being violently

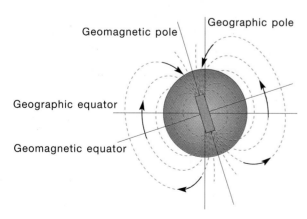

**Figure 6–18**   Magnetic lines of force for a simple dipole model of the earth's magnetic field.

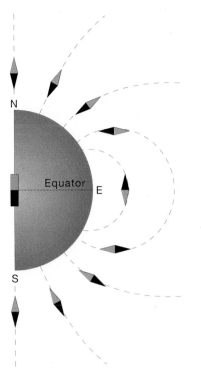

**Figure 6–19** A freely suspended compass needle aligns itself in the direction of the earth's magnetic field. The inclination of the needle will vary from horizontal at the equator to vertical at the poles.

shaken. The vibrating, shaking atoms are unable to line up until more cooling occurs. Finally, when the Curie temperature is reached (578°C for magnetite), enough energy has been lost to allow the atoms to come into alignment. They will stay aligned at still lower temperatures because the earth's field is not strong enough to alter the alignment already frozen into the minerals.

Igneous rocks are not the only kinds of earth materials that can acquire remanent magnetism. In lakes and seas that receive sediments from the erosion of nearby land areas, detrital grains of magnetite settle slowly through the water and rotate so that their directions of magnetization parallel the earth's magnetic field. They may continue to move into alignment while the sediment is still wet and uncompacted, but once the sediment is cemented or compacted, the depositional remanent magnetism is locked in.

Over the past two decades, geophysicists have been measuring and accumulating paleomagnetic data for all the major divisions of geologic time. Their results are partly responsible for the recent revival of interest in continental drift hypotheses. For example, when ancient pole positions were located on maps, it appeared that they were in different positions relative to a particular continent at different periods of time in the geologic past. Either the poles had moved relative to stationary continents, or the poles had remained in fixed positions while the continents shifted about. If the poles were wandering and the continents "stayed put," then a geophysicist working on the paleomagnetism of Ordovician rocks in France, for example, should arrive at the same location for the Ordovician poles as a geophysicist doing similar work on Ordovician rocks from the United States. In short, the paleomagnetically determined pole positions for a particular age should be the same for all continents. On the other hand, if the continents had moved and the poles were fixed, then we should find that pole positions for a particular geologic time would be different for different continents. The data suggest that this latter situation is the more valid.

Another way to view the data of paleomagnetism is to examine what are called **apparent polar wandering curves**. The word *apparent* in the expression is necessary, for polar wandering curves alone might imply that the poles wander. As we have just noted, this is highly unlikely. The apparent polar wandering curves are merely lines on a map connecting ancient pole positions relative to a specific continent for various times during the geologic past. As shown in Figure 6–20, the curves for North America and Europe meet in recent time at the present North Pole. This means that the paleomagnetic data from recently formed rocks from both continents indicate the same pole position. A plot of the more ancient poles results in two similarly shaped but increasingly divergent curves. If this divergence resulted from a drifting apart of Europe and North America, then one should be able to reverse the movements mentally and see if the curves do not come together. Indeed, the Paleozoic portions of the polar wandering curves can be brought into close accord if North America and its curve were to be slid eastward about 30° toward Europe. This sort of information gave new life to the old notion of continental drift. More important, it has become significant in the recent modification of that earlier concept—a modification known as **plate tectonics**.

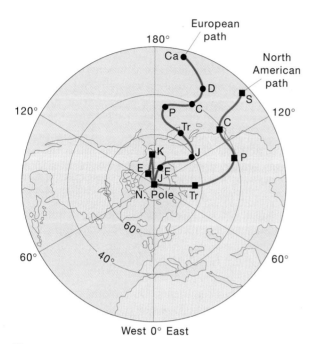

**Figure 6–20** Apparent paths of polar wandering for Europe and North America. A scatter of points have been averaged to a single point for each geologic period: Ca, Cambrian; S, Silurian; D, Devonian; C, Carboniferous; P, Permian; Tr, Triassic; J, Jurassic; K, Cretaceous; E, Eocene. *(After Bott, M. H. P.,* The Interior of the Earth: Structure and Processes. *New York: St. Martin's Press, 1971.)*

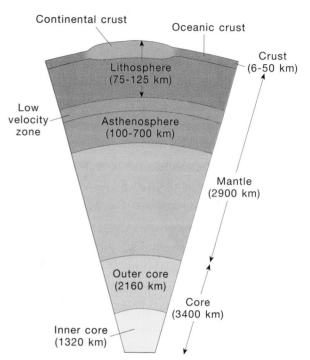

**Figure 6–21** Divisions of the earth's interior. (For clarity, the divisions have not been drawn to scale.)

## The Basic Concept of Plate Tectonics

When considered at the general level, plate tectonics is a remarkably simple concept. The **lithosphere**, or outer shell of the earth (Fig. 6–21), is constructed of 7 huge slabs and about 20 smaller plates that are squeezed in between them. The larger plates (Fig. 6–22) are approximately 75 to 125 km thick. Movement of the plates causes them to converge, diverge, or slide past one another, and this results in frequent earthquakes along plate margins. When the locations of earthquakes are plotted on the world map, they clearly define the boundaries of tectonic plates (Fig. 6–23).

A part of a plate containing a continent would have the configuration shown in Figure 6–24. Plates "float" upon a weak, partially molten region of the upper mantle called the **asthenosphere** (from the Greek *asthenos*, meaning "weak"). Geophysicists view the asthenosphere as a region of rock plasticity and

flowage. Its presence was first detected by Beno Gutenberg on the basis of changes in seismic wave velocities (as discussed in Chapter 5). It should be noted that the boundary between the lithosphere and the asthenosphere does not coincide with the crust–mantle boundary.

## Plate Boundaries and Sea Floor Spreading

Central to the idea of plate tectonics is the differential movement of lithospheric plates. For example, plates move apart at **divergent plate boundaries**, which may manifest themselves as mid-oceanic ridges complete with tensional ("pull apart") geologic structures. Indeed, the mid-Atlantic ridge approximates the line of separation between the Eurasian and African plate on one side and the American plate on the other (Fig. 6–25). As is to be expected, such a rending of the crust is accompanied by earthquakes and enormous outpourings of volcanic materials that are piled high to produce the ridge itself. The void between the separating plates is also filled with this molten rock, which rises from below the

**Figure 6-22** The earth's major tectonic plates. Arrows indicate general direction of movement of plates.

**Figure 6-23** World distribution of earthquakes. Notice the major earthquake belt that encircles the Pacific and another that extends eastward from the Mediterranean toward the Himalayas and East Indies. The midoceanic ridges are also the site of many earthquakes.

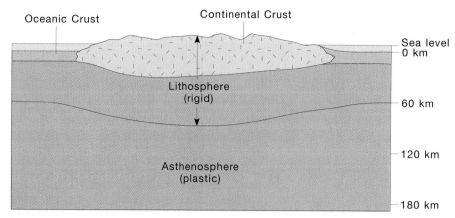

**Figure 6–24**   The outermost part of the earth consists of a strong, relatively rigid lithosphere, which overlies a weak, plastic asthenosphere. The lithosphere is capped by a thin crust beneath the oceans and a thicker continental crust elsewhere.

**Figure 6–25**   Locations of mid-oceanic ridges (red) and trenches (blue). Note fracture zones offsetting the ridges. Dashed lines in Africa represent the East African rift zone. *(After Isacks, B. et al. J. Geophys. Res. 73:5855–5899, 1968. Copyright, The American Geophysical Union.)*

lithosphere and solidifies in the fissure. Thus, new crust (that is, new sea floor) is added to the **trailing edge** of each separating plate as it moves slowly away from the mid-oceanic ridge (Figs. 6–26 and 6–27). The process has been appropriately named **sea floor spreading**. Zones of divergence may originate beneath continents, rupturing the overlying land mass and producing such rift features as the Red Sea and Gulf of Aden.

The axis of spreading is not a smoothly curving line. Rather, it is abruptly offset by numerous faults. These features are called **transform faults** and are an expected consequence of horizontal spreading of the sea floor along the earth's curved surface. Transform faults take their name from the fact that the fault is "transformed" into something different at its two ends. The relative motions of transform faults are shown in Figure 6–28. The ridge acts as a spreading

center that exists both to the north and to the south of the fault. The rate of relative movement on the opposite sides along fault segment X to X′ depends upon the rate of extrusion of new crust at the ridge. Because of the spreading of the sea floor outward from the ridge, the relative movement is opposite to that expected by ordinary fault movement. Thus, at first glance, the ridge-to-ridge transform fault (Fig. 6–28A) appears to be a left-lateral fault, but the actual movement along segment X–X′ is really right lateral. Notice that only the X–X′ segment shows movement of one side relative to the other. To the west of X and the east of X′ there is little or no relative movement.

If plates are receding from one another at one boundary, they may be expected to collide or slide past other plates at other boundaries. Thus, in addition to the divergent plate boundaries that occur

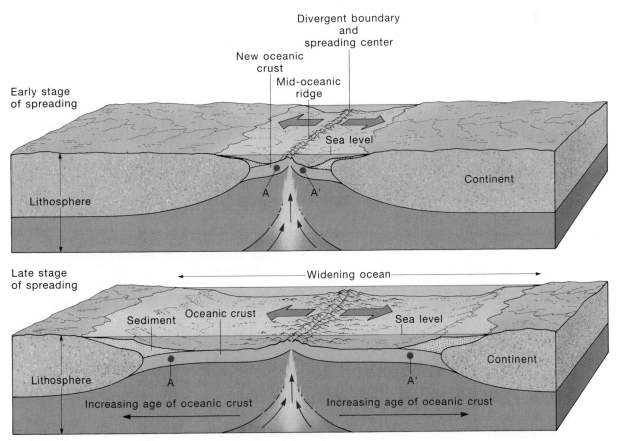

**Figure 6–26** Sea floor spreading. As the rift widens, basaltic lavas rise to fill the space and solidify to become a new part of the trailing edge of a tectonic plate. A and A′ are reference points.

**Figure 6–27**   Disruption of the supercontinent Pangea and dispersal of its continental fragments. Divergent zones indicated by red lines. (A) General conditions at the end of the Triassic, about 180 million years ago. (B) Position of the continents about 135 million years ago. (C) The continental positions at the end of the Cretaceous, about 66 million years ago. *(Based on maps by Dietz, R. S. and Holden, J. C., J. Geophys. Res. 75, 1970.)*

**Figure 6–28**   Three types of transform faults. (A) Ridge–ridge transform. (B) Ridge–trench. (C) Trench–trench.

along mid-oceanic ridges, there are **convergent** and **shear boundaries**. Convergent plate boundaries develop when two plates move toward one another and collide. As one might guess, these convergent junctions are characterized by a high frequency of earthquakes. In addition, they are thought to be the zones along which folded mountain ranges or deep sea trenches may develop (Fig. 6–29). The structural configuration of the convergent boundary is likely to vary according to the rate of spreading and whether the leading edges of the plates are composed of oce-

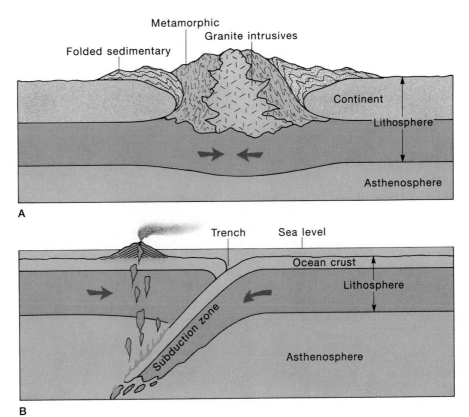

**Figure 6–29** Two types of convergent plate boundaries. (A) Convergence of two plates, both bearing continents. (B) Convergence of two plates, both bearing oceanic crust.

anic or continental crust. Geophysicists speculate that when the plates collide, one slab may slip and plunge below the other, producing what is called a **subduction zone**. The sediments and other rocks of this plunging plate are pulled downward (subducted), melted at depth, and much later, rise to become incorporated into the materials of the upper mantle and crust. In some instances, the silicate melts provide the lavas for chains of volcanoes.

An example of a shear plate boundary is the well-known San Andreas fault in California. Along this great fault, the Pacific plate moves laterally against the American plate (Fig. 6–30). Shear plate boundaries are decidedly earthquake prone, but are less likely to develop intense igneous activity. They are the active segments of transform faults along which no new surface is formed or old surface consumed.

### Crustal Behavior at Plate Boundaries

We have noted that there are three basic kinds of plate boundaries: convergent, divergent, and shear.

With regard to the convergent boundaries, there are also three kinds of tectonic behavior, depending on the kinds of crust constituting the margins of the plates that converge. If the leading edge of a plate is composed of continental crust and it collides with an opposing plate of similar composition, the result would be a folded mountain range in which are developed igneous rocks of granitic composition (see Fig. 6–29A). Because continental plate margins are too light and buoyant to be carried down into the asthenosphere, subduction does not occur in this type of collision. Instead, the crust at the plate margins is deformed and may detach itself from deeper zones, and slabs of crust from one plate may be thrust under the other. The zone of convergence between the two plates, recognized by the severity of folding, faulting, and intrusive activity, is termed the **suture zone**.

A dramatic example of an encounter between two continents occurred during the convergence of India and Eurasia (Fig. 6–31). The mighty Himalayas are the spectacular result of this collision. The

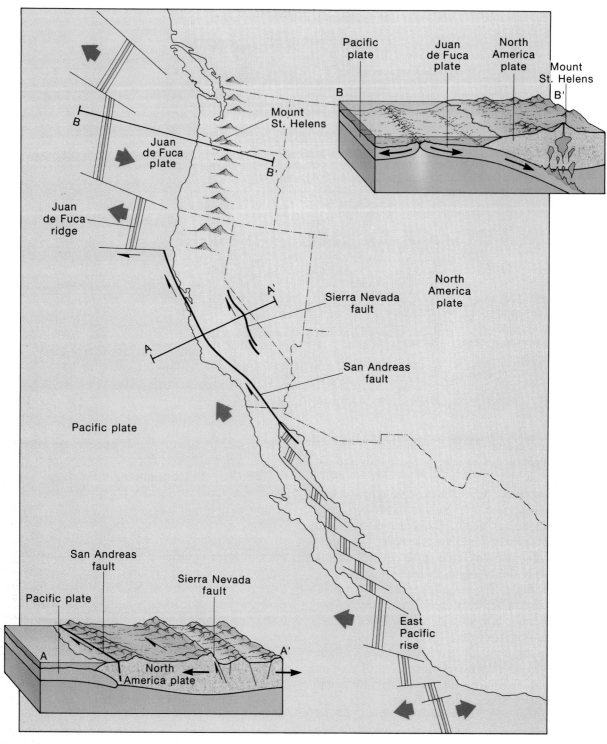

**Figure 6–30**   The juncture of the North American and the Pacific tectonic plates. The double lines are spreading centers. Note the trace of the San Andreas fault. To the north in Oregon and Washington, the small Juan de Fuca plate plunges beneath the North American continent to form the Cascades. *(Courtesy of U.S. Geological Survey.)*

along faults associated with convergence and evidence provided by photographs taken by satellites, indicate that this is not true. The plates may continue to converge, albeit at a slower rate, long after making contact with one another. Some of the continued movement is the result of crustal shortening that occurs whenever a flat span of crust is crumpled into a smaller area. The amount of crustal shortening resulting from the India–Eurasia convergence has been estimated at about 1500 km. Geologists can account for 500 to 1000 km of that amount in the folds and thrust faults of the suture zone. The remaining 500 to 1000 km was apparently dissipated along major east–west trending lateral faults in China and Mongolia (Fig. 6–32). Thus, plate convergence may continue long after initial closure by lateral release of plate movement along strike-slip faults (lateral faults) in the region peripheral to the suture zone. Today geologists are finding evidence of similar strike-slip faulting in older continent–continent convergences such as those that produced the Appalachians and Urals.

The second kind of convergence at plate boundaries involves the meeting of two plates that both have oceanic crust at their converging margins (Fig. 6–29B). Although the rate of plate movement in an ocean–ocean convergence will affect the kinds of structures produced, it is likely that such locations will develop deep sea trenches with bordering volcanic arcs such as those of the southwestern Pacific.

Finally, there is the third possibility. It involves the collision of a continental (granitic) plate boundary with an oceanic (basaltic) one. The result of such a collision might be a deep sea trench located offshore from an associated range of mountains. Numerous volcanoes would be expected to develop, and the lavas pouring from their eruptions would probably be a compositional blend of granite and basalt. The Andes are such a mountain range, and the igneous rock *andesite,* named for its prevalence in the Andes, may represent such a silicate blend.

Regions of ocean–continent convergence (Fig. 6–33) are characterized by rather distinctive rock assemblages and geologic structure. As we have seen, the convergence of two lithospheric slabs results in subduction of the ocean plate, whereas the more buoyant continental plate maintains its position at the surface but experiences intense deformation, metamorphism, and melting. A great mountain range begins to take form as a result of all of this

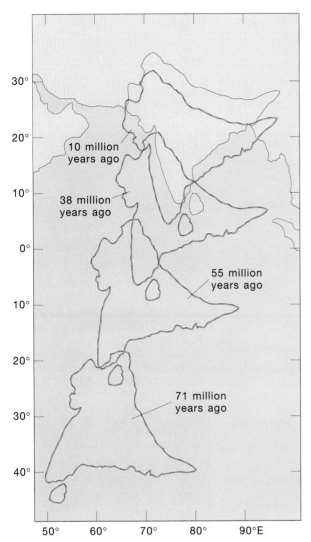

**Figure 6–31** The northward migration of India relative to Eurasia, showing the position of the continent 71 million years ago (Cretaceous), 55 million years ago (early Eocene), 33 million years ago (Oligocene), and 10 million years ago (late Miocene). The configuration of the northern boundary of India is conjectural. *(From Molnar, P. and Tapponnier, P. Cenozoic tectonics of Asia: Effects of a continental collision.* Science *189(4201): 419–426, Copyright 1975, AAAS.)*

study of structures produced during this convergence has provided some interesting insights into the history of continent–continent encounters. It was at one time assumed that once a convergence had occurred, motion between opposing plates would abruptly cease. However, analysis of movement

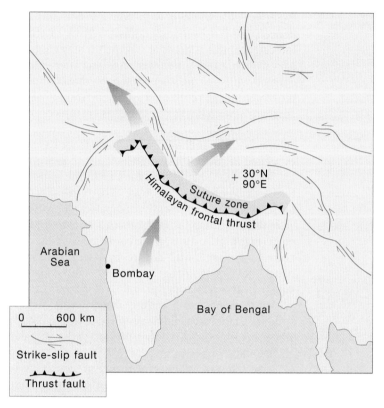

**Figure 6–32** Major lateral (strike-slip) faults above and adjacent to the Indian suture zone. These faults may represent lateral transfer of some of the movement involved in the India–Eurasia convergence. Geologically recent movement along the faults implies that the plate movement involved in the convergence is still in progress. *(Fault locations from Molnar, P. and Tapponnier, P. Cenozoic tectonics of Asia: Effects of a continental collision.* Science *189(4201):419–426, Copyright 1975, AAAS.)*

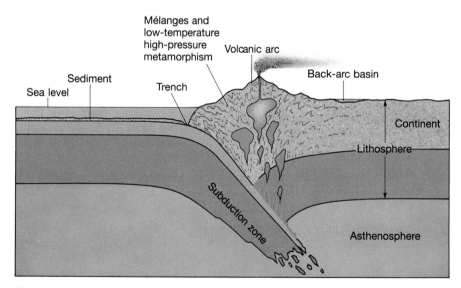

**Figure 6–33** Convergence of the continental part of a plate with the oceanic part of another plate. The leading edge of the continental plate is crumpled, whereas the oceanic plate buckles downward, creating a trench offshore. The situation generalized here is similar to that off the west coast of South America where the Nazca plate plunges beneath the South American plate.

dynamic activity. At the same time, sediments and submarine volcanic rocks along the subduction zone are squeezed, sheared, and shoved into a gigantic, chaotic medley of complexly disturbed rocks called a **mélange** (see Fig. 12–20). Within the mélange one finds a distinctive assemblage of deep-sea sediments containing microfossils, submarine lavas, and serpentinized peridotite possibly derived from the upper mantle. Altogether, these rocks constitute an ophiolite suite (Fig. 6–34). The **ophiolite suites** are splinters of the oceanic plate that were scraped off the upper part of the descending plate and inserted into the crushed forward edge of the continent. Thus, ophiolites mark the zone of contact between colliding continental and oceanic plates. Another

clue to the presence of such zones is a distinctive kind of metamorphic rock containing blue amphiboles. These rocks are called **blue schists.** They form at high pressures but relatively low temperatures. This rather unusual combination of conditions is characteristic of subduction zones where the relatively cool oceanic plate plunges rapidly into deep zones of high pressure.

## What Drives It All?

If we accept the fact that plates of lithosphere move across the surface of the globe, the next question concerns how they are moved. The propelling mechanism is thought by some scientists to consist of large thermal convection cells produced as mantle material is heated from below, expands, becomes less dense, and slowly rises. On encountering the lithosphere, the flow would diverge laterally, dragging along the overlying slab of lithosphere. Mantle material upwelling along the line of separating currents would join the trailing edges on either side. As the current moved horizontally, it would steadily lose its heat and become denser. Ultimately, it would encounter a similar current moving in an opposing direction, and both viscous streams would descend to be reheated and shunted toward a region of upwelling. Above the descending flow one would find subduction zones and deep sea trenches, whereas divergent zones and spreading centers would mark the locations of the ascending flow.

Several models of convection cell dimensions and distributions have been proposed (Fig. 6–35). One of these confines all convection movement to the relatively thin asthenosphere (Fig. 6–35A), with all of the remaining mantle serving as the heat source. Another suggestion is that convection cells span the entire mantle, from the top of the core to the base of the lithosphere (Fig. 6–35B). Yet a third model favors a stacked convection system with concentric layers of convection cells. Each of these concepts has components that are consistent with existing data, but it is presently too early to state which will ultimately provide the best interpretation. It is also possible that the configuration of convection currents has changed throughout the course of earth history. Initially, there may have been only one large convecting system that swept together scumlike slabs of continental crust into a single land mass. Later

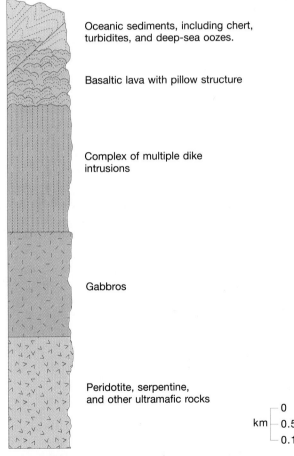

Oceanic sediments, including chert, turbidites, and deep-sea oozes.

Basaltic lava with pillow structure

Complex of multiple dike intrusions

Gabbros

Peridotite, serpentine, and other ultramafic rocks

km — 0
— 0.5
— 0.1

**Figure 6–34** Idealized section of an ophiolite suite. Ophiolites are thought to be splinters of the ocean floor squeezed into the continental margin during plate convergence.

A

B

C

D

**Figure 6–35** Models proposed for the plate tectonics driving force. (A) Convection model in which the convection cells are confined to the upper part of the mantle. (B) The convection cells are large and circulate through the entire mantle. (C) The thermal plume model. (D) Sometimes called the push–pull model, plates are pushed laterally at spreading centers and pulled by the plunging, cool, dense leading segment of the plate. A fifth model might be imagined, in which concentric layers of stacked convection cells exist, with the uppermost cells providing the force to move tectonic plates.

systems may have been characterized by larger numbers of convection cells located closer to the base of the lithosphere. A concept such as this may account for an apparent increase in continental fragmentations in the Mesozoic compared with earlier eras. This idea was originally advanced by the geophysicist S. K. Runkorn, who postulated that growth of the core of the earth and consequent shrinkage of the mantle may have brought on the changes in convection cell characteristics. Later information, however, suggests that after the core formed early in the earth's history, it barely increased in size. This does not, of course, preclude changes in the convection cell patterns for other reasons.

As a mechanism for moving plates of lithosphere, thermal convection may be the best hypothesis thus far advanced. This does not mean that this mechanism should be regarded as an established fact. As yet there has not been an entirely satisfactory way to test the concept, although such currents may be physically plausible, and there is some tenuous evidence that they exist. For example, heat flows from the earth's interior at a greater rate along midoceanic ridges than from adjacent abyssal plains. Yet analyses of hundreds of measurements by geophysicists at Columbia University suggest that although heat flow along the mid-Atlantic ridge is 20 percent greater than on the adjacent floors of the ocean basins, it is still not great enough to move the sea floor at the rates suggested by paleomagnetic studies. Another problem relates to the layering in the upper mantle that has been detected in seismologic studies. The mixing that accompanies convectional overturn would seem to preclude such layering.

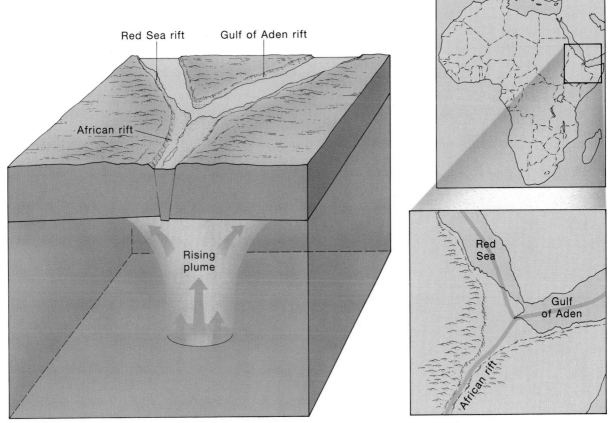

**Figure 6–36** Rising plumes of hot mantle may cause severe rifts, often forming 120° angles with one another. An example is the Afar triangle, shown at the south end of the Red Sea on the small map of Africa.

The search for a mechanism to drive lithospheric plates has produced another convection-related model. In this *thermal plume model* (Fig. 6–35C), mantle material does not circulate in great rolls but rather rises from near the core–mantle boundary in a manner suggestive of the shape and motion of a thundercloud. Proponents of this model suggest there may be as many as 20 of these thermal plumes, each a few hundred kilometers in diameter. According to the model, when a plume nears the lithosphere, it spreads laterally, doming surface zones of the earth and moving them along in the directions of radial flow. The center of the Afar triangle in Ethiopia (Fig. 6–36) has been suggested as the site of a plume that flowed upward and outward, carrying the Arabian, African, and Somali plates away from the center of the triangle. Geophysicists have suggested that other such triple rift systems in the world have had a similar origin.

Faced with the uncertainties in all existing theories for plate-moving forces, one should keep an open mind about other, perhaps less popular, hypotheses. Among other mechanisms proposed are "slab-pull" and "ridge-push" (Fig. 6–35D). Slab-pull is thought to operate at the subduction zone, where the subducting oceanic plate, being colder and denser than the surrounding mantle, sinks actively through the less dense mantle, "pulling" the rest of the slab along as it does so. The ridge-push mechanism, which may operate independently of or along with slab-pull, results from the fact that the spreading centers stand high on the ocean floor and have low-density roots. This causes them to spread out on either side of the mid-oceanic ridge and transmit this "push" to the tectonic plate. Other hypotheses suggest rafting of plates in response to stresses induced by differential rotation between the core and mantle, which might then be transmitted to the crust. In the years ahead, studies of rock behavior under high temperature and pressure, geophysical probings of

the mantle, and careful field studies will reveal which of the proposed mechanisms can be most confidently accepted.

## Tests of Plate Tectonics

We have examined the various lines of evidence supporting Wegener's notion of continental drift. Nearly

**Figure 6–37**   Magnetic field produced by the rocks on the floor of the northeast Pacific Ocean. Note the symmetry with respect to the ridges (the red areas). The ages of the rocks increase away from the ridges. The color bands are magnetically positive. *(After Raff, A. D. and Mason, R. G., Magnetic survey of the West Coast of North America, 40° N–52° N. latitude. Bull. Geol. Soc. Am., 72:1267–1270, 1961.)*

all of these clues also support the newer concepts embodied in the theory of plate tectonics. The earlier clues were based mostly on evidence found on land. The new theory, with its keystone concept of sea floor spreading, was developed from evidence gleaned from the sea floor.

Slightly before the time when geophysicists such as Harry Hess of Princeton University began formulating ideas of sea floor spreading, other scientists were puzzling over findings related to paleomagnetism. The new data were obtained from sensitive magnetometers that were being carried back and forth across the oceans by research vessels. These instruments were able to detect not only the earth's main geomagnetic field, but also local magnetic disturbances or magnetic anomalies frozen into the rocks along the sea floor. Maps were produced that exhibited linear anomaly bands of high and low field magnetic intensities parallel to the west coast of North America (Fig. 6–37). Surveys of magnetic field strength on traverses across the mid-Atlantic ridge revealed a similar pattern of symmetrically distributed belts of anomalies (Fig. 6–38). Directly over the ridge, the earth's magnetic field was 1 percent stronger than expected. Adjacent to this zone the field was somewhat weaker than would have been

predicted, and the changes from weaker to stronger and back again often occurred in distances of only a few score miles.

In 1963, F. J. Vine, a research student at Cambridge University, and a senior colleague, Drummond Matthews, suggested that these variations in intensity were caused by reversals in the polarity of the earth's magnetic field. The magnetometers towed behind the research vessels provided measurements that were the sum of the earth's present magnetic field strength and the paleomagnetism frozen in the crustal rocks of the ocean floor. If the paleomagnetic polarity was opposite in sign to that of the earth's present magnetic field, the sum would be less than the present magnetic field strength. This would indicate that the crust over which the ship was passing had reversed paleomagnetic polarity. Conversely, where the paleomagnetic polarity of the sea floor basalts was the same as that of the present magnetic field, the sum would be greater, and a normal polarity would be indicated.

Since 1963, geophysicists have learned that these irregularly occurring reversals have not been at all infrequent. During the past 70 million years, the earth's magnetic field has seen many episodes when the polarity was opposite to that of today (Fig. 6–39).

**Figure 6–38** The magnetic field produced by the ocean floor rocks over the Reykjanes ridge near Iceland. The outlined areas represent predominantly normal magnetization. The ages of the rocks increase away from the ridge. *(After Heirtzler, J. R., et al., Magnetic anomalies over the Reykjanes ridge,* Deep Sea Res., *13:427–443, 1966; and Vine, F. J., Magnetic anomalies associated with mid-ocean ridges, in Phinney, R. A. (ed.),* The History of the Earth's Crust, A Symposium, *Princeton, NJ: Princeton Univ. Press, 1968.)*

62°N

60°N

Ridge axis

Reykjanes Ridge Magnetics

30°W     25°W

| RADIOMETRIC TIME SCALE MY | GEOMAGNETIC POLARITY SCALE | EPOCH |
|---|---|---|
| | | PLEISTOCENE |
| | | PLIOCENE |
| 5 | | |
| 10 | | MIOCENE |
| 15 | | |
| 20 | | |
| 25 | | |
| 30 | | OLIGOCENE |
| 35 | | |
| 40 | | EOCENE |
| 45 | | |
| 50 | | |
| 55 | | PALEOCENE |
| 60 | | |
| 65 | | |
| 70 | | CRETACEOUS |

**Figure 6–39** Reversals of the earth's magnetic field during the past 70 million years. The field was "normal," as today, during intervals shown in black. *(Modified from Heirtzler, J. R., et al. J. Geophys. Res. 73:2119–2136, 1968. Copyright, The American Geophysical Union.)*

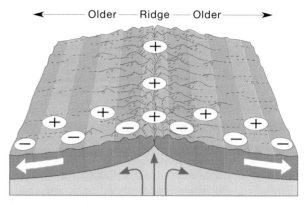

Older — Ridge — Older

**Figure 6–40** The normal (+) and reversed (−) magnetizations of the sea floor. Note the symmetry of the magnetizations with respect to the ridge. *(From McCormick, J. M., and Thiruvathukal, J. V., Elements of Oceanography, 2nd ed. Philadelphia: Saunders College Publishing, 1981.)*

These changes in polarity were incorporated into the remanent magnetism of the lavas extruded at the mid-oceanic ridges. The lava would acquire the magnetic polarity present at the time of extrusion and then would move out laterally, as has been previously described. If during that time the earth's polarity was as it is today, the stripe is said to represent "normal" polarity. If the earth's normal polarity reversed, the band of extruded lavas that followed behind the previous band would acquire "reverse" polarity as it cooled past the Curie temperature. As the process repeated itself through time, the result would be symmetric, mirror image patterns of normal and reverse bands on either side of spreading centers such as the mid-oceanic ridges (Fig. 6–40).

The stripes discovered by Vine and Matthews showed that Harry Hess was correct when he suggested in the early 1960s that the sea floor moves. At that time, however, there seemed to be no feasible way to determine the age of large areas of the ocean floor and thereby ascertain the rate of sea floor movement. Cores of the basaltic sea floor were not suitable for radiometric dating because basalt is altered, when hot, by contact with sea water. By lucky coincidence, a method for determining the age of the stripes was provided through the work of A. Cox, R. R. Doell, and G. B. Dalrymple. These geophysicists were investigating the magnetic properties of lava flows on the continents. They were able to accurately identify periodic reversals of the earth's magnetic field imprinted in superimposed layers of basalt (Fig. 6–41). Many of the layers in these superpositional sequences could be dated by radiometric methods. With such data, the time sequence of magnetic reversals for the past 5 million years was determined. The final step was to correlate these dates obtained from

land basalts to the rocks of the sea floor. Confirmation for the ages of some of the oceanic basalts was obtained from the study of fossils in overlying sediments.

With knowledge of the age of particular normal or reverse magnetic stripes, it was now possible to calculate rates of sea floor spreading and sometimes to reconstruct the former positions of continents. Figure 6–42 illustrates how this can be done. For example, if one wishes to know the distance between the eastern coast of the United States and the northeastern coast of Africa about 81 million years ago, one brings together the traces of the two 81-million-year-old magnetic stripes, being careful to move the sea floor parallel to the transform faults, which indicate the direction of movement.

The velocity of plate movement is not uniform around the world. Plates that include large continents tend to move slowly. Their velocity relative to the underlying mantle rarely exceeds 2 cm per year. Plates that are largely devoid of large continents have average velocities of between 6 and 9 cm per year. Figure 6–43 provides a summary of the rates and directions of sea floor spreading as determined by analyses of displacements along transform faults, magnetic anomalies, and other geophysical data.

### Evidence of Sea Floor Spreading from Oceanic Sediment

If newly formed crust joined the ocean floor at spreading centers and then moved outward to either side, then the sediments, dated by the fossil planktonic organisms they contain, could be no older than the surface on which they came to rest. Near a mid-oceanic ridge, the sediments directly over basalt should be relatively young. Samples of the first sedimentary layer above the basalt but progressively farther away from the spreading center should be progressively older (Fig. 6–44).

In a succession of cruises that began in 1969, the American drilling ship *Glomar Challenger* had collected ample evidence to prove that the previously stated conclusions on the age distribution of sediments are correct. The studies carried out by scientists associated with the drilling project confirmed earlier assumptions that there were no sediments older than about 200 million years on the sea floor, that the sediments on the sea floor were relatively

**Figure 6–41** The imprint of normal light tan and reversed dark tan magnetic polarity in a composite section of layered basalts from many continental localities. The magnetic reversal time scale is shown in the center column, and at the left are the *magnetic epochs* named after famous investigators of magnetic effects. The magnetic epochs reflect a predominance of one kind of polarity. Within the larger magnetic epochs are *magnetic events* of shorter duration. *(Modified from Cox, A., Geomagnetic reversals, Science, 163:237–245, 1969.)*

A

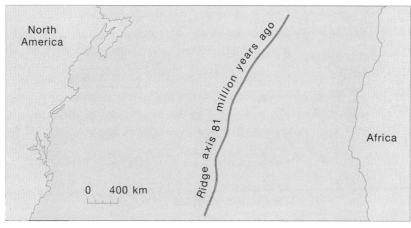

B

**Figure 6–42**  A method for determining the paleogeographic relations of continents. (A) The present North Atlantic with ages plotted for some of the magnetic stripes. To see the location of North America relative to Africa, say, 81 million years ago, the 81-million-year-old bands are brought together. (B) The result is the view of the much narrower North Atlantic of 81 million years ago. *(Modified from Pitman III, W. C., and Talwani, M., Geol. Soc. Am. Bull. 83:619–644, 1972.)*

thin, and that they became thinnest closer to the mid-oceanic ridges. Spreading provided a logical reason for these observations. A given area of oceanic plate surface was simply not in existence long enough to accumulate a thick section of sediments over a long span of geologic time. The sea floor clays and oozes were conveyed to subduction zones, dragged back down into the mantle, and (as suggested by rock compositional studies) resorbed. Thus, the entire world ocean is virtually swept free of deep sea sediments every 200 to 300 million years.

As an example, sea floor drilling and sampling has thus far been unable to locate sediments in the South Atlantic any older than 150 million years. This knowledge is related to the paleomagnetic calculations that suggests that South America and Africa have been moving apart from one another at a rate of about 4 cm per year. If this rate has been fairly constant, then extrapolating back in time, one may estimate that the two continents should have been in contact about 150 million years ago during the Jurassic. After that time, the two fragments (Africa and

**Figure 6–43** Rates and directions of sea floor spreading. *(Results from Minster, J. B., and Jordon, T. H. J. Geophys. Res. 83:5331–5354, 1978.)*

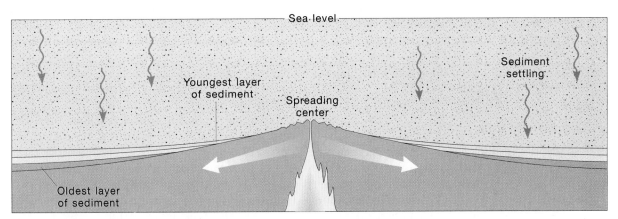

**Figure 6–44** As a result of sea floor spreading, sediments near the mid-oceanic ridges are youngest, and sediments *directly above the basaltic crust* but progressively farther away from the ridge are sequentially older.

---

**COMMENTARY**

## Rates of Plate Movement

On examination of the rates at which tectonic plates move away from midoceanic ridges at various places around the earth, it is immediately apparent that the rate of movement is greater at some places than at others. One reason for this relates to movement on a spherical surface. If the earth were flat, as some viewed it before the voyages of Columbus, then all locations along a similarly flat plate would be moving away from a midoceanic ridge at about the same velocity. On a sphere, however, the plates are curved. They begin to separate at a location termed a pole of rotation and form an ever-widening split. The effect is somewhat like pulling apart the outer dry layer of an onion. The split in the skin would be widest where it was being pulled apart, and the gap would narrow to a point corresponding to the pole of rotation. On the earth, plate movement away from the midoceanic ridge near the pole of rotation would be slower because of the lesser distance traversed over a given time interval. Conversely, far from the pole of rotation, plates diverge across a greater distance for the same time interval and hence have greater velocity. One can observe this effect in the North Atlantic, where spreading rates increase from about .75 cm per year near Iceland to over six times that amount near the equator. A corollary of this effect is that the width of new oceanic crust also increases with distance from the pole of rotation.

In addition to changes in plate velocity that are a consequence of the earth's spherical shape, it appears that large expanses of continental crust incorporated into a plate can decrease its velocity. The Pacific and Nazca Plates do not carry a heavy load of crust. These plates have higher velocities than the American and Eurasian plates which are burdened with large continents.

---

South America) moved apart as oceanic basalts filled the gap left behind. In this way, only sediments younger than the date of fragmentation would be present, since there was not an oceanic basin of deposition in existence prior to that time. Similar reconstructions have been made from sea floor samples and cores elsewhere around the world.

### Measurements from Space

The space program has provided yet another method for measuring the movement of tectonic plates. This new method uses laser devices, which emit an intense beam of light of definite wavelength. The beam is capable of traveling across an enormous distance without becoming dispersed. During the Apollo missions, astronauts placed three clusters of special reflectors on the moon. Laser beams from earth are "fired" toward the moon, and by recording the time required for the signal to travel to the moon and return to earth, one can determine the distance between the laser device and the lunar reflector. The method is accurate to within only 3 cm. Distances from the moon to stations on two moving tectonic plates will, of course, change over a period of time, and such measurements can be used to calculate how fast continents are moving toward or away from one another.

Beaming lasers to artificial satellites from two widely spaced earth stations can also be used to find the distance between two points on earth. The Lageos satellite, which circles the earth at an altitude of 5800 km, is currently being used to reflect laser beams back to their source on earth. By accurately measuring the time required for a laser pulse to travel to the satellite and return, the position of the earthbound station is ascertained. A similar measurement made from a second station (perhaps on a different continent) fixes the distance between the two stations. Similar measurements made after a given lapse of time are used to determine if the distance between the two stations has increased or decreased and at what rate.

A rather different technique that shows great promise is called Very Long Baseline Interferometry. In this method, several stations at widely spaced locations on earth monitor the radio signals from distant quasars (very large redshift, intragalactic, luminous objects). The differences in the arrival times of these signals determine the distance between the earth receiving stations, and changes in these distances are then noted for given increments of time, as in the laser methods. Calculations based on monitoring stations in the United States and Europe indicate that the North American and Eurasian tectonic plates are moving apart at a rate of about 1.9 cm per year. This figure compares very well with values based on the spacing of magnetic stripes on the floor of the north Atlantic Ocean.

## Seismologic Evidence for Sea Floor Spreading

The evidence for the actual movement of lithospheric plates appears to be fairly direct and susceptible to several methods of testing. It is more difficult to prove that plates of the lithosphere are dragged back down into the mantle and resorbed. The best indications that subduction actually does occur come from the study of earthquakes generated along the seismically active colliding edges of plates.

Well known among these studies are the impressive investigations of Hugo Benioff, a seismologist and designer of improved seismographs. Benioff and subsequent investigators compiled records of earthquake foci that occurred along presumed plate boundaries. The charting of the data showed that the deeper earthquake foci occur along a narrow zone that was tilted at an angle of 45° under the adjacent island arc or continent (Fig. 6–45). This inclined seismic shear zone was traced to depths as great as 700 km (435 miles). It was named the **Benioff seismic zone** after its discoverer. These inclined earthquake zones are believed to define the positions of the subducted plates where they plunge into the mantle beneath the overriding plate. The earthquakes are the result of fracturing and subsequent rupture as the cool, brittle lithospheric plate descends into the hot mantle.

**Figure 6–45** Vertical cross-section showing spatial distribution of earthquakes along a line perpendicular to the Tonga trench and volcanic island arc. The Tonga trench (a on inset map) lies just to the north of the Kermodec trench (b). (Simplified from Isacks, B., Oliver, J., and Sykes, R. J. Geophys. Res. 73:5855, 1968.)

In addition to the deep and intermediate focus earthquakes along the Benioff zone, there are intense and often destructive earthquakes at shallow depths where the edges of the two rigid plates press against one another. The 1964 Alaskan earthquake originated at this shallow level. Such quakes are thought to occur along the shear plane between the subducting ocean lithosphere and the continental lithosphere. Some may result from normal faulting as portions of the lithosphere are arched by the collision.

## Gravity Plays a Role

Gravity measured over deep sea trenches is characteristically lower than that measured in areas adjacent to the trenches. Geophysicists refer to such phenomena as negative gravity anomalies. A **gravity anomaly** is the difference between the observed value of gravity at any point on the earth and the computed theoretic value. **Negative gravity anoma-**

**lies** occur where there is an excess of low-density rock beneath the surface. The very strong negative gravity anomalies extending along the margins of the deep sea trenches can mean only that such belts are underlain by rocks much lighter than those at depth on either side. Because the zone of lower gravity values is narrow in most places, it is believed to mark trends where lighter rocks dip steeply into the denser mantle. Geophysicists assume that the less dense rocks of the negative zone must be held down by some force to prevent them from floating upward to a level appropriate to their density. That force might be provided by a descending convection current, or the plate may be pushed by rising magma along mid-oceanic ridges until it encounters an opposing plate and plunges into the asthenosphere.

## Hot Spots

Anyone examining the superb topographic maps (see Fig. 1–5) of the sea floor that have been produced over the past decade cannot help but notice the chains of volcanic islands and seamounts (submerged volcanoes) that rise in alignment on the sea floor. Because these volcanoes occur at great distances from plate margins, they must have originated differently from the volcanoes associated with mid-oceanic ridges or deep sea trenches. Their striking alignment has been explained as being a consequence of sea floor spreading. According to this notion, intraoceanic volcanoes develop over a "hot spot" in the asthenosphere. The hot spot is fixed in location and may be a manifestation of one of the deep plumes of upwelling mantle rock described earlier. Lava from the plume may work its way to the surface to erupt as a volcano on the sea floor. As the sea floor moves (at rates as high as 10 cm per year in the Pacific), volcanoes form over the hot spot and expire as they are conveyed away. New volcanoes form at the original location. The process may be repeated indefinitely, resulting in a great linear succession of volcanoes that may extend for thousands of kilometers in the direction that the sea floor has

**Figure 6–46** Bend in trend of the Hawaiian Island–Emperor Seamount chain was probably caused by change in direction of movement of the Pacific tectonic plate. *(From Watkins, J. S., Bottino, M. L., and Morisawa, M. Our Geological Environment Philadelphia: W. B. Saunders Company, 1975.)*

**Figure 6–47** Locations of some of the major hot spots around the earth. *(Courtesy of Tom Crough, Department of Geologic and Geophysical Sciences, Princeton University.)*

moved. The Hawaiian volcanic chain is believed to have formed in this manner from a single source of lava over which the Pacific plate has passed on a northwesterly course. In support of this concept are radioactive dates obtained from rocks of volcanoes that clearly indicate that those farthest from the source are the oldest.

At the western end of the Hawaiian Islands is a string of submerged peaks called the Emperor seamounts (Fig. 6–46). These submerged volcanoes trend in a more northerly direction than the Hawaiian Islands. However, both the seamounts and the islands are part of a single chain that has been bent as a result of a change in the direction of plate movement. Such an interpretation is supported by age determinations. The oldest of the Hawaiian Islands (near the bend) was formed about 40 million years ago. The seamounts continue the age sequence to-

ward the end of the chain, where the peaks are about 80 million years old. Thus, 40 million years ago the Pacific plate changed course and put a kink in the Hawaiian chain.

The ocean around Hawaii is not the only part of the globe having hot spots. As indicated in Figure 6–47, hot spots are widely dispersed and occur beneath both continental and oceanic crust. Yellowstone National Park (Fig. 6–48) is over a hot spot, which, unlike the Hawaiian Islands, is in the interior of a continent.

## Lost Continents and Alien Terranes

We are accustomed to thinking of continental crust in terms of large land masses such as North America or Eurasia. There are, however, many relatively small patches of continental crust scattered about on the

**Figure 6–48** Mammoth Geyser, Yellowstone Park, Wyoming. Yellowstone is over a hot spot. Surface waters percolating through a system of deep fractures reach the hot rocks below and erupt in columns of hot water and steam.

lithosphere. As long ago as 1915, Alfred Wegener described the Seychelles Bank (Fig. 6–49) in the Indian Ocean as a small continental fragment that had broken away from Africa. The higher parts of the Seychelles Bank project above sea level as islands, but many other such small patches of continental

**Figure 6–49** Location map of the Seychelles Bank.

crust are totally submerged. Geologists use the term **microcontinents** for these bits of continental crust that are surrounded by oceanic crust. They are recognized by their granitic composition, by the velocity with which compressional seismic waves traverse them (6.0 to 6.4 km/second), by their general elevation above the oceanic crust, and by their comparatively quiet seismic nature.

It is apparent that microcontinents are small pieces of larger continents that have experienced fragmentation. As these smaller pieces of continental crust are moved along by sea floor spreading, they may ultimately converge upon the subduction zone at the margin of a large continent. Because they are composed of relatively low-density rock and hence are buoyant, they are a difficult bite for the subduction zone to swallow. Their buoyancy prevents their being carried down into the mantle and assimilated. Indeed the small patch of crust may become incorporated into the crumpled margin of the larger continent as an exotic block, or so-called **suspect terrane.**

It is interesting that geologists found evidence of microcontinents long before the present theory for their origin was formulated. While mapping in Precambrian shields, they came across areas that were incongruous in structure, age, fossil content, lithology, and paleomagnetic orientation when compared to the surrounding geology. It was as if these areas

COMMENTARY COMMENTARY COMMENTARY COMMENTARY COMMENTARY
COMMENTARY COMMENTARY COMMENTARY COMMENTARY COMMENTARY COMMEN
COMMENTARY COMMENTARY COMMENTARY COMMENTARY COMMENTARY
COMMENTARY COMMENTARY COMMENTARY COMMENTARY COMMENTARY COMMEN

## COMMENTARY

### *The Missing Hawaiian Volcano*

In this chapter we have described how the Hawaiian chain of volcanic islands originated by passage of the Pacific Plate over a relatively stable hot spot. In addition to the general alignment of the Hawaiian Islands and Emperor Seamounts, however, one can also observe that the volcanoes located along the southern segment of the chain occur sequentially along two lines, as shown in the accompanying figure. Along each line, the volcanoes are fairly evenly spaced, except for a large gap in the western line between the volcanoes Kahoolawe and Hualalai. Geologists have been concerned with this extra space that seemed to disrupt an otherwise remarkably uniform pattern. Those concerns, however, were put to rest recently when new bathymetric data and geochemical analyses of lava dredged from the sea floor demonstrated the existence of a submerged volcano within what had been assumed was a gap in the chain. The missing volcano was about 70 km long by about 30 km wide, and it had a height of 4 km above the abyssal sea floor. Geologists named the volcano Mahukona after the closest settlement to the volcano on the island of Hawaii.

The question of the gap in the southern Hawaiian chain had been solved, but two questions remained. Why were the volcanoes in a paired sequence, and why did Mahukona become extinct before building itself upward to form an island? In laboratory experiments that model ascending hot spot plumes, the plumes bifurcate as they rise, forming two distinct branches. This may have occurred in the ascending plume that generated the volcanoes of the southern Hawaiian chain, so as to produce two

Southern Hawaiian Islands with paired sequence of volcanoes.

parallel lines of volcanoes about 35 km apart. One of these volcanoes, Mahukona, did not reach sea level possibly because the branch feeding the western line was smaller or became depleted, causing Mahukona to become dormant before it reached sea level. Other volcanoes on the western line, such as Lanai and Koolau, although they reached sea level, tend to be smaller than volcanoes in the eastern line. These are simple models to explain the characteristics of the Hawaiian chains, and they may be modified with time.

**Reference**
Garcia, M.O., Kurz, M.D., and Muenow, D.W. 1990. *Geology* 18:1111–1114.

were small, self-contained, isolated geologic provinces. Often their boundaries were marked by major faults. Geologists designated these areas as **allochthonous terranes** to indicate that they had not originated in the places where they now rested. Allochthonous terranes have been recognized on every major land mass, with well-studied examples in the northeastern Soviet Union, in the Appalachians, and in many parts of western North America (Fig. 6–50).

In particular, Alaska appears to be largely constructed of allochthonous terranes.

If splinters of continents can be transported on the spreading sea floor, so can pieces of oceanic crust. Particularly in the Cordilleran mountain belt of North America, one finds allochthonous terranes that were apparently microplates of ocean crust containing volcanoes, seamounts, segments of island arcs, and other features of the ocean floor. All of

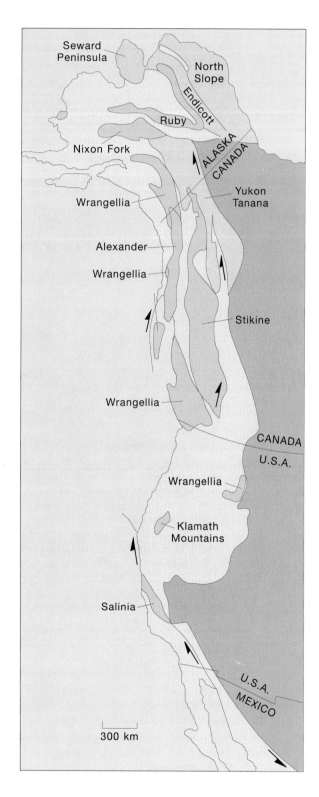

**Figure 6–50** The larger allochthonous continental terranes in western North America that contain Paleozoic or older rocks. At some time prior to reaching their present locations, these terranes were continental fragments embedded in oceanic crust or microcontinents. Those colored green probably originated as parts of continents other than North America, whereas the pink blocks are possibly displaced parts of the North American continent. *(From Ben-Avraham, Z., Am. Sci. 69:298, 1981.)*

these features were carried to the western margin of North America like passengers on a huge conveyor belt. As the plate that bore them plunged downward at the subduction zone located along the continental margin, volcanoes, seamounts, and the other features were scraped off the descending plate and plastered onto the continental margin. The result was a vast collage of accreted oceanic microplates interspersed with similarly transported microcontinents. It appears that about 50 such allochthonous or exotic terranes exist in the Cordillera, all lying west of the edge of the continent as it existed about 200 million years ago. Clearly this process of successive additions of oceanic and continental microplates can significantly increase the rate at which a continent is able to grow. It is also apparent that active continental margins such as those bordering the Pacific Ocean are likely to grow faster than passive margins because of microplate accretion.

Unraveling the orogenic history of mountain ranges containing multiple allochthonous terranes is an enormously complex task requiring the cooperation of geologists, geophysicists, and paleontologists. Such a collaboration can often yield dramatic results. Recently, paleontologists and geologists working in the Wallowa Mountains of Oregon discovered a massive coral reef of Triassic age. The reef, however, clearly did not belong where it was found. It rested on volcanic rocks whose paleomagnetic orientation indicated that they had solidified from lavas at a latitude considerably closer to the equator. Further investigation indicated that the fossil organisms in the reef were identical to those found in Triassic strata of the Austrian and German Alps. These strata were deposited prior to Alpine mountain building in a seaway called the Tethys that extended from the present east coast of Japan into the Mediterranean region. The similarity between the fossils of Oregon and those of the Alps strongly suggests that the reefs now in Oregon actually grew around the margins of volcanic islands in the Tethys Sea and, in the subsequent 220 million years or so, were transported thousands of kilometers eastward until they collided with the North American continent. By that time, of course, the reef organisms had died, but their fossilized skeletons remained as the reef was accreted onto the western margin of North America.

Another somewhat implausible explanation for the occurrence in Oregon of a Triassic fauna from the Tethys Sea would be the migration of the free-swimming larval stages of reef organisms from island to island across the Pacific until they reached North America. This would require a remarkably continuous string of islands, for larval stages are generally relatively short-lived.

## Life on a Traveling Crust

Of all the tectonic features discussed on previous pages, continents are the easiest to locate in the past. Their shifting positions and changing configurations had far-reaching effects upon the earth's inhabitants. The series of maps depicted in Figure 6–12 shows the continents in various stages of fragmentation since the breakup of Pangea, the supercontinent of Permian–Triassic time that Wegener championed. (Geologists now know that this land mass was preceded by an earlier supercontinent termed Proto-Pangea that fragmented and was reassembled.)

The American geologists J. W. Valentine and E. M. Moores have devised an interesting model that provides a means by which one can more easily imagine the effects on life of such continental fragmentations and reassemblies. Imagine the earth as having a single supercontinent reminiscent of Pangea centered on the South Pole (rather like Fig. 6–51). In this configuration, the entire coastline of the land mass would be at about the same latitude and thus would have rather similar climatic characteristics. The bottom-dwelling (benthic) shallow-water marine invertebrates that are so frequent in the fossil record would have had ease of migration around the entire supercontinent, because there would have been few climatic barriers. The result would have

been a single so-called **shelf province**. Within that province there might have been differences in the topography or composition of the sea floor that might have influenced the kinds of biologic communities that would have lived in particular locations. However, because of the general uniformity of conditions and ease of migration and interaction, global biologic diversity would probably have been relatively low.

A single supercontinent would also provide terrestrial vertebrates with relative ease of wide-spread migration and interaction. This is likely to have caused the development of cosmopolitan land faunas with fewer families and orders of animals than would have been the case if there had been many smaller isolated land areas. Paleomagnetic and paleontologic evidence indicates that although Pangea had begun to divide in the Triassic Era, the rifts did not become effective barriers to dispersal of terrestrial animals until very late in the Cretaceous and early Cenozoic (Fig. 6–51G and H). Thus, during most of the 200 million years of the Mesozoic "Age of Reptiles," dispersal was a relatively simple matter. In contrast, Cenozoic separations became increasingly more extreme, and many terrestrial segments were soon in positions of relative isolation. During this "Age of Mammals," the earlier supercontinent divided into eight smaller units: North America, Greenland, Eurasia, Africa, India, Australia, South America, and Antarctica. These smaller continents were still further divided by inland seas. Mammals evolving on each continent underwent many separate and different evolutionary radiations and developed distinctive populations. The result was a greater diversity of land vertebrates than the earth had ever hosted. The Age of Mammals gave rise to over 30 taxonomic orders of mammals in only 65 million years. The previous Age of Reptiles, which lasted over three times as long, saw the evolution of only 20 reptilian orders.

The Cenozoic world is somewhat analogous to the continential configuration depicted in Figure 6–51H. The continents in the model are exceptionally fragmented and widely dispersed. The benthic marine invertebrates that inhabit continental shelves of continents near the equator would probably develop a very high diversity of species because of the general environmental stability there. Continents in higher latitudes would have larger numbers of individuals but relatively less diversity because of the greater

**Figure 6–51** Continental fragmentation and the diversity of life show a clear relationship from Cambrian to recent time. *(After Valentine, J. W., and Moores, E. M., J. Geol. 80(2):167–184, 1972. Copyright © 1972 by the University of Chicago.)*

rigor of the poleward environments. In terms of the entire earth, there would be a greater variety of habitats, a larger number of individual centers of biologic radiations, and a corresponding greater increase in global diversity. Today we live in a fragmented world, and life is indeed diverse.

Professors Valentine and Moores are among several earth scientists who have suggested that unusual episodes of extinctions as well as changes in diversity can be related to plate tectonics. For example, when continents come together to form a supercontinent (Fig. 6–51E), there would be a trend toward extremes of climate, elimination of habitats, regression

of inland seas, and loss of barriers to migration. Scores of species would probably perish because of their inability to adjust to these changes, and the fossil record would suggest that an episode of mass extinction had occurred. The survivors and their progeny would probably develop into a rather cosmopolitan world fauna.

We have noted that fragmentation provides multiple centers of biologic diversification. This may well be the reason for the increased variety of life in the geologic periods following the breakup of Proto-Pangea and Pangea. The fragmentation episodes may have been accompanied by worldwide changes in sea level (eustacy) resulting from the growth of mid-oceanic ridges and the displacement of water that ensued. This is not a wild guess, for it has been calculated that if such ridges today were to vanish, sea level would be lowered approximately 650 meters. During these times of rising sea level, marine waters would spill over onto the low-lying regions of continents, creating vast inland seas ideal for the proliferation of marine invertebrates. These transgressing inland seas would also have a moderating effect on temperature, humidity, and seasonality. The result would be the increase in the number of families of animals depicted in Figure 6–51 for those segments of geologic time in which continents were many and scattered.

The study of the relationship between organisms and their environments is enormously complex. Small changes in faunal elements at lower levels in food pyramids profoundly affect higher levels, often on a global scale. To understand the many variables involved is difficult. To comprehend them from clues in rocks hundreds of millions of years old is far more complex. However, as paleontologists continue to reexamine the data in light of the new theories, they find that the effects of plate tectonics on all the earth's former inhabitants have been profound.

## SUMMARY

We have seen how the many notions of drifting continents and sea floor spreading have been combined into the modern theory of plate tectonics. As with any new theory, there is much that remains to be tested and explained. Nevertheless, the central idea is convincing and has great appeal because of the way it serves to unify and relate geologic phenomena. What is believed today will surely be modified in the years to come. This summary therefore represents only a brief review of progress to date.

According to the views embodied in plate tectonics, the crust of the earth and part of the upper mantle compose a brittle shell called the lithosphere. The lithosphere is broken into a number of plates that override a soft plastic layer in the mantle called the asthenosphere. Heat from within the earth may create convection currents in the mantle, and such currents may provide a propelling force to move the plates. However, sliding of plates in response to gravity may also prove to be a mechanism for their movement.

Materials for the lithospheric plates originate along fracture zones typified by mid-oceanic ridges. Along these tensional features, basaltic lavas rise and become incorporated into the trailing edges of the plates, simultaneously taking on the magnetism of the earth's field as they crystallize. In recent geologic time, this process has provided a record of geomagnetic reversals that in turn permit geophysicists to determine the rates of sea floor spreading.

Provided they do not have continents on opposing edges, when plates collide, one of them may slide beneath the other and enter the mantle at a steep angle, to be remelted at depth. The zone of collision between plates may be marked by systems of volcanoes, deep sea trenches, and great mountain ranges. The continental crustal segments of the lithosphere ride passively on the plates. They have a lower density than does oceanic crust and do not sink into subduction zones. This explains why continents are older than the ocean floors, and why, when continents collide, they produce mountain ranges rather than trenches. Where continents straddle zones of divergent movements in the asthenosphere, the land mass may break and produce rift features such as the Red Sea and Gulf of Aden in the Afar triangle. These fracture zones may then widen further, resulting in the formation of new tracts of ocean floor.

Aside from the suturing of one large land mass onto another, a continent may grow in either of two ways, both involving mountain building. One process involves orogenic compression and metamorphism of sediments that had accumulated in marginal depositional basins. The second is by accretion of microcontinents and oceanic microplates bearing volcanoes, sea floor plateaus, and seamounts. These bits and pieces of crust are carried by sea floor spreading to marginal subduction zones, where they are accreted onto the edge of the continent as allochthonous terranes. The Cordillera of North America appears to be largely composed of a complex collage of such terranes, each having its own distinctive geology that can often be traced to distant source regions.

Many once poorly understood and isolated problems appear to have found an explanation in the theory of plate tectonics. It has provided an interrelated network of answers for such questions as how continents grow, why basins of sedimentation experience upheaval to form mountains, what causes the majority of the world's earthquakes and volcanoes, what determines the locations of major mineral deposits, and why the history of life on earth is interrupted by episodes of crisis or of variation in biologic diversity.

## QUESTIONS FOR REVIEW

1. What are the principal categories of faults? Which kinds of faults result primarily from tension in the earth's crust and which come from compression?

2. What are folds? What are the principal kinds of folds? Are tracts of closely folded rocks the result of compressional or tensional forces? How are the topographic ridges of the Appalachian Mountains related to folds?

3. In chronologic sequence, list the geologic events that can be inferred from viewing an angular unconformity separating marine sedimentary rocks.

4. What large-scale features of the sea floor seem to have developed in response to crustal tension? What sea floor features seem related to crustal compression?

5. What is remanent magnetism? What is its origin, and how is it used in finding ancient pole positions?

6. How do the data of paleomagnetism support the concept of moving continents?

7. How has the detection of reversals in the earth's geomagnetic field been used as support for the concept of sea floor spreading?

8. How does the age distribution of chains of intraoceanic volcanoes enhance the concept of sea floor spreading?

9. What relationship might thermal convection cells in the mantle have to regions of crustal tension and compression?

10. Why are the most ancient rocks on earth found on the continents, whereas only relatively young rocks are found on the ocean floors?

11. Compile a list of items that Alfred Wegener might have used to convince a sceptic of the validity of the theory of continental drift.

12. Do mid-oceanic volcanoes have lavas of granitic or basaltic composition? What is the composition of lavas in continental mountain ranges adjacent to deep sea trenches? How might one account for these compositions?

13. According to plate tectonic theory, how might the Himalayan Mountains have originated? The San Andreas fault? The Dead and Red seas?

14. According to the theory of plate tectonics, where is new material added to the sea floor, and where is older material consumed?

15. A moving tectonic plate must logically have sides, a leading edge, and a trailing edge. Where are the leading and trailing edges of the North American plate located?

16. Why are the paths of polar wandering different for North America compared with Europe? Why is "polar wandering" a rather deceiving expression?

17. What evidence would one seek to support an interpretation that a particular area in the Appalachian Mountains was an allochthonous terrane? How might an allochthonous terrane derived from a microcontinent differ from one derived from a portion of a former volcanic island arc?

## TERMS TO REMEMBER

| | | | |
|---|---|---|---|
| allochthonous terranes | apparent polar | basin | continental drift |
| angular unconformity | wandering curves | Benioff seismic zone | convergent boundary |
| anticline | asthenosphere | blue schist | Curie temperature |

disconformity
divergent plate
  boundary
dome
fault
fold
foot wall
*Glossopteris* flora
Gondwanaland
hanging wall
lateral fault

Laurasia
lithosphere
mélange
*Mesosaurus*
microcontinents
monocline
negative gravity anomaly
nonconformity
normal fault
ophiolite suite

paleomagnetism
Pangea
Panthalassa
paraconformity
plate tectonics
Proto-Pangea
remanent magnetism
reverse fault
sea floor spreading
shear boundary

shelf province
subduction zone
suspect terrane
suture zone
syncline
thrust fault
tillite
trailing edge
transform fault
unconformity

## SUPPLEMENTAL READINGS AND REFERENCES

Ben-Avraham, Z. 1981. The Movement of Continents. *Am. Sci.* 69:291–299.

Cocks, L. R. M. (ed.). 1981. *The Evolving Earth* Cambridge, England: Cambridge Univ. Press.

Continents adrift and continents aground. 1976. *Readings from Scientific American.* San Francisco: W. H. Freeman Co.

Cox, A. (ed.). 1973. *Plate Tectonics and Geomagnetic Reversals.* San Francisco: W. H. Freeman Co.

Dietz, R. S., and Holden, J. 1970. Reconstruction of Pangaea: Break-up and dispersion of continents, Permian to recent. *J. Geophys. Res.* 75:4939–4956.

Glen, W. 1982. *The Road to Jaramillo.* Stanford, CA: Stanford Univ. Press.

Hallam, A. 1973. *A Revolution in the Earth Sciences.* New York: Oxford University Press.

Jones, P. L., Cox, A., and Beck, M. 1982. The growth of western North America. *Sci. Am.* 247:78–128.

Molnar, P., and Tapponnier, P. 1975. Cenozoic tectonics of Asia. *Science* 189:419–426.

Pitman, W. C. III, and Talwani, M. 1972. Sea-floor spreading in the North Atlantic. *Geol. Soc. Amer. Bull.* 83:619–646.

Saleeby, J. B. 1983. Accretionary tectonics of the North American Cordillera. *Ann. Rev. Earth Planet. Sci.,* 15:45–73.

Schwarzbach, M. 1986. *Alfred Wegener, the Father of Continental Drift.* Madison, WI: Science Tech Publishers.

Valentine, J. W., and Moores, E. M. Plate tectonics and the history of life in the oceans. *Sci. Am.* 230:80–89.

Van Andel, T. H. 1985. *New Views of an Old Planet.* Cambridge, England: Cambridge Univ. Press.

Wegener, A. 1929 (1966 translation). *The Origin of the Continents and Oceans.* New York: Dover Publications.

Weyman, D. 1981. *Tectonic Processes.* London: Allen & Unwin, Inc.

# The Oldest Rocks

*Gaea, mother goddess of the Earth,*
*guards her secrets carefully, and with*
*increasing jealousy as they recede in time.*

Preston Cloud, *Oasis in Space*
W. W. Norton, 1988

## Beneath the Cambrian

When eighteenth century geologists first began to describe local sequences of strata, they sometimes encountered a primary or basement complex of metamorphic and igneous rocks that lay beneath fossil-bearing sedimentary strata. These older mostly crystalline rocks came to be known as **Precambrian**, a derivative of the term **Cambrian** proposed for overlying younger rocks by Adam Sedgwick in 1835. To most of Sedgwick's contemporaries, correlating and even crudely estimating the age of the apparently unfossiliferous, altered, and deformed Precambrian rocks seemed an impossible task. Nevertheless, geologists recognized that Precambrian terranes were of great importance. They represented an enormous block of geologic time, formed the very cores of continents, and contained rich deposits of iron and other metals. For these reasons, it was inevitable that a few intrepid geologists would devote their lives to deciphering the complex relationships of Precambrian rocks.

One of these pioneers of Precambrian geology was Sir William Logan of the Canadian Geological survey. In the middle 1800s, Logan was able to associate groups of Precambrian rocks according to their superpositional and cross-cutting relationships. To substantiate these relationships, he inferred that those rocks that had experienced the most metamorphism were likely to be the oldest. Modern geologic investigations clearly show that Logan frequently erred in attempting to relate degree of metamorphism to age. Older rocks may escape metamorphism, and younger ones may be radically metamorphosed. Yet in southeastern Canada where Logan mapped, one could find an older terrane of gneissic rocks as well as younger sequences of less altered metamorphic and sedimentary rocks. Because of this, it seemed reasonable to think of Precambrian time as divisible into an older **Archean Eon** and a younger **Proterozoic Eon.** All of the remainder of geologic time up to the present day was designated the **Phanerozoic Eon.**

Today, all geologists agree that the only reliable basis for correlation of unfossiliferous Precambrian rocks is through isotopic dating techniques. Only after the absolute ages of rocks in a particular region are known can they be placed in correct chronologic order. Radiometric dating of Precambrian rocks has provided a basis for mapping Precambrian terranes and for deciphering their geologic history. The hundreds of dates already obtained permit a tentative calibration of the Precambrian geologic time table (Table 7–1). The beginning of the Proterozoic is placed at about 2.5 billion years ago (Fig. 7–1). The Archean is generally considered to have begun about 3.8 billion years ago, a date that approximates the age of oldest known crustal rocks.

For the earlier interval that began with the birth of the planet, the term **Hadean Eon** is often used. During the Hadean, the earth was still very hot, con-

Table 7–1  Time Divisions for the Precambrian

| Time in Billions of Years | Time Divisions Followed in This Book[*] | United States Geological Survey | Canada (Stockwell, 1964) | |
|---|---|---|---|---|
| | | | | 0.57 |
| | Late Proterozoic | Precambrian Z | | Hadrynian |
| | | 0.8 | | 0.88 |
| | 0.9 | | | |
| 1.0 | Middle Proterozoic | Precambrian Y | Proterozoic | Helikian |
| 1.5 | | | | |
| | 1.6 | 1.6 | | 1.64 |
| 2.0 | Early Proterozoic | Precambrian X | | Aphebian |
| | | | | 2.39 |
| 2.5 | 2.5 | 2.5 | | |
| | Late Archean | | | |
| 3.0 | 3.0 | | | |
| | Middle Archean | Precambrian W | | Archean |
| 3.5 | 3.4 | | | |
| | Early Archean | | | |
| | 3.8 | | | |
| 4.0 | Hadean | | | |
| 4.5 | | | | |
| | 4.6 | | | |

[*]After Harrison and Peterman, 1982

vection was vigorous, and steaming vents and volcanoes blistered the planet's cratered surface. As described in Chapter 4, it was during the Hadean that the earth's iron core separated from the silicate mantle and during which the earth was enveloped in an atmosphere devoid of free oxygen, but rich in carbon dioxide, water, and other volcanic gases. It was also a time during which the earth, its moon, and other planets in the solar system were being vigorously pelted by meteorites and asteroids. Although this episode of bombardment continued for several hundred million years, it has left little direct evidence on the present-day earth. Meteorites and craters formed by Hadean meteorite showers were destroyed by

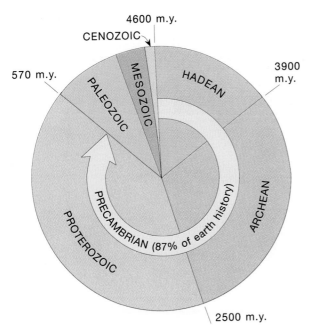

**Figure 7–1** Proportions of geologic time encompassed by the Precambrian and its divisions, the Hadean, Archean, and Proterozoic eons.

later episodes of melting and erosion. However, proof of the bombardment is clearly evident on the crater-scarred surface of the moon. Rocks from the lunar highlands are about 4.0 billion years old. At that time, there may have been sufficient heat generated by the vast infall of meteorites to cause melting and consequent resetting of radiometric clocks. Radiometric clocks on the Hadean earth would have been reset in a similar way.

## Shields and Cratons

Although exposures of Precambrian rocks may occur in the deep canyons of some mountains and plateaus, the most extensive exposures are found in broadly upwarped, geologically stable regions of continents called **shields**. Every continent has one or more shields (Fig. 7–2). Extending across 3 million square miles of northern North America is the great Canadian shield. Because continental glaciers of the last ice age have stripped away the cover of soil and

**Figure 7–2** Major areas of exposed Precambrian rocks (shown in yellow). Trends of mobile belts are indicated by the red lines.

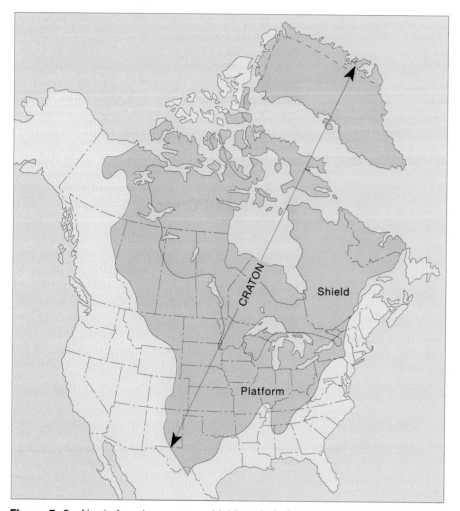

**Figure 7–3** North American craton, shield, and platform.

debris, Precambrian rocks of the Canadian shield are extensively exposed at the surface.

To the south and west of the Canadian shield, Precambrian rocks are covered by sedimentary sequences of Phanerozoic age. These younger rocks are mostly flat lying or only gently folded. Such stable regions where basement rocks are covered by relatively thin blankets of sedimentary strata are called **platforms.** The platform of a continent, together with its shield, constitutes that continent's **craton** (Fig. 7–3).

On the basis of differences in the trends of faults and folds, the style of folding, and the ages of component rocks, the Canadian shield has been divided into a number of **Precambrian provinces** (Fig. 7–4). The boundaries of these provinces are often marked by abrupt truncations in structural lineations, or they may be represented by bands of severely deformed rocks of former orogenic belts. In North America, seven Precambrian provinces are recognized: the Superior, Wyoming, Slave, Nain, Hearne, Rae, and Grenville. These provinces were once separate cratonic elements that have been consolidated to form the larger North American craton. They are bound together by belts of deformed, metamorphosed, and intruded rocks that mark the location of collision of the various cratonic elements. For this reason, the belts of deformed rocks are called **collisional orogens.** It has been estimated that all of the Archean and Proterozoic elements of the Canadian shield were consolidated by about 1.8 to 1.9 billion years ago.

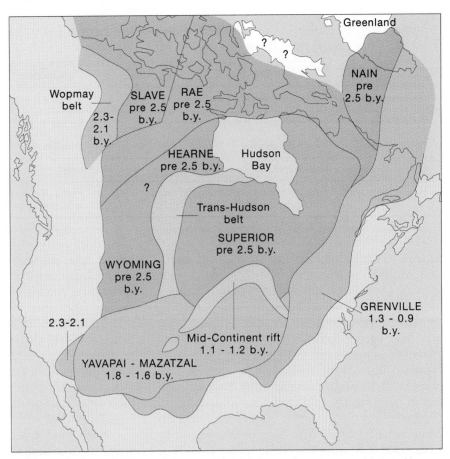

**Figure 7–4** Precambrian provinces of North America. Archean provinces (those greater than 2.5 billion years old) are in red. Proterozoic provinces are in green. The Trans-Hudson belt is a zone along which rifting and then closure occurred so as to weld the Superior province to the Wyoming province. The Hearne and Wyoming provinces may actually be a single province. A major episode of rifting that occurred 1.2 to 1.0 billion years ago is represented by the mid-continent rift zone. The Wopmay belt was developed by rifting followed by closure of an ocean basin. *(Adapted from Hoffman, P. F., Precambrian geology and tectonic history of North America, in Geology of North America, Vol. A, Ch. 16. Boulder, Colorado: Geological Society of America, 1989.)*

# The Archean Crust

## *Origin of the Oceanic Crust*

As described in Chapter 5, the earth's original crust was not granitic. It was a **mafic** crust, meaning that it was dominated by dark iron and magnesium silicates. This mafic crust was derived from an even more mafic (hence, **ultramafic**) underlying mantle. To understand the origin of the mafic crust, we must imagine the earth during its Hadean infancy, when it seethed with heat generated from the decay of short-lived radioactive isotopes, from gravitational forces,

and from the infall of meteorites. Even after the final stages of accretion, there was ample heat to partially or totally melt the upper mantle. As a result of the melting, an extensive magma ocean may have covered the earth's surface. When surface patches of the magma ocean cooled, they solidified as ultramafic rocks called **komatiites.** Komatiites form at temperatures even greater than the 1100°C required to produce basalt. Thus, they reflect the higher temperature gradients that prevailed during the late Hadean.

Even without total melting of the upper mantle to produce a magma ocean, it is probable that extensive partial melting in this zone produced sufficient quantities of magma to form an early crust. As this

Table 7–2 Summary of the Characteristics of the Earth's Early Oceanic and Continental Crust

| | Oceanic Crust | Continental Crust |
|---|---|---|
| First appearance | About 4.5 million years ago | About 4.0 million years ago |
| Where formed | Ocean ridges (spreading centers) | Subduction zones |
| Composition | Komatiite–basalt | Tonalite–granodiorite |
| Lateral extent | Widespread | Local |
| How generated | Partial melting of ultramafic rocks in upper mantle | Partial melting of wet mafic rocks in descending slabs |

From Condie, K. C., 1989, Origin of the Earth's Crust. *Paleogeography, Paleoclimatology, Paleoecology* 75:57–81, with permission.

was occurring, high mantle temperatures and associated vigorous convection probably resulted in some form of plate movement and recycling of surface rocks. Indeed, because komatiites are somewhat denser than hot, partly molten upper mantle, there may have been a tendency for slabs of primordial komatiitic oceanic crust to sink as they were being conveyed laterally by underlying convection currents. With the passage of time, with continued cooling of the mantle, and with partial melting of ultramafic rocks in the upper mantle, komatiitic additions to the surface may have been replaced by mafic (basaltic) magmas. These may have been added to the expanding oceanic crust at spreading centers (Table 7–2).

## Origin of Continental Crust

In contrast to the mafic nature of the oceanic crust, the continental crust is rich in feldspars, quartz, and muscovite. It is called a **felsic** crust. One of the oldest known patches of felsic crust was recently discovered at a location about 370 km (230 miles) north of the town of Yellowknife in Canada's Northwest Territories. Radiometric dating of zircon grains within these felsic rocks indicates that they are 3.96 billion years old (Fig. 7–5). These old continental rocks are tonalite gneisses. They were formed by metamorphism of **tonalite,** a variety of the igneous rock diorite that contains at least 10 percent quartz. The Isua region of southwestern Greenland also contains very old tonalite gneisses (3.8 billion years old) (Fig. 7–6). The Greenland outcrops are more extensive in area and contain fragments of ultramafic rocks that may be remnants of the earth's surface that was present

before there was any felsic crust. Other patches of extremely old felsic crust have been found in Enderby Land, Antarctica (3.9 billion years old), and in the Limpopo region of South Africa (3.8 billion years old). In addition, in western Australia, 4.0- to 4.3-billion-year-old detrital zircon grains have been found in Archean sedimentary rocks. Zircon is a relatively common accessory mineral in felsic igneous rocks. The Australian zircons had been eroded from such a rock, transported, and deposited in the fluvial sandstones from which they were extracted.

When we compare the ages obtained from the samples just described to the 4.6-billion-year age of our planet, it is apparent that the earth had acquired patches of continental, tonalitic, and granodioritic crust a few hundred million years after its formation, or about 4.0 to 4.3 billion years ago. But how did these patches form and grow? Many geologists believe they originated at subduction zones where slabs of sediment-covered mafic crust plunged into the mantle, there to be partially melted so as to release felsic components. Because of their lower density, felsic melts ascended to form tonalites and granodiorites. The more mafic materials remained behind in the mantle. By about 3.0 billion years ago, granites formed by a similar mechanism involving partial melting of tonalitic, granodioritic, and sedimentary rock materials.

It is likely that the early Archean segments of tonalitic crust were small, perhaps less than about 500 km (310 miles) in diameter. These initial patches grew larger by several mechanisms that continue to operate today. One such mechanism may have involved the coming together of two tectonic plates bearing felsic crust at their encountering mar-

A

B

**Figure 7–5**   (A) Photomicrograph of one of the 3.96-billion-year-old zircon grains extracted from the Acasta Gneiss, Slave province, Northwest Territories of Canada. The grain is 0.5 mm long. Its polished surface has been etched with acid so as to highlight crystal growth zones. Numbers refer to points selected for analysis. (B) Exposure of the zircon-bearing Acasta Gneiss at the discovery site. *(Courtesy of Sam A. Bowring.)*

**Figure 7–6**   Archean tonalite gneiss from the Isua region of southwestern Greenland. *(Courtesy of R. F. Dymek.)*

gins. Compressive deformation and suturing may have occurred so as to form a larger expanse of continental crust (see Fig. 6-29A). Growth may also have occurred as island arcs and microcontinents were added to the host continent along subduction zones. A third method may have involved the welding of prisms of sediment onto continental margins. This process begins with the weathering and erosion of preexisting rocks and deposition of the resulting sediment in the ocean adjacent to an upland area. Next, the thick accumulation of sediment is subjected to orogeny caused by an encounter with a converging tectonic plate. The sediment in the prism would have roughly the same bulk composition as granitic igneous rocks. Such material would be converted to felsic crystalline crustal rocks if subjected to intense metamorphism and melting. The repetition of such cycles of deposition and orogeny, along with suturing events and the incorporation of microcontinents, could account for continuing growth of continental crust. As indicated by the known distribution of 3.0- to 2.5-billion-year-old felsic rocks, very large continents had been assembled by late Archean time.

### Granulites and Greenstones

In general, Archean rocks of cratons can be grouped into two major rock associations. The first is the **granulite** association. It is composed largely of gneisses derived from strongly heated and deformed tonalites, granodiorites, and granites, as well as layered intrusive gabbroic rocks called **anorthosites**. Few, if any, clues about the nature of parent materials can be found in these severely metamorphosed granulite associations.

The second general group of Archean cratonic rocks are **greenstones**. Greenstone associations usually occur in roughly troughlike or synclinal belts (Fig. 7–7). They are prominent features of Archean terranes of all continents. The Abitibi belt of the Superior province is the largest uninterrupted greenstone trend in the Canadian shield (see Fig. 7–7). Like most other greenstone belts, it is composed of basaltic, andesitic, and rhyolitic volcanic rocks along with metamorphosed sediments derived by weathering and erosion of the volcanics. Lavas of greenstone belts exhibit pillow structures that indicate they were extruded under water (Fig. 7–8).

The most intriguing characteristic of greenstone belts is the roughly sequential transition of rock types from highly mafic near the bottom to felsic

**Figure 7–7** Greenstone belts of the Superior province. *(After A. M. Goodwin,* Proc. Geol. Assoc. of Canada, *19:1–14, 1968.)*

Hudson Bay

Granites and gneisses

Greenstone belts

Abitibi belt

0  500 km

**Figure 7–8** Exposure of greenstone showing well-developed pillow structures, near Lake of the Woods, Ontario, Canada. The original basalts were extruded under water during late Archean time. During the Pleistocene Ice Age, glaciers beveled the surface of the exposures and left behind the telltale scratches (glacial striations). *(Courtesy of K. Schultz.)*

**Figure 7–9** Generalized cross-section through two greenstone belts. Note their syclinal form and the sequence of rock types from ultrabasic near the bottom to felsic near the top. A late event in the history of the belt is the intrusion of granites. Ultramafic basal layers are particularly characteristic of greenstone belts in Australia and South Africa, but do not occur in the Archean of Canada.

near the top (Figs. 7–9 and 7–10). In many, the basal rock masses are komatiites. Overlying basalts are somewhat less mafic and include low-grade metamorphic minerals such as chlorite and hornblende. These green minerals impart the color for which greenstones are named. Proceeding upward in a typical greenstone succession, one finds felsic volcanics above mafic rocks, which are overlain in turn by shales, graywackes, conglomerates, and less frequently, by banded iron formations. The entire greenstone sequence is often surrounded by granitic intrusive rocks. In the Superior province of Canada, the granitic rocks were emplaced during a deformational event that occurred about 2.6 to 2.7 billion years ago. That event, called the **Kenoran orogeny**, marks the close of the Archean Eon in Canada.

## Archean Protocontinents

The upwardly changing sequence from ultramafic or mafic to felsic rocks in greenstone belts records the development of felsic crustal materials from older, more mafic source rocks. The sedimentary rocks that occur most commonly near the top of the greenstone sequence provide clues to the nature of the earliest patches of Archean continental crust—the so-called **protocontinents.** Among the sediments are elongate layers of coarse conglomerates containing pebbles of granite. Thus, felsic source rocks were already present at the earth's surface. Graywackes and dark shales present in the greenstone sedimentary sequence appear to have been deposited in deep water

environments adjacent to protocontinental coasts. In the greenstone sedimentary association, we do not find blankets of well-sorted, ripple-marked, cross-bedded sandstones. The existence of broad lowlands that could be flooded by shallow interior seas is therefore unlikely. Nor is there any evidence for wide marginal shelf environments. This absence of blanket sandstones and the presence of graywackes and elon-

**Figure 7–10** A representative greenstone belt stratigraphic sequence from the Precambrian shield of western Australia. (Adapted from D. R. Lowe, Ann. Rev. Earth Planet. Sci. 8:145–167, 1980.)

gate bodies of conglomerates suggest that Archean protocontinents were steep-sided, relatively small, narrow land masses with rugged shorelines.

## Archean Tectonics

The relationship between greenstone belts and surrounding granulite terranes is currently being actively investigated. Although no single model has emerged that is overwhelmingly accepted, many geologists favor a plate tectonics model. As discussed earlier, the plate tectonics mechanism is capable of providing emerging protocontinents with a supply of mantle-derived materials that could be recycled and partially melted to form additions to continental crust. It is probable that during the Archean, plate tectonics would have operated at a more dynamic level, as compared to the Phanerozoic, because of the greater amount of heat still present in the mantle. As

compared to Phanerozoic conditions, volcanism was far more extensive and rates of convection were far greater. The sequence of events that may have provided for both the granulite and greenstone associations would have begun with the development of volcanic arcs adjacent to subducting plates (Fig. 7–11). Lavas and pyroclastics of the arcs were weathered and eroded to provide sediment to the adjacent trench. In the subduction zones, this wet sediment was carried downward along with oceanic crust and was heated. Partial melting of the mix of materials produced increasingly more felsic magmas, and these worked their way upward and engulfed earlier volcanics. During or shortly after these events, compressional deformation and metamorphism altered the rocks to form the granulite terranes.

Whereas volcanic arcs were the sites for the birth of granulite associations, the back-arc basins were the probable locations for the development of green-

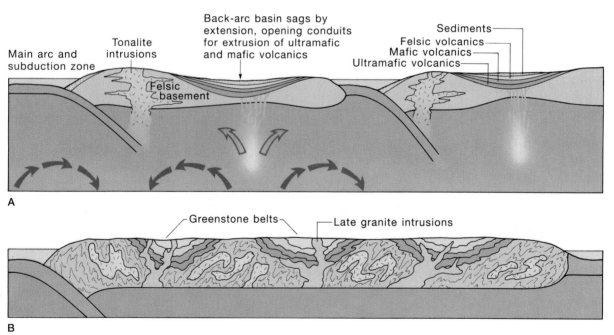

**Figure 7–11** Plate tectonics model for the development of greenstone belts and growth of continental crust. (A) Plates are in motion, driven by convection cells in the upper mantle. Subduction provides for the emplacement of wedges of oceanic crust and for mixing and melting to provide tonalite intrusions. Behind the main arc, the back arc sags by extension and the greenstone volcanic sequence is extruded. (B) Compression has occurred to create the greenstone belts with their synclinal form and also to aggregate small continental patches into a larger continental mass. Later, granites are intruded in and around greenstone belts. *(Simplified from a model proposed by B. F. Windley, 1977, The Evolving Continents.)*

stone belts. In the early stages of that development, ultramafic to mafic lavas ascended through rifts in the back-arcs to form the lower layers of the greenstone sequences. Later extrusions included increasingly more felsic lavas, pyroclastics, and sediments derived from the erosion of adjacent uplands. The final stage of development involved compression of the back-arcs so as to form the synclinal structures characteristic of greenstone belts. Compression may have occurred simultaneously with the magmatic activity that produced the granitic intrusions that surround the lower levels of greenstone belts.

## Archean Sediments

Southern Africa's Barberton Mountain Land and a terrane in western Australia called the Pilbara block contain excellent records of Archean sedimentation. Here one finds large cratons, rather than the small protocontinents discussed earlier. These cratons were fully formed and stable as early as 3.0 billion years ago. Because they have been stable since that early time, sedimentary rocks within them have often escaped severe metamorphism that might otherwise have obscured their environments of deposition. In southern Africa, some of these sedimentary rocks contain gold and have therefore been intensely investigated.

In the greenstone belts of Barberton Mountain Land, the geologic column begins in its lower part with ultramafic and mafic lava flows (Fig. 7–12), followed by pillowed oceanic basalts. These rocks give way upward to felsic lava flows and then graywackes and claystones probably derived from the erosion of volcanic mountains and cindercones. Pyroclastic beds capped by chert layers also occur. The chert probably formed from the silica of volcanic ash. The general sequence from ultramafic or mafic volcanics to more felsic volcanics to sediments is sometimes repeated, and with each cycle the younger rocks of the sequence are more felsic. Sandstones containing quartz, orthoclase, and granitic detritus are found near the top of the geologic column. These detrital sediments indicate the presence of nearby granitic source areas exposed during episodes of cratonic uplift. By the time these highly felsic clastic sediments were being deposited, a full range of depositional environments had developed. Deep water continental slope and rise environments are recorded by

**Figure 7–12** Stratigraphic sequence in a greenstone belt in Barberton Mountain Land, South Africa. *(After Lowe, D. R., Ann. Rev. Earth Planet. Sci. 8:145–167, 1980.)*

the graded turbidites and mudstones of the Fig Tree Group in Barberton Mountain Land. The overlying Moodies Group includes quartz sandstones and shales that were spread across shallow marine shelves, tidal flats, and deltas.

Intertidal clastic sedimentary rocks of the Pongola Group are above the Moodies rocks, and these are overlain by a remarkable sequence of river-borne conglomerates, sandstones, and shales known as the Witwatersrand Supergroup. The Witwatersrand is notable for several reasons. First, its areal extent (about 103,000 km$^2$, or 40,000 sq. miles), thickness (8 to 10 km, or 5 to 6.3 miles), and nonmarine environment of deposition (fluvial, deltaic, and lacustrine) clearly denote that very large cratons were in existence by about 2.7 billion years ago. Second, the Witwatersrand contains placer gold. Indeed, since the discovery of that gold in 1883, half the world's supply of the precious metal has come from the fluvial and deltaic conglomerates of the Witwatersrand.

# Life of the Archean

The splendid diversity of animals and plants on our planet is a consequence of billions of years of chemical and biological evolution that began during the Precambrian. We would like to have fossil documentation of the earliest events in the progression from nonliving to living things, but that kind of direct evidence is lacking. Theories for the origin of life are based on our understanding of present-day biology, coupled with evidence obtained from the study of Archean rocks, meteorites, and planetary compositions.

## *The Beginning of Life*

### From Simple to Complex

Systematic progressions from simple to complex structures are a manifest characteristic of both living and nonliving systems. Awareness of such progressions has strongly influenced theories on the origin of life. Subatomic particles such as neutrons, protons, and electrons are built into atoms. Atoms in turn combine to form molecules, and molecules may be joined to form sheets, chains, and a variety of complex molecular structures. Perhaps in a somewhat similar fashion, the atoms of elements essential to organisms (carbon, oxygen, hydrogen, nitrogen, sulfur, and phosphorus) are built into organic molecules. At some stage in the progression, simpler molecules might have added components that would gradually transform them into the complex organic

molecules essential to all living things. An analogy might be the manner in which small molecules of nitrogenous bases combine with sugar and phosphorous compounds to form the large deoxyribonucleic (DNA) molecule (see Fig. 3–11). In living organisms, DNA is found chiefly in the nucleus of the cell. It functions in the transfer of genetic characteristics and in protein synthesis. Some theorists suggest that such large and complex molecules might have organized themselves into bodies capable of performing specific functions. Such bodies in unicellular organisms are called *organelles*. Organelles might then have combined to form larger entities that grew, metabolized, reproduced, mutated, and produced mutations. These are the attributes of all living things. Could a progression even remotely similar to this have occurred in the early years of the Archean? We do not know for sure, but the concept is at least feasible.

### Preliminary Considerations

The oldest fossil evidence of life is found in 3.5-billion-year-old rocks of Australia's Pilbara shield. Living things, however, probably existed even before the ones that left these traces, as indicated by carbon found in 3.8-billion-year-old banded iron deposits in southern Greenland. The carbon extracted from these ancient sedimentary rocks has $^{13}$C to $^{14}$C ratios similar to those in present-day organisms.

As noted in Chapter 5, the earth accreted from a dust cloud about 4.6 billion years ago, and an early atmosphere was subsequently generated by volcanic outgassing. Presumably, the gases emitted were similar to those in volcanic eruptions today in that water and carbon dioxide were dominant, with lesser amounts of nitrogen, hydrogen, hydrogen sulfide, sulfur dioxide, carbon monoxide, methane, and ammonia. Note the absence of free oxygen in this list. The planet's earliest organisms must have been anaerobic, that is, they did not require oxygen for respiration. The environment in which they lived lacked a protective atmospheric shield of ozone, which is derived from oxygen and which absorbs ultraviolet radiation from the sun. Perhaps early life escaped these lethal rays by living beneath protective ledges of rock or at greater water depths. Some biologists insist that in the early stages of the development of life, ultraviolet radiation may not have been a hindrance, but rather served to impel molecules to interact and form complex associations.

The carbon, hydrogen, oxygen, nitrogen, phosphorus, and sulfur necessary to produce life are among the most abundant elements in the solar system. They were undoubtedly present on the primitive earth. What had to be accomplished was the enormously intricate organization of those elements into the molecular components of an organism. There are four essential components of life. The first component is protein. **Proteins** are essentially strings of comparatively simple organic molecules called *amino acids*. Proteins act as building materials and as compounds that assist in chemical reactions within the organism. The second of the basic components is **nucleic acids**, such as DNA, mentioned earlier, and ribonucleic acid (RNA). Organic phosphorous compounds provide a third component of life. They serve to transform light or chemical fuel into the energy required for cell activities. An important fourth ingredient for life is some sort of container, such as a cell membrane. The enclosing membrane provides a relatively isolated chemical system within the cell and keeps the various components in close proximity so that they can interact.

From the previous description, it is apparent that amino acids have an important role in the development of the larger and more complex molecules. They are the building blocks of proteins. Two environmental circumstances in the early years of earth history may have been important in the natural synthesis of amino acids. Prior to the accumulation of the ozone layer in the earth's upper atmosphere, ultraviolet rays bathed the earth's surface. As demonstrated in experiments, ultraviolet radiation is capable of separating the atoms in mixtures of water, ammonia, and hydrocarbons and of recombining those atoms into amino acids. A second form of energy capable of accomplishing this feat is electrical discharge in the form of lightning. Either together or separately, lightning and ultraviolet radiation may have stimulated the production of amino acids in the air, in tidal pools, in the upper levels of the oceans, and wherever suitable environmental conditions existed.

## Test Tube Amino Acids and a Step Beyond

Scientists in the mid-nineteenth century had succeeded in manufacturing some relatively simple organic compounds in the laboratory. However, it was not until 1953 that the laboratory synthesis of amino acids and other molecules of roughly similar com-

plexity was announced. Stanley Miller, at the suggestion of Harold Urey, performed the now famous experiment. He infused an atmosphere at that time thought to be like that of the earth's earliest atmosphere into an apparatus similar to the one shown in Figure 7–13. It was a methane, ammonia, hydrogen, and water vapor atmosphere. As the mixture was circulated through the glass tubes, sparks of electricity (simulated lightning) were discharged into the mixture. At the end of only 8 days, the condensed water in the apparatus had become turbid and deep red. Analysis of the crimson liquid showed that it con-

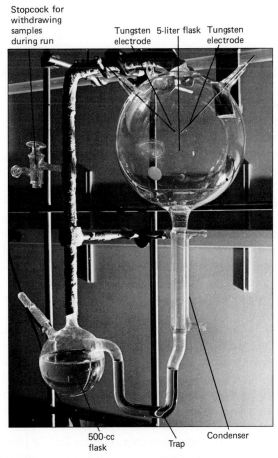

Stopcock for withdrawing samples during run

Tungsten electrode    5-liter flask    Tungsten electrode

500-cc flask    Trap    Condenser

**Figure 7–13**  Stanley Miller and Harold Urey used an apparatus similar to this to replicate what they believed were the conditions of early earth. An electric spark was produced in the upper right flask to simulate lightning. The gases present in the flask reacted together, forming a number of basic organic compounds. *(Dr. Stanley L. Miller, University of California, San Diego.)*

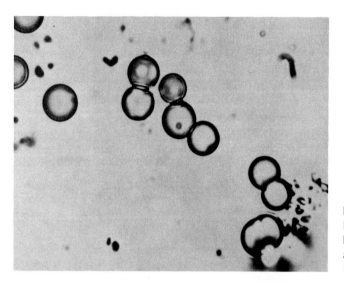

**Figure 7–14**  Proteinoid microspheres formed in hot aqueous solutions of proteinoids that have been permited to cool slowly. The spheres are about 1 to 2 $\mu$m in diameter. *(Photo by Sidney W. Fox)*

tained a bonanza of amino acids as well as somewhat more complicated organic compounds that enter into the composition of all living things. In additional experiments by other biochemists, it was shown that similar organic compounds could also be produced from gases (carbon dioxide, nitrogen, and water vapor) of the preoxygenic atmosphere. The main requirement for the success of the experiments seemed to be the lack, or near absence, of free oxygen. To the experimenters, it now seemed almost inevitable that amino acids would have developed in the earth's prelife environment. Because amino acids are relatively stable, they probably increased gradually to levels of abundance that would enhance their abilities to join together into more complex molecules.

In order to come together and form proteinlike molecules, amino acids must lose water. This loss can be accomplished by heating concentrations of amino acids to temperatures of at least 140°C. Volcanic activity on the primitive crust would be capable of providing such temperatures. However, the biochemist S. W. Fox discovered that the reaction also occurred at temperatures as low as 70°C if phosphoric acid was present. Fox and his co-workers were able to produce proteinlike chains from a mixture of 18 common amino acids. He called these structures **proteinoids** and reasoned that billions of years ago they were the transitional structures leading to true proteins. This is not extravagant conjecture, for Fox was able to find proteinoids similar to those he created in his laboratory among the lavas and cinders adjacent to the vents of Hawaiian volcanoes. Apparently, amino acids formed in the volcanic vapors and were combined into proteinoids by the heat of escaping gases.

Hot, aqueous solutions of proteinoids will, on cooling, form into tiny spheres that show many characteristics common to living cells. These **microspheres** (Fig. 7–14), as they are called, have a filmlike outer wall; are capable of osmotic swelling and shrinking; exhibit budding, as do yeast organisms; and can be observed to divide into "daughter" microspheres. They occasionally aggregate linearly to form filaments, as in some bacteria, and they exhibit a streaming movement of internal particles similar to that observed in living cells.

Although complete long-chain nucleic acids have not yet been experimentally produced under prelife conditions, short stretches of ordered sequences of nucleic acid components have now been produced in the laboratory. Indeed, in 1976, Har Gobind Khorana and his associates at the Massachusetts Institute of Technology announced that they had made a functioning artificial gene—one of thousands in the DNA spiral molecule of the bacterium *Escherichia coli.*

## Speculations about Earliest Life

There are several good reasons to believe that the earliest organisms originated in the sea. The sea contains the salts needed for health and growth. The waters of the oceans serve as universal solvents capable of dissolving a great variety of organic compounds, and currents in the oceans ceaselessly circulate and mix these compounds. Such constant motion would have favored frequent collisions of the vital molecules and would thereby have increased the probability of their combining into larger bodies. One can imagine the larger bodies adding components, increasing in complexity, and ascending in a thousand infinitesimally small steps from nonliving to transitional things of a biologic netherworld to the first living organisms. Such a sequence could not occur in an environment like that existing on land today. Oxygen and present-day microbial predators would destroy the delicate structures. However, life originated before there were organisms to cause decay and before there was sufficient free oxygen to be troublesome.

Largely from our knowledge of paleontology and biology, it is believed that the first living things were microscopic in size and were unicellular. It is likely that these earliest forms of life would not have evolved a means of manufacturing their own food, but rather assimilated small aggregates of organic molecules that were also present in the surrounding medium. Some, undoubtedly, would have consumed even their developing contemporaries. Today, organisms with this type of nutritional mechanism are termed **heterotrophs.** The food gathered by the ancestral heterotrophs was externally digested by excreted enzymes before being converted to the energy required for vital function. In the absence of free oxygen, there was only one way to accomplish this conversion—by **fermentation.** There are many variations of the fermentation process. The most familiar reaction involves the fermentation of sugar by yeast. It is a process by which organisms are able to disassemble organic molecules, rearrange their parts, and derive energy for life functions. A very simple fermentation reaction can be written as follows:

$$C_6H_{12}O_6 \rightarrow 2\ CO_2 + 2\ C_2H_5OH + Energy$$

Glucose    Carbon    Alcohol
dioxide

Animal cells are also able to ferment sugar in a reaction that yields lactic acid rather than alcohol.

Eventually, the original fermentation organisms would have begun to experience difficult times. By consuming the organic compounds of their environ-

ment, they would eventually have created a food shortage; this scarcity, in turn, might have caused selective pressures for evolutionary change. At some point prior to the depletion of the food supply, organisms evolved the ability to synthesize their needs from simple inorganic substances. These were the first autotrophic organisms. Unlike the heterotrophs, **autotrophs** were able to manufacture their own food. The organisms that had developed this remarkable ability saved themselves from starvation. Their evolution proceeded in diverse directions. Some manufactured their food from carbon dioxide and hydrogen sulfide and were the probable ancestors of today's sulfur bacteria. Others employed the life scheme of modern nitrifying bacteria by using ammonia as a source of energy and matter.

However, more significant than either of these kinds of autotrophs were the **photoautotrophs** (Fig. 7–15), which were capable of carrying on photosynthesis. Their gift to life was the unique capability of dissociating carbon dioxide into carbon and free oxygen. The carbon was combined with other elements to permit growth, and the oxygen escaped to prepare the environment for the next important step in the evolution of primitive organisms. In simplified form, the reaction for photosynthesis can be written as follows:

$$6\ CO_2 + 6\ H_2O \xrightarrow{\text{Sunlight}} C_6H_{12}O_6 + 6\ O_2$$

**Figure 7–15** Photoautotrophs. These organisms known as *Nostoc,* use light energy to power the synthesis of organic compounds from carbon dioxide and water. *(Sinclair Stammers/Science Photo Library/Photo Researchers, Inc.)*

With the multiplication of the photoautotrophs, billions upon billions of tiny living oxygen generators began to change the primeval nonoxygenic atmosphere to an oxygenic one. Fortunately, the change was probably gradual, for if oxygen had accumulated too rapidly, it would have been lethal to developing early microorganisms. Some sort of oxygen acceptors were needed to act as safety valves and to prevent too rapid a buildup of the gas. Iron in rocks of the continental crust provided suitable oxygen acceptors. Eventually, organisms evolved oxygen-mediating enzymes that permitted them to cope with the new atmosphere.

After the surficial iron on the earth's surface had combined with its capacity of oxygen, the gas began to accumulate in the atmosphere and hydrosphere. Solar radiation acted upon atmospheric oxygen to convert part of it to ozone, and ozone in turn formed an effective shield against harmful ultraviolet radiation. Still-primitive and vulnerable life was thereby protected and could expand into environments that formerly had not been able to harbor life. The stage was set for the appearance of aerobic organisms.

**Aerobic organisms** use oxygen to convert their food into energy. The reaction, which can be considered a form of cold combustion, provides far more energy in relation to food consumed than does the fermentation reaction. This surplus of energy was an important factor in the evolution of more complex forms of life.

## Prokaryotes, Eukaryotes, and Symbiosis

There is no unequivocal evidence to date the transition from chemical or prebiotic evolution to organic evolution. The most accurate statement that can be made in this regard is that the transition occurred at some time prior to 3.5 billion years ago. In rocks of that age, paleontologists have discovered the earliest fossil evidence of life. These oldest fossils belong to a category of microorganisms called **prokaryotes** (Fig. 7–16A). Prokaryotes lack definite membrane-bounded organelles and also do not have a membrane-bounded nucleus in which genetic material is neatly arranged into discrete chromosomes. Modern prokaryotes do possess cell walls, and most are able to move about. The cyanobacteria (formerly called blue-green algae) are prokaryotes that are capable of photosynthesis (see Fig. 7–15).

Prokaryotes are asexual and thereby restricted in the level of variability they can attain. In sexual reproduction, as noted in Chapter 3, there is a union of gametes (egg and sperm) to form the nucleus of a single cell, the zygote. The formation of the zygote results in recombination of the parental chromosomes and leads to a multitude of gene combinations among the gametes that give rise to the next generation. In the asexually reproducing prokaryotes, in contrast, a cell divides from the parent and becomes an independent individual containing the identical chromosome complement of the parent cell. Unless mutation intervenes, the number and kinds of chromosomes in individuals produced by asexual reproduction are exactly the same as those in the parent. Thus, the possibilities for variation are more limited than in sexually reproducing organisms. It is probably for this reason that prokaryotes have shown little evolutionary change through more than 2 billion years of earth history. Nevertheless, such organisms represent an important early step in the history of primordial life.

Evolution proceeded from the prokaryotes to organisms with a definite nuclear wall, well-defined chromosomes, and the capacity for sexual reproduction. These more advanced forms were **eukaryotes** (Fig. 7–16A). Unlike prokaryotes, eukaryotes contain organelles such as chloroplasts (which convert sunlight into energy) and mitochondria (which metabolize carbohydrates and fatty acids to carbon dioxide and water, releasing energy-rich phosphate compounds in the process).

Biologists believe that the organelles in eukaryotic cells were once independent microorganisms that entered other cells and then established symbiotic relationships with the primary cell. For example, an anaerobic, heterotrophic prokaryote such as a fermentative bacterium might have engulfed a respiratory prokaryote and thereby would have an internal consumer of the oxygen that might otherwise have threatened its existence. A nonmotile organism might acquire mobility by forming a symbiotic association with a whiplike organism like a spirochete. The resulting cell would appear to have a flagellum for locomotion. Natural selection would clearly favor such advantageous symbiotic relationships.

With the development of sexually reproducing eukaryotes, genetic variations could be passed from parent to offspring in a great variety of new combina-

**Figure 7–16**  (A) The basic differences between prokaryotes and eukaryotes. (B) An electron micrograph of the prokaryote *Bacillus subtilus,* a living bacterium that is about to complete cell division (magnification ×15,000). *(Figure (B) from Dr. A. Ryter.)*

tions. This development was truly momentous, for it led to a dramatic increase in the rate of evolution and was ultimately responsible for the evolution of complex multicellular animals.

## The Fossil Record for the Archean

Because of the prevalence of prokaryotic microorganisms in certain Archean sedimentary rocks, the eon might well be called the age of prokaryotes. Among the more widespread indications of these primitive organisms are fossils known as stromatolites. **Stromatolites** are laminar, organic sedimentary structures formed by the trapping of sedimentary particles and precipitation of calcium carbonate in response to the metabolic activities and growth of matlike colonies of cyanobacteria and fewer numbers of other prokaryotes. The study of present-day stromatolites indicates they develop when fine particles of calcium carbonate settle on the sticky, filamentous surface of cyanobacterial mats to form thin layers (laminae) (Fig. 7–17). The cyanobacterial colonies then grow through the layer and form another surface for the collection of additional fine sediment. Repetition of this process along with precipitation of calcium carbonate produces a succession of laminae. The resulting structures can be cabbagelike, columnar, tabular, or branching in shape.

Stromatolites are more common in Proterozoic rocks than Archean rocks. This may be a reflection of a rarity of the warm shelf environments conducive to the growth of marine stromatolites. The oldest stromatolites occur in 3.5-billion-year-old rocks of the Warrawoona Group of Australia's Pilbara shield. The black chert beds in the Warrawoona Group also contain a variety of filamentous fossil prokaryotes (Fig. 7–18). Somewhat younger Archean stromatolites are found in the 3.0-billion-year-old Pongola Group of southern Africa and the 2.8-billion-year-old Bulawayan Group of Australia. The $^{12}C$ to $^{13}C$ ratios in the Bulawayan rocks provide evidence of biologic fixation of carbon dioxide by photosynthetic organisms.

Some of the earliest discoveries of Archean microfossils were the result of careful study of thin sections and electron micrographs prepared from rocks

COMMENTARY COMMENTARY COMMENTARY COMMENTARY
MMENTARY COMMENTARY COMMENTARY COMMENTARY COMMEN
COMMENTARY COMMENTARY COMMENTARY COMMENTARY
MENTARY COMMENTARY **COMMENTARY** COMMENTARY COMMEN

## The Endosymbiotic Theory for the Origin of Eukaryotes

The origin of eukaryotic organisms from prokaryotic ancestors is termed the endosymbiotic theory. The accompanying figure illustrates the steps involved in this important evolutionary theory. At the top of the figure is the original prokaryotic host cell. The host ingests, but does not digest, aerobic bacterial endosymbionts. These survive and reproduce within the host cell so that future generations of the host organism also contain these endosymbionts. At the same time, the host cell provides a safe environment, raw materials, and nutrients. Eventually, the host and endosymbionts develop a mutalistic relationship, and the originally ingested cells lose their ability to function outside the cell. The endosymbionts that were originally photosynthetic bacteria become organelles called chloroplasts and perform the function of photosynthesis. The aerobic bacteria that had been ingested ultimately become mitochondria and function in oxidative metabolism. They are the powerhouses of the cell.

Evidence in support of the endosymbiotic origin of eukaryotes includes observations that mitochondria and chloroplasts possess some of their own genetic apparatus, and can, on a limited level, synthesize protein. They can also divide independently of the cell in which they reside. The double membrane enveloping mitochondria and chloroplasts is believed to have formed by invagination of the host cell's outer membrane.

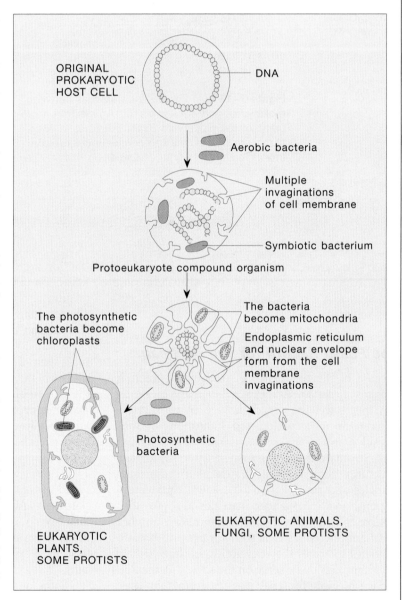

ORIGINAL PROKARYOTIC HOST CELL

DNA

Aerobic bacteria

Multiple invaginations of cell membrane

Symbiotic bacterium

Protoeukaryote compound organism

The bacteria become mitochondria

Endoplasmic reticulum and nuclear envelope form from the cell membrane invaginations

The photosynthetic bacteria become chloroplasts

Photosynthetic bacteria

EUKARYOTIC PLANTS, SOME PROTISTS

EUKARYOTIC ANIMALS, FUNGI, SOME PROTISTS

A                                                                                                           B

**Figure 7–17**  (A) Modern columnar stromatolites growing in the intertidal zone of Shark Bay, Australia. Metabolic activities of colonial marine cyanobacteria result in the formation of these structures. Fine particles of calcium carbonate settle between the tiny filaments of the matlike colonies and are bound with a mesh of organic matter. Successive additional layers result in the laminations that are the most distinctive characteristic of stromatolites. (B) Fossil stromatolites from Precambrian rocks exposed in southern Africa. *(Photo (A) courtesy of J. Ross, and (B) courtesy of J. W. Schopf.)*

**Figure 7–18**  Filamentous prokaryotic microfossils from 3.5-billion-year-old black cherts of the Archean Warra-woona Group, Pilbara shield of western Australia. *(Courtesy of J. W. Schopf and B. M. Packer.)*

**Figure 7–19** Electron micrographs of rod-shaped *Eobacterium isolatum,* about 0.6 micron in length. Specimen at lower right has been interpreted as a transverse section through a cell. *(Courtesy of J. W. Schopf.)*

of the southern African Fig Tree Group. These rocks are about 3.1 billion years old and consist of variously colored cherts, slates, ironstones, and hard sandstones. Elso S. Barghoorn and J. William Schopf were among the first paleontologists to study the fossils of the Fig Tree Group. They were able to find a number of tiny, double-walled, rod-shaped structures (Fig. 7–19) that had a striking resemblance to modern bacteria. Barghoorn and Schopf named their find *Eobacterium isolatum*—the "isolated dawn bacteria." In their search for larger fossils, the scientists prepared thin sections of the chert and examined these with the optical microscope. The examination disclosed numerous spheroidal bodies very similar to certain cyanobacteria. Because the samples were collected near the town of Barberton, Barghoorn and Schopf named the fossils *Archaeosphaeroides barbertonensis* (Fig. 7–20). The filaments of organic matter and hydrocarbons of organic origin found in Fig Tree rocks are clear evidence of the presence of abundant microscopic life in the Archean. Many beds within the Fig Tree contain so much organic carbon that they are black or dark gray in color. On analysis, the dark layers are found to contain organic compounds regarded as breakdown products of chlorophyll, indicating that photosynthesis was in operation at least 3.0 billion years ago.

## Mineral Wealth of the Archean

Archean rocks are rich in metallic ore deposits, including ores of gold, iron, copper, chromium, and nickel. Some of these metals were concentrated at the same time that their host rocks were formed, whereas others were introduced at a later date when ore-bearing hot solutions cooled and precipitated

**Figure 7–20** Spheroidal algalike microfossils (*Archaeosphaeroides barbertonensis*) in black chert of the early Precambrian Fig Tree Series. These fossils are 3.1 billion years old. The scale bars are in microns (1 micron is equal to 0.001 mm). *(Courtesy of J. W. Schopf.)*

their dissolved metals in veins and cavities. Gold is the metal particularly characteristic of the Archean. It occurs in Archean rocks in Africa, North America, Asia, South America, and Australia. Iron, although important locally, is not as abundant in Archean rocks as it is in Proterozoic rocks. Much of the copper, along with zinc and iron, occurs as sulfides that appear to have formed when sea water penetrated hot submarine lavas, reacted with them to extract the metals, and then, on cooling, precipitated the metallic sulfides on the floors of Archean seas. Ores of this type are called *volcanogenic massive sulfide deposits*. They have been observed forming today on the sea floor along the East Pacific rise.

## SUMMARY

The long span of time in the history of the earth that began 3.8 billion years ago after the Hadean and ended 2.5 billion years ago is known as the Archean. Rocks of the Archean are most extensively exposed for study in the broadly upwarped, geologically stable regions of continents known as Precambrian shields. The shields can be divided into a number of Precambrian provinces that tend to be distinctive in age and structural characteristics. The margins of provinces may be marked by belts of deformed rocks called collisional orogens.

The earth developed its crust during the Archean. Originally, that crust was mafic in composition and was probably formed of extrusions originating from a partially or totally molten underlying mantle. From this original mafic crust, patches of continental crust of a more felsic composition were formed by partial melting of mafic rocks and by passing the weathered products of original mafic terranes through cycles of tectonic and orogenic activity involving melting and metamorphism.

Because of high heat flow from the earth's interior during the early history of the Archean, it is likely that the crust was frequently disrupted and that continents were small. By late in the Archean, however, large continents had developed by accretion of microcontinents and consolidation of marginal depositional tracts during postdepositional orogenies.

Archean rocks can be grouped into two major rock associations. The first is the granulite association, which consists of gneisses derived from granodioritic and gabbroic rocks. The second is the greenstone association, which typically occurs in synclinal belts within the more extensive granulite terranes. Greenstones consist of mafic or ultramafic volcanics overlain by felsic volcanics and are capped by sedimentary rocks. In theory, both greenstone belts and granulite terranes may have developed in the course of Archean tectonics, in which granulite associations formed along subduction zones adjacent to volcanic arcs and greenstone belts developed in back-arc basins. Archean sedimentary rocks are found chiefly in greenstone belts and consist largely of graywackes, shales, claystones, conglomerates, and cherts.

The Archean was the time when life originated on earth. There is, of course, no primary evidence of that event in Archean rocks. However, biochemists have shown experimentally that, in an environment largely devoid of free oxygen but containing carbon, hydrogen, nitrogen, and water in which inorganic salts are dissolved, complex organic molecules can be formed when the mixture is activated by ultraviolet light or lightning-induced electrical discharges. These compounds and their derivatives are found in every living organism. They are the molecular basis of all organic structures, heritable information, and energy. A long episode of Precambrian chemical evolution eventually led to the formation of amino acids, sugars, and nucleic acids. These materials somehow acquired the organization and behavior of replicating living systems. The first living cells might have subsisted on the chemical energy they derived by consuming other cells and molecules. They were heterotrophs and were followed by cellular organisms (autotrophs) that could use the energy of the sun through photosynthesis. Free oxygen accumulated as a by-product of photosynthesis, making possible the evolution of organisms that use oxygen to burn their food for energy.

Archean fossils have been found in rocks about 3.5 billion years old and sporadically in younger rocks of Archean age. The organisms preserved are all prokaryotes. They consist of cyanobacteria and other bacteria often associated with the distinctive laminated structures known as stromatolites.

Rocks of the Archean are important sources of such metals as gold, iron, copper, chromium, and nickel. Gold is particularly characteristic of late Archean rocks, whereas the less precious metals occur in volcanogenic massive sulfide deposits that are about 3 billion years old.

## QUESTIONS FOR REVIEW

1. What is a Precambrian shield? Distinguish between a shield and a craton, and between a shield and a platform.

2. Why are Archean rocks generally more difficult to correlate than rocks of the Phanerozoic? What is the method used to date and correlate most Archean rocks?

3. Differentiate between the terms *mafic* and *felsic*. What would be an example of a felsic rock? Of a mafic rock? Name two felsic and two mafic minerals.

4. How might felsic continental crust have been derived from the mafic rocks that originally constituted the earth's crust?

5. What are komatiites? Where might they be found in some greenstone belts? What do they indicate about the temperature gradients present near the surface of the earth at the time they formed?

6. Describe the structural configuration and general vertical sequence of rock types in a greenstone belt. In a plate tectonic scenario, where might greenstone belts form and how would they develop?

7. What is the economic significance of the Archean Witwatersrand rocks of Africa? What kind of sedimentary rocks occur in the Witwatersrand Supergroup, and what was their environment of deposition?

8. What role might symbiosis have played in the evolution of eukaryotic organisms from prokaryotes?

9. What was the mixture that was used in Stanley Miller's famous experiment for artificial production of amino acids? Why was oxygen excluded?

10. In general, what is the inferred sequence of events that led to the origin of life?

11. What evidence indicates that oxygen was being produced by photosynthesis about 3 billion years ago?

12. Differentiate between a prokaryotic organism and a eukaryotic organism; between a heterotroph and an autotroph; and between anaerobic and aerobic organisms.

13. Where, and in what type of rock, do the oldest known fossils occur? What is the general nature of these remains?

14. What is the importance of amino acids in the molecular evolution and origin of life?

15. What are stromatolites? How do they originate and grow?

## TERMS TO REMEMBER

| | | | |
|---|---|---|---|
| aerobic organism | fermentation | mafic | prokaryote |
| anaerobic organism | granulite association | microspheres | proteins |
| Archean Eon | greenstone belt | nucleic acids | protocontinents |
| autotroph | Hadean Eon | photoautotrophs | shield |
| craton | heterotroph | photosynthesis | stromatolites |
| cyanobacteria | Kenoran orogeny | platform | tonalite |
| eukaryote | komatiites | Precambrian provinces | ultramafic |
| felsic | | | |

# SUPPLEMENTAL READINGS AND REFERENCES

Barghoorn, E. S., and Schopf, J. W. 1966. Micro-organisms three billion years old from the Precambrian of South Africa. *Science* 152(3723):758–763.

Bowring, S. A., and Compston, I. S. Wms. W. 1989. 3.96-Ga gneisses from the Slave Province, Northwest Territories, Canada. *Geology* 17:971–975

Fox, S. W., and Dose, K. 1972. *Molecular Evolution and the Origin of Life*. San Francisco: W. H. Freeman and Co.

Kershaw, S. 1990. Evolution of the Earth's atmosphere and its geological impact. *Geology Today* 6(2):55–60

Lowe, D. R. 1980. Archean sedimentation. *Ann Rev. Earth Planetary Sciences* 8:145–167.

Margolis, L. 1981. *Symbiosis in Cell Evolution*. San Francisco: W. H. Freeman and Co.

Nisbet, E. G. 1987. *The Young Earth*. Winchester, Massachusetts: Allen and Unwin, Inc.

# SAMUEL BOWRING

## INTERVIEW

Samuel Bowring was born in 1953 in Portsmouth, New Hampshire. His undergraduate education was completed at the University of New Hampshire where he received his B.S. degree with a major in geology in 1976. In 1979, he completed his M.S. degree in geology at the New Mexico Institute of Mining and Technology. His doctorate was obtained in 1985 from the University of Kansas where he completed a research project on uranium–lead zircon geochronology of the Wopmay orogen in the Northwest Territories of Canada. Subsequently, Bowring served on the faculty of the Earth and Planetary Sciences Department of Washington University. In 1991, he accepted his current position on the faculty of the Massachusetts Institute of Technology.

As a strong advocate for integration of field studies with experimental work, Bowring has been actively engaged in mapping projects in the southwestern United States and in geologically complex regions of northwestern Canada. An area of primary and continuing interest for him is the evolution of the earth's crust during the Archean and Proterozoic eras.

Bowring is a member of the Geological Society of America, the American Geophysical Union, and the New Mexico Geological Society. He has reviewed scientific papers for most of the major geologic journals and has been the author or co-author of over 30 articles relating to crustal evolution, the history of orogenic belts, radiometric geochronology of Precambrian rocks, and the geology of specific localities. At Washington University he taught undergraduate courses in physical geology, environmental geology, structural geology, volcanology, Precambrian geology, and tectonics.

**What originally got you interested in geology as an undergraduate?**

I became interested in geology after taking a course in physical geology as a freshman. I had always loved the outdoors and was an avid hiker and fisherman. Geology seemed to tie so many aspects of the earth together, that it really got my interest. I also had the good fortune of being involved in several research projects as an undergraduate, including a summer spent evaluating the uranium potential of north-central New Hampshire and a project using techniques of shallow exploration geophysics to find favorable sites for locating municipal wells in southeastern New Hampshire.

INTERVIEW INTERVIEW INTERVIEW INTERVIEW
VIEW INTERVIEW INTERVIEW INTERVIEW INTERVIEW INTER
INTERVIEW INTERVIEW INTERVIEW INTERVIEW INTERVIEW
VIEW INTERVIEW INTERVIEW INTERVIEW INTERVIEW INTER
INTERVIEW INTERVIEW INTERVIEW INTERVIEW INTERVIEW
VIEW INTERVIEW INTERVIEW INTERVIEW INTER

**Tell us about your research specialties and how you chose those specialties over other fields within the earth sciences.**

My research specialty is the study of the evolution of continental crust and lithosphere. In particular, I am interested in how continental lithosphere is formed, preserved, and destroyed. The best way to do this is to study the evolution of orogenic belts. In orogenic belts one can investigate the mechanisms by which continental crust develops by both tectonic and magmatic processes. I use an integrated approach that involves field mapping as well as laboratory work. One of the most important laboratory methods I use is the U-Pb zircon method of dating. This method allows very precise measurement and calibration of the geologic time scale. In many cases it is possible to determine the age of Precambrian rocks with an uncertainty of only 5 million years. The growing use of the U-Pb dating method in zircon has revolutionized our understanding of the growth and development of continental crust. In particular it has allowed us to examine the Precambrian rock record where fossils do not allow any insights into relative or absolute age control. In the last 10 years, it has become apparent that the timing between Precambrian orogenic episodes is not much different from that in Cenozoic examples.

I was drawn to the study of the evolution of continental crust and lithosphere when I began to study Precambrian rocks in the southwestern United States at graduate school. As Precambrian rocks lack fossils, precise geochronology becomes an important way to study the evolution of crust. One of the biggest questions in the evolution of the earth is whether there has been secular change. We know that the earth was considerably hotter early in its history, but we don't know how this extra heat was dissipated. Some have sug-

gested that the extra heat was lost by faster rates of plate tectonic activity. To evaluate this hypothesis and the question of secular change, we must study Precambrian orogenic belts and compare their evolution to Cenozoic examples. Another big question in the study of crustal evolution is how and when the earth differentiated into an enriched sialic continental crust and a depleted mantle and, once formed, how much of the sialic crust is recycled back into the mantle. Various models have been proposed that range from development of all of the sialic crust very early in earth history to the earliest crust being basaltic with sialic crust developing slowly through earth history. Clearly one of the biggest obstacles in understanding the early history of the earth is the paucity of preserved rocks in the age range from 4.6 to 3.8 billion years. Another big unknown in the first billion years of earth history is the role of meteorite bombardment in disrupting and recycling the earth's earliest crust.

One of the best ways to approach the question of crustal growth is to use isotopic studies to constrain the relative amounts of older crust and mantle-derived magmas involved in forming crust at different times in earth history. Time and time again as I began to consider the problems of crustal evolution, it became apparent that field mapping coupled with precise dating and isotopic tracer studies of Precambrian rocks was an extremely powerful approach toward constraining these problems.

**Let's talk about one of your recent discoveries of the earth's oldest known rock, dating back 3.96 billion years. What is the significance of this discovery? How was this date obtained?**

The significance of the oldest rock (3.96 billion years) is twofold. First, it shows us that contrary to many predictions, there is crust older than 3.8 billion years and its presence provides us with an important constraint on the destruction of earth's earliest crust by meteorite bombardment. Secondly, because the rocks are "evolved," it seems likely that they were derived in part from even older crust. Neodymium isotopic data from these rocks suggest that this older crust could be as old as 4.1 billion years.

So the Acasta Gneiss is the first chance we have had to investigate approximately 4.0-billion-year-old terrestrial rocks and to gain insights into their chemistry. Initial results suggest that these rocks are no different than many younger rocks, and in fact, one of the rocks is granitic in composition and appears chemically and petrographically similar to younger rocks in the region.

### What advice would you offer an undergraduate who is thinking of becoming a geology major? A professional geologist?

I tell undergraduate students who come to ask me about geology that the science offers many possibilities that range from understanding how the earth has evolved over geologic time to understanding how to find, exploit, and preserve our natural resources. As many wars are fought over natural resources, it is obviously a very important

Geologist Samuel A. Bowring examining the 3.96 billion-year-old Acasta Gneiss, Northwest Territories, Canada. *(Photo courtesy of Paul Hoffman.)*

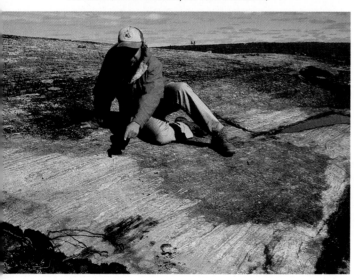

aspect of geology. In addition, a good knowledge of geology is essential for someone who works with environmental issues. Most importantly, geology offers the opportunity to integrate the fields of biology, chemistry, physics, and mathematics toward a better understanding of how this and other planets have evolved.

### Tell us more about your current or most recent research and your plans for future investigations.

My current research is centered around four major projects. First, I am continuing to map and date the Acasta Gneiss of the Slave Province, NWT, Canada. The terrane is apparently quite large and very complex so there is a lot of work to be done in understanding its evolution. As I have already described, these rocks provide the only example to date of rocks that are nearly 4.0 billion years old and with evidence for involvement of even older crust. As the nearly 4.0-billion-year-old rocks were discovered almost by accident, we are quite hopeful that we will find even older rocks in the region, as suggested by the neodymium isotope systematics.

Second, adjacent to the Slave Province is a Proterozoic (1.9-billion-year-old) orogenic belt called the Wopmay orogen. I have worked there for the past ten years, and it is often cited as one of the best examples of a Precambrian orogenic belt that can be explained in terms of plate tectonics. Working there has been a fantastic chance to see how scientific research advances. Each year as mapping and laboratory work progresses, we develop models for the orogenic belt and then test the models, find the flaws, and develop new models. We are continuing a large effort to understand its evolution and feel that it may be one of the best exposed and accessible collisional orogenic belts in the world.

Third, I am also working on understanding how the Precambrian rocks of the southwestern United States were added to the North American continent 1.8 to 1.6 billion years ago. This involves field mapping and dating of the rocks exposed in Arizona and New Mexico. This is an exciting place to study the processes of crustal growth and evolution. It appears as though much of the crust and lithosphere in this region is "juvenile." This means that the rocks were originally formed on oceanic lithosphere without the influence of older continental crust and were amalgamated into continental crust. Exactly how fragments of juvenile crust and lithosphere are transformed into stable continental crust is an important problem in geology and geophysics. We know that continental crust is underlain by thick lithospheric mantle that serves to stabilize the crust. Exactly when and how this lithosphere forms is not well understood. We are attempting to better constrain this by studying the history of crustal growth and stabilization in this region. To do this, we must study the structure, metamorphism, geochronology, and isotope geochemistry of these rocks.

Finally, I have also recently initiated a large collaborative project with scientists from the United States and the Soviet Union to study the geologic evolution of the Baikal rift in Siberia, one of the largest rift zones in the world. A continental rift zone is generally characterized by upwelling asthenosphere beneath continental lithosphere. This is a tremendous opportunity to study the effect of a preexisting boundary in the lithosphere on the rifting process and on the geochemistry of basalts erupted in the rift. The Baikal rift developed along a suture zone between the Archean Siberian Platform and a collage of Proterozoic crust. It has often been linked to the collision between India and Eurasia, but the geochronology of the rift-related magmatism is poorly known. By studying the rift-related magmatism and the basement rocks through which they were erupted, we hope to gain insights into the three-dimensional nature of this major suture zone and to investigate the effects of juxtaposing thick Archean lithosphere with thinner Proterozoic lithosphere.

## What good is an understanding of earth science to someone who does not plan to be a geologist?

A better understanding of the earth is essential for all students because, if nothing else, it offers explanations and insights into natural processes that we often take for granted. In any case, geology will play a major role in our survival as a species as we change the global weather patterns and deplete the earth of its natural resources.

## Has your work changed any of the traditional views of Precambrian geologic history? If so, how?

I think my work has helped to change the traditional view of Precambrian geology that existed 10 to 15 years ago. It has helped because it is part of a renewed interest and excitement in studying the evolution of the earth's crust and lithosphere. Plate tectonic activity has destroyed all oceanic lithosphere older than about 250 million years; thus, our only clues to tectonic processes and the geochemical evolution of the earth are contained in the continents. The Precambrian rock record provides us with the ultimate test of any model for the thermal and geochemical evolution of the earth. In addition, as more of the Precambrian is examined, the traditional view of 20 to 30 years ago that most of the Precambrian is a hopeless chaotic mass of gneisses has been changed. Wopmay orogen certainly has gone a long way in showing how spectacular Precambrian orogenic belts can be.

Precise geochronology has revolutionized our understanding of the origin of continental crust and lithosphere. Twenty years ago Precambrian orogenesis was viewed as a very protracted process that took place over hundreds of millions of years, as geochronologic studies had to rely mostly on K-Ar and Rb-Sr ages which are very susceptible to resetting during later metamorphism. With the recent proliferation of U-Pb zircon dating, we can see that Precambrian orogenic belts such as the Wopmay orogen develop with remarkable similarity in timing to Cenozoic orogens. As much more geochronologic data accumulates from Precambrian orogenic belts, it may be possible to begin to address the question of whether plate tectonic processes operated at a faster rate in the Precambrian.

The discovery of the 3.96-billion-year-old Acasta Gneiss has changed many people's view of earth's earliest history. Since the discovery of the almost 3.8-billion-year-old rocks in Greenland, I think many people assumed these were the oldest rocks that would ever be discovered. The Acasta Gneiss has generated new interest in finding and studying the oldest fragments of terrestrial rocks so that we can better understand the earth's earliest history.

Early Proterozoic graywackes with graded bedding interstratified with shaly limestone, deposited in the aulacogen east of Great Slave Lake, Northwest Territories
of Canada. *(Courtesy of P. Hoffman.)*

# The Proterozoic Eon

*Where these ancient formations have been intensely deformed by mountain-building forces or by later igneous intrusion, and then profoundly eroded, they are like sheaves of a manuscript that have been torn and scattered, leaving only "internal evidence" as a clue to the order in which the recovered parts should be fitted together.*

Charles Schuchert and Carl O. Dunbar, 1937

## Proterozoic Continental Growth

The Proterozoic Eon began 2.5 billion years ago and ended only 570 million years ago. This enormous period of time comprises 42 percent of earth history. The designation of the beginning of the Proterozoic at 2.5 billion years ago is somewhat arbitrary, but it is used because this is about the time when a more modern style of plate tectonics and sedimentation began to be prevalent. Because Proterozoic rocks are less altered, they are easier to study and interpret than rocks of the Archean. Difficulties persist, however, for they are not as abundantly fossiliferous as many Phanerozoic strata.

As described in the preceding chapter, a number of distinct cratonic elements, designated the Superior, Slave, Hearne, Rae, Wyoming, and Nain provinces, had developed in North America during the Archean. These Archean elements were welded together to form a large continent called **Laurentia** during the early Proterozoic. The welding occurred along belts of crustal compression, mountain building, and metamorphism called **orogens** (Fig. 8–1). This process of aggregation and consolidation of the once separate cratonic elements was completed by about 1.7 billion years ago. During the remainder of the Proterozoic, Laurentia continued to grow as a result of accretionary events. Accretion involves additions of crustal materials to continental margins. The additions may consist of sedimentary rocks de-

posited along the once passive margins of continents which may later be compressed and welded onto the continent. Accretion may instead involve the incorporation of small blocks of crustal material carried to subduction zones adjacent to continents by sea floor spreading.

The processes of accretion and aggregation of Archean crustal elements were in operation worldwide during the Proterozoic and resulted in the formation of several very large cratons. On and around these cratons, tectonic events occurred and environments of deposition developed that closely resemble those of the succeeding Phanerozoic Eon. By Proterozoic time, wide marginal shelves and epeiric seas existed in which shallow water clastic sediments and carbonate deposits could accumulate. On the whole, such sediments were rare in the Archean. It is also evident that the Proterozoic was characterized by substantial amounts of lateral plate motion and accompanying subduction, rifting, and sea floor spreading.

## Early Proterozoic Events in North America

### The Wopmay Orogen

In the Canadian shield, accretionary early Proterozoic rocks occur primarily in areas that circumvent the older Archean provinces. One such area lies

**Figure 8–1** Expanded and generalized relationships of Proterozoic orogens to Archean provinces. The Wopmay, Penokean, Labrador, and Mazatzal orogens developed by accretion of sedimentary prisms laid down along the margins of Archean cratons. The other orogens are collisional in nature. *(From Hoffman, P. 1989. Precambrian geology and tectonic history of North America, Ch. 16, in Geology of North America. Boulder, CO: Geological Society of America.)*

along the western margin of the Slave province in Canada's Northwest Territories. Here, in a belt of deformation, igneous intrusion, and metamorphism called the **Wopmay orogen** (Fig. 8–2), there is evidence of the opening of an ocean basin, sedimentation along resulting new continental margins, and closure of the ocean basin through plate tectonic processes. This sequence of events, well documented in the Phanerozoic, is called a **Wilson cycle**, after J. Tuzo Wilson, one of the pioneers of the plate tectonics theory.

In the region of the Wopmay orogen, evidence for the initial opening of an ocean basin consists of numerous normal (tensional) faults (Fig. 8–3) which created downfaulted or rift valley basins in which accumulated alluvial fan and fluvial deposits. Lavas flowed upward through the fault systems and became interlayered with many of these sediments.

Continued opening produced an oceanic tract which slowly widened. As the ocean widened, the western edge of the Slave province became a passive continental margin on which two parallel zones of deposition developed. An eastern zone next to the Slave platform received coastal plain deposits that graded westward (seaward) to shallow marine quartz sandstones. These sandstones were subsequently metamorphosed and are now quartzites. The quartzites pass upward into massive stromatolitic dolomites of the Rocknest Formation (Fig. 8–4). To the west of these shallow continental shelf deposits, deeper water turbidites of the continental rise were deposited.

After a period of relatively uneventful sedimentation along the Wopmay tract, the more dynamic episode of the Wilson cycle involving ocean closure was initiated. As a result of westward subduction of

**Figure 8–2**  Tectonic elements in and around the Wopmay orogen, Northwest Territories, Canada *(From Hoffman, P. 1973. Evolution of an early Proterozoic continental margin: The Coronation geosyncline and associated aulacogens of the northwestern Canadian shield.* Phil. Trans. Royal Soc. London, A. *273:547–581.)*

**Figure 8–3**  Relationship of rock units following early Proterozoic opening of an ocean basin along the western margin of the Slave craton. *(Simplified from Hoffman, P. 1989.* Geology of North America, *vol. A:447–512. Boulder, CO: Geological Society of America.)*

A

B

**Figure 8–4**   (A) Dolomite and interbedded shales of the Rocknest Formation, Northwest Territories, Canada. (B) This oblique aerial view impressively illustrates the folding of these dolomitic rocks that resulted from east–west crustal shortening and accretion that occurred during the orogenic phase of the development of the Wopmay orogen. *(Courtesy of P. Hoffman.)*

the leading edge of the Slave plate beneath a microplate approaching from the west, the continental shelf buckled downward, creating a deep elongated area of subsidence that ended carbonate deposition and served as a basin for the accumulation of deep water clastics. When the foredeep had filled,

deltaic and fluvial sands were deposited above the marine clastics. Subsequently, the region was subjected to additional folding and faulting caused by plate collisions and accompanying east–west crustal shortening.

These events in the Wopmay orogen closely parallel those that occurred during the late Paleozoic evolution of the Appalachian Mountains. The history of both tracts began with deposition of alluvial deposits in fault-bound basins followed by opening of an oceanic tract; progressed to the deposition of continental shelf and rise sediments along passive continental margins; and ended with ocean closure, deformation, and deposition of nonmarine clastic sediments derived from the erosion of mountains produced during the closure episode.

## Aulacogens

Another tectonic event of the Slave province that resembles Phanerozoic tectonics is the development of aulacogens. An **aulacogen** is an inactive rift of a radiating three-rift system that develops over an area of crust that is bulging upward, possibly because of an underlying mantle plume. As the bulge forms, the stretched crust breaks to form three rift zones that radiate from a common center. Two of these rifts may widen and become continental margins, whereas the third trends inland away from the continental margin, becomes the "failed arm" of the three-pronged system, and fills with sediment. Two aulacogens have been detected along the western side of the Slave province (see Fig. 8–2). The Athapuscow aulacogen extends northeastward from Great Slave Lake, and the Bathhurst aulacogen trends southeastward from Coronation Gulf.

## The Trans-Hudson Orogen

The Wopmay is just one of many orogens that developed along the margins of Archean provinces during the Proterozoic. Another is the **Trans-Hudson orogen** which separates the Superior province from provinces to the north and west (see Fig. 7–4). Rocks of the Trans-Hudson belt record a historical sequence of initial rifting, opening of an oceanic tract, and subsequent closure along a subduction zone. This closure, with the accompanying severe folding, metamorphism, and intrusive igneous activity, welded the Superior plate to the Hearne and Wyoming plates lying to the west.

## An Early Proterozoic Ice Age

In the region north of Lake Huron, rocks of the Proterozoic are called **Huronian.** They consist mainly of coarse clastics that were eroded from older crystalline rocks. In addition, the Huronian Supergroup includes the **Gowganda Formation,** a rock unit that is notable because of its conglomerates and laminated mudstones of probable glacial origin. Laminations in the mudstones appear to represent regularly repeated summer and winter layers of sediment that are called *varves.* As described in Chapter 4, varved mudstones often form in meltwater lakes that are marginal to ice sheets. The varved sediments in the Gowganda Formation alternate with tillites (consolidated rocks composed of unsorted glacial debris) indicating periodic advances of ice into the marginal lakes. Cobbles and boulders in the tillite have been scratched and faceted, providing evidence of the abrasive action of the ice mass as it moved across the underlying bedrock. The Gowganda Formation is intruded by igneous rocks that have been dated as 2.1 billion years old. They rest on a basement of older crystalline rocks that are 2.6 billion years old. Hence, in North America, glaciation occurred at some time between these two dates. Glacial conditions and accompanying cooler climates were widespread at this time, for rocks similar to the Gowganda are recognized in Finland, southern Africa, and India. This early Proterozoic episode of glaciation may have been the first our planet experienced, for the higher thermal regimes of the Archean would have made the accumulation of sufficient ice to form continental glaciers unlikely.

## The Animikie Group

Lower Proterozoic rocks surrounding the western shores of Lake Superior include those of the **Animikie Group.** This group of formations is world famous for the bonanza iron ores they contain. The ore minerals are oxides of iron and thus record the presence of free oxygen in the earth's atmosphere. By implication, the oxygen-generating process of photosynthesis was probably in vigorous operation by this time. Coarse sandstones and conglomerates deposited in shallow water lie near the base of the Animikie Group. These rocks are overlain by cyclic successions of chert, cherty limestones, shales, and banded iron formations (known by the acronym BIF). Banded iron formations are also known in Archean sequences, but these are not nearly as extensive as those of the Proterozoic (Fig. 8–5). Some of the Canadian deposits are over 1000 meters thick and extend over 100 km. Within the banded iron sequence, there is a formation known as the **Gunflint Chert** which contains an interesting assemblage of cyanobacteria and other prokaryotic organisms.

A

B

**Figure 8–5**  Banded iron formations. (A) This banded iron formation is exposed at Jasper Knob in Michigan's Upper Peninsula. (B) Banded iron formations occur worldwide in Proterozoic rocks, as suggested by these at Wadi Kareim, Egypt. *(Photograph (B) courtesy of D. Bhattacharyya.)*

**Figure 8–6** The Labrador fold belt depicted in a Landsat image made north of Quebec. The belt was once a chain of mountains raised by the collision of two continents about 1.8 to 1.7 billion years ago. Since then, the mountains have eroded, exposing the deeper metamorphic and igneous rocks that were deformed into these folds by the collision. *(Courtesy of NASA.)*

## The Labrador Trough

East of the Superior province, one finds extensive outcrops of rocks that, like those of the Wopmay orogen, were deposited on the continental shelf, slope, and a rise of a craton. These rocks occur along a curving, elongate tract known as the Labrador Trough. On the western side of the tract, quartz sandstones, dolomites, and iron formations lie above fluvial and alluvial deposits of fault-bound basins. Pillow lavas, tholeiitic basalts, mafic intrusives, and graywackes occupy an adjacent zone to the east. The eastern zone was subjected to intense folding (Fig. 8–6), metamorphism, and westward thrust

faulting during an orogenic event named the **Hudsonian orogeny.** The Hudsonian orogeny serves as the event that separates early Proterozoic from middle Proterozoic geologic history.

## Middle Proterozoic Events

### Keweenawan Sequence

Among the rocks deposited or emplaced during the middle Proterozoic were those of the Keweenawan sequence (Fig. 8–7). These rocks rest upon either

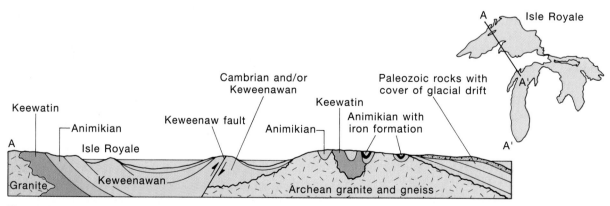

**Figure 8–7** Geologic cross-section across Lake Superior. *(Adapted from Michigan Geological Survey,* Bedrock of Michigan, *1968.)*

crystalline basement or Animikian strata. They extend from the Lake Superior region southward beneath a cover of Phanerozoic rocks for hundreds of kilometers. Keweenawan rocks consist of clean quartz sandstones and conglomerates as well as basaltic volcanics. The lava flows are well known for their content of native copper. Holes originally formed as gas bubbles in the extruding lava provided some of the voids in which the copper was deposited. In addition, copper fills small joints and pore spaces in associated conglomerates.

The Keweenawan lavas accumulated to a thickness of several thousand meters. Even with so great an outpouring, however, much of the supply of mafic magma was not tapped. It remained beneath the surface where it crystallized to form the 12,000-meter-thick and 160-km-long mass of the Duluth Gabbro. As noted earlier, basalts are rock types more characteristic of oceanic areas. When such a large quantity of mafic material comes to the surface within the central stable region of a continent, it often signals the presence of a rift zone along which a continent is going to break apart and admit an ocean. The tract where the break begins typically develops tensional faults along which mafic magma rises to the surface to form Keweenawan-like accumulations. Evidence from gravity and magnetic surveys, as well as samples from deep drill holes, indicates that the rift zones associated with Keweenawan volcanism developed about 1.2 to 1.0 billion years ago, and extended from Lake Superior southward beneath covering sedimentary rocks into Kansas. If this system of rifts had been extended to the edge of the craton, an ocean tract would have formed within the rift system and the eastern United States would have drifted away. Rifting ceased, however, before such a separation could occur.

## The Grenville Province

The Grenville province of eastern Canada is the youngest middle Proterozoic region of the Canadian shield and the last to experience a major orogeny. Exposures of Grenville rocks extend from the Atlantic coast of Labrador to Lake Huron. This, however, is only part of their true extent, for they continue beneath covering Phanerozoic rocks down the eastward side of the United States and westward into Texas (see Fig. 7–4). In the United States, the study of Grenville rocks is difficult, for they have been sub-

jected to the destructive processes that accompanied the building of the Appalachian Mountains during the Paleozoic.

Typical Grenville rocks consist of carbonates and sandstones (Fig. 8–8) that have been metamorphosed and intruded by igneous bodies. In eastern Canada, these rocks have been compressed into overturned folds having northeasterly axes. Deformation of Grenville sediments occurred 1.2 to 1.0 billion years ago during the **Grenville orogeny.** Two models have been proposed for the cause of this great orogenic event. The first is a plate tectonics model that began with sedimentation along the passive eastern margin of Laurentia. The sedimentation phase was followed by mountain building resulting from collision of another continental block against eastern Laurentia. The alternative model depicts the orogeny as occurring within the continental plate rather than at its margin. In this proposal, a region of subsidence is thought to have formed by ductile stretching of the continental crust over an underlying mantle plume. The depression formed by the stretching became the basin in which Grenville sediments were deposited. Subsequently, the area of

**Figure 8–8** Folded sandstones and carbonates of the late Proterozoic Grenville Group, Belmont Township, Ontario, Canada. *(Courtesy of the Geological Survey of Canada.)*

stretching shifted laterally, and sediments in the former basin were squeezed against the adjacent cooler and more rigid crust. It is not yet known which of these models is the most valid, but the majority of opinion seems to favor the plate tectonics model.

## The Late Proterozoic

The final segment of Proterozoic time is represented in North America by rocks that were deposited in basins and shelf areas surrounding the continent. Most of these sediments were later deformed during Paleozoic orogenic events. In eastern North America, one such event involved rifting and the filling of rift valleys with lava flows and coarse clastics. The sea entered the rift valleys as the beginning of a widening proto-Atlantic ocean, and a new continental margin was established east of the Grenville orogen. The stage was then set for the deposition of the first Paleozoic shelf sediments.

Also in upper Proterozoic rocks, one can find exposures of unsorted boulder beds that are considered by many geologists to be glacial deposits. Such deposits are known in Utah, Nevada, western Canada, Alaska, Greenland, South America, Scandinavia, and Africa. At every locality, the cobbles and boulders exhibit the telltale striations and faceted surfaces that attest to transport by glaciers. This late Proterozoic ice age lasted about 240 million years. It was not the first ice age for the Proterozoic. As described earlier, the Gowganda tillites and correlative formations on other continents reflect widespread glaciation slightly more than 2 billion years ago. Nor would the late Proterozoic glaciations be the last the earth would witness, for glacial ice advanced again over major portions of the continents during the Ordovician and Silurian, near the end of the Paleozoic, and again during the Quaternary Period of the Cenozoic Era (Fig. 8–9). Each of these major episodes of glaciation contained shorter intervals during which the ice sheets alternately advanced and regressed. The causes for these oscillations are a matter of great interest to geologists and will be discussed in Chapter 14.

In our standard geologic time scale, the late Proterozoic is followed by the Cambrian Period of the Paleozoic Era. Nineteenth century geologists re-

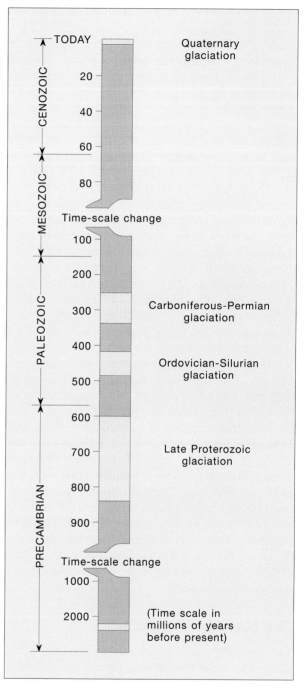

**Figure 8–9**  In its long history, the earth has experienced several major episodes of widespread continental glaciation. Each of these major episodes contained shorter intervals of advance and retreat of the glaciers.

garded strata that contained the first occurrence of fossils as Cambrian in age. They, of course, had no knowledge of the many animal fossils recently discovered in upper Proterozoic rocks, a few of which are even shell-bearing. These discoveries of the last few decades necessitate a more limited definition of the boundary between the Proterozoic and Phanerozoic. Most geologists now agree that the boundary should be defined by the first occurrence of abundant shelled invertebrates, including archaeocyathids. This definition excludes the relatively rare occurrences of tiny shelled tubular fossils found in continents of the Southern Hemisphere. Also, by insisting that the fossils be shell-bearing, one can exclude strata bearing impressions of soft-bodied animals, such as occur sporadically in latest Proterozoic rocks.

## Proterozoic Rocks South of the Canadian Shield

Although outcrops are not as extensive as in Canada, Precambrian rocks also occur south and west of the Canadian shield. Particularly impressive successions are exposed in the Rocky Mountains and Colorado Plateau regions. These rocks have had a complex history that began over 2.5 billion years ago with the evolution of an Archean terrane composed of strongly deformed and metamorphosed granitic rocks. Largely on the basis of the study of remnant patches of volcanic rocks and greenstones in Wyoming, it appears likely that this old Archean mass collided with one or more island arc terranes about 1.7 or 1.8 billion years ago. The line of collision is believed to be marked by a shear zone in southern Wyoming and western Colorado characterized by severely crushed and brecciated rocks (Fig. 8–10). Following this orogenic event, there was an extensive episode of magmatism during the middle Proterozoic. Plutons were emplaced in a broad belt across North America from California to Labrador. The 1.4- to 1.5-billion-year-old plutons are composed primarily of anorthosites and granites. They are *anorogenic,* meaning that they are not associated with an orogenic cycle. This was the first time in the earth's history that extensive anorogenic magmatism occurred. It is likely that the precondition for this widespread magmatic activity was the existence of a broad region of stable continental crust.

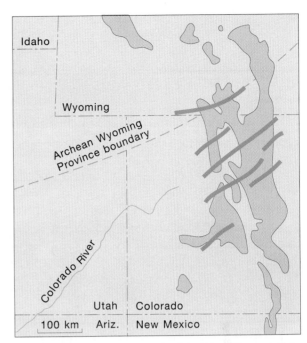

**Figure 8–10**   Shear zones (red) in Wyoming and Colorado developed when the Archean craton collided with island arc terranes during the early Proterozoic. Areas of lower Proterozoic outcrops are shown in green. *(Adapted from Karlstrom, K. E., and Bowing, S. A. 1988. Jour. Geology 96:561–571.)*

The next event south of the Canadian shield was a widespread episode of rift development about 0.85 to 1.4 billion years ago. Large fault-controlled depressions were formed in which thick sequences of late Precambrian sediments accumulated. They include the Uinta Series of central Utah, the Pahrump Group of southeastern California, and the **Belt Supergroup** of Montana, Idaho, and British Columbia. Because of the scenic features with which they are associated, the Belt rocks are of particular interest. In places these rocks are over 12,000 meters thick. Especially impressive are the massive cliff-forming limestones that can be viewed at Waterton Lakes and Glacier National Parks (Fig. 8–11). Although exceptionally thick, the Belt rocks display features such as ripple marks and algal structures that indicate they were deposited in relatively shallow water along the passive western margin of North America.

The last event in the late Precambrian history of the western United States was the development of a north–south trending passive margin. At this time the North American plate was moving away from a

**Figure 8–11**   Undeformed, horizontal limestones of the Belt Supergroup comprise the upper third of Chief Mountain, located just outside of Glacier National Park, Montana. The lower two-thirds of the mountain consists of deformed Cretaceous beds. A thrust fault separates the two and represents the surface over which the Proterozoic beds were pushed eastward over the weaker underlying Cretaceous strata. The fault is known as the Lewis thrust.

spreading center located somewhere to the west, and a passive margin marked the continent's trailing edge. The continental shelf was the site of the shallower zone. A westward-thickening wedge of latest Precambrian and Cambrian sediments was deposited on this shelf. The distribution of these sediments was not at all affected by the fault-controlled basins of the earlier episode of deposition.

Precambrian rocks of the Grand Canyon region consist of two distinct units. The lower and older unit is the Vishnu Schist, and the upper unit is the Grand Canyon Series. The Vishnu is a complex body of metamorphosed sediments and gneisses that have been intensely folded and invaded by granites. The granitic intrusions that cut into the Vishnu (and correlative rocks of the southwestern United States) were emplaced 1300 to 1400 million years ago as part of a deformational event named the **Mazatzal orogeny**. A splendid example of a nonconformity occurs at the contact between the Vishnu Schist and the overlying Grand Canyon Supergroup (Fig. 8–12). The latter unit is late Proterozoic in age and is itself unconformably overlain by Paleozoic rocks. The Grand Canyon Supergroup correlates to the Belt Supergroup. It consists mostly of clastic rocks (sandstones, siltstones, and shales) that accumulated in a troughlike basin that extended into the craton. The Chuar Group of the Grand Canyon Supergroup contains small, circular carbonaceous structures believed to be the fossil remains of algal spheres.

## Precambrian Events Outside North America

The most significant global event of late Proterozoic time was the gathering of Precambrian continents to form a large supercontinent named **Proto-Pangea** (to distinguish it from the late Paleozoic supercontinent known as Pangea). In the course of the coming together of continents, it is likely that a Southern Hemisphere land mass collided with the eastern margin of North America. Many geologists believe such a collision was a component of the Grenville orogeny. Similar collisions occurred around the globe, but this marriage of continents was brief.

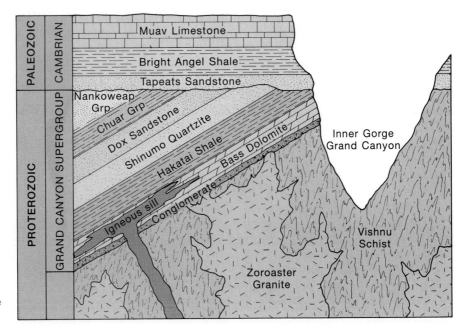

**Figure 8–12** Vishnu Schist, Grand Canyon Supergroup, and other rocks in the Grand Canyon of the Colorado River.

Shortly after coming together, rifts developed along which North America and Greenland moved away from the other elements of Proto-Pangea. The ocean flooded into the rift zone and then slowly broadened until an oceanic tract was formed called the **Iapetus Ocean**.

## The Baltic Shield

Precambrian rocks are found in many of Europe's mountain ranges. However, the most extensive exposures form the surface of the Baltic shield (Fig. 8–13). The northwestern edge of the shield is bounded

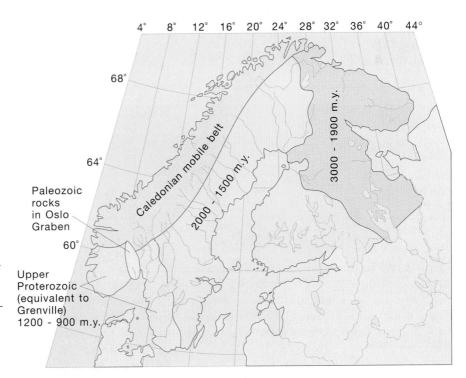

**Figure 8–13** Region over which Precambrian rocks outcrop in the Baltic shield. Oldest (Archean) rocks are found in the far west, and youngest at the southwestern tips of Norway and Sweden. The Caledonian mobile belt is composed mostly of Paleozoic rocks.

by an early Paleozoic orogenic belt. In general, there is an east–west trend in the age of rocks composing the Baltic shield. The oldest rocks are of Archean age and are exposed on the eastern side. These are followed by younger Proterozoic rocks in the central region of the shield. The youngest Precambrian rocks are found at the southwestern ends of Norway and Sweden. These rocks are approximately equivalent in age to the Grenville sequence in North America and may have been emplaced at about the time that western Europe began to rift away from Laurentia. A large downfaulted block (graben) near Oslo contains an interesting sequence of fossiliferous Paleozoic strata.

The extremely complex Precambrian history of the Baltic shield has been studied intensively by Swedish and Finnish geologists, who have recognized four great sedimentary and volcanic cycles separated by tectonic movements and batholithic intrusions. The rocks representing the two oldest cycles are mostly schists and gneisses that were once graywackes and volcanics. Conglomerates, ripple-marked quartz sandstones, dolomites, greenstones, slates, and iron ores characterize the third cycle. The most recent rocks escaped severe metamorphism and consist mostly of sandstones and lavas intruded by granites.

## The Angaran Shield

The Angaran shield (Fig. 8–14) is more difficult to study than either the Canadian or the Baltic shield. These Precambrian rocks are not exposed over a vast region. Rather, they are revealed as small exposed patches that elsewhere are covered by younger sedimentary rocks of the great Siberian Platform. Covered or not, however, the shield is important as the very nucleus of Asia. To the east and south of this region lay the mobile belts that were later to be fashioned into the Ural and Himalayan mountains. Not unlike the Precambrian of other shields, the Angaran has an older complex of gneisses, schists, and granites that provides evidence of a long and complicated Archean history. There is also a Proterozoic sequence of sedimentary rocks, volcanics, and granitic intrusives that dates from about 1.6 billion years ago. Russian geologists have named these rocks the

**Figure 8–14** Areas of Precambrian outcrop in the Angaran shield. Note that vast areas of Precambrian rocks are covered by younger rocks. *(Data from* Geologic Map of U.S.S.R., *1960.)*

*Rhiphaen rocks.* They include sandstones and limestones of shallow seas. Within the limestones one finds massive reefs of *stromatolites,* the laminar moundlike or fingerlike structures produced by algae that grew in profusion along carbonate banks of Precambrian seas (see Fig. 7–17).

For the very latest Proterozoic rocks, Russian geologists propose the term **Vendian System**. In the Ukraine where it is best exposed, the Vendian incorporates strata from the bottom of the lowest upper Proterozoic glacial deposits up to the base of the Cambrian. Rocks of the Vendian are notable for the burrows and impressions of soft-bodied metazoans (multicellular animals) they contain.

## Shields of South America

The number three seems to characterize the Precambrian geology of South America. There are *three* shields: the Guianan, the Brazilian, and the Patagonian (Fig. 8–15); also, the Precambrian rocks of South America are usually grouped into *three* divisions: lower, middle, and upper Precambrian. As might be expected, gneisses and schists form the oldest (lower) sequence. The middle Precambrian consists, in large part, of metamorphosed sediments, volcanics, and intrusions that appear to be the product of accretionary evolution. Metamorphic effects are less profound in the late Precambrian. Slates and phyllites are still in evidence, however, along with quartzites, conglomerates, volcanic flows, and ash beds. Because of a covering of younger sediments over parts of the basement as well as the dense forests that cover many regions, the Precambrian rocks of South America are still not as well known as those of the relatively barren shields of Canada and Scandinavia.

## Africa

Precambrian rocks occur over half the surface of Africa and elsewhere lie beneath a veneer of Paleozoic and Mesozoic rocks. Indeed, Africa appears to be essentially a vast platform of abutting shield segments. It is mostly a very stable continent, but along its eastern edge are the remarkable fault structures that produced the African rift valleys. Africa's Precambrian rocks are renowned for their treasure of minerals; exploration for gold, diamonds, copper, uranium, and chromium has resulted in improved understanding of the continent's complex geology.

**Figure 8–15** Regions of exposed Precambrian rocks in South America are shown in light red. The Andean mobile belt is colored green. Yellow areas consist of relatively undisturbed Phanerozoic strata. *(After Umgrove, J. H. F. The Pulse of the Earth. Nijholt: The Hague, 1947.)*

As described in Chapter 7, the Archean of Africa includes thick sections of komatiites and mafic lavas that often appear to be immersed in a sea of granitic intrusions. These igneous rocks constitute the Archean basement of South Africa. In most areas of South Africa, a nonconformity separates the lavas and batholithic granitic masses from the younger sequences of the lower Proterozoic.

By about 1.9 to 1.1 billion years ago, four stable segments of the African craton had been formed (Fig. 8–16). One large segment forms much of the western African bulge, two large elements were established in central Africa, and one formed in southern Africa. Belts of metamorphosed granitic crust formed between these crustal segments during the late Proterozoic and welded them into a unified shield. The orogenic activity accompanying the welding of crustal elements continued until about

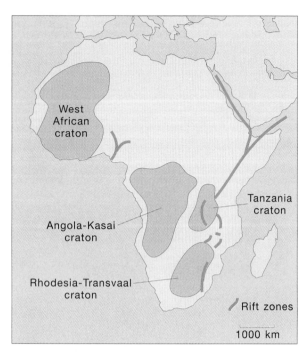

**Figure 8–16**   The four large cratonic segments of Africa.

400 million years ago and resulted in the continent taking on its present-day outlines.

## India

From a geologic point of view, India can be separated into two distinctly different regions. To the north is the Himalayan orogenic belt that developed long after the Precambrian, whereas to the south lies the Indian shield. It is now believed that the shield was brought into juxtaposition with the Himalayan region at a comparatively recent date by northward drifting of what was once an Indian "island continent."

As is the case in North America, geologists have been able to recognize a number of separate provinces within the shield that can be distinguished by their radiometrically determined ages and distinctive structural characteristics (Fig. 8–17). Five of these provinces are now recognized, and each is believed to be the craton of a separate continental fragment. Orogenic belts and zones of compressional faulting serve to join the cratons together. The joining of component cratons was apparently complete by about 1.5 billion years ago (middle Proterozoic). Subsequently, the shield was remarkably stable, until in Late Cretaceous time when huge volumes of Dec-

can basaltic lavas were extruded along numerous fissures. The outpouring of lava may have been in response to the separation of India from the Seychelles Islands. The relative tranquility of India was again disturbed by its collision with Asia only 40 million years ago.

## Australia

Australia is composed of Phanerozoic orogenic belts along the eastern side of the continent and consists of shield and platform elements over most of the central and western parts of the continent. The largest area of exposed Archean rocks is in the Yigarn shield in western Australia (Fig. 8–18). North of the Yigarn shield lies the smaller Pilbara shield. Both of these Archean elements contain the greenstone belts and granulite expanses that typify Archean geology on other continents. At least two great episodes of accretion and mountain building are represented in these Archean rocks.

**Figure 8–17**   Precambrian provinces of India. *(After Nagvi, S. M. and Rogers, J. J. W. Precambrian Geology of India. Oxford: Oxford Univ. Press, 1985.)*

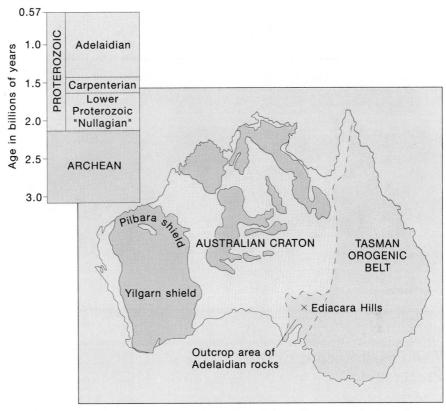

**Figure 8–18**  Central and western Australia consists of a vast cratonic region composed of Precambrian rocks (green) exposed as large patches wherever the cover of mostly horizontal Phanerozoic rocks is absent. The eastern third of the continent contains folded Paleozoic strata of the Tasman orogenic belt. Insert indicates Australian Precambrian time classifications. *(Generalized from geologic map of Australia, Department of National Development, Canberra, 1958.)*

The Proterozoic of Australia includes a thick, widespread sequence of relatively unaltered sedimentary rocks that span a time interval of over 1.5 billion years. Such rocks permit direct comparisons with modern sedimentation processes and provide many clues to paleoclimatic conditions in this part of the Proterozoic world. Included within the sedimentary sequence are extensive shallow marine deposits containing abundant stromatolites. Banded iron deposits of about the same age as those of the Canadian Animikian also occur. Near the top of the Proterozoic sequence are boulder beds deposited by glaciers, varved mudstones of marginal glacial lakes, and glacially striated pavements. Marine sediments slightly younger that the glacial deposits include the Rawnsley Quartzite (a metamorphosed sandstone), well known by geologists because of the impressions of soft-bodied animals it contains.

## Antarctica

Beneath the frozen surface of Antarctica lies a Precambrian shield that was poorly understood until the advent of modern techniques for geophysical exploration. Over the past four decades, geologists have had an opportunity to examine the bedrock exposed around the margins of the huge Antarctic ice sheet and to visit mountain peaks (nunataks) that pierce the ice cover. Most of the rounded eastern half of Antarctica consists of Precambrian rocks. Archean rocks occupy the eastern margin of the continent, and Proterozoic rocks occur across much of the remainder of the shield. The western part of Antarctica consists of fold belts of Late Cretaceous and younger sequences. These mountainous tracts show a marked similarity to those of the Andean orogenic belt of western South America.

**Figure 8–19**   An early Proterozoic colony of the cyanobacterium *Eoentophysalis* photographed in a thin section of a Belcher Group stromatolitic chert, Belcher Islands, Canada. *(Courtesy of H. J. Hofmann.)*

## The Fossil Record of the Proterozoic

Life of the early Proterozoic was not significantly different from that of the preceding late Archean. Blue-green scums of photosynthetic cyanobacteria (Fig. 8–19) probably floated on seas and lakes, and anaerobic prokaryotes multiplied in environments deficient in oxygen. Stromatolites, which were relatively uncommon in the Archean, proliferated in the early Proterozoic (Fig. 8–20) and continued to flourish throughout the remainder of the Phanerozoic Eon. Very likely, the expansion of stromatolites was enhanced by the greater number of epeiric sea and shallow shelf environments evident in the Proterozoic.

By middle Proterozoic time, evolution had progressed to the point where eukaryotes begin to appear in the fossil record. You may recall that single-celled eukaryotes contain a true nucleus enclosed within a nuclear membrane and have well-defined chromosomes and cell organelles. Early single-celled eukaryotes gave rise to multicellular forms and were the probable ancestors of the first unicellular animals. Very likely, the expansion of eukaryotes began about 1.4 billion years ago. It was followed in late Proterozoic time by the early evolution of multicellular animal life.

**Figure 8–20**   Two-billion-year-old columnar stromatolites from the Kona Dolomite near Marquette, Michigan. *(Courtesy of J. W. Schopf.)*

## COMMENTARY

### Heliotropic Stromatolites

Stromatolites are among the most abundant kinds of fossils found in Proterozoic rocks. Today, as during the Proterozoic, the photosynthetic cyanophyta that construct stromatolites depend on sunlight for survival and growth. In a study of modern stromatolites, it has been observed that certain species form columnar laminated growths that are inclined toward the sun so as to gather the maximum amount of light on their upper surfaces. This tendency among organisms to grow toward the sun is called **heliotropism**. Stromatolites, however, may also exhibit growth responses to currents and rising sea level, and thus must be carefully examined for evidence that a preferred growth orientation is related to sunlight. Stanley Awramic and James Vanyo (paleobiologist and astronomer, respectively) have studied 850 million year old stromatolites from Australia's Bitter Springs Formation for evidence of heliotropism. They believe they found such evidence in species of *Anabaria juvensis*. At the locality studied in central Australia, these stromatolites curve upward in a distinct sine wave form that appears to record growth that followed the seasonal change in the position of the sun above the shallow sea in which the stromatolitic cyanobacteria lived. In their elongate sinuous form, each sine wave represented a year of growth. The stromatolites grew by the addition of a layer or lamina each day, and thus by counting the laminae in the length of a single sine wave, one can obtain the number of days in the year. The results indicated that there were 435 days in the year 850 million years ago when the stromatolites of the Bitter Springs Formation were actively responding to Proterozoic sunlight.

## Proterozoic Prokaryotes

Unicellular prokaryotes were the characteristic organisms of the Archean as well as the early Proterozoic. Fossil remains of these relatively primitive organisms were first detected in 1965 by the American paleontologists Elso S. Barghoorn and J. William Schopf in a now-famous rock unit called the **Gunflint Chert.** This rock unit is exposed along the northwestern margin of Lake Superior. Radiometric dating methods provide an age of 1.9 billion years for the Gunflint Chert. The formation contains an abundant and varied flora of so-called thread bacteria and nostocalean cyanobacteria. Unbranched filamentous forms, some of which are septate, have been given the name *Gunflintia* (Fig. 8–21). More finely septate forms, such as *Animikiea* (Fig. 8–22) are remarkably similar to such living algae as *Oscillatoria* and *Lyngbya*. Other Gunflint fossils, such as *Eoastrion* (literally, the "dawn star"), resemble living iron- and magnesium-reducing bacteria. *Kakabekia* and *Eosphaera* do not resemble any living microorganism, and their classification is uncertain. What does seem certain is that there were an abundance of photosynthetic organisms in the early Proterozoic, all actively producing oxygen and thereby altering the composition of the atmosphere. Not only do the Gunflint fossils resemble living photosynthetic organisms, but their host rock contains organic compounds regarded as the breakdown products of chlorophyll.

## The Rise of Eukaryotes

The evolution of eukaryotes surely qualifies as one of the major events in the history of life. Eukaryotes possess the potential for sexual reproduction, and this provides enormously greater possibilities for evolutionary change. Unfortunately, the fossil record for the earliest eukaryotes is somewhat ambiguous. This is not surprising when one considers that the first eukaryotes were microscopic unicells whose identifying characteristics (enclosed nucleus, organelles, and so on) are rarely preserved. There is, however, one important clue to the identification of a

**Figure 8–21** Fossil remains of microorganisms from the Gunflint Chert. The three specimens across the top with umbrellalike crowns are *Kakabekia umbellata*. The three subspherical fossils are species of *Huroniospora*. The filamentous microorganisms with cells separated by septa are species of *Gunflintia*. (*Photographs provided by J. W. Schopf, courtesy of J. W. Schopf and Elso S. Barghoorn.*)

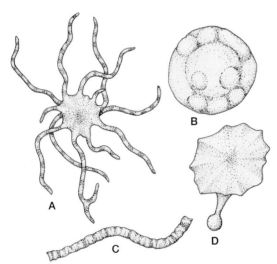

**Figure 8–22** (A) *Eoastrion,* (B) *Eosphaera,* (C) *Animikiea,* and (D) *Kakabekia* from the Gunflint Chert. All specimens are drawn to the same scale. *Eosphaera* is about 30 microns in diameter.

fossil as a eukaryote. It is based on the size of cells. Living spherical prokaryotic cells rarely exceed 20 microns in diameter, whereas eukaryotic cells are nearly always larger than 60 microns. Such larger (and hence probable eukaryotic) cells begin to appear in the fossil record by about 1.4 to 1.6 billion years ago.

Another clue to the presence of eukaryotes in the Proterozoic consists of fossils known as acritarchs. **Acritarchs** are unicellular, spherical microfossils with resistant, single-layered walls (Fig. 8–23). The walls may be smooth or variously ornamented with spines, ridges, or papillae. Most paleontologists believe acritarchs were planktonic algae that grew their thick coats during a resting stage in their life cycle. Some (but not all) resemble the resting stage of unicellular, biflagellate, marine algae known as dinoflagellates. Like dinoflagellates, acritarchs range in size from 20 to 120 $\mu$m (microns) and are therefore within the usual size range of eukaryotes.

A

B

C

**Figure 8–23**   Acritarchs. (A) The striated acritarch *Kildinosphaera lophostriata* from the Chuar Group of Arizona (see Figure 8–12). (B) A spiny acritarch named *Vandalosphaeridium walcotti,* also from the Chuar Group. (C) A spiny acritarch named *Skiagia scottica* from the Early Cambrian (about 560 million years old) of northern Greenland, depicted here for comparison with the late Proterozoic forms. *(Courtesy of G. Vidal.)*

Fossil microorganisms believed to be acritarchs are known from rocks as old as 1.6 billion years, but they are rare in rocks older than 0.9 billion years. Subsequently, they became more common and more varied, so that there is interest in using them as guide fossils for upper Proterozoic strata. Their period of expansion, however, was rather short-lived, for cooler climates accompanying late Proterozoic glaciation caused widespread extinctions. Only a few of the simple smooth-walled forms survived into the Paleozoic.

The oldest known acritarchs were recovered from middle Proterozoic rocks of the Ural Mountains. Similar remains occur in the 1.0-billion-year-old Beck Springs Dolomite of southeastern California. Other presumed eukaryotes are known from the Bitter Springs Formation of central Australia. Some of the fossil cells in the Bitter Springs exhibit ghostly internal structures that may be remnants of enclosed nuclei or organelles (Fig. 8–24).

## Dawn Animals

The most animal-like of the unicellular eukaryotes are protozoans. They are members of the Kingdom Protista, and they obtain energy in the animal-like heterotrophic manner by ingesting other cells. Amoebas and ciliates are living examples of protozoans. Protozoan fossils are common in many Phanerozoic rocks, largely because they had evolved shells and skeletal structures by that time, allowing them to

**Figure 8–24**   Microfossils from the Bitter Springs Formation of Australia arranged by J. W. Schopf in a sequence presumed to represent stages of cell division. It has been suggested that the microorganism (*Glenobotrydion*) may have been a species of green algae. The letter "n" indicates nuclear material. *(Courtesy of J. W. Schopf, from photographs used during Proceedings of XXIV International Geological Congress, Sec. 1, 1972.)*

**Figure 8–25**   Fossil metazoans from the Conception Group, Avalon Peninsula, Newfoundland. These early metazoans are similar to those of the Ediacaran fauna of Australia. The rocks in which they occur are thought to be late Proterozoic in age. The elongate forms have been interpreted as a form of colonial soft coral, and the circular fossil has been called a jellyfish. There are, however, other interpretations. *(Courtesy of B. Stinchcomb.)*

be more easily preserved. Most Proterozoic protozoans were naked, soft-bodied creatures, with little chance of preservation. The exception are tiny vase-shaped microfossils about 800 million years old from such widely separated localities as Spitsbergen and Arizona.

## The Ediacaran Fauna

The late Proterozoic fossil record for larger, multicellular animals is considerably better than that for protozoans. Indeed, since the 1960s, important discoveries of large, multicellular late Proterozoic animals have been made in Australia, Africa, South America, Russia, China, Europe, Iran, and North America (Fig. 8–25). The fossils consist of impressions in sedimentary rocks of animals that are clearly metazoans. **Metazoans** are multicellular animals that possess more than one kind of cell and have their cells organized into tissues and organs.

The best known fossils of late Proterozoic metazoans were first found in 1946 as well-preserved impressions in what were once the sands of an ancient beach. These sands now comprise the Rawnsley Quartzite of the Pound Subgroup. Exposures of the fossil-bearing rock occur in Ediacara

**Figure 8–26** Three members of the Ediacaran fauna. (A) *Pseudohizostomites,* a wormlike form of uncertain affinity. (B) *Tribrachidium,* an unusual discoidal form that appears to have no living relatives. (C) *Parvancorina,* possibly some form of arthropod. *(Specimen (A) is an imprint in the Rawnsley Quartzite; (B) and (C) are plaster molds made from the original fossils. All specimens courtesy of M. F. Glaessner.)*

**Figure 8–27** Impression of a soft-bodied discoidal fossil in the Ediacaran Rawnsley Quartzite of southern Australia. This organism has been interpreted as a jellyfish and named *Cyclomedusa*. Some paleontologists, however, believe it is unrelated to any living organisms. *(Courtesy of B. N. Runnegar.)*

Hills in the Flinders Range of southern Australia (see Fig. 8–18). Thousands of impressions of large soft-bodied animals have been recovered from these quartzites (Fig. 8–26). According to their shape, they fall into three groups. There are, for example, discoidal forms such as *Cyclomedusa* (Fig. 8–27) that were initially thought to be jellyfish. Another circular form, *Tribrachidium* (Fig. 8–26B), seems to have no modern counterpart and may be a member of an extinct phylum.

The second group of Ediacaran specimens can be called frond fossils. They resemble the living soft corals informally called sea pens (Fig. 8–28). Sea pens look rather like fronds of ferns, except that tiny coral polyps are aligned along the branchlets. The polyps capture and consume tiny organisms that float by. Frond fossils similar to those in the Rawnsley Quartzite are also known from Africa, Russia, and England. In those from England (*Charniodiscus*), the frond is attached to a basal, concentrically ringed disk that apparently served to hold the organism to the sea floor. The disks are frequently found separated from the fronds, which suggests that many of the discoidal fossils common in Ediacaran faunas around the world may be anchoring structures of frond fossils and not jellyfish.

The third group of Ediacaran fossils are ovate to elongate in form and were originally regarded as

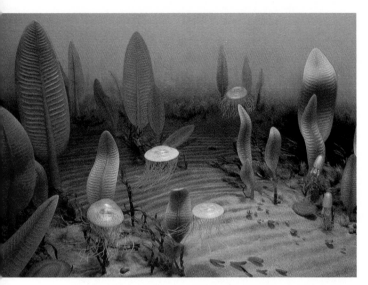

**Figure 8–28** Diorama of the sea floor in which lived Ediacaran metazoans. The large frondlike organisms are interpreted here as soft corals known today as sea pens. Silvery jellyfish are seen swimming about. On the floor of the sea one can find *Parvancorina* and elongate wormlike creatures. *(U.S. Natural History Museum, Smithsonian Institution.)*

**Figure 8–29** An exceptionally well preserved specimen of *Dickinsonia costata* in the Ediacaran Rawnsley Quartzite of southern Australia. This fossil has been interpreted as a segmented worm. *(Courtesy of B. N. Runnegar.)*

10 mm

**Figure 8–30** *Spriggina floundersi* from the Ediacaran Rawnsley Quartzite of southern Australia.

impressions made by flatworms and annelid worms. Typical of these fossils is *Dickinsonia* (Fig. 8–29), which attained lengths of up to a meter, and *Spriggina* (Fig. 8–30), a more slender animal with a distinctive crescent-shaped structure at its anterior end.

There is an interesting controversy currently taking place with regard to Ediacaran fossils. Most have long been interpreted as Proterozoic members of such existing phyla as the Cnidaria (which includes jellyfish and corals) and Annelida. Indeed, many paleontologists still retain this opinion. However, that view of the affinities of Ediacaran organisms has been challenged by the innovative German paleontologist Adolf Seilacher. Seilacher believes the resemblance between living sea pens and frond fossils is only superficial. In support of his beliefs, he notes that the branchlets in the frond fossils are fused together and do not have passages through which water currents might pass. Living sea pens have such openings, and this permits the polyps on the branchlets to efficiently gather food particles in the passing flow of water. With regard to the discoidal fossils thought to be jellyfish, Seilacher notes that living jellyfish have radial structures at their centers and concentric structures around the periphery. This arrangement is opposite to that found in the discoidal Ediacaran fossils. Finally, the rather superficial resemblance of *Spriggina* and *Dickinsonia* to worms may have little significance, for there is no evidence in the fossils of organs that are essential to modern-day worms, such as mouths, guts, and anuses.

The Ediacaran animals may be fundamentally different from cnidarians and annelids in yet another way. Cnidarians and worms are rarely preserved in sandy deposits. Preservation of soft, delicate tissue is unlikely in such coarse sediment. Yet the Ediacaran animals have left distinct impressions in the enclosing rock. For this to occur, they must have had firm and tough outer coverings. The ribbed and grooved appearance of many Ediacaran impressions suggests to Seilacher that these animals had an exterior construction suggestive of an air mattress and that this construction provided the firmness needed to make an impression in sand.

Seilacher and his American colleague Mark McMenamin believe that Ediacaran animals lived with symbiotic algae in their tissues as do modern corals. This relationship would have provided the animals with a way to derive nutrition from the photosynthetic activity of the algae. Perhaps also, the broad, thin shapes of many of these animals gave them sufficient surface area to allow for diffusion respiration, which would be essential to animals that had not yet evolved the complex circulatory, digestive, and respiratory systems of thicker animals.

The dissimilarities of Ediacaran creatures to animals existing today and their own unique characteristics support a view that they should not be placed in existing phyla. Seilacher suggests placing them in a separate taxonomic category for which he proposed the name Vendoza (after Vendian, the final period of the Proterozoic in Russia). There are, however, many paleontologists that still believe most of the Ediacaran animals are early members of existing phyla. The resolution of this controversy may have to await the discovery of fossils that provide more anatomic information.

The Ediacaran fauna survived for about 50 million years following their appearance about 630 million years ago. The fauna is important as a record of the earth's first evolutionary radiation of multicellular animals. Although their record is still too patchy to trace phylogenetic relationships, certain Ediacaran animals were probably ancestral to Paleozoic invertebrates.

At the present time there is only one known genus of a shell-bearing animal of Ediacaran age. That genus, first discovered in upper Proterozoic rocks of the Nama Group of Namibia, Africa, is named *Cloudina* after the American geologist Preston Cloud. *Cloudina* has also been reported in Brazil and other continents of the Southern Hemisphere. The

**Figure 8–31** *Cloudina,* the earliest known calcium carbonate shell-bearing fossils. *Cloudina* was first described from the Proterozoic Nama Group of Namibia. It is believed to be a tube-dwelling annelid worm.

fossils are tubular, calcium carbonate shells only a few centimeters long (Fig. 8–31). *Cloudina* has been interpreted as a tube-dwelling annelid worm.

As noted in Chapter 3, fossils need not consist only of petrifactions, impressions, molds, or casts (so-called body fossils). Trails, burrows, and other **trace fossils** also provide important clues to ancient life. Trace fossils of burrowing metazoans have been found in upper Proterozoic rocks of Australia (Fig. 8–32), Russia, England, and North America. In every locality, they occur in rocks deposited after the late Proterozoic episode of glaciation. It is interesting

**Figure 8–32** Trace fossil (horizontal trail) of an Ediacaran metazoan. The trace is in the Rawnsley Quartzite of southern Australia. *(Courtesy of B. N. Runnegar.)*

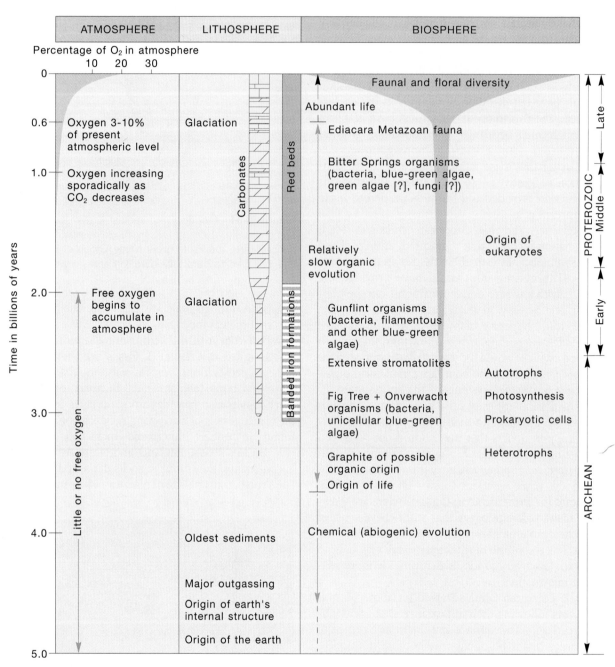

**Figure 8–33** Correlation of major events in the history of the biosphere, lithosphere, and atmosphere.

that the Proterozoic traces consist mostly of relatively simple, shallow burrows, whereas those of the overlying Cambrian are more complex, diverse, and numerous. One sees a similar increase in complexity, diversity, and abundance of metazoan body fossils as one passes from the Proterozoic into the Cambrian.

The sudden appearance of Ediacaran fossils near the end of the Proterozoic and the absence of such metazoans in older rocks may be related to the accumulation of sufficient free oxygen in the atmosphere to permit oxidative metabolism in organisms. On the other hand, the Ediacaran life may have evolved more gradually from earlier small and naked forms that were essentially incapable of leaving a fossil record. Perhaps the ancestral metazoans lived in "oxygen oases" in which marine plants were concentrated. After the atmospheric oxygen content had reached about 1 percent of the present level, these as yet undiscovered animals would have been free to leave their oases and spread widely in the seas. According to this view, the evolutionary development of metazoans may not have been abrupt, but their dispersal may have been rapid after suitable conditions became prevalent.

## The Changing Proterozoic Environment

The earth's early atmosphere and hydrosphere were largely devoid of free oxygen. Then about 2 billion years ago, the atmospheric oxygen began to accumulate because of increased plant activity (Fig. 8–33). One result of the oxygen buildup was the accumulation on land of considerable amounts of ferric iron oxide, which stained terrestrial sediments a rust-red color. Such sedimentary rocks are known as **red beds** and are considered a valid indication of the advent of an oxygenic environment. Of course, the oxygen level probably rose slowly and very likely did not approach 10 percent of present atmospheric levels of free $O_2$ until the Cambrian Period.

In general, Proterozoic rocks provide evidence for a wide range of climatic conditions, but there is no indication that these climates were especially unique in comparison with those of the Phanerozoic eras. Thick limestones and dolomites with reeflike algal colonies were deposited along the equator, where warm, tropical conditions prevailed much as they do today. During the early Proterozoic, the equator lay close to the northern border of North America. Precambrian evaporite deposits in eastern Canada and in Australia suggest that conditions were periodically rather arid during the middle Proterozoic. In the middle and low latitudes, climates were more severe, as indicated by consolidated deposits of glacial debris (tillites) and glacially striated basement rocks. As previously discussed, the best known of these poorly sorted, thick, boulderlike deposits is the Gowganda Formation.

## Mineral Wealth of the Proterozoic

Like the Archean, Proterozoic rocks have yielded important quantities of metals, including iron, nickel, silver, chromium, and uranium. Iron is the most notable of these ores in terms of tonnages that have been mined. The world's largest group of iron mines is in the Canadian shield. Of these, the most famous are the mines that process sedimentary ores in the Lake Superior region. The ore deposits were formed in local areas where iron-bearing sedimentary rocks were altered and enriched by removal of nonmetallic constituents. Other major sources of iron occur in Sweden, the Ukraine, South America, and South Africa.

Enormous amounts of copper were once recovered from Proterozoic rocks along the Keweenaw Peninsula of Lake Superior; however, today the easily accessible ores are almost exhausted. The abandoned mines are a testimony of the ultimate fate of even the richest ore deposits. Today, huge tonnages of copper are mined from Proterozoic rocks in Zambia, Africa.

Important deposits of lead and zinc occur in the Proterozoic rocks of Broken Hills, New South Wales (Australia), and Sullivan, British Columbia (Canada). A Proterozoic gabbro intrusion near Sudbury, Ontario, has provided nearly 70 percent of the world's nickel.

## SUMMARY

By the beginning of the Proterozoic about 2.5 billion years ago, many of the smaller cratonic elements of the Archean had come together to form a large craton. Lines of contact between the once separate cratonic elements were thus marked by orogens. In North America, the large continent known as Laurentia was formed by this welding of smaller crustal segments. Laurentia and other large cratons of the Proterozoic experienced further growth largely as a result of accretion of sedimentary prisms along continental margins, or additions of microplates. Proterozoic rocks in several parts of the earth provide evidence that they were involved in a modern style of plate tectonics.

The Wopmay orogen of Canada's Northwest Territories is a well-studied example of Proterozoic sedimentation and tectonics. The Wopmay contains a sequence of sedimentary deposits reflecting the opening and closing of an ocean basin. Similar events occurred in the Trans-Hudson and Grenville orogens of Canada. Aulacogens, the failed rifts of three-armed rift systems, are also evident in Proterozoic depositional basins.

Rocks of the Proterozoic contain the oldest evidence of continental glaciers. Two episodes of glaciation occurred, each documented by ice-deposited conglomeratic beds (tillites), varved mudstones of meltwater lakes, and rocks that have been scratched and faceted by glacial action. The Gowganda Formation contains glacial deposits of an early Proterozoic ice age. The second episode of glaciation occurred near the end of the late Proterozoic.

Banded iron deposits are notable for two reasons. They are an important mineral resource, and they reflect the build-up of sufficient oxygen in the atmosphere to oxidize iron at the earth's surface. The development of banded iron formations is followed by deposition of red beds beginning about 2.0 billion years ago.

The Keweenawan lavas of the middle Proterozoic reflect an episode of severe rifting in the North American mid-continental region. The extensive systems of faults associated with rifting provided conduits for the extrusion of Keweenawan lavas. The rifts, however, did not extend to the edge of the continent and did not evolve into an opening oceanic tract. Near the end of the middle Proterozoic, sedimentary rocks that had accumulated along the east-

ern margin of North America were compressed and intruded in a mountain-building event known as the Grenville orogeny. South of the Canadian shield, Proterozoic rocks are exposed in several of the Cordilleran ranges. Among these rocks, the Belt Supergroup was deposited along the rifted western margin of North America.

Life at the beginning of the Proterozoic resembled that of the late Archean. It consisted of prokaryotes such as cyanobacteria. Stromatolites became increasingly abundant during the Proterozoic. By the middle Proterozoic, the first eukaryotes appeared and subsequently diversified. Among the most common Proterozoic fossils of eukaryotic organisms are acritarchs: tiny spherical structures thought to represent the resting phase of planktonic, eukaryotic microorganisms.

The final and most dramatic evolutionary event of the Proterozoic was the appearance of metazoans (multicellular animals) near the end of the eon. The fossils of these animals are known from various localities around the world, but were first discovered in Ediacara Hills in southern Australia. The fossils are impressions in a former sandstone. They are generally large, discoidal, frondlike, or elongate animals that have traditionally been considered early members of such existing phyla as the Cnidaria or Annelida. Another view is that they are so dissimilar to existing phyla that they should be given a separate taxonomic designation, for which the name Vendoza has been proposed.

The Proterozoic seems to have had climates generally similar to those of the Phanerozoic. They ranged from warm tropical or subtropical conditions suggested by the presence of thick, stromatolitic carbonate sequences, to the cooler conditions that accompanied the two episodes of continental glaciation.

## QUESTIONS FOR REVIEW

1. With regard to the nature of the sedimentary rock record, the style of tectonics, and the fossil record, how does the Proterozoic differ from the Archean?

2. Describe the sequence of events recorded by the rocks of the Wopmay orogen. Where in the Cana-

dian shield is there evidence of a generally similar sequence of events?

3. What are aulacogens? How do they develop?

4. What evidence is there for Proterozoic episodes of continental glaciation? When did these glacia-

tions occur? Why is it unlikely that continental glaciations would have occurred in the Archean?

5. What is the environmental significance of the occurrence of banded iron formations in Proterozoic rocks around the world?

6. What kind of tectonic activity was the underlying cause of the massive extrusions of basaltic Keweenawan lavas in the Lake Superior region?

7. What metallic ores are extracted from Proterozoic rocks in Zambia, Africa, Ontario (Sudbury), and the region just north of Lake Superior?

8. When do eukaryotes first appear in the fossil record? How do these organisms differ from prokaryotes? (See Chapter 7.)

9. Describe the associations of fossils found in the Gunflint Chert, the Beck Springs Dolomite, and the Rawnsley Quartzite.

10. What are metazoans? What is the earliest known occurrence of metazoans? In relation to shape, what are the three general groups of Ediacaran metazoans?

11. What evidence suggests that Ediacaran "jellyfish" may not really be jellyfish, that Ediacaran frond fossils may not be soft corals, and that Ediacaran "annelids" may not really be annelids?

## TERMS TO REMEMBER

| | | | |
|---|---|---|---|
| acritarchs | Gowganda Formation | Laurentia | red beds |
| Animikie Group | Grenville orogeny | Mazatzal orogeny | Trans-Hudson orogen |
| aulacogen | Gunflint Chert | metazoan | Vendian System |
| banded iron formation | Hudsonian orogeny | orogen | Wilson cycle |
| Belt Supergroup | Iapetus Ocean | Protopangea | Wopmay orogen |
| Ediacaran fauna | | | |

## SUPPLEMENTAL READINGS AND REFERENCES

Diver, W. L., and Peat, C. L. 1979. On the interpretation and classification of Precambrian organic-walled microfossils. *Geology* 7:401–404.

Glaessner, M. F. 1984. *The Dawn of Animal Life*. Cambridge, England: Cambridge Univ. Press.

Hoffman, P. 1989. Precambrian geology and tectonic history of North America, in *Geology of North America,* Vol. A, pp. 447–511. Boulder, CO: Geological Society of America.

Hoffman, P. 1973. Evolution of an early Proterozoic continental margin: The Coronation geosyncline and associated aulacogens of the northwestern Canadian shield. *Phil. Trans. Royal Soc. London* 273:547–581.

Margulis, L. 1982. *Early Life*. New York: Van Nostrand–Reinhold Co.

McMenamin, M. A. S., and McMenamin, D. L. S. 1989. *The Emergence of Animals*. New York: Columbia Univ. Press.

Morris, S. C. 1990. Late Precambrian and Cambrian soft-bodied faunas. *Ann. Rev. Earth and Planetary Sci.* 18:101–122.

Morris, S. C. 1987. The search for the Precambrian–Cambrian boundary. *Scientific American* 75:157–167.

Vidal, G. 1973. The oldest eukaryotic cells. *Scientific American* 250:48–57.

The Tapeats Sandstone exposed in the vertical walls of Deer Creek Canyon, Grand Canyon, Arizona. The Tapeats is composed mostly of sand grains deposited in a nearshore environment of the shallow sea that transgressed over a large area of the southwestern United States during the Cambrian Period. It rests on erosionally truncated Protozeroic rocks, and grades upward and eastward into the Bright Angel Shale, which in turn grades into the Muav Limestone. *(Courtesy of N. Potter Jr., and the American Geological Institute.)*

# Early Paleozoic Events

*As these are stratified rocks, they must have been formed from the debris of still older rocks forming the land at that time. But as they are the oldest known rocks, we know nothing of the position of the land from which they were formed.*

Joseph LeConte, 1901

## Introduction

In many parts of the world, Phanerozoic rocks have yielded their secrets more readily than those of the older Archean and Proterozoic. They are often more accessible, less altered, and more fossiliferous. In Chapter 4 we noted that the Phanerozoic contains three eras: the Paleozoic, Mesozoic, and Cenozoic. We will first consider the geologic history of the Paleozoic Era, dividing it for convenience into informal early and late segments. The Cambrian, Ordovician, and Silurian periods constitute the early Paleozoic, which lasted about 190 million years. In the subsequent 180 million years, the Devonian, Mississippian, Pennsylvanian, and Permian periods ran their course. Each of the two approximate halves of the Paleozoic were characterized by general similarities in the historical sequence of events. For example, the early Paleozoic began on most continents with gradual marine invasions of low-lying interior regions. These wide epeiric seas moderated climate and provided habitats for a multitude of marine creatures. Near the end of the early Paleozoic, conditions on the continents became more unstable. Inland seas began an oscillating retreat back into the major ocean basins. Increasing thicknesses of terrestrial sediments and volcanic rocks appear in the stratigraphic sequence, and contorted strata attest to the growth of mountain ranges. It was a time when lands predominated, just as the preceding interval had been a time when seas covered much of the earth. Gradually, however, the lands were reduced, and the seas returned as time passed from early Paleozoic into late Paleozoic.

After we have examined the total physical history of the Phanerozoic, it will be apparent that long periods of gradual sedimentation were punctuated by intervals of severe change involving earth movements and mountain building. Proto-Pangea had begun to break apart during the late Proterozoic, and the return of the errant lands to form Pangea was associated with orogenies that affected most continents that now surround the Atlantic Ocean (Fig. 9–1). In North America, these events include the Taconic, Acadian, and Allegheny orogenies, whereas in Europe, the names Caledonian orogeny and Hercynian orogeny are applied to these same Paleozoic mountain-building events.

## Lands, Seas, and Orogenic Belts

### Clues to Paleogeography

Because of the many processes that have ceaselessly worn away or altered the older parts of the geologic record, it is difficult to reconstruct details of early Paleozoic geography. From paleomagnetic studies we have learned that during the Cambrian most of

| | |
|---|---|
| ▮ Early Paleozoic mountain belts | ▮ Middle and Late Paleozoic mountain belts |

**Figure 9–1** Pangea with mountain belts formed during the Paleozoic.

North America lay along the paleoequator, which was aligned nearly perpendicular to present parallels of latitude from north-central Mexico to Ellesmere Island in the Arctic (Fig. 9–2). The Ordovician equator extended slightly more northeastward, from what is now Baja California to Greenland and thence across the British Isles and central Europe. These paleoequatorial positions are substantiated by paleoclimatic indicators. Salt and gypsum accumulations, for example, are found in Ordovician rocks of Arctic Canada at a paleolatitude of about 10° from the paleoequator. Great thicknesses of carbonate rocks bearing fossils of marine invertebrates are located within 30° of the paleomagnetically determined equator.

Although Proto-Pangea had begun to break apart near the end of the Proterozoic, paleomagnetic and paleontologic evidence indicates that the continents were still relatively close together at the beginning of the Paleozoic. They appear to have been clustered in the Southern Hemisphere. The Gondwanaland region was turned about 180° from its present orientation, whereas what is now North America, Greenland, and western Europe lay somewhat east of South America. By late in the Cambrian, the oceanic tracts separating the continents had widened, and the rifted passive margins of the separating land masses became collecting sites for continental

shelf and rise sediments. Subsequent closure deformed these sediments and raised them into mountain ranges.

Figure 6–51 (in Chapter 6) illustrates in simplified diagrams the sequence of events as they are postulated by the earth scientists James Valentine and Eldridge Moores. These configurations show Proto-Pangea divided into four continents by Late Cambrian and Early Ordovician time. Then, during Late Ordovician and Silurian time, Europe west of the Uralian trough pushed into North America, forming a chain of ancient mountains called the Caledonides and welding the two continental segments together. The next major collision occurred about 300 million years ago when the Northern Hemisphere continents joined Gondwanaland, forming the ancient Hercynian Mountains. These mountains have since been largely eroded away and covered with younger strata over much of Europe. Finally, near the end of the Paleozoic, Asia was joined to Europe. The sediments caught and crushed in the closing ocean formed the Ural Mountains. Pangea was now fully constructed.

## The Continental Framework

As we consider the geologic history of the early Paleozoic, it becomes apparent that events of the more stable interiors of land masses differed from those of the often more dynamic continental margins.

### Platforms

In Chapter 7, we noted that continents can be described in terms of cratons and mobile belts or orogens. A craton, you may recall, is the relatively stable part of a continent consisting of a Precambrian shield and the buried extension of that shield known as a **platform**. In North America, the platform is constructed of flat-lying or only gently deformed Phanerozoic strata consisting of wave-washed sands and carbonates deposited in shallow seas that periodically flooded continental regions of low relief. Here and there, these strata are gently warped into broad synclines, basins, domes, and arches (Fig. 9–3). The resulting tilt to the strata is so slight that it is usually expressed in feet per mile rather than degrees. In the course of geologic history, the arches and domes stood as low islands or barely awash submarine banks.

Domes and arches seem to have developed in response to vertically directed forces quite unlike

**Figure 9–2**  Paleogeographic and tectonic elements of North America during the Cambrian Period, showing position of the Cambrian paleoequator.

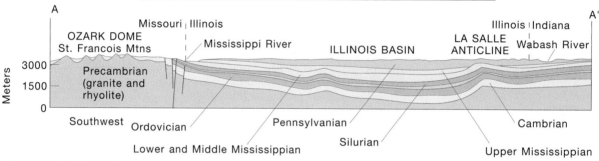

**Figure 9–3** Map of the central platform of the United States showing major basins and domes. Structural section below the map crosses part of the Ozark dome and the Illinois basin. The basins and domes developed at different times during the Phanerozoic.

those that formed the compressional folds of mountain belts. Whether domes or basins, structures of the platform can be recognized by their pattern of outcropping rocks. Erosional truncation of domes exposes older rocks near the centers and younger rocks around the peripheries. Sequences of strata over arches and domes tend to be thinner. Also, because these structures were periodically above sea level, they characteristically exhibit a greater frequency of erosional unconformities. Basins, in contrast, were more persistently covered by inland seas, have fewer unconformities, and developed greater thickness of sedimentary rocks. In erosionally truncated basins, younger rocks are centrally located and older strata occur successively farther toward the peripheries (Fig. 9–4).

## Orogenic Belts

The North American craton is bounded on four sides by great elongate tracts that have been the sites of intense deformation, igneous activity, and earthquakes. One or more of such orogenic belts are present on all continents (Fig. 9–5). Most are located along present or past margins of continents, such as the North American Cordilleran belt. As described in Chapter 6, the passive margin of a continent may experience an early stage in which thick sequences of sediments accumulate along the continental shelf and rise. This stage may be followed by subduction of oceanic lithosphere along continental margins and may terminate in continent-to-continent collisions, as in the classic Wilson cycle. As we review the his-

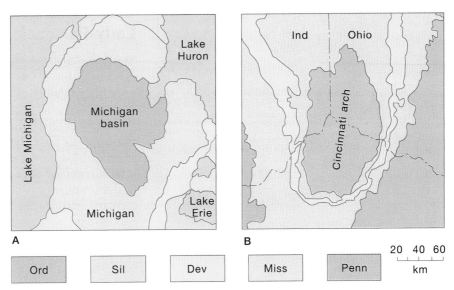

A

| Ord | Sil | Dev | Miss | Penn |

20  40  60
km

**Figure 9–4**  (A) In an erosionally truncated basin such as the Michigan basin, the youngest beds are centrally located. (B) In a domelike structure such as the Cincinnati arch, the oldest beds are located in the center.

tory of eastern North America later in this chapter, we will describe the geologic evidence for this sequence of events along the Appalachian orogen.

## The Base of the Cambrian

At one time, recognition of the boundary between the Cambrian and the underlying Precambrian seemed relatively simple. The base of the Cambrian

System could be identified by the first occurrence of shell-bearing multicellular animals. Among these, marine arthropods known as trilobites served as the principal marker fossils. A problem became evident in the 1970s, however, when a distinctive group of shelly fossils were found *beneath* the lowest occurrence of strata containing the first trilobites. These animals constructed their shells of calcium phosphate and were very small, which accounts for their not being discovered earlier. They include early mem-

**Figure 9–5**  Cratons and mobile belts of North America and Europe.

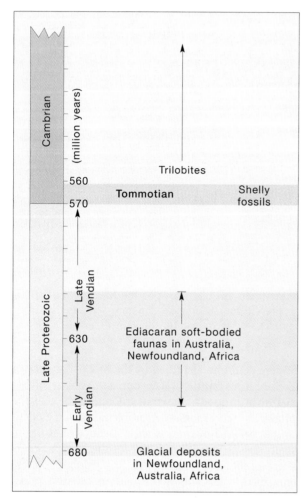

**Figure 9–6** Location of the Tommotian Stage in the geologic time scale. The preferred boundary between the Protozeroic and Phanerozoic is placed at the base of the Tommotian, close to the horizon at which diverse shell-bearing metazoans make their appearance.

bers of such existing phyla as Porifera, Brachiopoda, and Mollusca, as well as tiny tubes and cones of uncertain association. Although originally discovered in Siberia, these pre-trilobite shelled organisms have also been found recently in many other continents. Their discovery has resulted in the designation of a new stage, the **Tommotian Stage**, for rocks at the very bottom of the Cambrian System (Fig. 9–6). The Tommotian fauna has thus become the marker for the base of the Cambrian. Above the Tommotian, rocks contain trilobites (and other invertebrates), and below are upper Proterozoic rocks that, if fossiliferous, contain only soft-bodied organisms.

# Early Paleozoic History

The beginning of the Paleozoic Era was a geologically calm time for North America, as if the continent rested from the rigors of orogeny and severe climate that had occurred near the end of the preceding era. We know little of the Paleozoic ocean basins, for those old sea floors have long since been consumed at subduction zones. However, oceans did exist and gradually spilled out of their basins onto low regions of the continents. Although a large part of the sedimentary record has been removed by erosion, enough remains to indicate that practically all of the Canadian shield was inundated at times. Initially, the dominant deposits were sands and clays derived from the weathering and erosion of igneous and metamorphic rocks of the shield. Later, as these quartz- and clay-producing terranes were reduced, limestones and dolomites became increasingly more prevalent. These rocks contain abundant remains of lime-secreting marine organisms and indicate that shallow seas were common throughout much of the equatorial region of the earth during the early Paleozoic. Indeed, the advances and retreats of these shallow seas were the most apparent events in the early Paleozoic history of the continental interiors.

## *Events on the Craton*

It is convenient to discuss the geologic history of the continental mass that now constitutes North America in terms of its cratonic and orogenic regions. It is also advantageous to subdivide the early Paleozoic of the craton on the basis of advances (transgressions) and retreats (regressions) on the inland or epeiric seas. Regressions exposed the old sea floors to erosion and produced extensive unconformities that are used as stratigraphic markers.

In Chapter 4 we discussed the largely European basis for defining the geologic periods. Although the cratonic rocks of North America can be correlated with those of Europe on the basis of fossils, most of the stratigraphic breaks in deposition do not correspond to those that define the European period boundaries. Indeed, if the relative time scale had been worked out in North America instead of Europe, there would probably have been only two periods in the early Paleozoic rather than three. The American stratigrapher Laurence Sloss has suggested a workable remedy for this disparity. He suggested that the Paleozoic rocks of the North American craton be divided into "sequences" of deposition. For the Paleozoic, he named these the **Sauk, Tippecanoe, Kaskaskia,** and **Absaroka** sequences. As shown in Table 9–1, the boundaries of these sequences are marked by widespread unconformities.

The Sauk and Tippecanoe are early Paleozoic sequences, whereas the Kaskaskia and Absaroka are late Paleozoic.

During the earliest years of the Sauk sequence, the seas were largely confined to the continental

### Table 9–1   Cratonic Sequences of North America.[*]

| Geologic Time | | Cratonic Sequences (◄ Center of craton    Margin of craton ►) | Orogenic Events | Biological Events | Ice Ages |
|---|---|---|---|---|---|
| Cenozoic | | Tejas | Himalayan Alpine Laramide | Age of mammals — Massive extinctions — | |
| — 65 mya — | | | | | |
| Mesozoic | Cretaceous | Zuni | Sevier — Nevadan — | First flowering plants; Climax dinosaurs and ammonites | |
| | Jurassic | | | First birds; Abundant dinosaurs and ammonites | |
| | Triassic | | Sonoma | First dinosaurs; First mammals; Abundant cycads | |
| — 248 mya — | | | | Massive extinctions (including trilobites); Mammal-like reptiles | |
| Late Paleozoic | Permian | Absaroka | | | |
| | Pennsylvanian | | Alleghenian | Great coal forests; Conifers; First reptiles | |
| | Mississippian | Kaskaskia | — Antler — | Abundant amphibians and sharks; Scale trees; Seed ferns | |
| | Devonian | | Acadian– Caledonian | Extinctions; First insects; First amphibians; First forests; First sharks | |
| — 408 mya — | | | | | |
| Early Paleozoic | Silurian | Tippecanoe | | First jawed fishes; First air-breathing arthropods | |
| | Ordovician | | Taconic | Extinctions; First land plants; Expansion of marine shelled invertebrates | |
| | Cambrian | Sauk | | First fishes; Abundant shell-bearing marine invertebrates; Trilobites | |
| — 570 mya — | | | | | |
| Late Proterozoic | | | | Rise of the metozoans | |

[*]The green areas represent sequences of strata. They are separated by major unconformities, indicated in yellow. Note that the rock record is most complete near cratonic margins just as the time spans represented by unconformities are greatest near the center of the craton. Major biologic, orogenic, and glacial events are added for reference. (Cratonic sequence model after Sloss, L. L. 1965. *Bull. Geol. Soc. Amer.* 74:93–114.)

**Figure 9–7** The Upper Cambrian Galesville Sandstone exposed in a quarry near Portage, Wisconsin. The mature sandstone is being quarried for use in the manufacture of glass and in premixed cement.

Sandstone and siltstone

Shale

Carbonates

Sandstone, siltstone, and volcanics

Edge of thrust belt or strongly folded strata

0    Zero edge

**Figure 9–8** Upper Cambrian lithofacies map. *(Simplified and adapted from* Stratigraphic Atlas of North America and Central America. *Shell Oil Company, Exploration Department.)*

margins (continental shelves and rises), so that most of the craton was exposed and undergoing erosion. No doubt it was a bleak and barren scene, for vascular land plants had not yet evolved. Uninhibited by protective vegetative growth, erosional forces gullied and dissected the surface of the land. Over a span of at least 50 million years, the Precambrian crystalline rocks were deeply weathered and must have formed a thick, sandy "soil." Eventually, marine waters spilled out of the marginal basins and began a slow encroachment over the eroded and weathered surface of the central craton.

The craton was not a level, monotonous plain during Sauk time, but had distinct upland areas composed of Precambrian igneous and metamorphic rocks. During times of marine transgression, these uplands became islands in early Paleozoic seas and provided detrital sediments to surrounding areas. The absence of marine sediments over these upland tracts provides evidence of their existence and extent. One of the largest of the highland regions was the transcontinental arch (see Fig. 9–2), which extended southwestward across the craton from Ontario to Mexico.

By Late Cambrian time, seas extended across the southern half of the craton from Montana to New York. A vast apron of clean sand (Fig. 9–7) was spread across the sea floor for many miles behind the advancing shoreline (Fig. 9–8). This sandy aspect, or facies, of Cambrian deposition was replaced toward the south by carbonates. Here the waters were warm, clear, and largely uncontaminated by clays and silts from the distant shield. Marine algae flourished and contributed to the precipitation of calcium carbonate. Invertebrates, although present, did not contribute to the volume of sediment to the degree they did in later periods.

In the Cordilleran region, the earliest deposits were sands, which graded westward into finer clastics and carbonates. An excellent place to study this Sauk transgression is along the walls of the Grand Canyon of the Colorado River. In this region one finds the Lower Cambrian Tapeats Sandstone (see chapter opening photograph) as an initial strandline deposit above the old Precambrian surface. The Tapeats can be traced both laterally and upward into the Bright Angel Shale, which was deposited in a more offshore environment (Fig. 9–9). Next is the

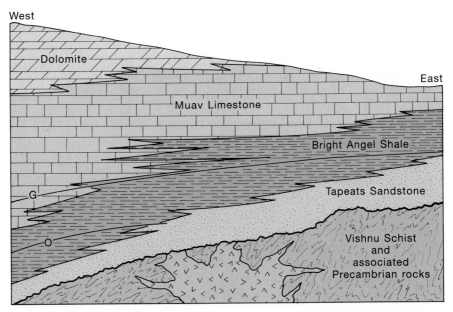

**Figure 9–9**  East–west section of Cambrian strata exposed along the Grand Canyon. The line labeled ''O'' indicates the top of the Lower Cambrian *Olenellus* trilobite zone. The line labeled ''G'' indicates the location of the middle Cambrian *Glossopleura* trilobite zone. The section is approximately 200 km long, and the section along the western margin is about 600 meters thick. *(Adapted from McKee, E. D. Cambrian Stratigraphy of the Grand Canyon Region. Washington, Carnegie Institute, 1945, Publ. 563.)*

**Figure 9–10** Lower Tippecanoe lithofacies map and cross-section along line A to A'. In the centrally located white area, Tippecanoe strata have been removed by erosion. Note the extensive unconformity between the Sauk and the Tippecanoe that is depicted on the cross-section.

Muav Limestone, which originated in a still more seaward environment. As the sea continued its eastward transgression, the early deposits of the Tapeats were covered by clays of the Bright Angel Formation, and the Bright Angel was in turn covered by limy deposits of the Muav Limestone. Together these formations form a typical transgressive sequence, recognized by coarse deposits near the base of the section and increasingly finer (and more offshore) sediments near the top.

Cambrian rocks of the Grand Canyon region not only provide a glimpse of the areal variation in depositional environments as deduced from changing lithologic patterns, but they also illustrate that partic-

ular formations are not usually the same age everywhere they occur. Detailed mapping of the Tapeats, Bright Angel, and Muav units combined with careful tracing and correlation of trilobite occurrences indicate that deposition of the highest facies (Muav) had already begun in the west before deposition of the lowest facies (Tapeats) had stopped in the east. Correlations based on the guide fossils provided reliable evidence that the Bright Angel sediments were largely Early Cambrian in age in California and mostly Middle Cambrian in age in the Grand Canyon National Park area. This example illustrates the **principle of temporal transgression**, which stipulates that sediments deposited by advancing or retreating seas are not necessarily of correlative geologic age throughout their areal extent.

The episode of carbonate deposition that was so characteristic of the southern craton continued into the Early Ordovician almost without interruption. Then, near the end of the Early Ordovician, the seas regressed, leaving behind a landscape underlain by limestones that was subjected to deep subaerial erosion. That erosion produced a widespread unconformity that geologists use as the boundary between the Sauk and the Tippecanoe sequences (Fig. 9–10).

The second major transgression to affect the platform occurred when the Tippecanoe sea flooded the region vacated as the Sauk sea regressed. Again the initial deposits were great blankets of clean quartz sands. Perhaps the most famous of these Tippecanoe sandstones is the St. Peter Sandstone (Fig. 9–11), which is nearly pure quartz and is thus prized for use in glass manufacturing. Such exceptionally pure sandstones are usually not developed in a single cycle of erosion, transportation, and deposition. They are the products of chemical and mechanical processes acting on still older sandstones. In the St. Peter Formation, waves and currents of the transgressing Tippecanoe sea reworked Upper Cambrian and Lower Ordovician sandstones and spread the resulting blanket of clean sand over an area of nearly 7500 sq km. As described in Chapter 2, sandstones can be described by their texture and composition as either mature or immature. These textural and compositional traits are, of course, related to the processes that produced the sandstone. Sandstones with a high proportion of chemically unstable minerals and angular, poorly sorted grains are considered immature, whereas aggregates of well-rounded, well-sorted, highly stable minerals (such as quartz) are considered mature. The St. Peter Sandstone is so pure as to be geologically unusual and perhaps should be termed an ultramature sandstone.

The sandstone depositional phase of the Tippecanoe was followed by the development of extensive limestones, often containing calcareous remains of brachiopods, bryozoans, echinoderms, mollusks, corals, and algae (Fig. 9–12). Some of these carbonates were chemical precipitates, many were fossil fragment limestones (bioclastic limestones), and some were great organic reefs. Frequently, the deposited calcite underwent chemical substitution of some of the calcium by magnesium and in the process was converted to the carbonate rock dolomite.

**Figure 9–11** The St. Peter Sandstone is the massive bed directly beneath the overhanging ledge of horizontally bedded dolomite of the Joachim Formation. Both formations are Middle Ordovician in age. The St. Peter is the initial deposit of the transgressing Tippecanoe sea. The brown stains are iron oxide. This exposure is 30 miles southwest of St. Louis, Missouri.

**Figure 9–12** Fossiliferous limestone deposited in the Tippecanoe sea. This specimen, from near Danville, Missouri, is about 20 cm in longest dimension. The most abundant fossils are brachiopods. On close examination, one can also find many twiglike branches of colonial bryozoans and echinoderm parts.

**Figure 9-13** Block diagram and stratigraphic section of Niagara Falls. The Lockport Dolomite forms the resistant lip of the falls. The rocks dip gently to the south in this area, and where harder dolomite layers such as the Lockport Dolomite intersect the surface, they form a line of bluffs known as the Niagaran Escarpment. *(Map after E. T. Raisz.)*

In the region east of the Mississippi River, the dolomites and limestones were gradually supplanted by shales. As we shall see, these clays are the peripheral sediments of the Queenston clastic wedge which lay far to the east. The geologic section exposed at Niagara Falls is a classic locality to examine these rocks (Figs. 9-13 and 9-14).

Near the close of the Tippecanoe sequence, landlocked, reef-fringed basins developed in the Great Lakes region (Fig. 9-15). During the Silurian,

**Figure 9–14** Niagara Falls, formed where the Niagara River flows from Lake Erie into Lake Ontario. The classic section of the American Silurian System is exposed along the walls of the gorge below the falls. *(Guido Alberto Rossi/The Image Bank.)*

evaporation of these basins caused the precipitation of salt and gypsum on an immense scale. These deposits clearly indicate the existence of excessively

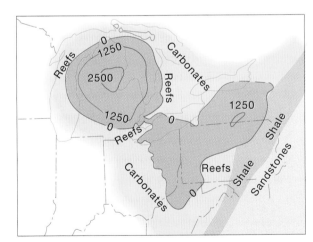

**Figure 9–15** Isopach map showing thickness and area of evaporite basins during the Late Silurian. Areas of evaporite precipitation were surrounded by carbonate banks and reefs. Gaps in the reefs and banks provided channelways for replenishment of the basins with normal sea water, thus replacing water lost by evaporation. *(Adapted from Alling, H. L. and Biggs, L. I. Bull. Amer. Assoc. Petrol. Geol. 45:515–547, 1961.)*

arid conditions. In the salt-bearing Salina Group, for example, evaporites have an aggregate thickness of 750 meters. The precipitation of this amount of salt and gypsum, if theoretically confined to a single continuous episode of evaporation, would require a column of sea water 1000 kilometers tall. Because such a column of water has never existed on earth, it is likely that the evaporating basins had connections to the ocean through which replenishment could occur. One setting in which this might have occurred would be in a basin that had its opening to the sea restricted by a raised sill (or possibly also by a submerged bar). Evaporation within the basin would have produced heavy brines which would have sunk to the bottom and been prevented from escaping because of the sill or bar (Fig. 9–16). This type of basin is called a **barred basin**.

## The Cordilleran Region

During much of early Paleozoic time, large areas of the Cordilleran region of western North America lay beneath sea level. A coastal plain and shallow marine shelf extended westward from the Transcontinental arch. This relatively level expanse was bordered on the west by deep marine basins adjacent to the western continental margin. Many of the marine sedi-

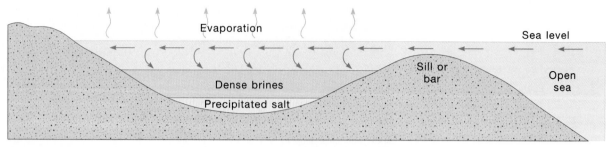

**Figure 9–16** Cross-section illustrating a model for the deposition of evaporites in a barred basin. Sea water is able to flow more or less continuously into the basin over the partially submerged barrier. It is evaporated to form dense brines. Because of their density, the brines sink and are thus unable to return to the open sea because of the barrier. When the brine has become sufficiently concentrated, salts are precipitated.

mentary basins of the Cordillera appear to have originated during the Proterozoic when North America separated from another continental block that drifted westward, rotated counterclockwise, and was ultimately incorporated into the Siberian platform (Fig. 9–17). As a result of the separation, a passive margin was developed along the western border of North America. Evidence for this hypothesis consists of similarities in Proterozoic rocks of the two now widely separated segments, as well as correspondence in the alignment of structural trends when the

two blocks are placed in proper juxtaposition. In addition, there is geophysical evidence for the presence of a failed rift or aulocogen in Southern California. As is characteristic of aulocogens, this one is perpendicular to the possible line of separation between the two land masses. One notes further that the Belt Supergroup of Montana, Idaho, and British Columbia, as well as the Uinta Series of Utah and the Pahrump Series of California, were all deposited in fault-controlled basins formed during an episode of Proterozoic rifting.

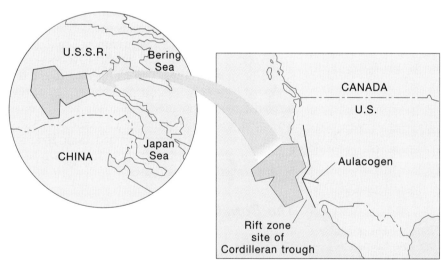

**Figure 9–17** Conceptual drawing indicating the development of a passive margin trough along the western edge of North America by detachment of a land mass that ultimately was incorporated into the Siberian Platform of Russia. (*Concept from Sears, J. W. and Price, R. A. Geology 6:227–270, 1978.*)

Some of the grandest sections of Cambrian rocks in the world are exposed in the Canadian Rockies of British Columbia and Alberta. The formations have been erosionally sculptured into some of Canada's most magnificent scenery (Fig. 9–18). Lower Cambrian rocks include ripple-marked quartz sandstones that were probably derived from the Canadian shield. By Middle Cambrian time, the seas had transgressed farther eastward, and shales and carbonates became more prevalent. An interesting section of these Middle Cambrian rocks is exposed along the slopes of Kicking Horse Pass in British Columbia near the border with Alberta. One of the units in this section, the Burgess Shale, has excited the keen interest of paleontologists around the world because it contains abundant remains of Middle Cambrian soft-bodied animals, most of which have never been discovered elsewhere. We will discuss these fossils in Chapter 11, along with other organisms of the early Paleozoic.

Patterns of sedimentation in the Cordilleran region during the Ordovician and Silurian indicate that the western edge of the continent was no longer a passive margin, as was the case in the late Proterozoic and earliest Cambrian. The Pacific plate was now moving against North America, and a subduction zone with an associated volcanic chain had formed (Fig. 9–19). A thick subduction zone complex, consisting of graywackes and volcanics was laid down in the trench above the subduction zone, whereas east of the volcanoes, siliceous black shales and bedded cherts accumulated. The shales are noted for their rich content of fossil colonial organ-

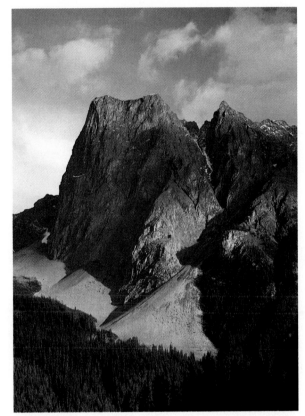

**Figure 9–18** Mount Burgess in Yoho National Park, British Columbia. The mountain is on the east side of Burgess Pass about 3 km southeast of the Burgess Shale fossil quarry where a diverse fauna of Cambrian metazoans was discovered in 1909 by Charles Walcott. The rocks exposed along the flanks of Mount Burgess consist primarily of Cambrian shales. *(Courtesy of the Image Bank/Wm. A. Logan.)*

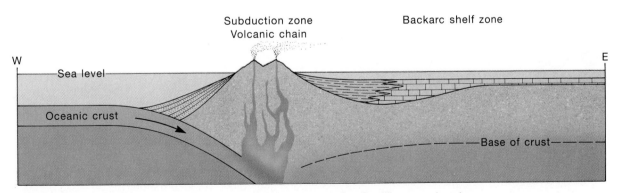

**Figure 9–19** Interpretive cross-section of conditions across the Cordilleran region during early Paleozoic time.

**Figure 9–20** Branches (stipes) of the graptolite *Diplograptus*. *Diplograptus* is common in dark shales of Ordovician age in both Europe and North America.

isms called graptolites (Fig. 9–20). Graptolites are so prevalent in lower Paleozoic rocks that the sediments have been referred to as the "graptolite facies." The associated bedded cherts may have derived their silica by dissolution of particles of volcanic ash and the tiny shells of siliceous marine microorganisms.

Eastward of the subduction zone complex in the Cordillera, one finds a thick sequence of fossiliferous carbonates, shales, and sandstones (Fig. 9–21). These are the rocks of the continental shelf. Because

many of the carbonates contain abundant fossils of marine invertebrates, they are sometimes dubbed the "shelly facies."

Silurian deposits of the Cordillera are generally similar to those of the Ordovician. Sediments of the subduction complex can be recognized in southeastern Alaska, California, and Nevada, where they occur in great thrust sheets. Nevada also contains exposures of Silurian carbonates deposited in shallower continental shelf areas.

## Deposition in the Far North

The cold and sparsely inhabited northern part of North America is a difficult region in which to carry on geologic field studies. Thus, this region has received less attention than more hospitable parts of the continent. To add to the difficulty, lower Paleozoic strata are often sparse, covered with younger rocks, and deformed or metamorphosed. Nevertheless, there is evidence that the northern edge of the continent formed late in the late Proterozoic when a continental mass split off from the North American craton. The rifting and displacement was probably part of the general dismemberment of Proto-Pangea occurring at this time. By Ordovician time, conditions had stabilized enough to permit the development of a carbonate platform on the continental shelf. To the north, the shelf dropped off abruptly, providing a deep-water environment in which turbidites were trapped. The Silurian section in this region is one of the thickest in the world. It contains extensive reef deposits, some of which stood more

**Figure 9–21** Ordovician sedimentation in the Montana region. Isopach lines indicate thickness of the Ordovician time–rock unit. Deep-water sediments reach thicknesses of 9500 ft (2900 meters) in Idaho and 10,000 ft (3080 meters) in British Columbia. *(After Sloss, L. L. Bull. Am. Assoc. Petr. Geol. 34:423–445, 1950.)*

than 200 meters above the sea floor. There are thick deposits of evaporites as well. These rocks imply warm conditions, an interpretation stengthened by paleomagnetic data indicating this presently frigid region lay within 15° of the early Paleozoic equator.

## The Appalachian Belt

Bordering the interior platform of North America on the east from Newfoundland to Georgia is the Appalachian orogenic belt. (Its southwestward extension is called the Ouachita orogenic belt.) The Appalachian belt contains the record of three major orogenic events and many minor disturbances. Study of the rocks deformed during these events has been the basis not only for classic theories of mountain building, but also for new ideas that view the Appalachian belt as a collage of microcontinents and other suspect terranes accreted to the eastern edge of North America by plate convergence.

The Appalachian depositional basin (see Fig. 9–2) can be rather neatly divided into a western belt composed of shales, limestones, and sandstones, and an eastern belt composed largely of graywackes, volcanics, and siliceous shales that are often highly metamorphosed and intruded by granite masses of batholithic proportions. Today these rocks underly the Blue Ridge and Piedmont Appalachian provinces, whereas rocks of the western belt occur beneath the Valley and Ridge province and Appalachian plateaus (Fig. 9–22).

At the beginning of the Paleozoic, the eastern margin of North America was a passive margin, much like it is today. Relatively shallow-water deposits of the shelly facies were being spread across the continental shelf, while deeper water sediments accumulated farther seaward along the continental rise. The initial deposits along the shelf were largely quartz sandstones and shales derived from the stable interior, which was still largely emergent in earliest Cambrian time. Gradually, however, the seas spread inland and sandy strandlines migrated westward. The early sandstone deposits of the shelf were covered by a thick sequence of carbonate rocks that formed a shallow bank or platform extending from Newfoundland to Alabama. Mudcracks and stromatolites attest to the deposition of these carbonates (now dolomites) at or near sea level.

This rather quiet depositional scenario of the Cambrian and Early Ordovician changed dramati-

**Figure 9–22** Physiographic provinces of the eastern United States.

cally during the Middle Ordovician (Fig. 9–23). At that time, carbonate sedimentation ceased, the carbonate platform was downwarped, and huge volumes of graptolitic black shales and immature sands spread westward over the dolomites. The Ordovician section in New England contains such coarse detritus, as well as volcanic pyroclastics and interbedded lava flows that indicate a subduction zone with an

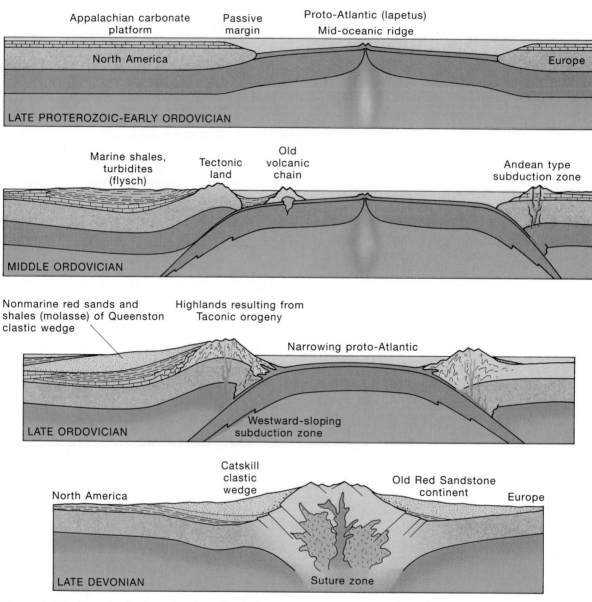

**Figure 9–23** Hypothesis depicting the role of plate movements in the evolution of the northern Appalachians. Cambrian and Lower Ordovician carbonates were deposited on the passive margin of North America. With the development of plates with opposing movements in the Ordovician, highland areas were produced that shed clastics into adjacent basins. Devonian clastic sediments had as their source the mountains produced by the closure of the Iapetus Ocean and convergence of North America and Europe. *(Based in part on concepts expressed in Bird, J. M. and Dewey, J. F. Bull. Amer. Jour. Sci. 81:1031–1059, 1972; Hatcher, R. D. Jr. Amer. J. Sci. 274:135–147, 1972.)*

accompanying volcanic chain had formed along the eastern margin of North America. Apparently, the Proto-Atlantic (or Iapetus) Ocean that had opened during the Proterozoic and Cambrian was beginning to close again. That closure resulted in the subduc-

tion–volcanic arc complex of the Paleozoic and ultimately carried the continents that were to compose Pangea into contact with each other.

The pulses of orogenic activity that had begun in the Early Ordovician were followed by several more

**Figure 9–24** These Ordovician shale beds exposed at Black Point, Port au Port, Newfoundland, are intensely folded as a result of the Taconic orogeny of Late Ordovician time. *(Courtesy of the Geological Society of Canada, Ottawa.)*

intense deformational events in Middle and Late Ordovician that constitute the **Taconic orogeny.** The Taconic orogeny was caused by the collision of another tectonic plate, probably part of ancestral western Europe, with the eastern coast of North America. Later orogenic episodes resulted from further compression between the opposing plates. The effects of the Taconic orogeny are most apparent in the northern part of the Appalachian belt. Caught in the vise between closing lithospheric plates, sediments were crushed (Fig. 9–24), metamorphosed, and thrust northwestward along a great fault. In this fault, named **Logan's thrust** (Fig. 9–25), continental rise sediments have been shoved over and across

**Figure 9–25** Logan's line. As a result of folding and thrust faulting, the New England region of North America was compressed by several hundred kilometers. Cambrian and Ordovician sediments originally deposited near the present coastline are found along the Hudson River and Lake Champlain as remnants of great thrust sheets. The boundary between the thrust sheets and the relatively less disturbed strata is named Logan's line after Sir Joshua Logan, a pioneering Canadian geologist who first described its significance. (Teeth on fault line are on the overthrust block.)

**Figure 9–26** Great wedges of clastic sediments spread westward as a result of erosion of mountain belts developed during the early Paleozoic. *(After Hadley, J. B. In Tectonics of the Southern Appalachians. V.P.I., Dept. Geol. Sci. Memoir 1, 1964.)*

some 48 km of shelf and shield rocks. Today we are able to view the remnants of this activity in the Taconic Mountains of New York. Ash beds, now weathered to a clay called bentonite, attest to the violence of volcanism. Masses of granite now exposed in the Piedmont help to record the great pressures and heat to which the sediments of the subduction zone complex were subjected. But even if the igneous rocks had never been found, geologists would know mountains had formed, for the great apron of sandstones and shales that outcrop across Pennsylvania, Ohio, New York, and West Virginia must have had their source in the rising Taconic ranges (Fig. 9–26).

From a feather edge in Ohio, this barren wedge of rust-red terrestrial clastics called the **Queenston**

**clastic wedge** (Fig. 9–27) becomes increasingly coarser and thicker toward the ancient source areas to the east (Figs. 9–28 and 9–29). Streams flowing from those mountainous source areas were laden with sand, silt, and clay. As they emerged from the ranges, they spread their load of detritus in the form of huge deltaic systems which grew and prograded westward into the basin, often covering earlier shallow marine deposits. The deltaic aspect of many of these deposits accounts for the alternate name of Queenston delta for the Queenston clastic wedge. It has been estimated that over 600,000 km$^3$ of rock were eroded to produce the enormous volume of sediment in the Queenston clastic wedge. This indicates that the Taconic ranges may have exceeded 4000 meters (13,100 feet) in elevation.

**Figure 9–27**  Paleographic map of Ordovician North America.

**Figure 9–28** Isopach map illustrating regional variation in thickness of Upper Ordovician sedimentary rocks in Pennsylvania and adjoining states. *(After Kay, M.* Geol. Soc. Am. Mem. *48, 1951.)*

During the Silurian, the locus of most intense orogenic activity seems to have shifted northeastward into the Caledonian belt. Meanwhile, erosion of the Taconic highlands continued. Early Silurian beds are often coarsely clastic, as evidenced by the Shawangunk Conglomerate of New York (Fig. 9–30). These pebbly strata give way upward and laterally to sandstones such as those found in the Cataract Group of Niagara Falls. Silurian iron-bearing sedimentary deposits accumulated in the southern

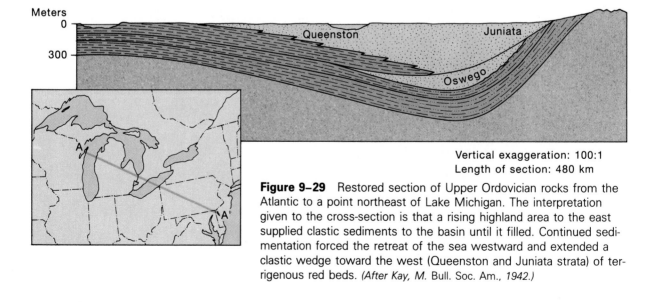

Vertical exaggeration: 100:1
Length of section: 480 km

**Figure 9–29** Restored section of Upper Ordovician rocks from the Atlantic to a point northeast of Lake Michigan. The interpretation given to the cross-section is that a rising highland area to the east supplied clastic sediments to the basin until it filled. Continued sedimentation forced the retreat of the sea westward and extended a clastic wedge toward the west (Queenston and Juniata strata) of terrigenous red beds. *(After Kay, M.* Bull. Soc. Am., *1942.)*

**Figure 9–30**  Paleographic map of Silurian North America.

part of the Appalachian region. The greatest development of this sedimentary iron ore is in central Alabama, where there are also coal deposits of Pennsylvanian age. The coal is used to manufacture coke, which is needed in the process of smelting. Limestone, used as blast furnace flux, is also available nearby. The fortunate occurrence of iron ore, coal, and limestone in the same area accounts for the formerly important steel industry of Birmingham, Alabama (Fig. 9–31).

Extending across the southern margin of the North American craton is the Ouachita–Marathon depositional trough (see Fig. 9–30). Although over 1500 km long, only about 300 km of its folded strata are exposed. Additional information about the distribution of early Paleozoic rocks within the belt is derived mostly from oil well drilling and geophysical surveys.

Altogether, nearly 10,000 meters of Paleozoic sediment fill the Ouachita–Marathon trough. However, of this thickness, only about 1600 meters were deposited in the Early Silurian. This indicates more

rapid subsidence and deposition during the late Paleozoic. Unfossiliferous metamorphosed graywackes, quartzites, and cherty beds of probable Early Cambrian age are the oldest Paleozoic sediments known from the Ouachita belt and indicate deposition along the continental rise. The next oldest rocks are Middle Cambrian igneous bodies that are 525 to 535 million years old. These are overlain unconformably by basal sandstones and fossiliferous limestones of platform facies. Ordovician and Silurian deposits in the Ouachita region can be differentiated into distinct graptolitic facies and shelly facies of the continental shelf. The deeper water deposits lay to the south, whereas the shallower shelf sediments were deposited along the edge of the craton. Northward, the early Paleozoic section thins as it becomes part of the interior platform. Paleozoic rocks of the Ouachita belt are noted for their unusually siliceous and cherty character. Very likely, the abundant silica was derived from the submarine weathering of volcanic ash ejected by volcanoes that lay to the south of the Ouachita trough.

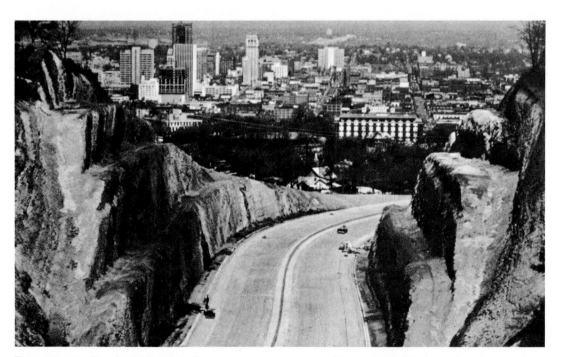

**Figure 9–31** The Red Mountain Expressway road cut near Birmingham, Alabama. Iron ore and underlying limestone are exposed in the cut. Coal to fuel Alabama's iron industry crops out nearby. *(From LeMoreaux, P. E. and Simpson, T. A. Geotimes, 17(8):10–11, 1970.)*

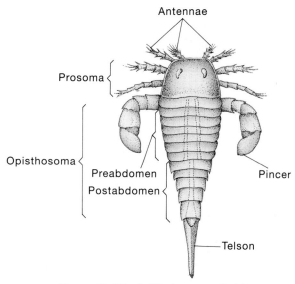

**Figure 9–32** A Silurian eurypterid (shown at one-fourth normal size).

## The Caledonides

Scotland's ancient name, still used in poetry, is Caledonia. Eduard Suess used the name for the Scottish remnants of an early Paleozoic mountain range, the **Caledonides.** It is also the basis for the name of the Caledonian orogenic belt which extends along the northwestern border of Europe (see Fig. 9–5).

The Caledonian and Appalachian belts have had a generally similar history. This is not surprising, for both are really part of one greater Appalachian–Caledonian system. Both elongate depositional sites evolved as a result of a cycle of ocean expansion and contraction, such as is schematically illustrated in Figure 9–23. The cycle began with a late Precambrian to Middle Ordovician episode of spreading, as the Iapetus Ocean widened to admit new oceanic crust along a spreading center. On the continental shelves and rises of the separating blocks, thick lenses of sediment accumulated. This depositional phase is marked by the development of two distinct facies. The graptolite facies consists of more than 6000 meters of volcanics, graywackes, and shales. Graptolites are prevalent in this facies and are used for subdividing the Ordovician and Silurian into biostratigraphic zones. A shelly facies with clean sandstones and fossiliferous limestones identifies the shallower or continental shelf facies. Here and there along the margins of the Caledonian trough there are deposits of Upper Silurian freshwater shales that are noteworthy for their fossil remains of early fishes and strange arthropods known as eurypterids (Fig. 9–32).

The closure of the Iapetus Ocean and crumpling of the Caledonian marine basins began in Middle Ordovician time, when subduction zones developed along the margins of the formerly separating continents. This event is recognized from the volcanic rocks that occur in the Canadian Maritime Provinces, northwestern England, northeastern Greenland, and Norway. Little by little, the Iapetus Ocean closed until the opposing continental margins converged in a culminating mountain-building event termed the **Caledonian orogeny.** The orogeny reached its climax in Late Silurian to earliest Devonian time. It was most intense in Norway, where Precambrian and lower Paleozoic sedimentary rocks are drastically metamorphosed and thrust-faulted. Southwestward, in the British Isles, the effects were not as severe, although mountainous terrains also developed there.

South of the unruly Caledonian belt there is evidence of a land mass that during Late Silurian and Devonian received sandy detritus from the growing Caledonides. This region, on which was deposited clays and sands of the Old Red Sandstone, is often referred to as the Old Red Continent (Fig. 9–33). The deposits here are in the form of a thick, clastic wedge very similar to the Queenston clastic wedge

A

B

C

**Figure 9–33** Paleographic maps of Europe during the (A) Cambrian, (B) Ordovician–Silurian, and (C) Devonian. *(After Brinkman, R. Geologic Evolution of Europe. Stuttgart:Enke; New York: Hafner Publishing Co., 1969.)*

on the opposite side of the Iapetus Ocean. In fact, the two clastic wedges are approximately mirror images of one another (see Fig. 9–23).

Rocks of the Caledonian belt are well exposed not only in the British Isles, but also across northeastern Greenland and Spitzbergen. The geologic succession in the Caledonides of eastern Greenland begins with a thick (10,000-meter) sequence of Precambrian rocks. The uppermost unit in the Precambrian sequence contains tillites, indicating an episode of glaciation just prior to the beginning of the Paleozoic. In eastern Greenland, Cambrian deposition began with sandstones followed by shales, dolomites, and limestones. Carbonates continued to be prevalent in the Early Ordovician, but thereafter sedimentary patterns began to change in response to the closing of the Iapetus Ocean as it was manifested in the Caledonian orogeny. Erosion of the metamorphosed and folded rocks led to the accumulation of the continental Old Red Sandstone facies in basins adjacent to the growing mountains.

## The Uralian Seaway

The elongate seaway in which lower Paleozoic rocks of the Urals were deposited extended southward from the Russian Arctic along the course of the present Ural Mountains. The seaway separated the Russian platform on the west from the Angaran shield on the east. The western side of the Uralian belt contains early Paleozoic shelf carbonates, sandstones, and shales that thin toward the west and pinch out against the Russian platform (Fig. 9–34). Deep marine conditions characterized the eastern side of the Uralian tract. Here we find turbidites, shales, and volcanics.

Although the Uralian orogenic phase was not to come until the late Paleozoic, folded strata, unconformities, and mafic intrusions attest to plate convergence and subduction in the region. The history of the Urals seems to have followed an ocean expansion and contraction cycle (that is, a Wilson cycle) in some ways similar to that of the Caledonian belt.

**Figure 9–34** The Uralian seaway.
*(Adapted from Read, H. H. and Watson, J.* Introduction to Geology. *New York, Halstead Press, 1975.)*

Paleomagnetic studies indicate that the Russian and Siberian platforms were separated by a wide oceanic tract during the early Paleozoic. During the Silurian and Devonian, subduction zones developed between the two opposing continental masses. As the oceanic crust was being subducted, the intervening ocean narrowed, and the continents began to converge upon one another (see Middle Silurian in Fig. (9–36). By the end of the Paleozoic, and possibly continuing into the Triassic, the former depositional belt experienced its orogenic climax as the two continental blocks collided (see early Late Permian in Fig. 10–1).

## Asia Begins to Take Form

Asia east of the Urals is a very ancient and complex terrane formed from the convergence and accretion of several crustal blocks. The sutures along which the continents collided are marked by ophiolite belts, narrow zones of high-pressure metamorphism, and volcanic rocks that characterize crustal tracts above subduction zones that descend at continental margins. Some geologists believe as many as nine separate crustal blocks or microcontinents may have combined to form Asia. In the early Paleozoic, faunal

COMMENTARY COMMENTARY COMMENTARY COMMENTARY COMMENTARY
MMENTARY COMMENTARY COMMENTARY COMMENTARY COMMENTARY COMMEN
COMMENTARY COMMENTARY COMMENTARY COMMENTARY COMMENTARY
MMENTARY COMMENTARY **COMMENTARY** COMMENTARY COMMEN

## The Big Freeze in North Africa

The rafting about of continents during the early Paleozoic resulted in climatic changes that provide some striking contrasts with conditions as they exist today. The Sahara region, which presently is largely a harsh, dry, often warm desert, was once a frigid, stormy landscape covered by a vast continental glacier. As can be seen in Figure 9–36, Africa during the Middle Ordovician lay far south of the equator and was nearly upside-down relative to its present orientation. By the end of the Ordovician, the continent had moved still farther to the south, to a location where the South Pole lay in what is now a barren area near the border between western Algeria and Mauritania. In and around this region, the evidence for an ancient episode of glaciation consists of glacial striations and grooves in flat-lying early Paleozoic sandstones that can be followed for kilometers. In addition, there are extensive tillites,

outwash plains, moraines, and meltwater channels that extend across thousands of square kilometers. The striations produced by the glacier indicate that the ice moved from south to north down the gentle slope of the Precambrian Shield.

Such an enormous buildup of continental ice must surely have lowered global sea level and may even have resulted in cooler temperatures in ocean regions far to the north. Such eustatic and climatic changes would have had a pervasive effect on the biosphere. Paleontologists have long recognized that the end of the Ordovician was marked by extinctions of large numbers of families of bryozoans, tabulate corals, brachiopods, sponges, nautiloid cephalopods, and crinoids. Their demise may have resulted from a general cooling of tropical seas, or possibly from loss of the epeiric and marginal shallow seas as sea level was lowered.

and paleomagnetic evidence suggests that most of these blocks were separated by considerable widths of oceanic crust. By the Middle Ordovician, narrowing of these oceans and convergence of the continental crustal blocks was in progress, although major episodes of collisional orogenesis did not occur in most regions until late in the Paleozoic and in the Triassic and Jurassic.

## Way Down South

Unlike the stratigraphic record of the early Paleozoic for North America and Europe, the record for the Southern Hemisphere land masses is relatively poor. This may be the result of the continents being prevailingly above sea level, which would favor erosion rather than deposition. However, there is a fairly complete record of sedimentation along the north–south trending Tasman belt of Australia (Figs. 8–18 and 9–35). Sedimentation began there just before the Paleozoic. These uppermost Proterozoic rocks are mostly sandstones that grade into overlying Lower Cambrian formations without a significant erosional unconformity. The relatively quiet period represented by these sediments was followed by a full-scale orogeny in the Late Cambrian. Thick graywackes and volcanics record the orogenic dis-

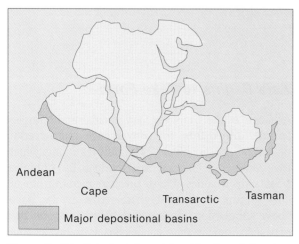

**Figure 9–35** Principal depositional basins of Southern Hemisphere continents during early Paleozoic time.

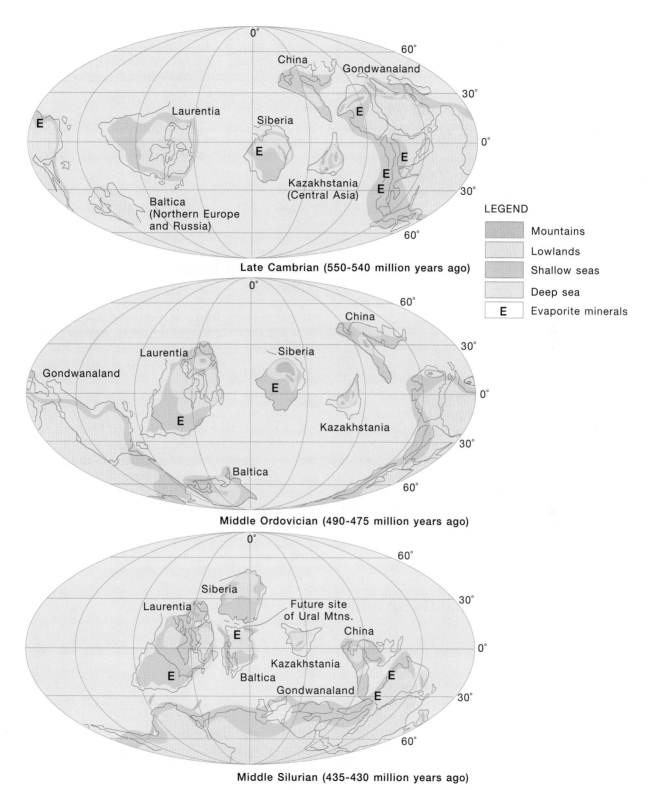

**Late Cambrian (550-540 million years ago)**

**Middle Ordovician (490-475 million years ago)**

**Middle Silurian (435-430 million years ago)**

LEGEND
- Mountains
- Lowlands
- Shallow seas
- Deep sea
- E  Evaporite minerals

**Figure 9–36**  Global paleogeography of the early Paleozoic. *(From Scotese, C. R., Bambach, R. K., and Barton, C., et al. J. Geol. 87(3):217–277, 1979.)*

turbances. During the Ordovician, a broad-shelf sea spread across central Australia and received deposits of well-sorted sandstones and limestones.

The African record of the early Paleozoic is far more obscure than that of Australia. The most complete sequences of strata are located north of the equator, where one finds Cambrian through Silurian rocks that were laid down in shallow embayments of a seaway that was the precursor of the Tethys basin. Africa south of the equator was persistently above sea level and was being eroded during most of the early Paleozoic. Strata of this age are indeed uncommon.

One feature in the geologic record of the northern part of Africa is rather astonishing. There is remarkable evidence of an Ordovician episode of glaciation in the area that is now the Sahara Desert. One can only surmise that this part of Africa was close to the Ordovician South Pole (as indicated by paleomagnetic data).

Lower Paleozoic rocks in South America are also poorly known and difficult to interpret. Strata deposited along the Andean tract have been greatly deformed and frequently obliterated by Mesozoic orogenic activity. Elsewhere, the Paleozoic sequence lies buried beneath an immense pile of sediment spread eastward from the eroding Andes Mountains. We do, however, have scattered bits of the record. We know that Cambrian and Ordovician seas occupied the western margin of South America and that graptolitic shales and shelly limestones were deposited there. Platform deposits were also laid down in shallow basins within the present Amazon region. In Ordovician rocks of Argentina, we find boulder beds and faceted cobbles produced by glaciers.

## Early Paleozoic Global Geography

We are so accustomed to the shapes and locations of today's continents that the earth's early Paleozoic geography seems alien to us. Geologists have attempted to reconstruct the positions of continents during this ancient time by combining all relevant paleomagnetic, paleoclimatologic, biogeographic, and tectonic information. This synthesis of data has resulted in global paleogeographic maps such as those provided in Figure 9–36. These maps indicate that by Late Cambrian time, the major land areas were positioned at low latitudes as rather isolated continents in a well-interconnected world ocean. The main continental blocks at this time were

Kazakhstania (a piece of crust now located in central Asia north of Iran), Siberia, Laurentia, Baltica, and Gondwanaland. There was no land north or south of 60° latitude, and the poles were therefore in the middle of oceanic areas. During the Ordovician and Silurian, the most dramatic change in global geography was the movement of Gondwanaland from its equatorial position southward to a polar location. Baltica (northern Europe and Russia), meanwhile, gradually circled and set a collision course toward eastern Laurentia. The actual collision took place in the Devonian as part of the Acadian orogeny and the assembly of Pangea.

## Aspects of Early Paleozoic Climates

During the early Paleozoic, the earth had a north and south pole, was encircled by an equator, and must have been characterized by latitudinal and topographic variations in climate, just as such variations exist today. As we have seen, during episodes of great marine transgressions and low-lying terrains, climates were probably somewhat milder; during those periods when continents stood high and great ranges diverted atmospheric circulation patterns, climates were more diverse and extreme, much as is the case today. However, there were also factors that would have made those climates of long ago rather unique. During the early Paleozoic, the earth turned faster and the days were shorter. Tidal effects were stronger, and there was no green covering of vegetation to better absorb the sun's radiation. There may even have been changes in solar radiation or atmospheric composition that left no unambiguous clues. The total climatic effects of these possibilities can only be surmised.

At the very beginning of the Cambrian, the climate was probably somewhat cooler than the average for the entire early Paleozoic. The earth was still recovering from a great ice age that had culminated near the close of the Precambrian. Soon, however, climates became warmer. Broad, shallow seas (epeiric seas) slowly began to encroach upon the continental interiors. Paleogeographic reconstructions such as those shown in Figure 9–36 suggest that North America, Europe, and even Antarctica may have lain astride or near the equator during the early Paleozoic. These geophysical data are compatible with the depositional record of thick limestones and extensive reefs, which today form only in the warmer regions of the oceans.

However, some evaporites and fossiliferous limestones have been found in rather high paleolatitudes, indicating that worldwide climates may also have been somewhat warmer than average for a time. In the Ordovician stratigraphic sequences of northern Canada, for example, thick deposits of gypsum and salt are found at paleolatitudes (latitudes determined by paleomagnetic studies) of about 10°S. The North American reefs of Ordovician age generally fall within 30° of the paleoequator, whereas the richly fossiliferous carbonate rocks of the craton would have been within about 40° of the paleoequator. The study of cross-bedding in sandstones suggests that eastern North America lay in a zone of trade winds that blew from northeast to southwest with respect to the present directional grid. These winds thus moved in a direction opposite to those prevalent in eastern North America today.

That there were severe as well as equable climates during the early Paleozoic is indicated by extensive Late Ordovician glacial deposits in the region of the present Sahara Desert. These deposits are sufficiently widespread that they must have been emplaced by continental glaciers and certainly must have been accompanied by a lowering of mean annual temperatures at middle and high latitudes. Study of the distribution of the glacial deposits, directions of glacial striations, and paleomagnetic data indicates that in Ordovician time the north African part of Gondwanaland was positioned astride the South Pole.

During the Silurian, there were latitudinal variations in climate somewhat similar to those of today. Glacial deposits of this age are found only in higher locations, usually in excess of 65° from the paleoequator. Coral reefs, evaporites, and desert sandstones are found within 40° of the Silurian equator. Certainly there were regions of marked aridity during the Silurian. Epeiric seas became relatively less extensive as the Silurian drew to a close. We can speculate that the Iapetus Ocean formed by the division of that hypothetic continent Proto-Pangea had already begun closing and that the Caledonian crunch was about to take place. The curtain was ready to rise on the world of the late Paleozoic.

## Mineral Resources of the Lower Paleozoic

The lower Paleozoic systems contain several kinds of earth materials having economic importance. Among these are building stones, such as slates that were once extensively quarried in New Jersey and Pennsylvania and marbles derived from Ordovician limestones that are still quarried in New England. The St. Peter Sandstone has been and continues to be an important source of clean quartz for use in the manufacture of glass. As noted earlier, salt is also an important mineral resource of the lower Paleozoic (Silurian). In New York state, salt recovery and marketing is a major industry. The method used in recovery involves pumping water down wells into salt-bearing formations. The water dissolves the salt, which is then pumped to the surface to be reprecipitated in evaporating pans and refined.

The most notable ore deposits of the lower Paleozoic are sedimentary iron ores that occur in Newfoundland and in a band extending southward from New York into Alabama. At Birmingham, these iron ore beds were once the basis for a thriving steel industry. However, most of the mines are now closed. Nevertheless, a large amount of ore remains, and it may become economically feasible to reopen mines at some future date. Petroleum has been recovered from Ordovician and Silurian rocks in Ohio, Oklahoma, and Texas, as well as in western Europe.

## SUMMARY

It is generally agreed that the Paleozoic Era began at about the same time that shell-bearing multicellular animals made their appearance. It also began at a time when the Proto-Pangean supercontinent had begun to pull apart, allowing the incursion of oceanic tracts between the separating landmasses. The Iapetus Ocean was an early Paleozoic proto-Atlantic ocean that was widening at the beginning of the

Cambrian, but was to close during later Paleozoic time as a result of conversion of tectonic plates bearing North America, western Europe, and Gondwanaland.

Most of North America at the beginning of the Paleozoic was above sea level and experiencing erosion. The initial sites of marine deposition were along the rifted continental margins. Gradually, however, the seas spilled out of these marginal basins onto the continental platform. The sediments deposited on the floors of epeiric seas that covered the platform record cycles of transgression and regression. The thicker sections are preserved in basins, and the thinner sequences with many unconformities developed over domes. The Sauk and Tippecanoe cratonic sequences record the transgressive–regressive cycles of the early Paleozoic. For much of the early Paleozoic, the Transcontinental arch was a positive area providing sediment to surrounding seas.

By Ordovician time, the relative calm that characterized the Cambrian was interrupted by a plate convergence and mountain-building episode called the Taconic orogeny. Periodic crustal disturbances associated with this orogeny raised mountains along the eastern margin of North America. Rapid erosion of these uplands produced huge volumes of detritus that were spread westward by streams and deposited in a great apron of detrital sediment called the Queenston clastic wedge. Similar mountain-building activity was also underway in northwestern Europe,

where the Caledonian depositional basin was compressed, intruded, and uplifted in the early stages of the Caledonian orogeny. Both the Taconic and Acadian orogenies were the result of plate convergence associated with the narrowing of the Iapetus Ocean.

The initial deposits of the early Paleozoic in western North America were continental shelf and rise deposits of the rifted continental margin. This early episode of relatively quiet sedimentation was interrupted during the Ordovician and Silurian by movement of the Pacific plate against the western edge of North America. The result of this movement was the development of a subduction zone and chain of volcanic islands along the western margin of the continent.

In general, the range of climatic conditions during the early Paleozoic was not significantly different from that of the recent geologic past. Cold climates are indicated by the presence of Ordovician glacial deposits in Southern Hemisphere continents. Warmer conditions prevailed in regions where reefs were extensive and richly fossiliferous limestones were deposited. Aridity is indicated by the immensely thick sections of evaporites in the region around the Great Lakes. Although the range of climates may not have been greatly different from that of today, there were many nonclimatic differences. In the early Paleozoic the days were shorter, tides were stronger, and until Silurian time, the lands had no cover of green plants.

## QUESTIONS FOR REVIEW

1. What are the names and locations of the major orogenic events of the early Paleozoic, and when did they occur?

2. Domes and basins are characteristic structural features of the platform region of North America and other continents. Define each and explain how they differ in thickness of strata and unconformities. How might one distinguish between these two structures on a geologic map?

3. Geologists have suggested that erosion by wind and water proceeded more vigorously during the Cambrian and Ordovician than today. What might have been the reason?

4. What is the Tommotian Stage of the Cambrian System? How do multicellular animals found as fossils below the Tommotian rocks differ from those above?

5. What geologic evidence indicates that the Transcontinental arch was above sea level during most of the early Paleozoic?

6. Why are rock units such as the Tapeats Sandstone and the Muav Limestone not precisely the same age at different localities?

7. What is a barred basin? In arid coastal regions, how might such a basin provide for the precipitation of evaporites?

8. What is a clastic wedge? Are clastic wedges

primarily marine or nonmarine? Under what geologic circumstances do clastic wedges form?

9. When and where did the Queenston clastic wedge develop? With what orogeny is it associated?

10. It is probable that tides were higher and days were shorter during the early Paleozoic. Assuming this is true, what would be the reason for such conditions?

## TERMS TO REMEMBER

barred basin
Caledonian orogeny
clastic wedge
Iapetus Ocean

Logan's thrust
platform
principle of temporal
    transgression

Proto-Pangea
Queenston clastic wedge
    (formerly Queenston
    delta)

Sauk cratonic sequence
Tippecanoe cratonic
    sequence
Tommotian Stage

## SUPPLEMENTAL READINGS AND REFERENCES

Bambach, R. K., Scotese, C. R., and Ziegler, A. M. 1980. Before Pangaea: The geographies of the Paleozoic world. *Amer. Scientist* 68(1):26–38.

Fairbridge, R. W. 1970. Ice age in the Sahara. *Geotimes* 15(6):18.

Frazier, W. J., and Schwimmer, D. R. 1987 *Regional Stratigraphy of North America*. New York: Plenum Press.

Hamilton, W. 1970. The Uralides and the motion of the Russian and Siberian platforms. *Bull. Geol. Soc. Amer.* 81:2553–2576.

Kummel, B. 1970. *History of the Earth* (2nd ed.). San Francisco: W. H. Freeman and Co.

Matthews, S. C., and Missarzhevsky, V. V. 1975. Small shelly fossils of late Precambrian and early Cambrian age: A review of recent work. *J. Geol. Soc. London* 131:289–304.

Sloss, S. L. 1963. Sequences in the cratonic interior of North America. *Geol. Soc. Amer. Bull.* 74:93–111.

Stearn, C. W., Carroll, R. L., and Clark, T. H. 1979. *The Geological Evolution of North America* (3rd ed.). New York: Ronald Press.

Valentine, J. W., and Moores, E. M. 1972. Global tectonics and the fossil record. *J. Geol.* 80(2):167–184.

Steeply dipping beds of the Fountain Formation form the so-called flatirons over-
looking Boulder, Colorado. The Fountain Formation consists of red arkosic sandstones, conglomerates, and mudstones deposited during late Pennsylvanian and
early Permian time. The source of these clastics was the Ancestral Rocky Mountains that lay to the west. The Fountain Formation was tilted upward to form the
dramatic ridges and hogbacks of the Colorado Front Range during the orogeny
that produced the modern Rocky Mountains. The view is northward, with the High

# Late Paleozoic Events

**10**

A continuous ocean spreads over the space now
occupied by the British islands: in the tract covered
by the green fields and brown moors of our own
country, the bottom, for a hundred yards downwards,
is composed of the debris of rolled pebbles and
coarse sand intermingled, long since consolidated into
the lower member of the Old Red Sandstone; the
upper surface is composed of banks of sand, mud,
and clay; and the sea, swarming with animal life,
flows over all.

Hugh Miller, *The Old Red Sandstone*, 1858

## Introduction

The Devonian, Mississippian, Pennsylvanian, and Permian are the geologic periods of the late Paleozoic in North America. Elsewhere, the Mississippian and Pennsylvanian are combined into the Carboniferous Period. The late Paleozoic was a time when most of the separate land masses of earlier periods were assembled into the great supercontinent of Pangea. The process of gathering the scattered continents had actually begun earlier. In the Silurian, for example, the ocean tract between North America (Laurentia) and northern Europe (Baltica) gradually narrowed as the two continents came together. The Silurian Caledonian orogeny was the result of the collision that continued into the Devonian of eastern North America as the **Acadian orogeny**. North America was now sutured to Europe, and the combined land mass has been named Laurussia (Fig. 10–1).

After the Devonian, the process of assembling Pangea continued as the plate-bearing Gondwanaland made its way northward at the expense of the intervening oceanic area. By late in the Carboniferous, Gondwanaland had come into contact with Laurussia, generating the Hercynian orogeny of central Europe and the Allegheny orogeny of eastern North America. It may seem startling that mountains in the southern Appalachian region were the result of a collision with northwestern Africa. Similarly, the now deeply eroded and largely buried mountains of the Ouachita orogenic belt were produced by collision with the South American segment of Gondwanaland. The part of South America that is now Venezuela apparently formed the southern margin of that late Paleozoic Ouachita mountain system.

The clustering of once separate continents to form Pangea was nearly complete by the Late Permian. Great mountain systems identified the suture zones where continents had come together. An enormous ocean called Panthalassa spanned the globe across nearly 300° of longitude.

In addition to continental collisions and recurrent orogenic activity, the late Paleozoic was a time of diverse sedimentation, progress in organic evolution, and diverse climatic conditions. During this time there was widespread colonization of the land by both large land plants and vertebrates. Amphibians and reptiles, along with spore-bearing trees and seed ferns, were especially evident in the landscapes of the late Paleozoic. Near the end of the era, conifers and reptiles more tolerant of drier, cooler climates became abundant. Marine invertebrates thrived in shallow epicontinental seas that were especially extensive

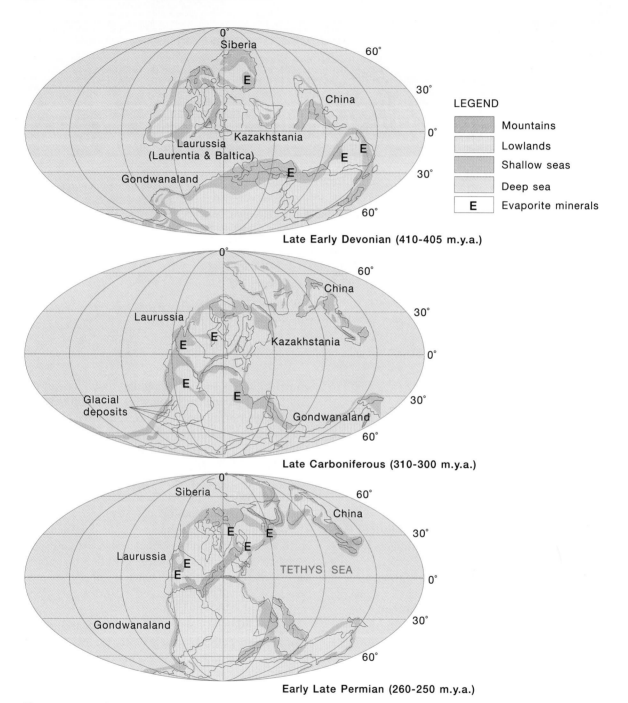

**Figure 10–1** Global paleogeography of the late Paleozoic. *(From Scotese, C. R., Bambach, R. K., Barton, C., Van Der Voo, R., and Ziegler, A. M. J. Geol. 87(3):217–277, 1979.)*

in North America during the Mississippian. In the central interior of the United States, limestones deposited in these inland seas are over 700 meters thick and are among the most extensive sheets of limestone anywhere in the world. Adjacent to the orogenic belts and in regions where the epicontinen-

**Figure 10–2**  The distribution of glacial tillites (blue triangles), coal (red circles), and evaporites (irregular green areas) during the Permian, about 250 million years ago. *(From Dewey, G. E., Ramsey, T. S., and Smith, A. G. J. Geol. 82(5):539, 1974.)*

tal seas were drained as a result of uplift of the land or lowering of sea level, sediment-laden streams deposited their loads of sand and mud.

These nonmarine sediments contain the fossil plants and animals that aid in inferring climatic conditions on the continents. The accumulation of abundant vegetation in swampy areas provided the raw materials from which the famous coal deposits of Pennsylvanian time were formed. Now consolidated sands of desert dunes and layers of evaporites attest to arid conditions at many localities around the globe (Fig. 10–2). Tillites and glacial striations on bedrock provide evidence of glaciation. This is not unexpected, for the more southerly parts of Gondwanaland were located near the pole. Evidence of Late Carboniferous or Permian ice sheets can be found on the now separated landmasses of South America, Africa, Australia (Fig. 10–3), Antarctica, and India.

Decades before geologists had an understanding of plate tectonics, the ancient tillite deposits on these continents were cited as evidence for continental drift.

## Lands, Seas, and Orogenic Belts

### *The Craton of North America*

In Chapter 9, we noted that the last event of the early Paleozoic cratonic region had been the withdrawal of inland seas. Many early Paleozoic formations were subjected to erosion following the regression of seas, and this exposed older and deeper formations in those areas that had experienced uparching. Gradual flooding of that early Paleozoic erosional surface was the first event of the late Paleozoic. The new inland

**Figure 10–3** Glacial striations formed in Early Permian time on Proterozoic quartzite in Australia. The ice flowed to the right, parallel to the pen. Note the downflow smoothing and crescent-shaped "chatter" marks with their concave sides facing upflow (upper right). *(Courtesy of Warren Hamilton, U.S. Geological Survey.)*

sea is termed the **Kaskaskia sea** (Fig. 10–4), and in it were deposited marine rocks of the Kaskaskia sequence (see Table 9–1). Except for minor regressions, the sea persisted over extensive areas of the craton until the end of the Mississippian Period. Erosion of the layers of sediment deposited on the Kaskaskia sea floor began, only to be interrupted by the advance of a second sea, the **Absaroka sea.**

## The Kaskaskia Sequence

Along the eastern side of the North American craton, Kaskaskia sedimentation consists initially of clean quartz sands. The most famous of these sandy formations is the Oriskany Sandstone of New York and Pennsylvania (Fig. 10–5). Because of its purity, Oriskany Sandstone is extensively used in making glass.

Although quartz is by far the dominant mineral in the Oriskany, the formation also contains a very small percentage of other silicate materials, which, like quartz, are resistant to weathering and erosion but are heavier than quartz. These so-called *heavy minerals* can be used to determine the kinds of source rocks and source areas from which clastic

sediments were derived. After study of the heavy minerals from scattered exposures of Oriskany sandstones, it became apparent that there are two distinctly different assemblages of heavy minerals. In the area south of New York, the formation contains exceedingly small amounts of only the most stable heavy minerals: tourmaline, zircon, and rutile. The grains in these sandstones are exceptionally well rounded and worn, and less stable heavy minerals are lacking. This part of the Oriskany is considered to have derived from older clastic sedimentary units to the east and north. In contrast, the Oriskany in New York State contains a heavy mineral assemblage of relatively unstable pyroxene, amphibole, biotite, and garnet. Garnet is ordinarily derived from metamorphic source rocks, whereas the others are usual components of a variety of igneous and metamorphic rocks. The grains are unaltered. One may infer, therefore, that the sands composing the Oriskany of the New York area may have had an igneous or metamorphic source area and were not recycled from older sandstones.

The Oriskany Sandstone is the initial deposit of a transgressing sea, as is indicated by the way the

**Figure 10–4**    Paleogeography of North America during the Devonian Period.

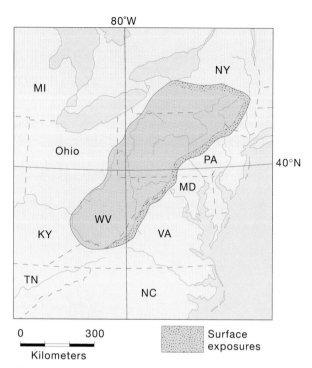

**Figure 10–5** Areal extent of the Oriskany Sandstone. Surface exposures occur along the northern and eastern borders of the area shown in green.

sands overlap older strata. As this marine advance continued, limy sediments were deposited over the Oriskany, and corals began to build reef structures. In areas where water circulation was restricted, salt and gypsum were deposited. During the Middle and Late Devonian of the region east of the Mississippi Valley, carbonate sedimentation gave way to shales. The change to clastic deposition was a consequence of mountain building associated with the Acadian orogeny in the Appalachians. As will be described shortly, highlands formed during this orogeny were rapidly eroded, and clastics were transported westward to form an extensive apron of sediments that are coarser and thicker near their eastern source area and give way to thin but widespread black shales in the east-central region of the United States.

In the far western part of the craton, Middle and Upper Devonian rocks are largely limestones, although there are shales as well. In a depressed area known as the Williston basin (from South Dakota and Montana northward into Canada), extensive reefs were developed. Arid conditions and restricted circulation in reef-enclosed basins resulted in the deposition of impressive thicknesses of gypsum and salt. The reefs provided permeable structures into which petroleum migrated, resulting in some of Canada's richest oil fields.

During the Late Devonian, mountains existed along the eastern margin of North America. A clastic wedge was built along the western side of the mountains. Conglomerates and sandstones were the characteristic deposits of the wedge, but farther to the west, suspended muds were carried into the shallow sea that covered the platform. These muds were deposited as a thin, but extraordinarily widespread sheet of dark shale. There are many local names for the shale, but the formal name most commonly used is **Chattanooga Shale.** Because the shale is both widespread and easy to recognize, it has served as an excellent marker in regional correlation. It also contains uranium and has been radiometrically determined to be 350 million years old.

In addition to its uranium content, the Chattanooga is rich in finely disseminated pyrite and organic matter. It is black or dark gray in color and contains few, if any, fossils of bottom-dwelling invertebrates. These characteristics indicate that the Chattanooga accumulated in stagnant, oxygen-deficient waters. Today, such environments are much more limited in size. It is difficult to conceive how a sea occupying an area as vast as the depositional area of the Chattanooga could sustain such oxygen-deficient conditions. One interesting suggestion is that phytoplankton proliferated catastrophically, consuming and thereby depleting the oxygen supply. Without the oxygen required for decay of organic matter, carbon would accumulate and contribute to the black color characteristic of the Chattanooga Shale.

During the passage from Devonian to Mississippian time, highland source areas that provided the Chattanooga Shale were reduced, and the quantity of muddy sediment decreased. Carbonates then became the most abundant and widespread kind of sediment in the epeiric seas of the platform. Cherty limestones (see Fig. 1–47), shelly limestones, limestones composed of the remains of billions of crinoids (Fig. 10–6) and other invertebrates, and limestones containing myriads of oöids (Fig. 10–7) formed extensive beds across the central and western parts of the craton (Fig. 10–8). You may recall from Chapter 1 that oöids are spherical grains formed by the precipitation of calcium carbonate around a nucleus such as a tiny particle of calcium carbonate. They form in

A

**Figure 10–6** (A) Specimen of crinoidal limestone (Burlington Formation) in which crinoid fragments (mostly stem plates) stand out in relief because of weathering. (B) Reconstruction of crinoids growing on the floor of the Kaskaskia (Mississippian) sea. *(Latter photo courtesy of the U.S. National Museum.)*

B

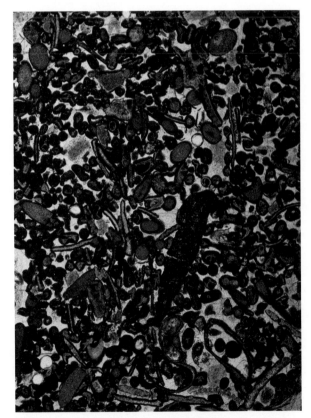

**Figure 10–7** Photomicrograph of Salem Limestone (Mississippian). The Salem consists primarily of fossil debris and oöids cemented with clear calcite. It is extensively quarried in southern Indiana and used as trim and building stone in public buildings throughout the United States. Width of section 2.5 mm.

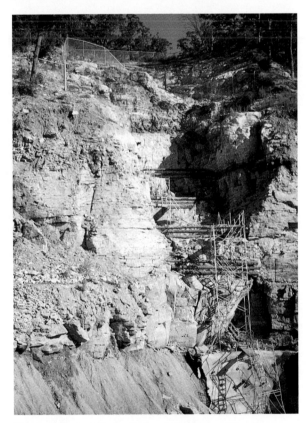

**Figure 10–8** Mississippian limestones (Chouteau and Burlington formations) overlying gray shale (Hannibal Shale) exposed during excavation for the Clarence Cannon Dam in northeastern Missouri. The Burlington Formation is noted for its rich content of crinoid remains. It also contains abundant chert nodules.

**Figure 10–9** Paleogeography of North America during the Mississippian Period.

**Figure 10–10**  Cross-bedding in sandstones of the Upper Mississippian Tar Springs Formation, southern Illinois. *(Photograph courtesy of J. C. Brice.)*

clear, shallow sea water where currents exist to move particles about on the sea floor. Salinity somewhat greater than normal appears to enhance the formation of oöids.

The sea in which these crinoidal, oölitic, and other types of limestones were deposited was the most extensive that North America had experienced since Ordovician time (Fig. 10–9). The widespread blanket of carbonates deposited in this sea is often referred to as the great Mississippian lime bank. The deposits record the last great Paleozoic flooding of the North American craton.

In Late Mississippian time, as the Kaskaskia seas began their final withdrawal, large quantities of detrital sediments and thin, scattered limestones were deposited. Today, these rocks are reservoirs for petroleum in Illinois and therefore have been studied extensively. Detailed maps of thicknesses of some of the sandstone units show that they are thickest along branching, sinuous trends that suggest old stream valleys developed on the former sea floor. Studies of grain size, cross-bedding (Fig. 10–10), and current-produced sedimentary structures strongly indicate that the detritus was derived from the northern Appalachians and was transported southwestward across the central interior.

## The Absaroka Sequence

When the Kaskaskia seas had finally left the craton at the end of Mississippian time, the exposed terrain

was subjected to erosion that resulted in one of the most widespread regional unconformities in the world. Not only was erosion extensive in areas, but it also eroded away entire systems of older rocks over arches and domes. The unconformity provides a criterion for separating those strata equivalent to the Carboniferous of Europe into the Mississippian and Pennsylvanian systems. It is appropriate to use these two names in North America, not only because of the unconformity but also because the rocks above the erosional hiatus differ markedly from those below. Indeed, the overlying Pennsylvanian strata are a consequence of quite different tectonic circumstances.

It was not until near the beginning of Middle Pennsylvanian time that the seas were able to encroach onto the long exposed central region of the craton. The deposits of this seaway are those of the Absaroka sequence. In general, the Pennsylvanian section of rocks near the eastern highlands are thicker, and virtually all are continental sandstones, shales, and coal beds (Fig. 10–11). This eastern section of Pennsylvanian rocks gradually thins away from the Appalachian belt and changes from predominantly terrestrial to about half marine rocks and half nonmarine rocks. Still farther west, the Pennsylvanian outcrops are largely marine limestones, sandstones, and shales.

One of the most notable aspects of Pennsylvanian sedimentation in the middle and eastern states is the repetitive alternation of marine and nonmarine strata. A group of strata that records the variations in

**Figure 10–11**    Paleogeography of North America during the Pennsylvanian Period.

depositional environments caused by a single advance and retreat of the sea is called a **cyclothem**. A typical cyclothem in the Pennsylvanian of Illinois contains ten units (Fig. 10–12). Units one through five are continental deposits, the uppermost of which is a coal bed (Fig. 10–13). The strata deposited above the coal bed represent an advance of the sea over an old forested area.

Cyclothems are notable not only for their cyclicity, but because they can be correlated over great distances. Those of the Appalachian region, for example, can be correlated to cyclothems in western Kansas. In Missouri and Kansas, at least 50 cyclothems are recognized within a section only 750 meters thick, and some of these extend across thousands of square kilometers.

It is apparent that cyclothems are the result of repetitive and widespread advance and retreat of seas. But what was the cause of such oscillations? One explanation involves periodic regional subsidence of the land to a level slightly below sea level, so that marginal seas could spill onto swampy lowlands. A relatively short time later, subsidence might cease, and sediments could be built above sea level to extend the shoreline seaward and reestablish continental conditions. Alternatively, the reestablishment of dry lands may have resulted from temporary regional uplifts. Finally, worldwide (eustatic) changes in sea level related to continental glaciation might have produced repetitive marine invasions and re-

**Figure 10–12** An ideal coal-bearing cyclothem showing the typical sequence of layers. Many cyclothems do not contain all ten units, as in this illustration of an idealized sequence. Some units may not have been deposited because changes from marine to nonmarine conditions may have been abrupt, and/or units may have been removed by erosion following marine regressions. The number 8 bed usually represents maximum inundation and, correlated with the same bed elsewhere, provides an important correlative stratigraphic horizon.

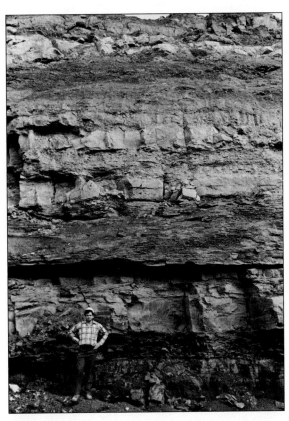

**Figure 10–13** Part of an Illinois cyclothem. The lowermost layer is the coal seam (cyclothem bed 5), followed upward by shale (bed 6) near the geologist's hand, limestone (bed 7), shale (bed 8), another limestone (bed 9), and the upper shale (bed 10). Part of another sequence caps the exposure. This cyclothem is part of the Carbondale Formation. (*Photograph courtesy of D. L. Reinertsen and the Illinois Geological Survey.*)

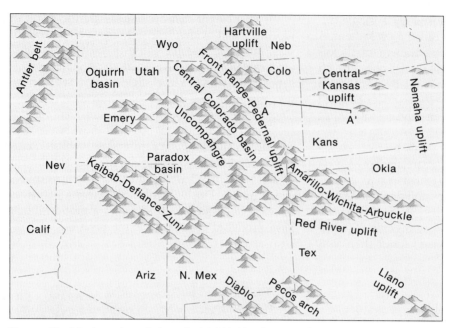

**Figure 10–14** Location of the principal highland areas of the southwestern part of the craton during Pennsylvanian time.

gressions. We have ample evidence that large regions of Gondwanaland were covered with glacial ice from Late Mississippian through Permian time. When the ice sheet grew in size, sea level would be lowered because of water removed from the ocean and precipitated as snow onto the continent. During warmer episodes, meltwater returning to the ocean would raise sea level. These sea level oscillations could result in the shifting back and forth of shorelines that produced cyclothems. Additional support for this idea comes from recognition of cyclothems of the same age on continents other than North America, as this indicates that the cause of the cyclicity was global.

Ordinarily, cratonic areas of continents are characterized by stability. The southwestern part of the North American craton provides an exception to this general rule, for during the Pennsylvanian this was a region of mountain building. The resulting highlands are generally called the Colorado mountains (also called the ancestral Rockies) and the Oklahoma mountains. These mountains and related uplifts appear to have resulted from movements of crustal blocks along large, nearly vertical reverse faults. The Colorado mountains included a range that extended north–south across central Colorado (the Front Range–Pedernal uplifts) and a segment curving from Colorado into eastern Utah (the Uncompahgre uplift). The central Colorado basin lay between these highland areas (Fig. 10–14). A separate range, the Zuni–Fort Defiance uplift, extended across northeastern Arizona on the southwestern perimeter of the Paradox basin. To the east of the Colorado mountains lay the southeastward-trending Oklahoma mountains. Eroded stumps of this once rugged range form today's greatly reduced Arbuckle and Wichita mountains. Remains of the Amarillo mountains are now buried beneath younger rocks and are known principally from exploratory drilling for oil.

Judging from the tremendous volume of sediments eroded from the Colorado mountains, it is likely that they attained heights in excess of 1000 meters. They also were probably subjected to repeated episodes of uplift, continuing in some places into the Permian. Erosion of these highlands eventually exposed their Precambrian igneous and metamorphic cores. As erosion and weathering continued, great wedge-shaped deposits of red arkosic sediments were spread onto adjacent and intervening basins (Fig. 10–15). A small part of this massive accumulation of clastic sediment is dramatically exposed in the Red Rocks Amphitheatre just a few

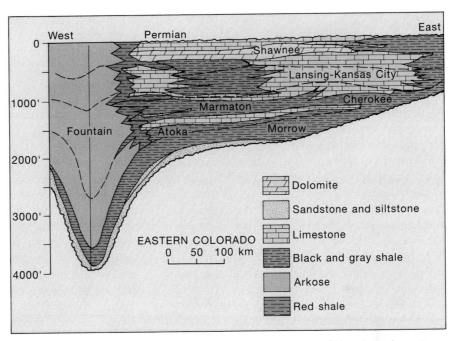

**Figure 10–15**  Pennsylvanian cross-section across eastern Colorado and western Kansas showing great accumulation of coarse arkosic sandstones east of the Colorado mountains. Line of section is along A–A′ in Figure 10–14. *(From* Stratigraphic Atlas of North and Central America. *Shell Oil Company, Exploration Department.)*

miles west of Denver (Fig. 10–16) and the flatirons near Boulder, Colorado (see chapter opening illustration).

Geologists have long puzzled over this episode of deformation of the craton, but plate tectonics has recently provided a hypothesis for this unusual event. It seems likely that the collision of Gondwanaland with North America along the site of the Ouachita orogenic belt generated stress in the bordering area of the craton to the north. Crustal adjustments to relieve these stresses resulted in the deformation that produced the highlands and associated basins.

The basins that lay adjacent to the ancient Colorado mountains are important in working out the geologic history of the southwestern craton because of the excellent stratigraphic record they contain. The Paradox basin is especially interesting. It was inundated by the Absaroka sea in Early Pennsylvanian time. The initial Pennsylvanian deposits, largely shales, were deposited over a karst topography developed on Mississippian limestones. By Middle

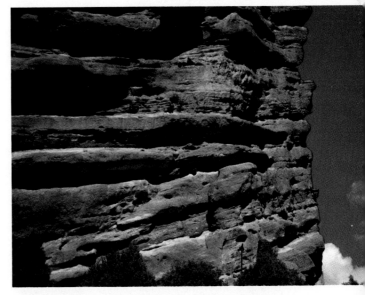

**Figure 10–16**  Coarse arkosic red beds of the Pennsylvanian Fountain Formation exposed around the famous Red Rocks Amphitheatre, a few miles southwest of Denver.

Pennsylvanian time, the western access of the Absaroka sea to the Paradox basin had become restricted, and thick beds of salt, gypsum, and anhydrite were deposited. Fossiliferous and oölitic limestones developed around the periphery of the basin, and patch reefs grew abundantly along the western side. The association of porous reefs and lagoonal deposits resulted in several sites suitable for the later entrapment of petroleum.

Near the end of the Pennsylvanian, the Paradox basin was filled to above sea level by arkosic sediments shed from the recently uplifted Uncompahgre highlands. Also at this time, the Absaroka sea, which had begun its transgression at the beginning of the Pennsylvanian, began a slow and irregular regression near the end of the same period. The withdrawal was still incomplete in Early Permian time, so that marine sediments continued to be deposited in a rather narrow zone from Nebraska to western Texas. Fossiliferous limestones characterized these inland seas, although near highlands in Colorado, Texas, and

Oklahoma, coarse clastics accumulated to sufficient thicknesses to bury surrounding uplands.

The deposits along what was the eastern edge of the seaway have been eroded away, but at several places along the western side, one can observe the change from richly fossiliferous beds below to barren shales, red beds, and evaporites above. The thick and extensive salt beds of Kansas provide testimony to the gradual restriction and evaporation of Permian seas in the central United States. The last Permian marine conditions occurred in the western part of Texas and southeastern New Mexico, where a remarkable sequence of interrelated lagoon, reef, and open-basin sediments were deposited (Figs. 10–17 and 10–18). In this region, several irregularly subsiding basins developed between shallowly submerged platforms. Dark-colored limestones, shales, and sandstones were deposited in the deep basins, whereas massive reefs formed along the basin edges. Behind the reefs, in what must have been the shallow waters of extensive lagoons, the deposits are thin

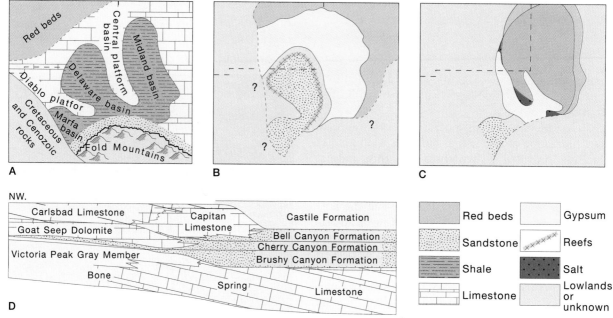

**Figure 10–17** The Permian of North America can be divided into four stages. The oldest is the Wolfcampian, which is followed by Leonardian, Guadalupian, and Ochoan. A, B, and C show the paleogeography and nature of sedimentation in West Texas during the (A) Wolfcampian, (B) Guadalupian and (C) Ochoan. (D) A simplified cross-section of Leonardian and Guadalupian sediments of the Guadalupe Mountains indicates the relationship of the reef to the other facies. *(After many sources, but primarily King, P. B.,* U.S. Geological Survey Professional Paper *215, 1948.)*

**Figure 10–18**   Generalized paleogeographic map for the Permian Period.

**Figure 10–19** Example of a facies change in rocks at the southern end of the Guadalupe Mountains about 100 miles east of El Paso, Texas. The prominent cliff is called El Capitan. It is composed of a Permian limestone reef deposit, rich in corals and associated marine invertebrate fossils. Behind the reef there once existed lagoons with abnormally high salinity, as suggested by evaporites, dolomites, and virtually unfossiliferous limestones. These lagoonal beds are exposed along the ridge on the right side of the photograph. Thus, a massive reef limestone facies lies adjacent to a lagoonal facies. To the south of the reef lay a marine basin in which normal marine sediments were deposited. *(Photograph courtesy of the National Park Service.)*

limestones, evaporites, and red beds. Late in the Permian, the connections of these basins to the south became so severely restricted that the waters gradually evaporated, leaving behind great thicknesses of gypsum and salt.

Much paleoenvironmental information has been obtained from study of the West Texas Permian rocks. From the lack of medium- and coarse-grained clastics, one may assume that surrounding regions were low-lying. The gypsum and salt suggest a warm, dry climate in which basins were periodically replenished with sea water and experienced relatively rapid evaporation. Careful mapping of the rock units has helped to establish an estimated depth of about 500 meters for the deeper basins and only a few meters for the intrabasinal platforms. The basin deposits are dark in color and rich in organic carbon as a likely consequence of accumulation under stagnant, oxygen-poor conditions. Perhaps upwelling of these deeper waters may have provided a bonanza of nutrients on which phytoplankton and reef-forming algae thrived. Along with the algae, the reefs contain the skeletal remains of over 250 species of marine invertebrates. Today, because of their relatively greater resistance to erosion, these ancient reefs form

the steep El Capitan promontory in the Guadalupe Mountains of West Texas and New Mexico (Fig. 10–19). Here one can examine the forereef composed of broken reef debris that formed a sort of talus deposit caused by the pounding of waves along the southeastern side.

## The Eastern Margin of North America

During the latter half of the Paleozoic Era, the great Appalachian and Ouachita belts experienced their culminating and most intense episodes of orogenesis. The crumpling of these former depositional tracts was a consequence of the reassembly of the formerly separated continents into Pangea. The northern part of the Appalachian–Caledonian belt had taken the shock of a collision with Europe during the Silurian and Devonian. That collision was the cause of the Caledonian orogeny in Europe and the Acadian orogeny in northeastern North America. As indicated in Figure 10–1, the southern half of the elongate Appalachian depositional basin as well as the Ouachita trend were crushed into mountainous tracts when they were struck by the northern margin of Gondwanaland (South America and northwestern Africa)

## Reefs

Reefs are of great interest to geologists for many reasons. They comprise immense masses of carbonate material generated entirely by marine invertebrates and algae. Reefs are often well preserved, and therefore provide intact representations of complex ancient and recent biological communities from which a wealth of ecological information can be obtained. Petroleum geologists are aware that ancient reefs serve as excellent traps for petroleum, and therefore carefully map their locations and trends.

Reefs are shallow water, wave-resistant buildups of calcium carbonate formed around a rigid framework of skeletal material. For most of the Phanerozoic, that framework has been built by corals, although the families of corals that construct reefs has varied through time as various groups evolved or became extinct. Before there were abundant reef-building corals, calcareous algae and bryozoa were the principal frame builders. The calcareous skeletons of other reef-dwelling organisms such as crinoids, brachiopods, mollusks, sponges, and algae add mass to the reefs

and help to trap interstitial accumulations of sand and lime mud.

In most reefs, such as the spectacular Permian Capitan Reef of western Texas (see Figs. 10–17 and 10–19), one can recognize three major reef components or facies. The first is the **reef core** itself, which is formed of organisms that have built the reef upward from the shallow sea floor to sea level. Waves constantly crash along the front or windward side of the reef core. In response to this kind of vigorous wave action, organisms living along the reef front form low-growing or encrusting skeletal structures. Waves continuously break off pieces of this part of the reef and the resulting debris forms a steeply dipping apron of debris that resembles a submarine talus deposit (see accompanying figure). This part of the reef is referred to as the **fore-reef facies**. A more sheltered area known as the **back-reef facies** is located landward of the reef core. Here carbonate sands, muds, and oöids accumulate among more erect growing frame builders. The back-reef is also known for the extraordinary diversity of its biological community.

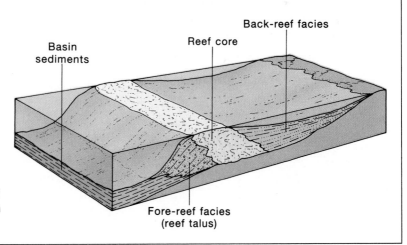

Position of reef core, back-reef facies, and fore-reef facies in a typical reef.

during the Late Carboniferous and Permian. This great encounter between Laurussia and Gondwanaland was the cause of the Allegheny orogeny. As described in the previous section, the collision from the south not only raised the old basins of sedimentation, but also transmitted stresses into the interior, causing deep-seated deformations such as those that produced the Colorado and Oklahoma mountains.

The effects of the Devonian deformation, called the Acadian orogeny, are clearly seen in a belt extending from Newfoundland to West Virginia. Here one finds thick, folded sequences of turbidites interspersed with rhyolitic volcanic rocks and granitic intrusions. The intensity of the compression that affected these rocks is reflected in their metamorphic minerals, which indicate that mineralization occurred at temperatures exceeding 500°C and pressures equivalent to burial under 15,000 meters of rock. The overall result of the Acadian orogeny was to demolish forever the marine depositional basin and establish in its place mountainous areas in which erosion was the prevalent geologic process. Here and there in isolated basins among the mountains, Devonian nonmarine sediments were deposited. How-

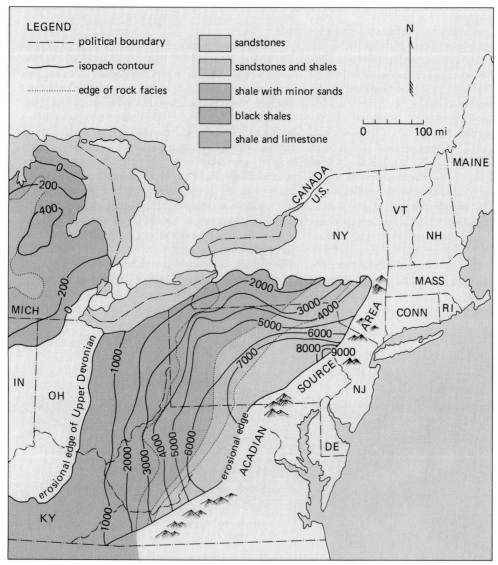

**Figure 10–20** Isopach and lithofacies map of Upper Devonian sedimentary rocks in the northeastern United States. Thicknesses are given in feet. *(Modified from Sevon, W. D. Geol. Soc. Am. Special Paper 201:79–90, 1985; and Ayrton, W. G. Pa. Geol. Survey Rpt. 39(4):3–6, 1963.)*

Red beds
Sandstones
Siltstones and shales
Shale
Limestone

Line of section

**Figure 10–21**   East–west section across the Devonian Catskill clastic wedge, New York. Note that continental red beds interfinger with nearshore marine sandstones, and these in turn grade toward the west into offshore siltstones and shales. Continental deposits prograded upon the sea, pushing the shoreline progressively westward. *(Based on several classic studies by Chadwick, G. H., and Cooper, G. A. completed between 1924 and 1942.)*

ever, the greatest volume of erosional detritus was spread outward from the highlands as a great wedge of terrigenous sediment called the Catskill clastic wedge (Figs. 10–20 and 10–21).

## The Catskill Clastic Wedge

For many reasons, the Devonian rocks of the **Catskill clastic wedge** (also called the Catskill delta) have excited the interest of geologists for well over a century. This apron of sediments provides an ideal area for the examination of facies from a varied assortment of both marine and nonmarine depositional environments (Fig. 10–22). Catskill facies characteristically exhibit rapid lateral changes from sandstones to shales. Such relationships form traps for petroleum and are the reason for the thousands of wells drilled into Catskill strata. The flat slabs of red sandstones have also provided vast tonnages of flagstones for buildings in eastern cities. But for historical geologists, the most pervasive reason for studying Catskill rocks is for the clues they provide to the locations and time of occurrence of phases of the Acadian orogeny. Most of these clues come from the study of subsidiary clastic wedges within the larger Catskill complex. The locations of these wedges indicate that the Acadian orogeny was caused by the oblique convergence of a displaced continental fragment named the **Avalon terrane** against the irregular eastern margin of the North American craton (Fig. 10–23). Pulses of orogenic activity apparently occurred progressively as the westward-moving Avalon terrane encountered parts of the cratonic margin that projected eastward as broad promontories. The first orogenic pulse occurred during the early Middle Devonian, when the Avalon terrane collided with a promontory in the present vicinity of the St. Lawrence valley. Orogenic pulses followed in succession throughout the remainder of the Devonian as the Avalon terrane encountered the more southerly promontories of regions now occupied by the central and southern Appalachians.

In regard to Catskill facies, the older term, Catskill delta, is not entirely appropriate. These sediments were not deposited as large deltas of one or

A

B

C

D

**Figure 10–22**  Sedimentary rocks of the Catskill clastic wedge. (A) Catskill Formation sandstone filling an abandoned stream channel (scale divisions are 10 cm). (B) Stream deposited (fluvial) sandstones and mudstones exposed north of Trout Run, Pennsylvania (scale divisions are 50 cm). (C) Fluvial channel sandstones grading upward to floodplain mudstones, east of Haines Falls, New York. (D) Tidal flat deposits near Altoona, Pennsylvania. *(Courtesy of W. D. Sevon.)*

two major streams. Study of the directional properties of the fluvial sandstones indicates deposition from many small streams, all flowing westward out of the Acadian highlands (Fig. 10–24). In many areas, marine processes reworked the sediment supplied by streams so quickly that even small deltas did not develop. Thus, for describing these facies, the expression *Catskill clastic wedge* is more appropriate.

Unlike the earlier Queenston rocks described in Chapter 9, the Catskill sediments were laid down at a time when land plants were abundant (Fig. 10–25) and provided a green mantle for the alluvial plains and hills. Fossil remains of these plants are most frequently encountered in the sediments deposited along former stream valleys. They indicate a tropical climate.

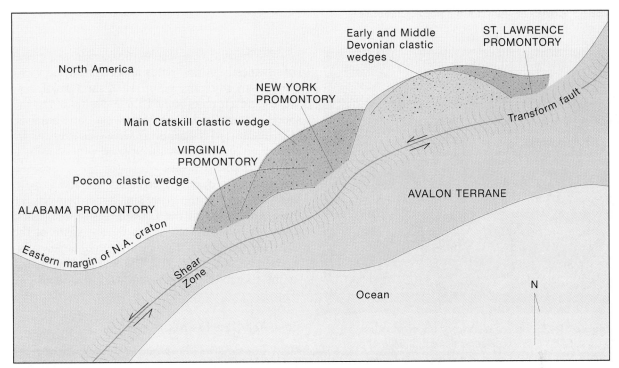

**Figure 10–23** An interpretation of the interaction between the Avalon terrane and the eastern margin of the North American craton during the late Paleozoic. In the position shown, the Avalon terrane has collided with the St. Lawrence promontory and the Virginia promontory. (The Alabama promontory is not converged upon until Mississippian time.) Because of the oblique direction of the convergence, the locations of collisions and accompanying deformations on the North American side of the collision boundary shift southwestward in time, encountering successive promontories as they do so. There is a corresponding southwesterly shift in clastic wedges. *(Adapted from Ettensohn, E. R. Geol. Soc. Am. Special Paper 201:39–49, 1985.)*

**Figure 10–24** A panoramic view of the Catskill clastic wedge as seen from a location above south-central Pennsylvania. Shoreline trends for about 300 km in a northeastern direction. *(Modified from Woodrow, D. L. Geol. Soc. Am. Special Paper 201:51–63, 1985.)*

**Figure 10–25** Fossil plant root traces in mudstones of the Catskill clastic wedge (Oneonta Formation), near Unadilla, New York. *(Courtesy of W. D. Sevon.)*

Nonmarine Catskill sedimentary rocks are dominated by sandstones and shales in which the contained iron is deeply oxidized. The oxide takes the form of the mineral hematite and causes the reddish brown hues so characteristic of Catskill nonmarine rocks. The majority of these so-called red beds represent deposits of braided or meandering streams.

The marine components of Catskill sediments consist mostly of fine sands and clays carried westward to an adjoining basin called the Catskill sea. Waves and currents shifted these clastics along the shoreline, forming lagoonal areas, bars, and beaches. More basinward deeper areas were characterized by deposition of dark clays interspersed with occasional sandy turbidite deposits.

Geologists in both North America and western Europe recognize the many similarities between the nonmarine rocks of the Catskill clastic wedge and those that were spread across Europe south of the Caledonian orogeny. This vast region of largely alluvial deposition is named the **Old Red continent** after its most famous formation, the Old Red Sandstone (Fig. 10–26).

## Post-Acadian Events

Mississippian strata crop out in the Appalachian region from Pennsylvania to Alabama. Nonmarine shales and sandstones predominate, indicating that erosion of the mountainous tracts developed during the Acadian orogeny was still in progress. Some of the finer clastics were spread westward onto the craton to form the vast deposits of Lower Mississippian black shales mentioned earlier. Particularly coarse sediments, including conglomerates, were deposited as part of the **Pocono Group**. Pocono sandstones

**Figure 10–26** The Old Red Sandstone exposed in sea cliffs at St. Ann's Head, Pembrokshire, England. *(Courtesy of Institute of Geological Sciences, London.)*

(Fig. 10–27) form some of the resistant ridges of the Appalachian Mountains in Pennsylvania. Westward, the Pocono section thins and changes imperceptibly into marine siltstones and shales. Evidently, the depositional framework consisted of another great clastic wedge complex of alluvial plains that sloped westward and merged into deltas that were being built outward into the epicontinental seas. The plains and deltas, standing only slightly above sea level, were backed by the rising mountains of the Appalachian fold belt. A large part of the coarser clastics had settled before reaching the southern portions of the epicontinental seas where marine limestones are the most prevalent rocks.

Pennsylvanian rocks of the Appalachians are characterized by cross-bedded sandstones and gray shales that were deposited by rivers or within lakes and swamps. Coal seams are, of course, prevalent in the Pennsylvanian System of the eastern United States and reflect the luxuriant growths of mangrove-like forests that clothed the lands. This was an ideal environment for coal formation. Vegetation that accumulated in the poorly drained swampy areas was frequently inundated and killed off. Immersed in water or covered with muck, the dead plant material was protected from rapid oxidation; however, it was attacked by anaerobic bacteria. These organisms broke down the plant tissues, extracted the oxygen, and released hydrogen. What remained was a fibrous sludge with a high content of carbon. Later, such peatlike layers were covered with additional sediments—usually siltstones and shales—and then compressed and slowly converted to coal.

The culminating deformational event in the southern Appalachians has been termed the **Allegheny orogeny** (also called the Appalachian orogeny). This great episode of mountain building probably began during the Pennsylvanian Period and continued throughout the remainder of the Paleozoic. It affected a belt that extended for over 1600 km, from southern New York to central Alabama. Like the earlier Acadian orogeny, the Allegheny deformations were the result of closure of an oceanic tract (the Proto-Atlantic, or Iapetus) and the convergence of continents. Whereas the Acadian even had been governed by convergence of North America and Europe the Allegheny orogeny occurred when northwestern Africa came into contact with southeastern North America (see Fig. 10–1).

The results of the Allegheny orogeny were profound and included Permian compression of early continental shelf and rise sediments as well as strata

**Figure 10–27**  Exposure of the Pocono Group (Mississippian) in a highway cut in western Pennsylvania. The section includes thick, cross-bedded fluvial sandstones as well as a few dark, coaly shales. The cut is across the axis of a syncline that was formed during the Allegheny orogeny. *(Courtesy of J. D. Glaser.)*

**A**

Section A-A'

Section B-B'

**B**

**Figure 10–28** (A) Map of physiographic provinces of the Appalachian region. (B) Two cross-sections showing the structural relationships of the Appalachian Mountains. *(Photo (A) from* Tectonic Map of the United States, *Geological Society of America, 1944; (B) from King, P. B. Bull. Am. Assoc. Petr. Geol., 1950.)*

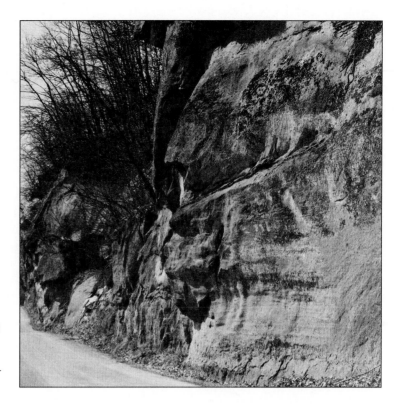

**Figure 10–29** The Waynesburg Sandstone of the Dunkard Series exposed along the Ohio River in Meigs County, Ohio. *(Photograph courtesy of the Ohio Division of Geological Survey.)*

deposited along the bordering tract of the craton. The great folds now visible in the Ridge and Valley Province were developed during this orogeny (Fig. 10–28). Less visible at the surface but no less impressive are enormous thrust faults formed along the east side of the southern Appalachians. The folds are asymmetrically overturned toward the northwest, and the fault surfaces are inclined southeastward, suggesting that the entire region was moved forceably against the central craton. Not unexpectedly, erosion of the rising mountains produced another great clastic blanket of nonmarine sediments. These mostly continental reddish sandstones and gritty shales compose the **Dunkard Series** (Fig. 10–29) and **Monongahela Series** of Permian–Pennsylvanian age.

## The Ouachita Deformation

The deformation of the Ouachita orogenic belt was caused by the collision of the northern margin of Gondwanaland (northern South America and perhaps part of northwestern Africa) with the southeastern margin of the North American craton (see Fig. 10–1). That deformation began rather late in the Paleozoic, for the rock record indicates that from Early Devonian to Late Mississippian time the region experienced slow deposition interrupted by only minor disturbances. Carbonates predominated in the more northerly shelf zone, whereas cherty rocks known as **novaculites** accumulated in the deeper marine areas (Fig. 10–30).

Novaculites are hard, even-textured, siliceous rocks composed mostly of microcrystalline quartz. They are formed from bedded cherts that have been subjected to heat and pressure. Arkansas novaculite is used as a whetstone in sharpening steel tools. The pace of sedimentation increased dramatically toward the end of the Mississippian, when a thickness of over 8000 meters of graywackes and shales was spread into the depositional basin. The flood of clastic deposition continued into the Pennsylvanian, forming a great wedge of sediment that thickened and became coarser toward the south, where mountain ranges were being formed (Fig. 10–31). Radiometric dating of now deeply buried basement rocks from the Gulf Coastal states indicates that these rocks were metamorphosed during the late Paleozoic

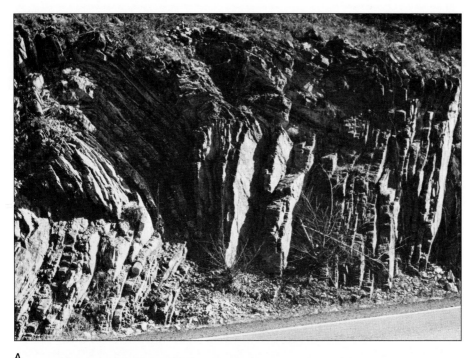

A

B

**Figure 10–30** (A) Tightly folded beds of Arkansas novaculite. (B) Close-up view of complex small folds in the novaculite beds. *(Photograph courtesy of C. G. Stone and the Arkansas Geological Commission.)*

**Figure 10–31** Geologic section of Pennsylvania rocks across Oklahoma and Texas, showing the thick wedge of sediment shed from mountain ranges to the south. *(From Stratigraphic Atlas of North and Central America. Shell Oil Company, Exploration Department.)*

and were the likely source for the Pennsylvanian clastics spread into the belt. These coarse sediments document several pulses of orogenesis that ultimately produced mountains along the entire southern border of North America. In the faulting that accompanied the intense folding, strata of the continental rise were thrust northward onto the rocks of the shelf. By Permian time, stability returned, and strata of this final Paleozoic period are relatively undisturbed.

Since the time of active mountain building, erosion has leveled most of the highlands, leaving only the Ouachita Mountains of Arkansas and Oklahoma and the Marathon Mountains of southwest Texas as remnants of once lofty ranges. Although the Ouachita belt has been traced over a distance of nearly 2000 km, only about 400 km are exposed. Thus, the actual configuration of the belt has been determined by the examination of millions of well samples and other data obtained during drilling activities associated with petroleum exploration.

## Sedimentation and Orogeny in the West

The late Paleozoic history of the western or Cordilleran belt was almost as lively as that of the Appalachian. You may recall from Chapter 9 that during the early Paleozoic, a passive margin existed along the western side of the North American craton. The familiar belts of continental shelf, slope, and rise sediments were deposited along that relatively quiet margin. More dynamic conditions began in the Devonian, when subduction of oceanic lithosphere beneath the western margin of the continent was initiated. This was the beginning of a disturbance known as the **Antler orogeny**. During the Antler orogeny, a volcanic island arc converged on the continent, crushing the sediment in the intervening basin. The convergence was accompanied by thrust faulting on a massive scale. The effects are clearly seen in the Roberts Mountain thrust fault of Nevada and at other locations northward into Idaho (Fig.

**Figure 10–32** Location of the Antler orogenic belt.

10–32). One can recognize continental rise and slope deposits that have been thrust as much as 80 km over shallow water sediments of the former continental shelf.

The Antler orogeny, which had begun in the Late Devonian, continued actively into the Mississippian and Pennsylvanian. Highlands created by the orogeny provided detrital sediment that was transported into adjacent basins (Fig. 10–33). Especially thick sequences of Pennsylvanian and Permian shelf sediments accumulated in the area now occupied by the Wasatch and Oquirrh mountains (Fig. 10–34). In the latter area, the Oquirrh Group is over 9000 meters thick.

Mississippian and Pennsylvanian deposits west of the Antler highlands include a great volume of coarse detritus and volcanic rocks. These materials were swept eastward from a volcanic arc source area that lay along the western side of North America (Fig. 10–35). Over 2000 meters of sandstones, shales, lavas, and ash beds are found in the Klamath Mountains of northern California. Volcanic rocks in western Idaho and British Columbia attest to a continuation of vigorous volcanism from the Mississippian through the Permian. Crustal deformation along the west side of the Cordilleran belt is indicated by an extensive area of angular unconformities between Permian and Triassic sequences. These Permian–Triassic disturbances of the Cordilleran region have been named the **Cassiar orogeny** in British Columbia and the **Sonoma orogeny** in southwestern United States. Like the earlier Antler orogeny, the Sonoma event was probably caused by the collision of an eastward-moving island arc against the North American continental margin in west-central Nevada. Oceanic rocks, and remnants of the arc were thrust into the edge of the continent and became part of North America.

Permian conditions in the shelf area east of the Antler uplift were quieter than those to the west. The region was occupied by a shallow sea during much of the Permian Period. Within that sea several large platform deposits accumulated. One of these was the Kaibab Limestone, an imposing formation that forms the vertical cliffs along the rim of the Grand Canyon. Beneath and eastward of the Kaibab Limestone, there are red beds that reflect deposition on coastal mud flats and floodplains. Sand dunes were prevalent nearby, as indicated by the massive, extensively cross-bedded Coconino Sandstone.

While the Kaibab was being deposited in the southwest, a relatively deep marine basin was developing to the north in the general area now occupied by Wyoming, Montana, and Idaho. In this basin,

**Figure 10–33** An interpretation of conditions in the Cordilleran orogenic belt in Early Mississippian time, shortly after the Antler orogeny. *(Based on diagrams by Poole, G. F. Soc. Economic Paleontologists and Mineralogists Special Publication 22:58–82, 1974.)*

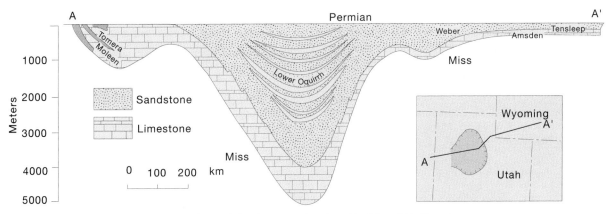

**Figure 10–34** Section across the Oquirrh basin. *(From* Stratigraphic Atlas of North and Central America. *Shell Oil Company, Exploration Department.)*

**Figure 10–35** The Mississippian Bragdon Formation of the eastern Klamath Mountains of California. The Bragdon is composed of siltstones, sandstones, and felsic ash shed from a volcanic island arc that formed next to North America during late Paleozoic time. *(Courtesy of M. Miller.)*

sediments of the Phosphoria Formation were deposited. Although the Phosphoria includes beds of cherts, sandstones, and mudstones, it takes its name from its many layers of dark phosphatic shales, phosphatic limestones, and phosphorites. **Phosphorite**, a dark gray concretionary variety of calcium phosphate, is mined from the formation and used in the manufacture of fertilizers and other chemical products. The unusual concentration of phosphates may have been produced by upwelling of phosphorus-rich sea water from deep parts of the basin. The metabolic activities of microorganisms may have assisted in the precipitation of phosphate salts.

## The Northern Seaway

The late Paleozoic record of the Northern or Franklinian seaway in northern Canada is initiated by Devonian shales and limestones. Coral reefs are abundantly preserved in the Devonian System. The existence of these reefs is not surprising, for the region lay at about 15° north latitude during the Devonian and temperatures were much warmer than today. Following deposition of the Devonian marine rocks, an extensive mountain system was developed across northern Canada during an event called the **Ellesmere orogeny.** Erosion of the mountains provided immense volumes of detrital sediments to adjacent lowlands. One such depositional site, named the Sverdrup basin, formed late in the Mississippian and served as a trap for a nearly continuous sequence of strata ranging in age from Late Mississippian to early Cenozoic. The Late Mississippian beds include plant-bearing deltaic shales and sandstones, red beds, and coal seams. By Pennsylvanian time, the mountains had been reduced, and hence the supply of detritus was diminished. The basin, however, continued to subside and became a relatively deep-marine environment of deposition. In this marine basin, thick sequences of Pennsylvanian and Permian sandstones, limestones, and evaporites were deposited.

## Europe During the Late Paleozoic

During the late Paleozoic, Europe was bordered by the Uralian seaway on the east and the Hercynian on the south. Along the northern margin of Europe, the Caledonian orogeny had created the vast land area of the Old Red continent. Uplands provided a source for clastic sediments, which were swept out into numerous basins to accumulate to thicknesses exceeding 10,000 meters. Judging from the many interlayers of ash and lava, volcanic activity was frequent on the Old Red continent. Fossil fishes and sedimentary evidence indicate that Europe's climate was tropical and possibly semiarid at the time.

South of the Caledonian mountains, the continental deposits gradually thin and grade into marine shales and limestones (see Fig. 9–33). For a time, quiet prevailed in western Europe. However, in the Late Devonian and Early Mississippian, the depositional sites were intensely folded, metamorphosed, and intruded by granites as continents began or continued the collisions that were to produce Pangea. The late Paleozoic orogenic event has been named the **Hercynian orogeny,** and its result was a great range of mountains across southern Europe. The Hercynian orogeny occurred at approximately the same time as the Allegheny orogeny in North America. For the most part, the eroded stumps of the Hercynian mountains are now buried, but here and there patches of the younger covering rocks have been eroded away, revealing the intensely folded, faulted, and intruded older rocks that lie beneath.

From the Hercynian uplands, gravel, sand, and mud were carried down into basins and coastal environments. The clastic deposits were quickly clothed in dense tropical forests. Burial and slow alteration of vegetative debris from these forests provided the material for coal formation in the great European coal basins. Plant fossils found in these coal seams are of the tropical kind and differ from the more temperate Gondwana flora of the Southern Hemisphere.

Although the greatest amount of Hercynian deformation occurred toward the end of the Early Carboniferous (Mississippian) in Europe, spasms of unrest continued in both the Late Carboniferous and the Permian. These later episodes of folding correlate with similar deformations in the southern Appalachian and Ouachita orogenic belts of North America.

No less important than the Hercynian basins in Europe was the lengthy Uralian tract, which extended along a belt now occupied by the Ural Mountains. The Uralian basins were already in existence at the beginning of the Paleozoic Era. Interestingly, there are discrepancies between the polar wandering curves for Paleozoic rocks from the cratons on either side of the Urals. The discrepancies indicate that the Russian and Siberian platforms were widely separated during early Paleozoic time and that they began to converge in the middle part of the era and ultimately collided by the end of the Paleozoic (see Fig. 10–1). Indeed, the western Urals consist largely of oceanic material scraped off against the edges of the converging plates. The collision resulted in the formation of a great mountain system along the entire Uralian orogenic belt and unified ancestral Siberia with eastern Europe.

In central Europe and parts of Russia, there were temporary Late Permian marine incursions that precipitated evaporites. The famous German potassium salts are a product of one of these inland seas, which has been named the **Zechstein sea** (Fig. 10–36). Apparently, aridity was as characteristic of western Europe during the Permian as it was of Texas and New Mexico.

## Gondwanaland During the Late Paleozoic

During the late Paleozoic, the great land mass of Gondwanaland remained fairly intact. It moved across the South Pole and more fully entered the side of the earth on which Laurussia was located. In its northward migration, Gondwanaland caused the closure of the ocean that separated it from Laurussia,

**Figure 10–36**   Outcrop area of sedimentary rocks deposited in the Zechstein sea. *(After Brinkman, R. Geological Evolution of Europe. Translation from German by Sanders, J. E. and Enke, F. Stuttgart: Verlag; and New York: Hafner Publishing Co., 1969.)*

causing the Hercynian orogeny of Europe and the Allegheny orogeny of North America. Orogenic activity associated with subduction zones was also evident in the late Paleozoic history of Gondwanaland, particularly along the Andean belt of South America and the Tasman belt of eastern Australia.

The most dramatic paleoclimatologic event of Gondwanaland's late Paleozoic history was the growth of vast continental glaciers (see Fig. 10–2). Extensive layers of tillite (Fig. 10–37) and the scour marks of glaciers (see Fig. 10–3) have been found at hundreds of locations in South America, South Africa, Antarctica, and India. There are indications of at least four and possibly more glacial advances, suggesting a pattern of cyclic glaciation not unlike that experienced by North America and Europe less than 100,000 years ago. The orientation of striations chiseled into bedrock by the moving ice suggests the glaciers moved northward from centers of accumulation in southwestern Africa and eastern Antarctica. During the warmer interglacial stages and in outlying less frigid areas, *Glossopteris* and other plants tolerant of the cool, damp climates grew in profusion and provided the materials for thick seams of coal.

In time the ice receded, and Gondwanaland's vast cratonic areas became sites for deposition of Permian nonmarine red beds and shales. As we shall see in the next chapter, some of these sediments contain the fossil remains of the ancestors of the earth's first mammals.

**Figure 10–37**   Upper Carboniferous tillite at base of slope (beneath hammer) composed of Precambrian basalt. Tillite is consolidated and unsorted glacial debris. Kimberley, South Africa. *(Courtesy of Warren Hamilton, U.S. Geological Survey.)*

**Figure 10–38**   Approximate relationship of continents to the equator during the Carboniferous (Mississippian–Pennsylvania).

## Climates of the Late Paleozoic

The main climatic zones of the late Paleozoic tended to parallel latitudinal lines, just as zones do today. Of course, the continents were located differently, so that the South Pole was in South Africa and the North Pole was over open ocean. The late Paleozoic equator trended northeastward across Canada and southward across Europe (Fig. 10–38). Coal beds, evaporites, coral reefs, and dune-deposited red beds developed within 40° of the paleoequator. Today, we cannot help but be startled to find the fossils of tropical amphibians and plants at locations within the Arctic circle. As already noted, cooler climates prevailed in Gondwanaland because of its late Paleozoic proximity to the South Pole, and by Permian time the southern continents were in the grip of a major ice age. To the north, however, warmer conditions prevailed. Fossil plants in Pennsylvanian coal seams and shales reflect tropical conditions, as do coral reefs developed in areas of marine deposition. Evaporites associated with red beds suggest arid conditions in some areas. It is likely that the northern regions of Pangea lay within the zone of northeasterly trade winds. Much as today, as these winds traveled westward and encountered mountains, they would have been forced to ascend, form clouds, and produce precipitation on the windward side of the ranges. On reaching the far or lee side of the mountains, much of the moisture in the air would have been lost, and arid conditions would be the expected

result, at least until moisture was again replenished as winds continued westward over epeiric seas.

## Mineral Products of the Late Paleozoic

Although a great variety of mineral deposits were formed during the late Paleozoic, the fossil fuels (coal, oil, and gas) are particularly significant. Coal occurs in all post-Devonian systems. In Northern Hemisphere continents, it is particularly characteristic of the Late Carboniferous (Pennsylvanian). Thick deposits of Pennsylvanian coal occur in the Appalachians, the Illinois basin (Fig. 10–39), and the in-

A

B

**Figure 10–39**   (A) A remote-controlled continuous mining machine with a flooded bed scrubber for dust control is removing coal in a West Virginia underground mine. (B) Large stripping shovel in the pit of an Illinois surface mine. *(Courtesy of Peabody Coal Co.)*

dustrial heartland of Europe. Single sequences of Pennsylvanian strata may include many coal beds, as in West Virginia, where 117 different layers have been named. In the anthracite district of western Pennsylvania, orogenic compression has partially metamorphosed coal into an exceptionally high-carbon, low-volatile variety that is prized for its industrial uses. Permian coal seams are found in China, Russia, India, South Africa, and Australia. The coal industry in North America, which had deteriorated for the last few decades, is currently experiencing an upsurge as a consequence of decreasing supplies of petroleum and natural gas.

Commercial quantities of oil and gas are frequently found in upper Paleozoic strata. Devonian reefs within the Williston basin of Alberta and Montana have been exceptionally productive reservoir rocks for petroleum. Devonian petroleum has also been produced in the Appalachians. Indeed, in 1859, the first U.S. oil well was drilled into a Devonian sandstone. (Oil was struck at a depth of only 20 meters.) Carboniferous formations of the Rocky Mountains, midcontinent (Fig. 10–40), and Appalachians also contain oil reservoirs. However, wells drilled into reefs and sandstones of the Permian basin of West Texas have yielded the greatest amount of oil from upper Paleozoic formations of the western United States. Oil trapped in upper Paleozoic strata beneath the North Sea is now being actively utilized to supply the needs of Europe.

Arid, warm climatic conditions, particularly in northern continents during the late Paleozoic, provided a suitable environment for deposition of sodium and potassium salts. In Late Permian time, enormous amounts of phosphates were also deposited as part of the Phosphoria Formation, which is exposed in Montana, Idaho, Utah, and Wyoming. The Phosphoria shales are extensively quarried for phosphate, which is sold as an important plant food.

**Figure 10–40**   Drilling for oil in upper Paleozoic rocks of eastern Kansas.

Mountain building, with its attendant intrusive and volcanic igneous activity, is nearly always accompanied by the emplacement of metallic ores. The Hercynian and Allegheny orogenies of the late Paleozoic generated ores of tin, copper, silver, gold, zinc, lead, and platinum. Deposits of all the precious metals, as well as copper, zinc, and lead, are found in the Urals of Russia and in China, Japan, Burma, and Malaya. Tin, tungsten, bismuth, and gold are mined in Australia and New Zealand. It is self-evident that late Paleozoic rock sequences, with their stores of both metallic and nonmetallic economic minerals and their content of fossil fuels, are vitally important to the welfare of modern civilizations.

## SUMMARY

During the late Paleozoic (Devonian through Permian), most of the major land masses of earlier Paleozoic periods were assembled into the supercontinent Pangea. Whereas the process of bringing continents into contact resulted in mountain building along their converging margins, the interiors were affected primarily by broad epeirogenic movements and associated advances and withdrawals of epicontinental seas. In North America, these shifts in shorelines are reflected in the marine and nonmarine sediments that compose the Kaskaskia and Absaroka cratonic sequences. Kaskaskia rocks are predominantly fossiliferous limestones, with sandstones and shales increasing in volume toward the eastern and southern depositional areas. As the Kaskaskia seas withdrew, deltaic and fluvial deposits spread across the old sea floor. The regression resulted in a great regional unconformity that marks the boundary between Mississippian and Pennsylvanian systems in North America. The most distinctive feature of the Pennsylvanian sediments is their cyclic nature. In the mid-continent, they consist of alternate marine and nonmarine groupings. To the east, the Pennsylvanian sediments were largely terrestrial, whereas marine deposition prevailed in the western part of the craton. Near the end of the late Paleozoic, the epicontinental seas gradually withdrew, and evaporite and red bed sequences developed in the Permian basins of New Mexico and West Texas.

The unrest that marked the Appalachian–Caledonian belt during the early Paleozoic continued into the Devonian with the Acadian orogeny and reached its climax in the great Allegheny orogeny. Those events are thought to have been manifestations of the closing of the Iapetus Ocean as Pangea was assembled. The compressive forces of continental collision raised great mountain ranges along the eastern and southern margins of North America.

In the west, crustal disturbances during the Mississippian uplifted a long and narrow range of mountains from Arizona to Idaho. Thick deposits of sand and gravel attest to the former existence of these ranges. There was further activity during the Pennsylvanian when uplands still visible were formed in Texas, Arkansas, and Oklahoma. A chain of mountains (the Colorado mountains) also developed across Colorado. These Colorado mountains were the source of sandstones that today form the famous red rocks of the Garden of the Gods along the Front Range of the Rockies.

The Old Red continent was the dominant feature of Europe at the beginning of the late Paleozoic. For a time, relative stability prevailed in Europe, and then much of the continent was thrown into an episode of folding, volcanism, and intrusion as it collided with Gondwanaland. These disturbances were part of the Hercynian orogeny. The collision of Gondwanaland with Laurussia was not only the cause of the Hercynian orogeny in central Europe, but was also the cause of the Allegheny orogeny along the eastern and southern margin of North America.

Gondwanaland was ringed by orogenic belts during the late Paleozoic, and the Andean and Tasman segments experienced late Paleozoic orogenesis. The central regions were occasionally covered with inland seas in which were deposited fossiliferous limestones and shales. During the Permian, when the northern land masses were relatively warm and sometimes arid, Gondwanaland appears to have been cooler. The South Pole for the Permian was located in what is now South Africa, and all of the now separated Gondwanaland continents exhibit the scars of widespread glaciations.

## QUESTIONS FOR REVIEW

1. How are the orogenies that produced the Appalachian and Ouachita mountains related to plate movements during the late Paleozoic?

2. What relationship might the Allegheny orogeny have had to the Hercynian orogeny in Europe?

3. Why is it logical to divide the late Paleozoic into two North American cratonic sequences—the Kaskaskia and Absaroka?

4. Where was the Old Red continent located? When did it come into existence? What sort of sediments were deposited on it?

5. What is the paleoenvironmental significance of the extensive red sandstones, salts, and gypsum deposits of the Permian? Where are the more extensive deposits of Permian evaporites located?

6. What is the geologic evidence for the occurrence of a geographically extensive episode of continental glaciation in the Southern Hemisphere near the end of the Paleozoic?

7. What is a cyclothem? What may have been the cause of this type of cyclic deposition during the Pennsylvanian?

8. When and where were the Colorado mountains ("ancestral Rockies") developed?

9. Geologists have suggested that erosion on land by water and wind probably proceeded more rapidly during the early Paleozoic than during the remainder of the era. Can you suggest why this might be a valid inference?

10. What geologic interpretation is given to great "clastic wedges" such as the Catskill clastic wedge and the Pocono Group?

11. Name three metallic and three nonmetallic mineral resources that are extracted from upper Paleozoic rocks.

## TERMS TO REMEMBER

| | | | |
|---|---|---|---|
| Absaroka sea | Cassiar orogeny | Ellesmere orogeny | oöids |
| Acadian orogeny | Catskill clastic wedge | Hercynian orogeny | Pocono Group |
| Allegheny orogeny | Chattanooga Shale | Kaskaskia sea | phosphorite |
| Antler orogeny | cyclothem | Monongahela Series | Sonoma orogeny |
| Avalon terrane | Dunkard Series | novaculites | |

## SUPPLEMENTAL READINGS AND REFERENCES

Bambach, R. K., Scotese, C. R., and Ziegler, A. M. 1980. Before Pangaea: The geographies of the Paleozoic world. *Amer. Scientist* 68(1):26–28

Bird, J. M., and Dewey, J. F. 1970. Lithosphere plate–continental margin tectonics and the evolution of the Appalachian orogeny. *Bull. Geol. Soc. Am.* 81:1031–1060.

Frazier, W. J., and Schwimmer, D. R. 1987. *Regional Stratigraphy of North America.* New York: Plenum Press.

Hamilton, W., and Krinsley, D. 1967. Upper Paleozoic glacial deposits of South Africa and southern Australia. *Bull. Geol. Soc. Am.* 82:1581.

Pelletier, B. R. 1958. Pocono palecurrents in Pennsylvania and Maryland. *Bull. Geol. Soc. Am.* 69:1033.

Reed, H. H., and Watson, J. 1975. *Introduction to Geology,* Vol. 2, Part II, *Later Stages of Earth History.* New York: Wiley Interscience.

Rodgers, J. 1970. *The Tectonics of the Appalachians.* New York: Wiley Interscience.

Woodrow, D. L., and Sevon, W. D. (eds.) 1985. The Catskill Delta. *Geol. Soc. Am. Special Paper* 201.

Diorama of a Silurian reef reconstructed from fossil invertebrates commonly found in Silurian strata of the central United States. The reef community includes both tabulate and colonial rugose corals, bryozoans, brachiopods, gastropods, blastoids, cystoids, crinoids, and marine algae. How many of these groups can you identify in the diorama? (Courtesy of the Milwaukee Public Museum.)

# Life of the Paleozoic

*The race of man shall perish, but the eyes*
*Of trilobites eternal be in stone,*
*And seem to stare about in mild surprise*
*At changes greater than they have yet known.*

T. A. Conrad

## Introduction

The rich fossil record of the Paleozoic stands in startling contrast to that of the earlier Precambrian. We have noted that the Precambrian was a world of bacteria, algae, and fungi. Unquestioned multicellular animals are found only in uppermost Proterozoic strata, such as those at Ediacara Hills in Australia. Except for a few tiny tubular creatures, latest Precambrian animals had not acquired the ability to secrete shells.

In passing from late Proterozoic to Early Cambrian time, the record for life is much improved. Scattered around the shallow margins of the continents, isolated groups of invertebrates had established themselves. Soon, epeiric seas began to creep across the cratons, providing as they did so a multitude of habitats and opportunities for the diversification and expansion of the phyla.

Certainly the improvement in the fossil record evident in Cambrian strata can be attributed to the spread of shell-building abilities. During the Cambrian, shell-building brachiopods and trilobites were abundant, but the most rapid expansion and diversification of shell-bearing groups began later, in Early Ordovician time. This proliferation in shell-building was probably related to the advantages afforded by shells in providing protection and support for soft tissues. Because there is little fossil evidence of numerous predatory animals until about Late Cambrian, support may have initially been the more important function. However, with the advent of abundant and vigorous carnivores (such as the predatory cephalopods) during the Late Cambrian, active predation may have provided the selective pressures that would have favored rapid evolutionary expansion of protective exoskeletons. The result was the multitude of shelled invertebrates that have populated the oceans down to the present day.

## Plants of the Paleozoic

The history of plants has an obscure beginning among the bacteria and algae of the Precambrian. Indeed, the Precambrian sedimentary iron ores are presumed to have been produced by bacterial activity. However, as has been noted, fossils of such tiny, soft-bodied creatures are very rare.

An exception to the generally poor fossil record of the earliest plants is provided by stromatolites, which are abundant in upper Proterozoic rocks and continue to be found in limestones of all younger ages. During the Cambrian Period, stromatolitic reefs were widespread (Fig. 11–1), but became more restricted in the Ordovician, perhaps because of the rise of marine invertebrates that grazed on the algae responsible for producing stromatolites.

**Figure 11–1** Hummocky stromatolites of Cambrian age exposed along the banks of the Black River in Missouri.

Another relatively common group of early Paleozoic marine plant fossils are the **receptaculids** (Fig. 11–2). Receptaculids are lime-secreting green algae of the Family Dasycladacea. In general appearance, they remind one of the seed-bearing central area of a large sunflower; this has caused amateur fossil collectors to designate them "sunflower corals." They are, however, neither sunflowers nor corals. Nor are they sponges, a designation once considered

**Figure 11–2** *Receptaculites,* a receptaculid from the Ordovician Kimmswick Formation of Missouri. Sometimes called sunflower corals, Receptaculids are the fossil remains of a type of green algae and not in any way related to corals.

because of their spicule-like rods and circulatory passages. Although most frequently found in Ordovician rocks, members of this group also occur sparsely in Silurian and Devonian strata.

## Invasion of the Lands

The appearance of our landscape today owes much to its verdant covering of forests and grasslands. At the beginning of the Paleozoic, however, there were no trees clothing the continents. Most living organisms were confined to the seas. It is likely that some species of algae invaded freshwater lakes and streams, and some may even have evolved means of moisture control independent of their environment. These algae moved into damp areas on land and began the evolutionary journey toward complex land plants. The algal group most likely to have begun this progression were probably the **chlorophytes** or green algae. A relationship between chlorophytes and land plants is indicated by the presence of the same kind of green pigment in both groups and by the fact that both produce the same kind of carbohydrate during photosynthesis. Also, some chlorophytes have become adapted to fresh water—a necessary accomplishment for the vegetative invasion of the continents. Today some chlorophytes can even survive in moist areas on land.

The vegetative invasion of the continents required the evolution of a vascular system. Vascular plants have tubes and vessels that convey fluids from one part of the plant to another. The importance of this system for a land plant is apparent when we remember that surficial moisture above ground is an undependable source of water in most environments. There is nearly always a greater persistence of moisture in the pore spaces of soil. But no light for photosynthesis is available underground. The vascular system of a plant permits part of it to exist underground where there is water but no light and another part of it to grow where water supplies are uncertain but sunlight for photosynthesis is ensured.

The transition from aquatic plants to land plants was apparently difficult, for it was late in coming. The first *unquestioned* remains of vascular plants are found in rocks of Middle Silurian age. These are **psilophytes.** They were small plants (less than 30 cm tall), consisting of horizontal stalks or rhizomes that grew just beneath the surface of the

**Figure 11–3**  A psilophyte from the Lower Devonian of the Gaspé Peninsula.

ground in moist soil. Extending vertically from the rhizomes (Fig. 11–3) were short slender stems bearing smaller branches and spore sacs. Delicate strands of wood cells and a rudimentary vascular system have been recognized in psilophyte fossils.

The psilophytes paved the way for the evolution of large trees. Because of the evolution of wood, plants were able to stand tall against the pull of gravity and the force of strong winds. By late in the Devonian, there were forests of lofty, well-rooted, leafy trees (Fig. 11–4). One of these forests, located near Gilboa, New York, has left many fossils as ample evidence of its existence. Some of the Gilboa trees were over 7 meters tall. Impressive as these trees were, however, they were to be dwarfed by some of their Carboniferous descendents.

When one surveys the entire history of vascular plants, three major advances become apparent. Each involved the development of increasingly more effective reproductive systems. The first advance led to seedless, spore-bearing plants, such as those that were ubiquitous in the great coal-forming swamps of the Carboniferous. The second saw the evolution of seed-producing, pollinating, but nonflowering plants (gymnosperms). This was a late Paleozoic event. The evolution of plants with both seeds and flowers (angiosperms) came late in the Mesozoic Era.

Among the moisture-loving plants of the Carboniferous were the so-called scale trees, or **lycopsids**. Today, they are represented by their smaller

**Figure 11–4**  Restoration of a Middle Devonian forest in the eastern area of the United States. (A) An early lycopod, *Protolepidodendron*. (B) *Calamophyton*, an early form of the horsetail rush. (C) Early tree fern. *Eospermatopteris*. *(Courtesy of Field Museum of Natural History; painting by C. R. Knight.)*

**Figure 11–5**   Fossilized bark of the Carboniferous tree *Lepidodendron*. (Photograph courtesy of Wards Natural Science Establishment, Inc., Rochester, N.Y.)

survivors, the club mosses, of which the ground pine *Lycopodium* is a member. Small size was not particularly characteristic of the late Paleozoic lycopsids. The forked branches of *Lepidodendron* (Fig. 11–5) reached 30 meters into the sky. The elongate leaves of the scale trees emerged directly from the trunks and branches. After being released, they left a regular pattern of leaf scars. In *Lepidodendron,* the scars are arranged in diagonal spirals, whereas species of *Sigillaria* have leaf scars in vertical rows.

Another dominant group of plants that grew side by side with the Carboniferous lycopsids were the **sphenopsids.** Living sphenopsids include the scouring rushes and horsetails. Fossil sphenopsids, such as *Calamites* (Fig. 11–6) and *Annularia* (Fig. 11–7), possessed slender, unbranching, longitudinally ribbed stems with a thick core of pith and rings of leaves at each transverse joint. At the top, a cone bore the spores that would be scattered in the wind.

True ferns (Fig. 11–8) were also present in the coal forests. Many were tall enough to be classified as trees. Like the lycopsids and sphenopsids, they reproduced by means of spores carried in regular patterns on the undersides of the leaves.

Seed plants made their debut during the late Paleozoic. They probably arose from Devonian fernlike plants. These plants, appropriately dubbed "seed ferns," had fernlike leaves, but unlike true ferns they reproduced by means of seeds.

One of the most widely known seed fern groups is *Glossopteris*, which was restricted to the Southern Hemisphere during the Carboniferous and Permian (see Fig. 6–16). Most species of *Glossopteris* were plants with thick, tongue-shaped leaves. Because of certain anatomic traits and because of their association with glacial deposits, *Glossopteris* and associated

**Figure 11–6**   *Calamites,* a sphenopsid.

**Figure 11–7** *Annularia,* an abundant sphenopsid of Pennsylvanian age.

**Figure 11–8** *Pecopteris,* a true fern from the Pennsylvanian of Illinois. *(Note the penny for scale.)*

**Figure 11–9** Reconstruction of a cordaitean plant of the Carboniferous.

# Invertebrate Animals of the Paleozoic

plants are thought to have been adapted to cool climates.

Fossils of seed plants are present in the Northern Hemisphere also. Cordaites (primitive members of the conifer lineage) were abundant, some towering 50 meters (Fig. 11–9). Their branching limbs were crowned with clusters of large, straplike leaves. These and somewhat more modern cone-bearing conifer plants spread widely during the Permian, perhaps as a consequence of drier climatic conditions. The first ginkgoes made their appearance during the Permian. Today, only a single species, *Ginkgo biloba,* remains as a survivor of this once flourishing group.

People are accustomed to thinking of the ocean as the birthplace of life. Thus, it is no surprise to learn that the earth's earliest animals lived in the sea. The only fossils ever found of animals that lived during the end of the late Proterozoic, the Cambrian, and the Ordovician have been those of ocean dwellers. As is the case today, these animals varied in their choices of habitat as well as in the kinds of food they required. In the geologic past, as in the present, there were basically three modes of life. Organisms that could live on the sea floor or burrow in sea-floor sediment are termed *benthic,* those that floated are termed *planktic,* and those that swam are termed *nektic.* In addition, many animals in the sea spend a phase of their lives in one of these modes and then change to accommodate themselves to another. In fact, the mobile larvae of a multitude of otherwise sedentary (stationary) "critters" are dispersed in this way.

As indicated by the Ediacaran fossils described in Chapter 8, multicellular marine organisms were already present near the end of the Precambrian. Those organisms, however, were soft-bodied and are only infrequently preserved as imprints in once soft sediment. When animals had achieved the ability to secrete hard parts, their chances of being preserved improved dramatically. The oldest assemblage of fossils to accomplish this feat occurs in rocks of the latest Precambrian (Yudomian Stage) and earliest Cambrian (Tommotian Stage) of Siberia, Sweden, Antarctica, and England. The fossils are the remains of small, shelly animals that include snails, sponges (known from preserved spicules), and a variety of tube- and cap-shaped creatures whose biologic classification is still uncertain (Fig. 11–10). Unlike the majority of later shelly invertebrates that built their shells of calcium carbonate, these early shell-bearers

secreted calcium phosphate for their skeletons. As a result, they are found in slabs of limestone and dolomite when the rock is etched in weak acid.

Following the Tommotian Stage, rocks are clearly dominated by more familiar fossils such as trilobites and brachiopods. Their robust shells provided a fine fossil record. But there were also many soft-bodied creatures. Some of these comprise the Early Cambrian Chengjian fossils of China and the Middle Cambrian Burgess Shale fauna of Canada.

## Window into the Past: Fossils of the Burgess Shale

High on a ridge near Mount Wapta, British Columbia, there is an exposure of Burgess Shale that contains one of the most important faunas in the fossil record (Fig. 11–11). The fossils in the Burgess Shale are reduced to shiny, jetlike impressions on the bedding planes of the black rock (Fig. 11–12). Most of them are the remains of animals that lacked shells. Altogether, they form an extraordinary assemblage (Fig. 11–13) that includes four major groups of arthropods (trilobites, crustaceans, and members of the taxonomic groups that include scorpions and insects), as well as sponges, onycophorans (see Fig. 11–11), crinoids, sea cucumbers, chordates, and many species that defy placement in any known phyla.

The special significance of the Burgess Shale fauna lies in the perspective that it provides on life at the beginning of the Paleozoic. It reveals that the Cambrian seas teemed with diverse and complex animals and that marine life at the beginning of the Paleozoic had already reached a quite advanced stage of evolution. The discovery of the Burgess fossils further demonstrates that reconstructions of ancient communities based on the fossils of shelled animals (often all that are preserved) are likely to be seriously incomplete. Much like marine communities today, there were more animals with soft parts in the Burgess fauna than those having hard parts, and these soft-bodied animals were more varied than the shelled ones. Thus, the Burgess fauna is a more valid representation of life as it really was than is usually found for other fossil faunas.

The Burgess fauna is significant for yet another reason. Sharing the sea floor with many other unique and sometimes bizarre Burgess creatures were small,

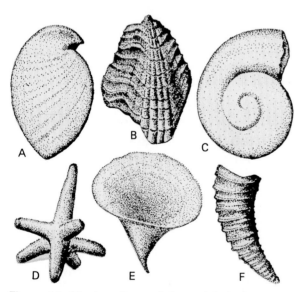

**Figure 11–10**  Late Precambrian and Early Cambrian shell-bearing fossils from Siberia. (a) *Anabarella*, × 20, a gastropod; (b) *Camenella*, × 18, affinity uncertain; (c) *Aldanella*, × 20, a gastropod; (d) sponge spicule, × 30; (e) *Fomitchella*, × 45, affinity uncertain; (f) *Lapworthella*, × 20, possibly a type of barnacle. *(After Matthews, S. L. and Missarzhevsky, V. V. J., Geol. Soc. London 131:289–304, 1975.)*

**Figure 11–11** (A) Cambrian stratigraphic section at (B) Kicking Horse Pass, British Columbia. According to a now famous story, this is where a slab of fossiliferous shale was kicked over by a pack horse so as to catch the eye of paleontologist Charles Walcott, who followed the trail of rock debris up the side of the mountain to discover the Burgess Shale beds with their rich fauna of Middle Cambrian fossils. Among these were (C) *Aysheaia*, an onycophoran believed to be intermediate in evolutionary position between segmented worms and arthropods. Among the thousands of specimens were trilobite-like arthropods such as (D) *Leanchoilia superlata* and (E) *Waptia fieldensis*.

**Figure 11–12** *Hyolithes carinatus*. This photograph illustrates the nature of preservation of Burgess Shale fossils. *Hyolithes* had a tapering shell surmounted by a lid or operculum which could be closed for protection. The lateral blades on either side may have served as props. *Hyolithes* is tentatively considered a mollusk. The length of the shell is 2 cm. *(Courtesy Smithsonian Institute; photograph by Chip Clark.)*

**379**

**Figure 11–13** The Burgess Shale diorama at the U.S. National Museum of Natural History. The reconstruction is based on actual fossil remains of organisms. It depicts a benthic marine community of the Middle Cambrian. A steep submarine escarpment forms the background. Slumping along this wall contributed to the preservation of the Burgess fauna by burying the organisms. Along the wall one can see green and pink vertical growths of two types of algae. The large purplish creatures that resemble stacks of automobile tires are sponges *(Vauxia).* The blue-colored animals are trilobites *(Olenoides).* The brown arthropods with distinct lateral eyes are named *Sidneyia.* Climbing out of the hollow on the sea floor are crustaceans called *Canadaspis.* The yellow animals swimming toward the right above the sea floor are *Waptia. Obabinia* has crawled out of the left side of the hollow. Burrowing worms are visible in the vertical cut at the bottom. *(Courtesy of the National Museum of Natural History; photograph by Chip Clark.)*

rod) and a dorsally situated nerve cord. Chordates in which the notochord is supplanted by a series of vertebrae are generally termed vertebrates. These earliest known chordates have been assigned the generic name *Pikaia* in honor of Mount Pika near the Burgess fauna discovery site. Fossils of *Pikaia* exhibit two features upon which their designation as chordates is based. The animals had a dorsal, longitudinal supportive structure known as a **notochord,** as do all other chordates at some stage in their life history. In addition, one can discern the outlines of muscles arranged in a series of **V**-shaped segments along the sides of the animal as is characteristic of the musculature in fish. These muscles, working with the rigid but flexible notochord, provided a sinuous, fishlike body motion required for swimming. Like *Pikaia,* humans are chordates. For this reason, we are keenly interested in this little animal that appears to be our earliest ancestor.

The discovery of the Burgess Shale fossil locality is an interesting story in itself. The initial discovery of a single fossil from the formation was made in August 1909 by Charles D. Walcott. Walcott's field crew included his wife, Marrella, who was an avid fossil collector, and his two sons. As they were riding along a trail south of Mount Wapta, Mrs. Walcott's horse dislodged a slab of black shale with a silvery impression on its surface. The slab caught Walcott's eye, and he dismounted to examine it closely. It contained the remains of a crustacean that Walcott later named *Marrella.* Unfortunately, the discovery came at the end of the field season, and exposures at higher elevations from which the rock had come were becoming snow covered. As a result, the Walcotts were forced to leave the site and return the following year, when they traced the fossil to its source in the Burgess Shale. Quarrying was begun and continued during the 1912, 1913, and 1917 field seasons.

Altogether, about 60,000 specimens were collected and stored in the U.S. National Museum of Natural History where Walcott served as secretary of the Smithsonian Institute. In the 1960s, these fossils were reexamined by Harry B. Whittington, who then joined paleontologists of the Geological Survey of Canada in reopening the quarry and assembling a second collection. Whittington devoted the next 15 years to an incisive and critical scrutiny of the Burgess Shale fauna. He was able to describe and depict the anatomy of the Burgess animals with great clar-

elongate animals that are currently believed to be the earliest known chordates. Chordates are animals that have, at some stage in their development or throughout their lives, a notochord (internally supportive ity. During the study, it became apparent that earlier interpretations of many Burgess animals as ancestors to existing taxonomic groups were incorrect and that Burgess fauna included creatures so different as to warrant placement in new phyla.

## Early Paleozoic Diversification

As indicated by the highly structured and diversified Burgess Shale community, as well as other communities known mostly from shelled animals, marine life during the early Paleozoic had evolved a variety of life styles and adapted to a multitude of habitats. There were **epifaunal** animals that lived on the surface of the sea floor, as well as **infaunal** creatures that burrowed beneath the surface. There were borers as well as burrowers, attached forms and mobile crawlers, swimmers, and floaters. Some were *filter-feeders* that strained tiny bits of organic matter or microorganisms from the water; others were *sediment-feeders* that passed the mud of the sea floor through their digestive tracts to extract the nutrients within. There were animals that grazed upon algae that covered parts of the ocean bottom, and carnivores that consumed these grazers. A host of scavengers processed organic debris and aided in keeping the seas suitable for life. Some classes of invertebrates maintained a single mode of life, whereas groups within other classes adapted to several different lifestyles. Snails, for example, included scavengers, herbivores, and carnivores. With the passage of geologic time, many evolutionary changes and extinctions were to occur among the invertebrates, but these changes tended to occur *within* the taxonomic groups that had been established during the early Paleozoic.

## Unicellular Animals

The postulated biochemical steps leading to the first unicellular animals were discussed in the previous chapter. The first unicellular organisms were plants, but among modern unicells are creatures that defy classification as either animal or plant. Fortunately, there are also some single-celled forms that can with confidence be identified as animals. Two groups of creatures in the last category that are also capable of building shells or supportive structures are the **foraminifers** and **radiolarians**.

Foraminifers build their tiny shells by adding chambers singly, in rows, in coils, or in spirals. Some species construct the shell, or *test,* of tiny particles of silt; others secrete tests composed of calcium carbonate. The test is characteristically provided with holes through which extend "tentacles" of protoplasm for feeding. It is from these holes, or *foramina,* in the test that foraminifers take their name. Foraminifers range from the Cambrian to the present, but they are rare and poorly preserved in lower Paleozoic strata. During these early stages of their history they were mostly simple, saclike, tubular, or loosely coiled benthic creatures with tests composed of sedimentary particles.

Rates of evolution increased for foraminifers during the Carboniferous when they became more abundant and varied. One reflection of their numerical increase is seen in the Mississippian Salem Limestone, which locally consists almost entirely of specimens of *Endothyra* (Fig. 11–14). The **fusulinids**—

**Figure 11–14** Specimens of the foraminifer *Endothyra* from the Salem Limestone of Mississippian age.

**Figure 11–15** Fusulinid limestone of Permian age from the Sierra Madre of western Texas. *(Courtesy of W. D. Hamilton.)*

In modern oceans, the tests of calcareous foraminifers rain down continuously on the sea floor, much like an unending limy snow. The accumulated debris forms a deep-sea sediment called globigerina ooze because of the prevalence of the remains of species of *Globigerina* and related planktonic foraminifers.

Radiolarians are also single-celled planktonic organisms that have been present on earth at least since the Paleozoic began. Like the foraminifers, they have threadlike pseudopodia that project from an ornate latticelike skeleton of opaline silica or a proteinaceous substance. In some regions of the oceans today, radiolarian skeletons accumulate to form deposits of radiolarian or siliceous ooze. Although radiolarians do occur in lower Paleozoic rocks (Fig. 11–16), they are rare and not yet useful for stratigraphic correlation. They are more abundant in Mesozoic and Cenozoic rocks and have been shown to be particularly useful in correlating Pleistocene deep-sea deposits. Of special interest is the observation that Pleistocene radiolarian extinctions appear to be associated with dates of reversal in the earth's magnetic field.

spindle-shaped descendants of endothyrids—proliferated during the Pennsylvanian and Permian (Fig. 11–15). As a group, fusulinids looked superficially similar, but individual species evolved complex and distinctive internal structures that permit their use as index fossils in many parts of the world.

## Cup Animals: Archaeocyathids

*Archaeocyatha* means "ancient cups." It is an appropriate name for a somewhat enigmatic group of Cambrian organisms that constructed conical or vase-shaped skeletons out of calcium carbonate (Fig. 11–17). **Archaeocyathids** hold two paleontologic records. They are the earliest abundant reef-building animals on earth, and they are members of a separate phylum that suffered extinction. Although quite abundant during the Early Cambrian, they had

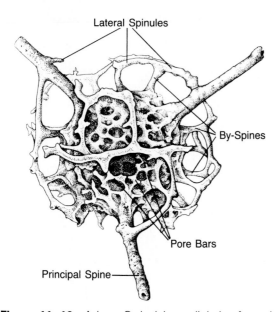

Lateral Spinules

By-Spines

Pore Bars

Principal Spine

**Figure 11–16** A Late Ordovician radiolarian from the Hanson Creek Formation of Nevada. It is unusual to find well-preserved radiolarians in rocks of the early Paleozoic. *(From Dunham, J. B., and Murphy, M. A. J. Paleo. 50(5):883, 1976.)*

(x1½)

(x1)

**Figure 11–17** An archaeocyathid.

entirely died out before the end of that period. Gregarious in habit, they carpeted the floors of the warm inland early Paleozoic seas. Low reefs formed primarily of archaeocyathids can be studied today in North America, Siberia, Antarctica, and Australia. Australian archaeocyathid reefs are often over 60 meters thick and extend horizontally in narrow bands for over 200 km.

## Pore-Bearers

Among the many stationary animals to colonize the early Paleozoic sea floor were the sponges. Sponges are members of the phylum **Porifera**. They appear to have evolved from colonial flagellated unicellular creatures and thus provide insight into how the transition from unicellular to multicellular animals may have occurred.

Sponges are a relatively conservative branch of invertebrates that have a long history. Cambrian representatives of all but one modern class of Porifera are known as fossils. Some fossil sponges, such as *Protospongia* from the Cambrian; *Astraeospongium, Microspongia,* and *Astylospongia* (Fig. 11–18) from the Silurian; and the siliceous sponge *Hydnoceras* from the Devonian are all well-known guide fossils (Fig. 11–19).

Sponges have always been predominantly marine creatures, although a few modern species live in fresh water. Spicules formed in the walls and cell layers of sponges are distinguishing characteristics of the phylum and provide both protection and support. The spicules may be composed of silica, calcium carbonate, or a proteinaceous material termed *spongin*. Naturally, the mineralized spicules are more commonly preserved and are frequently found by geologists examining rocks that have been disaggregated for study. Spicules are also important in the classification of sponges. For example, the Desmospongea consist entirely or in part of spongin, which may be reinforced with siliceous spicules. Hexactinellida develop siliceous spicules of distinctive shape, and Calcarea are characterized by spicules made of calcium carbonate. The Sclerospongea are a minor class of tropical marine sponges with skeletons composed of siliceous spicules, spongin fibers, and a basal layer of calcium carbonate.

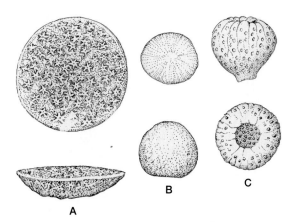

**Figure 11–18**  Early Paleozoic (Silurian) sponges. (A) *Astraeospongium* (it takes its name from the starlike spicules.) (B) *Microspongia*. (C) *Astylospongia*.

**Figure 11–19**  The Devonian siliceous sponge *Hydnoceras*. *(Photograph courtesy of J. Keith Rigby, Brigham Young University.)*

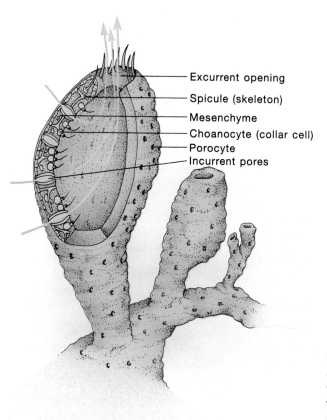

Excurrent opening
Spicule (skeleton)
Mesenchyme
Choanocyte (collar cell)
Porocyte
Incurrent pores

**Figure 11–20** Schematic diagram of a sponge having the simplest type of canal system. The path of water currents is indicated by arrows.

Although sponges vary greatly in size and shape, their basic structure (Fig. 11–20) is that of a highly perforated vase modified by folds and canals. The body is attached to the sea floor at the base, and there is an excurrent opening, or *osculum*, at the top. The wall consists of two layers of cells. Facing the internal space is a layer of collar cells (choanocytes), and on the outside is a protective wall of flat cells that resemble the bricks of a worn masonry pavement. Between these two layers one finds a gelatinous substance called *mesenchyme*. Here amoeboid cells go about the work of secreting the spicules. Sponges lack true organs. Water currents moving through the sponge are created by the beat of *flagellae*. These currents bring in suspended food particles, which are ingested by the collar cells. In a simple sponge, water enters through the pores, flows across sheets of choanocytes in the central cavity, and passes out through the osculum.

A group of early Paleozoic sclerosponges of particular interest because of their reef-building capabilities is the **stromatoporoids** (Fig. 11–21). These organisms constructed fibrous, calcareous skeletons of pillars and thin laminae that can nearly always be found in reef-associated carbonate rocks of the Silurian and Devonian. Apparently, stromatoporoids grew profusely and in close association with corals, brachiopods, and other invertebrate reef dwellers. Silurian coral–stromatoporoid reefs are well known

**Figure 11–21** Polished slab of limestone containing the stromatoporoid named *Stromatoporella,* from Devonian rocks of Ohio.

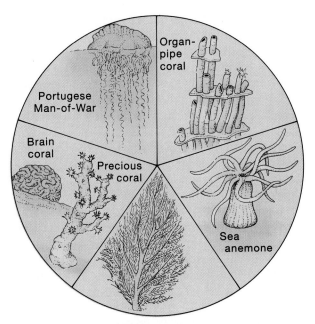

**Figure 11–22** Diversity among the cnidarians *(From Levin, H. L.* Life Through Time. *Dubuque, Iowa: William C. Brown Co., 1975.)*

to geologists primarily because of extensive study of reef exposures on Gotland Island off the Baltic coast of Sweden and in the region around and southwest of the Michigan basin. Reef development during the Devonian was similar to that in the Silurian, with

blanketlike, massive, and cylindric stromatoporoids forming a considerable part of the total mass of the reef structures.

## Corals and Other Cnidaria

Sea anemones, sea fans, jellyfish, the tiny *Hydra,* and reef-forming corals are all representatives of the phylum **Cnidaria**, which is known for the great diversity and beauty of its members (Fig. 11–22). The cnidarian body wall is composed of an outer layer of cells, the *ectoderm;* an inner layer, the *endoderm;* and a thin, noncellular intermediate layer, the *mesoglea.* In the endoderm are found primitive sensory cells, gland cells that secrete digestive enzymes, flagellated cells, and nutritive cells to absorb nutrients. A distinctive feature of cnidarians is the presence of stinging cells, which, when activated, can inject a paralyzing poison. These specialized cells, called cnidocytes, are a unique characteristic of all cnidaria.

Body form in cnidaria may be either polyp or medusa (Fig. 11–23). The medusoid form is seen in the jellyfish, which resembles an umbrella in shape. Jellyfish have a concave undersurface that contains a centrally located mouth; for this reason, it is designated the oral surface. In jellyfish such as the living *Aurelia,* which is common along the eastern shore of the United States, the mouth is surrounded by four oral arms. All jellyfish have tentacles, and in *Aurelia* these are located around the margin of the umbrella.

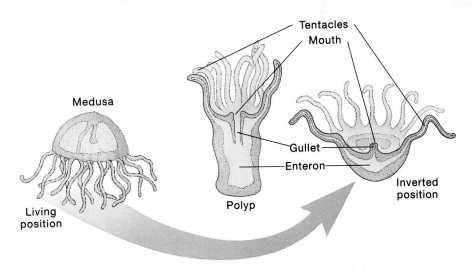

**Figure 11–23** Comparison of polyp and medusa forms in cnidarians.

**Figure 11-24** The common tabulate coral *Favosites.* An individual polyp resided in each of the tubes. Tiny pores visible in the walls of the tubes allowed for interconnections of tissues between adjacent polyps.

Most jellyfish swim by rhythmic contractions of the umbrella.

In the polyp form, as exemplified by *Hydra,* there is a circle of tentacles above the saclike body. Corals and sea anemones are among the frequently encountered cnidaria with this polyp form. In stony corals (class Anthozoa), the polyp secretes a calcareous cup, in which it lives. The animal may live alone or may combine with other individuals to form large colonies. The cup, or **theca,** may be divided by vertical plates called **septa,** which serve to separate layers of tissue and provide support. As the animal grows, it also secretes horizontal plates termed **tabulae.** Corals are identified according to the nature of their septa, tabulae, and other skeletal features. For example, after the development on an initial embryonic set of six protosepta, the Paleozoic **rugose corals,** or Rugosa, insert new septa at only four locations during growth. In other Paleozoic corals, septa are absent or poorly developed, so that tabulae are the most important features in classification. These are the Tabulata. **Tabulate corals** include many interesting colonial forms, such as the honeycomb and chain corals. The rugose corals and tabulate corals became extinct at the end of the Paleozoic and were followed in the Mesozoic by anthozoans of the order Sclerac-

tinia. In all scleractinian corals, septa are inserted between the mesenteries in multiples of six.

The fossil record for Cnidaria begins with fossil jellyfish impressions discovered in upper Precambrian rocks. The phylum is still poorly represented in Cambrian rocks. In the succeeding Ordovician the record improves dramatically, for the lime-secreting anthozoans begin to expand and diversify. The first stony corals were the tabulates, recognized by their simple, often clustered or aligned tubes divided horizontally by transverse tabula. These were tabulate corals, such as the common honeycomb coral *Favosites* (Fig. 11-24). Tabulate corals were the principal reef formers during the Silurian. They declined after the Silurian, but their role in reef building was assumed by the rugose corals. Both solitary and compound colonies of rugosids (Fig. 11-25) are abundant in the fossil record of the Devonian and Carboniferous, but this group declined and became extinct during the Late Permian.

Reefs built on the North American platform by Paleozoic corals were extensive and often exceeded 250 meters in total thickness. Most of these reefs lie beneath younger rocks and were discovered during oil explorations. Because of their porosity, these buried Paleozoic reefs became sites for the accumulation of petroleum.

## Moss Animals: Bryozoa

**Bryozoans** are minute, bilaterally symmetric animals that grow in colonies that frequently appear twiglike when viewed without the aid of a magnifier. The individuals, called *zooids,* are housed in a capsule, or *zooecium* (Fig. 11-26), that is often preserved as a result of its being calcified. The zooecia appear as pinpoint depressions on the outside of the colony, or *zoarium.* The zooid has a complete, **U**-shaped digestive tract with mouth surrounded by a tentacled feeding organ called the **lophophore.**

Today, there are more than 4000 living species of bryozoans, but nearly four times that number are known as fossils. Their earliest unquestioned occurrence is from Lower Ordovician strata of the Baltic, but they did not become abundant until Middle Ordovician and Silurian time. In rocks of these ages, their remains sometimes make up much of the bulk of entire formations. Like the corals, bryozoa contributed to the framework of reefs. One of the com-

A

B

C

D

**Figure 11–25** Devonian rugose corals. (A) The solitary horn coral *Zaphrenthis* with clearly visible radiating septa in the hornlike theca. (B) The compound (colonial) rugose coral *Lithostrotionella*. (C) A polished slab of the compound coral *Hexagonaria*. Water-worn fragments of this coral are found along the shore of Lake Michigan at Petoskey, Michigan, and this accounts for its being called Petoskey stone. Although not a rock, Petoskey stone is the designated state rock of Michigan. (D) Reconstruction of compound and solitary rugose corals on the floor of a Devonian epeiric sea. *(Diorama photograph courtesy of the National Museum of Natural History.)*

monest of all the early Paleozoic bryozoans was *Fistulipora,* different species of which develop massive, incrusting, or arborescent colonies. The genus *Hallopora* (Fig. 11–27A) is one of many branching twig bryozoans. Star-shaped patterns on the surface of *Constellaria* provided the inspiration for its generic name. Late Paleozoic bryozoans include varieties that constructed lacy, delicate, fanlike colonies (Fig. 11–27B). Associated with these so-called **fenestellid** colonies were the bizarre Mississippian corkscrew bryozoans known by the generic name *Archimedes* (Fig. 11–27C and D).

A

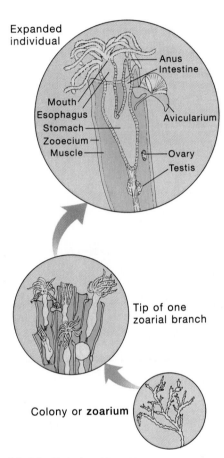

Expanded
individual

Anus
Intestine

Mouth
Esophagus
Stomach
Zooecium
Muscle

Avicularium

Ovary
Testis

Tip of one
zoarial branch

Colony or **zoarium**

**Figure 11–26**  Relationship of bryozoan individuals to the zoarium. *(From Levin, H. L.* Life Through Time. *Dubuque, Iowa: William C. Brown Co., 1975.)*

B

C

D

**Figure 11–27**  Paleozoic bryozoans. (A) The branching twig bryozoan *Hallopora* from the Ordovician of Kentucky. (B) *Fenestella,* a lacy bryozoan from Devonian limestones at the Falls of the Ohio River. (C) *Archimedes,* with part of the spirally encircling frond of lacy bryozoan colony attached and visible. (D) The central axis of *Archimedes.* ▶

**Figure 11–29** The inarticulate brachiopod *Obolus* from the Cambrian Conasauga Formation near Coosa, Georgia. The specimens are about 0.6 cm in diameter.

**Figure 11–28** Living positions of articulate and inarticulate brachiopods. (A) The articulate living brachiopod *Magellania* attached to the sea floor by pedicle and with lophophore barely visible through the gape in the valves. (B) Interior of brachial valve showing ciliated lophophore. (C) The inarticulate brachiopod *Lingula,* which excavates a tube in bottom sediment and lives within it. The pedicle secretes a mucus at the end that glues the animals to the tube.

## Brachiopods

**Brachiopods** are probably the most abundant, diverse, and useful fossils readily found in Paleozoic rocks. They are characterized by a pair of enclosing valves, which together constitute the shell of the animal (Fig. 11–28A). They resemble clams in this regard, but in symmetry of the valves and soft-part anatomy, brachiopods are quite different from pelecypods. Brachiopod valves are almost always symmetric on either side of the midline, and the two valves differ from each other in size and shape. The valves of a clam are right and left, whereas those of a brachiopod are dorsal and ventral. The valves may be variously ornamented with radial ridges or grooves, spines, nodes, and growth lines. Calcium carbonate is the usual hard tissue of brachiopods, although the valves of some families are composed of mixtures of chitin and calcium phosphate. This is particularly true of the group called **inarticulates** (Figs. 11–29 and 11–28C).

In the **articulate** brachiopods, the valves are hinged along the posterior margin and are prevented from slipping sideways by teeth and sockets. The less common inarticulates lack this definite hinge, are held together by muscles, and characteristically have simple spoon-shaped or circular valves. Although brachiopod larvae swim about freely, the adults are frequently anchored or cemented to objects on the sea floor by a fleshy stalk (*pedicle*) or by spines. Some simply rest on the sea floor. One of the more con-

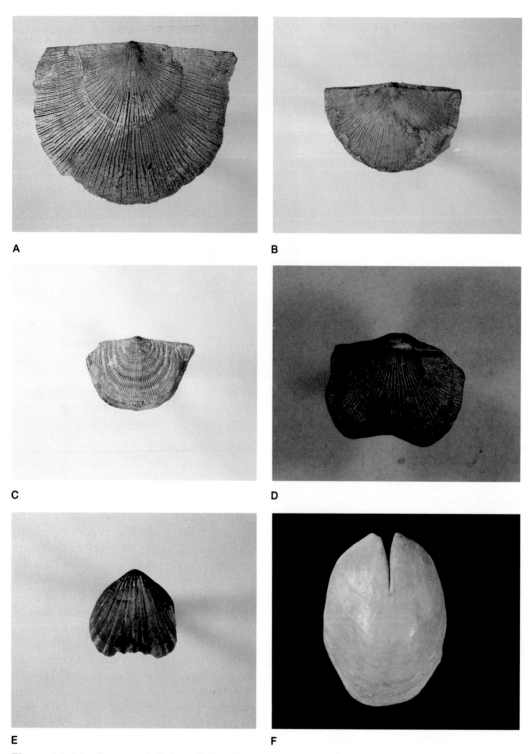

**Figure 11–30** Some early Paleozoic brachiopods. Along the top are Ordovician strophomenid brachiopods: (A) *Rafinesquina,* (B) *Strophomena,* and (C) *Leptaena.* Along the bottom are (D) *Hebertella,* an Orodovician orthid brachiopod; (E) *Lepidocyclus,* an Ordovician rhynchonellid; and (F) *Pentamerus,* a Silurian pentamerid ($\times\frac{1}{2}$).

A

B

**Figure 11–31** Devonian spiriferid brachiopods. (A) *Mucrospirifer.* (B) *Platyrachella.* (C) A spiriferid brachiopod with whell broken so as to reveal the internal spiral supports for the lophophore.

C

spicuous soft organs of brachiopods is the *lophophore,* a structure consisting of two ciliated, coiled tentacles whose function is to circulate the water between the valves, distribute oxygen, and remove carbon dioxide (see Fig. 11–28). Water currents generated by *cilia* on the lophophore move food particles toward the mouth and short digestive tract.

Brachiopods still live in the seas today, although in far fewer numbers than during the Paleozoic. Cambrian brachiopods were almost entirely the chitinous inarticulate types. These increased in diversity during the Ordovician, but then declined. A few species of only three families remain today.

The articulates also first appeared in the Cambrian, but became truly abundant during the succeeding Ordovician Period, when there was a great expansion of all sorts of shelled invertebrates. Across the floors of early Paleozoic epeiric seas, large and

small aggregates of these filter-feeders could be found. Today, their skeletons compose much of the volume of thick formations and provide the stratigrapher with essential markers for correlation. Among the articulate brachiopods that were particularly abundant during the early Paleozoic were the strophomenids, orthids, pentamerids, and rhynchonellids (Fig. 11–30). During the Devonian Period of the late Paleozoic, spiriferid brachiopods became particularly abundant. Spiriferids take their name from the calcified internal spirals that supported the lophophore (Fig. 11–31). The most characteristic brachiopods of the Carboniferous and Permian were the productids (Fig. 11–32). Productids were distinctively spinose (Fig. 11–33) with large, inflated ventral valves. They were so numerous during the Carboniferous that the period might well be dubbed the age of productids.

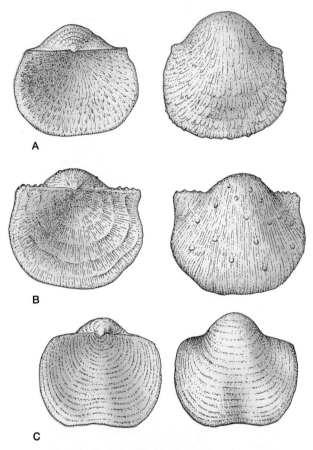

A

B

C

**Figure 11–32** The Pennsylvanian productid brachiopods (A) *Juresania,* (B) *Linoproductus,* and (C) *Echinoconchus. (From Guidebook Series 8, Illinois Geological Survey, 1970.)*

**Figure 11–33** Ventral (upper photograph) and side (lower photograph) views of the Permian spinose productid brachiopod *Marginifera ornata* from the Salt Range of West Pakistan. Valves (not including spines) are about 2 cm wide. *(Courtesy of R. E. Grant, U.S. Geological Survey.)*

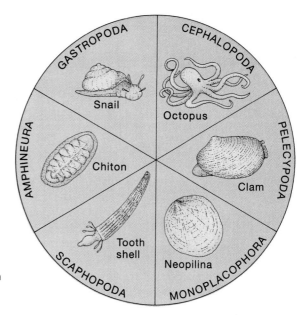

**Figure 11–34** Diversity among the members of the phylum Mollusca. *(From Levin, H. L. Life Through Time. Dubuque, Iowa: William C. Brown Co., 1975.)*

## Mollusks

A stroll along almost any seashore will provide evidence that mollusks are today's most familiar marine invertebrates. The shells of most members of this phylum have been readily preserved and provide enjoyment to collectors and conchologists today. Such well-known animals as snails, clams, chitons, tooth shells, and squid are included within the phylum **Mollusca** (Fig. 11–34). Although most mollusks possess shells, some, such as the slugs and octopods, do not. The various classes of Mollusca differ considerably in external appearance, yet they have fundamental similarities in their internal structure. There is a muscular portion of the body called the *foot* that functions primarily in locomotion. In cephalopod forms, the foot is modified into tentacles. A fleshy fold called the *mantle* secretes the shell. In aquatic mollusks, respiration is accomplished by means of gills. Well-developed organs for digestion, sensation, and circulation attest to the advanced stage of evolution that mollusks have attained.

Included within the phylum Mollusca are a group of marine animals that may be collectively referred to as placophorans. **Placophorans** are relatively primitive mollusks that have multiple paired gills and, in shelled forms, a creeping foot much like that seen in snails. The most familiar placophorans are the polyplacophorans, represented by chitons (Fig. 11–35), which possess eight overlapping plates

covering an ovoid, flattened body. Chitons are highly adapted for adhering to and grazing on rocks and shells. Their fossil record begins in the Cambrian and extends to the present day.

As indicated by their name, **monoplacophorans** are placophorans that have a single shell resembling a flattened or short cone. Their fossil record extends back to the Cambrian, but they disappeared from the fossil record in the Devonian. In 1952, however, ten

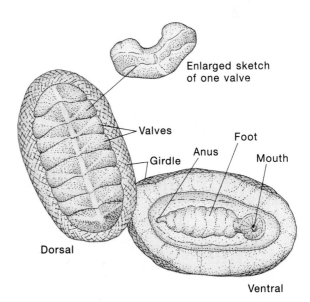

**Figure 11–35** The common Atlantic coast chiton.

**Figure 11-36** The "living fossil" *Neopilina ewingi.* (Magnification × ½.)

living specimens of a monoplacophoran were dredged from a deep ocean trench off the Pacific coast of Costa Rica. The specimens, later named *Neopilina* (Fig. 11–36), were of great interest to paleontologists, for they displayed a segmental arrangement of gills, muscles, and other organs that indicates mollusks branched off the invertebrate family tree at some point below the branch represented by segmented annelid worms and that the unsegmented condition seen in most mollusks is not primitive, but rather a secondary development. Many paleontologists regard the monoplacophorans as ancestral to more familiar mollusks such as pelecypods, gastropods, and cephalopods.

The **Pelecypoda** (also termed the Bivalvia) are a class of mollusks that includes clams, oysters, and mussels. Pelecypods are generally characterized by layered gills, a muscular foot (Fig. 11–37), bilobed mantle, and absence of a definite "head." Although members of the class made their first appearance in the Cambrian, they did not become notably abundant until Pennsylvanian and Permian when they populated the shallow seas of the time.

**Gastropods** first appear in Lower Cambrian strata. Earliest forms constructed small, depressed conical shells. During later Cambrian and Ordovician time, gastropods with the more familiar coiled conchs became commonplace (Fig. 11–38). It appears that coiling in a plane was not unusual initially, but was supplanted by spiral coiling. Gastropods of the Ordovician and Silurian had shapes similar to those of living species. One presumes that their soft parts were also similar, with distinct head, mouth, eyes, tentacles, and a vertically flattened foot to provide for gliding movement. By Pennsylvanian time, gastropods had become abundant and diverse. During this period, the oldest known air-breathing (pulmonate) gastropods appeared. The succeeding Permian Period ended with widespread extinctions among marine invertebrates, and many families of gastropods were decimated.

The **cephalopods** may very well be the most complex of all the mollusks. Today, this marine group is represented by the squid, cuttlefish, octopods, and the lovely chambered nautilus. The nautilus in particular provides us with important information about the soft anatomy and habits of a vast array of fossils known only by their preserved conchs. In the genus *Nautilus* (Fig. 11–39), one finds a bilaterally symmetric body, a prominent head with paired

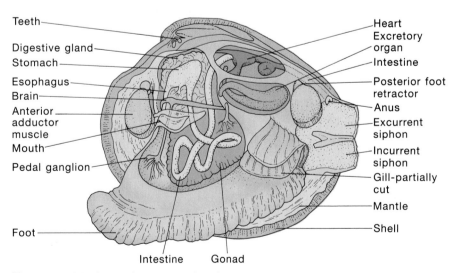

**Figure 11-37** Internal anatomy of a clam.

A                                              B

**Figure 11–38**  Two representative early Paleozoic gastropods. (A) *Taeniospira* from the Upper Cambrian Eminence Formation of Missouri. (B) *Clathospira* from the Ordovician Platteville Formation of Minnesota.

image-forming eyes, and tentacles developed on the forward portion of the foot. Water is forcefully ejected through the tubular "funnel" to provide swift, jet-propelled movement.

At first glance, the conchs of cephalopods resemble snail shells, but the resemblance is only superficial. Although we have mentioned an exception to this rule, the gastropod shell generally coils in a spiral, whereas the cephalopod conch characteristically coils in a plane. More importantly, the planispi-

rally coiled conch is divided into a series of chambers by transverse partitions, or septa. The bulk of the soft organs reside in the final chamber. Where the septa join the inner wall of the conch, *suture* lines are formed. These lines are enormously useful in the identification and classification of cephalopods. For example, cephalopods placed in the subclass **Nautiloidea** have straight or gently undulating sutures, whereas the **Ammonoidea** have more complex sutures.

A                                              B

**Figure 11–39**  *Nautilus,* a modern nautiloid cephalopod. (A) Conch of a nautiloid sawed in half to show large living chamber, septa, and septal necks through which the siphuncle passed. (B) Living animal photographed at a depth of about 300 meters. *(Photograph courtesy of W. Bruce Saunders.)*

A                                                B

**Figure 11–40**  Variation in conch shape among early Paleozoic nautiloid cephalopods. Both of these specimens are from the Silurian of Bohemia. (A) A sawed and
polished section of the straight conch of *Orthoceras potens* showing septa and
siphuncle. (B) Sawed and polished section of *Barrandeoceras,* exhibiting a coiled
form. Specimen (A) is 22.5 cm in length; (B) has a diameter of 18 cm.

The oldest fossils to be classified as cephalopods
were small, conical conchs discovered in Lower and
Middle Cambrian rocks of Europe. The class gradually increased in number and diversity and became
ubiquitous inhabitants of Ordovician and Silurian
seas. Indeed, by Silurian time, a great variety of conch
forms—from straight to tightly coiled (Fig. 11–40)—
had developed. Some of the elongate forms exceeded
4 meters in length. The first signs of a decline in
nautiloid populations can be detected in Silurian
strata. After Silurian time, the group continued to
dwindle until today only a single genus, *Nautilus,*
survives.

During the Devonian, the first ammonoid cephalopods appeared. These were the **goniatites,** characterized by angular and generally zig-zag sutures
without any additional crenulations (Fig. 11–41).
The goniatites persisted throughout the late Paleozoic and gave rise to the ceratites and ammonites of
the Mesozoic.

**Figure 11–41**  Goniatite ammonoid cephalopod exhibiting zig-zag sutures. Diameter is 4.6 cm.

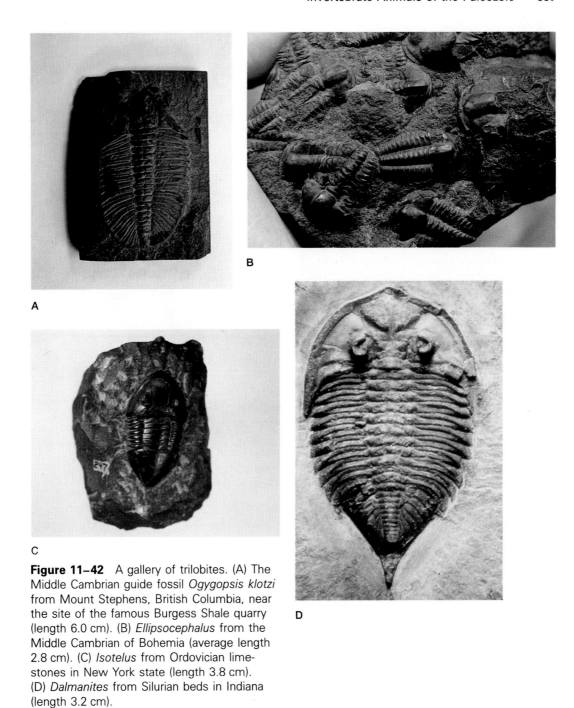

**Figure 11–42**  A gallery of trilobites. (A) The Middle Cambrian guide fossil *Ogygopsis klotzi* from Mount Stephens, British Columbia, near the site of the famous Burgess Shale quarry (length 6.0 cm). (B) *Ellipsocephalus* from the Middle Cambrian of Bohemia (average length 2.8 cm). (C) *Isotelus* from Ordovician limestones in New York state (length 3.8 cm). (D) *Dalmanites* from Silurian beds in Indiana (length 3.2 cm).

## *Arthropods*

The **arthropod** phylum is enormous. It includes such living animals as lobsters, spiders, insects, and a host of other animals that possess chitinous exterior skeletons, segmented bodies, paired and jointed appendages, and highly developed nervous system and sensory organs. Members of Arthropoda that have left a particularly significant fossil record are the **trilobites, ostracods,** and **eurypterids.**

Trilobites (Fig. 11–42) were swimming or crawling arthropods that take their name from divi-

## The Eyes of Trilobites

"The eyes of trilobites eternal be in stone, and seem to stare about in mild surprise at changes greater than they have yet known," wrote T. A. Conrad. Indeed, trilobites may have been the first animals to look out upon the world, for they possessed the most ancient visual systems known. The eyes of trilobites include both simple and compound types. A simple eye appears as a single, tiny lens, resembling a small node. A few receptor cells probably lay beneath each simple lens. Simple eyes are rare among trilobites. Far more abundant are compound eyes composed of a large number of individual visual bodies, each with its own biconvex lens. Each lens is composed of a single crystal of calcite. One of the properties of a clear crystal of calcite is that in certain orientations, objects viewed through the crystal produce a double image. Trilobites, however, were not troubled by double vision because the calcite lenses were oriented with the principal optic axis (the light path along which the double image does not occur) normal to the surface of the eye.

In some compound trilobite eyes, termed holochroal, the many tiny lenses are covered by a continuous, thin, transparent cornea. Beneath this smooth cover there may be as many as 15,000 lenses. In other trilobites, such as *Phacops* and *Calliops* (see Fig. 11–44 and accompanying figure), there are discrete individual lenses each covered by its own separate cornea, and separated from its neighbors by a cribwork of exoskeletal tissue. Such compound eyes are termed schizochroal.

Trilobite eyes can sometimes provide clues useful in determining the habits of certain species. Eyes along the anterior margin of the cephalon, for example, may indicate

The large schizochroal right eye of this well-preserved, silicified cephalon of the trilobite *Calliops*. Light directed toward the back of the specimen is emerging through some of the lensal pits.

active swimmers. Eyes located on the ventral side of the cephalon may indicate the trilobite was a surface-dweller. In the majority of trilobites, the eyes are located about midway on the cephalon in a position appropriate for an animal that crawled on the sea bottom or occasionally swam above it. Trilobites that either lacked eyes or were secondarily blind appear to have been adapted for burrowing in the soft sediment on the ocean floor.

sion of the dorsal surface into three longitudinal segments, or *lobes*. There are, for example, a central axial lobe and two lateral (pleural) lobes (Fig. 11–43). There was also a transverse differentiation of the shield into an anterior *cephalon,* a segmented *thorax,* and a posterior *pygidium*. The skeleton was composed of chitin strengthened by calcium carbonate in parts not requiring flexibility. As in many other arthropods, growth was accomplished by molting. Although many trilobites were sightless, the majority had either single-lens eyes or compound eyes composed of a large number of discrete visual bodies. The earliest trilobites had small pygidia and a large number of thoracic segments. These characteristics

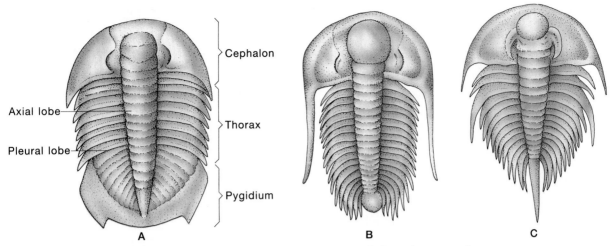

**Figure 11–43**   Three well-known Cambrian trilobites. (A) *Dikelocephalus minnesotensis* (Upper Cambrian), (B) *Paradoxides harlani* (Middle Cambrian), and (C) *Olenellus thompsoni* (Lower Cambrian).

have led paleontologists to speculate that trilobites may have evolved from annelid worms sometime during the Precambrian.

If one considers the entire fossil record for only the Cambrian, the trilobites are clearly the most abundant and diverse. More than 600 genera are known from the Cambrian. They first appear in rocks above those of the Tommotian Stage of the lowermost Cambrian.

The earliest trilobites apparently were bottom-dwelling, crawling scavengers and mud processors. Some preferred limy bottoms of cratonic and shelf areas, and others, such as *Paradoxides* of the Middle Cambrian, inhabited the muddy or silty floors of deep-water tracts. As a result of such environmental preferences, it has been possible to delineate trilobite faunal provinces in North America and Europe that suggest these two continents were once close together. (For example, the Lower Cambrian *Olenellus* assemblage is found in exposures in Pennsylvania, New York, Vermont, Newfoundland, Greenland, and Scotland.)

The optimum time for trilobites was reached in the Late Cambrian. After that they began to decline, perhaps in response to predation from cephalopods and fishes; however, they remained fairly abundant throughout the Ordovician and Silurian. Extinctions during the Middle and Late Devonian decimated the majority of trilobite groups, but the large-eyed group

known as proparians (Fig. 11–44) persisted and now serve as guide fossils. The largest trilobites ever found lived during the Devonian. One giant reached a length of over 70 cm. After the temporary increase in diversity during the Devonian, trilobites continued to wane until they became extinct near the close of the Paleozoic. Nevertheless, they were not at all biologic failures, for they had been important animals in the oceans for over 300 million years.

Anthropod companions to the trilobites in the Paleozoic seas were the small, bean-shaped ostracods (Fig. 11–45). At first glance, ostracods appear so different from trilobites as to cause one to question their classification within the same phylum. The os-

**Figure 11–44**   Side view of the trilobite *Phacops rana* showing the discrete visual bodies of the large compound eye. Length is 3.2 cm.

**Figure 11–45** Ostracods *(Leperditia fabulites)* in the Ordovician Plattin Limestone, Jefferson County, Missouri. Ostracods are the smooth, bean-shaped fossils. These average about 1 cm in length and are thus very large ostracods. Two strophomenid brachiopods are also present on the surface of the limestone.

**Figure 11–46** The fossil eurypterid *Eurypterus lacustris* from the Bertie Waterlime of Silurian age, Erie County, Pennsylvania. This well-preserved specimen is about 28 cm long. *(Courtesy of Ward's Natural Science Establishment.)*

tracods have a bivalved shell vaguely suggestive of some sort of tiny clam. However, this bivalved carapace encloses a segmented body from which extend seven pairs of jointed appendages. Adult animals are about 0.5 to 4 mm in length. The valves are composed of both chitin and calcium carbonate and are hinged along the dorsal margin.

Ostracods first appeared early in the Ordovician, and some limestones of this age are almost completely composed of their discarded carapaces. *Leperditia* and *Euprimitia* are common Ordovician genera. Ostracods continue in relative abundance to the present day. They occur in both marine and freshwater sediments. Because of their small size, they are brought to the surface in wells drilled for oil, and along with other microfossils, such as foraminifers and radiolaria, are used by exploration geologists in correlating the strata of oil fields.

Eurypterids (Fig. 11–46) are a group of arthropods that, because of their rarity, are less useful in stratigraphic studies than are either trilobites or ostracods. Nevertheless, they are among the most impressive of early Paleozoic marine invertebrates. Although many were of modest size, some were nearly 3 meters long and, had they survived, would have been suitable subjects for a Hollywood monster film. On their scorpion-like bodies were five pairs of appendages and a fearful-looking pair of pincers. They were also equipped with a venomous stinger. Eurypterids ranged across portions of the sea floor and brackish estuaries from Ordovician until Permian time, but were especially abundant during the Silurian and Devonian.

## Spiny-Skinned Animals

Just as modern seas abound with starfish, sea urchins, and sea lilies, so were the oceans of the early Paleozoic populated with members of the phylum

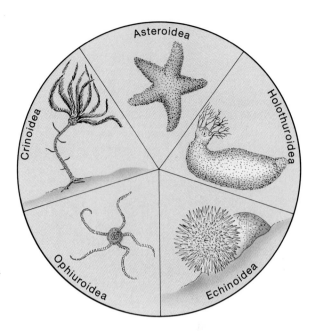

**Figure 11–47** Living echinoderms. *(From Levin, H. L.* Life Through Time. *Dubuque, Iowa: William C. Brown, 1975.)*

Echinodermata (Fig. 11–47). **Echinoderms** are animals with mostly five-way symmetry that masks an underlying primitive bilateral symmetry. They are well named "the spiny skinned," for spines are indeed present in many species. A unique characteristic of the phylum is the presence of a system of tubes—the water vascular system—which functions in respiration and locomotion (Fig. 11–48). Members of the phylum are exclusively marine, typically bottom-dwelling, and either attached to the sea floor or able to move about slowly.

Of considerable interest is the evolutionary relationship between echinoderms and vertebrates. There are so many allied embryologic parallelisms among primitive chordates and echinoderms that some zoologists speculate that both may have arisen from similar ancestral forms. For example, the larvae of echinoderms closely resemble those of the living protochordates called acorn worms. Also, in embryologic development, the mesodermal layer of cells, as well as certain other elements of the body, arises in the same way. Biochemistry has provided additional evidence for a relationship between echinoderms and chordates by revealing chemical similarities associated with muscle activity and oxygen-carrying pigments in the blood.

Among the many classes of the phylum Echinodermata, the Asteroidea (starfish), Ophiuroidea (brittle stars), Echinoidea (sea urchins), Edrioasteroidea, Crinoidea (sea lilies), Blastoidea, and Cystoidea are the most abundant and useful in geologic studies. The mostly attached and sessile crinoids, blastoids, and cystoids were particularly characteristic of the

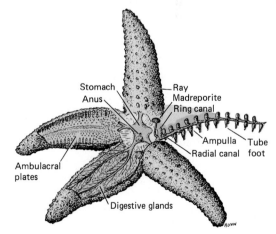

**Figure 11–48** Partially dissected starfish showing elements of the water vascular system and other organs.

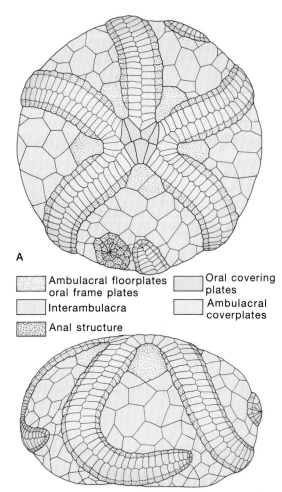

A

| Ambulacral floorplates oral frame plates | Oral covering plates |
| Interambulacra | Ambulacral coverplates |
| Anal structure | |

**Figure 11–49**   *Edrioaster bigsbyi,* a Middle Ordovician edrioasteroid. Specimen is 45 mm in diameter. (A) Oral surface. (B) Lateral view of the globoid fossil. *(From Bell, B. M. J. Paleo. 51(3):620, 1977.)*

**Figure 11–50**   The spiraled, spindle-shaped Early Cambrian echinoderm *Helicoplacus.* The specimen is about 2.6 cm long. *(After Durham, J. W., and Caster, K. E. Science 140:820–822, 1963.)*

Paleozoic Era, whereas the more mobile asteroids and echinoids were more frequent in later eras.

Echinoderms appear to have evolved late in the Proterozoic, although fossil remains of Proterozoic forms are few and often enigmatic. The Ediacaran fauna, for example, includes a globular fossil named *Arkarua* that has five rays on its surface and may be related to a group of echinoderms called edrioasteroids. Edrioasteroids (Figs. 11–49) are considered by many paleontologists to be ancestral to starfish and sea urchins.

By Early and Middle Cambrian time, several classes of rather peculiar echinoderms appeared, but this initial radiation was apparently not very successful. None of these became abundant, and three became extinct before the end of the Cambrian. One of the most bizarre of these early echinoderms was *Helicoplacus* (Fig. 11–50), a form having plates and food grooves arranged in a spiral around its spindle-shaped body.

The stemmed or stalked echinoderms first occur in Middle Cambrian strata, but do not become abundant until Ordovician and Silurian time. Stalked forms called **cystoids** (Fig. 11–51) are the most primitive among this group. The striking pentamerous (five-way) symmetry that is evident in most echinoderms is often less well developed in cystoids. Beginning students of paleontology often recognize cystoids by the characteristic pores that occur in rhomboid patterns on the plates of the calyx and compose part of the water vascular system. Although cystoids range from Cambrian to Late Devonian, they are chiefly found in Ordovician and Silurian rocks.

Unlike some of the cystoids, stalked echinoderms known as **blastoids** (Fig. 11–52) have a beautifully symmetric arrangement of plates. The five radial areas (ambulacral areas) are prominent, bear slender branches or brachioles along their margins,

**Figure 11–51**   A well-preserved specimen of the Silurian cystoid *Caryocrinites ornatus* from the Lockport Shale of New York *(From Sprinkle, J.* J. Paleo. *49(6):1062–1073, 1975.)*

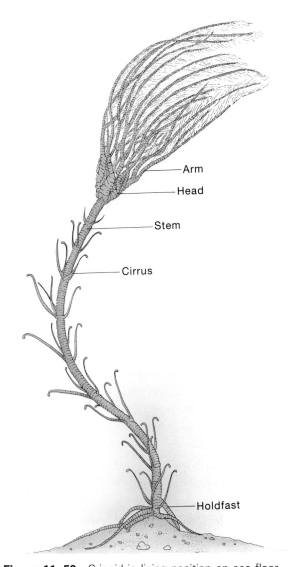

**Figure 11–53**   Crinoid in living position on sea floor.

**Figure 11–52**   Calyx of the blastoid *Pentremites symmetricus* from the Upper Mississippian near Floraville, Illinois. Height of calyx is 2.6 cm.

and have a well-developed and unique water transport system. The blastoids first appeared in Silurian time, expanded in the Mississippian, and declined to extinction in the Permian.

**Crinoids,** like most cystoids and blastoids, are composed of three main parts: the *calyx* (which contains the vital organs), the *arms,* and the *stem* with its rootlike *holdfast* (Fig. 11–53). The arms bear ciliated food grooves and, like the brachioles of blastoids, serve to move food particles toward the mouth. Crinoids are found in rocks that range in age from Or-

A                                    B                                    C

**Figure 11–54**    (A) Diorama depicting crinoids living on the floor of a Mississippian epeiric sea. (B) Calyx and arms of the Mississippian crinoid *Taxocrinus* (height 5.1 cm). (C) *Scytalocrinus* (height 4.5 cm). Both fossil crinoids are from the Keokuk Formation of Mississippian age, near Crawfordsville, Indiana. *(Diorama photograph courtesy of the National Museum of Natural History.)*

**Figure 11–55**    The large Mississippian echinoid *Melonechinus* from the St. Louis Limestone, St. Louis, Missouri. Average diameter of specimens is 12 cm. *(Photograph by John Simon.)*

dovician to Holocene. In some areas, Ordovician, Silurian, and Carboniferous rocks contain such great quantities of disaggregated plates of crinoids that they are named crinoidal limestones. So abundant are crinoid remains in Mississippian rocks that the period has been dubbed the age of crinoids (Fig. 11–54).

Although one group of crinoids survived the hard times at the end of the Paleozoic and gave rise to those still living today, most died out before the end of the Permian. The blastoids also died out before the end of the Paleozoic. Free-living echinoderms such as echinoids (Fig. 11–55) and starfish were locally abundant during the late Paleozoic, but not as numerous or diverse as they would become in post-Paleozoic time.

## Graptolites

Just as the early Paleozoic sea floors bustled with the activities of a myriad of invertebrates, so also did the overlying waters. However, many of these planktonic creatures lacked skeletons and so are poorly known. One group, the **graptolites** (Fig. 11–56), had preservable chitinous skeletons that housed colonies of tiny individual animals.

The graptolites made their appearance at the very end of the Cambrian but were rare until the

**Figure 11–56** Part of a stipe of the Ordovician graptolite *Orthograptus quadrimucronatus* (×15). The specimen has been bleached for better visualization of internal features. The cuplike thecae are clearly visible on either side. One can also see the inverted cone of the sicula. The nema would rise from the pointed end of the sicula. *(From Herr, S. R. J. Paleo. 45(4):628–632, 1971.)*

Ordovician when they became quite abundant and diverse. Unfortunately for paleontologists, they did not survive beyond the Mississippian, and the nature and functions of their soft organs can only be surmised. The structure of the exoskeleton is relatively simple. The tiny cups, or thecae, are arranged along a branch, or *stipe*. The stipes may be solitary or formed into a system of two or more branches. The entire colony is referred to as the *rhabdosome* and is supported or attached at one end by a thin filament—the *nema*. Where the lower end of the nema reaches the base of a stipe, there is a conical *sicula*, which may have served as the theca for the first individual. From the sicula, subsequent members of the colony add their thecae by budding. Although most graptolites were planktonic creatures and often developed flotational structures, some were apparently attached to objects by the threadlike nema.

Graptolites occur as streaks of flattened, carbonaceous matter in fine-grained rocks. Recently, however, uncompressed graptolites have been studied that reveal an unexpected relationship to primitive chordates. For example, both graptolites and the living protochordates known as pterobranches secreted tiny enclosed tubes, and in both, the unique structure of the enclosing wall of the sheath is so similar that a relationship is virtually certain.

During the early Paleozoic, graptolites were at times so abundant that they have been visualized as forming sargasso-like floating masses. They were apparently carried about by ocean currents and thus achieved a worldwide distribution, which has enhanced their use as index fossils. Their use, however, is limited in that they are seldom found preserved except in fine-grained, usually carbonaceous, shaly sediments. Perhaps these occurrences came about as masses of graptolite colonies floated into areas of toxic conditions, died, and settled to the sea floor. It is also possible that such accumulations resulted from fortuitously favorable preservation.

## Conodonts

One group of well-known but enigmatic microscopic fossils may have affinities to a prechordate animal. However, their true biologic relationships are still being hotly debated. These problematic fossils are **conodonts**. Some are platelike or cone-shaped (Fig. 11–57), and all are composed of calcium phosphate. Most conodonts are about 1 mm in length. At their base one finds cavities or attachment scars where they were presumably fastened to a supportive part of a parent organism. Conodonts occur mostly as single, disarticulated fossils, but occasionally paired groupings are discovered suggestive of a feeding or filtering mechanism.

Although their true natures are unknown, conodonts are still excellent guide fossils. They occur in a variety of marine sedimentary rocks and range from latest Proterozoic to Triassic age. There are now about 140 conodont-based stratigraphic zones. Their uncertain origins do not detract from their value in stratigraphy. Nevertheless, paleontologists are keenly interested in discovering their true affinities. Fortunately, the search has been narrowed by two or three discoveries of conodonts found in close association with soft remains of the animals that apparently contained them. Mississippian rocks near Edinburgh, Scotland, provided one of the most recent of these discoveries. The conodont-bearing fossil was an elongate (about 40 mm long), segmented animal with ray-supported fins. Conodonts were found clustered at the anterior end just behind a depression resembling the animal's mouth. The conodonts were ar-

**Figure 11–57**   Scanning electron micrographs of Ordovician conodonts. These simple, recurved, cone-shaped forms have a deeply excavated base resembling a pulp cavity. Specimens are about 1 mm long. *(Courtesy of D. Hearns.)*

ranged symmetrically in an apparatus-like grouping. The investigators are convinced they have found a true conodont animal, although once again its precise affinities are obscure.

## Continental Invertebrates

Because of the greater hazards of postmortem destruction of continental (terrestrial) invertebrates, their fossil record is not as complete as that for marine invertebrates. Nevertheless, as plants invaded the continents, animals were able to follow. Arthropods, with their sturdy legs and protective exoskeletons, were probably the marine animals best preadapted for coming ashore. Early invaders may have lived within the wet, nutritious plant debris washed up along the beaches of early Paleozoic seas. Their fossil record begins with possible millipede trace fossils in Ordovician rocks. During the Devonian, the record for arthropods improves, with discoveries of wingless insects, including mites, springtails, spiders, and centipedes. Carboniferous strata contain a slightly better record of insects, including giant dragonflies with wingspans of over 70 cm (2 feet). Cock-

roaches (Fig. 11–58) that reached lengths of 10 cm crept about among the rotting vegetation. Eurypterids, although not common, persisted throughout the late Paleozoic. Several species of advanced spi-

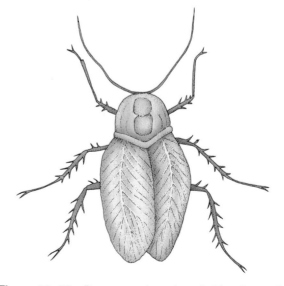

**Figure 11–58**   Reconstruction of a primitive Pennsylvanian cockroach. Length is 10 cm.

ders have been recovered from Devonian and Carboniferous strata in North America, Germany, and Great Britain. In Nova Scotia, an interesting collection of land snails has been collected from hollows within the preserved stumps of trees. Scorpions and centipedes have been found in abundance in a concretionary Pennsylvanian siltstone exposed along Mazon Creek in northern Illinois.

Of all the land invertebrates, it is apparent that the arthropods and gastropods have been the most persistently successful. Many paleontologists believe this may be because of attributes already evolved by their aquatic ancestors. As protection against desiccation, arthropods had evolved relatively impervious exoskeletons. Snails derived similar benefits from their shells. Both groups included very active animals with sufficient mobility to seek out food aggressively.

## Vertebrate Animals of the Paleozoic

### *The Rise of Fishes*

A momentous biologic event occurred in early Paleozoic time. It was the birth of the chordate line. In our earlier discussion of the Burgess Shale fauna, we briefly described chordates as animals that have (at least at some stage in their life history) a stiff, elongate supporting structure. In addition, all chordates have a dorsal, central nerve cord, gill slits, and blood that circulates forward in a main ventral vessel and backward in the dorsal. The supportive structure in primitive chordates is called a notochord and has been studied by generations of biology students in such animals as the lancelet *Branchiostoma* (Fig. 11–59). In taxonomic hierarchy, vertebrates are simply those chordates in which the notochord is supplemented or replaced by a series of cartilaginous or bony vertebrae.

As we have mentioned previously, the ancestors of the vertebrate lineage may lie somewhere among the echinoderms. The theoretic evolutionary progression that was to lead to vertebrates may have begun with sedentary, filter-feeding animals that had exposed cilia located along their arms, somewhat in the fashion of crinoids. From such a beginning there may have evolved filter-feeders with cilia brought inside the body in the form of gills. In a subsequent stage, the organisms may have become free-swimming, gilled animals, in superficial appearance not unlike small, simple fishes or frog tadpoles. *Pikaia*, from the Middle Cambrian Burgess Shale, appears to have been such an animal. *Pikaia* or animals like it

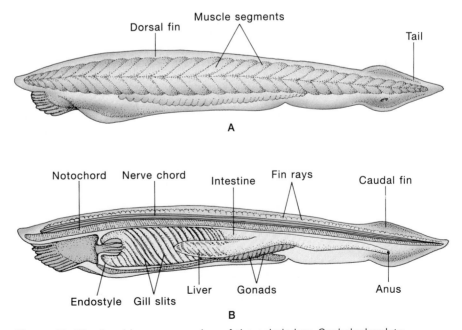

**Figure 11–59**  Amphioxus, a member of the subphylum Cephalochordata. (A) External view; (B) longitudinal section.

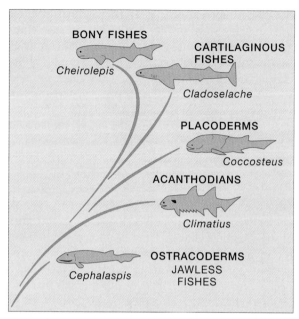

**Figure 11–60** The basic evolutionary radiation of fishes that began during the early Paleozoic and produced the sharks and bony fishes during the Devonian. *(From Colbert, E. H.* Evolution of the Vertebrates. *New York: John Wiley, 1969. With permission of the author, artist Lois Darling, and the publisher.)*

may be at the base of the evolutionary sequence leading to vertebrates. Presumably, we may call those earliest vertebrates fishes.

The oldest fossil remains of fishes consist of small pieces of external plates found in Upper Cambrian deposits of Wyoming. Their distinctive surface patterns and internal structure clearly indicate these plates were part of the body covering of armored jawless fishes. Somewhat younger scales and plates of jawless fishes are found in Ordovician rocks of Colorado, Spitzbergen, and Greenland.

The vertebrates that we loosely call fishes are actually divided into at least five distinct taxonomic classes (Fig. 11–60). There are the jawless fishes, or Agnatha; two groups of archaic jawed fishes, the Acanthodii and Placodermi; the cartilaginous fishes, or Chondrichthyes; and the familiar Osteichthyes with their highly developed bony skeletons. It is the first three of these categories—namely, the Agnatha, Acanthodii, and Placodermi—that frequented freshwater and saltwater bodies of the early Paleozoic.

The agnaths still exist today as the inelegant hagfish and lamprey. However, these specialized survivors are quite unlike the jawless fishes of the Ordovician and Silurian. Early Paleozoic agnaths are collectively termed **ostracoderms**. The name means "shell skin" and refers to the bony exterior that was a distinctive trait of many of these fishes. Ostracoderms comprised a rather diverse group that included the unarmored forms *Thelodus* and *Jamoytius,* as well as such armored creatures as *Pteraspis* (Fig. 11–61). The purpose of the bony armor that is the hallmark of many of the ostracoderms is still being debated. A widely held view is that the armor provided protection, although the fossil record shows few creatures of the same age and habitat that could have preyed on the ostracoderms. Nevertheless, it does not seem unreasonable that some of the smaller ostracoderms might have provided an occasional

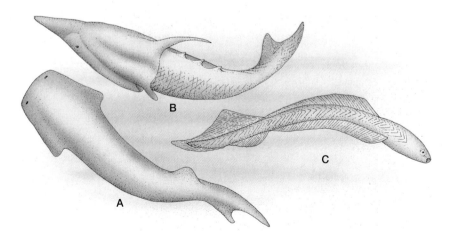

**Figure 11–61** The early Paleozoic ostracoderms, (A) *Thelodus,* (B) *Pteraspis,* and (C) *Jamoytius,* drawn to the same scale.

**Figure 11–62** The ostracoderm *Hemicyclaspis*. The creature was about 15 cm long. *(After Stensio, A. E. Br. Mus., 1932, p. 3, fig. 15.)*

The evolutionary development of the vertebrate jaw involved a remarkable transformation in which an older structure was modified to perform a role that was entirely different from its original function. Those older structures were supports called *gill arches* that were located between the *gill slits* (Fig. 11–63). As the mouth extended posteriorly, the first two of these arches were sequentially eliminated. The third set of gill arches was gradually modified

meal for a large cephalopod or eurypterid. Another theory stipulates that the dermal armor was not primarily for protection but rather was a device for storing seasonally available phosphorus. According to this idea, phosphates could be accumulated as calcium phosphate in the dermal layers during times of greater availability and then used during periods when supply was deficient. This cache of phosphorus may have been vital to early vertebrates for them to maintain a suitable level of muscular activity.

*Hemicyclaspis* (Fig. 11–62) is one of the most widely known of the ostracoderms. It is recognized by its large, semicircular head shield, on the top of which can be found four openings: two for the upward-looking eyes, a small pineal opening, and a single nostril. The pineal opening may have housed a light-sensitive third eye. The mouth was located ventrally in a position suitable for taking in food from the surface layer of soft sediment. Along the margin of the head shield were depressed areas that may have had a sensory function. Internally, the head contained a regular segmental arrangement of nerves, blood vessels, and gill pouches. These and other features of the soft internal anatomy have been discerned by careful dissection of exceptionally well-preserved specimens.

The ostracoderms continued into the Devonian, but did not survive beyond that period. For the most part they were small, rather sluggish creatures restricted to mud-straining or filter-feeding modes of life. They were to be gradually replaced by fishes that had developed bone-supported, movable jaws. The evolution of the jaw was no small accomplishment, for it enormously expanded the adaptive range of the vertebrates. Fishes with jaws were able to bite and grasp. These new abilities led to more varied and active ways of life and to new sources of food not available to the agnaths.

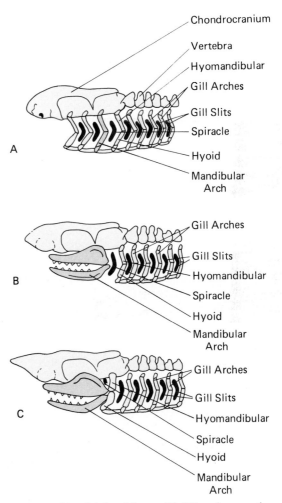

**Figure 11–63** Origin of jaws. (A) Gill arches as they might exist in a primitive jawless fish. (B) Early jawed fish with a complete gill slit behind jaws. (C) A jawed fish (like a shark) with the first gill slit reduced to a spiracle. *(Modified from Cockrum, E. L., and McCauley, W. J. Zoology. Philadelphia: W. B. Saunders, 1965.)*

**Figure 11–64**    The Early Devonian acanthodian fish *Climatius. (After Romer, A. S. Vertebrate Paleontology. Chicago: Univ. of Chicago Press, 1945.)*

into jaws. Classic anatomic studies have clearly shown the similarities in jaw architecture and neurology to elements of the gill arches and the nerves that serve these supports. In the earliest fishes, the upper jaw was attached to the skull by ligaments only. A full gill slit occurred immediately behind the jaw. In more advanced fishes, the upper jaw was anchored to the skull by the upper part of the next gill arch, the *hyomandibular*. The pair of gill slits that had once existed behind the jaw was squeezed into a smaller space and ultimately developed into the *spiracle* openings found in members of the shark family today. The spiracle was a structure destined to become part of the ear apparatus in higher vertebrates.

The oldest fossil remains of jawed fishes are found in nonmarine rocks of the Late Silurian. These fishes, called **acanthodians** (Fig. 11–64), became most numerous during the Devonian and then declined to extinction in the Permian. Acanthodians were archaic jawed fishes and were quite distinct from the great orders of modern fishes. Another group of archaic fishes includes the **placoderms**, or "plate-skinned" fishes. They too arose in the Late Silurian, expanded rapidly during the Devonian, and then in the latter part of that period began to decline and be replaced by the ascending sharks and bony fishes.

There was considerable variety among the placoderms. The most formidable of these plate-skinned

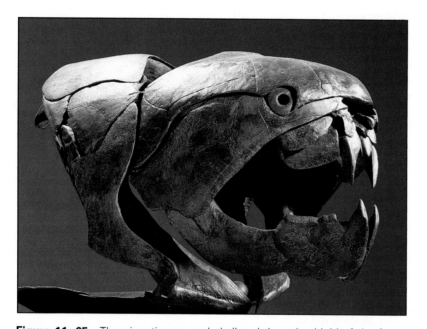

**Figure 11–65**    The gigantic armored skull and thoracic shield of the formidable Late Devonian placoderm fish known as *Dunkleosteus. Dunkleosteus* was over 10 meters (about 30 feet) long. The skull shown here is about 1 meter tall. It is equipped with large bony cutting plates that functioned as teeth. Each eye socket was protected by a ring of four plates, and a special joint at the rear of the skull permitted the head to be raised and thereby provided for an extra large bite. *Dunkleosteus* ruled the seas 350 million years ago. *(Courtesy National Museum of Natural History; photograph by Chip Clark.)*

**Figure 11–66** The Devonian antiarch fish *Pterichthyodes*. *(From Romer, A. S.* Vertebrate Paleontology. *Chicago: Univ. of Chicago Press, 1945, p. 54, fig. 38.)*

fishes was a carnivorous group called arthrodires. *Dunkleosteus* (Fig. 11–65) was a Devonian arthrodire whose length exceeded 9 meters and whose huge jaws could be opened exceptionally wide to engulf even the largest of available prey. Other placoderms, called antiarchs (Fig. 11–66), had the heavily armored form and mud-grubbing habits of their predecessors, the ostracoderms.

Among the Late Silurian acanthodians and placoderms were the ancestors of the bony and cartilaginous fishes that were to dominate the marine realm in succeeding periods. The ascendancy of these more modern kinds of fishes began with a veritable evolutionary explosion during the Devonian. Two important categories of fishes, the cartilagenous chondrichthyians and the bony osteichthyians, made their debut in the Devonian. Today, the cartilaginous fishes are represented by sharks, rays, and skates. Among the better known of the late Paleozoic sharks were species of *Cladoselache* (Fig. 11–67A). Remains

A

**Figure 11–67** Models of (A) the Devonian marine shark *Cladoselache,* and (B) the Pennsylvanian freshwater shark *Xenacanthus.*

B

**Figure 11–68** *Cheirolepis,* the ancestral bony fish that lived during the Devonian Period.

of this shark are frequently encountered in the Devonian shales that crop out on the south shore of Lake Erie. During the Late Carboniferous, a group represented by *Xenacanthus* (Fig. 11–67B) managed to penetrate the freshwater environment. A third group of Paleozoic cartilaginous fishes were the bradyodonts. These had flattened bodies like modern rays and blunt, rounded teeth for crushing shellfish. Apparently, modern sharks arose from cladoselachian ancestors but retained their archaic traits until the Jurassic.

Because of the role of bony fishes in the evolution of tetrapods (four-legged animals) and because they are the most numerous, varied, and successful of all aquatic vertebrates, their evolution is of particular importance. Bony fishes may be divided into two categories, namely the familiar ray-fin, or Actinopterygii, and the lobe-fins, or Sarcopterygii.

As implied by their name, ray-fin fishes lack a muscular base to their paired fins, which are thin structures supported by radiating bony rays. Unlike the Sarcopterygii, they do not possess paired nasal passages that open into the throat. The ray-fins began their evolution in Devonian lakes and streams and quickly expanded into the marine realm. They became the dominant fishes of the modern world.

The more primitive Devonian bony fishes are well represented by the genus *Cheirolepis* (Fig. 11–68). From such fishes as these evolved the more advanced bony fishes during the Mesozoic and Cenozoic.

The second category of bony fishes, the Sarcopterygii, are characterized by sturdy, fleshy lobe-fins and a pair of openings in the roof of the mouth that led to clearly visible external nostrils. Such fish were able to rise to the surface and take in air that was passed on to functional lungs. Lungs and fins do not seem to occur together in modern fishes, but in late Paleozoic fishes, the combination was not uncommon. Studies of living examples of Sarcopterygii indicate that lungs probably began their evolution as saclike bodies developed on the ventral side of the esophagus and then became enlarged and improved for the extraction of oxygen. (In modern fishes, the lung has been converted to a swim bladder, which aids in hydrostatic balance.)

Two major groups of lungfishes lived during the Devonian. They are designated the **Dipnoi** and **Crossopterygii**. The Dipnoi, represented in the Devonian by *Dipterus* (Fig. 11–69), were not on the evolutionary track that was to lead to tetrapods. They are, nevertheless, an interesting group, which includes the living freshwater lungfish of Australia, Africa, and South America. Their restricted presence south of the equator suggests that Gondwanaland was the probable center of dispersal for dipnoans. Dipnoi means "double breather." The name was suggested by the observation that living species are able to breathe by means of lungs during dry seasons. At such difficult times, they burrow into the mud before the water is gone. When the lake or stream is dry, they survive by using their accessory lungs, and when the waters return they switch to gill respiration.

Because of the arrangement of bones in their muscular fins (Fig. 11–70), the pattern of skull elements (Fig. 11–71), and the structure of their teeth (Fig. 11–72), fossil Crossopterygii are considered

**Figure 11–69** *Dipterus,* a Devonian lungfish.

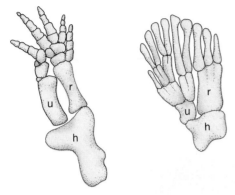

**Figure 11–70** Comparison of the limb bones of a crossopterygian fish (upper right) and an early amphibian. *(From Levin, H. L.* Life Through Time. *Dubuque, Iowa: William C. Brown, 1975.)*

the pelvic structure), articulated with the girdle bones and was followed by two bones, the ulna and radius, for the pectoral fins, and the tibia and fibula for the posterior fins.

Of course, the robust skeleton and sturdy limbs of the crossopterygians did not evolve because fishes had miraculously taken on a desire for life on land. These were adaptations for moving to and remaining in water during periods of drought. The air-gulping fishes found land a hostile environment and occasionally dragged themselves onto the land only as a means of reaching another pond or fresher body of water.

During the Devonian, two distinct branches of crossopterygians had evolved. One of these, the rhipidistians, led ultimately to amphibians, and the other led to saltwater fishes called coelacanths. Coelacanths were thought to have undergone extinction during the Late Cretaceous, but their survival down to the present day has been documented by several catches of the coelacanth *Latimeria* (Fig. 11–74) near Madagascar.

the ancestors of the amphibians. A rather advanced fish that exemplifies Devonian crossopterygians is *Eusthenopteron* (Fig. 11–73). In this genus, the paired fins were short and muscular. Internally, a single basal limb bone, the humerus (or femur for

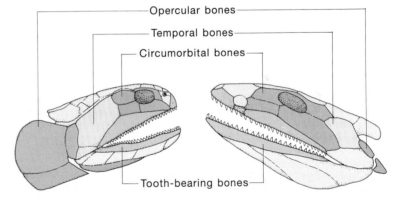

**Figure 11–71** Comparison of skulls and lower jaws of a crossopterygian (left) and the Devonian amphibian *Ichthyostega. (From Levin, H. L.* Life Through Time. *Dubuque, Iowa: William C. Brown, 1975.)*

Opercular bones
Temporal bones
Circumorbital bones
Tooth-bearing bones

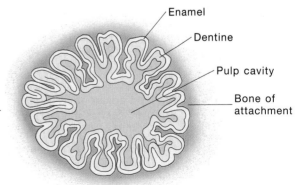

Enamel
Dentine
Pulp cavity
Bone of attachment

**Figure 11–72** Cross-section of a crossopterygian tooth clearly exhibits the distinctive pattern of infolded enamel. This is also a characteristic of the teeth of the amphibian descendents of crossopterygian fishes. *(From Levin, H. L.* Life Through Time. *Dubuque, Iowa: William C. Brown, 1975.)*

**Figure 11–73** The Devonian crossopterygian lungfish *Eusthenopteron* had sturdy fins whose structure foreshadowed that of four-footed land animals that were to be their descendents. *(St. Louis Science Center diorama.)*

**Figure 11–74** *Latimeria,* a surviving coelacanth living in the ocean near Madagascar and southeast Africa. *Latimeria* is a large fish, nearly 2 meters in length. *(From Levin, H. L. Life Through Time. Dubuque, Iowa: William C. Brown, 1975.)*

## *Amphibians*

It required tens of millions of generations to convert the crossopterygian fishes into animals that could live comfortably on solid ground (Fig. 11–75). Even so, the conversion was not complete, for amphibians continued to return to water to lay their fishlike, naked eggs. From these eggs came fishlike larvae, which, like fish, used gills for respiration.

A number of changes accompanied the shift to land dwelling. A three-chambered heart developed to route the blood more efficiently to and from the more efficient lungs. The limb and girdle bones were modified to overcome the constant drag of gravity and better hold the body above the ground. The spinal column, a simple structure in fishes, was transformed into a sturdy but flexible bridge of interlocking elements. To improve hearing in gaseous rather than fluid surroundings, the old hyomandibular bone, used in fishes to prop the braincase and upper jaw together, gradually evolved into an ear ossicle—the stapes. The fish spiracle (a vestigial gill slit) became the amphibian eustachian tube and middle ear. To complete the auditory apparatus, a tympanic membrane (ear drum) was developed across a prominent notch in the rear part of the amphibian skull.

The fossil record for amphibians begins with a group called **ichthyostegids** (Fig. 11–76). As suggested by their name, these creatures retained many features of their fish ancestors. The amphibians that followed the ichthyostegids fall mostly within a group collectively termed the **labyrinthodonts**. The labyrinthic wrinkling and folding of tooth enamel inherited from the rhipidistian Sarcopterygii provided the inspiration for the labyrinthodont name (see Fig. 11–72). During the Carboniferous, large numbers of labyrinthodonts wallowed in swamps and streams, eating insects, fish, and one another.

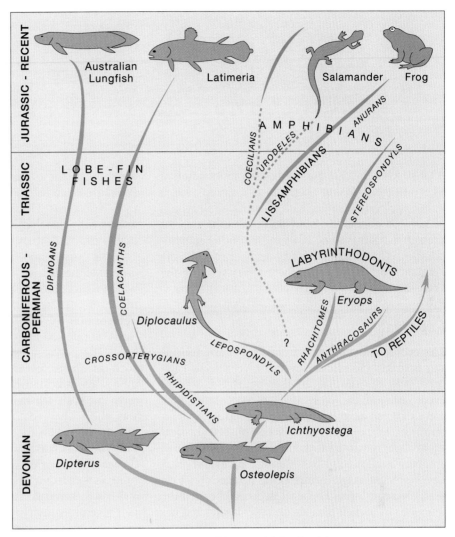

**Figure 11–75**  The evolution of amphibians and lobe-fin fishes. *(From Colbert, E. H. Evolution of the Vertebrates. New York: John Wiley, 1969. With permission of the author, artist Lois Darling, and the publisher.)*

**Figure 11–76**  The skeleton of *Ichthyostega* still retains the fishlike form of its crossopterygian ancestors. *(From Levin, H. L. Life Through Time. Dubuque, Iowa: William C. Brown, 1975.)*

**Figure 11–77**  The large Permian amphibian *Eryops*. Length is about 2 meters.

A labyrinthodont that exemplifies the culmination of their lineage is *Eryops* (Fig. 11–77). Eryops was a bulky, inelegant creature with a flattish skull typical of labyrinthodonts and bony nodules in the skin for protection against some of its more vicious contemporaries. The labyrinthodonts declined during the Permian, and only a relatively few survived into the Triassic.

## Reptiles

The evolutionary advance from fish to amphibians was no trivial biologic achievement, yet the late Paleozoic was the time of another equally significant event. Among the evolving land vertebrates were some that had developed a way to reproduce without returning to the water. They accomplished this feat by evolving enclosed eggs in which the embryonic animal was allowed to pass through larval and other developmental stages before being hatched in an essentially adult form. This enclosed egg liberated the tetrapods from their reliance on water bodies and has been hailed as a major milestone in the history of vertebrates. The animals that first evolved the so-called amniotic egg were amphibians. Evolutionary processes had provided them with a method for protecting developing young from predation and desiccation.

The oldest known remains of reptiles were discovered in Early Pennsylvanian swamp deposits of Nova Scotia. Named *Hylonomus*, these reptiles were only about 24 cm (9.5 inches) from head to tail and had a body form resembling that of a salamander. Their skeletal remains are found in holes within the stumps of Pennsylvanian scale trees. In slightly younger Late Pennsylvanian rocks of Kansas, additional early reptiles have been recovered from claystones and siltstones of former mudflats. The most common of these reptiles is called *Petrolacosaurus*. *Petrolacosaurus* had a longer neck and longer legs than *Hylonomus*. Details of its skull indicate that it was an early member of the lineage that would eventually lead to dinosaurs.

The small and relatively primitive reptilian groups of the Pennsylvanian provided the stock from which a succession of more specialized Permian reptiles evolved. The most spectacular of these were the **pelycosaurs**, several species of which sported erect "sails" supported by rodlike extensions of the vertebrae. The pelycosaurs were a varied group. Some, such as *Edaphosaurus*, were plant eaters, whereas others, as indicated by their great jaws and sharp teeth, ate flesh. *Dimetrodon* (Figs. 11–78 and 11–79) was one such predator. There have been many attempts to explain the function of the pelycosaurian sail. The most reasonable explanation is that the sail-fin acted in temperature regulation by serving sometimes as a collector of solar heat and at other times as a radiator giving off surplus heat. Because pelycosaurs had several mammalian skeletal characteristics, they are thought to be the early mammal-like reptilian group from which advanced mammal-like reptiles known as **therapsids** arose. In this regard, it is interesting that they were attempting a kind of body temperature control.

For paleontologists, therapsids are among the most fascinating of fossil vertebrates. They were widely dispersed during Permian and Triassic time. There is a fabulous record of these reptiles and their contemporaries in the Karoo basin of South Africa and more modest discoveries in Russia, South America, Australia, India, and Antarctica. Therapsids were predominantly small to moderate-sized animals that displayed at least the beginnings of several mammalian skeletal traits. There were fewer bones in the skull than generally found in reptile skulls, and there

**Figure 11–78**   Permian reptiles. The sailback reptiles with the larger skulls and teeth are *Dimetrodon. Edaphosaurus,* a herbivorous form, is just right of center. The smaller lizardlike reptiles are *Casea. (Courtesy of Chicago Field Museum of Natural History; painting by C. R. Knight.)*

was a mammal-like enlargement of the lower jaw bone (dentary) at the expense of more posterior elements of the jaw. A double ball-and-socket articulation had evolved between the skull and neck. Teeth showed a primitive but distinct differentiation into incisors, canines, and cheek teeth. The limbs were swung more directly into vertical alignment beneath the body, and the ribs were reduced in the neck and lumbar region for greater overall flexibility. These features are well developed in the Permian therapsid *Cynognathus* (Fig. 11–80). The therapsids continued from the Permian through the Triassic, but became

**Figure 11–79**   Mounted skeleton of the Permian "sail-reptile" *Dimetrodon gigas.* The tail was actually somewhat longer. *(Courtesy of the Smithsonian Institution.)*

**Figure 11–80** Mammal-like reptiles. The scene depicts three carnivorous forms *(Cynognathus)* about to attack a plant-eating reptile *(Kannemeyeria) (Courtesy of Chicago Field Museum of Natural History; painting by C. R. Knight.)*

extinct early in the Jurassic. However, before they died out completely, they gave origin to the early mammals.

## Extinctions at the Close of the Paleozoic

The fossil record indicates that there has been a persistent increase in the numbers of different species through geologic time. It also shows that this trend has been upset or even reversed at times as a result of environmental adversity. Extinctions of large numbers of previously successful animals and plants may have resulted from their inability to adapt to changing conditions. Early geologists were able to identify such times of biologic crisis and used extinctions to define further the termination of the Paleozoic and the Mesozoic eras.

As discussed in the previous chapter, there were drastic geographic changes as the Paleozoic drew to a close. Pangea was experiencing mountain building and regional uplift, there was widespread volcanic activity, and seas had withdrawn from the extensive tracts they once covered. Climates became more severe, and a protracted glacial age occurred in the south. There is evidence of marked seasonal changes and aridity in many areas of the globe. As one or another group of animals gradually diminished in response to these changes, they contributed to the demise of others farther along in the food chain. Then, as now, the balance in the biologic environment could be easily upset, causing far-reaching waves of extinction.

Among the invertebrate groups that became extinct near the end of the Paleozoic were the trilobites, eurypterids, rugose and tabulate corals, productid brachiopods, many families of bryozoans, and blastoids. Entire families of other invertebrate taxa were lost as well. The archaic jawed fishes (acanthodians) became extinct during the Permian, whereas amphibians, reptiles, and several groups of plants declined conspicuously. The misfortunes of the invertebrates may have been associated with the drastic reduction of the inland seas, whereas the amphibians

may have suffered in the competition with evolving reptiles. Among those reptiles were a few hardy groups that were able to meet the new challenges and become the founders of the reptilian communities that were to dominate the Mesozoic terrestrial scene.

## SUMMARY

The fossil record of the Paleozoic is immensely better than that of the preceding Proterozoic. Shell building is one important reason for this improvement in the fossil record of invertebrates.

The earliest Paleozoic plants were mainly bacteria and algae, with the most obvious larger fossils being stromatolites and, by Ordovician time, the receptaculids. By Middle Silurian, the first land plants, the psilophytes, made their appearance. It is probable that they evolved from chlorophytes. Psilophytes were followed in the Devonian and Carboniferous by lycopsids, sphenopsids, and seed ferns. In temperate Southern Hemisphere continents, a flora dominated by *Glossopteris* developed. Permian forests included conifers and ginkgoes.

The ability to secrete hard parts had already begun in earliest Cambrian time, and this adaptation led to a rapid radiation of marine invertebrates, with archaeocyathids, brachiopods, and trilobites becoming abundant. By Ordovician time, however, all major invertebrate phyla had become common. The best record of soft-bodied invertebrates are those of the Middle Cambrian Burgess Shale fauna, which includes many previously unknown arthropods, echinoderms, sponges, and cnidarians, as well as an animal (*Pikaia*) believed to be the earliest known chordate. Among the other important invertebrate groups of the Paleozoic were foraminifers (fusulinids expanded during the Pennsylvanian and Permian), sponges, corals (tabulate and rugose), bryozoa, mollusks, echinoderms (especially blastoids and crinoids), and graptolites. The enigmatic conodonts serve as excellent index fossils for Paleozoic rocks.

The first uncontested fossils of vertebrate animals are found in rocks of Ordovician age. They are remains of a group of fishes known as ostracoderms.

Archaic jawed fishes appear late in the Silurian, but do not begin to dominate the marine realm until Devonian time. This was also the time when cartilagenous (chondrichthyian) and bony (osteichthyan) fishes began their rise to dominance. A special group of bony fishes called crossopterygian arose from osteichthyan stock. These fishes could breathe air by means of accessory lungs and possessed muscular fins that provided short-distance overland locomotion. The bones within the fins of crossopterygians resembled those of primitive amphibians. Other skeletal traits clearly suggest that these Devonian fishes were the ancestors of the first amphibians—the ichthyostegids. From a start provided by ichthyostegids, amphibians called labyrinthodonts underwent a successful adaptive radiation that lasted until the close of the Triassic. Long before their demise, however, they provided a lineage from which evolved the reptiles.

The transition from amphibian to reptile entailed a significant breakthrough in evolution. It involved the development of the amniotic egg, a biologic device that liberated land vertebrates from the need to return to water bodies to reproduce. The first known reptiles are Pennsylvanian in age and include such forms as *Hylonomus* and *Petrolacosaurus*. During very late Pennsylvanian and Permian, reptiles underwent an elaborate evolutionary radiation that produced the fin-back reptiles and mammal-like reptiles, as well as several other major reptilian groups.

The end of the Paleozoic is marked by a major episode of extinction. Near the end of the Permian period, nearly half of the known families of animals disappeared. The decimation of marine animals was particularly dramatic and included the loss of fusulinids, spiny productid brachiopods, rugose corals, two orders of bryozoans, and many taxa of stemmed echinoderms, including blastoids. Among the vertebrates, over 70 percent of amphibian and reptilian families perished. Fortunately, a few groups survived the Permian crisis and were able to continue their evolution during the Triassic. The cause of the Permian biologic crisis is not definitely known. Many believe it was the result of a high stand of Permian continents, reduction of epeiric seas, mountain building, and sharp climatic changes resulting from all of these factors.

## QUESTIONS FOR REVIEW

1. What group of algae were the probable ancestors of the first land plants? What evidence supports this interpretation?

2. Why was the evolution of a vascular system critical to the conquest of the continents by land plants?

3. In terms of location, climatic preference, and reproduction, how did *Lepidodendron* differ from *Glossopteris?*

4. What does the fauna of the Burgess Shale reveal about the bias in the fossil record of early Paleozoic life?

5. What is the evolutionary significance of the Burgess Shale fauna named *Pikaia?*

6. To what geologic system would rocks containing the following fossils be assigned? (a) archaeocyathids, (b) fusulinids, (c) *Archimedes.*

7. The phylum Cnidaria was formerly termed Coelenterata. Why is Cnidaria an appropriate name for these animals?

8. How do the sutures of nautiloids differ from those of ammonoid goniatites?

9. What is the difference between the meaning of the terms *chordate* and *vertebrate?* What phylum of invertebrates has been considered ancestral to the earliest chordates? Why?

10. What are conodonts? What other animal group has hard tissue (tooth or bone) of similar composition?

11. From what previous anatomic structure was the vertebrate jaw derived?

12. List three characteristics of therapsid reptiles that indicate they were on the main line of evolution toward mammals.

## TERMS TO REMEMBER

| | | | |
|---|---|---|---|
| acanthodians | Dipnoi | labyrinthodont | Porifera |
| Ammonoidea | echinoderm | lophophore | psilophyte |
| archaeocyathid | edrioasteroid | lycopsid | radiolarian |
| arthropod | epifaunal | Mollusca | receptaculid |
| blastoid | eurypterid | monoplacophoran | rugose coral |
| brachiopod | fenestellid bryozoan | Nautiloidea | septa |
| cephalopod | foraminifer | notochord | sphenopsid |
| chlorophyte | fusulinid | ostracod | stromatoporoids |
| Cnidaria | gastropod | ostracoderm | tabulae |
| conodont | goniatite | pelycosaur | tabulate corals |
| crinoid | graptolite | pelecypod | theca |
| Crossopterygii | ichthyostegid | placoderm | therapsid |
| cystoid | infaunal | placophoran | trilobites |

## SUPPLEMENTAL READINGS AND REFERENCES

Clarkson, E. N. K. 1979. *Invertebrate Paleontology and Evolution*. London: Allen & Unwin.

Conway, M. S. 1987. The search for the Precambrian–Cambrian boundary. *Amer. Scientist* 75:157–167.

Cowen, R. 1990. *History of Life*. Boston: Blackwell Scientific Publications.

Gould, S. J. 1989. *Wonderful Life*. New York: W. W. Norton Co.

Stearn, C. W., and Carroll, R. L. 1989. *Paleontology: The Record of Life*. New York: John Wiley.

Whittington, H. B. 1985. *The Burgess Shale*. New Haven: Yale Univ. Press.

The Egyptian Temple, Capitol Reef National Monument, Utah. The massive rock at the top is Shinarump Formation, beneath which are banded beds of the Moenkopi Formation. *(Courtesy U.S. Geological Survey, L.C. Huff photograph.)*

# The Mesozoic Era

*The earth, from the time of the chalk to the present day, has been the theater of a series of changes as vast in their amount as they were slow in their progress. The area on which we stand has been first sea and then land for at least four alternations and has remained in each of these conditions for a period of great length.*

Thomas Huxley, 1868, *On a Piece of Chalk*

## Introduction

The extinction of Paleozoic animals and plants that occurred at the close of the Permian Period did not go unnoticed by the founders of geology. The fossil evidence for this time of crisis provided a natural boundary that was used to mark the end of the Paleozoic sequence. Overlying younger strata contained different and generally more progressive organisms, yet not as modern as those that live today. Thus, it seemed appropriate to formulate a middle chapter in the history of life. They named that chapter the Mesozoic Era. In its turn, the Mesozoic also ended in a biologic crisis that marks its separation from the youngest era of all—the Cenozoic.

The Mesozoic Era lasted an estimated 160 million years, ending approximately 65 million years ago. The three periods of the Mesozoic are unequal in duration. The Triassic lasted about 30 million years, the Jurassic 55 million years, and the Cretaceous about 75 million years.

During the lengthy span of the Mesozoic, many new families of plants and animals evolved and experienced often spectacular radiations. It was the era in which two new vertebrate classes, the birds and the mammals, first appeared. It was also a time in which the supercontinent Pangea, which had formed in the previous era by the joining of ancestral continents, was gradually dismembered. As the continents moved slowly apart, the oceanic rifts between them deepened and broadened. A process of fragmentation and drift had begun that would ultimately lead to the present physical geography of the planet.

## The Breakup of Pangea

Any discussion of the history of the Mesozoic Era must have Pangea as its starting point (see Fig. 10-2). The same forces that had drawn the plate segments together to form this great supercontinent were still in operation at the beginning of the Mesozoic; however, they now moved in different directions or had shifted their locations. In general, the dismemberment of Pangea seems to have occurred in four stages. The first stage began in the Triassic, with rifting and volcanism along normal fault systems. Events such as these result when a region of the lithosphere is subjected to tensional stresses. In this instance, the forces were associated with the separation of North America and Gondwanaland. As the rifting progressed, Mexico was decoupled from South America and the eastern border of North America parted from the Moroccan bulge of Africa (Fig. 12–1). Oceanic basalts were added to the sea floor of the newly formed and gradually widening

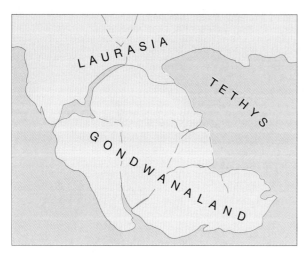

**Figure 12–1** Paleogeographic reconstruction of the world about 180 million years ago when the breakup of Pangea was beginning.

Atlantic Ocean. The largely tensional geologic structures, as well as the volcanic and clastic sedimentary rocks in the Triassic System of both the eastern United States and Morocco exhibit many striking similarities. These now widely separated regions were on opposite sides of the same axis of spreading and were affected by similar forces and events.

Recently, geologists have used a geophysical technique called *seismic reflection profiling* to locate the 300-million-year-old zone of convergence, or **suture zone,** formed by the convergence of Africa and North America. The zone, which trends roughly east–southeast across southern Georgia, was left behind when the two continents subsequently parted company about 190 million years ago. The discovery was particularly interesting to paleontologists, who had postulated the convergence and later separation of the two continents from fossil evidence.

Throughout Triassic time, Laurasia north of the Maritime Provinces remained intact. Sea floor spreading was apparently more rapid in the central and northerly region of the newly opening Atlantic. It also appears that South America maintained its hold on Africa throughout most of the Triassic.

The second stage in the breakup of Pangea involved the rifting and opening of narrow oceanic tracts between southern Africa and Antarctica. This rift extended northeastward between Africa and India and developed a branch that separated north-ward-bound India from the yet unseparated Antarctica–Australia land mass. The rifting was accompanied by outpouring of great volumes of basaltic lavas.

In stage three of the breakup of Pangea, the Atlantic rift began to extend itself northward, and clockwise rotation of Eurasia tended to close the eastern end of the Tethys Sea, a forerunner of the Mediterranean. By the end of Jurassic time, an incipient breach began to split South America away from Africa. The cleft apparently worked its way up from the south, creating a long seaway somewhat reminiscent of the present Red Sea. Australia and Antarctica remained intact, but India had moved well along on its long voyage to Laurasia. By Late Cretaceous time, about 70 million years ago, South America had completely separated from Africa, and Greenland began to separate from Europe. However, northeasternmost North America was still attached to Greenland and northern Europe.

The fourth stage in the dismemberment of Pangea was not a Mesozoic event. It occurred early in the Cenozoic, during which time the North Atlantic rift slowly penetrated northward until eventually the North American and Eurasian continents were completely separated. Also during this final era of earth history, Antarctica and Australia separated. This final separation of continents occurred only about 45 million years ago, with the total span of time needed for the fragmentation of Pangea being about 150 million years.

## The Mesozoic History of North America

### To the East and South

#### Triassic and Jurassic

At the beginning of the Mesozoic, conditions in the eastern part of North America were much the same as they had been near the end of the Paleozoic. The rugged Appalachian ranges that had been raised during the Allegheny orogeny were undergoing vigorous erosion. The coarse clastics derived from the uplands filled intermontane basins and other low areas during the Early and Middle Triassic. Then, during the Late Triassic and Early Jurassic, North America began to experience the tensional forces that precede actual separation of continents. As a result, a series of fault-bounded troughs developed along the eastern coast from Nova Scotia to North Carolina (Figs. 12–2 and 12–3). The downfaulted blocks received great thicknesses of poorly sorted arkosic red sandstones and shales. These sediments are nonmarine and indicate that insufficient separation had occurred during the Late Triassic to allow ingress of marine waters. Rocks deposited in the downfaulted basins constitute the **Newark Group** of Late Triassic and Early Jurassic age.

The poor sorting of the coarser Newark clastics and their high content of relatively unweathered feldspar grains indicate transportation and deposition by streams flowing swiftly down from the granitic highlands that bordered the fault basins (Fig. 12–4). In some places, drainage in the basins became impounded and lakes formed. These lake deposits contain the remains of freshwater crustaceans and fish. Ripple marks, mud cracks, raindrop impressions, and even the footprints of early members of the dinosaur line (Fig. 12–5) are frequently found on the lithified surfaces of Newark sediments. Lavas, rising along steeply dipping fault surfaces, flowed out upon the newly deposited sediment, and volcanoes spewed ash onto the surrounding hills and plains.

Three particularly extensive lava flows and an imposing sill are included within the Newark Group in the New Jersey–New York area. The exposed edge of a vertically jointed sill forms the well-known Palisades of the Hudson River (Fig. 12–6). Radiometric dates obtained for the Palisades basalt indicate that it solidified about 190 million years ago. Although most of the Newark beds are Triassic, recent spore and pollen studies suggest that Newark rocks in some areas accumulated during the Early Jurassic as well.

By Late Triassic time, the fault block mountains that had been produced as a result of rifting of Laurussia had been severely reduced by erosion. The topography was further reduced during the Jurassic and Early Cretaceous until only a broad, low-lying erosional surface remained. That old surface is called the Fall Line surface because its profile can best be seen along the Fall Line, which separates more resistant rocks of the Piedmont from the softer rocks of the Atlantic coastal plain. Beneath the Fall Line surface are Triassic faulted sediments and older crystalline rocks, and above it lie the more or less flat-lying

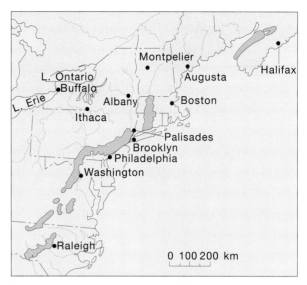

**Figure 12–2** Outcrop areas of Triassic rocks in eastern North America. Green areas show troughlike deposits of Late Triassic age.

Cretaceous sediments of the Atlantic coastal plain.

South of the old Appalachian–Ouachita tracts, a new depositional province, the Gulf of Mexico, began to take form. The initial deposits were mostly evaporites and do not lend themselves to radiometric dating. However, analyses of pollen from a sample of salt taken from this sequence suggest an age of Late Triassic. Evaporites continued to be deposited well into the Jurassic, indicating aridity in the Gulf region (Fig. 12–7). Apparently, the Gulf of Mexico was rather like a great evaporating basin, concentrating the waters of the Atlantic Ocean and precipitating salt and gypsum to thicknesses exceeding 1000 meters (3280 feet). The Jurassic evaporite beds are the source of the salt domes of the Gulf Coast. Salt domes are economically important structures associated with the entrapment of petroleum. When compressed by a heavy load of overlying strata, salt tends to flow plastically. As it moves upward in the direction of lesser pressures, the salt bends and faults overlying strata. The permeable, often faulted beds that slope downward away from the salt, or are arched over the dome, afford excellent sites for petroleum accumulation (see Fig. 1–48).

Evaporative conditions abated later in the Jurassic, and several hundred meters of normal marine

**Figure 12–3**  Generalized paleogeographic map for the Triassic of North America.

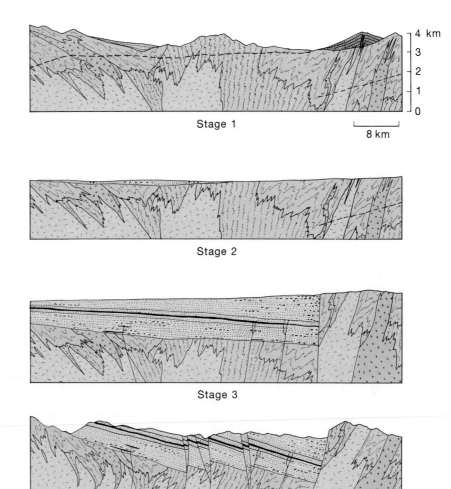

Stage 1

8 km

Stage 2

Stage 3

Stage 4

**Figure 12–4** Four stages in the Triassic history of the Connecticut Valley. (Stage 1) Erosion of the complex structures developed during the Allegheny orogeny. (Stage 2) Mountains have been eroded to a low plain, and Triassic sedimentation has begun. (Stage 3) Newark sediments and basaltic sills, flows, and dikes accumulate in troughs resulting from faulting. (Stage 4) In Late Triassic time, the area is broken into a complex of normal faults as part of the Palisades orogeny. *(Modified from the classic 1915 paper by J. Barrell, Central Connecticut in the Geologic Past, Conn. Geol. Nat. Hist. Bull., No. 23.)*

**Figure 12–5** Slab of Newark Group sandstone containing the cast of a footprint made by a three-toed Triassic reptile. The footprint is about 20 cm long. The linear ridges in the slab are casts of mudcracks.

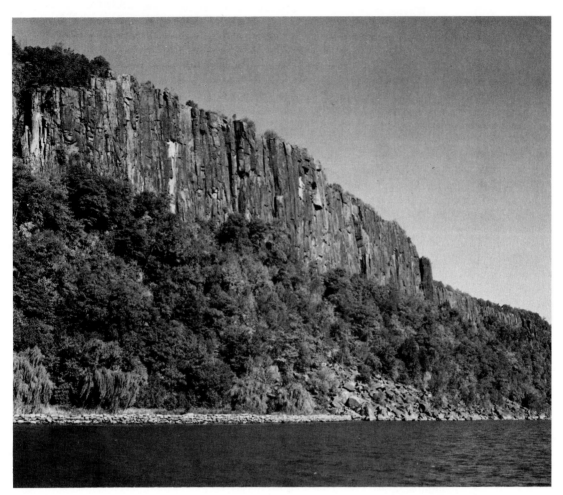

**Figure 12–6** Palisades of the Hudson River. (Photograph courtesy of Palisades Interstate Park Commission.)

limestones, limy muds, shales, and sandstones accumulated in the alternately transgressing and regressing seas of the Gulf embayment. Although there are surface exposures of these rocks where they extend into northeastern Mexico, in the United States they are deeply buried beneath a thick cover of Cretaceous and Cenozoic sediments (Fig. 12–8). Were it not for the drilling activities of oil companies, the nature of these rocks could only be inferred from geophysical data.

## Cretaceous

The Cretaceous was a time of great marine inundations of land masses, and the Atlantic and Gulf coastal regions received their full share of flooding.

Early in Cretaceous time, the Atlantic coastal plain, which had been experiencing erosion since the beginning of the Mesozoic, began to subside. At about the same time, the Appalachian belt that lay to the west was elevated. On the subsiding coastal plains, alternate layers of marine and deltaic terrestrial deposits accumulated and were gradually built into a great wedge of sediments that thickened seaward (Fig. 12–9). Today, the thinner eastern border of the wedge is narrowly exposed in New Jersey, Maryland, Virginia, and the Carolinas. But the greatest volume of Cretaceous sediments was deposited farther east, along the present continental shelf.

To the south, the Florida region was a shallow submarine bank during the Cretaceous. The oldest strata consist of limestones, but later in the period,

**Figure 12–7** Generalized paleogeographic map for the Jurassic of North America.

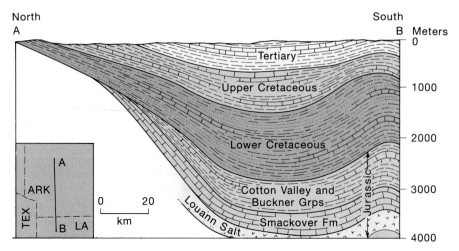

**Figure 12–8** Jurassic rocks in the Gulf Coast region are deeply buried beneath Tertiary and Cretaceous rocks. *(Adapted from* Stratigraphic Atlas of North and Central America, *Shell Oil Co.)*

terrigenous clastics were brought into the area by streams flowing from the southern Appalachians. These source areas were gradually worn down, and carbonate deposition prevailed.

The Cretaceous is noteworthy as a period of the Mesozoic in which carbonate reefs were particularly extensive in the warmer regions of both the Eastern and Western Hemispheres. Among the invertebrates that contributed their skeletal substance to these reefs, a group of pelecypods called **rudists** were immensely important. In many Cretaceous reefs, the shells of these creatures (Fig. 12–10) form the basic framework of the reef. Because of their high porosity and permeability, rudistid reefs are reservoir rocks

for some of the world's greatest Cretaceous oil accumulations.

Following the withdrawal of Jurassic seas, the region northwest of Florida consisted of low-lying coastal plains. However, early in the Cretaceous, these relatively level lands were invaded by marine waters moving northward from the ancestral Gulf of Mexico. Nearshore deposits such as sandstones are overlain by finer sedimentary rocks that are characteristic of deeper water and provide clear evidence of the northward migration of the shoreline. The advance of the Cretaceous sea was not uniform. An extensive regression occurred near the end of the first half of the period; however, flooding resumed in

**Figure 12–9** Section across the Atlantic coastal plain from the vicinity of the Fall Line to the edge of the continental shelf. Data based on deep drilling. *(From King, P. B.* The Tectonics of Middle North America. *Princeton N.J.: Princeton Univ. Press. 1951, p. 165.)*

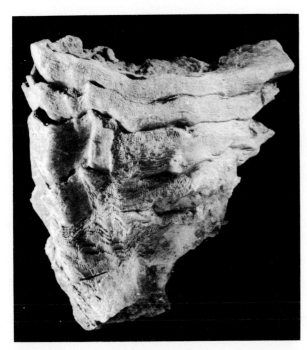

**Figure 12–10** *Eoradiolites davidsoni,* a Cretaceous rudistoid pelecypod from Texas. Rudistoids appeared in the Jurassic and proliferated during the Cretaceous when they were important reef-making organisms. They bore a close resemblance to some Paleozoic corals. This specimen is 8.6 cm tall.

Late Cretaceous time when a wide seaway occupied a tract from the Gulf of Mexico to the Arctic Ocean (Fig. 12–11). Among the Upper Cretaceous formations deposited in this inland sea, chalk was particularly prevalent. Chalk is a white, fine-grained, soft variety of limestone that is composed mainly of the microscopic calcareous platelets (called *coccoliths*) of golden brown algae. It is common among beds of Cretaceous age in many other parts of the world (Fig. 12–12). Indeed, the Cretaceous System takes its name from *creta,* the Latin word for chalk.

## To the West

As the eastern border of North America was experiencing the tensional effects caused by separation from Europe and Africa, compressional forces prevailed in the west. While the newly formed Atlantic Ocean was widening, North America moved westward, overriding the Pacific plate. Thus, deformation of western North America is related to events in the east. It has been shown, for example, that the pace of tectonic activity in the North American Cordillera was most intense during the time when sea floor spreading was most rapid in the Atlantic.

## Accretionary Tectonics

During the Mesozoic, a steeply dipping subduction zone existed along the western margin of North America. This was not a relatively simple Andean-type subduction zone, however, for the advancing Pacific plate carried considerably more than ocean basalts and sea floor sediments. Entire sections of volcanic arcs, fragments of distant continents (the microcontinents described earlier), and pieces of oceanic plateaus were carried to North America's western margin as well. These fragments, now incorporated into the Cordilleran belt, constitute **displaced** or **allochthonous terranes** described briefly in Chapter 6. The better documented of these terranes can be identified by their distinctive age, rock assemblages, and mineral resources that indicate the part of the world from which the displaced fragments came. Many of the fragments have one or more fault boundaries.

Altogether, more than 50 displaced terranes have been recognized in the Cordillera. They may constitute as much as 70 percent of the total Cordilleran region. Thus, the growth of North America has occurred not only by accretion of materials along subduction zones, but also by incorporation of huge crustal fragments formed elsewhere and subsequently conveyed to North America by sea floor spreading. Some of these fragments were evidently so large that they caused changes in subduction zone configuration. Others composed of low-density rocks were too buoyant for subduction. These were scraped off the subducting plate, splintered, and thrust onto the continental margin. In contrast to subduction, the process whereby one rock mass rides up and over another is call **obduction.** Stated differently, in subduction, one plate stays at a constant level and the other plate is forced downward under it, whereas in obduction, one plate remains at a constant level, while the other plate is forced upward over it.

The growth of a continent by the progressive incorporation of crustal fragments in the manner described above is termed **accretionary tectonics.** In an oversimplified analogy to accretionary tecton-

**Figure 12–11** Generalized paleogeographic map for the Cretaceous of North America.

B

**Figure 12–12**    (A) Sea cliffs composed of chalk along the Dorset coast of England. (B) Close-up view of the chalk. The dark nodules are flint, a gray or black variety of chert.

A

ics, one can imagine icebergs in an ocean being driven onshore by powerful winds. Each crashes into the coast and is then struck from behind by subsequent arrivals until a jumbled region of displaced icebergs is constructed. In geology, an analogous region of displaced terranes is called a **tectonic collage.**

## Triassic

The Cordilleran region as it existed during the Mesozoic can be divided into a western belt containing thick volcanic and siliceous deposits and a wide eastern tract adjacent to the more stable interior of the continent. Nearly 800 meters of clastic sediments and volcanics exposed in southwestern Nevada and southeastern California attest to the instability of the western zone. Geologists speculate that the belt may well have resembled the Indonesian island arc of today.

The initial orogenic event of the Cordilleran region is difficult to date precisely, although evidence for deformation at or near the Permian–Triassic boundary can be found from Alaska to Nevada. In the United States, the event has been named the **Sonoma orogeny.** Its effects can be studied in west-central Nevada. The Sonoma orogeny occurred when an eastward-moving volcanic arc, called the *Golconda arc,* collided with the Pacific margin of North America. At that time, North America was bordered by a westward-dipping subduction zone. As a result of the collision, the island arc and its adjacent accretionary wedge of sediment was welded onto the western edge of North America, adding as much as 300 km to the continent's east–west dimension. Oceanic rocks were thrust eastward onto the eroded structures of the old Antler orogen (Fig. 12–13). The larger of the thrust sheets moved scores of kilometers eastward and were themselves broken into many smaller internal faults, with each fault slice bringing a repeated stratigraphic sequence to the surface.

The Triassic rocks of the far western part of the Cordillera include great thicknesses of volcanics and graywackes presumably derived from the island arc. It is uncertain, however, whether these rocks were deposited where they are now found or whether they were part of a displaced terrane, called the **Sonoma terrane**, that actually originated in some unknown part of the Pacific.

Early Triassic rocks of the eastern Cordillera consist of sandstones and limestones of mostly shallow marine origin. The thickest section of these marine strata occurs in southeastern Idaho, where nearly 1000 meters of Lower Triassic sediments ac-

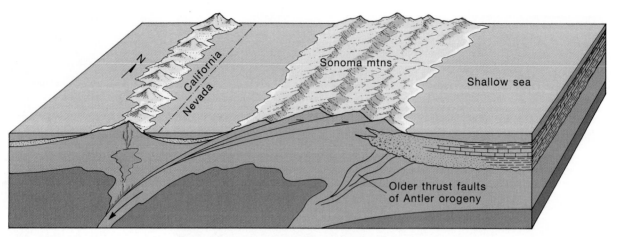

**Figure 12–13** Tectonic conditions and paleogeography along the Cordilleran region during the Late Permian and Early Triassic.

**Figure 12–14** Generalized geologic section of Upper Triassic and Lower Jurassic sedimentary rocks of central Utah.

cumulated. Eastward from the Cordillera, these marine beds interfinger with continental red beds. Although Triassic seas remained in Canada during the Middle Triassic, in the United States they regressed westward, leaving vast areas of the former sea floor subject to erosion.

Upper Triassic formations rest on the unconformity resulting from this erosion. The Upper Triassic series consists mostly of continental deposits transported by rivers flowing westward across an immense alluvial plain. There were also upland source areas in western and southern Nevada, Arizona, and California. For example, the lowermost strata consist of the sandy **Moenkopi Formation**, followed by the pebbly **Shinarump Formation** (Figs. 12–14 and 12–15), derived from uplifted neighboring areas in Arizona, western Colorado, and Idaho. Above the Shinarump lie the vividly colored shales, siltstones, and sandstones of the **Chinle** and **Kayenta** formations (Fig. 12–16). The sediments of these two formations were deposited in stream valleys and lakes. They alternate with sand dune accumulations of the **Navajo Sandstone** (Fig. 12–17) and **Wingate Sandstone** (Fig. 12–18). These formations are beautifully exposed in the walls of Zion Canyon in southern Utah. They display sweeping cross-bedding, such as is formed in sandy deposits transported by winds. (However, such so-called festoon cross-bedding is not always caused by winds, for submarine dunes may be very similar in form.)

The Painted Desert of Arizona is developed mostly in Chinle rocks. The formation is known

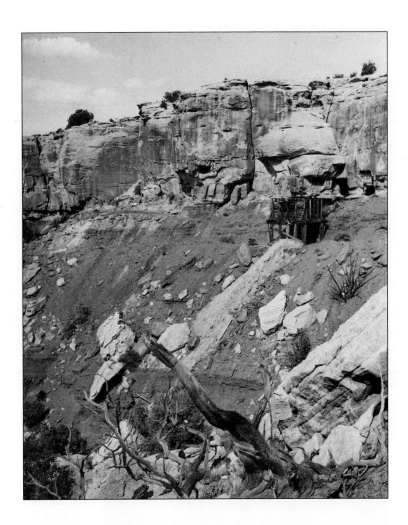

**Figure 12–15** Opening to small uranium mine in Shinarump Formation (massive bed), Colt Mesa Copper Mine, Utah. Slope debris covers Moenkopi Formation. *(Photograph courtesy of the Utah Geological and Mineral Survey.)*

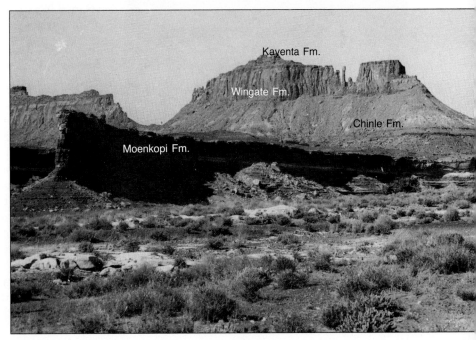

**Figure 12–16** A typical scene in the Orange Cliffs area of Utah. The high cliffs along the horizon are Wingate Sandstone. They are capped by beds of Kayenta visible near the top center. The slope immediately below the cliffs is eroded in the Chinle Formation. The ridge in the foreground is capped by the Lower Triassic Moenkopi Formation. *(Photograph courtesy of the Utah Geological and Mineral Survey.)*

Carmel Fm.

Navajo Fm.

Chinle Fm.

Shinarump

Moenkopi Fm.

**Figure 12–17** Panoramic view of Triassic and Jurassic formations in Zion National Park, Utah. The lowermost slope is formed by shales of the Moenkopi Formation. Above that slope is a prominent ledge of Shinarump sandstones. The next slope above the Shinarump is the one formed in Chinle Formation. The towering cliffs at the top are Navajo and Carmel formations. *(Photograph courtesy of the U.S. National Park Service.)*

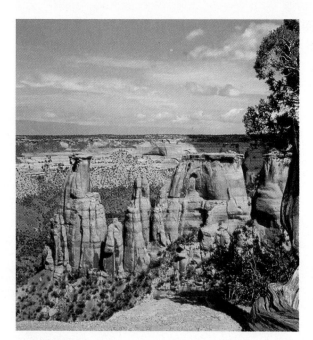

**Figure 12–18** The erosional features in the foreground are sculpted from Wingate Sandstone, beneath which lies the red Chinle Formation. The cliffs in the far background are also Wingate Sandstone. Colorado National Monument near Grand Junction, Colorado. *(Courtesy of R. J. Weimer.)*

throughout the world for the petrified logs of conifers it contains. Each year, thousands of tourists examine these logs, now turned to colorful agate, in Petrified Forest National Monument (Fig. 12–19). Apparently, during times of Triassic floods, the trees were left on sandbars or trapped in log jams and covered by sediment. Percolating solutions of underground water subsequently replaced the wood with silica.

## Jurassic to Early Tertiary Tectonics

Most of the orogenic activity in the Cordillera during the Mesozoic was a result of the continuing eastward subduction of oceanic lithosphere beneath the continental crust of the North American plate (see Fig. 12–21). That subduction varied in rate, in inclination, and, to a small degree, in direction. It resulted in eastward-shifting phases of deformation, which initially affected the far western part of the Cordillera, and then proceeded eastward to finally reach the margin of the craton.

The deformational and magmatic activity associated with the western tract is sometimes termed the **Nevadan orogeny**. During the Triassic, and increasing in frequency during the Jurassic and Cretaceous, graywackes, mudstones, cherts, and volcanics that

**Figure 12–19**   Petrified logs, Petrified National Monument, Arizona. The petrified logs and wood fragments are *Araucarioxylon*. They have been weathered from the Chinle Formation of Triassic age. *(Courtesy of Lehi F. Hintze.)*

**Figure 12–20**   Mélange that is within the Franciscan fold belt of northern California. The exposure is north of Crescent City, California. Width of photographed area is 2.4 meters.

had been swept into the subduction zone were severely folded, faulted, and metamorphosed. The intensely crumpled and altered rock sequences that were affected by compression between the converging plates are appropriately termed *mélange*, which means a jumble (Fig. 12–20). The Franciscan fold

**Figure 12–21** Greenstones of the Franciscan Formation exposed at Lime Point, Marin Peninsula, California. The west side of the Golden Gate Bridge tower rests on rocks of the Franciscan Formation. *(Courtesy of U.S. Geological Survey.)*

**Figure 12–22** Granodiorite of the Sierra Nevada batholith grandly exposed in Yosemite Valley, California.

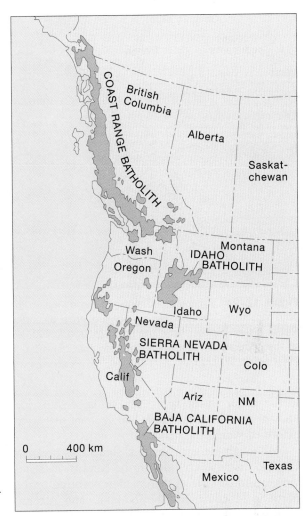

**Figure 12–23** Mesozoic batholiths in west–central North America.

belt of California (Fig. 12–21) provides a good example of a mélange. In addition to the deformation of the sedimentary rock sequences, enormous volumes of granodiorite were generated above the subduction zone and intruded overlying rocks repeatedly during the Jurassic and Cretaceous. The Sierra Nevada (Fig. 12–22), Idaho, and Coast Range batholiths are impressive evidence of the vast scale of this Nevadan magmatic activity (Fig. 12–23).

Somewhat before the Sierra Nevada batholith was emplaced during the Cretaceous, another series of deformations had begun east of the present day Sierra Nevada. This second phase of the tectonic development of the Cordillera primarily affected shallow water carbonates and terrigenous clastics deposited over a time span from Middle Jurassic to earliest Cenozoic. This new orogeny has been named the **Sevier**. During the Sevier orogeny, strata were sheared from underlying Precambrian rocks and broken along parallel planes of weakness to form multiple, imbricated, low-angle thrust faults (Fig. 12–24). The French word **décollement** (unsticking) has been used to describe this kind of structure, in which older rocks are thrust upon younger in multiple, nearly parallel slabs of crust. It has been estimated that the compressional structures produced during the Sevier orogeny resulted in over 100 km of crustal shortening in the Nevada–Utah region. In addition

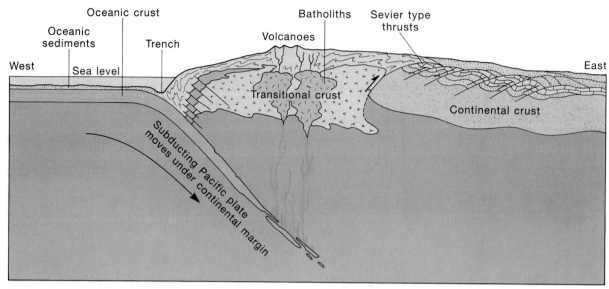

**Figure 12–24**   An advanced stage in the evolution of the North American Cordillera, with structures developing as a consequence of underthrusting of the continent by the Pacific oceanic plate. Note the multiple, imbricated, low-angle thrusts on the east side of the section. *(The diagram is simplified from Dewey, J. F., and Bird, J. M. J. Geophys. Res. 75(14):2638, 1970.)*

to the major thrust faults, several large folds are known, as well as intricately folded strata that were deformed within the major thrust units.

Although the term Sevier is sometimes reserved for deformational episodes in the Nevada–Utah region, a similar style of deformation occurred to the north in Montana, British Columbia, and Alberta. Each of the major ranges in this region is a fault block composed of Paleozoic strata of the shelf that have been thrust toward the east along westwardly dipping fault surfaces. The most famous of these faults is the Lewis thrust (Fig. 12–25), along which Proterozoic rocks of the Belt Supergroup were carried 65 km toward the east.

Magmatic activity along the far western edge of the North American plate had diminished near the end of the Cretaceous, and by early Tertiary time, much of the major thrusting of the marginal shelf region had subsided. Once again, deformational events shifted eastward to the cratonic region where the Rocky Mountains of New Mexico, Colorado, and Wyoming are now located. These more eastward disturbances are called the **Laramide orogeny.** Although thrust faults do occur in the region of Laramide activity, the more characteristic kind of structures are broadly arched domes, basins, and

anticlines. Strata composing the domes and anticlines are draped over central masses of Precambrian igneous and metamorphic rocks. In several instances, erosion has stripped away the cover of strata, exposing the central crystalline mass. Resistant layers of inclined beds that surround the central cores stand as mountains today. Many of these up-arched areas appear to have been produced by movements along underlying faults in the basement rocks. It has been suggested that these controlling structures were in turn developed by drag when the eastward-moving subducted oceanic lithosphere scraped along the sole of the cratonic margin of North America.

Most of the structures of the present Rocky Mountains are the result of Laramide phases of orogeny. The landscape we see today, however, is the result of repeated episodes of Cenozoic erosion and uplift. Erosion, acting on the geologic structures already present, sculpted the final scenic design.

## Jurassic Sedimentation

During the Jurassic, when the mélange was being formed in the subduction zone complex, depositional sites inland experienced a less dramatic kind

**Figure 12–25** Northeast–southwest cross-section showing the Lewis and Eldorado thrust faults at a location about 66 km south of Glacier National Park. Formations shown in green are Proterozoic in age. Tan color indicates Paleozoic and Mesozoic rocks. *(After Mudge, M. R., and Earhart, R. L. U.S. Geological Survey Prof. Paper, No. 1174, 1980.)*

of sedimentation. Early Jurassic deposits consist of clean sandstones, such as the **Navajo Sandstone**. Large-scale cross-bedding is well developed in the Navajo (Fig. 12–26). However, thin beds of fossiliferous limestone and evaporites occur locally and indicate that at least some parts of the formation were deposited in water. Indeed, recent interpretations suggest that much of the cross-bedding in the Navajo resulted from water currents that formed sand dunes on the floors of shallow marine areas. The Navajo and associated sand bodies were probably deposited in a nearshore environment, and it is

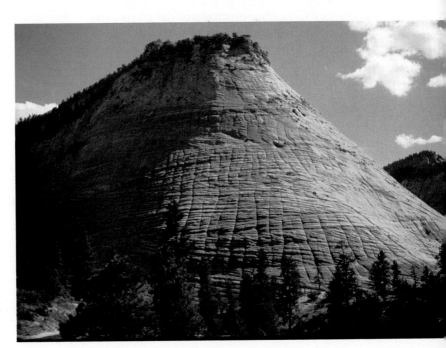

**Figure 12–26** Cross-stratification in the Navajo Sandstone, Checkerboard Mesa, Zion National Park, Utah. *(Courtesy of Lehi F. Hintze.)*

**Figure 12–27** Paleogeographic map for the Early Jurassic of the western United States, showing general extent of sea and land as well as paleolatitudes. *(From Stanley, K. O., Jordan, W. M., and Dott, R. H.* Bull. Am. Assoc. Petrol. Geol. *55(1):13, 1971.)*

**Figure 12–28** Region in the western North America inundated by the Middle Jurassic Sundance sea. (Land areas are shown in tan, marine areas in blue.)

likely that some of the deposits were part of a coastal dune environment (Fig. 12–27). They do not seem to have been laid down on the floors of a vast interior desert, as has often been postulated. Judging from studies of cross-bedding orientations, the source area for these clean sandstones was probably the Montana–Alberta craton. Somewhat older but similar quartz sandstones were recycled and spread southward into Wyoming and Utah, and then westward across Nevada.

Marine conditions became more widespread in the Middle Jurassic, when the entire west–central part of the continent was flooded by a wide seaway that extended well into central Utah (Fig. 12–28). This great embayment has been dubbed the **Sundance sea.** Within the Sundance sea were deposited the sands and silts of the Sundance Formation, famous for its fossil content of Jurassic marine reptiles. Sediment for the Sundance Formation and overlying younger Jurassic rock units was derived from the Cordilleran highlands that lay to the west. These highlands continued to grow throughout the Jurassic, ultimately extending from Central America to Alaska. They provided so great a volume of sediment

that the Sundance sea was gradually silted up, and what had been an inland sea was transformed into a vast, swampy plain across which meandering rivers built floodplains. These deposits compose the **Morrison Formation** (Fig. 12–29), which extends across millions of square kilometers of the American West. Enclosed within these floodplain deposits are the bones of more than 70 species of dinosaurs, including some of the largest land animals ever to have existed.

## Cretaceous Sedimentation

During the Cretaceous, the Pacific border region of North America was a land of lofty mountains formed during the orogenesis that had begun in the Jurassic. Erosion of these ranges brought sediments downward into adjoining rapidly subsiding basins, many of which were open to the Pacific Ocean. In some places, over 15,000 meters of volcanics and clastics

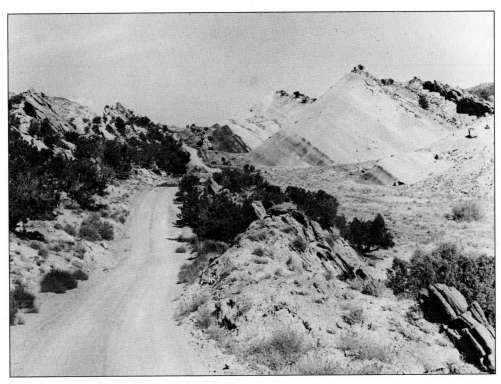

**Figure 12–29** Morrison Formation exposed along the east flank of Circle Cliffs, Garfield County, Utah. *(Photograph courtesy of the Utah Geological and Mineral Survey.)*

accumulated. Cycles of folding and intrusion occurred repeatedly, with rather exceptional deformations taking place during middle and then Late Cretaceous time.

In the more eastward regions of the Cordillera, the Cretaceous began with the advance of marine waters both northward from the ancestral Gulf of Mexico and southward from arctic Canada. These Early Cretaceous invasions did not meet, so that an area of dry land existed in Utah and Colorado. Their advance was reversed by a general regression that resulted in the unconformity used to separate the Cretaceous into an early and a late division. The flooding that followed in the Late Cretaceous was the greatest of the entire Mesozoic Era. This time, the embayment from the north joined with the southern seaway and effectively separated North America into two large islands.

Sedimentation in Cretaceous epicontinental seas was largely controlled by local conditions. Along the Gulf Coast, limestones and marls (clayey limestones) accumulated (Fig. 12–30). North of Texas, however, the great bulk of sediments consisted of terrigenous clastics supplied by streams flowing from the Cordilleran highlands.

In terms of plate tectonics, the depositional basin that received these sediments (as well as smaller amounts from the craton) can be termed a *back-arc basin* (Fig. 12–31). The back-arc basin of the western interior of North America was immense. It extended from the Arctic to the Gulf of California, and in places was over 1600 km wide. It was bounded on the west mostly by fold and thrust belts such as the Sevier, and on the east by the craton. Many of the rock formations in the basin yield commercial quantities of oil, natural gas, and coal. Exploration for these resources has provided hundreds of well logs and detailed surface studies and has permitted geologists to recognize and map Cretaceous rocks of piedmont, coastal plains, shorelines, marine shelf areas, and deep water tracts of the central basin. The picture that has emerged shows coarse, terrigenous clastics immediately adjacent to the Cordilleran highlands. These interfinger with coal-bearing conti-

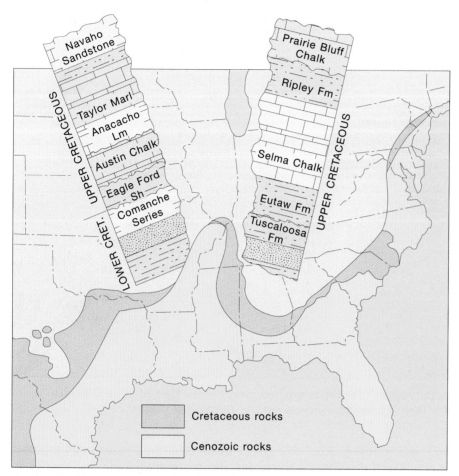

**Figure 12–30** Area of outcrop of Cretaceous limestone and marl rocks in the Atlantic and Gulf coastal plains, and generalized columnar sections of the Texas (left) and Alabama (right) Cretaceous.

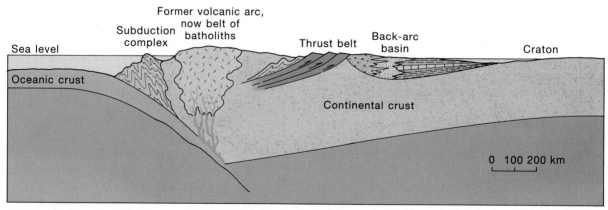

**Figure 12–31** Conceptual cross-section indicating major tectonic features present during the Cretaceous across the western United States.

Figure 12–32    Generalized cross-section illustrating the pattern of deposition of Cretaceous rocks in the back-arc basin of the western interior of the United States. *(Simplified from King, P. B.* Evolution of North America. *Princeton, N.J.: Princeton Univ. Press, 1977.)*

nental deposits that similarly interfinger with marine
rocks farther to the east (Fig. 12–32). The marine
and nonmarine facies shift repeatedly with the alternate advances and regressions of the Cretaceous sea.
As is normally true, the transgressive phase is
marked by sandstone beds, such as those of the Late
Cretaceous **Dakota Group** (Fig. 12–33). The red,
yellow, and brown layers of the Dakota are exposed
in many places along the eastern front of the Rocky
Mountains, where the inclined beds form prominent
ridges called *hogbacks* (Fig. 12–34). In parts of the
Great Plains, the Dakota sandstones are an important
source of underground water.

Cretaceous rocks of Wyoming and Colorado also
include extensive beds of a soft, plastic, light-colored
clayey rock called *bentonite*. Bentonite is composed
of clay minerals that are formed by the alteration of
volcanic ash. Volcanoes in the Idaho region during
the Late Jurassic and Cretaceous explosively ejected
tremendous volumes of ash that were carried into
adjacent states by westerly winds. The resulting ash
beds, subsequently converted to bentonites, represent single geologic events of short duration and can
be traced for great distances, even across changing
facies. Hence, bentonite beds are of great value in
time-stratigraphic correlation. Volcanic ash also pro-

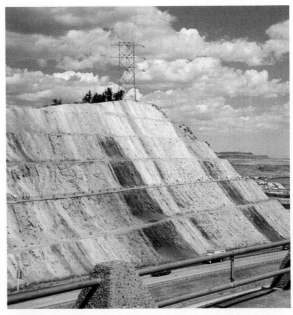

Figure 12–33    The north side of the impressive In-
terstate-70 road cut west of Denver, Colorado. The
rocks visible in this view are part of the Dakota
Group. They are dipping steeply toward the east
(right). Sandstones of the Dakota Group serve as an
important source of artesian groundwater beneath the
Great Plains. They have also yielded oil and gas in
eastern Colorado, Nebraska, and Wyoming.

**Figure 12–34** Prominent north–south trending hogback formed by the eastward-dipping Dakota Sandstone of Cretaceous age, a few miles south of Golden, Colorado. The Dakota Sandstone at this location is a beach deposit of the Cretaceous sea. Well-developed ripple marks and occasional dinosaur footprints are preserved in the sandstone. *(Courtesy of U.S. Geological Survey.)*

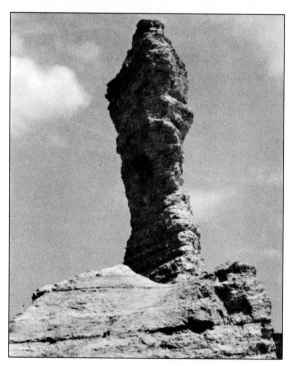

**Figure 12–35** An exposure of Niobrara chalk in Kansas sculpted by processes of erosion into a pinnacle. The pinnacle is about 7.5 meters tall. *(Photograph courtesy of the Kansas Geological Survey, photograph by Jim Enyeart.)*

vided a source for the silica that occurs in some Cretaceous siliceous shales and bedded cherts.

Cretaceous carbonate formations include soft, clayey limestones and chalky shales of the **Niobrara Formation** (Fig. 12–35). The Niobrara has yielded the remains of a variety of marine creatures, including enormous numbers of oysters, a large Cretaceous diving bird, (Fig. 12–36), marine reptiles, and the large flying reptile *Pteranodon*. Toward the end of the Cretaceous, the seas that supported these creatures began a slow withdrawal. This regression of the Cretaceous epicontinental sea from central North America was contemporaneous with the Laramide deformation that produced the Rocky Mountains. As the seas withdrew, coal-bearing deltaic and other continental sediments were gradually spread across the old sea floor.

Geologic field work in frigid polar climates is difficult, and this has limited our knowledge of the history of the northern margin of North America. Within the last two decades, however, exploration for petroleum has provided new information about Mesozoic basins of deposition in the far north. Mesozoic sediments accumulated mainly in two basins along the northern margin of the continent belt (Fig. 12–37): the Brooks basin of northern Alaska and the Sverdrup basin farther east. The Mesozoic record in

**Figure 12–36** Skeleton of the Cretaceous diving bird *Hesperornis*. This bird was over a meter tall when standing. Propulsion was provided by its large webbed feet. A notable primitive trait of this bird was the retention of teeth. *(Courtesy of the U.S. Natural History Museum.)*

the Brooks basin begins with Triassic clastics and limestones, some of which are fossiliferous. Jurassic and Cretaceous beds consist of volcanic rocks, sandstones, and shales.

Mesozoic formations of the Sverdrup basin include both marine and nonmarine sandstones and shales. Some of the formations contain petroleum. Sandstones and siltstones predominate within the Triassic system. Similar rocks, interspersed with coal beds, were deposited during the Jurassic. Volcanic rocks, continental clastics, red beds, and additional coal beds characterize the Cretaceous rock record.

## Eurasia and the Tethys Seaway

Extending eastward from Gibraltar and across southern Europe and Asia to the Pacific is the great Alpine–Himalayan mountain belt. Most of the rocks composing the ranges within the belt were laid down in an important depositional trough known as the **Tethys seaway**. Long after their deposition, these rocks were deformed into mountain ranges as northward-moving segments of Gondwanaland collided with Eurasia. Here and there, however, the waters

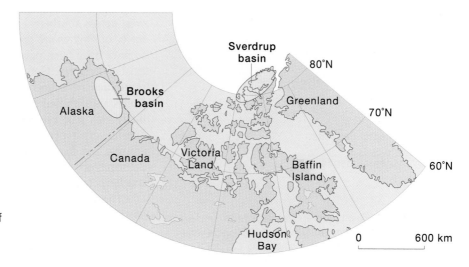

**Figure 12–37** Location of the Brooks and Sverdrup depositional basins.

remained and persist today as the Mediterranean, Caspian, and Black seas.

The geologic history of the Tethys seaway can be traced back to the Paleozoic. But unlike the Appalachians, for which most deposition and tectonic activity occurred during the Paleozoic, the Tethys experienced most of its eventful geologic history in the Mesozoic and Cenozoic.

During the Triassic Period, limestone deposition predominated within the Tethys (Fig. 12–38). Within this lime-depositing sea, pelecypods, cephalopods, crinoids, reef corals, and calcareous algae thrived and left a rich fossil record. Of particular importance for biostratigraphic correlation of Triassic strata are the abundant ammonoid cephalopods (see Chapter 13).

North of the Tethys, the development of a highland tract known as the Vindelician arch separated marine sedimentation to the south from quite different sedimentation in north-central Europe. In that northern region, the Triassic record begins with reddish, nonmarine clastic rocks that resemble those

deposited under arid conditions during the preceding Permian Period. These continental deposits are overlain by shoreline sandstones, marls, evaporites, and limestones deposited during a temporary marine invasion of central Europe. The sea did not remain long. Nor was it able to reach as far north as Great Britain, where a nonmarine sequence known as the New Red Sandstone was being deposited. As the sea withdrew, continental sedimentation much like that of the earliest Triassic resumed.

Marine conditions were far more widespread during the Jurassic than they had been during the Triassic. Shallow seas advanced from both the Tethys and the Atlantic and spread across Europe, leaving a rich sedimentary record in the basins that lay between the old Hercynian uplands. Although the predominant Jurassic sediment is limestone, there is often a gradation to fine clastics adjacent to the highlands. Eventually, marine conditions extended from the Tethys across Russia and into the Arctic Ocean. The invasion was short-lived, however, and the shallow seas gradually drained from the continent by the end of the Jurassic. Marine conditions persisted in the Tethys, and a complicated pattern of facies changes developed in response to block-faulting on the sea floor.

The Cretaceous is the most eventful geologic period of European Mesozoic history. During this time, the African plate was moving northward, narrowing the region occupied by the Tethys seaway. Along the southern margin of Europe, formerly quiet depositional tracts of the Tethys were subjected to powerful compressional forces. The record of this event is seen in complex overturned folds and thrust faults and in a zone of ophiolites that trend east–west across southern Europe. As described in Chapter 6, ophiolites are greenish igneous rocks believed to have once been part of the crust of the ocean that existed between Gondwanaland and Laurasia. The ophiolites are found in association with radiolarian cherts and lithified deep-sea oozes, which were very likely also part of that sea floor. Subduction of the oceanic crust beneath Eurasia as the African plate approached created the ophiolite zone as well as the deformation and volcanism that characterized the southern margin of Eurasia during the Cretaceous.

The Late Cretaceous is noteworthy not only for its mountain-building events but also as a time of extensive marine transgressions. More of Europe was inundated during the Late Cretaceous than at any

Continental deposits

Continental-marine mix

Marine

**Figure 12–38** Paleogeographic map of the Triassic in Europe. *(Modified from Brinkmann, R. Geologic Evolution of Europe. New York: Hafner Publishing Co., 1969.)*

COMMENTARY COMMENTARY COMMENTARY COMMENTARY
MMENTARY COMMENTARY COMMENTARY COMMENTARY COMMEN
COMMENTARY COMMENTARY COMMENTARY COMMENTARY
MMENTARY COMMENTARY **COMMENTARY** COMMENTARY COMMEN

# COMMENTARY

## *Chunnelling in the Cretaceous: The Channel Tunnel*

For nearly three centuries, visionaries have dreamed of constructing a tunnel between England and France. That dream became reality in October of 1990 when English and French tunnelers that had been working toward one another met deep beneath the floor of the English Channel. They had excavated and installed the lining of the first of a three-part tunnel system that is to be completed by the summer of 1993. The Channel Tunnel will be the longest undersea tunnel on earth. It will have two "running tunnels" flanking a central "service tunnel." High speed trains will carry passengers through the running tunnels from Folkestone, England (southwest of Dover) to France in 30 minutes.

About 85 percent of the "chunnel," as it has been dubbed, has been excavated in a Lower Cretaceous formation

known as the Chalk Marl. The term marl indicates the chalk contains clay, and the mixture of chalk and clay results in a rock that is relatively strong, stable, and impermeable. It is an excellent natural material in which to excavate a tunnel. In addition, throughout most of the tunnel route, the Chalk Marl is over 25 meters thick, and thus easily accomodates the 8 meter diameters of the running tunnels. The Chalk Marl is overlain by the Upper Cretaceous Gray and White Chalk formations. Beneath the Chalk Marl is the Gault Clay, a unit unsuitable for tunnelling because it is plastic, inherently weak, and thus likely to deform under the stresses of excavation.

Cross-section along the route of the Channel Tunnel and cross-section of the tunnel. The cross-section showing the channel route has large vertical exaggeration. (*Adapted from* Tunnelling '88, *Instit. Mining & Metallurgy.*)

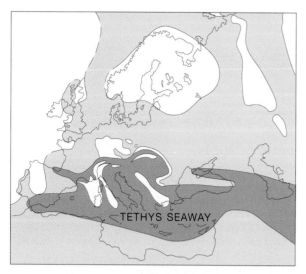

**Figure 12–39**  Areas covered by seas during the Cretaceous. *(Simplified from Brinkmann, R. Geologic Evolution of Europe. New York: Hafner Publishing Co., 1969.)*

data, there was no Bay of Biscay separating western France from Spain prior to the Late Jurassic. The Bay of Biscay opened as the Iberian Peninsula (Spain and Portugal) rotated about 35° from its former line of contact with France. The upheaval of the Pyrenees was another consequence of this rotation. The dating of the rotational event is confirmed by faults of Late Jurassic age on the northern border of the Bay of Biscay and by the fact that the oldest rocks in the bay are Late Cretaceous in age.

East of the Alpine region, the Tethys extended as a great loop past Burma and southeastward into Indonesia. Triassic and Jurassic deposits of the eastern Tethys consist of shales, fossiliferous limestones, and dolomites. During the Cretaceous, volcanic activity occurred throughout what is now the Himalayan region. Some areas seem to have subsided, and terrigenous rocks resembling those moved by submarine landslides or turbidity currents merged with marine volcanic tuffs, breccias, and radiolarian cherts.

other time since the Cambrian (Fig. 12–39). A great embayment from the Tethys worked its way northward until it ultimately joined with a similar southward encroachment from the region of the North Atlantic. One result of the linking of the two seaways was a rapid interchange of marine organisms from the two formerly separated regions. In the Tethys–Alpine region, block-faulting that had begun during Late Jurassic time continued until about the middle Cretaceous, when further compression from the south resulted in folding of the thick sequence of marine Tethys sedimentary rocks. The axes of these folds paralleled the east–west trend of the Tethys. Most of these great anticlines and synclines remained submerged, but the tops of some of the folds rose above the level of the waves, forming elongate islands along the north side of the narrowing Tethys seaway. Erosion of the anticlinal islands produced terrigenous clastics that were periodically swept downward by turbidity currents into the adjoining basins, to be deposited as thick sequences of interbedded sandstones and shales.

An event that determined the geographic configuration of the northern coastlines of Spain and France occurred in Europe between Late Jurassic and Late Cretaceous time. As indicated by paleomagnetic

## The Gondwanaland Continents

### Africa

Near the beginning of the Mesozoic, separations had already developed between the eastern coast of America and northwest Africa. However, as yet no rift existed between Africa and South America (see Fig. 12–1), and thus there was no South Atlantic Ocean. Most of Africa was relatively stable throughout the Mesozoic, with only relatively minor transgressions along the northern and eastern borders. During the Cretaceous, the Tethyan waters spread across what is today the northern tier of countries from Algeria to Egypt. Fine terrigenous muds and carbonates were the most prevalent deposits.

Although the Mesozoic rock record in the more interior parts of Africa is often poor, some regions yield particularly interesting geologic information. One such region is the Karoo basin at the southern end of Africa (Fig. 12–40). The Karoo was a continental basin that was formed late in the Carboniferous and received swamp, lake, and river deposits until late in the Triassic Period. Rocks of the **Karoo sequence** are known to paleontologists around the world for their wealth of fossilized mammal-like rep-

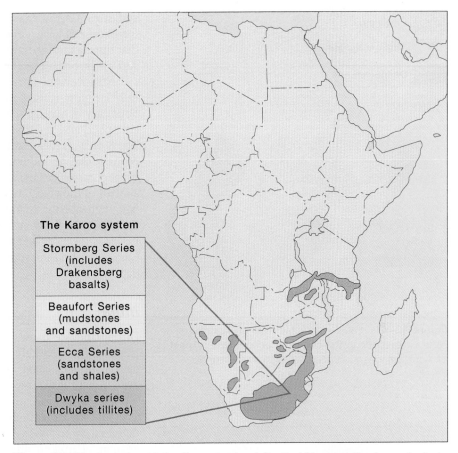

**Figure 12–40**  Location of the Karoo basin of South Africa and the four principal units of the Karoo system.

tiles. The fauna included large and small herbivores, diverse carnivores, and a range of insectivores and omnivores.

As the Triassic drew to a close, the reddish siltstones and sandstones of the Karoo sequence were covered by layer after layer of low-viscosity lava, which flowed onto the surface from fissures and, less commonly, from volcanoes to the southeast. The old Karoo landscape was buried beneath more than 1000 meters of basalt. Geologists speculate that these great outpourings of lava were associated with the pulling away of the Gondwanaland segments that once adjoined South Africa. Such fragmentation would very likely have caused severe fracturing of the foundations of the continents and provided multiple avenues for the upwelling of molten rock. The extrusions continued well into the Jurassic

Period. Contemporaneous lava floods and volcanism also occurred on the separating land masses of South America, Australia, and, somewhat later, Antarctica.

Africa experienced very little tectonic activity throughout the remainder of the Jurassic and Cretaceous. Marginal seas extended along the eastern edges of the continent. Periodic advance and retreat of these seas brought an alternation of marine and nonmarine beds.

### Australia and New Zealand

During the Triassic, Australia was still attached to Antarctica, and both continents were predominantly emergent. Terrestrial sandstones and shales resembling those of the Permian were deposited. Except for brief marginal embayments, deposition of lake

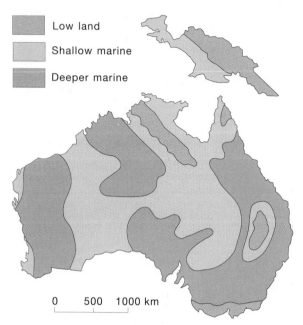

**Low land**

**Shallow marine**

**Deeper marine**

0    500   1000 km

**Figure 12–41** Paleogeographic map of Australia and New Guinea during the Early Cretaceous when most of Australia was inundated by shallow to slightly deeper seas.

Deccan Traps

Bombay ○

INDIA

0        400 km

**Figure 12–42** Present-day distribution of Deccan Trap basalts. Much of the original cover of lava has been lost to erosion, as indicated by the isolated remnants to the east of the large outcrop area.

and stream sediments continued to dominate until the middle of Early Cretaceous time (Fig. 12–41). At that time, a marine transgression brought sandstones and chalk beds into the interior.

New Zealand during the Mesozoic was geologically far more restless than Australia. It lay along the boundary between the southwestern part of the Pacific plate and the eastern edge of the Indian plate. The Mesozoic rock sequence includes turbidites, siliceous rocks, and abundant volcanics. Near the end of the Jurassic, a series of orogenic events occurred as a result of the collision and subduction of the Pacific plate beneath the island. In the course of the orogeny, mountains were erected and deep, unstable trenches developed. The activity continued into the Cenozoic and, intermittently, even to the present day.

## India

During the Mesozoic, India moved steadily northward on its remarkable voyage to Laurasia. The Indian continent itself remained tectonically stable, even though it was moving at a rapid rate toward its ultimate collision with Laurasia in Cenozoic time. The land mass experienced only relatively minor marginal incursions of the sea. Across the interior, erosion of upland areas spread terrigenous clastic sediments into the lowlands and plains. Some of these continental formations have yielded a wealth of dinosaur bones and superb fossils of plants. By Cretaceous time, India was nearing the Tethyan region. Here and there, the floor of the Tethyan trough was buckled into elongate ridges that were early indications of the coming tectonic collision. While these ridges were developing, the northwestern half of India was flooded with immense quantities of low-viscosity basaltic lava (Fig. 12–42). These now solidified lavas are flood basalts known as the **Deccan Traps**. Radiometric dates obtained from the basalts indicate that the outpourings continued from Cretaceous time well into the Cenozoic. It is likely that these basalt floods record the passage of India across a fixed hot spot in the mantle.

## South America

Much like Africa, South America stood well above sea level at the beginning of the Mesozoic. Along the western margin of the continent, two parallel zones of deposition were already established in Triassic time. The more western tract was characterized by turbidites, conglomerates, and siliceous sediments laid down in the deeper waters of the continental rise and slope. Eastward of these deep water facies, carbonates and shales were deposited across shelf zones. The approximate boundary between the two suites of rocks can now be traced along the central ranges of the Andes. In addition to these marine tracts, broad basins in the interior accumulated continental deposits. Particularly remarkable are the eolian and fluvial sandstones and siltstones that spread across the southeastern region of the continent during the Triassic. The beds contain a rich fauna of Triassic vertebrates. Lava flows of the same age as the Karoo rocks of Africa are also prevalent in the upper part of this stratigraphic sequence.

By Jurassic time, the initial narrow split between South America and Africa had widened into a configuration resembling the present-day Red Sea. New sea floor formed along the spreading center of the developing South Atlantic. The western side of the continent was at the leading edge of the westward-moving tectonic plate. It was opposed by the Pacific plate, the forward margin of which plunged beneath South America as an eastward-inclined subduction zone.

Near the end of the Jurassic, the Andean belt experienced deformation and volcanism. This activity appears to have begun in the south and then sporadically shifted with time toward the north. The climax of orogenic deformation occurred in Late Cretaceous and early Tertiary time and was accompanied by regional uplift that led to widespread withdrawal of the seas. Evaporites were deposited in scattered isolated basins that for a time retained marine waters.

The frequency and intensity of deformational and volcanic activity along the Andean mobile belt increased during the Cretaceous. In the subduction zone created by the movement of the Pacific plate against South America, oceanic crust and its cover of sea floor sediment was carried down to deep zones where melting occurred. The melts in turn worked their way back to the surface as great intrusions and outpourings of andesite. Deformation and igneous activity continued well into the Cenozoic and formed many of the geologic structures now seen in the towering peaks of the Andes.

## Antarctica

Antarctica was predominantly emergent throughout the Mesozoic. Eastern areas of the now frigid land mass were sites of continental deposition. Beds of volcanic ash and lava flows occur between these lake and stream deposits.

The nature of sedimentation was markedly different in western Antarctica. Outcrops along the Antarctic Peninsula consist of rocks that are very similar to those of equivalent age in the Andes. The strata of the Antarctic Peninsula most likely represent part of an orogenic belt that was continuous with the Andes to the north.

# Economic Resources

## Uranium Ores

Rocks of the Mesozoic System, like those of earlier eras, contain a wealth of important mineral resources. Notable among such resources are the nuclear fuels. In the United States, uranium ores are derived chiefly from continental Triassic and Jurassic rocks of New Mexico, Colorado, Utah, Wyoming, and Texas. The chief ore mineral is **carnotite** (Fig. 12–43), a yellow, earthy oxide of uranium that usually occurs within the pore spaces of sandstones, as encrustations, or as replacements of fossil wood. In some instances, petrified logs have provided amazing

**Figure 12–43** The bright yellow uranium mineral carnotite disseminated through and encrusting brown Shinarump sandstone. Specimen is 9 cm long.

concentrations of not only uranium but also vanadium and radium. These are striking exceptions, however, and most of the ores are of very low concentration.

There is presently some concern that the richer sources of fissionable uranium-235 in this country are likely to be exhausted by about the same time that our petroleum reserves are depleted. Among the alternate sources of nuclear energy now being studied are uranium-238 and thorium-232, which can be converted to fissionable isotopes by a process termed *breeding*. This will permit the use of the more abundant low-grade uranium ores. Of course, breeder reactors are not yet perfected and may not prove sufficiently safe or as enduring as other possible energy sources.

## Fossil Fuels

Until either the breeder reactor or devices to utilize solar energy are perfected, the nations of the world will continue to rely heavily on fossil fuels. Mesozoic rocks contain significant amounts of these critical resources. The Jurassic, for example, contains workable deposits of coal in Siberia, China, Australia, Tasmania, Spitzbergen, and North America. In North America, thick seams of Jurassic coal occur in British Columbia and Alberta. Cretaceous coal underlies more than 300,000 sq km of the Rocky Mountain region (Fig. 12–44). Much of this coal is now being vigorously mined, particularly because of its rela-

tively low content of environmentally offensive sulfur.

In addition to coal, Mesozoic rocks supply large quantities of oil and gas to energy-hungry industrial nations. The oil provinces of the Middle East and North Africa probably contain more oil than the combined reserves of all other countries. Middle East petroleum comes primarily from thick sections of Jurassic and Cretaceous sedimentary rocks that accumulated in the central part of the Tethys seaway. Other areas of petroleum production from Mesozoic rocks occur in the western United States (Fig. 12–45), Alaska, Arctic Canada, the Gulf Coastal states, western Venezuela, southeast Asia, beneath the North Sea, and beneath the eastern offshore area of Australia. Notwithstanding the size of some recent discoveries in rather forbidding parts of the world, it appears likely that the present rates of oil and gas consumption will cause exhaustion of the world's resources in less than a century. Therefore, oil and gas must be replaced by other energy sources in the near future, especially if we are to conserve these valuable materials for the chemical industry. The time is approaching when petroleum will be too valuable to burn.

## Metallic Mineral Deposits

Metalliferous deposits were formed widely throughout the active orogenic belts of the Mesozoic world. A variety of metals now found in the Rocky Moun-

**Figure 12–44** Coal beds in the Mesaverde Formation (Upper Cretaceous) exposed in a road cut on Highway 6-50 at Castlegate, Utah.

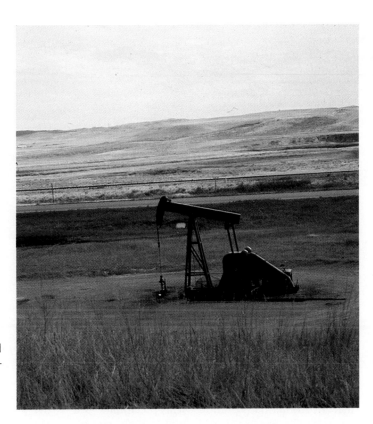

**Figure 12–45** Oil being pumped from Jurassic and Cretaceous formations in the Bighorn basin of Wyoming. *(Photograph courtesy of L. E. Davis.)*

tains, the Pacific coastal states, and British Columbia were emplaced during batholithic intrusions that accompanied Jurassic and Cretaceous mountain building. Among these ore deposits are the gold-bearing quartz veins known as the "Mother Lode." The California gold rush of 1849 was a consequence of the discovery of gold-bearing gravels eroded from the Mother Lode.

The copper, silver, and zinc veins of the Butte, Montana, and Coeur d'Alene, Idaho, mining district were emplaced as a result of Cretaceous igneous activity. One belt of copper-bearing rocks extends from Denver into northeastern Arizona. These deposits and similar ones in Utah occur in igneous rock having a porphyritic texture (large crystals of feldspar and quartz in a finer matrix) and hence are called *porphyry copper deposits* (Fig. 12–46). The copper minerals, principally copper and copper-iron sulfides, are disseminated through rock that has been pervasively fractured during episodes of explosive volcanic activity. The many fractures provided easy passage for hot, mineralizing solutions from which

the ore was precipitated. Porphyry copper deposits occur not only in western North America but also in South America, the Soviet Union, Iran, Yugoslavia, and the Phillipines.

Important nonmetallic deposits of the Mesozoic include sulfur and salt. Both are produced from the salt domes mentioned earlier (see Fig. 1–48). Diamonds are also obtained from Mesozoic igneous rocks. Siberia's diamond-bearing intrusions are believed to have penetrated the upper crust during the Triassic and Jurassic. Similar diamond pipes in Africa are probably of Cretaceous age.

## Plate Tectonics and Ore Deposits

In recent years, geologists have begun to see numerous correlations between patterns of mineral distribution and locations of present and former tectonic plate boundaries. Plate boundaries are likely sites for movements of hot, aqueous fluids that bear important metals in solutions. With changes in temperature or pressures, or on contact with reactive rocks,

**Figure 12-46** The open pit copper mine at Bingham Canyon Utah. The copper minerals at Bingham Canyon are disseminated throughout porphyritic igneous rock.

such hydrothermal solutions will precipitate ore minerals. The process may operate at both convergent and divergent plate boundaries. Both the Cordilleran and Andean mobile belts are examples of convergent boundaries. An example of hydrothermal mineralization at a divergent boundary is provided by the Red Sea, which has pools of very hot and exceptionally salty water along its bottom. The pools are rich in iron, manganese, zinc, and copper. The brines apparently have percolated upward through the young oceanic crust, dissolving the metals en route. Metals brought to the surface in this way in the past may have been conveyed in thin layers across immense tracts of the ocean floor by sea floor spreading. Sediments containing metallic ions extracted from the sea water itself may have enriched the accumulation. Ultimately, sea floor spreading would move the sediments and their contained metals into collision with another plate, providing an opportunity for their inclusion as ore bodies within the deformed belt.

Of course, the formation of ores in orogenic belts is a far more complicated process than this conceptual model suggests. Every ore body requires very special local conditions and events for its emplacement. Many ore deposits have no relationship to plate boundaries. However, in the case of the metallic deposits of North and South America, a relationship is likely, for both of these regions have overriding convergent boundaries with the plates of the Pacific.

## SUMMARY

Within the approximate 160 million years of the Mesozoic Era, the pattern of lands and seas was extensively altered as large segments of Pangea were separated. The breakup appears to have begun during the Triassic, with the splitting off of North America from Gondwanaland. By Late Jurassic time, rifts had developed between all of the Gondwanaland segments except Australia and Antarctica. These two continents, like northwestern North America, Greenland, and northwestern Europe, retained a hold on one another until early in the Cenozoic.

At the beginning of the Mesozoic, most of North America was emergent. In the east, crustal tension was expressed by the development of large-scale block-faulting and volcanism. In down-faulted valleys, arkosic sandstones, conglomerates, and lacustrine shales accumulated. While this was occurring, the ancestral Gulf of Mexico began to take form far to the south. During the Jurassic and Cretaceous, thick sequences of limestones, evaporites, sandstones, and calcareous shales accumulated along what are now the Gulf and Atlantic coastal states.

Continental deposits, particularly red beds, characterize the Triassic of the Rocky Mountain regions of North America. Marine conditions persisted along the Pacific margin, where volcanic rocks and graywackes accumulated to great thicknesses. During the Jurassic, this western belt was strongly deformed as North America began to override the subduction zone at the leading edge of the eastward-moving Pacific plate. The advancing Pacific plate carried large fragments of volcanic arcs, oceanic plateaus, and microcontinents to the western margin of North America. These crustal fragments became incorporated into the Cordillera as a tectonic collage of displaced terranes. Growth of a continental margin in this way is referred to as accretionary tectonics.

In general, deformation of the North American Cordillera during the Mesozoic began along the Pacific coast and moved progressively eastward, until by Late Cretaceous time, inland seas that had occupied the western interior of North America were displaced. Majestic mountain ranges stood in their place. Crustal unrest continued to affect the western states even during the early epochs of the Cenozoic Era.

The focus on Eurasian Mesozoic events was the great east–west trending Tethys mobile belt. The initial Mesozoic deposits of the western Tethys were predominantly carbonates. Block-faulting developed along parts of the belt during the Jurassic. This middle part of the Mesozoic witnessed marine invasions that completely spanned the continent from the Tethys to the Arctic. During the Cretaceous, the Tethys was affected by powerful compressional forces as the northward-moving oceanic plate bearing peninsular India moved toward and under the Eurasian plate. One result of these movements was the development of a series of anticlinal welts along the northern side of the Tethyan seaway. Rapid erosion of these elongated, uplifted tracts produced thick sequences of coarse clastics in the intervening troughs. Along the southern margins of the Tethys, conditions were more stable, and relatively undisturbed sequences of carbonates and sandstones were deposited.

Although Australia was predominantly emergent during the Mesozoic, an active tectonic zone existed across a tract now occupied by New Zealand. This zone originated in the late Paleozoic and reached its orogenic climax during the Jurassic. New Zealand's Cretaceous rocks were for the most part deposited in shallow marginal seas adjacent to the ranges uplifted during the previous period.

The Mesozoic history of the interior regions of Africa is read primarily from continental deposits. Among these, none is more famous than the beds of the Karoo basin of South Africa. The Karoo sequence includes both Permian and Triassic strata, some of which have yielded splendid collections of the abundant and diverse mammal-like reptiles that once inhabited southern Africa.

During the Mesozoic, a largely emergent peninsular India was progressing steadily toward its Eurasian destination. By Cretaceous time, it had approached sufficiently near to cause initial buckling in the eastern Tethys.

The Jurassic was an important geologic period for South America, which during this time began to separate from Africa. The separation continued during the remainder of the Mesozoic, and a new ocean, the South Atlantic, was born. The western margin of the continent formed the leading edge of the South American plate and was characterized by volcanic activity, folding, and intrusions that increased in frequency within the mobile belt throughout the Mesozoic.

The rocks of the Mesozoic have supplied the world with a variety of important economic resources. These include such fossil fuels as coal, petroleum, and natural gas, as well as nuclear fuels. Orogenic activity during the Mesozoic resulted in the emplacement of such critical metals as copper, zinc, chromite, gold, silver, lead, mercury, and many others. Theories of ore genesis now being formulated suggest that metallic ores emplaced during mountain building along continental margins may represent concentrates of materials on the sea floor that have been conveyed to orogenic belts by sea floor spreading. Once such materials are brought to the plate boundary, favorable physical and chemical conditions must exist locally to provide for concentration of the dispersed metals into ore deposits.

## QUESTIONS FOR REVIEW

1. What is the total duration in years of the Mesozoic Era? Which period of the Mesozoic was longest?

2. What geologic events immediately precede the separation of a land mass into two or more parts? Cite an example from Mesozoic geologic history.

3. Describe the general geologic conditions associated with the deposition of the Newark Group. Why did normal faulting rather than thrust-faulting occur in this area?

4. What was the probable source area for the sediments of the Morrison Formation of the Rocky Mountain region? What evidence indicates that the Morrison was a continental rather than a marine formation?

5. During which period of the Mesozoic were epicontinental seas most extensive? During which period were such incursions most limited?

6. What was the Tethys seaway? In which of today's mountain ranges might one go to examine the rocks deposited in the Tethys?

7. At what time in geologic history did the modern South Atlantic Ocean begin to form?

8. What nonmetallic mineral resources are derived from Mesozoic strata? What metallic ores occur?

9. What is the paleontologic importance of the Permian–Triassic strata of the Karoo basin of South Africa?

10. What kind of deformation is particularly characteristic of the Sevier orogeny?

11. Of all the Gondwanaland segments that separated from Pangea, which seems to have drifted over the greatest distance during the Mesozoic?

12. During what geologic period did most of the ranges of the Rocky Mountains undergo their major deformation? In what areas are thrust faults prevalent?

13. During what geologic period of the Mesozoic were large volumes of evaporites deposited in the Gulf Coast region? What relationships, if any, do these deposits have to oil entrapment?

## TERMS TO REMEMBER

| | | | |
|---|---|---|---|
| accretionary tectonics | displaced terrane | Nevadan orogeny | Shinarump Formation |
| allochthonous terranes | Karoo sequence | Newark Group | Sonoma orogeny |
| carnotite | Kayenta Formation | Niobrara Formation | Sundance sea |
| Chinle Formation | Laramide orogeny | obduction | suture zone |
| Dakota Group | Moenkopi Formation | ophiolites | tectonic collage |
| Deccan Traps | Morrison Formation | rudists | Tethys seaway |
| décollement | Navajo Sandstone | Sevier orogeny | Wingate Sandstone |

# SUPPLEMENTAL READINGS AND REFERENCES

Ager, D. V. 1980. *The Geology of Europe*. New York: John Wiley.

Armstrong, R. L. 1968. The Sevier orogenic belt in Nevada and Utah. *Bull. Geol. Soc. Am.* 79:429–452.

Dietz, R., and Holden, J. C. 1971. The breakup of Pangaea. In *Continents Adrift: Readings from Scientific American*. San Francisco: W. H. Freeman and Co.

Greenwood, P. H. (ed.). 1981. *The Evolving Earth*. Cambridge, England: Cambridge Press.

Jones, D. L., Cox, A., Coney, P., and Beck, M. 1982. The growth of western North America. *Sci. Am.* 247(5):70–128.

Manspeizer, W., Putter, J. H., and Cousminer, H. L. 1978. Separation of Morocco and eastern North America: A Triassic–Liassic stratigraphic record. *Bull. Geol. Soc. Am.* 89:901–920.

Reynolds, M. W., and Dolly, E. D. (eds.). 1983. *Mesozoic Paleogeography of the West-Central U.S.* Rocky Mountain Symposium. Society of Economic Paleontologists and Mineralogists, Denver, Colorado.

Rona, P. A. 1974. Plate tectonics and mineral resources. In *Planet Earth: Readings from Scientific American*. San Francisco: W. H. Freeman and Co., pp. 170–180.

Saleeby, J. B. 1983. Accretionary tectonics of the North American Cordillera. *Ann. Rev. Earth Planet Sci.* 15:45–73.

# BURRELL CLARK BURCHFIEL

## *INTERVIEW*

*Burrell Clark Burchfiel was born in 1934 in Stockton, California. He studied at Stanford University (B.Sc., 1957; M.S., 1958) and Yale University (Ph.D., 1961) where he was awarded the Silliman Prize. After receiving his Ph.D., Dr. Burchfiel joined the faculty of Rice University where he taught for 16 years before joining the faculty at the Massachusetts Institute of Technology. He is currently Schlumberger Professor of Geology at MIT. His areas of interest include structural geology, regional geology, plate tectonics,*

*and orogenesis (the process by which mountain belts are formed).*

*Burchfiel's current research areas include the investigation of the processes of orogenesis and their relationship to plate interactions and intraplate deformation. His major investigations are in five areas: (1) the Cordilleran orogenic belt of western North America, (2) the Alpine belt of eastern Europe, (3) the Caledonian belt of Scandinavia, (4) the Andean belt in Peru and Bolivia, and (5) north-central China and Tibet. His research focuses on processes of intracontinental deformation and is designed to examine orogenic belts and other regions of significance within the continents that are of various ages, tectonic settings, and hence at different levels of exposure.*

*Burchfiel has been a visiting professor at the University of Southern California, the University of Houston's Continuing Education Program, and Australian National University. He has served as a consultant to Exxon, ARCO, and Mobil Oil Company. Currently, he is a Fellow of the National Academy of Sciences, the American Academy of Arts and Sciences, the Geological Society of America, the American Geophysical Union, and the European Union of Geologists. He is also Chairman of the Cooperative Exchange Program, National Academy of Science/Academy Sciences U.S.S.R. in Earth Sciences, and serves as a member on the Commission on Geosciences, Environment and Resources of the National Research Council. He is Co-Editor of both the symposium volume on "Tectonics of Eastern Asia and Western Pacific Continental Margin"— Special Issue of Tectonophysics, and the "Decade of North American Geology"–Cordilleran volume.*

**What originally got you interested in geology?**

I originally became interested in geology during high school. I have always been fond of camping, hiking, and the outdoors, and geology offered me

the possibility of combining these things with a career. While in high school and college, I worked during the summers as a roughneck on oil rigs to earn money for college, and during this period I came into contact with numerous aspects of the

INTERVIEW INTERVIEW INTERVIEW INTERVIEW INTERVIEW
VIEW INTERVIEW INTERVIEW INTERVIEW INTERVIEW INTER
INTERVIEW INTERVIEW INTERVIEW INTERVIEW INTERVIEW
VIEW INTERVIEW INTERVIEW INTERVIEW INTERVIEW INTER
INTERVIEW INTERVIEW INTERVIEW INTERVIEW INTERVIEW
VIEW INTERVIEW INTERVIEW INTERVIEW INTERVIEW INTER

oil business and with geologists. This further enhanced my interest in geology. My undergraduate education at Stanford University was a very rewarding experience, and the faculty in the geology department really kindled my interest into a lifelong profession. It was my contact and relationships with the faculty that completed my desire to become a geologist.

**Tell us about your research specialties in plate tectonics and geophysics and how you chose those specialties over other fields within the earth sciences.**

I began as a sedimentologist because of my interest in the oil industry, and I took a Master's Degree in that subject. I became interested in structural geology and tectonics during my last year at Stanford and during my graduate work at Yale. My interests drifted into tectonics because it is a field that requires an integration of most of the disciplines of earth science. I found the study of global processes of mountain building fascinating, and this area of research also provided an opportunity to work outdoors in the high country.

**How have our concepts of continental evolution and ocean basin modification changed over the past several decades?**

Our geologic understanding of the western United States (and other mountain belts as well) has evolved in three stages since I have been working there. First was a stage of detailed geologic mapping of small areas, until a sufficient number of these maps had been completed so that they could be combined into maps of the entire western United States. I entered into this stage, which continues today, just before the beginning of the second stage. The vast background of information

accumulated by previous generations of geologists was integrated and used in the interpretation of the entire West. Second was a stage that coincided with the discovery and development of plate tectonics, in which the evolution of western mountain systems was interpreted as part of single dynamic systems active at any one time. This required geologists to place the various aspects of the evolution of mountains—the deformation, volcanism, metamorphism, and sedimentation—into a single integrated picture. This stage of investigation of the western United States is still active and evolving and will probably continue for generations to come.

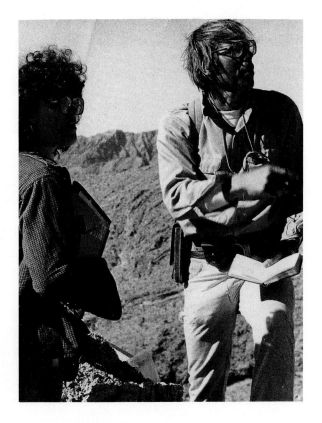

Third was a stage that began perhaps only a decade or so ago and focuses on a variety of processes that occur during mountain-building. This third stage requires a global view. Earth scientists must be aware of the global range of plate tectonic processes so that they can be applied to the mountain belts of the western United States. Active or young plate interactions must be examined because it is from these areas that processes can best be understood. More information on the contemporaniety of geologic features can be obtained from studying active and young mountain belts than from studying older belts. The older belts, however, are useful in looking at deeper parts of the crustal section because they have been more deeply eroded. They provide an opportunity to examine the results of deeper crustal processes first hand. In active or young mountain belts, the effects of these processes can only be imaged by remote sensing. First stage type studies will continue to provide basic information and set boundary conditions for second and third stage studies, which will occupy our efforts for a long time into the future.

**What advice would you offer an undergraduate who is thinking of becoming a geology major? A professional geologist?**

My advice to an undergraduate thinking of becoming a geology major would be to sample several different areas in the earth sciences to see if there is something that they find really interesting. I changed my direction once before I found what really interested me. I know others that have changed more than once before they found their niche. I would also suggest that during undergraduate work, students continually evaluate whether they really like the earth sciences and, if so, the special area in which they would like to specialize. I think it would be terrible to spend a lot of time training for a profession that I didn't like. Earth science offers such a breadth of opportunity that I would be surprised that someone couldn't find an area that he or she finds interesting.

For the undergraduate who has decided to pursue a professional career in geology, I would advise taking as much mathematics, physics, chemistry, and computer science as can be worked into the curriculum. The field of earth science has become more process-oriented, and such research requires a solid understanding of the basic sciences. Geology has clearly broadened from its traditional areas of interest and should take the lead in areas more oriented toward society, such as the environment, hazards, energy, and resources. These areas will only become more important in the future.

## Tell us more about your current or most recent research and its relationship to future investigations by yourself and others in your field.

My current research has shifted more toward trying to understand the processes of mountain building. This has led me to work more in the young and active mountain belts, but not to the neglect of the older belts. Several of my recent research efforts have been to look at active and Quaternary deformation in various parts of the world. I hope that in the future, these studies can be integrated into an understanding of the broad dynamic systems active in these mountain belts today. One area that I spend more and more time in is China. It contains perhaps the broadest and most active region of intracontinental deformation in the world today. It is a huge area, but one in which the mountain-building processes can be well studied because deformation and related features have evolved very rapidly. The Mediterranean region and the Andes are other similar areas where I have worked recently or am now working.

## What good is an understanding of earth science to someone who does not plan to be a geologist?

I think an understanding of earth science is useful for a variety of reasons. Two are obvious. Our society will depend in the future on more informed judgments about the earth: our environment, hazards, energy, and resources. One needs only to look at the problems we have created for ourselves today to see that changes must be made. The more informed the individual is about these earth science problems, the better they will be able to legislate or support the proper efforts to solve them. A second point is that a better understanding of the features of the earth permits a fuller enjoyment of our surroundings. Everyone takes trips, and it is always more interesting when we know something about what we are looking at. There is great beauty on the earth, and we have an opportunity to more fully enjoy it when we understand the origin of features that make our planet such a pleasant place on which to live.

## Has your work changed any of the traditional views of plate tectonics? If so, how?

One always hopes that he has contributed to advancing the science, but it is impossible to take singular credit for many of the ideas that I have been associated with. I have been very fortunate to have worked with some marvelous people, and clearly they deserve as much credit as I do, if not more. The works that I am most pleased about have been the syntheses of western U.S. geology that I have published with my colleague Greg Davis at U.S.C.; geologic studies in China with Peter Molnar and my wife, Leigh Royden; and studies in eastern Europe with my wife and our many eastern European colleagues. These studies include developing concepts about treating mountain building within the framework of large, single, dynamic plate boundary systems; the influence of gravity in mountain building; the character of regional crustal extension; and large-scale intracrustal decoupling at both convergent and divergent plate boundaries. I am also extremely proud of the ideas developed by my students over the years and proud to watch their careers flourish as they develop into leaders within their profession. It is a great experience to see how fast and how far the earth sciences have come in the past 30 years, and I look forward to seeing how far it will advance in the near future.

Animals and plants of the Cretaceous shallow marine environment. This reconstruction is based on fossils recovered from Cretaceous strata in western Tennessee. On the left is a large ammonite cephalopod with a coiled conch. For scale, it is about 36 cm in diameter. Another coiled cephalopod ornamented with nodes on its conch can be seen at the bottom center. Many straight-shelled ammonites dominate the scene. Notice also the cork-screw ammonite on the right, the shrublike algae, and the numerous clams and snails on the sandy sea floor. *(Courtesy of National Museum of Natural History, Smithsonian Institution.)*

*Tyrannosaurs, enormous bipedal caricatures of men, would stalk mindlessly across the sites of future cities and go their way down into the dark of geologic time.*

Loren Eiseley
*The Immense Journey,* 1959

## Introduction

The biosphere is the world of life. In this chapter, we direct our attention to the biosphere of the Mesozoic Era. The great nineteenth century naturalist Louis Agassiz called the Mesozoic the *Age of Reptiles.* It was a clearly appropriate designation, for reptiles thrived on land, in the seas, and even in the air. Yet, with all our fascination with the dinosaurs and other often bizarre reptiles of the Mesozoic, we should not forget that it was during this era that birds and mammals also began their evolution. In the world of plants, grasses and modern trees bearing seeds, nuts, and fruits expanded across the continents. As today, the distribution, evolution, and abundance of all these many forms of life were dependent upon climate. Climate, in turn, was affected by the changing locations of continents, major marine transgressions and regressions, and the formation of great mountain systems.

## Mesozoic Climates

Cool climates seem to have characterized many continental areas during the final days of the Paleozoic Era. The vastness of the Pangea supercontinent, the upheaval of mountains, general uplifts, and with-drawal of inland seas in many regions were contributing causes of the generally cooler conditions. Gradually, however, the climate warmed, and glaciers in Africa, Australia, Argentina, and India began to melt as these continents drifted away from the South Pole. In general, during most of the Jurassic and Cretaceous, climates for most regions were warm and equable.

During the Triassic, the continents were still tightly clustered. The paleoequator extended from central Mexico across the northern bulge of Africa (Fig. 13–1). As noted in Chapter 12, the Triassic was a time of general emergence of the continents. Mountains, thrust upward at the end of the Paleozoic, inhibited the flow of moist air into the more centrally located regions, causing widespread aridity. Evaporites, dune sandstones, and red beds accumulated at both high and low latitudes and attest to relatively dry and warm conditions.

Reconstructions based in part on paleomagnetic studies suggest that during the Jurassic, the continents were at the approximate latitudinal positions they occupy today. Marine waters extended northward into a great trough formed by the opening of the Atlantic Ocean, and in many places shallow inland seas spilled out of the deep basins onto the continents. An arm of the Pacific Ocean extended westward as the great Tethys seaway. It is not unreasonable to assume that warm, westward-flow-

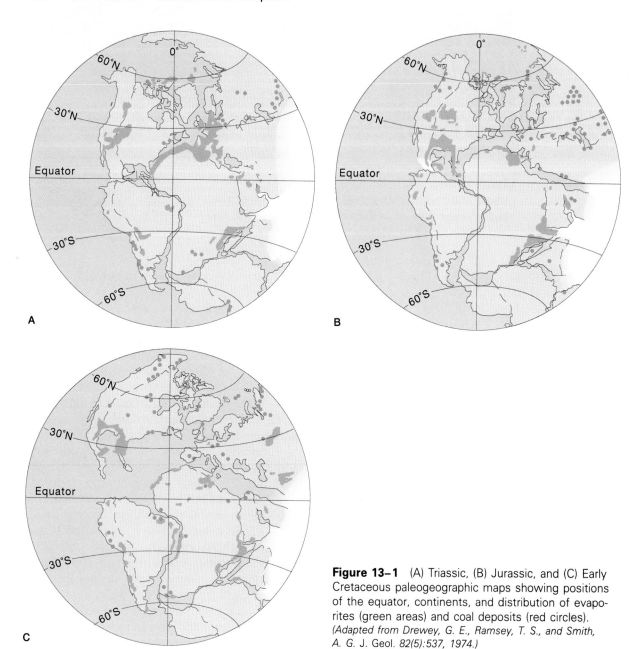

**Figure 13–1** (A) Triassic, (B) Jurassic, and (C) Early Cretaceous paleogeographic maps showing positions of the equator, continents, and distribution of evaporites (green areas) and coal deposits (red circles). *(Adapted from Drewey, G. E., Ramsey, T. S., and Smith, A. G. J. Geol. 82(5):537, 1974.)*

ing equatorial currents may have penetrated far into the Tethys. These same equatorial currents, deflected by the east coast of Pangea, were shunted to the north along coastal Asia and to the south along northeastern Africa and India. To complete the cycle, the cooled currents may have returned to the equator along the west side of the Americas. The presumed ocean and wind currents and the rather extensive coverage of seas, both upon and adjacent to continents, brought mild climates to many regions during the Jurassic. There also is evidence, however, of aridity and climatic patterns characterized by monsoons. Glacial deposits of this age are not known, and coal beds occur in Antarctica, India, China, and Canada. Paleobotanical evidence suggests that tropical conditions prevailed over regions that are now temperate.

During most of the Cretaceous, climatic conditions were a continuation of the generally warm and stable conditions that had prevailed during the Jurassic. A remarkably homogeneous flora spread around the world, with subtropical plant families thriving in latitudes 70° from the equator. Coal beds formed on nearly every continent and even at very high latitudes. However, such pleasant conditions did not persist, and toward the end of the Cretaceous, the climatic pendulum began a slow backward swing toward more rigorous conditions. The Late Cretaceous South Pole was centrally located in Antarctica, and the North Pole was located at the north edge of Ellesmere Island. Europe and North America had moved somewhat farther north, and the widespread inland seas had begun to recede. The worldwide regressions were accompanied in some regions by major orogenic disturbances.

In the Late Cretaceous seas, the golden-brown algae known as coccolithophorids spread in enormous profusion. Their calcium carbonate platelets (see Fig. 1–46) accumulated on the floors of the inland seas and were converted to thick layers of chalk. They are estimated to have been so abundant in the Late Cretaceous that their photosynthetic activity may have produced a carbon dioxide shortage in the atmosphere. Because carbon dioxide provides the earth with a warming "greenhouse effect," it is at least theoretically possible that the great blooms of plankton may have contributed, along with other environmental conditions, to a temporary chilling of the Late Cretaceous world.

There are several lines of evidence that favor the interpretation of a terminal Cretaceous cooling episode. Terrestrial plants provide some clues. The tropical cycads underwent a sharp reduction, and ferns declined in both North America and Eurasia. Hardier plants such as conifers and angiosperms extended their realms. Evidence comes also from paleotemperature studies based on the oxygen isotope method described in Chapter 3. The oxygen isotope ratios obtained from open ocean planktonic calcareous organisms indicate a decline in ocean temperatures beginning about 80 million years ago. If there was indeed a dip in worldwide mean annual temperatures, it might have had a deleterious effect on plant and animal life. For this reason, some paleontologists speculate about its relationship to the extinctions that occurred at the end of the Mesozoic.

# Mesozoic Flora

The existence of animal life on earth is ultimately dependent on plant life. This generalization was as valid during the Mesozoic as it is today. Then, as now, plants made up the broad base of the food pyramid. Their nutritious starches, oils, and sugars made possible the evolution and continuing existence of animals. Plants are a fundamental part of the earth's essentially self-sustaining ecologic system. The operation of the system is dependent upon oxygen and carbon dioxide. Animal respiration provides the carbon dioxide needed for plant photosynthesis, whereas plants supply—as a byproduct of photosynthesis—the oxygen needed by animals. In the geologic past, variations in plant productivity may have caused corresponding changes in the amount of carbon dioxide and oxygen in the atmosphere. Such variations may have favored the evolution of some organisms over others and may even have been responsible for the demise of particular groups. As an example, in 1984 geologist Dewey M. McLean suggested that the pronounced extinctions among planktonic foraminifera and coccolithophorids at the end of the Cretaceous Period may have been the result of rapid release of carbon dioxide derived from massive outpourings of lavas, such as occurred during the extrusion of the Deccan lavas in India. The surplus carbon dioxide may have then caused chemical changes in the upper 45 to 100 meters of the ocean that might have been lethal to microplankton. Without these organisms, extinctions of animals higher in the food chain would have been inevitable.

## *Marine Plants*

Because plants that live in the oceans are suspended in water, they do not require the vascular and support systems that characterize land plants. Most marine plants are therefore unicellular, although they may grow in impressive colonies and aggregates. The major groups of marine plants are part of that vast realm of floating organisms called plankton; because they are plants, they are further designated as **phytoplankton**. The geologic record of the important fossil groups of phytoplankton is shown in Figure 13–2. As indicated on the chart, phytoplankton that did not secrete mineralized coverings pre-

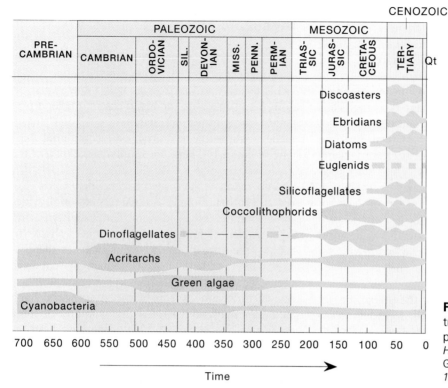

**Figure 13–2** Geologic distribution and abundances of phytoplankton. *(From Tappan, H., and Loeblich, A. R., Jr. Geol. Soc. Am. Special Paper 127:257, 1970.)*

dominated in the pre-Mesozoic eras. These include both cyanobacteria and green algae, as well as a group of cellulose-covered unicellular algae of uncertain affinity that are loosely called acritarchs. Of more immediate importance to us are the coccolithophorids, dinoflagellates, silicoflagellates, and diatoms, which were the more abundant phytoplankton groups of the Mesozoic. The expansion of the coccolithophorids and dinoflagellates really began in the Early Jurassic.

Dinoflagellates are frequently encountered as fossils and are important aids in Mesozoic and Cenozoic stratigraphy. From the Jurassic on, they were among the primary producers in the marine food chain. Dinoflagellates (Fig. 13–3) are unicellular organisms having a cellulose cell wall like that in pollen. For propulsion, the cell wall is equipped with two flagella: one is longitudinal and whiplike, and the other is transverse and ribbonlike. During their life cycle, dinoflagellates develop a motile planktonic form and a cyst phase that is formed within the motile organism. The cyst form is extremely resistant to decay, and, in fact, only dinoflagellate cysts are known as fossils.

The coccolithophorids also began their expansion during the Early Jurassic. Unlike the dinoflagellates, these calcium carbonate–secreting organisms have a splendid fossil record. Their abundant remains have formed many of the extensive coccolith limestones of the Mesozoic and early Cenozoic. Today, they are frequently present in the deep-sea sediment known as calcareous ooze. The coccolithophorid organism is one of several varieties of unicellular golden-brown algae. These algae deposit calcium carbonate internally on an organic matrix and construct tiny, shieldlike structures called **coccoliths.** Once formed, the coccoliths move to the surface of the cell and form a calcareous armor (Fig. 13–4). Coccoliths measure only from 3 to 15 $\mu$m in diameter (1 $\mu$m equals 0.001 mm). However, with the aid of the electron microscope, it is possible to discern their intricate construction. With such magnification, coccoliths are seen to consist of one, or sometimes two, superimposed elliptical plates. These plates are concave on one surface so that they can fit closely around the outside of the spherical cell. Each disc or plate is itself composed of still smaller crystalloids that may be triangular, rhombic, or variously

**Figure 13–3**  Fossil dinoflagellate cyst, *Prionodinium alveolatum,* from the Cretaceous of Alaska. *(From Leffingwell, H. A., and Morgan, R. P. J. Paleontol. 51(2):292, 1977.)*

shaped by the organism. The crystalloids are uniformly arranged, usually in a circular, radial, or spiral plan that is often astonishing in its beauty and precision (see Fig. 1–46). Because they are frequently fossilized, have experienced frequent evolutionary changes through time, and are readily dispersed by oceanic currents, coccoliths have become extremely useful in stratigraphic correlation.

The earliest known silicoflagellates (Fig. 13–5) and diatoms appeared in the middle Cretaceous.

Along with other phytoplankton, they experienced a decline at the end of the Cretaceous, and then all groups expanded again into the early epochs of the Cenozoic. The silicoflagellates and diatoms, along with the coccolithophorids, are members of the phylum Chrysophyta.

As suggested by their somewhat ponderous name, the silicoflagellates are flagella-bearing algae that secrete a siliceous internal skeleton. Radiating spines characterize most genera, whereas others are roughly stellate. Most range in size from 20 to 100 $\mu$m in diameter.

Like the silicoflagellates, **diatoms** also secrete siliceous coverings (Fig. 13–6). The covering is called a *frustule*. It is usually composed of an upper

**Figure 13–4**  Scanning electron micrograph of a coccosphere covered with coccoliths. *(From McCormick, J. M., and Thiruvathukal, J. V.,* Elements of Oceanography, *2nd ed. Philadelphia: Saunders College Publishing, 1981. Photography courtesy of S. Honjo.)*

**Figure 13–5**  Three Mesozoic silicoflagellates. (A) *Corbisema* (Cretaceous–Eocene). (B) *Lyramula* (Cretaceous). (C) *Vallacerta* (Cretaceous).

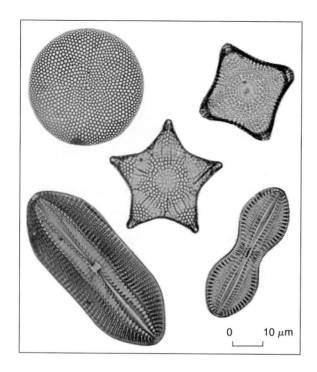

**Figure 13–6** Diatoms. These modern diatoms were collected off the coast of Crete. *(Photographs courtesy of Naja Mikkelsen, Scripps Institution of Oceanography.)*

A

B

**Figure 13–7** (A) Restoration of the cycadophyte *Williamsonia* from the Jurassic of India. (B) A cycad growing in South Africa. Note the immense seed cones. *(Part (A) from Sahni, B. 1932. Mem. Geol. Survey India, Paleontology Indica 20(3):1–19; part (B) W. H. Hodge/ Peter Arnold, Inc.)*

and lower part that fit together like a lid on a box. The test may be circular, cylindric, triangular, or a variety of other shapes and is usually quite beautiful. Today, marine diatoms are prevalent in the cooler regions of the oceans. In the past, a proliferation of diatoms was often associated with volcanic activity. Apparently, the silica supplied to sea water as fine volcanic ash stimulated diatom productivity.

## Terrestrial Plants

Although a vertebrate paleontologist might refer to the Mesozoic as the Age of Reptiles, paleobotanists might well argue that the term Age of Cycads would be equally appropriate. The **cycads**, more formally known as the class Cycadophyta (Fig. 13–7), are seed plants in which true flowers have not been developed. Jurassic cycads included tall trees with rough, columnar branches marked by the leaf bases of earlier growths and by crowns of leathery pinnate leaves. Actually, the Cycadophyta include two related groups: the cycadeoids (fossil cycads) and the Cycadales (true cycads). Although the adult plants in both of these groups were superficially similar, they differed in reproductive structures and in other anatomic details. Cycads experienced a marked decline in the Late Cretaceous, and only a few have survived to the present time. One such survivor is the sago palm, often used as a large house plant.

In Chapter 11, it was noted that there were three important episodes in the evolution of land plants. The first stage led to the development of the spore-bearing, leafy, treelike plants. The second involved the evolution of such nonflowering, pollinating seed plants as cycads, ginkgoes (Fig. 13–8), seed ferns, and conifers. As indicated in Figure 3–47, all but the seed ferns have living representatives, and the conifers are today widespread and highly successful. The third important episode in plant history is marked by the advent of species having enclosed seeds and flowers. Such plants are known as **angiosperms.** Angiosperm-like pollen grains obtained from rocks of Early Cretaceous age provide the earliest evidence of the existence of flowering plants. By middle Cretaceous time, angiosperms were widespread. Forested areas included stands of birch, maple, walnut, beech, sassafras (Fig. 13–9), poplar, and willow trees. Before the Cretaceous came to a close, angiosperms had surpassed the nonflowering plants in both abundance and diversity. Flowering trees, shrubs, and

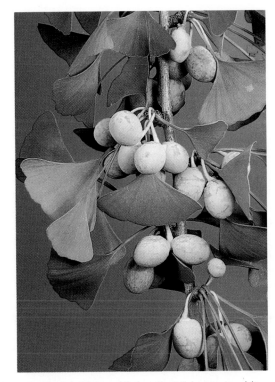

**Figure 13–8** *Ginkgo biloba,* the ginkgo or maidenhair tree (so-called because the leaves resemble those of the maidenhair fern). Note the naked, fleshy seeds. The ginkgo represents the oldest genus of living trees. Fossils over 200 million years old are nearly identical to the living forms. *(W. H. Hodge/Peter Arnold, Inc.)*

**Figure 13–9** Fossil leaf of a Cretaceous sassafras tree from the Dakota Sandstone, Elsworth, Kansas. The slab is about 12.5 cm (5 inches) wide.

vines expanded across the lands and, except for the absence of grasses, gave the landscape a modern appearance.

Either directly or indirectly, the evolution of Mesozoic insects, birds, reptiles, and mammals was influenced by this remarkable floral revolution. Angiosperms produced a variety of seeds, nuts, and fruits needed for the survival of the plant embryo, but which also provided for the survival of animals that used these plant products as food.

The relationship between insects and flowering plants is well known. The angiosperms, by encouraging insect visits, are able to use insects as delivery agents for pollen. This provides a far more efficient means of pollen dispersal than the more random wind pollination, which requires each plant to disperse tens of millions of pollen grains rather than a few dozen. The insect pollinators are encouraged to visit particular flowers to obtain the nectar and pollen needed for food. Having had success with a flower of a certain form, color, and scent, they move off to find others of the same kind to repeat the favorable experience. In this way, pollen is transported on the hair, legs, and bodies of insects from plant to plant of the same species. The selective competition for efficient pollinators has induced a constantly changing range in variations among both plants and insects. In the angiosperms, the need for each plant to be recognizably different resulted in a spectacular floral variety that has persisted from the Cretaceous down to the present.

## Mesozoic Invertebrates

### Terrestrial Invertebrates

Although paleontologists have assembled an enormous body of information about the marine invertebrates of the Mesozoic, relatively little is known about continental groups. The pulmonate, or air-breathing, snails are rarely found. Freshwater clams and snails are more commonly fossilized. Freshwater crustaceans, including nonmarine species of ostracodes, are frequently found in Mesozoic lake bed deposits. It is reasonable to assume that many varieties of worms existed, but they left few traces. The spiders, millipedes, scorpions, and centipedes that had been abundant in Carboniferous forests were undoubtedly also present in the Mesozoic, al-

though their remains are rare. Among the insects, flies, mosquitoes, caddis flies, earwigs, wasps, bees, and ants are known from rocks as old as Jurassic. Many of the best fossils are collected from the Solnhofen Limestone, an unusual Jurassic formation in Bavaria. Even the Solnhofen collection is not truly representative, however, because most of the fossils are of larger species. Smaller insects are more difficult to find and are not as readily preserved. The fossil record of Cretaceous insects is also sparse. In-

A

B

**Figure 13–10** (A) The pelecypod *Gryphaea* is found in shallow marine Jurassic and Cretaceous strata around the world. It is a member of the oyster family and is characterized by a large left valve that is arched up and over the smaller right valve. (B) Fossils of *Gryphaea* exposed by weathering of the Tununk Shale Member of the Mancos Shale of Cretaceous age. *(Tununk Shale photograph courtesy of Lehi Hintze.)*

sects preserved in amber (fossilized tree sap) of Cretaceous age are known from New Jersey, Arkansas, Canada, and Alaska. Specimens include bees, wasps, ants, beetles, flies, and mosquitoes. Butterflies, moths, termites, and fleas occur in rocks of the early Cenozoic, but have not yet been found in rocks as old as Cretaceous.

## Marine Invertebrates

At the end of the Paleozoic, many families of marine invertebrates either declined or became extinct. The Mesozoic resurgence of marine organisms was somewhat tardy, as indicated by the limited nature of Early Triassic faunas. However, after a few groups had become well established, marine life expanded dramatically. The pelecypods (bivalves) became increasingly prolific from Middle Triassic on and eventually surpassed the brachiopods in colonization of the sea floor. Among the most successful pelecypods were the oysters, represented by such genera as *Gryphaea* (Fig. 13–10) and *Exogyra* (Fig. 13–11). Some members of the oyster group became giants of their kind. Other pelecypods grew conical shells that strikingly resembled the horn corals of the Paleozoic.

In these forms, called **rudists** (Fig. 13–12), the left valve formed a small lid that closed the open end of the cone. Rudists, which are known from the Jurassic and Cretaceous, formed impressive reefs.

In those parts of the Mesozoic marine environment where the water was warm, relatively clear, and shallow, corals proliferated. During the Late Jurassic, for example, the Tethyan seaway was the site of major coralline evolution and reef building. Corals of the Mesozoic are called **scleractinians**. Scleractinians are divided into two important groups: the hermatypic scleractinians that built reefs and the ahermatypic or nonreef scleractinians. As is true today, hermatypic scleractinians had rather restrictive environmental requirements. Most could exist only in clear water of normal salinity no deeper than about 50 meters, with temperatures no lower than about 20°C. One reason hermatypic corals require shallow water is that they have a symbiotic relationship with an algae that lives within the coral polyp and is dependent on sunlight.

**Figure 13–11** *Exogyra arietina* from the Cretaceous Del Rio Clay of Texas. *Exogyra* differs from *Gryphaea* in having the beak of the larger left valve twisted to the side so as not to overhang the right valve. Both genera occur in enormous numbers in shell banks of Cretaceous age along the Gulf and Atlantic coastal plains.

**Figure 13–12** The Late Cretaceous rudist pelecypod *Coralliochama orcutti* from Baja California. (A) Cluster of specimens from the reef deposit. (B and C) Two single specimens. *(From Marincovich, L., Jr. J. Paleontol. 49(1):212–223, 1975. Used with author's permission.)*

**Figure 13–13** Large, well-preserved specimen of the crinoid *Pentacrinus subangularis* from Lower Jurassic strata near Holzmaden, Germany. The extended arms span about 1.2 meters. *(Washington University collection, photograph by John Simon.)*

The great reefs of stony corals offered food and shelter to a host of other kinds of oceanic life. Persisting groups of brachiopods clung to the reef structures, as did pelecypods, algae, bryozoans, and other sedentary creatures. Gastropods grazed ceaselessly along the reef structures, while crabs and shrimp scuttled about seeking food in the recesses and cavernous hollows of the reefs. In the quieter lagoonal areas behind the reefs and on the floors of the epicontinental seas, starfish, sea urchins, and crinoids thrived (Fig. 13–13).

Those cousins of the crinoids, the echinoids, became far more diverse and abundant in the Mesozoic than they had ever been in the preceding era. Some Lower Cretaceous formations in the Gulf Coast states contain prodigious remains of these spiny creatures. Collectors in Europe prize the silicified echinoids obtained from Cretaceous chalk beds. The regular sea urchins were especially numerous. In these forms, the symmetry is fivefold, and the shell, or **test**, is nearly spherical (Fig. 13–14A). The regular forms were overtaken by the irregular echinoids during the Cretaceous (Fig. 13–14B). These are mostly flattened bilateral sea urchins that live as burrowers in the sediment of the sea floor. Starfish and ophiuroids (Fig. 13–15) were also abundant in Mesozoic seas.

A        B

**Figure 13–14** (A) The regular fossil echinoid *Cidaris* from Jurassic marine strata of Germany. (B) *Hemiaster,* a fossil irregular echinoid from the Cretaceous of Mississippi. Specimens are about 6 cm in diameter.

A                                                    B

**Figure 13–15**  Two members of the echinoderm class Stelleroidea. (A) An ophiuroid (brittle star or serpent star) from the Triassic of England. Slab is about 10 cm wide. (B) An imprint of an asteroid (starfish) in a California Cretaceous sandstone. The starfish is about 15 cm across.

If Poseidon had been a paleontologist, he would probably have designated the Mesozoic as the Age of Ammonoids. The name would be suitable, not only because these cephalopods were so abundant, but also because they are useful in worldwide correlation. Zones developed on the basis of ammonoid guide fossils permit correlation of Mesozoic time–rock units with a level of precision surpassing even that of radiometric techniques. You may recall that two orders of cephalopods arose during the Paleozoic Era: the Nautiloidea, having relatively smooth sutures, and the Ammonoidea, having wrinkled sutures. **Sutures** of cephalopods are lines formed on the inside of the conch where the edge of each chamber's partition or **septum** meets the inner wall. Wrinkled sutures are simply a reflection of septa that, like the edges of a pie crust, are fluted. Based on the complexity of the suture patterns (Fig. 13–16), **ammonoids** can be subdivided into three groups: goniatites (see Fig. 11–41), which lived from Silurian to Permian time; ceratites, which were abundant in Permian and Triassic marine areas (Fig. 13–17); and ammonites (Fig. 13–18). Ammonites, although represented in all three Mesozoic periods, were most prolific during the Jurassic and Cretaceous.

Knowledge of the exact suture pattern of ammonoid cephalopods is necessary for their identification and hence their use in correlation. The function of the septal fluting that produced the suture patterns provides an interesting subject for speculation. Most paleontologists favor the theory that, like the corrugated steel panels used in buildings, fluted septa provided greater strength. Comparison studies based on the living cephalopod *Nautilus* have revealed that the gas-filled chambers exert only a slight outward pressure, whereas the water pressure on the outside of the conch wall is considerable. The septal fluting may have helped the animal to withstand the differences in pressure. Proponents of this concept call attention to the fact that ammonoid conchs (unlike nautiloid conchs) tend to thin toward the adult chambers. Apparently, to compensate for that thinner and presumably weaker conch wall, the septa in many species became more closely spaced and more intricately fluted.

The greater variety of Mesozoic ammonoids is an indication of their success in adapting to a variety of marine environments. They seem to have expanded not only within the shallow epicontinental seas but also in the open oceans. During the Cretaceous, many ammonites became aberrant in shape, so that the normally planispiral forms were joined by species with open spirals, straightened conchs, and even some that coiled in a helicoid fashion, like that of a

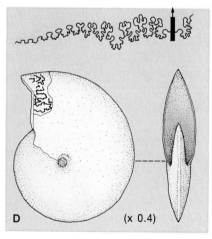

◀ **Figure 13–16** Sutures of cephalopods. (A) nautiloid cephalopod (arrow at midventral line pointing toward conch opening.) (B) Ammonoid cephalopod with goniatitic sutures. (C) Ammonoid with ceratitic sutures. (D) Ammonoid cephalopod with ammonitic sutures. *(From Twenhofel, W. H. and Shrock, R. R. Invertebrate Paleontology. New York: McGraw-Hill Book Co., 1935. Copyright © 1935 McGraw-Hill Book Co.)*

**Figure 13–17** An ammonoid cephalopod displaying ceratitic sutures (note the tiny serrations on the lobes of the sutures). This Triassic specimen was recovered from the Upper Muschelkalk strata of Germany. *(Photograph by E. Holdener.)*

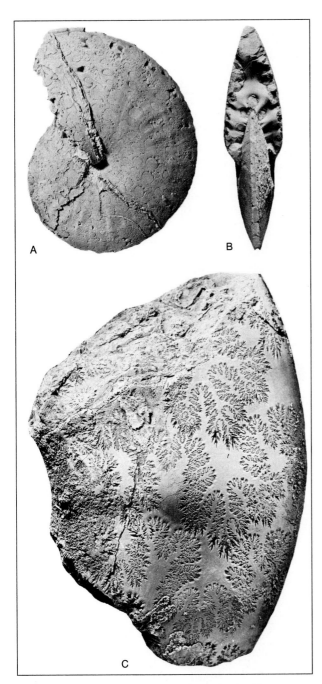

**Figure 13–18** Cretaceous ammonoid cephalopods from the Mancos Shale of New Mexico. (A) *Hoplitoides sandovalensis* (vertical diameter of 9.3 cm). (B) Apertural view of another individual of same species as in part (A). (C) Part of outer whorl of *Tragodesmoceras socorroense* (maximum height of 16 cm). *(Photograph courtesy of Cobban, W. A., U.S. Geological Survey.)*

snail. However, near the end of the Cretaceous, the entire diverse assemblage began to decline and, rather mysteriously, became extinct by the end of the era. Only their close relatives, the nautiloids, managed to survive.

Another group of cephalopods that became particularly common during the Mesozoic were the squidlike **belemnites** (see Fig. 3–41). The belemnite conch was inside the animal and resembled a cigar in shape. The pointed end was at the rear of the animal, and the forward part was chambered. Most belemnite conchs were less than a half meter in size, but some attained lengths of over 2 meters. A few remarkable specimens from Germany are preserved as thin films of carbon and clearly show the 10 tentacles and body form. X-ray study of these fossils has revealed details of internal anatomy as well. Much like the modern-day squid, the belemnites were probably able to make rapid reverse dashes by jetting water out of a funnel located at the anterior end.

The belemnites were highly successful during the Jurassic and Cretaceous. Triassic belemnites may very well have been the ancestors to the squids, which were also numerous during the Jurassic and Cretaceous. Octopods, because they lack a conch, are inadequately recorded in the Mesozoic. However, their presence is affirmed by an imprint of an octopus found in strata of Late Cretaceous age from Lebanon.

Marine gastropods were also abundant during the Mesozoic. Many are found in sediments that represent old beach deposits. Then, as now, cap-shaped limpets grazed slowly across wave-washed boulders while a variety of snails with tall, helicoid conchs crawled about on the surfaces of shallow reefs. For the most part, the gastropod fauna had a decidedly modern appearance and included many colorful and often beautiful forms that have present-day relatives.

Modern types of marine crustaceans, such as crayfish, lobsters, crabs, shrimp, and ostracodes, were abundant by Jurassic time. At some localities, barnacles grew in profusion on reefs and wave-washed rocks.

Among the single-celled protozoan animals that crawled or floated about the Mesozoic seas were the radiolarians (Fig. 13–19) and the foraminifers (Fig. 13–20). As noted in Chapter 11, both of these groups appeared in the Paleozoic. Radiolarians make their open, delicate, spinose skeletons from opaline

**Figure 13–19** Scanning electron micrographs of Jurassic radiolaria from the Coast Ranges of California. Top row, left to right: *Paronaella elegans, Crucella sanfilippoae,* and *Emiluvia antiqua.* Bottom row, left to right: *Tripocyclia blakei, Parvicingula santabarbaraensis,* and *Parvicingula hsui. (From Pessagno, E. A., Jr.* Micropaleontology *23(1):56–113, 1977, selected from Plates 1–12.)*

**Figure 13–20** A living planktonic foraminifer similar to those that became abundant during the Cretaceous and Cenozoic. The fine rays are cytoplasm that is extruded through the pores of the tiny calcium carbonate shell. The shell is approximately 85 $\mu$m in diameter. *(Manfred Kage/Peter Arnold, Inc.)*

silica. In some regions today, radiolarian and diatom skeletal remains accumulate on the sea floor to form extensive deposits of siliceous ooze and in the past have contributed to the formation of chert beds.

The tests of foraminifers are generally more durable than those of radiolarians. The *forams*, as they are often called, left an imposing Mesozoic fossil record and one of great importance in stratigraphic correlation. They are especially important in petroleum exploration. Because of their small size and strong tests, large numbers of foraminifers can be obtained unbroken from the small pieces of rock recovered while drilling for oil. They are then used in tracing stratigraphic units from well to well. Forams are also sensitive indicators of water temperature and salinity and have provided data useful in reconstructing ancient environmental conditions.

Foraminifers were only meagerly represented in the Triassic, but began to proliferate in the Jurassic and Cretaceous. Their expansion continued well into the Cenozoic. Nearly all were bottom-dwelling species until Cretaceous time, when plankton groups began their colonization of the upper levels of the ocean in prodigious numbers. Among the planktic foraminifers, such genera as *Rotalipora* and *Globotruncana* (Fig. 13–21) contributed their empty tests to the calcareous sea floor sedimentary beds, some of which were later lithified into Cretaceous chalks and marls.

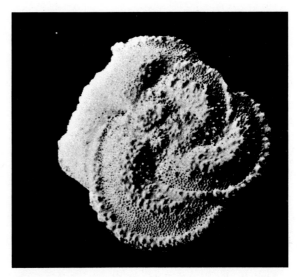

**Figure 13–21**   The Cretaceous planktic foraminifer *Globotruncana* (magnification ×100).

# Mesozoic Vertebrates

## *Reptile Clans*

The reptiles of the Mesozoic were a remarkably diverse group. So diverse that a simple classification will facilitate their study. One such classification is based on the position and number of openings on the sides of reptilian skulls, behind the eye orbits. Some reptiles had two such temporal openings, others only one, and still others lacked this feature altogether (Fig. 13–22). Those with two temporal openings can be placed in the Subclass **Diapsida**. The dinosaurs, flying reptiles or pterosaurs, and all living reptiles except the turtles are diapsids (Fig. 13–23). (Diapsids originated far back in the Pennsylvanian with a primitive group called eosuchians.) Reptiles having only a single temporal opening are designated either **Synapsida** if the opening is low on the skull (below the squamosal and postorbital bones) or

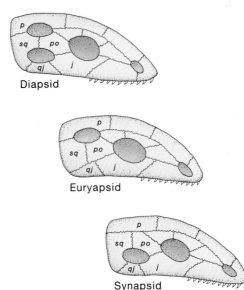

**Figure 13–22**   Reptile skull types. (*p*, parietal; *sq*, squamosal; *po*, postorbital; *j*, jugal; *qj* quadratojugal.)

**Euryapsida** if it is high on the skull (above the squamosal and postorbital bones). Those reptiles such as turtles that lack a temporal opening constitute the **Anapsida**.

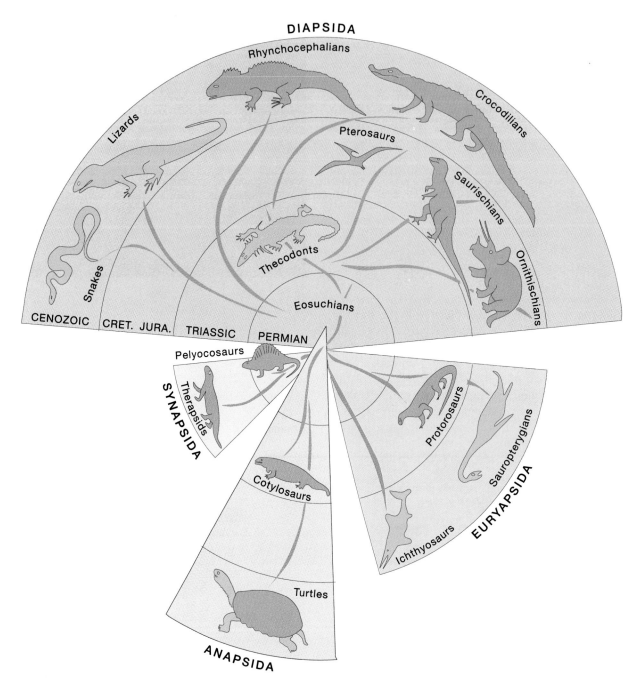

**Figure 13–23** Evolution and general classification of reptiles. In this classification, reptiles are grouped according to the position and number of temporal openings in the skull. Anapsida have no openings, Diapsida have two, Synapsida have one located low on the skull, and Euryapsida have one located higher on the skull. *(From Colbert, E. H. Evolution of the Vertebrates. New York: John Wiley & Sons, 1969. Used with permission of the author, artist Lois Darling, and publisher.)*

## *The Triassic Transition*

The general unrest, broad uplifts, and upheavals that occurred during the Carboniferous and Permian periods caused regressions of epicontinental seas, resulted in a variety of continental environments, and generally provided the environmental stimulus needed to maintain the spread and diversification of land vertebrates. Although marine faunas changed rather abruptly in passing from the Paleozoic to the Mesozoic, there was considerable continuity among land animals. The labyrinthodont amphibians continued into the Triassic before becoming extinct, and the mammal-like reptiles were also able to cross the era boundary. The most progressive of the mammal-like reptiles, the ictidosaurs, succeeded their Permian precursors to become contemporaries of primitive Triassic mammals.

Many new reptile groups appeared in the Triassic. Among these were the ancestors of the first turtles. Well-preserved turtles have been recovered from Upper Triassic rocks in Germany. Although they appear similar to modern turtles, many of these early forms retained teeth in their jaws. The Triassic was also the time during which many lineages of marine reptiles appeared. The rhynchocephalians, represented today by the tuatara of New Zealand, also were abundant. Most interesting of all, however, were reptiles known as **archosaurs**. The Archosauria are a large and important group of vertebrates that include crocodiles, the extinct flying reptiles, dinosaurs, and thecodonts. The thecodonts have a distinguished place in vertebrate history, for they are the ancestors of the dinosaurs.

### The Thecodonts

Early **thecodonts**, as exemplified by *Hesperosuchus* (Fig. 13–24) were small, agile, lightly constructed reptiles with long tails and short forelimbs. They had

**Figure 13–24** *Hesperosuchus* from the Triassic of the southwestern United States. Adult *Hesperosuchus* was about four feet long.

already developed the unique habit of walking relatively erectly on their hind legs. This bipedal mode was an important innovation. Bipedalism permitted thecodonts to move about more speedily than their sprawling ancestors. Because their forelimbs were not used for support, they could be employed for catching prey; even more important, they could be modified for flight. Thus, the thecodonts were the ancestors not only of dinosaurs but of flying reptiles as well.

Not all thecodonts were nimble, bipedal sprinters; some reverted to a four-footed stance and evolved into either armored land carnivores or large crocodile-like aquatic reptiles called **phytosaurs** (Fig. 13–25). Occasionally in the history of life, initially unlike organisms from separate lineages gradually become more and more similar in form. The once distinctly different groups, in fact, change over

**Figure 13–25** *Rutiodon*, a Triassic phytosaur.

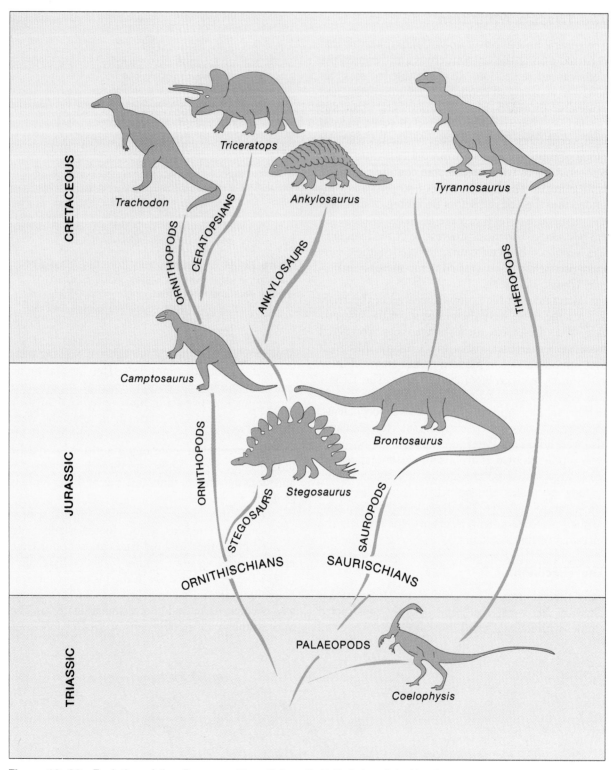

**Figure 13–26**    Evolution of the dinosaurs. *(From Colbert, E. H.* Evolution of the Vertebrates. *New York: John Wiley & Sons, 1969. Used with permission of the author, artist Lois Darling, and publisher.)*

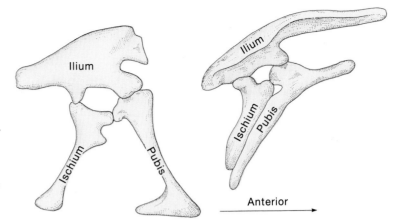

**Figure 13–27**  Basis for the division of dinosaurs into two groups. On the left is the arrangement of pelvic bones in the *Saurischia;* on the right, the arrangement in *Ornithischia*. These views show one side only; the bones are duplicated on the other side.

many generations so that they are better adapted to a particular environment. The evolutionary process responsible for the trend toward similarity in form is called **convergence**. Phytosaurs and crocodiles are good examples of evolutionary convergence. Indeed, the most visible distinction between the two groups is the position of the nostrils, which are at the end of the snout in crocodiles but were just in front of the eyes in phytosaurs.

## The Dinosaurs

Of all the reptiles that have ever lived on this planet, few are more fascinating than the dinosaurs (Fig. 13–26). Dinosaurs are the most awesome and familiar of prehistoric beasts. These headliners of the Age of Reptiles compose not one order, but two, each having evolved separately from the thecodonts. The two orders are the Saurischia (lizard-hipped) and the Ornithischia (bird-hipped). As suggested by these names, the arrangement of bones in the hip region provides the criterion for the twofold classification. The reptile pelvis is composed of three bones on each side. The uppermost bone is the ilium, which is firmly clamped to the spinal column. The bone extending downward and slightly backward is the ischium. Forward of the ischium is the pubis. In the saurischians, the arrangement of the three pelvic bones is triradiate, as it was in their thecodont ancestors. However, in the ornithiscians the pubis is swung downward and backward so that it is parallel to the ischium, as in birds (Fig. 13–27).

The earliest dinosaurs were nearly all saurischians. Most were relatively light, nimble, carnivorous bipeds known from fossils discovered in Triassic beds of both South and North America as well as China. *Coelophysis* (Fig. 13–28), for example, was a small, hollow-boned early saurischian found in the Chinle Formation of New Mexico. These birdlike reptiles, called coelurosaurs, continued into the Ju-

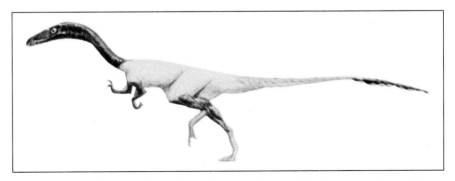

**Figure 13–28**  The Late Triassic carnivorous dinosaur *Coelophysis*. This reptile was about 2.5 meters long. *From A Gallery of Dinosaurs and Other Early Reptiles by David Peters. Alfred A. Knopf, Inc. Copyright © 1989, with permission.)*

**Figure 13–29**  *Ornithomimus,* an ostrich-like Cretaceous dinosaur that was capable of swift running. *(From Osborn, H. F. Bull. Am. Mus. Nat. Hist. 35:733–777, 1917.)*

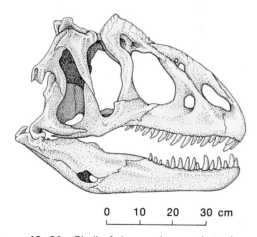

0    10    20    30 cm

**Figure 13–30**  Skull of the carnivorous Late Jurassic theropod *Allosaurus.* Note the long jaws and dagger-like teeth which *Allosaurus* used effectively in dealing with its prey of other reptiles. The skull consisted largely of a framework of sturdy bony arches to which powerful muscles were attached. *(From Osborn, H. F. Bull. Am. Mus. Nat. Hist. 35:733–771, 1917.)*

rassic and Cretaceous. *Ornithomimus* (Fig. 13–29) must have looked very much like an ostrich, with its long neck, toothless jaws, and small head. Because it lacked teeth, paleontologists speculate that it lived on the eggs laid by its contemporaries.

The larger carnivorous saurischians (including coelurosaurs) are called **theropods.** Of more dramatic dimensions than the coelurosaurs were the large carnosaurians such as *Allosaurus* (Fig. 13–30) of the Jurassic and the Cretaceous dinosaurs *Deinonychus* (Fig. 13–31) and *Tyrannosaurus* (Fig. 13–32). The last-named beasts attained lengths of over 13 meters and weighed in excess of 4 metric tons. Carnosaur hindlimbs were robust and muscular. Great curved claws for tearing flesh protruded from each of three toes, whereas a nearly functionless fourth toe bore a smaller claw. The forelimbs were small but powerful. As befits an animal that must kill with its jaws and teeth, the head of the carnosaur was large. Doubly serrated teeth, up to 6 inches long,

**Figure 13–31**  The Cretaceous dinosaur *Deinonychus.* This reptile was about 8 feet long. It possessed a large skull, and the margins of the jaws were set with serrated, saber-like teeth.

**Figure 13–32**  *Tyrannosaurus,* a Late Cretaceous theropod.

lined the powerful jaws. These were truly spectacular predators.

The saurischian group included herbivorous sauropods as well as flesh-eating theropods. The ancestry of the sauropods can be traced to the Late Triassic protosauropod known as *Plateosaurus* (Fig. 13–33). From this smaller, partially bipedal form, the more typical giant sauropods evolved at the beginning of the Jurassic and survived right up to the end of the Cretaceous. They are the animals that people first think of when they hear the word "dinosaur." The best-known sauropods were enormous, long-necked, long-tailed beasts that had returned to the four-legged stance to support their tremendous bulk. A well-known Jurassic representative whose remains have been found in the Morrison Formation

**Figure 13–33** *Plateosaurus,* the Late Triassic ancestor of the giant sauropods.

*Diplodocus* (Fig. 13–35), it had a neck and tail that were unusually long even for a sauropod.

For many years paleonotologists speculated that, even with their pillar-like legs, these largest of land animals could not have supported their own weight continuously. It was therefore surmised that they dwelt in the buoyant waters of lakes and streams. However, there is little evidence to support this theory. It is likely that sauropods roamed the Jurassic forests and uplands much like modern elephants.

Large size afforded certain advantages to the sauropods. Predators often avoid encounters with huge animals. In addition, great size in reptiles also serves to slow changes in body temperature. The ratio of surface area to mass for an animal decreases as size increases. Consequently, a large animal has a proportionately smaller surface for heat loss, and just as a large pot of water loses its heat more slowly than a small pot, so does a large animal lose its heat more slowly than a small animal.

The other major dinosaur line, the Ornithischia, evolved near the end of the Triassic and thrived throughout the remaining Mesozoic. Ornithischians were plant eaters. The teeth in the forward part of the jaws were replaced by a beak suitable for cropping vegetation. The group included both quadrupedal and bipedal varieties, with the bipedal condition considered more primitive. Even the most advanced quadruped ornithischians had such short forelimbs that their descent from bipedal forms seems certain.

of Colorado is *Apatosaurus* (Fig. 13–34). This favorite of schoolchildren (formerly known as *Brontosaurus*) measured almost 20 meters in length and weighed over 3 metric tons. Yet *Apatosaurus* was a relative lightweight when compared to *Supersaurus* and *Ultrasaurus* whose weights probably exceeded 80 tons. *Seismosaurus,* the earth shaker, was over 35 meters (115 feet) long. Like its smaller relative

**Figure 13–34** The enormous Jurassic sauropod *Apatosaurus* (formerly called *Brontosaurus*).

**Figure 13–35** Mounted skeleton of the lengthy Jurassic sauropod *Diplodocus*. From head to tail, *Diplodocus* measured 27 meters (88 feet) long. *(Courtesy of the National Museum of Natural History.)*

The bipedal group of ornithischians is known as **ornithopods.** Their evolutionary history began in the Triassic with relatively small species that lived primarily on dry land. A representative large Jurassic ornithopod is *Camptosaurus* (Fig. 13–36). This was a bipedal dinosaur of medium size with a heavy tail, short forelimbs, and long hind legs. The articulation of the jaw was arranged to bring the teeth together at the same time, an arrangement frequently seen in herbivores down through the ages. Leaves and stems were cropped by the forward, beaklike part of the jaws and passed backward to the cheek teeth for chopping and chewing. From camptosaurid-like ancestors, the larger Cretaceous ornithopods developed. Among these was *Iguanodon* (Fig. 13–37), one of the first dinosaurs to be scientifically described. *Iguanodon,* sometimes called the thumbs-up dinosaur because of the horny spike that substituted for a thumb, is thought to have been a gregarious animal that may have moved about in herds. Evidence for

**Figure 13–36** *Camptosaurus,* a small to medium-sized (2 to 3 meters long) herbivorous dinosaur. Although there were no teeth in the beaklike forward part of the jaw, the remainder of the jaw was equipped with a tight mosaic of row upon row of teeth suitable for grinding plant food.

# Changing Views of the Dinosaurs

Until a few decades ago, paleontologists engaged in the study of dinosaurs devoted most of their efforts to the discovery and description of dinosaur skeletal remains. Assumptions about a dinosaur's behavior were often based on comparisons with crocodiles, large land turtles, or creatures like the Komodo dragon (a giant lizard) of Indonesia. As a result, most dinosaurs were seen as cold-blooded, solitary, and weak-minded. This view of dinosaurs began to change in 1969, when John H. Ostrom presented evidence that some dinosaurs were very likely warm blooded. Other paleontologists quickly joined Ostrom with additional new arguments for endothermy in dinosaurs, as well as fresh ideas about dinosaur behavior, physiology, and ecology. Among these investigators was a tall, laconic dinosaur hunter named John (Jack) R. Horner.

In 1978, Horner was shown a fossil site in the Cretaceous Two Medicine Formation of western Montana. On close examination, it became evident that the fossil location included a hatchery of hadrosaurian dinosaurs, complete with nests, clutches of eggs, hadrosaur embryos, and nestlings. There was evidence that hadrosaur babies were nutured by their parents, that they lived within the social structure of large herds, that they were warm-blooded, and that, in general, they behaved more like birds than living reptiles.

Excavations within the Two Medicine Formation revealed that hadrosaurs had hollowed out bowl-shaped nests in soft fluvial sediments and that within each nest they had laid about 20 eggs in neatly arranged circles. Impressions in the sediment suggest that the eggs were covered with decaying vegetation so that fermentation would provide warmth. Nests were about 7.5 meters (24.6 feet) apart from one another, a distance approximating the length of the adult parent. Horner found two lines of evidence suggesting that the parent hadrosaurs, named *Maiasaura* (from the Greek word for good mother lizard), nurtured their young. He observed that many of the nests contained the bones of juveniles that were about a meter long. Thus, babies that were only about 30 cm long at the time they hatched, stayed in their nests where food was brought to them until they had grown sufficiently to fend for themselves. In addition, the teeth of these juveniles exhibited distinct signs of wear, suggesting they had hatched earlier and had been feeding for some time while still in the

Reconstruction of a nest of hadrosaur eggs from Cretaceous beds in western Montana. *(Courtesy of the Museum of the Rockies, Montana State University, with permission.)*

nest. The interpretation that newly hatched babies stayed in the nest until they tripled their size also supported the interpretation that *Maiasaura* was warm-blooded. A baby crocodile would require three years to triple its size. It is inconceivable that *Maiasaura* babies would have stayed in their nests three years. Like today's ostriches, they probably attained their meter-long size in just a few months. Such rapid growth occurs only in endothermic vertebrates.

In 1984, the hunt for dinosaur remains in western Montana provided Horner and his field party with their best evidence that *Maiasaura* lived and traveled together in enormous herds. The evidence consisted of a bed of rock containing nearly 30 million bones that represent about 10,000 individual *Maiasaura*. The bone bed is about 1¼ miles long by ¼ mile wide. A layer of volcanic ash occurs above the bone bed. Its presence suggests that the *Maiasaura* herd was killed by suffocating ash, lethal gases, or possibly mud flows associated with a nearby volcanic eruption. Such volcanic events were commonplace in the Rockies during the Cretaceous.

**Reference**
Horner, J.R. 1988. *Digging Dinosaurs.* New York, Workman Publishing.

**Figure 13–37** *Iguanodon* (in rear) was a herbivorous ornithischian dinosaur that lived during the Early Cretaceous. Approaching from the right is the carnivorous dinosaur *Baryonx*. The boy in the painting is for scale. *(From* A Gallery of Dinosaurs and Other Early Reptiles *by David Peters. Alfred A. Knopf, Inc. Copyright © 1989, with permission.)*

this comes from a Belgian coal mine where 29 of these individuals were found together as a result of having fallen into an ancient fissure.

By Cretaceous time, ornithopods had moved into a variety of terrestrial environments. A particularly successful group were the trachodonts or hadrosaurs. Some of the hadrosaurs appear to have been aquatic and inhabited lakes and streams. The evidence for this consists of skin impressions showing that the feet were webbed. The vertically flattened tail of these reptiles may have been used in swimming. Other hadrosaurs, however, seem to have lived in drier areas of coastal plains and floodplains. Discoveries of nesting sites in what were the foothills of the ancestral Rocky Mountains indicate they may have moved to higher ground for nesting.

In hadrosaurs, the forward part of the face was broad, flat, and toothless, sometimes resembling the bill of a duck (Fig. 13–38), hence, the name duckbill dinosaurs for members of this group. Behind the toothless forward part of the jaw were lines of lozenge-shaped teeth that appear to have been well adapted for chewing coarse vegetation.

An interesting peculiarity of some hadrosaurs was the development of bony skull crests containing tubular extensions of the nasal passages (Figs. 13–39 and 13–40). The function of these bizarre structures is of special interest to paleontologists. It has been

**Figure 13–38** Mounted skeletons of two Cretaceous duckbill dinosaurs (trachodonts) on display at the American Museum of Natural History in New York.

A

B

**Figure 13–39** (A) Skull of the hadrosaur *Lambeosaurus* with its peculiar hatchet-shaped crest. (B) *Corythosaurus* had a helmet-shaped crest. These crests may have functioned as vocal resonators for dinosaur bellowing. The skulls are approximately 0.75 meters (30 inches) long. *(Courtesy of the National Museum of Natural History.)*

suggested recently that the cranial crests were used to catch the attention of sexual partners of the same species and could be further employed as vocal resonators for promoting an individual's success in obtaining a breeding partner.

Not all ornithopods had skull crests, and in some the crests lacked nasal tubes. One crestless ornithopod that certainly qualifies as a legitimate "bone head" was *Pachycephalosaurus* (Fig. 13–41). The skull in pachycephalosaurians consisted mostly of solid bone with only a small space to accommodate the unimpressive brain.

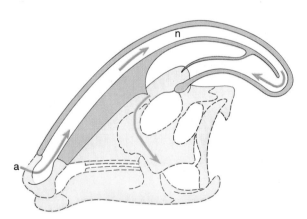

**Figure 13–40** Internal structure of the skull crest of *Parasaurolophus cyrtocristatus*. The superficial bone of the left side has been removed to expose the left nasal passage *(n)*. Air enters nostrils at *a*, moves up and around partition in crest, and from there moves down and back to internal openings in the palate. *(From Hopson, J. A. Paleobiology 1:24, 1975.)*

**Figure 13–41** Reconstruction of the head of *Pachycephalosaurus,* the "bone-head" dinosaur.

**Figure 13–42** *Stegosaurus.* Restudy of the fossil remains of this dinosaur indicate that the plates along the back were arranged in a single line rather than in two alternating rows, as has been frequently depicted.

The best known of the quadrupedal ornithischians are the **stegosaurs,** (Fig. 13–42). Stegosaurs had two pairs of heavy spikes mounted on the tail that were used for defense. However, their more identifying feature was the row of pentagonal plates that stood upright along the back. Scientists have debated the purpose of these plates for many years. One theory that is currently being considered is that the plates functioned in the regulation of body temperature by serving as body heat dissipaters. Indeed, the arrangement, size, and shape of the plates, and their probable rich supply of blood vessels, would favor the thermoregulatory suggestion, although the plates may have served as camouflage or protection as well.

During the Cretaceous, stegosaurs were succeeded by the heavily armored **ankylosaurs.** These bulky, squat ornithiscians were completely covered by closely fitted bony plates that covered the entire length of their 6-meter backsides. The head was small, and also armored. Small, weak teeth were present in the jaws of some ankylosaurs, but *Nodosaurus* (Fig. 13–43) was completely toothless.

The fourth group of quadrupedal ornithischians are the **ceratopsians.** These beasts take their name from horns that grew on the face of all but the earli-

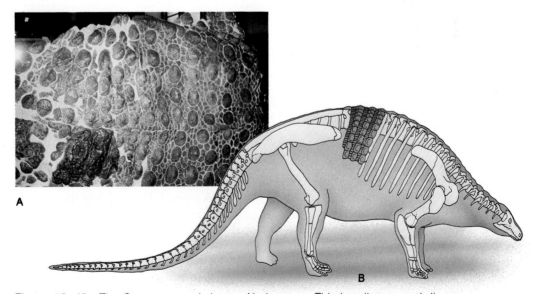

**Figure 13–43** The Cretaceous ankylosaur *Nodosaurus.* This heavily armored dinosaur was about 5 meters long. Part of its armored carapace is shown in the photograph. *(Drawing from Lull, R. S. Amer. Jour. Sci. 5th ser. 1:97–126, 1921.)*

A

B

**Figure 13–44**    The Cretaceous ceratopsian dinosaur *Styracosaurus*. To an advancing enemy, *Styracosaurus* presented a many-horned threat. *(From A Gallery of Dinosaurs and Other Early Reptiles by David Peters. Alfred A. Knopf, Inc. Copyright © 1989, with permission.)*

est forms. Typically, ceratopsians possessed a median horn just above the nostrils, and in some species an additional pair projected from the forehead. The head was quite large in proportion to the body and displayed a shieldlike bony frill at the back of the skull roof that served as protection for the neck region and as a place of attachment for powerful jaw and neck muscles. All ceratopsians possessed jaws that had the shape of a parrot's beak (Fig. 13–44). Judging from the scars that mark the shield bones of some ceratopsians, they were often attacked by the great carnosaurs. No doubt they frequently emerged the victor. During the Late Cretaceous, ceratopsins moved eastward from Asia across the land connection to North America and inhabited the region that lay to the west of the epicontinental sea.

**Were Dinosaurs Warm-Blooded?**    Since the late eighteenth century when dinosaurs were first studied scientifically, they have been regarded as reptiles and hence ectothermic or cold-blooded. Ectothermic animals have little or no ability to maintain a uniform body temperature by physiologic processes. Some ectotherms, however, may regulate their body temperature to a certain degree by seeking either sun or shade in response to temperature needs. In living reptiles, the pineal gland may play a role in directing this type of behavior. In extinct reptiles, certain ana-

tomical features such as the sail in *Dimetrodon* or the plates on the back of *Stegosaurus* may have served to catch the sun's rays or dissipate body heat.

In contrast to largely ectothermic animals, endothermic animals such as mammals and birds maintain a constant body temperature by physiologic production of heat internally and the radiation of excess heat away from the body. Like other animals, mammals produce heat by oxidizing food. However, when body temperature rises, special physiologic mechanisms regulated by the hypothalamus (part of the brain) help to dissipate the heat. These mechanisms include expansion of the blood vessels in the skin, perspiring, or (in furry animals) panting. When temperatures fall, other mechanisms—such as restriction of blood vessels in the skin and shivering—minimize heat loss.

In 1968, paleontologist Robert Bakker boldly proposed that dinosaurs were warm blooded, or endothermic, and thus could no longer be classified as reptiles. Bakker knew that birds and dinosaurs are similar in general anatomical design. If, like birds, dinosaurs were truly endothermic animals, then it would appear logical to restructure the classification of vertebrates by removing dinosaurs from the class Reptilia and erecting a new taxonomic class that would include both dinosaurs and birds. The class Aves would be eliminated. In the new classification, there would be four classes of terrestrial vertebrates:

the class Amphibia (ectothermic amphibians), the class Reptilia (ectothermic reptiles such as lizards and snakes), the class Dinosauria (endothermic dinosaurs and birds); and the class Mammalia. In a very broad interpretation of this proposed reclassification of vertebrates, dinosaurs still "live" today—as birds.

Bakker supported his theory of dinosaurian endothermy with several interesting lines of evidence. One relates to the way dinosaurs stood and walked. Today's lizards and salamanders are ectotherms and most have a sprawling stance, with their limbs directed more or less to the side. In contrast, mammals and birds stand and walk more erect. The limbs are held directly beneath the body. Dinosaur stance resembles that of mammals and birds, and hence, by correlation, dinosaurs were endothermic. But is this an entirely valid correlation? Dinosaur posture may be only an evolutionary solution for supporting their enormous weight.

A second observation considered by some to favor dinosaur endothermy relates to the microscopic structure of bones. Dinosaur bones are richly vascular, like the bones of mammals. Most reptiles have less vascular bones, indicating a poorer supply of blood to the bone tissue. The correlation, however, is not absolute. Some of the bones in living crocodiles have considerable bone vascularity, and crocodiles are primarily ectothermic. Perhaps the high amount of vascular bone in dinosaurs evolved in response to requirements related to growth rates or size and was not related to endothermy.

Remains of dinosaurs have been discovered at Jurassic and Cretaceous paleolatitudes that would normally have had temperate or even cool climates. Their occurrence at such northern locations provides yet another argument in favor of their being warm blooded.

Among the many additional arguments for dinosaur endothermy is one that seeks to make a correlation between predator to prey proportions in mammals (endotherms) as opposed to living reptiles (ectotherms). Today's endothermic communities consist of about 3 percent predators and 97 percent plant-eating prey. It clearly takes a lot of food to fuel the energy requirements of an endothermic predator such as a lion or wolf. In the ectothermic community, 33 percent of the animals are predators and 66 percent are prey. Determining the proportion of predators to prey among dinosaurs is rather tricky because the fossil record cannot be as precise as data obtained from living communities. At the present time, the evidence is contradictory. Early studies indicated that the proportion of predators to prey among dinosaurs was rather like that seen in endothermic communities, but more recent studies provide ratios more like that of ectothermic communities.

By now it should be apparent that the question of dinosaur warm-bloodedness has not been resolved. We can expect the great dinosaur debate to continue into the coming decade. However, even if it is demonstrated that dinosaurs were not warm blooded and were truly reptiles, this does not mean that some did not achieve a measure of body temperature regulation. Their large size provided the means for achieving this feat. As noted earlier, size determines the rate at which an animal loses heat to the outside environment. Large animals have less surface area per unit volume, so they lose heat very slowly and can maintain a more or less consistent internal temperature for a long time, especially in an environment where cold weather is not a factor.

## Aerial Reptiles of the Mesozoic

Again and again in the history of life, the descendants of a small group of animals that were initially adapted to a narrow range of ecologic conditions have, by means of evolutionary processes, radiated into peripheral environments. As the new groups diverged from their ancestral lineage, they changed in ways that made them better suited to their new surroundings. This process, known as **adaptive radiation**, is well demonstrated by the Mesozoic reptiles. The adaptive radiation originated with the stem reptiles of the Late Carboniferous and ultimately produced the enormous variety of large and small, herbivorous and carnivorous, dry land and aquatic animals of the Mesozoic. The radiation did not end with terrestrial vertebrates, however, for during the Mesozoic, reptiles invaded the marine environment and even managed to overcome the pull of gravity for short periods of time.

The first reptiles to attempt to conquer the air were not active flyers, but rather were gliders similar to present-day flying lizards and so-called flying squirrels. For such animals, gliding provided an easy

**Figure 13–45** Two flying reptiles from Soviet Central Asia. (A) *Sharovipteryx* is a Triassic form that is particularly interesting because it appears to represent an intermediate stage between primitive forms and more highly adapted flyers of the Jurassic and Cretaceous (B). The larger form is *Sordes pilosus* of Late Jurassic age. The hairy covering of *Sordes pilosus,* clearly visible in the fine limestone that enclosed the fossil, provides evidence that these creatures were warm blooded.

way to move from tree tops to the ground or from branch to branch. The earliest of such gliders were discovered in rocks of Permian age. They have been given the name *Coelurosauravus.* The skin membranes that served as wings for *Coelurosauravus* were supported along the sides of the body by greatly elongated ribs. Rib-supported wings are not unusual in gliding reptiles. The Triassic lizard *Icarosaurus* (named after Icarus, the ill-fated son of Daedalus) also evolved a similar support for its wings. It is also a characteristic of the present-day *Draco* lizard of the Orient.

A distinct different structure for flying appeared in the Triassic reptile *Longisquama.* The scales in this bizarre creature were as long as its body, and along the forward edge of its forelimbs, they were nearly seven times as long. The long scales could be spread out to either side to make an effective wing. Paleontologists believe that feathers could have evolved from such scales.

The Triassic beds that contained the remains of *Longisquama* also yielded a reptile that was not a passive glider but an active flyer. Its name is *Sharovipteryx.* Because this reptile could truly fly and maneuver in the air, it is considered the first known **pterosaur.** In *Sharovipteryx,* the skin membranes extended between the elbows and the knees and from the rear legs to the tail (Fig. 13–45A).

A discovery from Jurassic beds in the Soviet Union confirmed the generally accepted theory that active, wing-flapping reptilian flyers must have been able to maintain a constant high internal temperature. Without that kind of metabolism, cold air would have reduced their power of exertion and they would have stalled. The actual evidence was found in a well-preserved pterosaur that had a covering of soft hair. Appropriately, it was named *Sordes pilosus* (Fig. 13–45B), meaning the hairy devil. No doubt,

**Figure 13–46** The Jurassic flying reptile *Rhamphorhynchus. Rhamphorhynchus* was a small reptile with a wingspan of about 50 cm. Its most distinctive trait was its long tail, which had a rudder-shaped membrane at its end.

**Figure 13–47** Restored and mounted skeleton of the great flying reptile *Pteranodon ingens* from Cretaceous chalk deposits of Kansas. *Pteranodon* had a wing span of over 7 meters.

the hair provided the insulation required for an endothermic animal.

The pterosaurs may appear to us as ugly, graceless creatures, but their existence from Late Triassic until Late Cretaceous attests to their adaptive success. The Jurassic and Cretaceous pterosaurs that are most familiar typically had rather large heads and eyes and long jaws, which in most forms were lined with thin, slanted teeth. The bones of the fourth finger were lengthened to help support the wing, whereas the next three fingers were of ordinary length and terminated in claws. The wing was a sail made of skin stretched between the elongate digit, the sides of the body, and the rear limbs. There were two general groups of pterosaurs: the more primitive were the rhamphorhynchoids (Fig. 13–46), which had long tails, and the more advanced were tailless pterodactyloids. The latter group is exemplified by *Pteranodon* (Fig. 13–47), species of which had an astonishing wingspan of over 7 meters. The body of *Pteranodon* was about the size of that of a goose. The skeleton was lightly constructed, as is fitting for an aerial vertebrate. The animals probably soared along much like oversized sea birds, snapping up various sea creatures in their toothless jaws. Relative to body size, pterosaurs had somewhat larger brains than some of their land-dwelling relatives. Perhaps this was a result of the higher level of nervous control and coordination needed for flight.

The first prize for large size among pterosaurs must certainly go to *Quetzalcoatlus northropi* (Fig. 13–48) from Upper Cretaceous beds of western Texas. This giant, named after an Aztec god that took the form of a feathered serpent, had an estimated wing-span of 15.5 meters (50.8 feet). *Quetzalcoatlus northropi* ranks as the largest known flying vertebrate ever to soar over the earth's surface.

**Figure 13–48** The giant flying reptile *Quetzalcoatlus* discovered in Upper Cretaceous strata of West Texas. This reptile had a wingspan of 15.5 meters. The flying reptile *Pterodaustro* stands in front of *Quetzalcoatlus*. *(From* A Gallery of Dinosaurs and Other Early Reptiles *by David Peters. Alfred A. Knopf, Inc. Copyright © 1989, with permission.)*

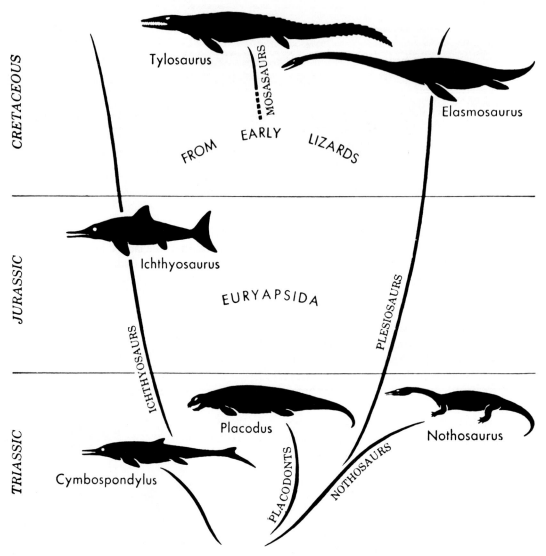

**Figure 13–49** Evolutionary relationships of major groups of marine reptiles of the Mesozoic. *(From Colbert, E. H. Evolution of the Vertebrates. New York: John Wiley & Sons, 1969. Used with permission of the author, artist Lois Darling, and publisher.)*

## A Return to the Sea

The marine habitat is one in which the archosaurs were not notably successful. Only one archosaurian group—the sea crocodiles—were able to invade the oceanic environment. Other reptilian groups, such as the ichthyosaurs, plesiosaurs, mosasaurs, and sea turtles, however, were very successful in their return to the productive seas of the Mesozoic (Fig. 13–49). Many fed upon ammonoids, sharks, and bony fishes

**Figure 13–50** The skull of *Placodus,* a bulky, mollusk-eating, paddle-limbed marine reptile of Triassic age. The forward teeth were modified for plucking shellfish from the sea floor, and the cheek teeth for crushing their shells.

that had preceded them in populating the seas. During the Cretaceous, the most modern of the bony fish, the teleosts, made their appearance.

Not unexpectedly, the invasion of the marine environment required many modifications in form and function. Paddle-shaped limbs and streamlined bodies evolved to allow efficient movement through the water. Because these reptiles were unable to abandon air breathing and reconvert to gills, their lungs were modified for greater efficiency. In those sea-going reptiles that were unable to lay their eggs ashore, reproductive adaptations provided for birth of the young while at sea.

Marine reptiles that had paddle-shaped limbs for locomotion were already present during the Triassic Period. One group, the **nothosaurs,** were just beginning to take on adaptations that would be perfected in their likely descendants, the plesiosaurs. The nothosaurs were joined in the Triassic by a group of mollusk-eating, flippered reptiles known as **placodonts** (Fig. 13–50). These bulky animals had distinctive pavement-type teeth in the jaws and palate, which they used for crushing the shells of the marine invertebrates upon which they fed.

By far the best known of the paddle swimmers were the **plesiosaurs** (Fig. 13–51). Their earliest remains are found in Jurassic strata. Sometimes nicknamed "swan lizards," they had short, broad bodies and large, many-boned flippers. In some species the neck was extraordinarily long and was terminated by a smallish head. Slender, curved teeth, suitable for ensnaring fish, lined the jaws. *Elasmosaurus,* a well-known Cretaceous plesiosaur, attained an overall length in excess of 12 meters.

In addition to the long-necked types, there were many plesiosaurs characterized by short necks and large heads. It is likely that the short-necked plesiosaurs were aggressive divers. *Kronosaurus,* a giant, short-necked form from the Lower Cretaceous of Australia, had a skull size that probably exceeds that of any known reptile. It was 3 meters long.

The most fishlike in form and habit of all the marine reptiles were the Triassic and Jurassic **ichthyosaurs** (Fig. 13–52). In many ways, they were the reptilian counterparts of the toothed whales of our present-day oceans. Ichthyosaurs had fishlike tails in

**Figure 13–51** The large Cretaceous plesiosaur *Elasmosaurus.*

**Figure 13–52**   Restoration of an ichthyosaur, a marine reptile similar to a modern porpoise in form and habits.

which vertebrae extended downward into the lower lobe, boneless dorsal fins to help prevent sideslip and roll, and paddle limbs for steering and braking. The head was a suitably pointed entering wedge for cutting rapidly through the water. A ring of bony plates surrounded the large eyes and may have helped protect them against water pressure. Clearly, these were active predators with good vision and the ability to move swiftly through the water.

The **mosasaurs,** a group of giant monster lizards, were a highly successful group of Cretaceous sea dwellers. A typical mosasaur (Fig. 13–53) had an elongate body and porpoise-like flippers. The creatures propelled themselves through the water by the sculling action of their long, vertically flattened tails and the rhythmic undulations of their long bodies. The lower jaw had an extra hinge at midlength (Fig. 13–54), which greatly increased its flexibility and gape. Mosasaurs were primarily fish eaters, but some frequently dined on large mollusks. Conchs of cephalopods have been found with puncture marks that precisely match the dental pattern of their mosasaurian foes.

Perhaps less spectacular than the mosasaurs, but far more persevering, were the sea turtles. In this group, also, we find a trend toward increased size. The Cretaceous turtle *Archelon* (see Fig. 13–52), for example, reached a length of nearly 4 meters. As an adaptation to their aquatic habitat, the carapace of the marine turtle was greatly reduced and the limbs were modified into broad paddles.

As briefly indicated earlier, the marine crocodiles (Fig. 13–55) were the only archosaurian group that entered the sea. They became relatively common during the Jurassic, but only a few remained by Early Cretaceous. It is likely that they did not fare well in competition with the mosasaurs.

## The Birds

From the time of Darwin, naturalists have been aware of the structural similarities between birds and

**Figure 13–53**   *Tylosaurus,* the giant marine lizard or mosasaur of the Cretaceous, prepares to make a meal of *Archelon,* a huge marine turtle. *Pteranodon* soars above the waves seeking its food of fish and cephalopods. *(Courtesy of the National Museum of Natural History.)*

**Figure 13–54** Mounted skeleton of a Cretaceous mosasaur. These giant marine lizards attained lengths of over 9 meters (30 feet). *(Courtesy of the National Museum of Natural History.)*

reptiles. These similarities have prompted the statement that birds are only glorified reptiles that have gained wings and feathers and lost their teeth. However, such statements depreciate the marvelous attainments of birds, not the least of which are their superior powers of flight and high level of endothermy. Both of these attributes are related to the evolution of feathers. Developmentally, feathers are homologous with reptilian scales. In their earliest development, they may not have functioned in flight, but instead for insulation, camouflage, or display.

There is little doubt that birds evolved from small Triassic theropods that were already birdlike in their bipedal stance and in the structure of their forelimbs and hindlimbs, shoulder girdle, and skull. Earlier in this chapter we noted a proposed new classification of vertebrates in which the close relationship between dinosaurs and birds is indicated by the inclusion of both groups within the same taxonomic class (the Dinosauria).

The first, undisputed fossil bird to be discovered, *Archaeopteryx,* is the perfect evolutionary link between small, bipedal theropods and modern birds. *Archaeopteryx* (Fig. 13–56) was about the size of a crow. With the exception of its distinctly fossilized feathers (arranged on the forelimbs as in modern flying birds), its skeletal features were largely reptilian. The jaws bore teeth, and the creature had a long, lizard-like tail that bore feathers. Unlike the wings of modern birds, in which the bones of the digits coalesce for greater strength, the primitive wings of *Archaeopteryx* retained claw-bearing free fingers for climbing and grasping. The light-weight sternum lacked a keel, indicating that the heavy muscles needed for sustained flight were lacking. Because of this, some paleontologists believe *Archaeopteryx* was a structurally primitive evolutionary side branch and that it did not give rise to more advanced fliers. Support for this view came in 1986 with the discovery of two crow-sized skeletons of birds that not only were 75 million years older than *Archaeopteryx,* but also possessed keel-like breast bones. Although feather impressions were not found with the skeletons, the tiny bones of the forearms possessed a series of nodes that may have been the sites for the attach-

**Figure 13–55** Toothy skull of the Jurassic marine crocodile *Geosaurus.* Length of skull is about 45 cm.

A                                    B

**Figure 13–56**    The Jurassic bird *Archaeopteryx.* (A) The skeleton of *Archaeopteryx* in the Solnhofen Limestone of Germany. The Solnhofen beds formed from lime mud deposited on the floor of a tropical lagoon. The specimen resides in the Berlin Museum of Natural History. *(John D. Cunningham/Visuals Unlimited.)* (B) Restoration of *Archaeopteryx.* *(From a painting by Rudolph Freund, Courtesy of the Carnegie Museum of Natural History.)*

ment of feathers. This seemingly more advanced vertebrate has been given the name *Protoavis.*

Small, delicate, hollow-boned animals are not readily preserved, and the fossil record for birds—especially Mesozoic birds—is not good. The Cretaceous provides only a partial glimpse of bird evolution. Toothed birds otherwise resembling terns and gulls are occasionally found in the Cretaceous deposits of the inland chalky seas. Some Cretaceous birds, such as *Hesperornis,* became excellent swimmers. They retained feathers and other birdlike characteristics but lost their wings and relied upon their webbed feet for swimming (see Fig. 12–36). Marine sediments provided the rapid burial needed to preserve these aquatic birds. On land, preservation of Mesozoic birds was rare indeed, and their fossil record is quite inadequate.

## The Mammalian Vanguard

While the Mesozoic reptiles were having their heyday, small, furry animals were scurrying about in the undergrowth and unwittingly awaiting their day of supremacy. These shrewlike creatures were the primitive mammals. On the basis of rare and often minuscule remains, they are known from all three systems of the Mesozoic. Among the earliest of the mammals were the **morganucodonts** (Fig. 13–57), jaws, teeth, and skull fragments of which have been recovered in Upper Triassic rocks of south Wales. It is evident from these rather scrappy remains that morganucodonts still retained many vestiges of reptilian structures. However, the articulation of the jaw to the skull was mammalian, and the lower jaw was functionally a single bone, as in mammals. New dis-

**Figure 13–57** Restoration of *Morganucodon,* an early mammal from the Late Triassic of Wales.

coveries of early mammals in regions other than England are currently being studied. One of these from the Triassic of Texas may be about 10 million years older than *Morganucodon.*

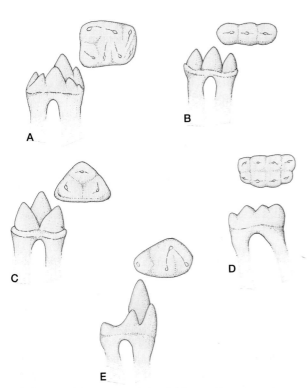

**Figure 13–58** Molar teeth of Mesozoic mammals. Side views of lower molars (as viewed from inside the mouth) and top views of the oral surfaces. (A) Docodont. (B) Triconodont. (C) Symmetrodont. (D) Multituberculate. (E) Eupantothere.

There were at least six additional groups of Mesozoic mammals. Each is recognized primarily on the basis of tooth morphology. Among those known from the Jurassic and Cretaceous are the docodonts, triconodonts, symmetrodonts, multituberculates, and eupantotherians (Fig. 13–58). The **docodonts** had multicusped molar teeth, which suggests that they may have been the stock from which present-day monotremes evolved. Very likely, they fed upon insects, as did many of these primitive mammals. The **triconodonts** can be recognized by cheek teeth in which three cusps are aligned in a row. Some were as large as cats and may have preyed upon smaller vertebrates. **Symmetrodonts** had molars constructed on a more or less triangular plan. As suggested by their name, **multituberculates** had teeth with many cusps. They may have been the first entirely herbivorous mammals. Their chisel-like incisors and the gap between the incisors and the cheek teeth gave them a decidedly rodent-like appearance (Fig. 13–59) and indicate that they probably had rodent-like habits as well.

**Figure 13–59** The rodent-like multituberculate *Taeniolabis.*

From an evolutionary point of view, the **eupantotheres** are the most important of the mammalian vanguard, for in the Late Cretaceous they gave rise to the marsupials and placentals. Marsupials, for example, have an asymmetrically triangular pattern of cusps that was evidently inherited from the pantotheres. In general, pantotheres were small, ratlike animals with long, slender, toothy jaws. It is likely that they preyed not only upon insects but also upon small lizards and mammals.

For the mammals, the Mesozoic was a time of evolutionary experimentation. For 100 million years, they effectively and unobtrusively lived among the great reptiles while simultaneously improving their nervous, circulatory, and reproductive anatomy. Equipped with an exceptionally reliable system for control of body temperature, they were able to thrive in cold as well as warm climates. As the reptile population declined near the end of the era, mammals quickly expanded into the many habitats vacated by the saurians.

# A Second Time of Crisis

Just as the end of the Paleozoic was a time of crisis for animal life, so also was the conclusion of the Mesozoic. Primarily on land but also at sea, extinction overtook many seemingly secure groups of vertebrates and invertebrates. In the seas, the ichthyosaurs, plesiosaurs, and mosasaurs perished. The ammonoid cephalopods and their close relatives, the belemnites, as well as the rudistid pelecypods, disappeared. Entire families of echinoids, bryozoans, planktonic foraminifers, and calcareous phytoplankton became extinct. On land the most noticeable losses were among the great clans of reptiles. Gone forever were the magnificent dinosaurs and soaring pterosaurs. Turtles, snakes, lizards, and crocodiles, and the New Zealand reptile *Sphenodon* (the tuatara lizard) are the only reptiles that survived the great biologic crash (see Fig. 13–23). Altogether, in the Late Cretaceous, extinctions eliminated about one-fourth of all known families of animals.

The question of what caused the decimation in animal life at the end of the Mesozoic continues to intrigue paleontologists. Scores of theories, some scientific and many preposterous, have been offered. Those that have the most credibility attempt to explain simultaneous extinctions of both marine and terrestrial animals and seek a single or related sequence of events as a cause. In general, the theories tend to fall into two broad categories. The first of these categories relies on some sort of extraterrestrial interference, such as an encounter with an asteroid or comet. Such events are considered catastrophic, in that their effects are concentrated within a relatively short span of time. The second group of hypotheses proposes that the cause of extinctions was events that occurred right here on earth.

## *Extraterrestrial Causes of Extinctions*

### Asteroid Impact

Since the time that geologists first became aware of the extinctions that mark the Cretaceous–Tertiary boundary, there have been attempts to place the blame for those extinctions on meteorites, comets, or lethal cosmic radiation. Tangible evidence to support those ideas, however, was lacking. This situation changed in 1977 when geologist Walter Alvarez discovered a thin layer of clay at the Cretaceous–Tertiary boundary outside of the town of Gubbio, Italy (Fig. 13–60). Alvarez sent samples of the clay to his father Luis, a physicist, who had the clay analyzed. The result of the analysis was startling. The samples contained approximately 30 times more of the metallic element iridium than is normal for the earth's crustal rocks. Where could this high concentration of iridium have come from? Iridium is probably present in the earth's core and perhaps the mantle, but how could the metal from so deep a source find its way into a clay layer at the Cretaceous–Tertiary boundary? While volcanism is a possibility, iridium also occurs in extraterrestrial objects such as asteroids (or their fragments, meteorites). The father and son Alvarez team favored this extraterrestrial origin for the iridium in the clay layer and proposed that an iridium-bearing asteroid had crashed into the earth at the end of the Cretaceous. The explosive, shattering blow from the huge body (presumed to be over 10 km in diameter) would have thrown dense clouds of iridium-bearing dust and other impact ejecta into the atmosphere. Mixed and transported

**Figure 13–60**    The coin marks the location of the iridium-rich layer of clay that separates Cretaceous and Tertiary rocks near Gubbio, Italy. The gray Cretaceous limestone below the coin contains abundant fossil coccolithophores, but few of these phytoplankton remain in the Tertiary beds above the iridium-rich clay layer. *(Lawrence Berkeley Laboratory, University of California.)*

by atmospheric circulation, the dust might have formed a lethal shroud around the planet, blocking the sun's rays, and thereby causing the demise of marine and land plants upon which all other forms of life ultimately depend. As the dust settled, it would have formed the iridium-rich clay layers found at Gubbio and subsequently in Denmark, Spain, New Zealand, North America, Austria, Haiti, the Soviet Union, and in sediment layers beneath the Atlantic and Pacific oceans (Fig. 13–61).

In addition to the iridium-rich clay layer found at various sites around the world, there are other kinds of evidence for the impact of a large extraterrestrial body at the end of the Cretaceous. One of these is the widespread occurrence of **shocked quartz** (Fig. 13–62) at the Cretaceous–Tertiary boundary. These mineral grains are recognized by distinctive parallel sets of microscopic planes (called *shock lamellae*) that were produced when high-pressure shock waves, such as those emanating from the impact of a large meteorite, travel through quartz-bearing rocks. Often in the same stratum containing grains of shocked quartz one also finds tiny glassy

spherules thought to represent droplets of molten rock thrown into the atmosphere during the impact event. Finally, sediments at the Cretaceous–Tertiary boundary often include a layer of soot that may be the residue of vegetation burned during widespread fires caused by extraterrestrial impact.

A meteorite or comet that explodes on striking the earth is called a **bolide**. If a bolide struck our planet about 65 million years ago, one would hope to find traces of the crater it produced. Recently, a buried impact structure of appropriate age and size has been discovered in the Yucatan Peninsula of Mexico. This feature, as well as other probable impact craters in India, the United States, and Siberia, are currently being investigated.

On first examination, the bolide-impact hypothesis seems a tidy way to account for the extinction of dinosaurs and many of their animal and plant contemporaries. Like all hypotheses, however, it had to stand the test of careful scrutiny by the scientific community. Were the Alvarezes correct in assuming the iridium was derived from an impacting asteroid? And if indeed such an impact did occur, did it cause

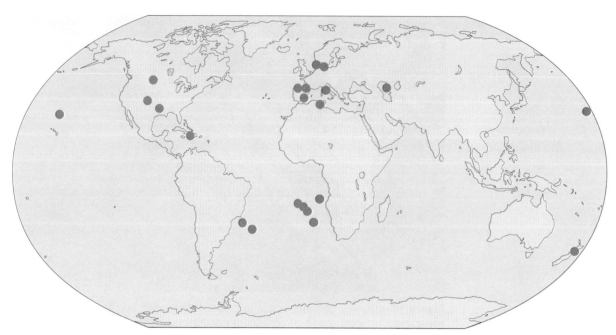

**Figure 13–61** Occurrences of the iridium-rich sediment layer at the Cretaceous–Tertiary boundary. *(From Alvarez, W., et al. Geol. Soc. Amer. Special Paper 190:305–315, 1990.)*

**Figure 13–62** A shocked quartz grain from the Cretaceous–Tertiary boundary layer. Note the parallel intersecting sets of microscopic planes called shock lamellae. These intersecting lamellae are similar to those observed in quartz from rock fragments clearly known to have been subjected to meteorite impact. *(Courtesy of Jim S. Alexopoulos.)*

the extinctions or merely contribute to terrestrial processes already in operation?

In regard to the first question, geologists have found evidence that iridium in clays like that at Gubbio may have its source in the earth's mantle, from which it can move to the surface by way of conduits and blast into the atmosphere as iridium-rich volcanic ash and dust. Volcanism was indeed very prevalent during the late stages of the Cretaceous.

If we assume the impact hypothesis is correct, then it is still difficult to explain why some animals perished from the proposed blast, yet small mammals, birds, lizards, crocodiles, bony fishes, many deciduous plants, and certain mollusks survived the hard times and proliferated during the Cenozoic.

## Planet X

Stimulated by the idea of an extraterrestrial cause for extinctions, scientists have provided several interesting alternatives to the asteroid-impact hypothesis. Blasts generated by comets encountering the earth, for example, could also be disastrous for life. The arrival of one or more large comets would heat the atmosphere to intolerable levels and possibly provide a rain of toxic cyanide to doom life in the seas. Scientists promoting this hypothesis have been encouraged by evidence that extinctions, less drastic in effect than the ones at the Cretaceous–Tertiary boundary, seem to have occurred on earth every 26 million years or so. This very periodicity would seem to favor some sort of regular astronomic interference, in contrast to the randomness of asteroid impact. One hypothesis to account for the periodicity proposes the existence of a tenth planet, dubbed Planet X, in the outer reaches of our solar system. Astronomers have been predicting the existence of such a planet since the late nineteenth century to explain perturbations in the movements of Neptune and Uranus. Calculations suggest that Planet X would require about 56 million years to complete its huge inclined and elliptical orbit around the sun. Twice during each orbital trip (every 28 million years) it would pass through a cloud of comets (called the Oort cloud) that lies beyond the orbit of Neptune. As it traveled through the field of comets, Planet X would probably dislodge some of those comets and send them hurtling to the earth, with devastating ef-

fect upon the biosphere. It is an interesting hypothesis that awaits evidence that can only come from continuing astronomic observation.

## Cosmic Rays

If there were no actual impacts of comets, asteroids, or large meteorites to cause extinctions at the end of the Cretaceous, might there still have been other more subtle but equally effective extraterrestrial causes? One culprit might have been an influx of abnormally high amounts of cosmic radiation. Such radiation might have destroyed organisms directly or severely impaired their reproductive capabilities. Most proponents of this idea suggest that reversals of the earth's magnetic field (a phenomenon known to occur) may have temporarily eliminated the planet's protection from harmful rays. The hypothesis, however, has some aspects that are difficult to defend. It does not explain why marine organisms, which could descend a few meters and escape radiation damage, were decimated at levels far in excess of those among unprotected land plants. Also, several known periods of polar reversals do not correlate to episodes of mass extinction. In addition, the Mesozoic extinctions spanned a time interval far greater than that encompassed by documented magnetic reversals.

Another possibility for radiation damage unrelated to magnetic reversals might have been the arrival of a blast of lethal rays from a nearby supernova. Supernovas are colossal stellar explosions that radiate about as much energy as 10 billion suns and can affect the surface of planets located as far as 100 light years away (1 light year is approximately 8,898,000,000,000 km). Astronomers estimate that, on the average, a supernova may occur within 50 light years of the earth every 70 million years. If such an event did take place, there seems little doubt that plant and animal communities would have been severely affected. For the present, however, we must await firm evidence that such a stellar explosion did occur about 65 million years ago.

## *Terrestrial Causes of Extinctions*

Several hypotheses that seek to explain the demise of plants and animals near the end of the Cretaceous propose that events on earth upset the ecologic balance between organisms and the environment to which they had become adapted. Recall from Chapter 12 that continents during the Cretaceous were extensively covered by shallow, warm, epicontinental seas. Many geologists believe these inundations were caused by displacement of ocean water when mid-oceanic ridges were raised as a consequence of accelerated rates of sea floor spreading. Whatever the cause, the warm inland seas that resulted were very favorable for marine life and helped to moderate and stabilize climates on the continents as well. Conditions favored the growth of living things, and plants and animals experienced remarkable increases in variety and abundance. But these favorable conditions were soon to change. Studies of the stratigraphic sequence across the Cretaceous–Tertiary boundary in Denmark indicate a global lowering of sea level at the end of the Mesozoic. Perhaps this change in sea level was caused by a slowing of rates of sea floor spreading. Whatever its cause, the change in sea level spelled disaster to animals and plants of shallow coastal areas, and especially to the phytoplankton on which the marine food web depends. Without the moderating effects of vast tropical seas, continental regions would also have experienced harsher climatic conditions and more extreme seasonality. Many families of organisms that had become adjusted to the previous environment might not have been able to adjust to the changes and, one by one, would have met their end. As is characteristic within the complex ecologic web, their demise would affect associated organisms, resulting in waves of extinctions among directly or indirectly dependent species.

There are many other variations on the general theme of environmental factors as a cause of extinctions. Some geologists dispute the evidence for a fall in sea level and invoke other events to put stress on the global ecosystem. For example, there is evidence that the Arctic Ocean was landlocked during the Cretaceous, with the result that it became a freshwater body. Near the end of the Cretaceous, tectonic plate movements broke the land barrier containing the Arctic water and the light, nonsaline water flowed southward over the denser, salty Atlantic. Marine plankton living in the vital upper zones of the ocean could probably not tolerate the change in water chemistry and normal marine circulation. A chain of ecologic disasters would have been the probable result.

Several hypotheses attempt to relate terminal Cretaceous extinctions to volcanism. As mentioned earlier, the late Mesozoic world experienced an abundance of volcanic activity. It is not unreasonable to hypothesize that the emission of large quantities of $CO_2$ from volcanic and fissure eruptions such as those that produced the extensive Deccan basalts of India may have caused climatic changes harmful to both terrestrial and marine organisms.

There are still many other hypotheses to account for the Late Cretaceous extinctions, but the examples provided here illustrate the complexity of the problem. The debate between those favoring sudden extraterrestrial causes and those supporting purely terrestrial hypotheses will surely continue for decades. Whatever the final outcome, it is a fact that hard times near the end of the Mesozoic doomed the dinosaurs, pterosaurs, ammonites, and over three-fourths of the known species of marine plankton. The extinctions, however, did not happen at the same time for all families, nor were they sudden occurrences. Indeed, families or groups of families died out sporadically over an interval of 0.5 to 5 million years. This fact alone is the most persuasive in the argument against a sudden extraterrestrial catastrophe. Perhaps the real cause of extinctions will be found in a combination of the two categories of hypotheses. The paleontologic evidence strongly indicates that extinctions were under way in the Late Cretaceous. The debilitated and diminished survivors of environmental hard times may have been dealt a coup de grace by an extraterrestrial event.

# SUMMARY

Climates of the Mesozoic were in general mild and equable except for occasional intervals of aridity and an episode of cooler conditions near the end of the era. In the widespread Mesozoic seas, coccoliths and diatoms flourished, as did such invertebrate groups as ammonoids, belemnites, oysters and other pelecypods, echinoderms, corals, and foraminifers. On land, seed ferns and conifers were common in Triassic and Jurassic forests. In the Cretaceous Period, the flowering plants expanded and with them a multitude of modern-looking insects. The changing composition of plant populations was matched by innovations among terrestrial vertebrates. The most dramatic of these changes involved the evolution of dinosaurs, pterosaurs, and birds from the small bipedal thecodonts of the Triassic.

The dinosaurs were the ruling reptiles of the Mesozoic. Both carnivorous and herbivorous varieties occupied a variety of habitats. Based on a basic difference in pelvic structure, two orders of dinosaurs, the Saurischia and Ornithischia, are recognized. Certain of the dinosaurs as well as the pterosaurs may have been endothermic, as indicated by bone structure, posture, and predatory–prey relationships. The reptilian dynasty extended to the oceans as well, where ichthyosaurs, plesiosaurs, and mosasaurs competed successfully with the most modern of fishes, the teleosts. The Mesozoic is also noteworthy as the era during which two new vertebrate classes appeared: mammals and birds. The birds, with true feathers for insulation and flight, evolved from small carnivorous dinosaurs (theropods). The oldest remains that are unquestionably those of a true bird are of *Archaeopteryx* from the Jurassic. *Protoavis,* from Triassic strata, is currently being evaluated as a possibly older bird.

Mammals made their debut during the Triassic. In general, most of these primitive mammals remained small and inconspicuous. One group, the eupantotheres, gave rise to marsupial and placental mammals during the Cretaceous.

Like the Paleozoic, the Mesozoic Era closed with an episode of extinctions. Many groups of both marine and terrestrial reptiles succumbed, as did the ammonoids and belemnites. Several of the major taxonomic classes survived, but entire families within those classes were exterminated. Still other groups of Mesozoic organisms experienced rapid evolutionary development and thereby were able to keep pace with environmental changes.

Geologists are not agreed on the cause of terminal Cretaceous extinctions. There is, however, an impressive body of evidence that a large meteorite or asteroid struck the earth about 66 million years ago, and the effects of the impact may have caused or contributed to the mass extinction. Others argue that the extinctions can be attributed to extensive volcanic activity, withdrawal of epeiric seas, and changes in climate.

# QUESTIONS FOR REVIEW

1. What are the coccoliths? In what way are they involved in the formation of chalk layers? When were they particularly abundant?

2. What are diatoms? How do they differ in composition and morphology from coccolithophorids?

3. In what way might prodigious phytoplankton productivity affect global climate and atmospheric composition?

4. How did the flora of terrestrial plants during the Cretaceous differ from Jurassic floras?

5. In general, how did the marine invertebrates of the Mesozoic Era differ from those of the preceding Paleozoic?

6. How did Mesozoic ammonoids differ from nautiloids? What attributes of the ammonoids result in their having special value as guide or index fossils?

7. What are foraminifers? What is there about these organisms that has resulted in their extensive use by petroleum geologists in the correlation of subsurface strata?

8. What clues might one hope to find in attempting to establish that a particular group of Mesozoic reptiles were endothermic (warm blooded)?

9. What two major classes of vertebrates appear for the first time during the Mesozoic Era?

10. What is meant by evolutionary convergence? Cite an example.

11. Formulate a single sentence description of the following Mesozoic animals.

a. Pterosaurs
b. Ichthyosaurs
c. Sauropods
d. Ornithischians
e. Teleosts
f. Theropods
g. Ornithopods
h. Mosasaurs
i. Morganucodonts
j. Saurischians

12. What is the evolutionary importance of the following Mesozoic vertebrates?
a. Thecodonts
b. *Archaeopteryx*
c. Eupantotheres

13. What reptilian groups managed to survive the wave of extinctions that occurred at the end of the Cretaceous Period?

14. What attributes already present in Jurassic and Cretaceous mammals contributed to their survival during the biologic crisis at the end of the Mesozoic?

15. What evidence at the boundary between the Cretaceous and Tertiary systems at many localities supports the asteroid impact hypothesis for the extinction of the dinosaurs? What evidence tends to dispute asteroid impact as a cause of dinosaur extinction?

## TERMS TO REMEMBER

adaptive radiation
ammonoids
Anapsida
angiosperm
ankylosaur
archosaurs
belemnites
biosphere
bolide
ceratopsian
coccoliths

convergence
cycads
Diapsida
diatoms
docodont
Euryapsida
ichthyosaur
mosasaur
morganucodont
multituberculate

nothosaur
ornithopod
pantotherian
phytoplankton
phytosaur
placodont
plesiosaur
pterosaur
rudist
scleractinians

septum (septa)
shocked quartz
stegosaur
suture
symmetrodont
Synapsida
test
thecodonts
theropod
triconodont

## SUPPLEMENTAL READINGS AND REFERENCES

Albritton, C. C., Jr. 1989. *Catastrophic Episodes in Earth History*. New York: Chapman and Hall.

Alvarez, L. W., Alvarez, W., Asaro, F., and Michel, H. V. 1980. Extraterrestrial cause for the Cretaceous–Tertiary extinction. *Science* 208:1095–1108.

Alvarez, W., Alvarez, L. W., Asaro, F., and Michel, H. V. 1982. Current status of the impact theory for terminal Cretaceous extinction, In *Geologic Implications of*

*Impacts of Large Asteroids and Comets on the Earth.* (eds. Silver, L. T., and Schultz, P. H.), *Geol. Soc. Amer. Special Paper* 190:305–315.

Bakker, R. T. 1986. *The Dinosaur Heresies*. New York: William Morrow.

Gartner, S., and McGuirk, P. 1979. Terminal Cretaceous extinction, scenario for catastrophe. *Science* 206:1272–1276.

Halstead, L. B. 1982. *The Search for the Past*. Garden City, New York: Doubleday & Co.

Horner, J. R., and Gorman, J. 1988. *Digging Dinosaurs*. New York: Workman.

Hsu, H. J. 1986. *The Great Dying*. New York: Harcourt Brace Jovanovich.

Langston, W., Jr. 1981. Pterosaurs. *Sci. Am.* 244(2):122–137.

McLean, D. M. 1985. Deccan Traps mantle degassing in the terminal Cretaceous marine extinctions. *Cretaceous Res.* 6:235–259.

McNeil, A. R. 1989. *Dynamics of Dinosaurs and other Extinct Giants*. New York: Columbia Univ. Press.

Russell, D. A. 1982. A paleontologic consensus on the extinction of the dinosaurs, In *Geologic Implications of Impacts of Large Asteroids and Comets on the Earth*. (eds. Silver, L. T., and Schultz, P. H.), *Geol. Soc. Amer. Special Paper* 190:401–405.

Weishample, D. B., Dodson, P., and Osmólska, H. 1990. *The Dinosauria*. Berkeley, California: Univ. Calif. Press.

The landscapes of the modern world acquired their present form during the Cenozoic Era. Even the spectacular scenery of the Colorado Plateau has been sculpted since about middle Tertiary time. This view of the Colorado River was taken from Dead Horse Pass, Utah. *(Courtesy of Lehi Hintze.)*

# The Cenozoic Era

*Many an aeon moulded earth before*
*her highest, man, was born.*
*Many an aeon too may pass when*
*earth is manless and forlorn.*
*Earth so huge and yet so bounded—*
*pools of salt and plots of land—*
*Shallow skin of green and azure—*
*chains of mountains, grains of sand!*

A. Tennyson, *Locksley Hall:*
*Sixty Years After,* 1866

## Introduction

The Cenozoic is the final era of geologic time. It is the era during which the continents and their landscapes acquired their present form. Many of the animals and plants of the present world have been shaped and modified by Cenozoic events.

There are two sets of subdivisions used for the Cenozoic. (Table 14–1). In a scheme long used by geologists, the era is separated into a Tertiary and a Quaternary Period. Many European geologists, however, prefer a time table that divides the era into **Paleogene** and **Neogene** periods, arguing that this represents a more equal division of the epochs and also a more natural way of dividing Cenozoic rocks of Europe. The Paris basin is the type area for most of the Cenozoic epochs. In that area, a major unconformity representing a marine regression does indeed serve nicely as a boundary between the Paleogene and Neogene systems. However, because predominant usage still seems to favor the old scheme it will be employed here.

The Cenozoic was and is a time of considerable tectonic plate motion and sea floor spreading. It has been estimated that approximately 50 percent of the present ocean floor has been renewed along mid-ocean ridges during the past 66 million years. Much

of this new ocean floor was emplaced in the expanding Atlantic and Indian oceans. As this widening progressed, the Americas moved westward. The area that is now California came into contact with the northward-moving Pacific plate and thereby produced the San Andreas fault system. South America moved firmly against the Andean trench and actually bent and displaced it. Orogenic and volcanic activity was vigorous along the western backbone of the Americas and resulted in the formation of the Isthmus of Panama that today links North and South America. The North Atlantic rift extended itself to the north, separating Greenland from Scandinavia and destroying the land connection between Europe and North America. Far to the south, Australia separated from Antarctica (Fig. 14–1) and moved northward to its present location. During the Cenozoic, a branch of the Indian Ocean rift split Arabia away from Africa and in the process created the Gulf of Aden and the Red Sea (Fig. 14–2). The most dramatic crustal events of the era, however, were the collisions of Africa and India with Eurasia. These magnificent smashups transformed the Tethys seaway into such lofty mountain ranges as the Alps and the Himalayas.

The interior regions of continents stood relatively high during the Cenozoic, and as a result, marine transgressions were limited. Climatic zones were

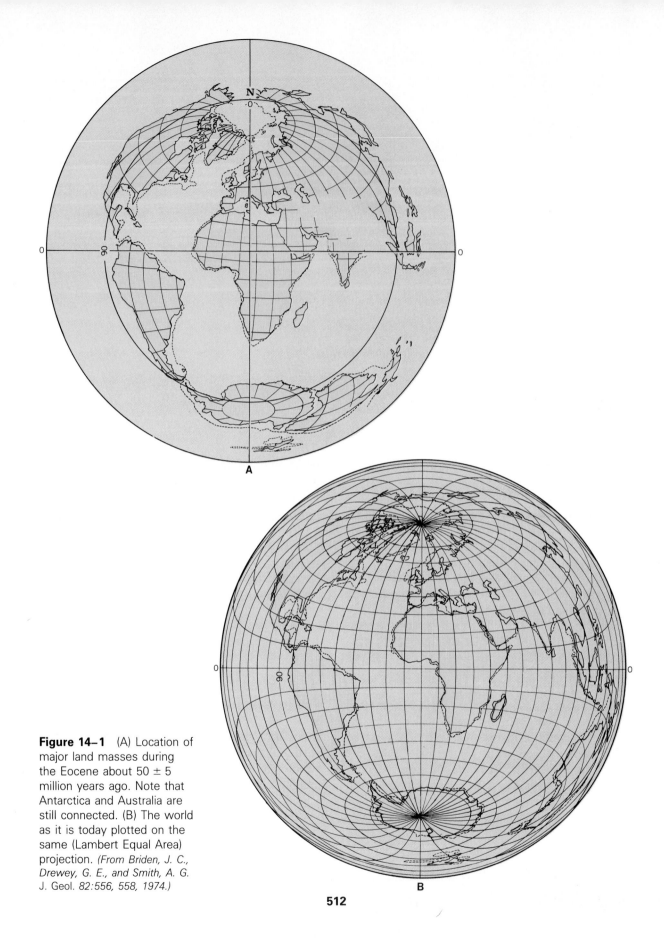

**Figure 14–1** (A) Location of major land masses during the Eocene about 50 ± 5 million years ago. Note that Antarctica and Australia are still connected. (B) The world as it is today plotted on the same (Lambert Equal Area) projection. *(From Briden, J. C., Drewey, G. E., and Smith, A. G. J. Geol. 82:556, 558, 1974.)*

| ERA | PERIOD | | EPOCH |
|---|---|---|---|
| CENOZOIC | QUATERNARY | NEOGENE | RECENT |
| | | | PLEISTOCENE |
| | | | PLIOCENE |
| | | | MIOCENE |
| | TERTIARY | PALEOGENE | OLIGOCENE |
| | | | EOCENE |
| | | | PALEOCENE |

**Table 14–1**  Geochronologic Terminology Used for Divisions of the Cenozoic Era

**Figure 14–2**  The Red Sea, shown here in a photograph taken from the Gemini spacecraft, formed about 30 million years ago when the Arabian Peninsula broke away from Africa. The seaway continues to widen today. The view is toward the south. The wedge of land in the lower left is the Sinai Peninsula, bordered on the right by the Gulf of Suez and Egypt, and on the left by the Gulf of Aqaba and Saudi Arabia. *(Courtesy of NASA.)*

more sharply defined than in earlier eras. A cooling trend that was to culminate in the Pleistocene Ice Age is clearly indicated by paleobotanical studies.

# Before the Ice Age

## Eastern United States

The eastern margin of North America experienced little orogenic activity during the Cenozoic. Erosion gradually continued in the Appalachians. Periodically, as the uplands were beveled by erosion, broad, gentle uplifts occurred. Streams, revitalized by the uplifts, sculpted a new generation of well-defined ridges and valleys. The most recent of the uplifts in the Appalachian belt was accompanied by gentle tilting of the Atlantic coastal plain and adjacent parts of the continental shelf. Terrigenous clastic sediments were brought out of the highlands by streams and deposited on the plains and, on occasion, were reworked by waves and currents of periodic marine transgressions. The Cenozoic section of marine rocks is thinner near the Appalachian source areas and becomes thicker and less clastic toward the region that lies offshore (Fig. 14–3). Southward, in the vicinity of Florida, terrigenous clastics were less available. Carbonate sediments accumulated to thicknesses of more than 2500 meters along a series of subsiding, elongate coralline platforms that were

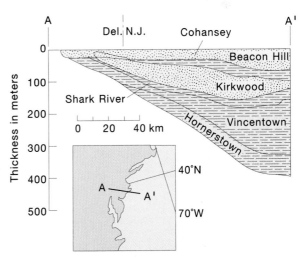

**Figure 14–3**  Cross-section of Tertiary strata across the New Jersey coastal plain. *(From Stratigraphic Atlas of North and Central America, courtesy of Shell Oil Company.)*

probably much like the Bahama Banks (Fig. 14–4). In the final ages of the Tertiary, uplift along the northern end of this tract raised the land area of Florida above the waves.

## Gulf Coast

The best record of Cenozoic strata in North America is found in the Gulf coastal plain (Fig. 14–5). Altogether, eight major transgressions and regressions are recorded in this region. The Paleocene transgression brought marine waters as far north as southern Illinois. Frequently, during marine regressions, nearshore deltaic sands were deposited above offshore shales. The resulting interfingering of permeable sands and impermeable clays provided ideal conditions for the eventual entrapment of oil and gas.

"A wedge of sediments that thickens seaward" is a particularly suitable description for the Cenozoic formations of the Gulf Coast. Geophysical measurements suggest that the thickness of Tertiary sediments beneath the northern border of the Gulf may exceed 10,000 meters. The area must have been subsiding rapidly as it received this great mass of sediment.

## Rocky Mountains and High Plains

While marine sedimentation was under way along the eastern coastal regions, terrestrial deposition pre-

vailed in the Rocky Mountain region. The exception was a single area of marine sedimentation during the Paleocene (Fig. 14–6). The strata recording the presence of this seaway are found in western North Dakota and consist of dark shales containing over 150 species of invertebrates. There is no evidence of a connection between this sea and either the Gulf of Mexico or the Arctic, and it is therefore believed to be a vestige of the more extensive Late Cretaceous epicontinental sea.

The Late Cretaceous and early Tertiary phases of deformation described in Chapter 12 were largely responsible for the major structural features of the Western Cordillera. However, the present topography of this region is due primarily to erosion following uplifts that began during the Miocene. As always, erosion acting upon the tilted and folded hard and soft layers was the final factor in shaping the landscape. Following Late Cretaceous orogenesis, erosional debris was trapped in lowlands and intermontane basins, and these terrestrial sediments provide the record of the earlier Tertiary epochs. Later uplifts and erosion resulted in the spectacular relief of the Rockies and caused the detritus of the older basins and newer uplands to be spread over the plains that lay to the east. The result was the creation of a vast apron of nonmarine Oligocene through Pliocene sands, shales, and lignites that underlies the western high plains. Beds of volcanic ash interspersed within

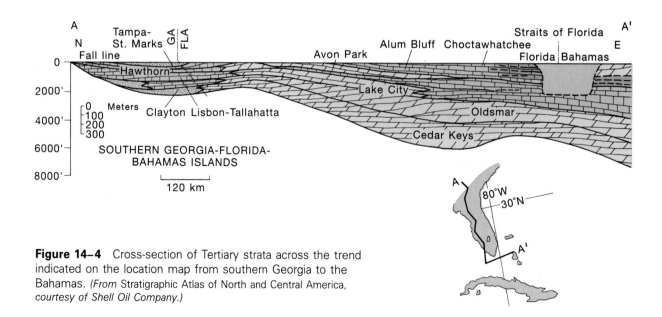

**Figure 14–4** Cross-section of Tertiary strata across the trend indicated on the location map from southern Georgia to the Bahamas. (From Stratigraphic Atlas of North and Central America, courtesy of Shell Oil Company.)

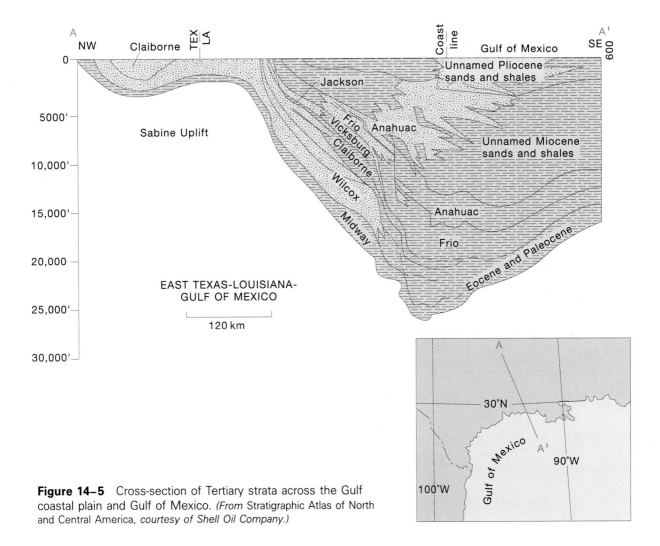

**Figure 14–5**  Cross-section of Tertiary strata across the Gulf coastal plain and Gulf of Mexico. *(From* Stratigraphic Atlas of North and Central America, *courtesy of Shell Oil Company.)*

the Tertiary sections of the Western Cordillera attest to periodic episodes of igneous pyrotechnics.

The lower Tertiary sedimentary rocks formed from the continental sediments deposited in intermontane basins consist of gray siltstones and sandstones, as well as carbonaceous shales, lignite, and coal. These rocks are well represented in the Fort Union Formation, a unit that can be recognized in several intermontane basins in the Rocky Mountain region. The Fort Union is approximately 1800 meters (5900 ft) thick and holds immense tonnages of low-sulfur coal in its lower levels. The coal beds are indicators of widespread swampy conditions in this region during Paleocene time.

Eocene deposits of the Rocky Mountains include such well-known formations as the Wasatch and

Green River formations. The sediments of the Wasatch Formation were deposited by streams, whereas the Green River is a lake deposit. Wasatch sedimentary rocks are often coarse and gravelly near the base of the formation, but at higher levels consist of finer clastics. These are mostly red sandstones and shales with interbedded olive-colored shales. The bright colors contrast markedly with the underlying drab gray shales of the Fort Union. At Bryce Canyon and Cedar Breaks in Utah, the attractively colored Wasatch beds are eroded into the spires and pinnacles of a magnificent badlands topography (Fig. 14–7).

Basins between the Uinta, Wind River, and Big Horn mountains received not only the erosional detritus from those mountains, but also the discharge of their streams. During the Eocene, that volume of

**Figure 14–6**    Paleogeographic map of North America during the early Tertiary.

**Figure 14–7** Cedar Breaks National Monument in Utah. Here the Eocene Wasatch Formation has been eroded into steep ravines, pinnacles, and razor-sharp divides.

stream water filled many of the basins and formed extensive shallow-water lakes. One lake, notable for its lacustrine deposits, formed in the Green River basin of southwestern Wyoming (Fig. 14–8) creating the Green River Formation, noted above (Fig. 14–9). Over 600 meters of freshwater limestones and fine, evenly laminated shales were deposited in this basin as it slowly subsided. The laminations are varves, each of which consists of a thin, dark winter layer and a lighter-colored, summer layer. By counting the varves, it has been determined that more than 6.5 million years were required to deposit the Green River sediments. Within the fine shales are well-preserved fishes (Fig. 14–10), insects, and plant

**Figure 14–8** Cenozoic basins of Colorado, Utah, and Wyoming containing important oil shale deposits. Map boundary is the Upper Colorado River drainage basin. *(Simplified from Rickert, D. A., Ulman, W. J., and Hampton, E. R., (eds.) 1979. Synthetic Fuels Development, U.S. Geologic Survey publication.)*

**Figure 14–9** The Green River Formation exposed in a road cut about 10 miles west of Soldier Summit, Wyoming. The exposure here consists of dark shales and thin, lighter colored limestones.

**Figure 14–10** Fossil remains of an Eocene freshwater fish (*Diplomystis*) from the Green River Formation, Wyoming.

fragments. In addition, the shales are rich in waxy hydrocarbons. For this reason, they are called **oil shales** and can be processed to yield up to 200 liters of petroleum per ton. The Wasatch Formation, also of Eocene age, underlies and interfingers laterally with the Green River beds. Several oil fields in Wyoming derive their oil from the Wasatch Formation.

The Uinta basin south of the Green River basin is notable because it is structurally the deepest of those in the Colorado Plateau. It received a particularly thick succession of Tertiary sediments (Fig. 14–11). These formations can still be observed in exposures in the interior of the basin where they have been only slightly affected by post-Eocene erosion. Strata along the northern edge of the basin are sharply upturned, forming hogbacks.

During the late Eocene and Oligocene, explosive volcanic activity blanketed the region that today includes Yellowstone National Park (Fig. 14–12) and the San Juan Mountains with layers of volcanic ash. For paleontologists, however, the more interesting rocks of this epoch are floodplain deposits of the White River Formation. This famous formation contains entire skeletons of Tertiary mammals in extraordinary number, variety, and excellence of preservation. If one recalls newsreels of recent floods, it is not difficult to understand how Oligocene mammals became buried in the sediment of overflowing streams. The clays, silts, and ash beds of the White River Formation are the sediments from which the Badlands of South Dakota have been sculptured (Fig. 14–13).

**Figure 14–11** North–south cross-section through the Uinta basin. Tertiary units are above the Mesaverde Group.

**Figure 14–12** Yellowstone Falls and canyon in Yellowstone National Park. Rocks exposed in the canyon walls consist of lower Tertiary lava flows and volcanic ash, often altered to bright colors by hydrothermal activity. *(Courtesy of L. E. Davis.)*

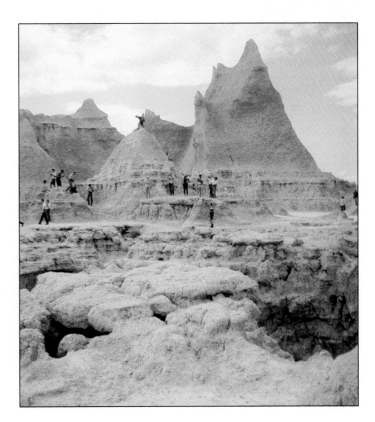

**Figure 14–13** Poorly consolidated, horizontal claystone and shale beds of the Oligocene White River Formation form the picturesque pinnacles of the South Dakota badlands. Lacking the protection afforded by vegetation, the soft sediments are susceptible to severe erosion during infrequent but torrential rains. *(Courtesy of R. Arvidson.)*

**Figure 14–14**   Fossil spider (upper left) and fossil insects from the Oligocene tuff beds near Florissant, Colorado. *(Tuff is consolidated volcanic ash.)*

Another particularly interesting occurrence of Oligocene beds is exposed near Florissant, Colorado. Explosive volcanic activity in this area produced a great deal of ash, which settled to the bottom of a neighboring lake, burying thousands of insects (Fig. 14–14), leaves, fish, and even a few birds. The Florissant flora, known not only from leaves but also from spores and pollen, indicates subhumid conditions for the region and elevations of between 300 and 900 meters.

The Miocene was a time of continued fluvial and lacustrine sedimentation in intermontane basins as well as on the plains east of the mountains. By Miocene time, climates had become somewhat cooler. There were expanding areas of grasslands populated by Miocene camels, horses, rhinoceroses, deer, and other grazing mammals. In the central and southern Rockies, Miocene formations include beds of vol-

canic ash as well as interspersed lava flows attesting to vigorous volcanic activity. The well-known gold deposits at Cripple Creek, Colorado, are mined from veins associated with a Miocene volcano. Regional uplift of the Rockies also began in the Miocene and was accompanied by increased rates of erosion. Great volumes of terrigenous detritus eroded from uplifted areas, filled intermontane basins, and spread eastward, where the detritus contributed to the construction of the Great Plains. Some of these crustal movements continued throughout the remaining epochs of the Cenozoic and brought some of the highest peaks of the Rockies to spectacular elevations.

As indicated earlier, the present Rocky Mountain topography is largely the result of the uplift and erosional sculpturing that began in the Miocene. Much of the erosional detritus was spread on top of the

**Figure 14–15** The jagged Teton Range has been cut by water and ice erosion from a crustal block uplifted along a nearly vertical fault. Movement along the fault began in the Pliocene and continued sporadically through the Pleistocene. *(Courtesy of Eva Moldovanyi.)*

White River deposits and equivalent beds over extensive areas of South Dakota and Nebraska. The general character of the sedimentary layers does not change in the overlying Pliocene. For example the Pliocene Ogallala Group consists of slightly consolidated sands, gravels, silts, and clays also derived from the erosion of the Rockies. Remains of plants and animals in some of these sediments indicate that the Pliocene was somewhat cooler and drier than the Miocene had been. This cooling was an early indication that a great ice age was about to begin.

At particular localities west of the Great Plains, Tertiary geologic events produced some of the most striking scenic attractions in the world. Normal fault-ing and volcanism accompanying late Tertiary uplifts were responsible for some of these features. In north-western Wyoming, the lofty Teton Range (Fig. 14–15) was elevated along great normal faults, with displacements of up to 6000 meters. The magnificent east face of the Tetons is a fault scarp that rises nearly 2.5 km to an elevation of over 4000 meters.

## Basin and Range Province

A larger scale physiographic feature that is the result of late Tertiary faulting is the Basin and Range Province of Nevada, Arizona, New Mexico, and southern California (Fig. 14–16). This region had been folded

**Figure 14–16** The California and Nevada Nopah Range of the Basin and Range Province.

Late Cenozoic sediment

**Figure 14–17** Block diagram showing general relationship of Basin and Range topography to normal faulting.

and overthrust during the Mesozoic and existed as a structural arch during the early Tertiary. Then, beginning in the Miocene, the arch subsided between great normal faults that developed on both the west and the east sides. Similar faults with general north–south alignment developed in the interior of the region. The uplifted blocks formed linear mountain ranges that became sources of sediment for the adjacent downdropped basins (Figs. 14–17 and 14–18).

Subsequent geologic history in the region is recorded in coarse terrigenous clastics washed out of the mountains. These Miocene to Holocene sediments frequently blocked the passes and caused lakes to develop. Miocene sediments often include the salty layers formed when these lakes evaporated.

The cause of all the large-scale tensional faulting is still being debated by geologists. On the basis of paleomagnetic evidence, some geologists believe that uparching and faulting in this region occurred when westward-moving North America overrode part of the oceanic plate and spreading center (East Pacific rise) that was being subducted along the coast of California. When the subducted spreading center reached the region beneath eastern Nevada and western Utah, it caused uplift and stretching of the crust in east and west directions. Recently, however, this idea has fallen into disfavor because of a lack of evidence of progressive eastward deformation that should accompany the passage of a spreading center beneath a continent.

As an alternative theory, many geologists now believe that the normal faulting in the Basin and Range is simply the way the crust adjusted to the change along the California coast when oblique shearing of the edge of the continent during the early Miocene replaced the earlier subduction zone. Yet another model proposes extension and uplift of the crust from the remnants of an oceanic plate that had been carried beneath the Basin and Range region by an earlier episode of subduction. When subduction ceased, the oceanic slab may have formed a partially molten buoyant mass that pressed upward against

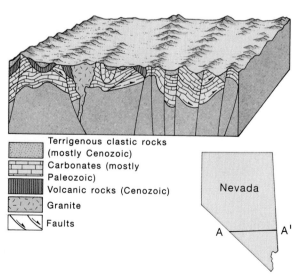

Terrigenous clastic rocks (mostly Cenozoic)

Carbonates (mostly Paleozoic)

Volcanic rocks (Cenozoic)

Granite

Faults

Nevada

A       A'

**Figure 14–18** Geologic section across the Basin and Range Province along line A—A' in southern Nevada.

the overlying crust and caused tensional faulting and escape of lava along fault and fracture zones. Finally, there is the possibility that the crustal extension and tensional faulting is related to convectional movements beneath the continental plate similar to those that cause the breaking apart of continents.

## Colorado Plateau

One of the most magnificent regions of uplift in the American West is the Colorado Plateau (Figs. 14–19 and 14–20). Somehow this block of crust had remained undeformed during the Mesozoic orogenies,

**Figure 14–19**  Canyon of the Dolores River, Colorado Plateau south of Grand Junction, Colorado.

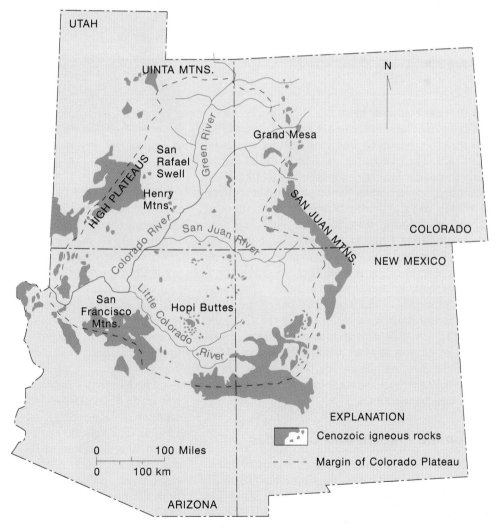

**Figure 14–20**  Location of volcanic and other centers of igneous activity of Cenozoic age in the Colorado Plateau and its periphery. *(From Hunt, C. B.,* Cenozoic Geology of the Colorado Plateau, *U.S. Geological Survey Professional Paper 279, 1956.)*

**Figure 14–21** Vertical aerial photograph of a large cinder cone in the San Francisco volcanic field of northern Arizona. The solidified flow issuing from the cone is 7 km long and more than 30 meters thick. *(Courtesy of the U.S. Geological Survey.)*

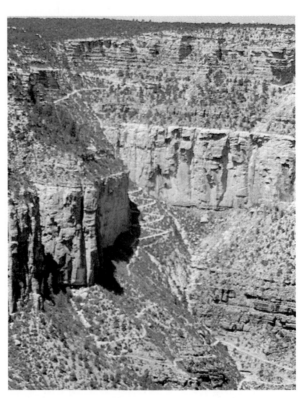

**Figure 14–22** The Grand Canyon of the Colorado River. *(Courtesy of L. E. Davis.)*

for its Paleozoic and Mesozoic rocks are relatively flat-lying. The region formed a buttress around the perimeter of which folding and faulting produced highlands. The plateau was repeatedly raised during early to middle Pliocene time (about 5 to 10 million years ago). Steep faults developed on the plateau during its rise and provided avenues for the upward escape of volcanic materials. The San Francisco Mountains near Flagstaff, Arizona, are a group of impressive recent volcanoes and cinder cones built above the level of the plateau (Fig. 14–21).

The best known feature resulting from the linked processes of uplift and erosion on the Colorado Plateau is the Grand Canyon of the Colorado River (Fig. 14–22). This awesome monument to the forces of erosion is more than 2600 meters (8500 feet) deep in some places. The river has penetrated through Phanerozoic strata into crystalline Precambrian basement rocks.

### Columbia Plateau and Cascades

Unlike the Colorado Plateau, which is constructed of layered sedimentary rocks, the Columbia Plateau in the northwestern corner of the United States has been built by volcanic activity (Fig. 14–23). During the late Tertiary and Quaternary, basaltic lavas erupted along deep fissures in the region. The liquid

rock spread out and buried more than 500,000 km$^2$ of existing topography under layer after layer of lava. In some places, these low-viscosity lavas flowed 170 km from their source. Their combined thickness exceeds 2800 meters.

West of the Columbia Plateau lies an uplifted belt that was also the site of extensive volcanic activity. Here, however, the outpourings of more viscous lavas resulted in the mountains of the Cascade Range. Volcanism in this region began about 4 million years ago.

The fact that it has continued up to the present was made dramatically obvious to Americans by the eruption of Mount St. Helens on May 18, 1980 (Fig. 14–24). The volcano exploded with a force equivalent to 50 million tons of TNT and with a roar heard over 300 km away. A great turbulent cloud of hot gases, steam, pulverized rock, and ash burst laterally from the north side of the mountain, followed by a vertical column of gas and ash that rose 20,000 meters into the atmosphere and began to drift slowly toward the east. An estimated 1 km$^3$ of airborne ash

**Figure 14–23** Lava flows exposed on either side of Summer Falls, Columbia Plateau. *(Photograph courtesy of U.S. Department of the Interior, Bureau of Reclamation, Columbia Basin Project.)*

**Figure 14–24** Aerial view of erupting Mount St. Helens volcano taken at about 11:30 A.M. on May 18, 1980. Clouds of steam and ash are being blown toward the northeast from the prominent plume, which reached about 20,000 meters in altitude. The white linear features at the base are logging roads, and the dark patches are stands of mature trees. The May 18 eruption produced an amount of ash roughly equivalent to that ejected during the A.D. 79 eruption of Mount Vesuvius that buried Pompeii. *(Photograph by Austin Post, courtesy of U.S. Geological Survey.)*

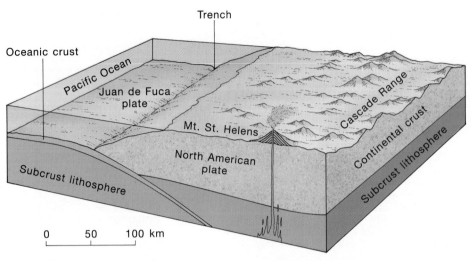

**Figure 14–25**   As the small Juan de Fuca plate plunges beneath Oregon and Washington, molten rock is generated and rises to supply the volcanoes of the Cascade Range.

and other rock debris from the explosion blocked out the sun's light and caused automatic street lamps to switch on in towns hundreds of kilometers downwind from the volcano. Turbulent hot masses of gas and ash surged down the sides of the mountain, flattening and scorching millions of trees. Meanwhile, the ash fell like a dismal gray snow, blanketing streets, dangerously burdening the roofs of buildings, choking the engines of vehicles, and covering the leaves of trees and crops. Hot gas and ash from the volcano melted part of Mount St. Helens snow and ice cap. The meltwater mixed with ash and formed huge mudflows that surged down the mountain slopes at speeds as great as 80 km per hour. A steaming turbulent mass of debris crashed through the once beautiful valley of the Toutle River and demolished 123 homes in the nearby town of Toutle. People camping or working near the volcano were killed by heat, gases, or burial under the downpour of ash. The May 18th eruption was clearly a major volcanic event. The fact that it was actually rather puny when compared with the famous eruptions of Krakatoa or Santorini was no consolation to the citizens of Washington as they viewed the devastated gray landscape, the hundreds of square kilometers of valuable trees toppled like matchsticks, buried roads, harbors choked with ash-sludge, and smothered fields of wheat.

The recent activity at Mount St. Helens and the older eruptions that gave us the volcanic peaks of the Cascades are surface manifestations of an ongoing collision between the American plate and the small Juan de Fuca plate of the eastern Pacific, which moves eastward on a collision course toward the coasts of Oregon and Washington. The Juan de Fuca plate plunges beneath the coastline, and molten rock generated as the plate moves downward rises to supply the lava of the volcanoes (Fig. 14–25).

Mount St. Helens is not the only famous volcanic mountain in the Cascades, nor is it the only peak having periodic eruptions. As recently as 1914, Mount Lassen extruded lava and ash for a year and then exploded violently on May 19, 1915, producing a **nuée ardente** (fiery cloud) that roared down the mountainside, destroying everything in its path. Other well-known Cascade peaks include Mount Baker, Mount Hood, and Mount Rainier. The last major eruption of Mount Rainier was about 2000 years ago, although minor disturbances occurred frequently in the late 1800s. Crater Lake in Oregon (Fig. 14–26) was formed when the top of a volcanic cone collapsed into its subsiding lava column to form a caldera.

### Sierra Nevada and California

The ranges of the Sierra Nevada lie south of the Cascades. The rocks of these majestic mountains were folded and intruded during the Jurassic Nevadan orogeny. For most of Tertiary time, the peaks were

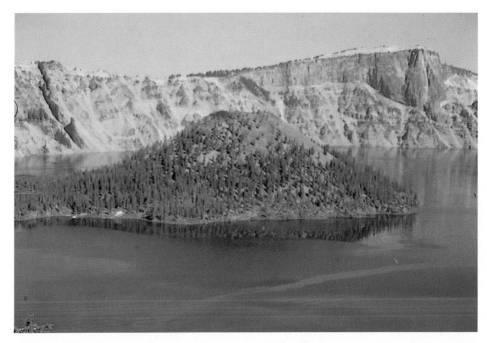

**Figure 14–26**   View of the west wall of Crater Lake, Oregon, with Wizard Island rising from the caldera. *(Courtesy of L. E. Davis.)*

steadily reduced by erosion until their granitic basement lay exposed at the surface. Then, in Pliocene time and continuing into the Pleistocene, the entire Sierra Nevada block was raised along normal faults bounding its eastern side and tilted westward (see Fig. 6–30). Its high eastern front was lifted an astonishing 4000 meters. The depressed western side formed the California trough. As these movements were under way, rejuvenated streams and powerful valley glaciers began to sculpture the magnificent landscapes of the present-day Sierras.

In the early epochs of the Cenozoic, the region west of the Sierras was most directly affected by subduction tectonics. As will be discussed shortly, however, this region underwent a change from subduction tectonics to strike-slip tectonics at about the beginning of Miocene time. Complex fault movements resulting from this shift in tectonic style created islands and intervening sedimentary basins in which fine-grained marine clastics, diatomites, and bedded cherts accumulated. Also during the Miocene, folding and uplifting resulted in marine regressions in many areas, and by the end of the Tertiary, seas were restricted to a narrow tract along the western edge of California. Ultimately, Quaternary uplifts caused the regression of this tract as well.

The Tertiary stratigraphic succession of California is impressive, but also imposing are sections of Tertiary rocks far to the north in Alaska and the arctic islands of Canada. Coal found associated with these strata, along with fossil spores and pollen, indicate that a temperate or even warm-temperature climate prevailed in these northern lands. Along the coastal mountains and Aleutian chain, Tertiary sands and shales are interspersed with layers of pyroclastics and lava flows. Then, as now, the area of present-day Alaska was tectonically active.

## West Coast Tectonics

As described in Chapter 12, the western edge of North America during most of the Cenozoic was the site of an eastward-dipping subduction zone. This subduction was in one way or another responsible for the batholiths, compressional structures, volcanism, and metamorphism that accompanied Mesozoic and early Tertiary orogenies. The oceanic plate that was being fed into the subduction zone has been named the **Farallon plate.** During the Cenozoic, the Farallon plate was being consumed at the subduction zone faster than it was receiving additions at its spreading center. As a result, most of the Farallon

EXPLANATION

| | | |
|---|---|---|
| Pacific plate | ⌐▲▲ Subduction zone | mya Million years ago |
| Farallon plate | Oceanic ridge | |
| North American plate | Transform fault | ⇧ Pacific plate motion relative to North American plate |

**Figure 14–27** Schematic model of the interaction of the Pacific, Farallon, and North American plates for six time intervals during the Cenozoic. Note how the Farallon plate was largely subducted by late Cenozoic time, leaving only remnants to the north (Juan de Fuca plate) and to the south (Cocos plate). The San Andreas and associated faults were caused by right-lateral movements beginning about 29 million years ago. *(Adapted from Nilsen, T. H. San Joaquin Geological Society Short Course No. 3, 1977. Data source from Atwater, T. Bull. Geol. Soc. Am. 81(12):3513–3536, 1970.)*

**Figure 14–28** Sequence of cross sections of California and its offshore area illustrating the subduction of the Farallon plate. In this model, the Pacific plate is considered fixed as the spreading center (East Pacific Rise) encounters the continental margin. *(After Atwater, T. Bull. Geol. Soc. Am. 81(12):3513–3536, 1970.)*

plate and part of the East Pacific rise that generated it was gobbled up at the subduction zone along the western edge of California (Fig. 14–27). Today, the small Juan de Fuca plate near Oregon and Washington and the Cocos plate off the coast of Mexico are the remnants of the once more extensive Farallon plate.

With the loss of the Farallon plate near California, the North American plate was brought into direct contact with the Pacific plate, and a new set of plate motions came into operation. Before making

contact with the west coast, the Pacific plate had been moving toward the northeast. As a result, when contact was made, the Pacific plate did not plunge under the continental margin but rather slipped along laterally, giving rise to the San Andreas fault. No longer did the California sector of the west coast have an Andean type of subduction zone (Fig. 14–28). Strike-slip movements now characterized the western margin of the United States.

An important result of the shearing and wrenching of the west coast was the tearing away of Baja

California from the mainland of Mexico about 5 million years ago. Because it was located west of the San Andreas fault system, Baja California was sheared from the American plate and now accompanies the northward movement of the Pacific plate.

## South America

Extensive deformation, metamorphism, and emplacement of granitic masses had characterized the Andean belt during the Cretaceous. Folding and volcanism continued to be widespread during the Cenozoic as well and was especially intense during the Miocene. Subsequently, the Andean highlands were eroded to low relief and then were uplifted in late Pliocene. Erosion of the uplifted surface is responsible for the present relief and topography of the Andes. Cenozoic instability in the Andes has been related to the continued movement of the floor of the South Pacific Ocean beneath South America. An oceanic trench off the Pacific coast and an eastwardly inclined seismic zone characterized by deep-focus earthquakes attest to this interpretation.

As a consequence of the frequent upheavals that beset the west side of South America, epicontinental seas were either marginal or limited. Sedimentation was primarily terrestrial, as silt, sand, and gravel eroded from the Andes was washed down into the Amazon and Orinoco basins and across the Pampas. Riftlike basins created by block faulting served as collecting sites for prodigious thicknesses of clastic sediments.

## The Tethyan Realm

The conversion of a major seaway separating Eurasia and Gondwanaland to a spectacular array of mountains and plateaus must be considered one of the greatest events of the Cenozoic Era. During most of the era, the Tethys seaway lay to the south of Europe. Its waters spilled onto the northern margin of Africa as an epicontinental sea. Disturbances prophetic of the impending tectonic upheavals began along the Tethys during the Eocene with large-scale folding and thrust-faulting. It was at approximately this time that the northward-moving African block first encountered the western underside of Europe and crumpled the strata that now compose the Pyrenees

and Atlas mountains. Then, with a sort of scissor-like movement, the Alpine region began to be squeezed. At first, marine conditions persisted between the emerging folds. Siliceous shales, cherts, and poorly sorted sandstones accumulated between elongate submarine banks.

By Oligocene time, compression from the south caused enormous recumbent folds to rise as mountain arcs out of the old seaway and to slide forward over the lands that once lay to the north of the Tethys. Great folds were cut along their undersides by thrust faults and pushed on top of one another as spectacular monuments to the forces involved in plate collisions. North of these contorted and rising structures lay a topographic depression that received the piedmont deposits eroded from the mountains. These terrestrial clastics, termed **molasse** by European geologists, resemble similar deposits swept eastward from the Rocky Mountains during the Cenozoic.

Even after the Oligocene, the compressions continued. During the Pliocene, further thrusts from the south carried the older folded belts northward over the molasse deposits and crumpled the Jura folds, which today form the northern front of the Alps. The thrusting was followed by spasmodic regional uplifts that continue even to the present day.

That part of the Tethys that extended far to the east of the Alps experienced its own paroxysm of mountain making during the Cenozoic. Volcanism, folding, thrusting, and emplacement of massive granitic intrusions began early in the era and increased markedly during the Miocene. Great elongate tracts of the former sea floor were squeezed into folds and thrust southward. Much of the early deformation occurred at or near sea level, but ultimately the sea floor section was forced up and above the level of the waves. A broad, subsiding trough formed along the northern edge of newly arrived peninsular India. On this lowland more than 5000 meters of continental sediments were deposited. These sediments contain an important fossil record of Cenozoic mammals and plants. In the two final epochs of the era, regional (epeirogenic) uplifts brought the plateaus and ranges to lofty elevations and caused the retreat of those marginal seas that still remained. The uplift may very well have been caused by the continued movement of the Indian block northward beneath the southern edge of the Asian plate.

**Figure 14–29**   The glaciated Swiss Alps near St. Moritz.

The history of the Tethys provides yet another remarkable example of how a region of the earth's crust can undergo a complete change-about in form. The once quiet seaway that bordered Eurasia on the south was transformed into a structurally complex region of rugged highlands that includes the Alps (Fig. 14–29), Atlas, Apennines, Carpathians, Caucasus, Pamirs, and Himalayas (Fig. 14–30). The Mediterranean, Black, and Caspian seas are the only surviving parts of the great Tethys gulf.

One event in the history of the Tethys was not directly related to orogeny in the region. As described in Chapter 1 under the heading "A Study of Mediterranean History," the Mediterranean once dried up and for a short time may have resembled the present Death Valley region of California. This extraordinary event occurred about 5.5 million years ago near the end of the Miocene. It has been named the **Messinian event**, in that the Messinian is the final stage of the Miocene in Italy. The Messinian event was an indirect result of a sudden but temporary expansion of the Antarctic ice sheet. Water needed to form the ice came from the ocean, and the loss in volume of ocean water lowered mean global sea level by as much as 50 meters. This decline left the Mediterranean isolated from other oceans. Without a way to replenish water lost by evaporation, the Mediterranean dried up. Evidence for this event comes primarily from an extensive evaporite bed encountered in cores that penetrated deep into the Mediterranean sea floor.

## North of the Tethys

Important geologic events were also occurring in regions north of the Tethys during the Cenozoic. In the early part of the era, lavas were extruded in Scotland, Ireland, Spitzbergen, Greenland, and Baffin Island.

**Figure 14–30**   Shivling Peak in the Himalayas. The mountain has been eroded by glaciers into a glacial horn. *(Courtesy of D. Bhattacharyya.)*

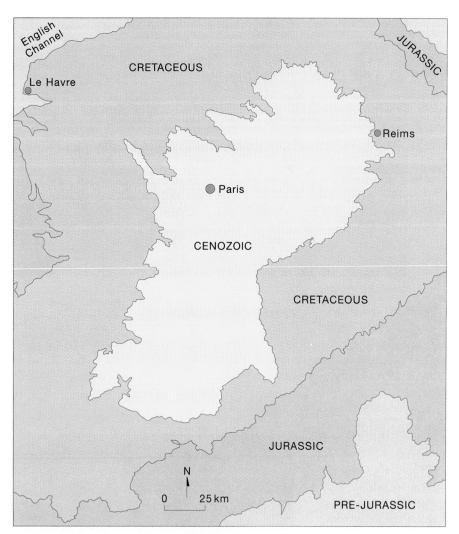

**Figure 14–31**  Geologic sketch map of the Paris Basin.

In Ireland, these lavas are vertically jointed and form the famous Giant's Causeway. The volcanic activity appears to have accompanied the separation of Greenland and Europe.

Across the English Channel in northeastern Europe there is a record of repeated marine incursions during the Paleocene, Eocene, and Oligocene. The most extensive of the transgressions occurred during the Oligocene. Like the earlier incursions, waters flooded toward the southeast from the North Sea. The rocks deposited during the alternate marine transgressions and regressions have been thoroughly studied in the Paris basin. The Paris basin is not only a sedimentary basin, but a structural basin as well. This is quickly apparent on a geologic map of the area (Fig. 14–31), which shows younger Tertiary deposits surrounded by Mesozoic rocks. After the Oligocene, general uplift prevented further marine incursions. Thus, the late Tertiary record is read largely from lake and stream deposits.

East of Europe in Mongolia and China, epeirogenic uplift occurred during the time that the Himalayas were being deformed. Lakes and swamps were plentiful across Asia, and some extended over

**Figure 14–32** (A) East African rift valleys and associated lakes. (B) Schematic illustration of the formation of rift valleys from the action of tensional forces in the crust. (C) Example of how compressional forces may also result in the formation of valleys. *(A, from Read, H. H., and Watson, J.* Introduction to Geology. *London: Macmillian & Co., 1975, p. 237, Figure 8.3.)*

vast regions. Meandering streams built wide floodplains. Important Cenozoic mammalian remains have been unearthed in these Asian fluvial deposits.

## Africa

The southern margin of the Tethys seaway formed the upper boundary of Africa during most of the Cenozoic. The region had a much quieter tectonic history than did the northern Tethys. In Libya and Egypt, the formations are flat or only moderately folded carbonates; the only severely folded mountain ranges are in western North Africa, which experienced pulses of orogeny throughout the Cenozoic.

The larger part of Africa that lay south of the Tethys was generally emergent during the Cenozoic. It was a time in which erosion prevailed. The most

conspicuous changes that occurred were those associated with the rift valleys and uplands along the east side of the continent. During the late Cenozoic, much of eastern Africa was arched upward nearly 3000 meters above sea level. Fracturing and faulting occurred across the crest of the arch. The famous **rift valleys** of east Africa formed as narrow slivers of crust that had slipped downward between great fault blocks. As the faulting continued, volcanoes developed along the fault trends. Mount Kenya and Mount Kilimanjaro are two of the most notable of these volcanoes, but there are many others as well. In addition to the volcanoes, a series of elongate lakes (Fig. 14–32) formed within the downfaulted blocks. Today, Lake Nyasa and Lake Tanganyika are splendid examples of these fault-controlled inland water bodies. This region of lakes, volcanoes, and fault-controlled ranges is well known for its exceptional scenic grandeur.

## The Western Pacific

Most of Australia was comparatively stable during the Cenozoic Era. Marine deposits were restricted to peripheral basins. Near the center of the continent, coal-bearing terrestrial clastics are found. However, the general lack of tectonic activity that prevailed over most of the continent was not characteristic of the extreme southeastern border, where Paleocene and Eocene volcanism was intense. This activity was apparently associated with the separation of Antarctica from Australia an estimated 50 or 60 million years ago (see Fig. 14–1).

The Cenozoic history of New Zealand and the region north of Australia was also marked by exceptional instability. Throughout the islands that extend from the Aleutians down through Japan, the Phillipines, Indonesia, and into New Zealand, folding, volcanism, and spectacular upheavals were common occurrences. Tropical rains and storms caused rapid erosion and deposition of coarse sediments. Deep sedimentary basins received the erosional detritus and frequently spread it seaward as vast deltas.

The enormous volumes of Cenozoic andesites and basalts found in New Zealand clearly indicate its participation in the instability. Indeed, a belt of active volcanoes persists today on the north island. Late in the Cenozoic, vertical movements raised a great fault segment an incredible 18 km along the east side of the north island.

The crustal activity that characterized the western Pacific during the Cenozoic has continued to the present. If one were to speculate about the most active region of the globe during the near geologic future, the sinuous arcs from the Kamachatka Peninsula through Japan and down into Indonesia (Fig. 14–33) would be likely candidates. (The Caribbean region might qualify as a second choice.)

## Antarctica

One of the most interesting of the inferences that have resulted from the study of Cenozoic rocks of Antarctica is that this now frigid land had a genial or semitropical climate throughout the early part of the era. It was not until the Miocene that snow and ice began to accumulate and spread across the land mass. Deep-marine clastic sediments and volcanics accumulated along the Pacific side of Antarctica. The instability of west Antarctica can be inferred to be the result of the eastward movement of an oceanic plate against and beneath the continent.

**Figure 14–33** Lightning strikes during the eruption of Galunggung volcano, West Java, Indonesia, December 3, 1982. *(Photo by Ruska Hadian, volcanology survey of Indonesia. Courtesy of the U.S. Geological Survey.)*

Areas covered with glacial
ice during the Pleistocene

**Figure 14–35**  Areal extent of major glaciers in Europe during the Pleistocene Epoch.

Existing glaciers

Land areas formerly
covered by glaciers

**Figure 14–34**  Maximal area of glacial coverage in North America. *(From the U.S. Geological Survey Pamphlet, The Great Ice Age, No. 0-357-128. U.S. Government Printing Office, 1969.)*

# The Great Pleistocene Ice Age

The final two epochs of the Cenozoic Era are the Pleistocene and Holocene (or Recent). They represent only about 2 million years of geologic time, but for humans they are an exceedingly significant interval. In the first place, it was during the Pleistocene that primates of our own species evolved and rose to a dominant position. Second, during the Pleistocene, more than 40 million km³ of snow and ice were dumped on about one-third of the land surface of the globe (Figs. 14–34 and 14–35). Such an extensive cover of ice and snow had profound effects not only upon the glaciated terrains themselves (Fig. 14–36), but also upon regions at great distances from the ice fronts. Climatic zones in the Northern Hemisphere were shifted southward, and arctic conditions prevailed across northern Europe and the United States. Mountains and highlands in the Cordilleran and Eurasian ranges were sculpted by spectacular moun-

**Figure 14–36**  Glaciated Canadian Shield north of Montreal, Canada. *(Courtesy of the Royal Canadian Air Force.)*

tain glaciers, rather like those depicted in Figure 14–37. While the snow and ice accumulated and spread in higher latitudes, rainfall increased in lower latitudes, with generally beneficial effects on plant and animal life. Even as late as the beginning of the Holocene, presently arid regions in north and east Africa were well watered, fertile, and populated by

**Figure 14–37** Oblique aerial view of glaciers in the Fairweather Mountains of Alaska. Note the debris flowlines. *(Courtesy of M. E. Dalechek, U.S. Geological Survey.)*

nomadic tribes. Peoples of middle and late Pleistocene time seemed to thrive by hunting along the fringes of the continental glaciers where game was abundant and the meat kept longer with less danger of spoilage. Animal furs provided warm clothing, and following the discovery of fire, the caves were warm against the arctic winds.

In addition to the glaciations, the Pleistocene was a time of recurring crustal unrest. Volcanoes were active in New Mexico, Arizona, Idaho, Mexico, Iceland, Spitzbergen, and the Pacific borders of both North and South America.

Pleistocene crustal uplifts occurred in the Tetons, the Sierra Nevadas, and the ranges of the central and northern Rocky Mountains. Pulses of uplift during the Pleistocene also characterized the Alps, the Himalayas, and the ranges that lay between. All of this crustal and climatic activity may suggest that the Pleistocene was a unique time in geologic history. However, this is not the case, for evidence suggests that widespread glaciations have also occurred in the Precambrian, late in the Ordovician; in the Permo-Carboniferous, and possibly during the Oligocene and Pliocene as well. Indeed, study of relatively accessible Pleistocene and Recent glacial deposits has greatly improved our ability to interpret the vestiges of more ancient glaciations. Carbon and oxygen isotope analyses of foraminifers from deep-sea cores have provided evidence of several episodes of ocean water cooling that may not necessarily have been associated with glaciation on land.

## Pleistocene and Holocene Chronology

It is a popular but biostratigraphically untenable conception that the Pleistocene Epoch began with a sudden worldwide onslaught of frigid climates and that the lower boundary of the Pleistocene Series is easily recognized by sedimentologic and paleontologic clues to that frigidity. However, as is the case with the other epochs of the Cenozoic, the Pleistocene was defined by Charles Lyell in 1839 according to the proportion of extinct to living species of mollusk shells in the layers of sediment. For example, strata containing 90 to 100 percent of present-day mollusk species were designated as belonging to the Pleistocene. Thus, the definition of what should be designated Pleistocene is straightforward, but it is no simple matter at all to find suitably fossiliferous sediments in various parts of the globe that can be confidently correlated to Lyell's type section in eastern Sicily.

At the present time, the most widely accepted time for the beginning of the Pleistocene is 1.8 million years ago. This date, however, does not coincide with the beginning of glaciation (as indicated by glacial sediments). Actual glaciation appears to have been initiated about 1 million years ago. Also, the

**Figure 14–38** Discoaster (*Discoaster challengeri*) seen through a light microscope at a magnification of ×1200.

low temperatures and abundant precipitation needed for extensive continental glaciation occurred at different times at different places.

For Pleistocene marine beds, fossils are indispensible for correlation. In some cases, a horizon correlative to the standard section can be recognized by the earliest appearance of certain cooler water mollusks and foraminifers. When studying cores of deep-sea sediments, one can frequently recognize the basal Pleistocene oozes by the extinction point of fossils called discoasters. **Discoasters** (Fig. 14–38) are calcareous, often star-shaped fossils believed to have been produced by golden brown algae related to coccoliths. In continental deposits, the fossil remains of the modern horse (*Equus*), the first true elephants, and particular species of other vertebrates are used to recognize the deposits of the lower Pleistocene.

Before the mid-1970s, geologists customarily thought of the Pleistocene as divisible into four distinct glacial stages, with intervening interglacial stages. Recent investigations indicate, however, that there may have been as many as 30 periods of severely frigid conditions over the past 3 million years. Widespread cold climates thus existed well before the beginning of the Pleistocene, and indeed the first subtle evidence of the coming cold began to appear 40 million years ago. Antarctica (Fig. 14–39) has been frozen into an ice age for at least the past 15 million years.

The end of the Pleistocene (and beginning of the Holocene) is considered to be the time of melting of the ice sheets to approximately their present extent and the concomitant rise in sea level. This would mean that the Pleistocene ended about 8000 years ago. However, there is some argument about the date of this boundary. Some geologists prefer to set the Pleistocene–Holocene boundary at the midpoint in warming of the oceans, in which case the Ice Age would have ended between 11,000 and 12,000 years ago.

Whatever date is confirmed for the end of the Pleistocene Epoch, it is evident from historical records and carbon-14 dating of old terminal moraines that cold spells have recurred periodically into the Holocene. Climates on our planet are in delicate balance with atmospheric, geographic, and astronomic variables. A change in any one of these variables is likely to affect climate. Meteorologists, for example, have found a very good correlation between periods of minimum sunspot activity and episodes of colder conditions on earth. One well-documented period of cooler and drier conditions in the Northern Hemisphere occurred between A.D. 1540 and 1890, when temperatures were often 2° to 4°F cooler than today. These four centuries encompass the so-called **little ice age**. During the little ice age, frigid conditions

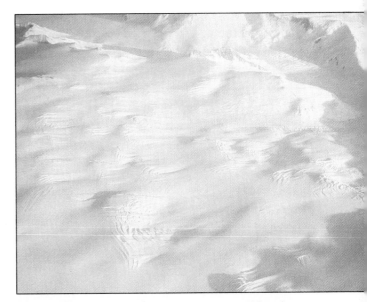

**Figure 14–39** The Antarctic ice sheet near Victoria Land, Antarctica. *(Courtesy P. B. Larson and the U.S. Geological Survey.)*

**Figure 14–40**
Cross-section of a terminal moraine, ground moraine, and outwash plain.

periodically extended across most of Europe and in the United States as far south as the Carolinas. Sea ice intruded far to the south of its present margin. The cold caused loss of harvests, with resulting famine, food riots, and warfare in Europe. In the middle of the fourteenth century, harsh conditions of the little ice age were a contributing cause of the demise of a once flourishing Norse colony in Greenland.

## Stratigraphy of Terrestrial Pleistocene Deposits

Glacial deposits are difficult to correlate. They often consist of chaotic mixtures of coarse detritus completely lacking in fossil remains. Moving glaciers may carry away even these materials, leaving only a bare surface of scoured and polished bedrock. However,

if the ice does not advance for a time, a terminal moraine may be deposited in front of the glacier (Fig. 14–40). As the ice melts, a widespread ground moraine is left behind. Moraines may consist of till, an unsorted mixture of particles from clay to boulder size dropped directly by the ice (Fig. 14–41). **Stratified drift**, on the other hand, has been washed by meltwater, is better sorted, and has many of the characteristics of stream deposits. In either case, however, the stratigrapher is faced with similar-looking masses of debris that are woefully deficient in distinguishing characteristics. As a result, several mutually substantiating criteria must be used in making correlations. The degree to which one sheet of glacial debris has been dissected by streams may indicate that it is older or younger than another sheet. The depth of oxidation and amount of chemical weathering may also provide some estimate of the relative age of

**Figure 14–41** Glacial till in the moraine of a valley glacier, Kenai Peninsula, Alaska *(Courtesy of U.S. Geological Survey).*

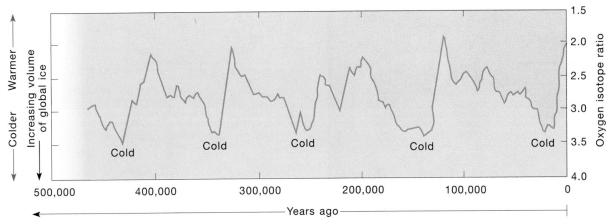

**Figure 14–42** Curve reflecting variations in the global volume of ice (and, indirectly, paleotemperatures) during the past 500,000 years. Data are from radiometric dating and isotope measurements of cores from the Indian Ocean. *(Data from Hays, J. D., and Shackleton, N. J. Science 194:1121–1132, 1976.)*

interglacial soils. Fossil pollen grains in thick exposures of bog or lake deposits often clearly reflect fluctuations of climate and can be used to mark times of glacial advance and retreat. Varved clays deposited in lakes near the glaciers can sometimes be correlated to similar sediments of other lakes and in addition may provide an estimate of the time required for deposition of the entire thickness of lake deposits. However, the most accurate means of dating and correlating Pleistocene sediments is to extract pieces of wood, bone, or peat and determine their age by radiocarbon dating techniques. Unfortunately, even this method has its limitations, the most significant of which is the relatively short half-life (5570 years) of carbon-14. This limits the method to materials less than about 100,000 years old and thus restricts the use of the carbon-14 technique to deposits of the last glacial stage.

## Pleistocene Deep-Sea Sediments

Sediments deposited on the ocean floor during the Pleistocene are easier to date and correlate than terrestrial deposits. The ocean basins are sites of a relatively continuous sedimentary record, the column of sediments contains abundant fossil remains, and the deposits can be dated by relating them to paleomagnetic data and to radiometric isotopes having short half-lives.

Continuous sections of deep-sea sediments can be obtained by means of piston coring devices. These tools provide cores more than 15 meters long. The cores can then be analyzed in various ways. One approach is to use the oxygen isotope ratios obtained from the shells of calcareous foraminifers within the cored sediment to determine the volume of ocean water that was stored at the time as glacial ice. Because of its greater mass, less oxygen-18 is evaporated and precipitated as snow. Hence, during an ice age when ice is accumulating globally, there will be relatively greater percentages of oxygen-18 relative to oxygen-16 in sea water and in the shells of marine invertebrates. Plotted against depth in a deep-sea core, these isotope ratios indicate variations in ice volume and temperature with time (Fig. 14–42). Indications of cooler conditions can then be correlated with declines in worldwide sea level.

Another method used in the correlation of Pleistocene marine sediments is to plot at each level in the core the relative abundance of foraminifers known to be especially sensitive to temperature. For example, the tropical species *Globorotalia menardii* (Fig. 14–43) is alternately present or absent within Pleistocene cores from the equatorial Atlantic. The absence of the species in part of a core is taken to indicate an episode of cooler climates and glaciation. Carbon-14 dates in the upper part of such cores permit one to determine the rates of sedimentation.

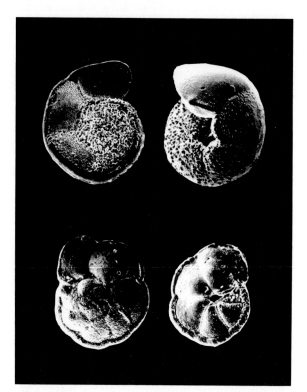

**Figure 14–43** Two well-known species of planktonic foraminifers used in correlating deep-sea sediments. At the top are views of both sides of *Globorotalia truncatulinoides*. Below are views of opposite sides of *Globorotalia menardii*. Magnification of all specimens ×50.

Another foraminifer, *Globorotalia truncatulinoides*, also permits one to recognize alternate cooling and warming of the oceans (Fig. 14–44). This species is trochospirally coiled, which means that it coils in a spiral with all the whorls visible on one side but only the last whorl visible on the other side. Individuals that coil to the right dominate in warmer water, whereas left-coiled individuals prefer colder water. During glacial advances, when ocean temperatures declined, the right-coiled populations of *G. truncatulinoides* in middle and low latitudes were replaced by populations in which left coiling predominated. The record of such changes is clearly apparent in the deep-sea cores.

Remanent magnetism in some deep-sea sediments is sometimes sufficiently strong and stable that a particular section of the core can be correlated to a magnetic reversal documented on land in volcanic rocks of known age. However, the analysis is likely to be in error if the original orientation of detrital

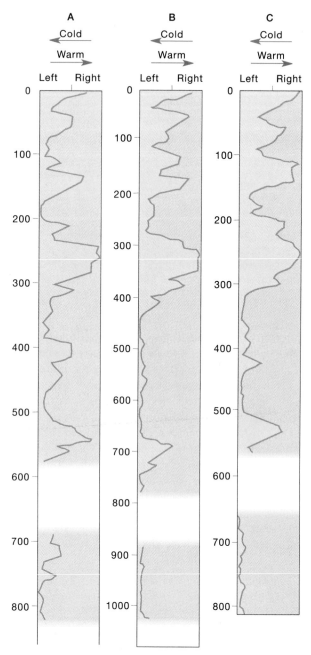

**Figure 14–44** A, B, and C are curves depicting the percentages of right- and left-coiled *Globorotalia truncatulinoides*. The scales run from 100 percent right-coiling forms at the right margins of each column to 100 percent left-coiling forms at the left margins. Left-coiling forms prefer colder waters. Numbers to the left of the columns are deep-sea core depths in centimeters. *(Adapted from Erickson, D. B., and Wollin, E. Science 162(3859):233.)*

magnetic grains has been disturbed by burrowing organisms.

## Effects of Pleistocene Glaciations

Earlier it was noted that during maximum glacial coverage, over 40 million km$^3$ of ice and snow lay upon the continents, equivalent to a tremendous amount of water; removal of that water from the oceans had a multitude of effects on the Pleistocene environment. It has been estimated that sea level may have dropped at least 75 meters during maximum ice coverage. Extensive tracts of the present continental shelves became dry land and were covered with forests and grasslands. The British Isles were joined to Europe, and a land bridge stretched from Siberia to Alaska. During interglacial stages, marine waters returned to the low coastal areas, drowning the flora and forcing terrestrial animals inland.

The glaciers had a direct impact on the erosion of lands and the creation of glacial land forms. The great weight of the ice depressed the crust of the earth over large parts of the glaciated area, in some places to a level of 200 to 300 meters below the preglacial position. With the removal of the last ice sheet, the downwarped areas of the crust gradually began to return to their former positions. The rebound is dramatically apparent in parts of the Baltic, the Arctic, and the Great Lakes region of North America, where former coastal features are now elevated high above sea level (Fig. 14–45).

| 80 or more meters | 60 to 79 meters | 40 to 59 meters | 20 to 39 meters | 0 to 19 meters |

**Figure 14–45** Postglacial uplift of North America determined by measuring elevation of marine sediments 6000 years old. *(Simplified and adapted from Andrews, J. T., The pattern and interpretation of restrained, postglacial and residual rebound in the area of Hudson Bay, in Earth Science Symposium on Hudson Bay, Ottawa, 1968, Canadian Geological Survey Paper 68–53, p. 53, 1969.)*

As the great continental glaciers advanced, they obliterated old drainage channels and caused streams to erode new channels. These dislocations are especially evident in the north-central United States. Prior to the Ice Age, the northern segment of the Missouri River drained northward into Hudson bay, and the northern part of the Ohio Rover flowed northeastward into the Gulf of St. Lawrence. The lower Ohio drained into a preglacial stream named the Teays River. Today, geologists recognize the former location of the Teays by a thick linear trend of sands and gravels. Those parts of the Missouri and Ohio rivers that once flowed toward the north were turned aside by the ice sheet and forced to flow along the fringe of the glacier until they found a southward outlet. The present trend of the Missouri and Ohio rivers approximates the margin of the most southerly advance of the ice. The equilibrium of streams was also affected, for with glacial advances there was a lowering of sea level and thus an increase in stream gradients near coastlines. A reverse effect occurred with wasting away of the ice sheets.

Prior to the Pleistocene, there were no Great Lakes in North America. The present floors of these water bodies were lowlands. Glaciers moved into these lowlands and scoured them deeper. As the glaciers retreated, their meltwaters collected in the vacated depressions (Fig. 14–46). Niagara Falls, between Lake Erie and Lake Ontario, came into existence when the retreating ice of the Wisconsin glacial stage uncovered an escarpment formed by southwardly tilted resistant strata. Water from the Niagara River tumbled over the edge of the escarpment, which is supported by the resistant Lockport

**Figure 14–46** Four stages in the development of the Great Lakes as the ice of the last glacial advance moved away. *(After Hough, J. L.,* Geology of the Great Lakes, *Urbana: Univ. of Illinois Press, 1958, figs. 56, 69, 73, and 74.)*

Limestone. Weak shales beneath the limestone are continually undermined, causing southward retreat of the falls (see Fig. 9–13).

Another large system of ice-damned lakes covered a vast area of North Dakota, Minnesota, Manitoba, and Saskatchewan. The largest of these lakes has been named Lake Agassiz in honor of Jean Louis Rodolphe Agassiz, the great French naturalist who initially insisted on the existence of the Ice Age. Today, rich wheatlands extend across what was once the floor of the lake.

Other lakes developed during the Pleistocene, occupying basins that were not near ice sheets. These were formed as a consequence of the greatly increased precipitation and runoff that characterized regions south of the glaciers. Such water bodies are called **pluvial lakes** (which comes from the Latin *pluvia,* meaning rain). Pluvial lakes were particularly numerous in the northern part of the Basin and Range Province of North America, where faulting produced more than 140 closed basins. So-called pluvial intervals, when lakes were most extensive, were generally synchronous with glacial stages, whereas during interglacial stages, many lakes shrank to small saline remnants or even dried out completely. Lake Bonneville in Utah was one such lake. It once covered more than 50,000 km² and was as deep as 300 meters in some places. Parts of Lake Bonneville persist today as Great Salt Lake, Utah Lake, and Sevier Lake.

A particularly spectacular event associated with the formation of Pleistocene lakes occurred in the northwestern corner of the United States. Lobes of the southwardly advancing ice sheet repeatedly blocked the Clark Fork River, and the impounded water formed a long, narrow lake extending diagonally across part of western Montana. The freshwater body, called Lake Missoula, contained an estimated 2000 km³ of water. With the recession of the glacier, the ice dam broke, and tremendous floods of water rushed out catastrophically across eastern Washington, causing severe erosion and depositing huge volumes of gravel, boulders, and cobbles (Fig. 14–47). The dissected region is appropriately termed the **channeled scablands** (Fig. 14–48).

The glacial conditions of the Pleistocene also had an effect upon soils. In many northern areas, fertile topsoil was stripped off the bedrock and transported to more southerly regions, which are now among the

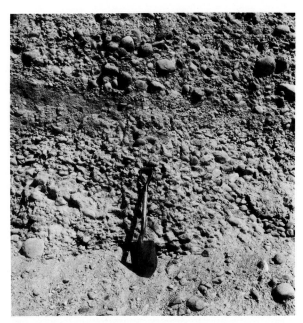

**Figure 14–47**  Conglomeratic scabland deposits exposed near Lyons Ferry, Washington. *(Courtesy of U.S. Geological Survey.)*

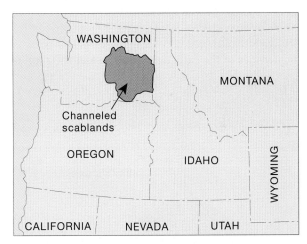

**Figure 14–48**  Location of the channeled scablands.

world's most productive farmlands. Because of the flow of dense, cold air coming off the glaciers, winds were strong and persistent. Fine-grained glacial sediments that had been spread across outwash plains and floodplains were picked up and transported by the wind and then deposited as thick layers of wind-blown silt called **loess.** Such deposits blanket large areas of the Missouri River valley, central Europe, and northern China.

## Cause of Pleistocene Climatic Conditions

The results of oxygen isotope research indicate that world climates grew progressively cooler from the Middle Cenozoic to the Pleistocene. The culmination of this trend was not a single sudden plunge into frigidity but rather an oscillation of glacial and interglacial stages. Any theory that adequately explains glaciation must consider not only the long-term decline in worldwide temperatures but the oscillations as well. Further, the theory must include reasons for the ideal combination of temperature and precipitation required for the buildup of continental glaciers. Although geologists, physicists, and meteorologists have speculated about the cause of the Ice Age for more than a hundred years, no undisputed cause has been found. Indeed, it seems likely that Pleistocene climatic conditions came upon us as a result of several simultaneous factors.

A widely accepted theory for the Ice Age was developed by the Yugoslavian mathematician Milutin Milankovitch (1879–1958). After 30 years of careful study, Milankovitch convincingly proposed that irregularities in the earth's movements and their influence on the amount of solar radiation received by the earth could account for glacial and interglacial stages of the Pleistocene. His calculations were based on three variables: the earth's axial tilt, precession, and orbital eccentricity. With regard to the first of these variables, Milankovitch recognized that the tilt of the earth's axis varies between about 22° and 24° over a period of about 41,000 years. This results in a corresponding variation in the seasonal length of days and in the amount of solar radiation received at higher latitudes. Precession, the second variable, refers to the way the axis of rotation moves slowly in a circle that is completed about every 26,000 years. The effect is equivalent to tilting a rapidly spinning top whose axis responds by describing a cone in space (Fig. 14–49). The third variable is the eccentricity of the earth's orbital path around the sun, which over an interval of about 100,000 years varies about 2 percent. As a result of this variation, the earth is at times closer or farther away from the sun.

According to Milankovitch's calculations, the combination of these variables periodically results in less solar radiation received at the top of the earth's atmosphere, and this might suffice to cause cooling and recurrent glaciations. The Milankovitch cycles correspond rather well to episodes of glaciation over the past 100,000 years. However, if the **Milankovitch effect** has been in existence for bil-

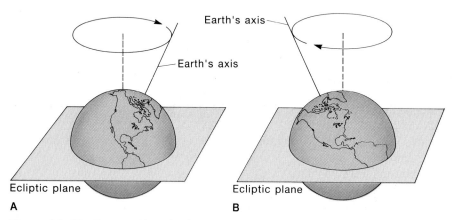

**Figure 14–49**  Two positions in the precession of the earth's axis. The period of a complete precession is 26,000 years. Thus, the two positions shown are separated by 13,000 years. *(From Pasachoff, J.* Contemporary Astronomy, *3rd ed. Philadelphia: Saunders College Publishing, 1985.)*

lions of years, why haven't there been Pleistocene-like glaciations continuously down through geologic time? Apparently, other factors must also be involved.

One such factor might be a variation in the amount of solar energy reflected from the earth back into space rather than being absorbed. The fraction of solar energy reflected back into space is termed the earth's **albedo.** At the present time, it is about 33 percent. Theorists suggest that by the beginning of Pleistocene time, when continents were fully emergent, and hence highly reflective, temperatures may have been lowered enough for the Milankovitch effect to begin to operate. This is not an unreasonable suggestion, for only a 1 percent lessening of retained solar energy could lead to as much as an 8°C drop in average surface temperatures. This would be sufficient to trigger a glacial advance if ample precipitation were available over continental areas. Still other geologists speculate that absorption of solar energy was hindered by cloud cover, volcanic ash, and dust in the atmosphere or fluctuations in carbon dioxide. A decrease in carbon dioxide content would cause a corresponding decrease in the warmth-gathering greenhouse effect, for example. On the other hand, it is also possible that a buildup of $CO_2$ might trigger glaciation, for as warming occurred, there might be more rapid evaporation and an increase in highly reflective cloud cover.

Yet another proposal, developed by the geophysicists M. Ewing and W. Donn, stresses the importance of ample precipitation in building continental glaciers and draws on the interrelationships of ocean currents and pole positions to supply the necessary moisture. Although there are areas of uncertainty in the Ewing and Donn theory, it accounts for both the cooling of the atmosphere that led to the glacial epoch and the oscillating advance and retreat of ice sheets. Several conditions may have had a bearing on the initiation of the Ice Age. One was the drifting of continental areas over or near the poles, where the cold would not be dissipated by the general movement of ocean waters. For example, when the north pole became established in the nearly enclosed Arctic Ocean basin, the heat exchange system of ocean currents was impaired, and polar regions became increasingly cooler. For a period of time, the Arctic Ocean supplied moist air for snowfall on surround-

ing land areas. Additional moisture shunted northward from the Atlantic and Pacific further nourished the ice sheet. Ultimately, the great blanket of ice and snow increased albedo to the point at which temperatures fell so low that the surface of the Arctic Ocean froze and evaporation declined. Deprived of precipitation, the ice sheets began to melt away. The climate warmed, and the Arctic Ocean became ice-free once again. The world entered an interglacial stage, but only temporarily, for once again the original conditions for glacial growth were established. Snow and ice began to accumulate once more, and another stage of glaciation commenced. If the theory is valid, we may be in an interglacial stage today, and climates will continue to oscillate between glacial and nonglacial until the relationship of the continents and poles is changed in some way.

Although scientists are not yet ready to formulate a complete and unified theory for the cause of the Pleistocene's multiple glaciations, it appears that they may be very near the answer. Few deny that the Milankovitch effect, albedo, ocean currents, and pole positions have played a role in creating Pleistocene climatic conditions. The problem is to determine the relative importance of these factors and how they relate to one another. If Pleistocene-like climatic cycles are fundamentally the result of changes in the earth's orbital geometry, one can make some interesting speculations about future climate. Calculations indicate, for example, that the long-term climatic trend over the next 20,000 years is toward extensive Northern Hemisphere glaciation and, of course, cooler conditions. What effect human activities, such as the burning of coal, oil, and gas, might have on this trend is uncertain.

## Cenozoic Climates

The worldwide cooling that culminated in the great Ice Age actually had begun during the early Cenozoic. However, the decrease in temperatures was not entirely uniform. There were warming trends during the late Paleocene and Eocene, as indicated by fossils of palm trees and crocodiles found in Minnesota, in Germany, and near London. Trees that are today characteristic of moist temperature zones thrived in

## Oil Shale

The Eocene oil shale (see Fig. 14-51) that occurs in parts of Wyoming, Utah, and Colorado is rich in an oil-yielding organic compound known as kerogen. When heated to 480°C, the kerogen in oil shale vaporizes. The vapor can then be condensed and will form a thick oil, which, when enriched with hydrogen, can be refined into gasoline and other products in much the same way as ordinary crude oil. The retort in which the shale is heated rather resembles a giant pressure cooker that fuels itself with the very gases generated during heating.

Many oil shales yield from one-half to three-quarters of a barrel (42 U.S. gallons) of oil per ton of rock. If one were to mine only the shale layers thicker than 10 meters, they would yield an impressive 540 billion barrels of oil. If this oil were to be produced during the next decade, it would appreciably reduce the United States' dependency on foreign crude oil. Production of oil from oil shale, however, requires the use of vast amounts of water, and also might damage the environment. Thus, Americans are faced with the question of how to balance the harmful effects resulting from the use of an important resource with the critical need for that resource.

Two environmental problems have prevented full-scale mining and retorting of oil shales. The first is the waste disposal problem. In the process of being crushed and retorted, the shale expands to occupy about 30 percent more volume than was present in the original rock. Geologists call this the popcorn effect. Where can this great volume of light, dusty material be placed? The second problem relates to air quality, for the processing of huge tonnages of oil shale is likely to release large amounts of dust into the atmosphere. Unfortunately, the shales occur in arid regions, and there is little water available for use in processing the rock so as to contain the dust and provide for revegetation of the land. Until these problems are solved, production of oil from oil shale will be slow. Perhaps this will work to our advantage, for we may need this oil for processed goods a century or two from now, when the internal combustion engine may be as rare as the horse and buggy is today.

---

Alaska, Spitzbergen, and Greenland. Coral reefs grew in regions 10° to 20° north of their present optimum habitat. Cooling resumed during the Oligocene, and coral reefs began a slow retreat toward the equator. On land, temperate and tropical forests were also displaced toward lower latitudes. Another pulse of warmer conditions occurred during the Miocene, after which climates grew steadily cooler, as if in anticipation of the approaching Pleistocene glaciations.

Although the most extensive glaciations of the Cenozoic were those of the Pleistocene, glaciation occurred on a more limited scale at other times during the Cenozoic. Deep-sea cores taken near Antarctica indicate than an ice sheet had already begun to form during the Eocene. In addition, Miocene glacial conditions are indicated both by terrestrial glacial deposits in Antarctica and ice-rafted glacial deposits in Miocene deep-sea cores from the Bering Sea. As noted earlier, the accumulation of Miocene glacial ice and resulting lowering of sea level contributed to the drying up of the Mediterranean Sea. Evidence for Pliocene glaciation has been recognized in the Sierra Nevada, Iceland, South America, and the Soviet Union.

## Mineral Resources of the Cenozoic

Most of the known accumulations of oil have been found in Tertiary reservoir rocks. In the United States, about half of all the oil ever discovered has been in Tertiary rocks. Paleocene petroleum reser-

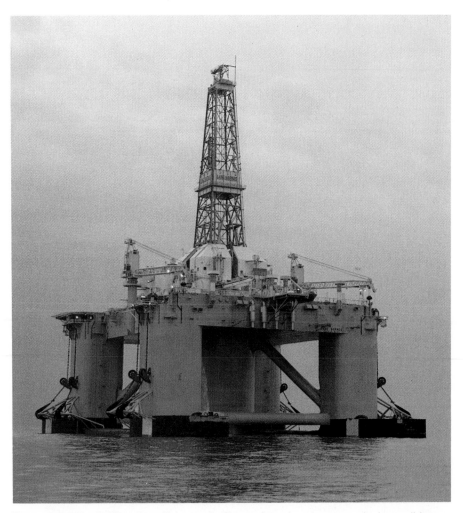

**Figure 14–50**   Drilling for offshore oil. Shown here is a huge semisubmersible drilling rig. *(Courtesy of Reading and Bates Drilling Company.)*

voirs occur in Libya and beneath the North Sea. Petroleum trapped in Eocene permeable limestones and sandstones is being pumped in Texas, Louisiana, Iraq, the Soviet Union, Pakistan, and Australia. Strata of Oligocene age yield oil in western Europe, Burma, California, and the Gulf Coast states of the United States. Miocene reservoir sandstones yield oil on every continent except Australia. They are extraordinarily productive in Saudi Arabia, Kuwait, California, Texas, and Louisiana. Offshore oil fields in the Gulf Coast (Fig. 14–50) and California tap oil held in rocks of Pliocene age. Finally, the largest reserves of oil shale in the world occur in the Eocene Green

River Formation of the western United States (Fig. 14–51).

Coal also occurs in Cenozoic rocks. Most Cenozoic coal is of the lignitic or subbituminous variety, but because of its low content of sulfur, it is extensively used. The coal beds of the Paleocene Fort Union Formation are mined in the Dakotas, Wyoming, and Montana. They represent the largest recoverable fossil fuel deposits in the United States. Scattered coal beds, mostly of Eocene age, are exploited along the Pacific coast of the United States.

Metals mined from Cenozoic rocks include placer gold gathered by dredging and hydraulic min-

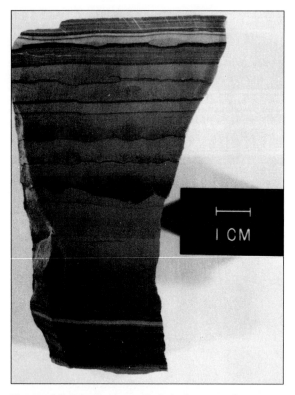

**Figure 14–51** Eocene oil shale from northeastern Colorado. *(Courtesy of U.S. Geological Survey.)*

ing of gravel deposits in California. The gold in these deposits has been eroded from older Jurassic gold-bearing quartz veins in the Sierra Nevada. About 60 percent of the world's supply of tin is also supplied by Cenozoic stream deposits. Manganese, an essential additive in the manufacture of steel, is mined extensively from lower Tertiary rocks of the Soviet Union. These deposits comprise 75 percent of the world's richest manganese ores. Another metal used in making certain types of steel is molybdenum. One deposit in Climax, Colorado, supplies 40 percent of the world's molybdenum. Other metals, such as copper, silver, lead, mercury, and zinc, are frequently found associated with Tertiary intrusive rocks of western North America and the Andes and with Tertiary orogenic belts along the western margins of the Pacific Ocean. Similar Cenozoic rocks were formed in southern Europe and Asia as a consequence of the deformation of the Tethys belt.

Nonmetallic materials from Cenozic rocks are also important resources. Diatomite (Fig. 14–52), the white, porous rock formed from the silica coverings (frustules) of diatoms, is mined in great quantities in California. Cenozoic building stone, clay, phosphates, sulfur, salt, and gypsum are quarried in North America, Europe, the Middle East, and Asia.

**Figure 14–52** Massive, jointed diatomite of the Bruneau Formation of middle Pleistocene age, Elmore County, Idaho. *(Courtesy of U.S. Geological Survey.)*

# SUMMARY

A compilation of the great historical events of the Cenozoic would certainly include the formation of the Alpine–Himalayan mountain system, the drift of continents to their present locations, the uplift of mountains around the perimeter of the Pacific, the development of an ice age, and the evolution of *Homo sapiens*. For most of the duration of the Cenozoic, the continents stood at relatively high elevations; epicontinental and marginal seas were of limited extent.

Along the east side of North America, the Appalachian chain was subjected to cycles of erosion and uplift that served to sculpt the surface to its present topography. Along the Atlantic and Gulf coastal plains, repeated transgressions and regressions reworked the clastics transported there from the interior. Florida was a shallow, limestone-forming, submarine bank until it was uplifted late in the era. In the Gulf of Mexico, Cenozoic deposits accumulated in thicknesses in excess of 9000 meters as the depositional basin was continuously downwarped.

In western North America, compressional forces that had produced many of the structures of the Cordillera ceased, and in their place vertical crustal adjustments occurred. The erosional debris worn from the highland areas filled the basins between ranges and was then spread eastward to form the clastic wedge of the Great Plains. Lakes formed in some basins and served as collecting sites for oil shales. Crustal movements late in the Tertiary caused the elevation of the Sierra Nevada along a great fault and produced the Basin and Range Province. Intense volcanic activity accompanied these crustal movements. Volcanism continues today along sectors of the west coast of North America that lie adjacent to subduction zones, such as in the Cascades, which are fed by lavas from the subducting Juan de Fuca plate. Also in the far west, the Farallon plate and its spreading center were overridden by the westward-moving continent. As a result, the formerly convergent plate boundary was changed to a shear boundary characterized by strike-slip, rather than subduction, tectonics. The development of the San Andreas fault system was a direct result of this change in the style of west coast tectonics.

In the Tethys region, there was intense deformation during the Cenozoic as major tectonic plates moved against one another. Orogenies that accompanied the compressions and uplifts along the southern margin of the Tethys region resulted in the Atlas Mountains, whereas along the northern border, the Alps, Carpathians, Apennines, Pyrenees, and Himalayas were constructed from the sediments that had accumulated in the former seaway. The Black, Caspian, and Mediterranean seas were left as remnants of the Tethys seaway. Orogenic activity was also prevalent over much of the Middle East and Far East and along the western borders of the Pacific. Africa witnessed Cenozoic marine deposition along its northern border. However, the most dramatic occurrences in Africa were the continuing development of the great rift valleys that extend along the eastern side of the continent.

It is apparent from fossil evidence and oxygen isotope studies that the earth's climate was on a somewhat fluctuating cooling trend during the Cenozoic. The culmination of that trend was the beginning of extensive glaciation about 1 million years ago. Variations in climate during the Pleistocene caused the vast ice sheets that had formed on the northern continents to alternately advance and recede. There were at least four major advances, with minor fluctuations before and within each. The ice, at times attaining thicknesses of 3000 meters, depressed the earth's crust, smoothed and rounded the landscape, disrupted drainage patterns, indirectly altered climates of adjacent regions, and was responsible for the development of numerous lakes, not the least of which are the Great Lakes. As the ice alternately accumulated and melted, sea level was caused to rise and fall with profound effects on low-lying coastal areas. It is a possibility that today we live in an interglacial stage and that some thousands of years from now the ice will again accumulate in Canada and northern Europe and advance southward.

Although a wealth of gold, tin, copper, silver, and other metals is found in deposits or associated with intrusions of Cenozoic rock, the greatest resource of the era has been petroleum. Indeed, most of the world's petroleum has been found in strata that range from 1 to 60 million years of age.

# QUESTIONS FOR REVIEW

1. Briefly describe the manner in which each of the major physiographic features below developed.
   - a. Mountains of the Basin and Range Province
   - b. Great Plains
   - c. Columbia and Snake River Plateau
   - d. Red Sea
   - e. Teton Range
   - f. Cascade Range

2. What Eurasian mountain ranges resulted from the compression and upheaval of large areas of the Tethys seaway?

3. What epochs of the Cenozic Era are included within the Tertiary? What epochs compose the Paleogene?

4. The Cenozoic stratigraphic record for the Gulf Coast region indicates that there were eight major transgressions and regressions. What characteristics of a stratigraphic column might indicate that it was deposited during a transgression? A regression?

5. What evidence indicates that the Gulf of Mexico experienced steady subsidence during the Cenozoic?

6. What is the economic importance of the Fort Union Formation? The Green River Formation?

7. In terms of plate tectonics, why did the western borders of the Americas experience folding and volcanism through the Cenozoic?

8. Account for the frequent juxtaposition of older, often chaotic, marine deposits lying *above* younger molasse deposits in areas of the Alpine–Himalayan belt.

9. What is the origin of such east African lakes as Nyasa and Tanganyika?

10. What is the difference in texture and mode of origin between stratified drift and till?

11. What are the advantages and limitations of the carbon-14 isotopic dating method for determining the age of Pleistocene deposits?

12. What three variables are the basis for the Milankovitch theory for the cause of the Ice Age? Why is it unlikely that the Milankovitch theory alone would provide an explanation for all former ice ages?

13. What is meant by *albedo?* What circumstances might cause an increase in albedo and possibly the initiation of an ice age?

14. Why were extensive deposits of loess formed in the Pleistocene, but not in earlier Cenozoic epochs? Most of the particles in loess are within the range from $\frac{1}{16}$ to $\frac{1}{32}$ mm, and very little sand or clay is present. How do you account for this characteristic of loess?

15. Why is the Great Salt Lake of Utah so salty?

16. What might be the effect upon humans if all the ice present today as continental ice sheets were to undergo relatively rapid melting?

17. In a 30-meter marine core that had penetrated the entire Pleistocene Series, how might you differentiate the sediment that had accumulated during an interglacial stage from that of a glacial stage?

18. What land bridge, important in the migration of Ice Age mammals, resulted from glacial lowering of sea level?

19. What is the explanation for the gradual rise in land elevations that has occurred within historic time around Hudson Bay, the Great Lakes, and the Baltic Sea?

20. What changes in movement of tectonic plates along the western border of the United States were responsible for the development of the San Andreas fault?

# TERMS TO REMEMBER

| | | | |
|---|---|---|---|
| albedo | little ice age | Neogene | pluvial lakes |
| axial tilt | loess | nuée ardente | precession |
| channeled scablands | Messinian event | oil shale | rift valley |
| discoaster | Milankovitch effect | orbital eccentricity | stratified drift |
| Farallon plate | molasse | Paleogene | till |

## SUPPLEMENTARY READINGS AND REFERENCES

Ager, D. V. 1980. *The Geology of Europe*. New York: Halstead Press.

Atwater, T. 1970. Implications of plate tectonics for the Cenozoic evolution of western North America. *Bull. Geol. Soc. Am.* 81:3513–3536.

Covey, C. 1984. The earth's orbit and the ice ages. *Sci. Am.* 250(2):58–77.

Curtis, B. F. (ed.). 1975. Cenozoic history of the southern Rocky Mountains. *Geol. Soc. Am. Memoir* 144:1–279.

Ewing, M., and Donn, W. L. 1966. A theory of ice ages III. *Science* 152:1706.

Hsu, K. J. 1972. When the Mediterranean dried up. *Sci. Am.* 227(6):27–46.

Imbrie, J., and Imbrie, K. P. 1979. *Ice Ages: Solving the Mystery*. Short Hills, N.J.: Enslow Publishers.

Match, C. L. 1976. *North America and the Great Ice Age*. New York: McGraw-Hill.

Van Andel, T. H. 1985. *New Views of an Old Planet*. Cambridge, England: Cambridge Univ. Press.

An Eocene insect preserved in amber. The insect is a member of the Order Diptera, which includes flies, mosquitoes, and gnats. Although preservation of insects is uncommon, it is likely that they approached modern levels of abundance and diversity early in the Cenozoic Era. *(Courtesy of W. Bruce Saunders.)*

# Life of the Cenozoic

*During the Cenozoic Era or the Age of Mammals, the vegetation was of modern cast and life conditions in the main were similar to those of today, although there is much evidence of a gradual elevation of nearly all lands with a consequent increase in aridity and diminution of moisture-loving vegetation.*

Richard Swann Lull, *Organic Evolution,* 1924

## Introduction

Because biological developments of the Cenozoic have occurred so recently and because Cenozoic fossils are topmost in the stratigraphic column, we have more facts about this era than about the far lengthier ones that preceded it. Armed with this more adequate data, paleontologists are better able to compare biological evolution with environmental and paleogeographic changes in the Cenozoic. Continental fragmentation clearly stimulated biological diversification and resulted in distinctive faunal radiations on separated land masses, as well as in climatically isolated marine basins. Among the many interesting evolutionary developments, none seem more fascinating than the changes experienced by the primates, which by late Tertiary time had produced species considered to be the ancestors of humans. In the Pleistocene Epoch, our own species, *Homo sapiens,* evolved.

## Plants

### Response of Mammals to the Spread of Prairies

Although they are the most recent group to evolve, the flowering plants are now the most widespread of all vascular land plants. Angiosperm floras did not explode upon the lands until mid-Cretaceous (Fig. 15–1). There were few spectacular floral innovations during the Cenozoic. Rather, this was a time of steady progress toward the development of today's complex plant populations. The Miocene is particularly noteworthy as the epoch during which grasses appeared and grassy plains and prairies spread widely over the lands. In response to the proliferation of this particular kind of forage, grazing mammals began their remarkable evolution.

Numerous evolutionary modifications among herbivorous mammals can be correlated to the development of extensive grasslands. Especially evident were changes in the dentition of grazing mammals. Grasses are abrasive materials. Many contain siliceous secretions, and because they grow close to the ground, they are often coated with fine particles of soil. To compensate for the wear that results from chewing grasses, the major groups of herbivores evolved high-crowned cheek teeth that continue to grow at the roots during part of the animals' lives. To provide space for these high-crowned teeth, the overall length of the face in front of the eyes increased. Enamel, the most resistant tooth material, became folded, so that when the tooth was worn, a complex system of enamel ridges was formed on the grinding surface (see Fig. 3–17). In general, incisors gradually aligned into a curved arc for nipping and chopping grasses.

In the wide open plains environment, it was more difficult to escape detection by predators. As a

553

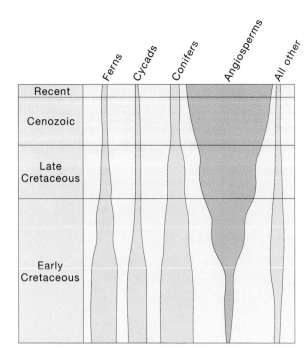

**Figure 15–1** Relative proportions of genera in Cretaceous to Recent floras.

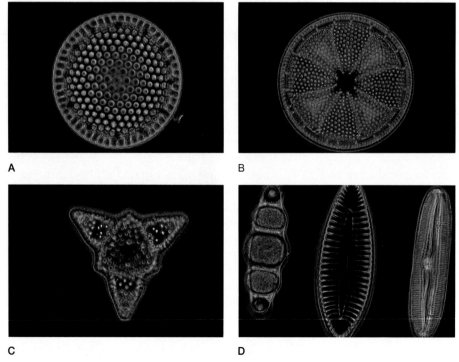

**Figure 15–2** Cenozoic diatoms exhibiting various frustule shapes, including discoidal, triangular, and spindle-shaped. (A) *Arachnoidiscus.* (B) *Actinoptychus.* (C) *Triceratum.* (D) From left to right, *Biddulphia, Surirella,* and *Pinnulria.*

**Figure 15–3** Dinoflagellate as seen with the aid of a scanning electron microscope. From Paleocene sedimentary rocks of Alabama. *(Courtesy of Standard Oil Company of California, photographer, W. Steinkraus.)*

prairie sod. Herbivores called ungulates evolved the four-chambered stomach in response to selection pressures favoring improved digestive mechanisms for breaking down the tough grassy materials. The response of mammals to the spread of grasses provides a fine example of how the environment influences the course of evolution.

### *Marine Phytoplankton*

As noted in Chapter 13, entire families of phytoplankton orders experienced extinction at the end of the Cretaceous. Only a few species in each major group survived and continued into the Tertiary. However, the survivors were able to take advantage of decreased competitive pressures and rapidly diversified. In general, peaks in diversity of species were reached in the Eocene and Miocene. A decrease in diversity has been recorded in the intervening Oligocene. Diatoms (Fig. 15–2), dinoflagellates (Fig. 15–3), and coccolithophores (Fig. 15–4) provided the most abundant populations of Cenozoic marine phytoplankton.

result, many herbivores evolved modifications that permitted speedy flight from their enemies. Limb and foot bones were lengthened, strengthened, and redesigned by the forces of selection to prevent strain-producing rotation and permit rapid fore-and-aft motion. To achieve greater speed, the ankle was elevated, and, like sprinters, animals ran on their toes. In many forms, side toes were gradually lost. Hoofs developed as a unique adaptation for protecting the toe bones while the animals ran across hard

## Invertebrates

The invertebrate animals of the Cenozoic seas included dense populations of foraminifers, radiolarians (Fig. 15–5), mollusks of all classes, corals,

**Figure 15–4** A coccolith of the coccolithophorid *Coccolithus* as seen with the aid of the transmission electron microscope. Maximum diameter is 16 $\mu$m.

**Figure 15–5**  Quaternary radiolarians from the tropical Pacific Ocean. *(Courtesy of Annika Sanfilippo, Scripps Institute of Oceanography.)*

**Figure 15–6**  Diversity of form in Cenozoic foraminifers. Magnification ×50.

bryozoans, and echinoids. The combined fauna had a decidedly modern aspect. Once successful groups such as ammonites and rudistid pelecypods were no longer present. No new major groups of invertebrates appeared during the Cenozoic.

## Protozoans

Foraminifers were enormously prolific and diverse during the Cenozoic (Fig. 15–6) and included large numbers of benthic as well as planktic forms (Fig.

15–7). Among the larger genera, coin-shaped nummulitic foraminifers thrived in the clear Tethys seaway, as well as in warm water areas of the western Atlantic. The tests of these organisms have accumulated to form thick beds of nummulitic limestone. The ancient Egyptians quarried this rock and used it to construct the Pyramids of Gizeh. Even the famous Sphinx was carved from a large residual block of nummulitic limestone. The incredible numbers and variety of foraminifers have permitted their extensive use in correlating Cenozoic strata in oil fields of the Gulf Coast, California, Venezuela, the East Indies, and the Near East.

## Corals

Corals (Figs. 15–8 and 15–9) grew extensively in the warmer waters of the Cenozoic oceans. Fossil solitary corals are commonly found in shallow-water deposits of early Tertiary age in Europe and the United States Gulf Coastal region. Reef corals were most extensively developed in parts of the Tethyan belt, West Indies, Caribbean, and Indo-Pacific regions. Careful comparison of coral species in Cenozoic rocks on either side of the Isthmus of Panama has given geologists clues to when the Atlantic and Pacific oceans were connected across this present-day barrier. Continuous deposition of reef limestones occurred throughout the Cenozoic in atolls of the open Pacific. **Atolls** are ringlike coral reefs that grow around a central lagoon. In 1842, Charles Darwin brilliantly discerned the relationship between atolls and volcanic islands. In his subsidence theory,

**Figure 15–7**  Present-day planktic foraminifers. (A) *Globorotalia tumida.* (B) *Globigerinoides conglobatus.* (C) *Globigerinoides rubra. (Courtesy of C. G. Adelseck, Jr., and W. H. Berger, Scripps Institute of Oceanography.)*

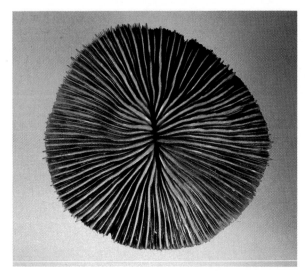

**Figure 15–8** *Fungia,* a solitary coral that takes its name from its resemblance to the underside of the cap of a mushroom. It has a geologic range from the Miocene to the present and is common today in the seas of the Indo-Pacific region. This specimen is about 9.0 cm in diameter.

A

B

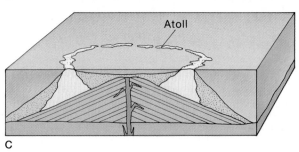

C

**Figure 15–10** Three stages in the development of an atoll. (A) In initial stage, a fringing reef develops around the shoreline of a volcanic island. (B) The island begins to subside, and corals build upward to stay in their optimal shallow water life zone. The result is the development of a barrier reef backed by lagoons. (C) As subsidence continues, the original land area has become inundated, and a circle of reefs and coralline islands remains.

**Figure 15–9** The living coral *Montastrea cavernosa,* with polyps extended for feeding. *(Courtesy of Charles Seaborn.)*

he proposed that, because of their great weight, volcanic islands subside slowly. Furthermore, such subsidence is usually slow enough to permit the corals that grow around the fringe of the sinking island to grow upward and maintain their optimum habitat near the ocean's surface. Eventually, the central is-

land would be submerged, but the encircling, living reef would persist (Fig. 15–10). Several deep borings into atolls have validated Darwin's theory. A boring drilled at Eniwetok island, an atoll in the Marshall islands of the west Pacific, encountered the basaltic summit of the volcano at depths of about 1200 meters. Eocene reefs lie directly above these igneous rocks.

It is likely that the subsidence associated with the development of some atolls is related to sea floor spreading rather than to the weight of the volcanic

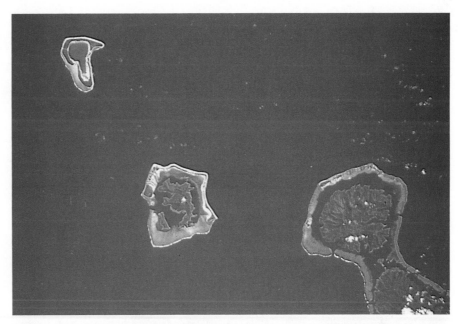

**Figure 15–11**  The Leeward Islands of the French Society Islands provide an illustration of the origin of coral reefs in deep oceans. At the bottom right, corals have constructed a fringing reef around the volcanic islands of Tahaa and Raiatea. Moving northward, the volcanoes are older and surrounded by barrier reefs. Bora Bora near the center of the photograph illustrates this phase of reef development. Tupai, visible in the top left of the photograph, is an atoll, and all that remains of the volcano around which the reef originally grew lies below the shallow floor of the central lagoon. *(Courtesy of L. E. Davis.)*

mass, as Darwin thought. Spreading begins along the crest of a mid-oceanic ridge, and as the plates move away from the ridge, they subside very slowly due to thermal contraction and isostatic adjustment. The subsidence is estimated at about 9 cm per year for the first million years. Thus, a volcano formed at the crest of the mid-oceanic ridge is carried down the slight incline of the ocean plate until it is completely submerged. Coral growth that is able to keep pace with the subsidence will result in the formation of barrier reefs and atolls (Fig. 15–11).

There is an alternate method for atoll formation. It is possible that corals might build **fringing reefs** around the perimeters of volcanic islands that had been erosionally truncated during a glacial period of low sea level. If sea level subsequently rose (as it did when the great Pleistocene ice sheets began to melt), then the corals would build the reef upward in order to stay at their optimum living depth. It seems not unreasonable that particular atolls may be primarily the result of subsidence of the island, whereas others may have been triggered by a rise in sea level.

## Mollusks

Cenozic shells of mollusks look very much like those likely to be found along coastlines today. Pelecypods and gastropods were the dominant classes of Cenozoic mollusks (Fig. 15–12). Their range of adaptation was truly outstanding. Arcoids, mytiloids, pectinoids, cardioids, veneroids, and oysters were particularly abundant pelecypods. Although the climax of cephalopod evolution had passed with the demise of the ammonites, nautiloids similar to the modern pearly nautilus lived in Tertiary seas, as they do today. Shell-less cephalopods, such as squid, octopi, and cuttlefish, were also well represented in the marine environment, although their fossil record is understandably sparse.

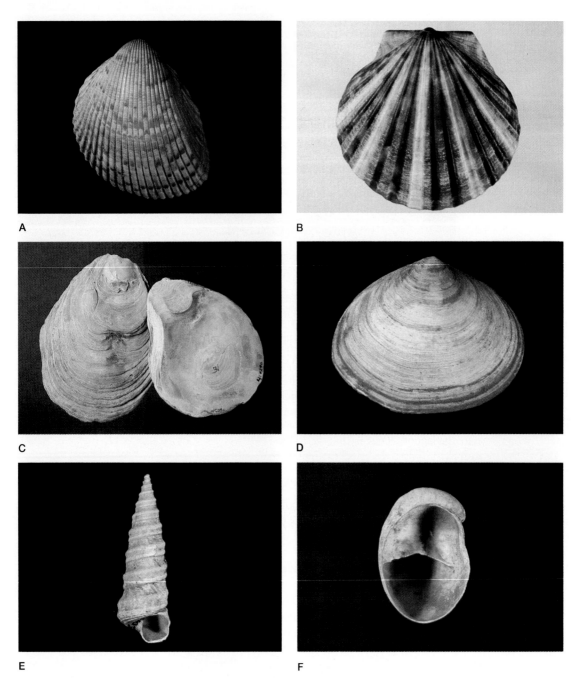

**Figure 15–12**  Common Cenozoic pelecypods (A, B, C, D) and gastropods (E, F). (A) *Cardium.* (B) *Pecten.* (C) *Ostrea.* (D) *Mya.* (E) *Turritella.* (F) *Crepidula,* a small marine snail having a horizontal shelf that extends across the posterior part of the shell.

**Figure 15–13**  The freshwater shrimp *Bechleja rostrata* from the Eocene Green River Formation. *(Photograph courtesy of Rodney M. Feldmann. From Feldmann, M. et al. J. Paleontol. 55(4):788–799, 1981.)*

## Echinoderms, Bryozoans, Crustaceans, and Brachiopods

Many other invertebrates have continued successfully through the Cenozoic. Echinoderms, mainly free-moving types, were particularly prolific. Members of the phylum Bryozoa are common in Tertiary rocks and are still very abundant in many parts of the ocean today. It was during this final era that the modern crustaceans became firmly established both in freshwater bodies (Fig. 15–13) and in the oceans. Brachiopods declined in abundance and diversity during the Cenozoic. Fewer than 60 genera survive today. They consist mostly of terebratulids (Fig. 15–14a), rhynchonellids, and such inarticulate brachiopods as *Lingula* (Fig. 15–14b).

**A**

**B**

**Figure 15–14**  (A) Cluster of present-day articulate terebratulid brachiopods, *Terebratulina septentrionalis.* (B) *Lingula,* a persistent primitive inarticulate brachiopod with a thin shell of proteinaceous material and a long, fleshy, muscular pendicle.

## Vertebrates

### *Fishes and Amphibians*

Bony fishes that evolved to the highest level of ossification and skeletal perfection are the **teleosts**. Teleosts have achieved an enormous range of adaptive radiation during the Cenozoic. That radiation has produced such varied forms as perches, bass, snappers, seahorses, sailfishes, barracudas, swordfishes, flounders, and others too numerous to mention. The Green River strata of Wyoming are well known for their content of beautifully preserved Eocene teleosts (see Fig. 14–10). Eocene strata at Monte Bolca in Italy also yield well-preserved teleost fossils (Fig. 15–15).

In addition to the bony fishes, the cartilagenous sharks were at least as common in the Tertiary as they are today. *Carcharodon,* an exceptionally large shark, had teeth as big as a man's hand and was more than 12 meters long.

Amphibians throughout the Cenozoic have resembled modern forms. All have been small-bodied, smooth-skinned creatures not at all like their Paleozoic ancestors. Frogs, toads, and salamanders were relatively abundant. The first frogs appeared during the Triassic, and by Jurassic time were already completely modern in appearance. Thus, they have continued almost unchanged for more than 200 million years.

### *Reptiles*

By the beginning of the Cenozoic, the dinosaurs had disappeared from the lands, as had the flying reptiles from the air and several groups of marine reptiles from the seas. The reptiles that have managed to survive and continue to the present include the tuatara, which inhabits islands off the coast of New Zealand, as well as turtles, crocodilians, lizards, and snakes (Fig. 15–16). The Tuatara, formally known as *Sphenodon,* resembles a large lizard. It is the sole survivor of a group of ancient reptiles known as rhynchocephalians that evolved and diversified during the Triassic.

Cenozoic turtles are the descendents of a lineage that can be traced back into the Late Permian. Turtles are readily recognized by everyone because of their distinctive adaptations. The most apparent of these adaptations is the shell. In turtles, the ribs have expanded differentially to form a broad *carapace.* On the underside, a bony growth called the *plastron* provides a similar covering. Both carapace and plastron are covered with a horny sheath. The jaws in turtles are toothless and covered by a beak that is used effectively in slicing through plants or meat.

Crocodilians also began their evolution during the Triassic and were very successful contemporaries of the dinosaurs throughout the Mesozoic. Modern crocodilians include the broad-snouted alligators, the narrow-snouted crocodiles, and the *very* narrow-snouted gavials.

Both lizards and snakes belong to an order of reptiles known as the Squamata. The Squamata are by far the most varied and numerous of living reptiles. Lizards are the ancestors of snakes. In fact, snakes are essentially modified lizards in which the limbs are lost, the skull bones modified into a highly flexible and mobile structure for engulfing prey, and the vertebrae and ribs greatly multiplied to provide the elongate form.

As evidence of their tetrapod ancestry, certain primitive snakes retain vestiges of rear limb and pelvic bones. Poisonous snakes evolved during the Miocene. They are characterized by specialized teeth or fangs for the injection of poison. One type of poison (neurotoxin) affects parts of the nervous system that control breathing and heart action, whereas a second

**Figure 15–15** The Eocene fish *Carangopsis analis* from Monte Bolca, Italy. Length is 20 cm.

A

B

C

**Figure 15–16**   Surviving rep-tiles. (A) monitor lizard, (B) hog-nose snakes emerging from their shells, (C) *Alligator missis-sippiensis,* (D) three-toed box turtle. *(B from Animals Animals © 1992 Zia Leszczynski; C from Ed Reschke; D from L. Stone/The Image Bank)*

D

type (hemotoxin) destroys red blood cells and causes disintegration of small blood vessels. It has been suggested that the evolution of poisonous snakes and the feeding behavior and skull structure of all snakes may be tied to the diversification of the Cenozoic mammals that served as prey.

## Birds

The Cenozoic record for birds is generally poor because they are rarely preserved. The rather fragmentary record indicates, however, that birds have been essentially modern in basic skeletal structure since the beginning of the Cenozoic. Distinctive skeletal features of birds include the fusion of bones of the "hand" to help support the wing, development of a vertical plate or keel on the sternum for attachment of the large muscles leading from the breast to the wings, and fusion of the pelvic girdle and vertebrae to provide rigidity during flight. Other characteristics include a body covering of feathers, light and porous bones, jaws in the form of a toothless horny beak, a four-chambered heart, and constant body temperature. The avian fauna is composed of a rich variety of families, including song birds such as robins, upland birds (pheasants), forest birds (owls), oceanic birds (albatrosses), wading birds (plovers), flightless aquatic birds (penguins), and flightless land birds (ostriches).

**Figure 15–17** *Diatryma* from the Eocene (Wasatch Formation) of Wyoming. This large, flightless bird was about 7 feet tall.

The fossil record for large terrestrial flightless birds is somewhat better than for small flying varieties. *Diatryma* (Fig. 15–17), an Eocene representative of this group, stood more than 2 meters tall. In New Zealand, huge moas lived until relatively recent time. Some were over 3 meters tall and laid eggs having a 2-gallon capacity. Perhaps the most famous of all Cenozoic ground birds was the dodo (*Didus ineptus*), which lived on the island of Mauritius east of Madagascar until about 1700 when they were exterminated by sailors searching for provisions. The moas of New Zealand were also exterminated by humans. Maori tribespeople killed them for food. The African ostrich, the South American rhea, and the emus and cassowaries of Australia are surviving flightless land birds.

## Mammals

During the Cenozoic, mammals came to dominate the earth in much the same way as reptiles had done so during the preceding Mesozoic era (Fig. 15–18). As we noted in Chapter 13, however, the evolution of mammalian traits had already begun among the therapsids of the Permo-Triassic. The Karoo beds of Africa, for example, contain bones of near-mammals that had almost made the transition from reptile to mammal.

Just as birds are easily recognized by the possession of feathers, so are mammals by their possession of hair. It is sometimes only present as a few whiskers, but it is present. Like feathers, hair functions as an insulating layer. Mammals are, of course, also recognized by mammary glands. However, neither body hair nor mammary glands are particularly helpful to the paleontologist, who must recognize mammalian remains on the basis of skeletal characteristics. Among these, the lower jaw is particularly useful. It consists of a single bone, the dentary, on either side. In reptiles and birds, there are always several. Another mammalian trait is the bony mechanism that conveys sound across the middle ear. In mammals, there is a chain of three little bones rather than a single one, as in lower tetrapods. Two of these ossicles, the incus and malleus, were derived from bones of the reptilian jaw. Unfortunately, these delicate ossicles are so small that they are rarely preserved. Typically, mammals have seven cervical, or neck, vertebrae, regardless of the length of the neck.

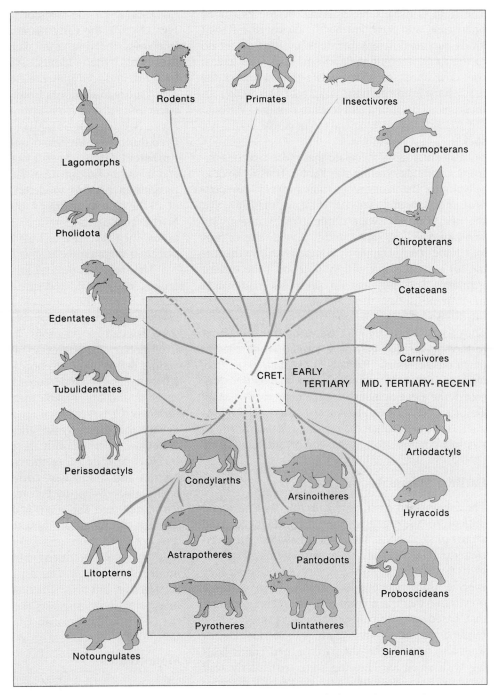

**Figure 15–18**  General relationships of the major orders of placental mammals. *(From Colbert, E. H.* Evolution of the Vertebrates. *New York: John Wiley, 1969. Used with permission of the author, artist Lois Darling, and publisher.)*

The skull is usually recognizable by the expanded brain case, and teeth are nearly always of different kinds, serving different functions in eating. As an aid to their endothermic metabolism, mammals have a well-developed secondary palate that separates the oral cavity from the nasal passages. This makes simultaneous breathing and feeding possible. Without the secondary palate, an infant mammal could not suckle.

Paleontologists speculate that more rigorous climatic conditions during the Permo-Triassic favored selection of the mammalian traits of warm-bloodedness and postnatal care. All evidence indicates that the first mammals were diminutive creatures, and small animals lose heat rapidly. A high rate of heat loss could have been partly compensated by insulating fur and an ample supply of food. If the earliest mammals laid eggs, as do primitive mammalian monotremes today, then the newly hatched young might also have had to cope with the heat problem. They were very likely kept warm by snuggling next to the soft fur that may have formed an incubation patch on the female. Perhaps at the same time, they were nourished by secretions from glands that preceded the development of true mammary glands. Unfortunately, fossils provide few clues to the reproductive characteristics of early mammals, and theories will continue to rely heavily on inference.

## Earliest Mammals

The mammalian fossil record begins with rare and difficult-to-find small jaw fragments, tiny teeth, and scraps of skull bone. The fossils, however, are sufficient to indicate that the earliest members of our own taxonomic class were very small, that they evolved from mammal-like reptiles such as those previously described from the late Paleozoic, and that, based on tooth patterns, they were insect eaters. Evidence from brain casts made of their inner cranial walls also suggests that the parts of their brains dealing with smell and hearing were particularly well developed, as in nocturnal animals. Perhaps the most significant and ironic fact about these tiny creatures is that they came on the scene at about the same time as the dinosaurs that they were destined to succeed. For 140 million years of dinosaur dominance, they quietly thrived as if waiting for just the right conditions for their own biological triumph.

Largely on the basis of tooth morphology (see Fig. 13–57), the early mammals are separated into categories bearing the names morganucodonts, docodonts, triconodonts, symmetrodonts, eupantotheres, and multituberculates. The oldest remains are of **morganucodonts** from Upper Triassic beds of Great Britain and from recent discoveries in South Africa (see Fig. 13–56). All of the groups just named lived during the Mesozoic, and the rodent-like **multituberculates** managed to cross the era boundary and live until Eocene time. The **docodonts** seem the possible ancestors of modern monotremes. However, of all the Mesozoic groups, the **eupantotheres** were of the greatest importance, for during the Cretaceous, they gave rise to both the marsupial and the placental mammals.

The stage was thus set for the varied mammalian groups that evolved and spread across seas and continents during the Cenozoic. The absence of dinosaurs and other groups of previously numerous reptiles favored a vigorous mammalian radiation. Evolutionary experimentation among Mesozoic ancestors had provided Cenozoic mammals with more efficient nervous and reproductive systems, greater speed and agility, reliable systems of body temperature control, larger brains, and levels of animal intelligence that far exceeded those of earlier classes of vertebrates. Armed with these attributes, mammals quickly expanded into the habitats vacated by the reptiles and found new pathways of adaptation as well (Table 15–1). By Paleocene time, 18 mammalian orders had appeared, and throughout the early Tertiary, there was an intricate interplay of appearances of new groups and extinctions of older groups that ultimately produced more than 30 taxonomic orders of mammals. The list is far too long to treat each group, but a brief review of some of the more interesting forms suggests the spectacular variety of the Cenozoic mammalian fauna.

## Monotremes

The most primitive of all living mammals are called **monotremes**. These relics of an older time still lay eggs in the reptilian manner. However, unlike reptiles, monotremes have primitive mammary glands and provide their newly hatched offspring with nourishment, if only for a brief period. Unfortunately, the fossil record for monotremes begins in

**Table 15–1    Evolutionary Modifications of Mammals for a Variety of Habitats***

| Habitat | Limbs | Teeth | Other Features |
|---|---|---|---|
| Primitive walking (ambulatory) | Limbs generalized; feet flat on ground | Incisors for grasping; canines for piercing; premolars for cutting; molars for grinding | Tongue, stomach, and intestine generalized; cecum small |
| Running herbivores (cursorial)<br>Horse and cow | Limbs elongate, toes elongate and reduced in number; hoof of horse is one toenail; in cow, two toenails | Incisors for nipping grass; cheek teeth heavy and ridged to withstand wear of grass | Stomach (cow) or intestine (horse) large and complex; cecum long |
| Running carnivores (cursorial)<br>Wolf, cheetah | Limbs elongate; often walk on toes; claws often long and sharp | Canines well-developed; premolars sharp for cutting flesh | Stomach large, intestine short |
| Heavy body<br>Elephant | Bones in limbs massive; flat jointed; toes in circle around pad | Cheek teeth massive and ridged for grinding plant materials | Nose elongated into trunk to reach food on ground or in trees |
| Tree-climbing (arborial)<br>Tree squirrel, monkey | Limbs elongate; wide range of motion; toes long, flexible | Cheek teeth usually smooth surfaced and flattened for crushing | Tail sometimes prehensile; diet usually fruit or vegetation |
| Diet of small, colonial insects<br>Anteaters, spiny echidna | Claws enlarged for digging after insects | Reduced or absent | Tongue long, sticky, and extensile |
| Burrowing (fossorial)<br>Moles, marsupial moles<br>Various rodents | Limbs short and stout; forefeet often enlarged, shovel-like; claws long and sharp | Teeth various, depending upon diet (vegetation or insects and worms) | Pinna of ears usually short or absent; fur short; tail short |
| Aquatic<br>Whales, seals, sea cows | Limbs shortened; feet enlarged and paddle shaped or absent | Often reduced to simple, conelike structures, sometimes absent | Tail often modified into swimming organ |
| Volant<br>Flying phalangers, flying squirrels | Limbs extendible to sides to hold gliding membrane outstretched | Various but usually show herbivorous adaptations | Tail often flattened for use in balance |
| Flying<br>Bats | Forelimbs and especially fingers of forelimbs elongate to support wing | Various for diets of insects, fruits and nectar, or other specific foods | Tongue, stomach, and intestine various, correlated with diet |

*From Cockrum, E. L., and McCauley, W. J. *Zoology*. Philadelphia: W. B. Saunders Co., 1965.

Pleistocene sediments, and we know nothing of their earlier evolution.

The platypus of Australia and Tasmania and two species of spiny anteaters of New Guinea and Australia (Fig. 15–19) are the only three species of monotremes still in existence.

## Marsupials

**Marsupials** are mammals that nurture their young in a special pouch, or marsupium (Fig. 15–20). Today, they are represented by such animals as kangaroos, wallabies, wombats, phalangers, bandicoots, koalas of Australia, and opossums of the New World (Fig. 15–21). Compared with the placentals, it is apparent that they are a dwindling group.

Marsupials had their greatest success in Australia and South America. Both continents were more or less isolated during most of the Tertiary, and marsupials were able to evolve without excessive competition from placentals. Although a few middle Tertiary Australian mammal sites have recently been found, the fossil record of Australian marsupials is not good until Pleistocene time. An older and more complete

**Figure 15–19**   Two present-day monotremes. (A) The duck-bill platypus of Australia and Tasmania. (B) The spiny anteater of Australia. *(A from E. R. Degginger.)*

**Figure 15–20**   The short-nosed bandicoot, a common marsupial of Australia. (A) Advanced offspring that have left the pouch return for nursing. (B) A fetal bandicoot attached to a teat in the mother's pouch.

record exists in South America where marsupials evolved from opossumlike ancestors and produced an array of mostly flesh-eating types. *Thylacosmilus* (Fig. 15–22), a large Pliocene form, was similar in appearance to the North American saber-toothed cats. South American marsupials fared poorly in competition with placental immigrants from the north when North and South America became connected in the early Pleistocene.

## Placental Mammals

Placental mammals appear during the Cretaceous as small, unspecialized insectivores. Modern moles are members of the order **Insectivora**, but the tiny shrew (Fig. 15–23) is more representative of the kind of animal from which other orders of Cenozoic placentals evolved. The descendants of the insectivores include the edentates, bats, primates, rodents,

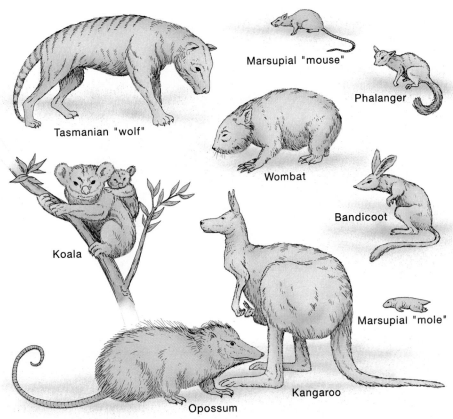

**Figure 15-21**   Adaptations of marsupials for various habitats is evident in their diversity. (The Tasmanian wolf is now thought to be extinct.)

**Figure 15-22**   *Thylacosmilus,* a Pliocene South American carnivorous marsupial comparable to the placental saber-toothed cats. The bladelike upper canine teeth were about 18 cm long and were protected when the jaws were closed by a deep flange of bone in the lower jaw.

**Figure 15-23**   The common tree shrew most resembles the ancient insectivores that gave rise to the primates. *(Warren Garst/Tom Stock & Associates.)*

**Figure 15–24** Restoration of a scene during the middle Pleistocene in Argentina. The heavily armored animals are glyptodonts. On the left is a giant ground sloth. *(Copyright © Chicago Museum of Natural History, painting by C. R. Knight, with permission.)*

flesh-eating mammals, a host of herbivores, and various marine mammals.

Armadillos, tree sloths, and South American anteaters are **edentates** that have not become extinct. Fossil species of armadillos have been found in rocks as old as Paleocene. Among the extinct edentates are the glyptodonts, which survived up until the Pleistocene. *Glyptodon* (Fig. 15–24) was a walking fortress with a spike-covered knob on its tail for bludgeoning pesky predators. Quite unlike the glyptodonts were the ungainly Cenozoic ground sloths. This group of edentates includes truly colossal Pleistocene beasts such as *Megatherium* (Fig. 15–25).

The rodents have been exceptionally successful Cenozoic mammals. Today, they probably outnumber all other mammals and have invaded every habitat of the earth. Their diversification has produced such burrowers as the marmot, the partially aquatic muskrat and beaver, the desert-dwelling jerboas and kangaroo rats, and the arboreal squirrels. The fossil record of rodents begins in the Paleocene with squirrel-like forms, such as *Paramys* (Fig. 15–26), and the horned rodent *Ceratogaulus*.

Beavers are well represented among fossil rodents. During the Pleistocene, beavers, along with many other mammals, grew to exceptional sizes. Some were as large as modern large bears. The other gnawers and nibblers of the Cenozoic that are not classified as rodents are the rabbits, hares, and pikas.

**Figure 15–25** The great ground sloth *Megatherium,* which lived during the Pleistocene and was nearly as large as a present-day elephant. This specimen is on display at the Chicago Field Museum.

**Figure 15–26** The Paleocene rodent *Paramys,* which resembled a large squirrel in size and form.

The fossil record for these **lagomorphs** begins with skeletal fragments of a primitive rabbit found in Paleocene beds of Mongolia.

Bats were also evolving during the Cenozoic. Their teeth have been discovered in Paleocene strata of France. A well-preserved skeleton was recovered from lower Eocene strata in Wyoming and indicates that by that time bats were already very similar to their living relatives. Bats are the only mammals to have achieved true flight. They have greatly elongated finger bones for the support of the membrane that forms the wing. Bats fly mostly at night when insect prey are abundant in the air. Although they occur in temperature zones, they are most abundant in the tropics. Vampire bats, which live in South America, feed on the blood of other mammals. Their front teeth are adapted to pierce the skin of their prey. When blood oozes from the wound, it is lapped up by the bat.

The history of flesh-eating placental mammals began in the Cretaceous with the advent of small, weasel-like animals called **creodonts** (Fig. 15–27). They were soon joined by members of the order **Carnivora**, which have common ancestry with the Creodonta. It appears that toward the end of Eocene time, with the coming of speedier and more progressive plant eaters, the creodonts gradually lost ground to carnivores. Most became extinct by the end of the Eocene, although one hyena-like group persisted until Pliocene time. The expansion of the Carnivora accelerated during the late Tertiary, producing a host of familiar flesh eaters, including bears, raccoons,

**Figure 15–27** The Eocene creodont, *Patriofelis. (Courtesy of the Smithsonian Institution.)*

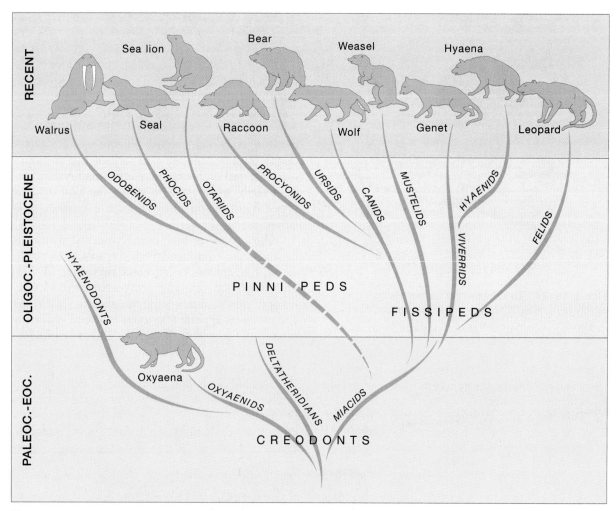

**Figure 15–28**   General relationships of major groups of flesh-eating placental mammals. *(From Colbert, E. H.* Evolution of the Vertebrates. *New York: John Wiley, 1969. Used with permission of the author, artist Lois Darling, and publisher.)*

weasels, genets, hyenas, dogs, and cats (Fig. 15–28). Wild dogs were doing very well on earth long before humans came along to make companions of them. The modern genus *Canis,* in the form of the dire wolf, was much in evidence during the Pleistocene. The most famous of the extinct cats are the so-called stabbing cats, exemplified by *Smilodon* (Fig. 15–29). This robust flesh eater preyed upon the larger herbivorous animals of the Pleistocene. A second line of cat evolution produced the biting cats, which resembled modern leopards and pumas. The skeletons of

these carnivores indicate they were strong, speedy, and agile predators.

As is true of carnivores today, Cenozoic carnivores were an essential element in the evolutionary process. To survive, they had to equal or better the speed and cunning of the herbivores on which they fed. The herbivores, in turn, responded to the carnivore threat by evolving adaptations for greater speed and defense. Then, as now, predators were not villains but necessary constituents of the total biological scheme. They were able to cull out the weak, de-

**Figure 15–29** The saber-toothed cat *Smilodon.*

formed, or sickly animals and thereby help to counteract the effects of degenerative mutation and to prevent overpopulation.

Not all the Carnivora of the Cenozoic were land dwellers. Some, such as seals, sea lions, and walruses, gathered their food in or at the edge of the sea. As indicated by their sharp, pointed teeth, early Cenozoic seals and sea lions ate fish as they do today. The walrus had (in addition to tusks) broad, flat teeth for crushing the shells of mollusks. Seals, sea lions, and walruses are known as **pinnepeds.** They probably descended from semiaquatic mammals somewhat similar to present-day otters, but the transitional forms are as yet undiscovered.

In adaptation to the marine environment, **cetaceans** (whales and porpoises) have made the most complete adjustment. Whales first appeared quite suddenly about 50 million years ago as already fully developed oceanic creatures. For this reason, paleon-

tologists suspect that cetaceans underwent an extraordinarily rapid evolution. The first whales are represented by *Basilosaurus* of the Eocene (Fig. 15–30). The structure of the skull in this toothed whale provides some evidence that the line was derived from early Cenozoic terrestrial flesh eaters.

Modern whales arose from the group that included *Basilosaurus* and divided into two lineages: the toothed whales and the whale-bone whales. Toothed whales, which made their debut during the Oligocene, today include porpoises, killer whales, and sperm whales like Melville's famous Moby Dick. The titanic blue whale, right whale, and Greenland whale are all representatives of the second cetacean group. These plankton-feeding giants first appeared in the Miocene. Instead of teeth, they possess ridges of hardened skin that extend downward in rows from the roof of the mouth. The ridges are fringed with hair that serves to entangle the tiny inverte-

**Figure 15–30** The Eocene whale *Basilosaurus.* The tendency toward increase in size was already evident in this whale, which was over 20 meters long.

**Figure 15-31** The early Tertiary archaic herbivorous mammal *Phenacodus* was a member of a group of primitive plant eaters called condylarths. Like most of the early Tertiary herbivores, *Phenacodus* walked upon all five toes in a method termed plantigrade. Speed was not essential to this forest-dwelling browser. *(Courtesy of the National Museum of Natural History, Smithsonian Institution.)*

**Figure 15-32** Eocene mammals. (A) *Uintatherium* is the large six-horned and tusked animal in the upper right. Other animals on this restoration are (B) the small, fleet rhinoceros *Hyrachus;* (C) *Trogosus,* a gnawing-toothed mammal; (D) *Mesonyx,* a hyenalike flesh eater; (E) *Stylinodon,* a gnawing-toothed mammal; (F) three early members of the horse lineage *(Orohippus);* (G) a saber-toothed mammal, *Machaeroides;* (H) *Patriofelis,* an early carnivore; and (I) *Palaeosyops,* an early titanothere. Restorations are based on skeletal remains from the middle Eocene Bridger Formation of Wyoming. *(Courtesy of the National Museum of Natural History, J. H. Matternes mural, with permission.)*

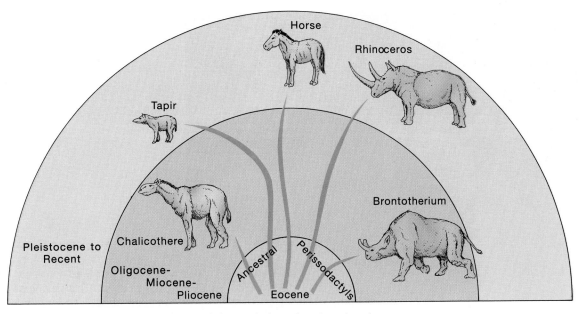

**Figure 15–33**   Generalized diagram of the evolution of perissodactyls.

brates on which these cetaceans feed—thus, the paradox of the largest of all animals feeding on some of the smallest. The great blue whales far exceed in size even the largest dinosaurs, for some have attained lengths of 30 meters and weights of over 135 metric tons.

The largest category of Cenozoic planteaters is that comprising the **ungulates**. Simply defined, ungulates are animals that walk upon hoofs and feed upon plants. The earliest ungulates were members of a group called condylarths. *Phenacodus* (Fig. 15–31), a representative condylarth, had simple primitive teeth and five toes, each terminated with a small hoof. The animal walked mostly in a flat-footed fashion, for these ungulates had not yet evolved the ability to walk or run on their toes for greater speed. This type of stance, termed **plantigrade**, was characteristic of early Tertiary ungulates, including the huge six-horned *Uintatherium* (Fig. 15–32). Primitive plantigrade and relatively small-brained plant eaters are often termed **archaic ungulates**.

The modern and more familiar ungulates fall into two categories: the **perissodactyls** and the **artiodactyls**. The perissodactyls include the modern horses, tapirs, and rhinoceroses and the extinct chalicotheres and titanotheres (Fig. 15–33). They seem to have originated from condylarth ancestors and reached the peak of their evolutionary history during the Miocene Epoch. Since that time, they have declined steadily.

Perissodactyls have certain distinctive characteristics. For example, the number of toes on each foot is usually odd, and the axis of the foot, along which the weight of the body is primarily supported, lies through the third, or middle, toe. There is a tendency toward reduction of the lateral toes. In the modern horse, only the single, central toe remains. They are clearly digitigrade, meaning that they run or walk on their toes in order to attain a longer stride and greater speed.

The oldest known perissodactyl, *Hyracotherium* (Fig. 15–34), has been found in upper Paleocene and Eocene strata of both North America and England. Its ancestors were probably condylarths. Familiarly known as eohippus, *Hyracotherium* has the distinction of being the earliest member of the horse family. This little "dawn horse" was not much larger than a fox. It had four hoofed toes on the front feet (Fig. 15–34) and three on the hind feet. The back curved like that of a condylarth, and the dentition was primitive, with canines still present and the premolar teeth not similar to the molars, as they are in more recent perissodactyls. In addition, the molars were bluntly cusped for browsing and had not yet developed the ridged oral surface and high crowns that characterize modern grazers.

**Figure 15–34** *Hyracotherium,* the small Eocene horse more popularly known as eohippus, the "dawn horse." This tiny horse was only about a half meter in length. *(Courtesy of the American Museum of Natural History.)*

From ancestors such as *Hyracotherium,* the horse family evolved, at least until the Miocene, in a rather straightforward manner (see Fig. 3–16). The skeletal remains clearly show progressive increases in size, length of legs, height of crowns of cheek teeth, and brain size. The curved back became straightened, and the middle toe was strengthened and emphasized at the expense of lateral toes (Fig. 15–35). The premolars came to resemble the molars, and their grinding surfaces developed increasingly complicated patterns of resistant enamel ridges. *Mesohippus,* an Oligocene horse about twice the height of

**Figure 15–35** Evolution of the lower foreleg in horses, beginning at far left with *Hyracotherium* and ending with the modern horse *Equus* at the far right. *(Courtesy of the National Museum of Natural History.)*

*Hyracotherium,* showed the beginnings of many of these trends but still retained relatively low-crowned teeth best adapted for browsing.

With the spread of grass-covered prairies during the Miocene, the family tree of horses became more complicated as more conservative species stayed behind in the forests while others took advantage of the open environments by grazing and running across the open plains. These prairie dwellers were the ancestors of the modern horse. *Merychippus,* a horse in the vanguard of this new group, ran on a single middle toe; the side toes were much reduced and of little use. The head was deeper and the face longer to accommodate the long cheek teeth.

In the Pliocene, two groups of horses arose from the general *Merychippus* base. The more conservative group, typified by *Hipparion,* retained three-toed feet. They found their way from North America into Europe and survived until the beginning of the Pleistocene. The more progressive branch of the horse family tree produced *Pliohippus* (Fig. 15–36), a completely one-toed form with lateral digits reduced to mere splints that were concealed beneath the skin. The descendants of *Pliohippus* also evolved along two separate branches. The horses of one branch, represented by *Hippidion,* migrated into South America over the Panamanian land bridge. The other branch led to *Equus,* the genus to which all modern horses belong.

During the Pleistocene, several different species of horses thrived. Some, such as *Equus occidentalis,* were the size of small ponies, but others, such as *Equus giganteus,* were larger than the great draft horses of today. They spread across the length of North America, Eurasia, and Africa. Only a few thousand years ago, horses suffered extinction in North America. (The continent was later restocked with the progeny of domestic horses that had escaped from the Spanish explorers in the sixteenth century.) The cause of the extinction of horses in North America is somewhat of a mystery. Some believe it was brought on by contagious disease, whereas others speculate that the cause was overkill by prehistoric human hunters.

Among the other surviving perissodactyls are the tapirs and rhinoceroses. Tapirs retain the primitive condition of four toes on the forefeet and three on the rear, as well as low-crowned teeth. They are

**Figure 15–36**   The Pliocene horse *Pliohippus.* Modern horses belonging to the genus *Equus* are descendents of *Pliohippus. (Courtesy of the National Museum of Natural History.)*

**Figure 15–37**  The giant Oligocene to early Miocene hornless rhinoceros *Baluchitherium.*

**Figure 15–38**  A restoration of Oligocene mammals based primarily on skeletal remains from the White River Formation of South Dakota and Nebraska. The flora is based on nearly contemporaneous plant fossils from the Florissant beds of Colorado. (A) *Trigonias,* an early rhinoceros. (B) *Mesohippus,* a three-toed horse. (C) *Aepinocodon,* a remote relative of the hippopotamus. (D) *Archaeotherium,* an entelodont. (E) *Protoceras,* a horned ruminant. (F) *Hyracodon,* small, fleet rhinoceros. (G) The giant titanothere *Brontotherium.* (H) Oreodonts named *Merycoidodon.* (I) *Hyaenodon,* an Oligocene carnivore. (J) *Poëbrotherium,* an ancestral camel. *(Courtesy of the National Museum of Natural History, J. H. Matternes mural, with permission.)*

forest-dwelling, leaf-eating animals whose fossil record begins in the Oligocene. More impressive to humans are the rhinoceroses, which began their evolution during the Eocene as small, swift-running creatures and eventually produced such giants as *Baluchitherium* (Fig. 15–37). *Baluchitherium* is the largest land mammal thus far discovered. It stood 5 meters tall at the shoulders.

The perissodactyls that did not survive into modern times were the **titanotheres** and **chalicotheres.** The most familiar titanothere of the Cenozoic was ponderous *Brontotherium* (Fig. 15–38G), readily remembered because of the pair of hornlike processes that grew over the snout. Chalicotheres differed from all other Cenozoic perissodactyls in having three claws rather than hoofs on their feet. These odd creatures (Fig. 15–39A) were rather similar to a horse in the appearance of the head and torso. The fore legs were longer than the hind legs, and the back sloped rearward. They lived from Eocene into the Pleistocene, at which time they became extinct.

The even-toed ungulates, or artiodactyls, have been far more successful than the perissodactyls in terms of survival, variety, and abundance (Fig. 15–40). Modern artiodactyls include pigs, deer, hippos, goats, sheep, cattle, and camels. The center of artiodactyl evolution was not North America, as had been the case with the odd-toed ungulates, but rather Eurasia and Africa. They were already present during the Eocene, and by late Tertiary time had clearly achieved numerical and varietal superiority over other herbivores.

Most artiodactyls have an even number of toes on each foot, either four or two. Most of the weight of the animal is carried on the two middle toes, which form an identical pair and thus a cloven, or split, hoof. Unlike advanced perissodactyls, the molar and premolar teeth are dissimilar. Among the Tertiary artiodactyls that became extinct, the **oreodonts** and **entelodonts** are particularly interesting. Oreodonts (see Fig. 15–38H) were short, stocky grazers that roamed the grassy plains of North America in enormous numbers. Entelodonts, some of

**Figure 15–39** Mural depicting an assemblage of early Miocene mammals. (A) The chalicothere *Moropus.* (B) The small artiodactyl *Merychyus.* (C) *Daphaenodon,* a large wolflike dog. (D) *Parahippus,* a three-toed horse. (E) *Syndyoceras,* an antelope-like animal. (F) *Dinohyus,* a giant, piglike entelodont. (G) *Oxydactylus,* a long-legged camel. (H) *Stenomylus,* a small camel. *(Courtesy of the National Museum of Natural History, J. H. Matternes mural, with permission.)*

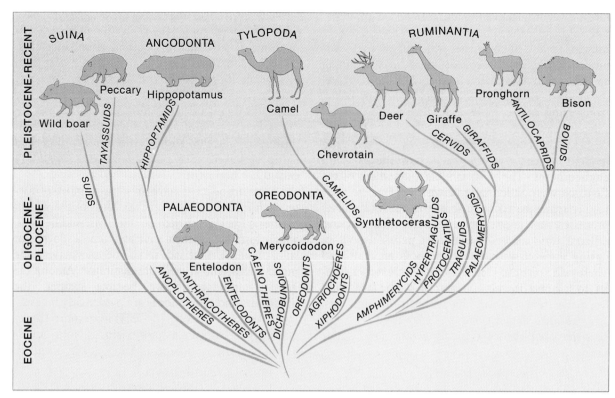

**Figure 15–40** The evolutionary radiation of the even-toed ungulates, or artiodactyls. *(From Colbert, E. H. Evolution of the Vertebrates. New York: John Wiley, 1969. Used with permission of the author, artist Lois Darling, and the publisher.)*

**Figure 15–41** *Dinohyus,* a giant piglike entelodont from the Miocene of America. The animal was about 3 meters long. *(Courtesy of the Denver Museum of Natural History.)*

which were as large as buffalo, were repulsive-looking, hoglike beasts. Their trademarks were curious bony processes that grew along the sides of the skull and jaws (Fig. 15–41 and see Fig. 15–39F).

The major radiation of artiodactyls began in the Eocene and ultimately produced the three major surviving categories: swine, camels, and ruminants. Of these, the swine family has probably remained the most primitive. They have kept their four toes, even though most of the weight is carried by the two middle digits. The swine group includes both pigs and the more lightly constructed and primarily South American peccaries. Hippopotami are the only modern amphibious artiodactyls. They are relatives of the pig family, having arisen from a group of Miocene piglike animals called anthracotheres.

People are often surprised to learn that much of the evolutionary development of the camel lineage—**Tylopoda**—occurred in North America. The geologic history of camels and llamas began in Eocene time with tiny creatures about the same size as a small goat. As they evolved through the Tertiary, they lost their side toes and increased the length of their legs and neck. The long-necked and long-limbed *Oxydactylus* (Fig. 15–42) of the Miocene probably browsed on leaves in much the same manner as the giraffe. By Pleistocene time, there were numerous modern-looking camels and llamas in North America. The llamas moved southward and took up residence in the highlands and plains of South America. Camels migrated across the Bering land bridge to Eurasia and Africa, where they established themselves in arid regions. Those that were left behind in North America mysteriously became extinct.

The **ruminants** are the most varied and abundant of modern-day artiodactyls. This group takes their name from the *rumen,* or the first of four compartments in their multichambered stomach. For the most part, they are cud chewers. The earliest ruminants were small, delicate, four-toed animals called **tragulids.** They are represented today by the skinny-legged and timid mouse deer of Africa and Asia. *Hypertragulus* (Fig. 15–43), whose remains are found in lower Oligocene strata, was such an animal. Later lines of tragulids sometimes developed peculiar assortments of horns, exemplified by *Synthetoceras* (Fig. 15–44B). This Pliocene tragulid had not only a pair of horns over the eyes but also a Y-shaped horn on the snout.

**Figure 15–42**    Skeleton of the Miocene camel, *Oxydactylus*. This small, graceful camel was only one of a varied group of camels that populated the grasslands of North America. *(Courtesy of the Denver Natural History Museum.)*

**Figure 15–43**    The primitive ruminant *Hypertragulus* from lower Oligocene beds of North America. The tiny animal stood only about 30 cm high at the shoulder. *(Courtesy of the Denver Museum of Natural History.)*

**Figure 15-44** A variety of early Pliocene mammals. (A) *Amebelodon,* the shovel-tusked mastodon. (B) *Synthetoceras.* (C) *Cranioceras.* (D) *Merycodus,* an extinct prong-horned antelope. (E) *Epigaulis,* a burrowing, horned rodent. (F) *Neohipparion,* a Pliocene horse. (G) The giant camel *Megatylopus* and smaller *Procamelus.* (H) *Prosthennops,* an extinct peccary. (I) The short-faced canid *Osteoborus.* *(Courtesy of the National Museum of Natural History, J. H. Matternes mural, with permission.)*

Ruminants such as sheep, cattle, giraffes, and deer are called **pecorans.** Deer are primarily browsers that have made the forests their principal habitat ever since their first appearance in the Oligocene. The early stages of deer evolution are exemplified by *Blastomeryx,* a small, hornless creature that lived in North America during the Miocene. Subsequent evolution involved increases in size and development of antlers. The culmination of this trend is represented by the Pleistocene *Megaloceros,* whose antlers measured more than 3 meters from tip to tip (Fig. 15–45).

The giraffe family probably branched off from the deer lineage sometime during the Miocene and became specialized in browsing on leaves of trees that grew rather sparsely in areas where most other herbivores were grazers. *Palaeotragus,* the ancestral giraffe, looked much like the modern okapi of Africa. From this general form, the obvious evolutionary trends emphasized leg and neck elongation.

Of course, the bovids—cattle, sheep, and goats—are presently the most numerous of ruminants. Miocene strata provide the earliest fossil record of bovids. The bison are particularly interesting to Americans. Seven species of bison lived in North America during the Pleistocene, some of which were truly giants with horns that measured 2 meters from end to end.

A discussion of Cenozoic mammals would not be complete without a brief mention of elephants and their older relatives, the mammoths and mastodons. Because these animals have a trunk, they are collectively termed **proboscideans**. The trunk is the elephant's principal means of bringing food to its mouth. Other animals as tall as elephants reach food on the ground easily because of their long necks. However, proboscideans, with short, muscular necks to support their massive heads, have evolved their own unique anatomic solution to food gathering. Paleontologists are able to follow the development of the trunk in early proboscideans by noting the position of the external nasal openings at the front of the skull. Those openings recede toward the rear of the skull in sequential stages of trunk development. Another elephantine trademark is the tusks, which evolved by elongation of the second pair of incisors.

The early fossil record of proboscideans includes a trunkless, tapir-like animal named *Moeritherium* (Fig. 15–46). Bones of this ancestral proboscidean have been found in upper Eocene and lower Oligocene beds near Lake Moeris in Egypt. It is likely that the moeritheres arose from early Tertiary condylarths. From an ancestry represented by the moeritheres, proboscideans separated into two branches. One led toward a group of Miocene and Pliocene animals called **dinotheres** (Fig. 15–47). The tusks of dinotheres were distinctive in that they were present only in the lower jaws and curved downward and backward, an orientation presumably useful for uprooting plants and digging for roots and tubers.

**Figure 15–45** *Megaloceros,* the giant Irish "elk," was actually a deer whose remains are frequently found in the peat bogs of Pleistocene age in Ireland. From tip to tip, the antlers were 3.6 meters in breadth.

**Figure 15–46** The first proboscideans (members of the group that includes elephants) were moeritheres, named after this typical genus *Moeritherium,* which lived in Egypt during the late Eocene.

**Figure 15–47** Skull of *Dinotherium,* a Miocene proboscidean. Length of skull is about 1.2 meters. This specimen is on display in the Sedgwick Museum of Cambridge University.

The other branch of the proboscidean family tree produced mastodons and elephants. *Palaeomastodon* in the Oligocene of North Africa, was probably representative of the group from which true mastodons evolved. The face was long, the trunk was short, and the second incisors in both the upper and the lower jaws had developed into tusks. By Miocene time, a larger proboscidean named *Gomphotherium* had found its way into North America by way of the Bering isthmus. Subsequent proboscidean evolution produced a variety of long-jawed mastodons. *Trilophodon* from the Pliocene had lower jaws almost 2 meters long. The most bizarre proboscidean was the "shovel-tusked" *Amebelodon* (see Fig. 15–44A) of the Pliocene. The tusks of the lower jaws in these beasts were flattened and formed two sides of a broad, scooplike ivory shovel. Mastodons with shorter jaws and longer trunks are best represented by species of *Mammut* (Fig. 15–48), which were common throughout North America during the great Ice Age and survived until comparatively recent times. *Mammut* was heavier but not as tall as the African elephants. It had the sharply crested molars typical of mastodons, a stocky build, and great curving tusks nearly 2 meters long.

True elephants and mammoths evolved from gomphotheres in the Old World. Their line of evolution can be traced backward to a Miocene animal named *Stegolophodon.* Next on the evolutionary scene

**Figure 15–48** Skeleton of *Mammut americanus,* the American mastodon, on display at the Los Angeles County Museum.

**COMMENTARY**

## How the Elephant Got Its Trunk

Among the many mammals that evolved during the Cenozoic, few inspire more interest than the proboscideans—mastodonts, mammoths, and elephants. Our interest seems to relate not just to their great size, but to their most distinctive feature, the trunk. The elephant's trunk is a highly muscularized, much elongated organ containing nostrils that extend downward through its entire length to the tip. The Indian elephant uses a one-finger process to pick up small objects with the tip of the trunk, while the African elephant uses a two-finger process.

The question, "how did the elephant get its trunk?" might be readily answered by a child on reading Rudyard Kipling's *The Elephant Child*. According to this story, an overly inquisitive elephant child tried to find out what the crocodile had for dinner. When the elephant child lowered his head along the bank of the great, gray-green, greasy Limpopo River, the jaws of the crocodile closed on his nose. There ensued a fierce struggle in which the "bi-colored-python-rock-snake" came to the aid of the poor elephant child, but in the tug-of-war, the elephant child's poor nose was stretched into the trunk we know today. Actually, the elephant achieved his wonderfully long and prehensile trunk by means of evolutionary processes that operated over many generations and resulted in an enormous elongation of the nose and upper lip. Scientists refer to such extraordinary growth of a body part as hypertrophy, and the elephant got its trunk by hypertrophy of the nose and upper lip. But why?

As one views the bones of proboscideans, beginning with the relatively small and primitive groups of the Eocene through the giant Pleistocene mastodonts and mammoths, several trends are apparent. One obvious trend was the increase in size, not only of the body, but of the huge proboscidean head. To support the weight of the ponderous head required a short, powerful neck. As the proboscideans grew taller, reaching food at ground level when equipped with a short neck would have been difficult. Selection paralleling that of the large head and short neck resulted in the development of the trunk. The elephant's trunk is a structural adaptation for feeding that could be used, not only for raising food from the ground, but for stripping leaves from branches and stuffing them into the mouth. Accompanying the evolution of the trunk in proboscideans, the lower jaw became shorter, allowing the trunk to hang downward. Tusks in the lower jaw that were present in older proboscideans, grew smaller and gradually disappeared.

was *Stegodon,* a Pliocene form with a short, tuskless lower jar and molar teeth already containing the numerous cross-ridges that characterize true elephants. Such teeth are more suitable for a diet of abrasive grasses. Further along in the evolution of the elephant, the trunk grew longer, the height of the skull increased enormously, and the jaws became so short that they could accommodate no more than eight grinding teeth at one time. As the molars of elephants gradually wear down, new teeth form in the posterior region of the jaws and push forward until the older, worn teeth break out. The last molars appear when the animals are 40 to 50 years old.

**Mammoth** is a term loosely applied to the Ice-Age elephants of North America, Europe, and Africa. They were a magnificent group of animals that included the famous woolly mammoths (Fig. 15–49) drawn by our own ancestors on the walls of their caves. The great imperial mammoth reached heights of 4.5 meters and ranged widely across California, Mexico, and Texas. The Columbian elephant had immense spiral tusks that in older individuals overlapped at the tips, thus becoming useless for digging purposes. Many of the late Pleistocene proboscideans were hunted by humans. Then, about 8000 years ago, all but two genera of elephants became extinct.

**Figure 15–49**    The woolly mammoth. In the late Pleistocene, these magnificent animals lived along the borders of the continental glaciers. Their remains have been found frozen in the tundra of northern Siberia. *(Courtesy and copyright of The Chicago Museum of Natural History; painting by C. R. Knight, with permission.)*

Their continued survival, as is the case with many wild animals, will depend upon our good judgment and that of the governments we support.

## Demise of the Pleistocene Giants

At the time of maximum continental glaciation (approximately 11,000 years ago), the Northern Hemisphere supported an abundant and varied fauna of large mammals, comparable to that which existed in Africa south of the Sahara several decades ago (Fig. 15–50). The fauna included giant beavers, mammoths, mastodonts, elk, most species of perissodactyls, many even-toed forms, and ground sloths. From all available evidence, most of these great beasts maintained their numbers quite well during the most severe episodes of glaciation, but experienced rapid decline and extinction in the period around 8000 years ago, when the climate became milder. What might have been the cause for the extermination of these large Pleistocene mammals? Explanations tend to fall into two categories. In one, theories attempt to explain the extinctions as being related to human overkill; a second interpretation seeks to relate the losses to climatically controlled environmental causes.

The overkill concept is based on the inference that early humans had developed the ability to hunt in highly organized social groups and skillfully bring down large numbers of big animals on open prairies and tundras. Human predators may have killed then, as they do now, in excess of their needs. Unlike such predators as wolves, early human hunters probably did not seek out the weak or sick to kill but brought down the best animals in the herds. By decimating particular species of herbivores, the other, non-human predators would have necessarily also suffered. In support of this idea, paleontologists note that most Pleistocene extinctions involved large terrestrial animals. Marine genera, protected by the sea from human predation, continued to thrive. Small mammals also survived; although frequently hunted, they were difficult to exterminate because of their more rapid breeding rate, their greater numbers, and the probability that they were killed only one at a time. The big, gregarious herbivores were easily accessible and reproduced more slowly, and entire herds could be driven off cliffs or into ravines, thus providing opportunities for slaughter of hundreds at the same time.

Another line of evidence favoring overkill stresses the observation that an early wave of extinction began in Africa and southern Asia, where predatory humans first became prevalent. The better known extinctions of the colder north did not occur until much later, when humans had moved into this region after having first devised the means to better clothe and shelter themselves.

The overkill concept has not, however, gone unchallenged. Many paleontologists feel that human

**Figure 15–50** An assemblage of animals that lived in Central Alaska about 12,000 years ago, during the late Pleistocene. Fossil remains of this period are abundant and indicate a fauna in which grazing animals predominated, with the remaining fauna composed of browsers and predators. *(Courtesy of the National Museum of Natural History, J. H. Matternes mural, with permission.)*

populations were too small and that nomadic tribes shifted too frequently to account for the extermination of entire species of mammals. Also, it cannot be demonstrated that the primitive peoples who live at stone age levels today have endangered the survival of species of animals they hunt. In some cases, animals that were probably not desirable for food, such as the cave lion, became extinct; more desirable forms, such as certain species of deer, also disappeared completely. Such arguments have led some paleontologists to favor theories that stipulate changes in climate operated selectively against large terrestrial animals.

It is possible that as the climate was warming up following the final retreat of the ice, parts of the temperate belt became excessively arid, producing severe stress on the large animals, which make the biggest demands on their environment. Also, North American herbivores could not freely move north and south over great distances to escape unfavorable conditions (as could, for example, those in Africa).

Unfortunately, theories that relate Pleistocene extinctions to climatic alteration are difficult to validate. Fossilized pollen (as clues to climate) cannot usually be precisely correlated to the demise of particular species of mammals. Whatever the cause of the exterminations, it is apparent that they coincide with the expansion of ancient human tribes and with a change from cool and moist to warmer and drier climates over large continental tracts.

## SUMMARY

The Cenozoic was a time of gradual approach to the conditions of the present. The flowering plants continued to expand and diversify, with the most important single occurrence being the Miocene proliferation of grasses as extensive prairies and savannahs. Among the marine invertebrates, large discoidal foraminifera such as nummulites were characteristic of areas in the Tethys seaway and the tropical west Atlantic. Smaller planktonic foraminifers proliferated and underwent a marked diversification during the Cenozoic. Gastropods and pelecypods were the dominant higher marine invertebrates. Reef corals grew extensively across tracts of shallow warm water in the Tethys belt, the Caribbean, the West Indies, and around the many volcanic islands of the Pacific.

The reptile dynasty had collapsed at the end of the Mesozoic. Only turtles, crocodiles, lizards, snakes, and rhynchocephalians lived through the Cenozoic and survived to modern times. Although the fossil record for birds is less than adequate, it is likely that most modern orders were present by Eocene time. Nearly all the continents seem to have had one or more species of large, flightless, flesh-eating birds during the Cenozoic.

With considerable justification, the Cenozoic has been called the Age of Mammals. During the Cenozoic, mammals expanded rapidly into the multitude of potential habitats vacated by the reptiles. From ancestral shrewlike creatures, either directly or indirectly, the higher orders of placental mammals evolved. In the vanguard were small-brained archaic ungulates, such as amblypods and condylarths, as well as primitive flesh-eating creodonts.

Following this early Cenozoic radiation, a second wave of placentals began to evolve in the middle part of the era, and most of the modern mammalian orders became established. This adaptive radiation was as remarkable as had been that of the earlier Mesozoic reptiles. Mammals in the form of bats conquered the air, whereas others, such as whales, seals, and walruses, returned to the sea, where the evolution of vertebrates had begun long ago.

The Carnivora diversified and became dependent upon the multitude of advanced ungulates that populated forests and grasslands. Ungulates included two great categories: the perissodactyls and artiodactyls. The former include horses, rhinoceroses, and tapirs, as well as the extinct titanotheres and chalicotheres. Among artiodactyls, or even-toed ungulates, the large, piglike entelodonts and smaller oreodonts failed to survive beyond the Oligocene. However, living artiodactyls include swine, camels, hippopotami, and a diverse host of ruminants. As all of these ungulates evolved during the Cenozoic, they underwent specialization of teeth for grazing and limbs for running across relatively open plains. Grass tends to wear down teeth rapidly. To compensate, grass eaters evolved teeth with complicated ridges of hard enamel and exceptionally high crowns.

Similar changes also occurred in members of the proboscidean line. The earliest proboscideans are recorded as fossils in Eocene beds of Egypt. Their radiation produced an impressive array of variously tusked and specialized mastodonts and mammoths. The remains of those great beasts are frequently found in Pleistocene bog deposits in the eastern United States, Alaska, and Siberia.

The Pleistocene was a time of splendid diversity among mammals. The fauna was probably more varied and impressive than that of Africa a century ago. The Pleistocene was an age of giants, in which colossal ground sloths, beavers over 2 meters tall, giant dire wolves, bison, and tall, rangy mammoths and woolly rhinoceroses ranged far and wide across the continents. One might think an assemblage of beasts that had proved their stamina against the difficulties of an ice age would persist for at least as long as the dinosaurs. Yet, widespread extermination and extinction was the dominant biological theme at the end of the Pleistocene. The cause of the extinctions is not known, although paleontologists suspect that, either directly or indirectly, the advent of humans may have had a bearing on the decline of some of these animals.

## QUESTIONS FOR REVIEW

1. What was the effect of geographic isolation on the evolution and distribution of Cenozoic mammals?

2. Which groups of marine phytoplankton proliferated during the Cenozoic?

3. How did the cephalopod faunas of the Mesozoic differ from those of the Cenozoic?

4. What depth of water is preferred by reef corals? What is the relationship of this preference to the origin of coral atolls?

5. Which group of bony fishes is particularly characteristic of the Cenozoic?

6. During what epoch of the Cenozoic did grasslands first become extensive? What effect did this have upon adaptations developed by herbivorous mammals?

7. What is the importance of the order Insectivora in the evolutionary history of placental mammals?

8. What general evolutionary changes are apparent in the Cenozoic history of horses? What environmental conditions stimulated these changes?

9. Discuss the premise that the adaptive radiation of Cenozoic mammals was at least as successful as that of Mesozoic reptiles.

10. What groups of reptiles managed to survive the exterminations of the late Mesozoic and survive today?

11. How might a paleontologist determine that a fossil lower jaw containing a full complement of teeth belonged to a mammal rather than a reptile?

12. What possible advantages may be inherent in the ruminant type of digestion? Name three living groups of ruminants.

13. How is it possible to trace the development of the trunk, or proboscis, in the skeletal remains of ancient proboscideans?

14. What is the evolutionary importance of the primitive Jurassic mammals known as eupantotheres?

15. Account for the reductions and extinctions in marsupialian groups in South America during the early Pleistocene.

## TERMS TO REMEMBER

| | | | |
|---|---|---|---|
| archaic ungulate | docodont | marsupial | plantigrade |
| artiodactyl | edentates | monotreme | proboscidean |
| atoll | entelodont | morganucodont | ruminant |
| Carnivora | eupantothere | multituberculate | teleost |
| cetacean | fringing reef | oreodont | titanothere |
| chalicothere | Insectivora | pecoran | tragulid |
| Creodonta | lagomorph | perissodactyl | Tylopoda |
| dinothere | mammoth | pinniped | ungulate |

## SUPPLEMENTARY READINGS AND REFERENCES

Carroll, R. L. 1988. *Vertebrate Paleontology and Evolution*. New York: W. H. Freeman and Co.

Colbert, E. H. 1980. *Evolution of the Vertebrates* (3d ed.). New York: John Wiley.

Halstead, L. B. 1982. *The Search for the Past*. Garden City, New York: Doubleday & Co.

Kürten, B. 1972. *The Age of Mammals*. New York: Columbia Univ. Press.

Match, C. L. 1976. *North America and the Great Ice Age*. New York: McGraw-Hill.

Martin, P. S., and Wright, H. E. 1967. *Pleistocene Extinctions*. New Haven, Conn.: Yale Univ. Press.

Savage, R. J. G., and Long, M. R. 1987. *Mammalian Evolution: An Illustrated Guide*. London: British Museum of Natural History.

The badly corroded skull of a Cro-Magnon human. As in present-day humans, the face of Cro-Magnon was vertical rather than projecting, and there was a prominent chin.

# Human Origins

*It walked on its hind feet, like
something out of the vanished Age
of Reptiles.
The mark of the trees was in its body
and hands.
It was venturing late into a world
dominated by fleet runners and swift
killers.
By all the biological laws this
gangling, ill-armed beast should
have perished, but you who read
these lines are its descendant.*

Loren Eiseley, 1971, *The Night Country.*
New York: Charles Scribner's Sons.

## Introduction

Evolution during the late epochs of the Cenozoic produced a mammal capable of shaping and controlling its own environment. That mammal, of course, was our own species, *Homo sapiens.* Unlike previous vertebrates, this remarkable creature profoundly changed the surface of the planet, modified the environment in ways both beneficial and destructive, and so manipulated the populations of other creatures as to change the entire biosphere. For these reasons, it is fitting that the final chapter in our history of the earth deal with the evolution of humankind.

## Primates

### Primate Characteristics

What traits should an animal have to qualify as a primate? It seems that this question cannot be answered easily, for primates have remained structur-ally generalized compared with most other orders of placental mammals. They retain the primitive number of five digits, have teeth that are not specialized for dealing with either grain or flesh, and have never developed hoofs, horns, trunks, or antlers. In the course of their evolution from shrewlike insectivores, their principal changes have involved progressive enlargement of the brain and certain modifications of the hand, foot, and thorax that were related to their life in the trees and their manner of obtaining food. These adaptations were not insignificant, however, for they enabled the human primate to shape a mode of life qualitatively different from that of any other animal.

The fictional pig Snowball in George Orwell's book *Animal Farm* remarked that "the distinguishing mark of man is the hand, the instrument with which he does all his mischief." Although this is clearly a biased point of view, it is correct in its assessment of the importance of the primate grasping hand with its opposable thumb (Fig. 16–1). This characteristic not only permitted primates a firm grip on their perches but also allowed them to grasp, release, and manipu-

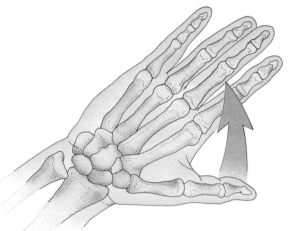

**Figure 16–1** The right hand of a human (palm up). The human hand is not used in locomotion and can be used to manipulate small objects between the fingers and thumb.

late food and other objects. The forearms retained their primitive mobility, in which rotation of the ulna and radius upon one another permitted the hands to be reversed in position.

The development of the grasping, mobile hand was accompanied by improvement in visual attributes. The eyes of primates became positioned toward the front of the face so that there was considerable overlap of both fields of vision, resulting in an improved ability to judge distances. It would seem that grasping hands and feet and good binocular vision are obvious adaptations for an animal that leaps from branch to branch and seeks its food from precarious boughs. Yet many such tree-dwelling animals as squirrels, civets, and opossums lack the short face, close-set eyes, and opposable digits and get along very well. For this reason, anthropologists have recently suggested that the visual attributes of primates originated as predatory adaptations that allowed early insectivorous primates to gauge accurately the distance to insect prey without movement of the head. By being able to grasp narrow supports securely with its feet, the animal was able to use both of its mobile hands to catch the prey. Claws, which are used by animals like squirrels in moving about on relatively wide branches, would not be as advantageous for climbing about on thin boughs and vines and might not provide a sufficiently secure hold for the quick catch of a moving insect. Thus, it seems

## Table 16–1 A Simplified Classification of the Order Primates

| Order | Suborder | Superfamily | Common Names of Representative Forms | |
|---|---|---|---|---|
| Primates | Prosimii | TUPAIOIDEA | Tree shrew | |
| | | LEMUROIDEA | Lemur | |
| | | LORISOIDEA | Bush baby, Slender loris | |
| | | TARSIOIDEA | Tarsier | |
| | Anthropoidea | CEBOIDEA | Howler monkey, Spider monkey Capuchin, Common marmoset, Pinche monkey | |
| | | CERCOPITHECOIDEA | Macaque, Baboon, Wanderloo, Common langur, Proboscis monkey | |
| | | | **Family** | |
| | | HOMINOIDEA | HYLOBATIDAE | Gibbon, Siamang |
| | | | PONGIDAE | Orangutan, Chimpanzee, Gorilla |
| | | | HOMINIDAE | Humans |

**Figure 16–2** The ring-tailed lemur of Madagascar. *(Frans Lanting/Minden Pictures.)*

that binocular vision and the grasping hand may not have been originally related to arboreal locomotion alone, but to the cautious, well-controlled, manual capture of visually located prey in the insect-rich canopy of tropical forests.

Other evolutionary modifications of primates were related to changes in the eyes and limbs. To protect the eyes, a postorbital bar evolved. As the eyes were positioned more closely together, the snout was reduced so that the face became flatter. In response to a brachiating habit (swinging from

**Figure 16–3** *Tarsius,* a small East Indian nocturnal, arboreal, prosimian with large, forwardly directed eyes. *(Gary Milburn/Tom Stack & Associates.)*

branches), forelimbs and hindlimbs diverged in form and function and an inadvertent predisposition toward upright posture developed.

## Modern Primates

The primates are divided into the suborders **Prosimii** which includes tree shrews, lorises, and tarsiers, and the suborder **Anthropoidea**, which includes monkeys, apes, and humans (Table 16–1). The Prosimii retain a greater number of primitive characters relative to the Anthropoidea. The tree shrews (see Fig. 15–23), for example, possess clawed feet and a long muzzle with eyes in the lateral positions. The lemurs (Fig. 16–2), which are confined largely to Madagascar, still have a long snout and lateral eyes, although they are positioned more to the front than those in the shrew. Some of the digits are clawed, whereas others are equipped with flattened nails. The East Indian tarsier (Fig. 16–3) has a relatively flat face. As befits a nocturnal animal, the eyes are large and positioned toward the front to provide stereoscopic vision. The fingers and toes terminate in nails rather than claws.

In the more progressive Anthropoidea, the trends initiated by prosimians are further developed. Monkeys are the more primitive members of the Anthropoidea. They are grouped into New World and Old World forms (Fig. 16–4). The *Ceboidea,* or New World monkeys, are an early branch and were not involved in the eventual evolution of humans. Included among the ceboids are the spider monkey

**Figure 16–4** Representative New World and Old World monkeys. Most New World monkeys have prehensile tails that can function almost as effectively as another limb. *(From Cockrum, E. L., et al. Zoology. Philadelphia: Saunders College Publishing, 1966.)*

and marmoset, as well as the familiar little capuchin, or organ-grinder monkey. New World monkeys can be identified by their flattish faces, widely separated nostrils, and prehensile tails. Most are small in comparison to their Old World cousins.

The more advanced Old World monkeys, or *Cercopithecoidea,* are widely distributed in tropical regions of Africa and Asia. They include the familiar macaque or rhesus monkey of laboratories and zoos, the Barbary ape of Gibraltar, langurs, baboons, and mandrills. In this group of monkeys, the nostrils are close together and directed downward (as in humans) and the tail is not prehensile.

The anthropoid apes (Fig. 16–5) are tail-less primates. Modern species probably evolved from the same ancestral stock that produced humans. However, the divergence probably took place back in the middle Tertiary. The more primitive branch of the tail-less apes is composed of the *Hylobatidae,* best

represented by those long-armed acrobats the gibbons. Orangutans, chimpanzees, and gorillas are grouped within the *Pongidae.* The red-haired orangutan, or "wild man" of Borneo and Sumatra, is considerably larger than the gibbon and has a larger brain as well. Chimpanzees and gorillas are similar in many ways and tend to be less pronounced in their arboreal adaptations. Gorillas are essentially ground dwellers and spend only a small fraction of their time in the trees.

## The Prosimian Vanguard

The fossil record for primates begins with the appearance of a creature named *Purgatorius,* known only from a few teeth discovered in the Hell Creek Formation at Purgatory Hill in Montana. These finds indicate that the earliest primates were contempo-

A

B

C

D

**Figure 16–5**   The apes. (A) Gibbons are extremely acrobatic and often move through the trees by brachiation. (B) Orangutans are solitary apes that seldom leave the protection of the trees. (C) A gorilla knuckle walking. (D) Chimpanzees live in groups and have a complex social behavior. *(A, E. R. Degginger; B, Tom McHugh/Photo Researchers, Inc.; C, John D. Cunningham/Visuals Unlimited; D, Animals Animals © 1988, Zig Leszozynski.)*

raries of the last of the dinosaurs, at least in the tropical latest Cretaceous environments of North America. The fossil record improves somewhat in the early epochs of the Cenozoic. *Plesiadapis* (Fig. 16–6), found in Paleocene beds of both the United States and France, has the distinction of being the only genus of primate (other than that of humans) that has inhabited both the Old World and the New

**Figure 16–6**   The Paleocene rodentlike prosimian *Plesiadapis*.

World. The presence of this **prosimian** on the now widely separated continents is one of many clues indicating that the continents were not yet completely separated by the widening Atlantic in the Paleocene. *Plesiadapis* was a rather distinctive and specialized primate and represents a sterile offshoot of the primate family tree. The incisors were rodentlike and were separated from the cheek teeth by a toothless gap or **diastema**. Fingers and toes terminated in claws rather than nails. The rodentlike characteristics and habits of the plesiadapiformes may have contributed to their extinction by late Eocene time. About that time, rodents were having their own radiation. Rodents are highly successful as gnawing, seedeating animals and are able to reproduce at a rate with which no primate can compete.

General trends in prosimian evolution during the Eocene Epoch (37 to 55 million years ago) involved reduction in the length of the muzzle, increase in brain size, shifting of the eye orbits to a more forward position, and development of a grasping big toe. These trends are evident in the fossil remains of *Cantius* from the Wind River basin of Wyoming. In the Wyoming stratigraphic sequence, one can trace the evolution of successive species of *Cantius* from lower to upper beds. Near the top of the section, a new genus, *Notharctus* (Fig. 16–7), occurs as the apparently direct descendent of *Cantius*. *Notharctus* has been known from other localities for over a century. In general form, it appears to be just the kind of lemurlike animal from which monkeys

and apes are derived. Prosimian populations during the late Paleocene and Eocene also included tarsiers. *Tetonius* was an Eocene form with tarsioid traits of closely spaced large eyes and a shortened muzzle.

Both tarsiers and lemurs were abundant and widely dispersed on Northern Hemisphere continents during the Eocene. However, with the advent of cooler Oligocene climates, they virtually deserted North America, whereas in the Eastern Hemisphere they were forced southward into the warmer latitudes of Asia, Africa, and the East Indies. Surviving prosimians are much reduced in variety and number. The prevailing opinion is that many have been replaced by the monkeys.

## The Early Anthropoids

The next step in the primate evolutionary advance was the appearance of the first **anthropoids**. Discoveries in the Fayum region of Egypt have provided a wealth of information about the early anthropoids.

**Figure 16–7**   The Eocene lemur *Notharctus*.

The fossils there occur in several upper Oligocene horizons and include well over 100 specimens. Although many of the skull fragments and teeth retain subtle vestiges of prosimian ancestry, none of the remains are from prosimians. They are fossils of primates that have reached what may be called the monkey stage of organization. One primate discovered at Fayum is *Aegyptopithecus,* a relatively robust arboreal primate with monkeylike limbs and tail, a brain larger than that of *Notharctus,* and eye orbits rotated to the front for stereoscopic vision. The fossil of the species *Aegyptopithecus zeuxis* is dated as 33 to 34 million years old, indicating that the prosimian–anthropoid transition had taken place by Oligocene time.

Fossils of the earliest known New World monkeys are known from upper Oligocene and lower Miocene strata of South America. Although they appear superficially similar to some Old World forms, New World monkeys evolved quite independently from prosimian ancestors and without genetic contact with their Old World cousins.

Because evolution is a continuing process, with each animal truly a transitional link between older and younger species, it is a difficult and often arbitrary task to assign the fragment of a fossil jaw or tooth to either the monkey or the ape category. One of many clues used in attempting to make this distinction is the pattern of cusps on certain molar teeth. In Old World monkeys, specific molars have four cusps. Among apes (and humans), these same molars have five cusps, with an intervening Y-shaped trough. Molars belonging to *Aegyptopithecus* display the Y-5 pattern (Fig. 16–8), which is one reason anthropologists consider the genus to be the possible very early ancestor of Miocene apes. It seems safe to state that the Oligocene Epoch may have witnessed not only the transition from prosimians to anthropoids, but also the differentiation of apes from monkeys.

During the Miocene, plate tectonics had a significant influence on the evolution of primates. Africarabia (Africa and Arabia) was drifting northward and ultimately collided with Eurasia. (The Arabian portion of the continent then rotated counterclockwise to eventually produce the Red Sea.) As a result of these movements, east–west circulation of tropical currents across the former Tethys seaway

was prevented, and East Africa became cooler and dryer. Extensive grass-covered savannas replaced former areas of dense forests. The evolution of Miocene primates was strongly affected by these changes.

Among the new players that came on the scene during the Miocene were a group called **dryomorphs** (formerly called dryopithecines). The dryomorphs take their name from *Dryopithecus fontani,* a species first discovered in France in 1856. The dryomorphs varied in size and appearance and include forms bearing different generic names from France, Spain, Greece, Hungary, Turkey, India, Pakistan, and Africa. Many anthropologists consider the important species *Proconsul africans* to be either a very early dryomorph or the immediate dryomorph ancestor. *Proconsul africans* was discovered by Mary and Louis Leakey in 1948 on an island in Lake Victoria, Kenya. The skull, jaws, and teeth of this small primate are decidedly apelike rather than monkey-

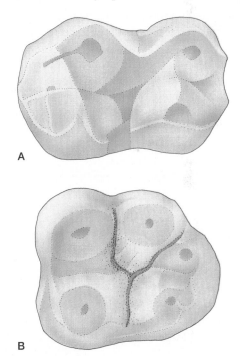

**Figure 16–8**    General pattern of cusps on the molars of Old World monkeys. (A) The lower molar of a baboon showing a cusp at each corner. (B) The lower molar of a chimpanzee showing the "lazy Y-5" pattern characterized by a Y-shaped depression that separates five cusps.

**Figure 16–9** Skull of *Proconsul* (a dryomorph) from Lake Victoria, Kenya.

like (Fig. 16–9). The slightly more advanced middle Miocene dryomorph descendents of *Proconsul* are probably ancestral to both modern African apes and the first homininids, the australopithecines. This means that *Proconsul* is distantly ancestral to humans as well.

The study of Miocene primates clearly indicates an absence of orderly sequential change along a single trend. The family tree of primates is a complex of parallel and diverging branches. For this reason, the attempt to trace the ascent of humans up through the many bifurcations and dead ends is an exciting but difficult task. Paleontologists must be ready to reinterpret the story each time new fossil material is uncovered.

## The Australopithecine Stage and the Emergence of Hominids

The story of the emergence of humans (hominids) begins with Raymond Dart's 1924 discovery of fossil remains of an immature primate in a limestone quarry in South Africa. Dart named the fossil *Australopithecus africanus* (Fig. 16–10). In succeeding years, many additional skeletal fragments of species of *Australopithecus* have been found in upper Pliocene or lower Pleistocene deposits of Africa, Java, and China. Fossil sites in East Africa have become increasingly important because of the veritable bonanza of hominid bones they have yielded. The new collections of fossils have given paleoanthropologists an unsurpassed record of human evolution over the past 3 million years. Some of the East African sites, such as Olduvai Gorge (Fig. 16–11), are now famous as a result of the lifelong research programs of Mary

**Figure 16–10** *Australopithecus africanus* from the Transvaal of South Africa. *(Photograph of Wenner-Gren Foundation replica by David G. Gantt.)*

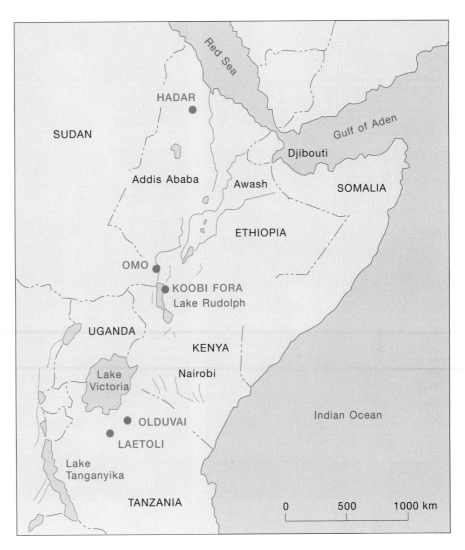

**Figure 16–11** Location of the Hadar, Laetoli, Olduvai, Omo, and Koobi Fora (East Lake Rudolf) paleontologic sites.

Leakey and the late Louis Leakey. Their efforts, and those of their son Richard, have provided fossils of relatively heavy-bodied, robust **australopithecines** as well as lighter, so-called gracile types. Among the latter are remains of creatures that may well belong within our own genus.

East Africa during the Cenozoic experienced a lively history of volcanic activity. As a result, many of the fossil sites have interspersed layers of volcanic ash and lava flows. This fact has been of great benefit to the paleoanthropologists, for samples of ash can often be dated by the potassium–argon method. Dates from two succeeding ash beds would then pro-

vide a close estimate of the age of fossils found in the intervening layers of sediment.

Among the oldest undisputed hominids of East Africa are those discovered at Laetoli in Tanzania and Hadar in Ethiopia (see Fig. 16–11). These localities have yielded remains of human ancestors that roamed Africa between 3 and 4 million years ago. Here have been found the remains of the oldest hominids. Among them, none is more famous than the skeleton of a young female discovered by Donald C. Johanson in 1974. The individual, informally dubbed "Lucy" (after a popular Beatles song, "Lucy in the Sky with Diamonds"), is the most complete

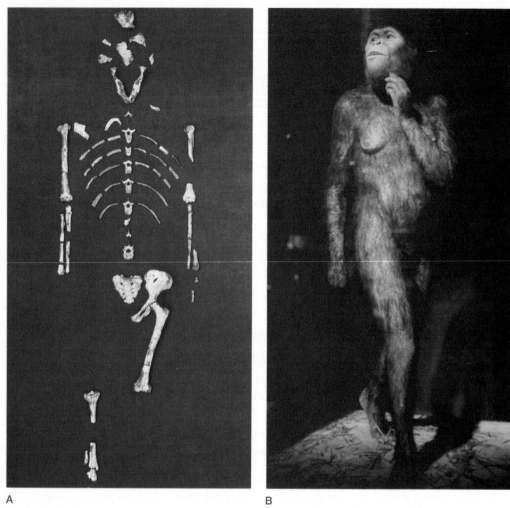

A    B

**Figure 16–12**    *Australopithecus afarensis,* informally known as "Lucy." Lucy was a young, erect walking hominid that lived in East Africa about 3.5 million years ago. (A) The specimen shown here is the oldest known, most complete skeleton of an erect walking human ancestor. (B) A reconstruction of Lucy on display at the St. Louis Zoo. *(Photograph of skeletal remains courtesy of the Cleveland Museum of Natural History, with permission.)*

Pliocene hominid thus far discovered (Fig. 16–12). Lucy, whose scientific name is *Australopithecus afarensis,* was well muscled and had powerful arms that were slightly longer relative to body size than our own. She was barely a meter (about 3.5 feet) tall. Her head was about the size of a softball and contained a brain comparable in size to that of a chimpanzee. The shape of the skull was more like that of an ape than a human, with jaws thrust forward and no chin.

Lucy was fully bipedal. The fact that she and her contemporaries walked erect is evident in the shape of her leg and pelvic bones. Further evidence of bipedalism came with the discovery of footprints in layers of volcanic ash (tuff) at Laetoli (Fig. 16–13). The footprints can be traced over a distance of 9 meters (about 30 feet) and were made by two contemporaries of Lucy as they walked side by side over a layer of soft, moist, volcanic ash. (Subsequently, another ashfall formed a protective seal over the footprints.)

**Figure 16–13** Footprints made by *Australopithecus afarensis* as they walked across wet ash scattered by a volcanic eruption over 3.5 million years ago. The footprints confirm skeletal evidence that the species had a fully erect stance. *(Peter Jones.)*

Measurements of the footprints and length of stride indicate that the hominids that made them were about 1.2 meters (about 4 feet) tall and walked in much the same way as we do today.

East African fossil sites have yielded the bones of as many as 65 individuals of *Australopithecus afarensis.* As indicated by their skeletal remains and other footprints in volcanic ash, these hominids lived in the presence of a varied fauna of other vertebrates. At Laetoli in Tanzania one finds hundreds of footprints of giraffes, rhinoceroses, antelopes, ostriches, hyenas, baboons, chalicotheres, and extinct horses.

Recently, fossils of *A. afarensis* that are 400,000 years older than the famous skeleton of Lucy were found in the Awash River valley east of Hadar in Ethiopia. The remains were found in tuff deposits that have been dated by potassium–argon and fission track methods as 4.0 million years old (middle Pliocene). As such, they represent the oldest hominid fossils known. The femur was that of a male, about 4.5 feet tall, who clearly had a bipedal posture and locomotion. The absence of any significant differences between the skull and other bones in this creature and those of Lucy suggests that, as a species,

*A. afarensis* was stable over a remarkable span of time. Some paleontologists see this as support for the punctuational pattern of evolution (see Chapter 3) rather than the more traditional view of evolution by slow and gradual steps.

Although debate about the precise evolutionary position of these fossils is currently lively, many paleontologists consider Lucy and her kin to be the direct ancestors of the genus *Homo,* as well as the ancestor of species of *Australopithecus* (Fig. 16–14).

Two other East African fossil localities that have yielded species of both *Homo* and *Australopithecus* are the Omo and Koobi Fora sites (see Fig. 16–11). The Omo digs are located along the Omo River in remote southwestern Ethiopia. Here one finds a nearly continuous section of sediments and volcanics that span 2.2 million years. The Koobi Fora locality is on the east side of Kenya's great alkaline Lake Rudolf (Fig. 16–15). It has been investigated by Richard Leakey and Glynn Isaac. Altogether, the Omo, Koobi Fora, Olduvai, Hadar, and Laetoli localities have given us sufficient skulls, jaws, teeth, and other bones to indicate that the australopithecines were not an entirely homogeneous group. However, certain characteristics appear in nearly all of the specimens unearthed. From the structure of their pelvic girdles, it is known that they stood upright in a fashion more humanlike than apelike. Australopithecine dentition was essentially human, although the teeth were more robust than in modern humans. In contrast to these humanlike characteristics, australopithecine cranial capacity was comparable to that of modern large apes and reached a volume of 600 or 700 cc (cubic centimeters). However, even with a brain only about half the size of *Homo sapiens,* they were able to make several kinds of crude implements from horn, teeth, and bone.

Recent investigations by anthropologists working in Kenya and Ethiopia have provided evidence that there may have been several humanlike lineages coexisting in Africa between 1 and 3 million years ago. One of these lineages included *Australopithecus africans* and the more recent *Australopithecus habilis* (some prefer *Homo habilis*) from Olduvai Gorge. The possible existence of a second lineage became apparent in 1972, when a team led by Richard Leakey unearthed a skull at the Koobi Fora site in beds that are about 2.0 million years old. The skull (Fig.

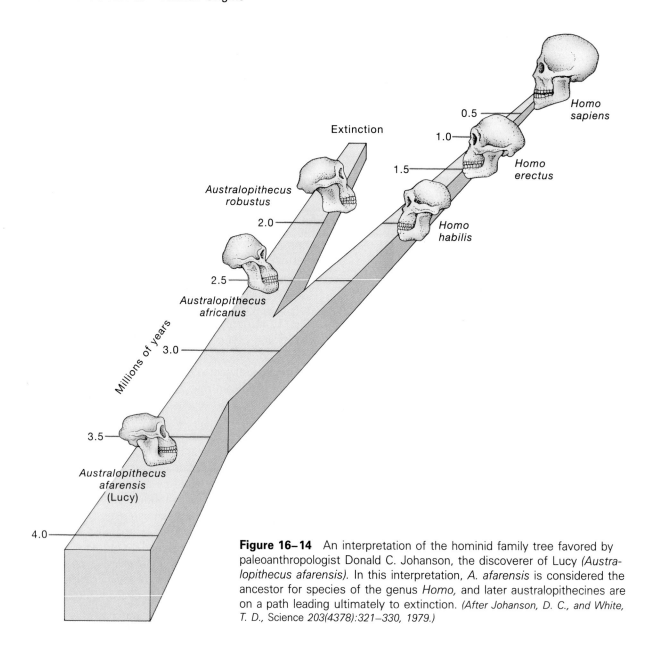

**Figure 16–14** An interpretation of the hominid family tree favored by paleoanthropologist Donald C. Johanson, the discoverer of Lucy *(Australopithecus afarensis)*. In this interpretation, *A. afarensis* is considered the ancestor for species of the genus *Homo*, and later australopithecines are on a path leading ultimately to extinction. *(After Johanson, D. C., and White, T. D., Science 203(4378):321–330, 1979.)*

16–16), assigned the code name KNM-ER-1470, had a cranial capacity of about 800 cc, much larger than that of any *Australopithecus* and close to the cranial capacity of later hominids already assigned to the genus *Homo*. Although KNM-ER-1470 may not have quite reached the *Homo* stage of evolution, it was tending in that direction. The fossil indicates that large-brained, humanlike creatures may have been present on earth much earlier than previously suspected.

Yet another group of primates at the australopithecine stage had smaller brains but were larger than the *Australopithecus africanus* line. This group is exemplified by *Paranthropus* and may have evolved into a still heavier creature appropriately named *Gigantopithecus*. Such species as *Australopithecus boisei* (once called *Zinjanthropus boisei*) and *Australopithecus robustus* (formerly *Paranthropus robustus*) (Fig. 16–17) apparently represent sterile side branches that failed to give rise to any known de-

**Figure 16–15** The Koobi Fora fossil site in East Africa where the skeletal remains of many australopithecines, as well as the skull designated KNM-ER-1470, were discovered. In addition to the hominid skeletal remains unearthed at Koobi Fora, paleontologists found pebble choppers and flake tools associated in deltaic deposits with the bones of a hippopotamus. The position of both the tools and hippo bones suggests that this was a butchering site. Apparently one day about 1.8 million years ago, some australopithecines came across the body of a hippopotamus that had recently died. They feasted on its carcass, using the sharp edges of flaked chert to get at the meat. *(Courtesy of Kay Behrensmeyer.)*

**Figure 16–16** Lateral view of a nearly complete skull, designated KNM-ER-1470, from the Koobi Fora fossil site in Kenya. The creature may be an early representative of the genus *Homo*.

**Figure 16–17** *Australopithecus robustus*. Because of the observation that forms such as this one had larger molars and premolars than did *Australopithecus africanus*, some paleoanthropologists believe that it was primarily a plant eater and was taxonomically distinct from *A. africanus*. Recent interpretations favor inclusion of these creatures under the generic name *Australopithecus*. This specimen was derived from Bed I on the south side of the main ravine at Olduvai Gorge. *(Photograph of Wenner-Gren Foundation replica by David G. Gantt.)*

**Figure 16–18**   Replica of *Homo erectus,* previously known as *Sinanthropus pekinensis. (Photograph of Wenner-Gren Foundation replica by David G. Gantt.)*

scendant. Careful study of the teeth of some of these larger australopithecines suggests that, unlike *A. africanus,* they may have been primarily plant eaters.

From the preceding discussion, it is apparent that paleoanthropologists have established a rather large number of names for various australopithecine groups that display relatively minor differences. Some experts do not support the recognition of so many different species and believe that many forms now designated as separate species may prove to be local races or subspecies.

## The *Homo Erectus* Stage

The next stage in hominid evolution is represented by *Homo erectus,* formerly known as *Pithecanthropus erectus* from Java and *Sinanthropus pekinensis* (Fig. 16–18) from China. Another important locality was added in 1963, when Louis Leakey discovered a skull cap of *Homo erectus* in Bed II of Olduvai Gorge (Fig. 16–19). Bed II had been dated by potassium–argon techniques and found to be about 750,000 years old. These finds indicate that *Homo erectus* was probably widely dispersed in Asia and Africa during the middle Pleistocene. *Homo erectus* has been considered the first true species of humans and is generally regarded as having evolved from a species of *Australopithecus.*

The fossil evidence already available indicates a certain amount of variability between different groups of *Homo erectus.* However, the variability was probably no greater than that which exists today among members of our own species.

The bones of the axial skeleton and limbs of *Homo erectus* were quite similar to those of modern humans. In this regard, these hominids were more advanced than the australopithecines. For example, the pelvic bones of *Australopithecus* indicate that, although they were fully bipedal, they walked with their feet turned outward in a sort of half-running, rolling gait. *Homo erectus,* on the other hand, was an excellent walker.

Although the postcranial skeleton of *Homo erectus* was quite modern, the skull was not. Cranial capacity ranged from about 775 cc in earlier forms to nearly 1300 cc in specimens of more recent age. Brain capacity in modern *Homo sapiens* ranges from 1200 to 1500 cc, and thus the brain size of *Homo erectus* overlaps at least the lowermost range of modern peoples. Pithecanthropines represent a stage in

**Figure 16–19** Upper portion of skull of a specimen of *Homo erectus* from Olduvai Gorge. This specimen was found associated with stone implements of the type known as Chellan. *(Photograph of Wenner-Gren Foundation replica by David G. Gantt.)*

the evolution of hominids during which relatively rapid increases in brain size had begun. No doubt the expansion of the brain involved the reshaping not only of the cranium but also of the birth canal. Selection would very likely have favored large pelvic inlets to accommodate fetuses with larger brains.

The skull of *Homo erectus* was massive and rather flat, with heavy supraorbital ridges over the eyes (see Fig. 16–18). The forehead sloped, and the jaws jutted forward at the tooth line in a condition termed **prognathous**. A definite jutting chin was lacking, and very likely the nose was broad and flat. These were rather primitive traits. However, except for their being somewhat more robust, the teeth and dental arcade in *Homo erectus* were essentially modern.

Evidence exists that *Homo erectus* made good use of his larger brain. From the bones of other animals found at living sites, it is clear that these hominids were good hunters. They were also skilled at making simple implements out of flint and chert (Fig. 16–20), such as axes and scrapers, the former being equipped with wooden handles. Some also appear to have engaged in cannibalism, perhaps not for food but as part of a ritual system. Anthropologists do not know if *Homo erectus* spoke a distinctive language, wore clothes, or built dwellings. There is vague evi-

dence that they hunted together in bands. What appear to be pithecanthropine slaughter sites have been found in Europe. Unfortunately, these sites did not yield a single bone of *Homo erectus*. A few *Homo erectus* fossil localities in Europe and China contain traces of carbon. However, this cannot be taken as indisputable evidence that *Homo erectus* had learned to use fire, for the carbon may have been the product of naturally occurring brush fires.

## Homo Sapiens

### The Neanderthals

From the *Homo erectus* stage of the middle Pleistocene (Fig. 16–21), it is only a short step to late Pleistocene *Homo neanderthalensis* (Fig. 16–22). The initial specimen of this early human was found in the Neander Valley near Dusseldorf, Germany. Subsequently, sufficient additional fossils have been found to indicate that the **Neanderthal** people ranged across the entire expanse of the Old World. With their heavy brow ridges and prognathous, chinless jaws, Neanderthals have become the very personification of the "cave man." Indeed, the face seems a brutish carryover from middle Pleistocene. However,

**Figure 16–20**  Progressive improvement in making tools from stone during the Pleistocene. The crude stone tools of the early Pleistocene were produced by australopithecines. *Homo erectus* produced the better shaped tools of the middle Pleistocene. The Upper Paleolithic tools included carefully chipped blades and points. The next stage, not shown here, would be the Neolithic, characterized by refined and polished tools of many kinds.

in other features, the neanderthaloids were quite modern. Below the neck, their skeletons matched our own, and their brain size equalled or exceeded that of present humans. Thus, the once popular depiction of Neanderthal as a bent-kneed, flat-footed, in-toed, bull-necked brute with a curved back is false. Re-examination of the skeleton on which the original restoration had been based revealed that it

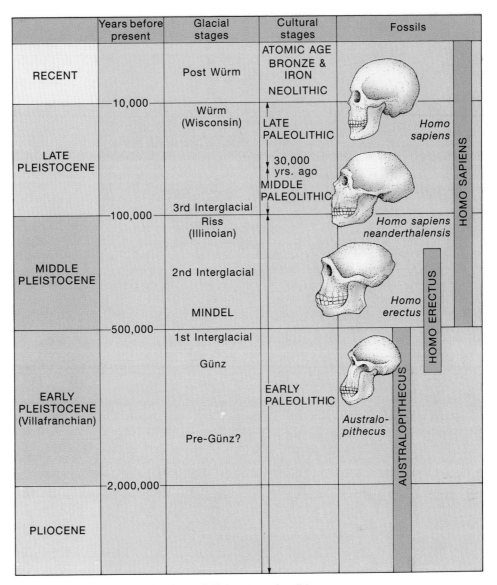

| Years before present | Glacial stages | Cultural stages | Fossils | | |
|---|---|---|---|---|---|
| **RECENT** | Post Würm | ATOMIC AGE BRONZE & IRON NEOLITHIC | | *Homo sapiens* | **HOMO SAPIENS** |
| —10,000— | Würm (Wisconsin) | LATE PALEOLITHIC | | | |
| **LATE PLEISTOCENE** | | ↓ 30,000 yrs. ago ↑ MIDDLE PALEOLITHIC | | *Homo sapiens neanderthalensis* | |
| —100,000— | 3rd Interglacial Riss (Illinoian) | | | | |
| **MIDDLE PLEISTOCENE** | 2nd Interglacial MINDEL | | | *Homo erectus* | **HOMO ERECTUS** |
| —500,000— | 1st Interglacial Günz | | | | |
| **EARLY PLEISTOCENE (Villafranchian)** | Pre-Günz? | EARLY PALEOLITHIC | | *Australo- pithecus* | **AUSTRALOPITHECUS** |
| —2,000,000— | | | | | |
| **PLIOCENE** | | | | | |

**Figure 16–21**  Chronologic chart of Pleistocene fossil humans.

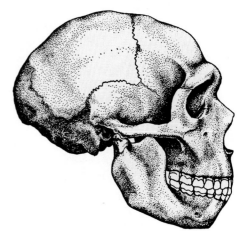

**Figure 16–22**  Right lateral view of the skull of *Homo neanderthalensis*.

607

## COMMENTARY

### Neanderthal Ritual

The Neanderthals lived from about 125,000 to about 40,000 years ago. Their culture included chipped flint tools, crude carvings, the use of fire, and evidence of deliberate burial of their dead in carefully prepared graves. Some of the burial sites contain indications that these ancient members of our own species may have had religious beliefs. The La Chapelle fossil site in southwestern France, for example, contains the body of a Neanderthal male placed in ritual position within a shallow grave with the leg of a bison on his chest. Flint tools were also placed in the grave, possibly in the belief that they could be used in an afterlife.

At the Ferrassie Neanderthal site in France, an apparent family cemetery was discovered within a rock shelter. Two adults, presumed to be a father and mother, are buried head to head. Nearby, three small children and a newborn infant are buried. Once again, flint flakes and bone splinters were placed in the grave of the adult male, and a heavy flat stone placed over his head and shoulders. Perhaps the stone was to protect the body, or possibly to restrain the deceased from returning to life.

That Neanderthals had a special attitude about death is also evident at a fossil site known as Teshik-Tash in Soviet Uzbekistan. Anthropologists working at Teshik-Tash uncovered an array of goat horns surrounding the buried body of a nine-year-old child.

Yet another particularly interesting Neanderthal fossil site is located in Shanidar, Iraq. The remains of nine Neanderthals have been uncovered at Shanidar, and four of these were deliberately buried. One of the burials contains an assortment of pollen grains indicating flowers had been placed in the grave. Although not related to funerary ritual, the Shanidar site includes the remains of a 30-year-old male who had been born with a crippled arm. The arm had been amputated below the elbow, and this hardy Neanderthal survived the surgery. The excessive wear on his teeth suggests that he used them to compensate for the missing limb.

**Figure 16–23** Reconstruction of a Neanderthal family group. *(Courtesy of the National Museum of Natural History.)*

was that of an elderly individual, between 60 and 70 years old, who had been severely afflicted with osteoarthritis.

Most of the so-called classic Neanderthals were sturdy people of small stature that had adapted to life in the cold climates near the edge of the ice sheet. The relatively short limbs and bulky torso may have been an advantage in helping them to conserve body heat. Neanderthals often lived in caves (Fig. 16–23) and successfully hunted many contemporary cold-tolerant mammals, including cave bears, mammoths, woolly rhinoceroses, reindeer, bison, and fierce ancestors of modern cattle known as aurochs. They were not at all devoid of culture and manufactured a variety of stone spear points, scrapers, borers, knives, and saw-edged tools. In addition, Neanderthals made ample use of fire and apparently could ignite one at will in the hearths they excavated in the floors of caves.

**Figure 16–24** Lateral view of the skull of Cro-Magnon *(Homo sapiens)*. Note the absence of heavy supraorbital ridges. *(Photograph of Wenner-Gren Foundation replica by David G. Gantt.)*

The use and control of fire had many advantages for Pleistocene hominids. It provided light in caves that otherwise would have been avoided because of the near helplessness of humans in deep darkness. It gave warmth to creatures that had evolved in the tropics but had moved northward into colder realms, and it gave protection at night from predators. During the frigid winters of the Ice Age, fire provided the means to thaw meat that had become quickly frozen after a kill. Conceivably, cooking may have been a simultaneous result of the thawing process. Cooked foods provided the added advantages of promoting easier digestion and destroying harmful microorganisms by heat.

In addition to their use of fire, there is evidence that Neanderthals constructed shelters of skins, sticks, and bones in areas where caves may not have been available. Perhaps as a hunting ritual, these people killed cave bears and stacked the skulls carefully in chests constructed of stones. That they also pondered the nature of death and believed in an afterlife is indicated by their custom of burying artifacts along with the dead.

### Cro-Magnon

About 34,000 years ago, during the fourth glacial stage, humans closely resembling modern Europeans moved into regions inhabited by the Neanderthals and in a short span of time completely replaced them, probably by tribal warfare and by competition for hunting grounds. The new breed, designated *Homo sapiens sapiens* but informally dubbed **Cro-Magnon** (Figs. 16–24 and 16–25), were mostly taller than their predecessors, had a more vertical

**Figure 16–25** Cro-Magnon hunters. *(Courtesy of the Field Museum, Chicago.)*

**Figure 16–26** Comparison of the skulls of (A) *Homo neanderthalensis;* (B) a neanderthal/Cro-Magnon transitional form from a rock shelter on the slope of Mount Carmel; and (C) Cro-Magnon. The Mount Carmel skull is both intermediate in form and age between neanderthal and Cro-Magnon.

brow, and had a decided projection to the chin. In short, Cro-Magnon's bones were modern, and anthropologists have recognized definite Cro-Magnon skull types among today's western and northern Europeans, as well as North Africans and native Canary Islanders.

The modern types of *Homo sapiens* did not appear suddenly on the evolutionary scene. Changes in dentition and supporting facial architecture were not an "overnight" occurrence. Transitional forms between Neanderthal and Cro-Magnon do exist (Fig. 16–26) and include fossil skulls from Palestine, South Africa, Germany, and Czechoslovakia. These forms vary among themselves but generally show less robust brow ridges, the beginnings of the modern chin, and less pronounced muscle markings than are evident in typical Neanderthal specimens. The transitional forms seem to have evolved outside Europe and, after they had reached the Cro-Magnon stage, migrated into Europe and Asia during a temporary regression of the ice sheets about 35,000 years ago.

The cultural traditions of the Neanderthals appear to have been continued and further developed by Cro-Magnon. The variety and perfection of stone and bone tools were increased. Finely crafted spear points, awls, needles, scrapers, and other tools are found in caves once inhabited by these people. These caves also retain splendid paintings and drawings on their walls and ceilings. Carvings and sculptures of obese women (Fig. 16–27), probably used in fertility rites, were produced from fragments of bone or ivory, as were small, elegant engravings and statues of mammoths and horses. Cro-Magnon people enjoyed wearing body ornaments and frequently fash-

ioned necklaces from pieces of ivory, shells, and teeth. Burial of the dead became a truly elaborate affair. Hunters were buried with their weapons and children with their ornaments. This apparent concern for an afterlife, and the sense of self-awareness that resulted in art and complex ritual, suggests that the beginning of the age of the philosopher had arrived.

Through most of his early history, *Homo sapiens* was a wandering hunter and gatherer of wild edible plants. However, about 10,000 to 15,000 years ago, near the beginning of the Holocene Epoch, tribes began to domesticate animals and cultivate plants. They learned to grind their tools to unprecedented perfection and to make utensils of fired clay. With more reliable sources of food, permanent settlements were developed, and individuals, spared the continuous demands of searching for food, were able to build and improve their cultures. Languages improved, and symbols developed into forms of writing. The era of recorded history had begun.

## Early Humans in the Americas

The fossil record for hominids in the New World is discontinuous and often ambiguous. More often than is desirable, anthropologists must trace human presence on the basis of fragments of tools or weapons rather than actual skeletal remains. These clues, however, suggest that humans wandered into Alaska during the late phase of the Wisconsin glacial stage, when the Bering Straits provided a dry land bridge. It is likely that tribal groups seeking new hunting areas then drifted southeastward across the less mountainous parts of Alaska, Canada, and western United

**Figure 16–27** Prehistoric art of late Pleistocene *Homo sapiens*. (A) Venus figure originally found in Austria ("The Venus of Willendorf"). (B) Thong-stropper used either to
work hide thongs or to straighten arrow shafts. The tool is made from an antler and is
intricately carved. *(Courtesy of Musée de l'Homme, Paris.)*

States. Precisely when these migrations occurred is
difficult to determine and is often hotly debated.
Very crude flints, tentatively reported to be artifacts,
have been found in alluvial fan sediments in California and may be up to 40,000 years old. Charred
bones of a dwarf mammoth believed to have been
deliberately roasted in a campfire have been found
on Santa Rosa Island off the coast of California.
Carbon-14 dating indicates that the bones are about
28,000 years old. At Tule Springs, Nevada, fire pits
determined to be 24,000 years old have been found
in questionable association with flaked flints. In
Mexico, an obsidian blade was found under a log
that was dated at about 23,000 years.

There are a number of more recent sites that
yield indirect evidence of **paleoindians**. Projectile
points are found in definite association with extinct
late Pleistocene elephants and bison in the western

United States. Sites near Clovis and Sandia, New
Mexico, have yielded the distinctive tools of two separate cultures that may have lasted from about
13,000 to 11,000 years ago. A somewhat younger
group of paleoindians lived from about 11,000 to
9000 years ago. This group has been named the Folsom Culture. Folsom people manufactured short,
finely flaked, and fluted points that were mounted
on shafts. These flints have been found in kill sites in
association with extinct species of bison.

The search for skeletal parts of early humans in
America has provided fewer discoveries than has the
search for their artifacts. However, there have been a
few significant finds. A skull unearthed near Laguna
Beach, California, has been dated by radiocarbon
techniques as between 15,680 and 18,620 years old.
Charred human bones from a cave on Marmes Ranch
in southeastern Washington state have been dated at

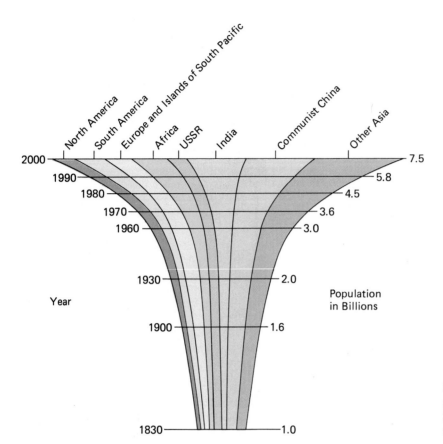

**Figure 16–28** World population growth projected to the year 2000. *(From U.S. Dept. of State information circular.)*

11,500 to 13,000 years old. A projectile point, a bone needle, and the roasted bones of rabbits, deer, and elk have been found in association with the remains of "**Marmes Man**." The oldest known human remains from South America consist of a portion of a jaw and teeth from Guitarrero Cave in northern Peru. These remains are believed to be about 12,600 years old. Other Pleistocene diggings in Florida, Mississippi, California, Texas, Venezuela, Chile, and Argentina clearly indicate the presence of humans in the Americas prior to 11,000 years ago.

## Population Growth

One of the many lessons of historical geology is that the advent of *Homo sapiens* represents only one recent and momentary event along the sinuous, branching, 500-million-year evolution of vertebrates. We humans are linked by similarities of structure and body chemistry to lungfish struggling out of stagnating Devonian pools, to the small, shrewlike precursors of the primate order, and to the axe-carrying hunters that wandered along the margins of great ice sheets. The descendants of those and other Pleistocene hunters have emerged as the dominant species of higher life presently on this planet. Like other animals, *Homo sapiens* has been shaped by the combined powers of genetic change and environment. However, our species has quickly become a pervasive force in modifying the very physical and biological environment from which it was shaped. Humanity has come to rely heavily on science to improve its lot, but has had great difficulty in finding ways to manage the resulting technology and to control the burgeoning problems arising from too few resources for too many humans.

About 10,000 years ago (a mere moment to a geologist), there were more than 6 million humans on the earth. By the year A.D. 1, the number had jumped 50 fold to 300 million. In 1970, the figure stood at 3.6 *billion*. If we project current rates of increase into the future, there will be 7.5 billion people on earth by the year 2000 (Fig. 16–28). Only a decade into the twenty-first century, global population is likely to exceed 10 billion. Can our fragile planet sustain so enormous a population? Few would doubt that the survival of *Homo sapiens* will depend upon our ability to control population growth and to solve the many often related resource and environmental problems. Should we fail to solve these problems, other animals may replace our species and add their own chapters to the history of life.

## SUMMARY

During the early epochs of the Cenozoic, the main groups of placental mammals were differentiated. One group, the probable descendants of primitive insectivores, took to the trees of dense early Cenozoic forests, where food was plentiful and natural enemies few. The earliest of these primates have been named prosimians, literally "pre-monkeys." They were widespread during the Paleocene and Eocene but decreased during the Oligocene Epoch. One reason for the decline among prosimians was the appearance and differentiation of monkeys and apes. Looking back over early Cenozoic primate evolution, we can see the beginnings of the transformation of snouty, squirrel-like prosimians into progressive anthropoids. The head was reshaped and rounded, the muzzle was lost, and the face was flattened. The eyes were moved forward to provide good binocular vision, not only for quickly estimating the distance of nearby insect prey but also for accurately gauging the distance to the next bough. Manual dexterity became important for hunting, feeding, and gripping branches. Forelimbs with clawed digits became arms and hands equipped with an opposable thumb and an improved sense of touch. Swinging from branch to branch by the arms paved the way toward the erect posture that would be exploited at a later stage in primate history.

During the Miocene, a group of apes appeared that were the probable ancestors of both modern apes and the hominid line. They were the dryomorphs (the dryopithecines of older nomenclature). *Proconsul africans* was a Miocene primate whose skull, face, and teeth make it a candidate for the subsequent apes, but whose limbs were not yet quite apelike. Other dryomorphs of the Miocene were more completely apelike and were widespread throughout Europe and Asia. In many ways the Miocene dryomorphs were the evolutionary products of changing late Cenozoic environments. Dense, junglelike forests that had cloaked entire regions of Africa were being replaced by grassy, parklike habitats called savannas. The semierect dryomorphs and later the australopithecines were quick to respond with a restructuring of the axial skeleton, pelvic girdle, and hind feet. With this improved, erect stability, the hands were now completely freed for manipulative purposes. These creatures probably took advantage of the protection inherent in unified group action and may have had the ability to communicate, at least to the same degree as do modern apes. All of these functions led ultimately to greater development of the cerebral cortex, the part of the brain associated with dexterity, memory, learning, and thought. Great reproductive advantage must have accrued to those species that made the best use of their brains. Although growth in brain size was slow until the end of the Pliocene, it accelerated phenomenally during the Pleistocene.

The australopithecines, or southern ape men, appeared on earth in the Pliocene about 4.0 million years ago. Their cranial capacity was about 600 cc, only about a third as large as that of humans. However, their dentition closely resembled that of modern people. Australopithecines were upright in stature and used crude tools. There is no evidence that they had discovered the use of fire. Structurally, the australopithecines were close to the grade of primate evolution that might warrant their being considered members of the genus *Homo*. In fact, they seem to have been either the ancestors or the contemporaries of larger brained creatures currently being designated as species of *Homo*.

The next stage of human evolution witnessed the appearance of *Homo erectus*, about 500,000 to 1 million years ago. These erect-walking primitive people had a brain capacity that ranged from 775 to 1300 cc. By middle Pleistocene time, *Homo erectus* has spread widely throughout Europe and Africa and had progressively improved his tool-making abilities.

The late Pleistocene was marked by the appearance of the very distinctive and famous Neanderthal people. Remains of the Neanderthals have been discovered in dozens of European sites, as well as in the Middle East and Africa. The Neanderthal skull was more massive than our own, had a less prominent forehead and chin, and had more prominent ridges above the eyes. The dozens of skulls already collected indicate that the Neanderthal peoples varied as widely in characteristics as do modern human populations. Neanderthals had brains as large as or larger than modern humans. Their tools were more finely constructed than those of *Homo erectus*, they fashioned objects for aesthetic purposes, they apparently had a religious concern about an afterlife, and they knew the use of fire.

Fire was of immense importance to Pleistocene humans. Because of the light that fire provides, the length of time during which they could effectively accomplish tasks was increased. Fire provided protection during the night and permitted *Homo erectus* to occupy cave shelters with less danger from great bears and other predators. (This alone has been a boon to anthropologists, in that it narrows down the number of places that are likely to have good fossils.) Fire also gave early humans the means to thaw and cook food. Its most important value, however, was that it provided warmth, thus enabling humans to extend their range into frigid latitudes.

Modern *Homo sapiens* appeared during the second interglacial stage of the Pleistocene. They continue down to the present day in the form of you and me. An early group, known as Cro-Magnon, replaced the Neanderthals in Europe and continued their cultural traditions. Even finer tools and weapons in bone and flint were manufactured by these peoples. They exhibited remarkable artistic skills, painting pictures of the animals vital to their survival on the walls of their caves and carving voluptuous Venus-type statues to encourage fertility. Humans came to the Americas at a rather late date. Artifacts and sparse skeletal remains indicate their presence somewhat over 20,000 years ago. Their spear heads are sometimes found among the remains of the extinct large mammals that they hunted.

## QUESTIONS FOR REVIEW

1. Early arboreal primates were characterized by close-set eyes and grasping extremities. What function, other than simply moving about in the trees, might these adaptations have performed?

2. In general, what modifications occur in the skulls of primates in the course of evolution from a tree shrewlike primate to *Homo sapiens*?

3. By what route and at what stage in the Pleistocene did *Homo sapiens* enter the New World?

4. During what geologic period do we find the earliest known remains of primates?

5. What physical characteristics of flint and chert result in their being useful in the manufacture of spear points and axes by early humans?

6. What is the evolutionary importance of the Miocene primate *Proconsul africanus*?

7. What advantage is there in a fossil site that consists of fossil-bearing sedimentary layers interspersed with layers of volcanic ash or lava flows?

8. What distinctly humanlike traits were possessed by *Australopithecus*? What apelike characteristics did this creature retain?

9. With the advent of *Homo erectus*, the trend toward increase in brain size was initiated. What related anatomical changes must have been synchronous with increases in cranial capacity?

10. What features of Cro-Magnon peoples differentiate them from Neanderthals?

## TERMS TO REMEMBER

anthropoid
australopithecine
Cro-Magnon

diastema
dryomorphs
"Marmes Man"

Neanderthal
paleoindian
pongid

*Proconsul*
prognathous
prosimian

## SUPPLEMENTAL READINGS AND REFERENCES

Brace, C. L. 1991. *The Stages of Human Evolution, 4th ed. Englewood Cliffs, New Jersey: Prentice Hall.*

Fleagle, J. G. 1988. *Primate Adaptation and Evolution.* San Diego: Academic Press.

Johanson, D. C., and Edey, M. A. 1981. *Lucy, the Beginnings of Humankind.* New York: Simon and Schuster.

Keyfitz, N. 1989. The growing human population. *Sci. Am.* 261(3).

Leakey, R. E. F. 1976. Hominids in Africa. *American Scientist* 64:174–178.

Lewin, R. 1984. *Human Evolution.* New York: W. H. Freeman and Co.

Pilbeam, D. 1972. *The Ascent of Man.* New York: Macmillan.

Reader, J. 1981. *Missing Links: The Hunt for Earliest Man.* Boston: Little, Brown & Co.

Rosen, S. I. 1974. *Introduction to the Primates.* Englewood Cliffs, N.J.: Prentice-Hall.

Simons, E. L. 1972. *Primate Evolution.* New York: Macmillan Co.

Collisions between planetesimals during the formative stage of the solar system.
(Painting by Don Dixon)

# Meteorites, Moons, and Planets

*Give me matter and motion, and I will construct the universe.*

René Descartes, 1640

17

## Introduction

In this book, we have traced much of the history of the earth since its origin about 4.6 billion years ago. We are aware that the earth is only one of nine planets that circle the sun and that each of these neighboring planets has an interesting history as well. It thus seems appropriate to briefly consider the characteristics of these other planets and the other extraterrestrial bodies that are the earth's nearby companions in space. In many instances, the study of other planets, their satellites, and meteorites has provided valuable new understanding of the origin and early development of the earth.

## Meteorites: Samples of the Solar System

Because meteorites are masses of mineral or rock that have reached the earth from space, they are of great importance to scientists interested in the origin and history of planets. They are the only objects from the universe beyond the moon and the earth that can currently be held in the hand and scrutinized in the laboratory. This research indicates that meteorites are fragments of asteroids formed when asteroids collided in space and were shattered. Many of these fragments were once part of meteors (Fig. 17–1), those rapidly moving, luminous bodies we refer to as shooting stars. Most of these particles burn up when they enter the earth's atmosphere, but if a portion survives its fall through the atmosphere and crashes into the earth, it is then termed a **meteorite**. About 500 meteorites at least as large as a baseball reach the earth's surface each year. Weathering and erosion on the earth have erased most of the craters made by their impacts, but about 70 partially preserved craters or clusters of craters remain.

The best preserved meteor crater in the world is Meteor Crater in Arizona (Fig. 17–2). The impact that produced Meteor Crater 50,000 years ago is estimated to have released energy equivalent to a 20-megaton nuclear bomb. In the past, much larger impacts have occurred. Meteor showers or the impacts of asteroids may even have contributed to episodes of mass extinctions on earth. There is evidence, for example, that an asteroid with a diameter of 10 km or greater struck the earth about 66 million years ago, lifting millions of tons of dust into the atmosphere. The dust blocked the sun's rays so that darkness and cold prevailed for several weeks, or possibly even several months. These effects could certainly have been catastrophic for life on earth and have been proposed as a major cause for the extinction of the dinosaurs (see Chapter 13).

**Figure 17–1** A meteor crossing the field of view of the Palomar telescope while it was taking a 15-minute exposure of the Comet Kobayashi-Berger-Milon, August 1975. *(Hale Observatory photography/John Huchra; from Pasachoff, J. M.* Contemporary Astronomy, *4th ed., Philadelphia: Saunders College Publishing, 1989.)*

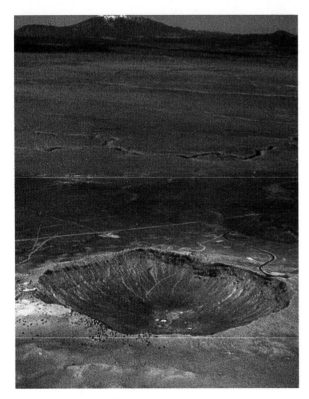

**Figure 17–2** Aerial view of Meteor Crater, formed by the impact of a large meteorite about 50,000 years ago. Conconino County, Arizona. *(Courtesy of W. B. Hamilton.)*

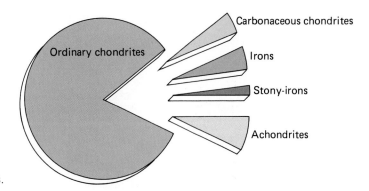

**Figure 17–3**  The major categories and proportions of meteorites.

Meteorites can be classified according to their composition as ordinary chondrites, carbonaceous chondrites, achondrites, irons, and stony-irons (Fig. 17–3). The most abundant of these types are the **ordinary chondrites** (Fig. 17–4), which are crystalline stony bodies composed of high-temperature ferromagnesian minerals. Ordinary chondrites can be dated by the uranium–lead, strontium–rubidium, and potassium–argon methods, and they are found to be about 4.6 billion years old. Many ordinary and carbonaceous chondrites contain spherical bodies called **chondrules** (see Fig. 17–4). Their spherical shape suggests that chondrules solidified from molten droplets splashed into space during collisions of objects swirling about in the solar nebula.

Less abundant than the ordinary chondrites are the **carbonaceous chondrites** (Fig. 17–5). In these meteorites, one finds the same abundance of metallic elements as in ordinary chondrites, with the addition

**Figure 17–4**  Stony meteorite of the type termed a chondrite. Chondrites contain small, rounded particles called chondrules composed of ferromagnesian minerals. In most cases, chondrules are only about 2 or 3 mm in diameter, but the one visible in the lower right area of this specimen is approximately 2 cm in diameter. *(Washington University collection.)*

**Figure 17–5**  A carbonaceous chondrite. This is a piece of a larger meteorite that apparently fragmented on entering the atmosphere above Mexico in 1869. The outer crust has been fused by heat. The white inclusions are minerals (refractory minerals) that withstand high temperatures without melting. Carbon imparts some of the dark color to the meteorite. *(Courtesy of T. J. Bernatowicz.)*

**Figure 17–6** Meteorite of the achondrite variety, found in Sioux County, Nebraska. *(Washington University collection.)*

of about 5 percent organic compounds, including inorganically produced amino acids. Suspended in the blackish earthy matrix of carbonaceous chondrites are both chondrules and irregular pieces of high-temperature minerals that may have condensed from a cooling vapor.

One of the most interesting findings about carbonaceous chondrites is that they contain nonvolatile elements in approximately the same proportion as those that occur in the sun. Indeed, if it were possible to extract some solar material, cool it down, and condense it, the condensate would be chemically similar to carbonaceous chondrites. This similarity suggests that carbonaceous chondrites are samples of primitive planetary material that formed when the sun formed and have never been remelted.

A small percentage of stony meteorites do not contain chondrules and therefore are termed **achondrites** (Fig. 17–6). Most of these have compositions similar to terrestrial basalts, and many exhibit angular broken grains. These fragmental textures indicate that achondrites may be the products of collisions of larger bodies.

The **iron meteorites** (Fig. 17–7) comprise about 6 percent of all observed falls of meteorites. Most iron meteorites are intergrowths of two varieties of iron–nickel alloy. The large size of the crystals and metallurgic calculations indicate that some iron–nickel meteorites cooled as slowly as 1°C per million years. Such a slow rate of cooling would be possible

only in objects at least as large as asteroids. Thus, the history of the iron meteorites probably involved a long episode of slow crystallization within a parent body sufficiently large to provide insulation for the hot interior, followed by violent disruption of that body by a collision.

The least abundant of all the categories of meteorites are the **stony-irons** (Fig. 17–8). As indicated by their name, they are composed of silicate minerals and iron–nickel metal in about equal proportions. The stony–iron meteorites are generally considered to represent fragments from the interfaces between the silicate and metal portions of asteroidal bodies. Some of them may have originated at the boundary between an iron core and the silicate mantle of a planetary body destroyed by collision.

# The Moon

## Moon Features

As satellites go, our moon is large in relation to the size of its parent planet. It has a diameter of over one-fourth that of the earth, and its density of about

**Figure 17–7** An iron meteorite. The meteorite has been cut, polished, and etched with acid to show an interlocking mosaic of the nickel-poor mineral kamacite, and nickel-rich mineral taenite. The distinctive interlocking texture is called Widmanstatten pattern. *(From the Washington University collection.)*

**Figure 17–8**  Stony-iron meteorite. The light-colored areas consist of iron–nickel, and the darker minerals are mostly olivine. *(Washington University collection)*.

$3.3 \text{ g/cm}^3$ is the same as that of the upper part of the earth's mantle. Gravity on the moon is insufficient to maintain an atmosphere. The moon and the earth revolve around their mutual center of gravity as a sort of "double planet." The moon orbits the earth and rotates on its axis at the same rate. As a result, we always see the same side of the moon and must depend on transmissions from space vehicles for images of the far side (Fig. 17–9). These photographs have revealed a densely cratered surface interrupted only by few and relatively small level areas.

Telescopic observation of the near side of the moon was first made in 1609 by a rather irascible professor of mathematics at the University of Padua. The mathematician was Galileo Galilei. With the use of his primitive telescope, Galileo was able to see that

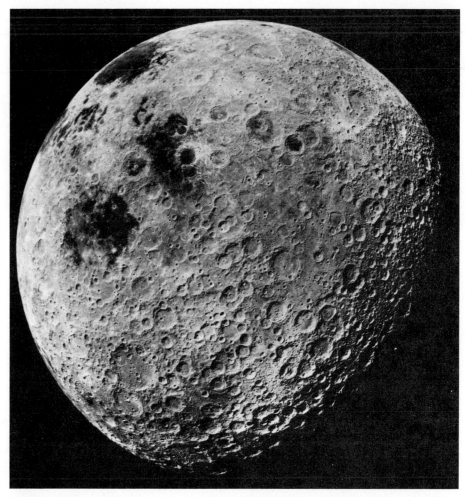

**Figure 17–9**  The far side of the moon as photographed by *Apollo 16*. *(Courtesy of NASA.)*

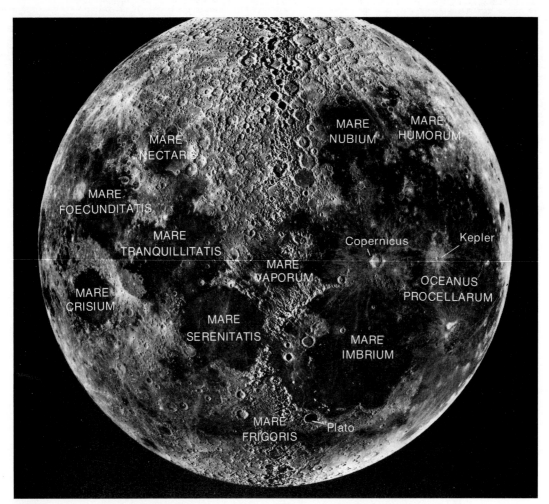

**Figure 17-10** The near side of the moon. To show the whole moon but still reveal the detail that does not show up well at full moon, the Lick Observatory has put together this composite of first and third quarters. Note the dark maria and the lighter, heavily cratered highlands. Two young craters, Copernicus and Kepler, can be seen to have rays of light (that is, lighter than the background) material emanating from them. This is a ground-based view of the moon, and it is shown inverted as it would be when viewed through a telescope. *(Lick Observatory photograph; from Pasachoff, J. M. Contemporary Astronomy, 4th ed. Philadelphia: Saunders College Publishing, 1989.)*

the surface of the moon was not smooth but "uneven, rough, full of cavities and prominences. . . ." He recognized "small spots," which we now know are craters, and dark patches that children imagine to be the eyes and mouth of the "man in the moon." These dark patches lie within large basins. Galileo named them **maria** ("seas") and incorrectly suggested they were filled with water. Today, we know the dark areas have been flooded, not with water, but with basaltic lavas. In many maria, the lava has been contained within depressions that are hundreds of kilometers across and 10 to 20 km deep. These depressions are termed **mare basins** (Figs. 17–10 and 17–11). Mare basins had to have developed prior to the extrusion of the dark lavas that form the maria, for these lavas are contained within the basins.

Galileo also provided the name for the lighter hued rougher terrains of the moon. He called these

**Figure 17–11**   A view of part of the Mare Tranquillitatis (Sea of Tranquillity), where astronauts aboard the *Apollo 11* spacecraft made their landing. *(Courtesy of NASA.)*

regions terrae, although modern space scientists refer to them as **lunar highlands.** The highlands are the oldest parts of the moon, having already formed long before the earliest mare lavas appeared. The highlands are heavily cratered regions reminiscent of a battlefield following a heavy artillery barrage. They provide stark evidence of the moon's early episodes of intense meteoritic bombardment.

On the moon, as on the earth, craters are roughly circular, steep-sided basins normally caused by either volcanic activity or meteoritic impact (Fig. 17–12). Although a very few lunar craters show questionable evidence of volcanic origin, the great majority are clearly the result of meteoritic impact. Unlike the earth, the moon has an insufficient atmosphere to burn up approaching meteorites, and thus the frequency of impact is high. Lunar geologists are able to determine the relative ages of certain areas on the moon's surface by the density of craters. Younger areas have fewer craters that do older ones. Another aid to recognizing differences in the age of lunar features is provided by the rays of material splashed radially outward from the crater at the time of im-

pact. Younger rays and other impact features cross and partially cover older features. Light-colored impact rays composed of finely crushed rock are exceptionally well developed in the large lunar crater known as Copernicus (Fig. 17–13). This familiar feature is about 90 km across and is rimmed by concentric ridges of hummocky material blasted out by the impact of the crater-forming body. Not all craters have rays as spectacular as those around Copernicus. Impact rays are sometimes obliterated by the rather continuous rain of tiny meteorites that strike the moon. For this reason, younger craters may still be adorned with impact rays, whereas these features have been lost around older craters.

There is ample evidence of former igneous activity on the moon. We have already noted the extensive lava floods of mare basins. In addition to these flood basalts, a few dome-shaped volcanoes measuring about 10 km in basal diameter have been observed. About 1 percent of the lunar craters may have resulted from volcanic activity. Meandering channelways occur on the surface of our satellite. Most of these are thought to have been produced by

**Figure 17–12** A view of the heavily cratered far side of the moon, photographed by the crew of the *Apollo 11* spacecraft. *(Courtesy of NASA.)*

**Figure 17–13** The crater Copernicus, as seen from an earth-based telescope, has rays of lighter colored material emanating from it. This light material was ejected radially when the meteorite that formed the crater struck the moon's surface. *(From Pasachoff, J. M.* Contemporary Astronomy, *4th ed., Philadelphia: Saunders College Publishing, 1989.)*

the turbulent flow of lava or by the collapse of lava tunnels, although the evidence is not conclusive. These winding features, are distinctly different from straight channels, which probably result from faulting. Tall volcanic peaks are not found on the moon. Perhaps this is because lunar lavas, like lavas associated with shield volcanoes on the earth, are highly fluid and spread laterally rather than piling up around a vent.

Are lunar volcanoes still active today? In 1958, Russian astronomers observed evidence of gas emission from the crater Alphonsus, and on two occasions in 1963, observers at the Lowell Observatory saw glowing red spots in and near the crater Aristarchus.

## Moon Rocks

Direct examination of samples returned from the moon (Fig. 17–14), as well as instrumental data gathered by orbiting spacecraft, indicate that there are three main kinds of rocks on the moon. One of these forms the floor of the smooth mare regions, whereas the other two are characteristic of the lunar highlands. The mare rocks are basalt, the finely crystalline, dark-colored igneous rock that was described in Chapter 1. Calcium plagioclase, pyroxene, olivine, and ilmenite are the principal minerals in basalt.

The rock types that occur in the lunar highlands differ from the mare basalts in both texture and composition. They are coarser (Fig. 17–15) and richer in calcium plagioclase. Most of them are aggregates of crushed rock mixed with material that appears to have been molten. Although several compositional varieties of highland rocks have been identified, in

A

B

C

**Figure 17–14** Moon rocks. (A) A photograph of a lunar basalt from the Sea of Tranquillity. This mare basalt was collected by *Apollo 15* astronauts. The lava solidified so rapidly that bubbles formed by escaping gases were trapped and formed vesicles. (B) A nonvesicular lunar basalt. (C) A photomicrograph of a thin section of lunar basalt composed of interlocking grains of augite (larger crystals), plagioclase (the zoned crystals), and olivine (center cluster of small crystals). The cubes are 2.5 cm square and indicate size. The area of the thin section is 3.7 mm. *(Courtesy of NASA.)*

**Figure 17–15**   Lunar rock from the lunar highlands. This coarsely crystalline rock is composed of approximately equal amounts of olivine (yellow grains) and plagioclase. *(Courtesy of NASA.)*

general they resemble the family of rocks on earth known as gabbros.

The samples brought to earth from the moon were all obtained from the blanket of rock fragments and dust that covers the moon's surface. Within this loose material, called **regolith**, it is possible to find fragments of rocks of all sizes, ranging from microscopic grains (Fig. 17–16) to huge boulders. Some of the boulders and cobbles consist of aggregates of smaller rock fragments welded together to form lunar breccias.

At Tranquility Base, site of the first moon landing (1969), the regolith was estimated to be 3 to 4 meters thick. It is believed to have formed as a result of meteorite impacts, each of which could dis-

lodge a mass of debris many hundred times greater than its own mass.

Radiometric dating of lunar rocks has been of great importance to the understanding of the sequence of events in the moon's history. Most mare basalts are between 3.2 and 3.8 billion years old and were extruded during the great lava floods that produced the mare basins (Fig. 17–17). Most lunar breccias were formed during impact events that occurred prior to these floodings. The oldest rocks on the moon were taken from the lunar highlands, and they have been found to be about 4.6 billion years old. This age represents the time that meteoritic impacts formed the rugged highland terrain and crushed and melted the rocks. Samples of moon rocks now being studied in the United States have been found to retain weak magnetism imparted to the rocks at the time they solidified. This indicates that at one time the core of the moon was liquid and that the moon had a significant magnetic field between about 4.2 and 3.2 billion years ago. The moon has no general magnetic field today.

## Lunar Processes

On earth there are processes capable of raising mountain ranges, as well as such opposing processes as erosion and weathering that are constantly at work wearing away the lands. Geologic processes also operate on the moon, but they are slower and of a different kind. The moon is a desolate world in which there is no air or surface water. Whenever gases or liquids escape to the lunar surface, they are dissi-

**Figure 17–16**   Lunar regolith (soil), which consists of impact fragments and tiny spheres of dark glass. Scale divisions at lower left are in millimeters. *(Courtesy of NASA.)*

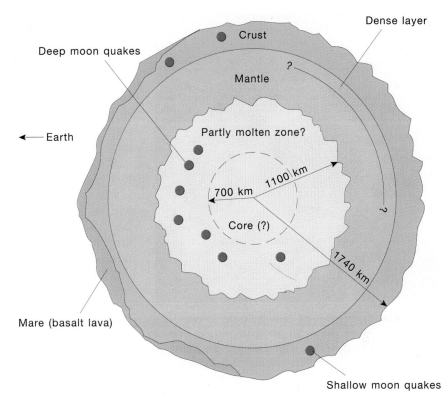

**Figure 17–17**  Cross-section through the moon showing an outer crust, an inner mantle, and an innermost zone that may still be partly molten. Within the innermost zone there may be a small metallic core (dashed circle.) The earth-facing side is relatively smooth and shows larger accumulations of the mare basin lavas. The far side is more rugged and has almost no lava. The drawing is not to scale, and the surface relief is exaggerated. Data for the drawing were provided by A. M. Dainty, N. R. Goins, and M. N. Toksoz. *(Courtesy of NASA. From French, B. M.,* What's New on the Moon?*)*

pated into space because the moon lacks sufficient gravitational attraction to hold them. There can be no stream-cut canyons and valleys, no glacial deposits, and no accumulations of wave-sorted sand. Rocks brought back to earth from the moon seem bright and newly formed and lack the discoloration common to weathered earth rocks. Yet, erosion does occur on the moon, primarily through the continuous bombardment of the lunar surface by large and small meteorites. The larger meteorites produce craters and rough terrain, but it is the rain of small meteorites and **micrometeorites** (those with diameters less than 1 mm) that subtly alter the moon's surface and are important in the formation of lunar soil. The upper or exposed surfaces of lunar rocks brought back to earth from the moon are riddled with tiny glass-lined pits produced by the impact of fast-moving micrometeorites. Geologists call these tiny depressions microcraters or **zap pits.** Most are less than 0.01 mm in diameter (Fig. 17–18). The micrometeorite barrage continues today, as was evident on examination of the Apollo spacecraft after its return to earth. Ten tiny microcraters were chipped into its windows.

Although gravity on the moon is only $\frac{1}{6}$ that on earth, it is nevertheless capable of moving lunar materials from high to low places. Whenever loose material on slopes is unable to resist the pull of gravity, it will break away and produce masses of slumped debris. Steplike slump masses are recognized as a common feature along the rims of some lunar craters. Elsewhere, photographs reveal tracks in lunar

**Figure 17–18** Photograph of a *zap pit* or microcrater. The tiny pit is only 20 $\mu$m (0.02 mm) in diameter and was photographed with a scanning electron microscope. *(Photo by F. Horz, courtesy of G. J. Taylor.)*

soils that record the rolling and sliding of large boulders down slopes.

## The Moon's Early History

Even with the explosion of scientific data from the space program, scientists are unable to support with confidence a single theory for the origin of the moon. As a result, several hypotheses are now being actively examined. The first of these stipulates that the moon accreted along with the earth as part of an integrated two-planet system. The second concept suggests that both bodies accreted independently and that the moon was then captured by the earth. The third hypothesis proposes that the moon broke off from the earth before either had completely solidified.

If both the earth and the moon formed together from the same part of the solar nebula, they should have similar compositions (as indicated in part by specific gravity). However, the average specific gravity of the earth is about 5.5, whereas that of the moon is only 3.3. This difference, together with the absence of a magnetic field on the moon, indicates that it lacks a large iron core. These considerations have caused skepticism about the double-planet concept for the origin of the moon.

The compositional difficulties seem to be resolved in a second hypothesis, in which the moon is considered to have formed in a distant part of the solar system but was subsequently captured when it approached the earth's gravitational field. Unfortunately, there are some severe problems relating to the mechanics of such a capture, and long-term measurements of the moon's orbit required for verification of the concept are presently lacking.

In 1898, George Darwin, the son of Charles Darwin, speculated that the moon was composed of a large mass of material that spun off the earth at an early time when its rotation was much faster than now. His idea was that the earth's centrifugal effect, together with the sun's tide-raising ability, first caused a huge tidal bulge on the earth. The bulge then separated to form the moon. When the moon's composition was found to be much like that of the earth's mantle, scientists speculated that the development of the earth had first proceeded to the point at which the earth's core of iron and other heavy elements had formed. After these heavy elements (which are substantially depleted in the moon's composition) had settled inward, a great bulge of mantle material formed, then pinched off, and began to orbit in widening circles around the earth. This interesting theory will continue to be examined over the next several years. It cannot be accepted yet, largely because of calculations indicating that if the moon spun off the mantle, it would subsequently be drawn back again. Geophysicists are also uncertain that there would have been sufficient force available to lift the mass of the moon from the earth and throw it into orbit.

As some scientists continue to debate the possibility that the moon was derived from the earth's mantle, a large number of investigators have come to favor the hypothesis that the moon, along with the other solar system bodies, was created about 4.6 billion years ago by the accumulation of matter and smaller bodies from the original dust cloud. Theoretically, the infall of small objects released so much energy that the outer part of the moon became molten to a depth of several hundred kilometers. At this time, a melt of lighter minerals (such as feldspar) floated to the surface of this vast ocean of molten rock and formed a crust. Heavier minerals such as pyroxene, olivine, and ilmenite settled to form an underlying layer about three times thicker than the crust. Then, about 200 or 300 million years later, great quantities of molten rock rose through fractures in the crust and hardened on the surface.

Next, the moon's primordial surface was subjected to a meteoritic barrage that lasted nearly 1.5 billion years and produced the scarred and cratered landscapes of the lunar highlands. The internal heat generated by the steady decay of such radioactive minerals as uranium and thorium began to melt portions of the moon again, especially at rather shallow depths of 100 to 250 km. The results of this melting event were massive extrusions of basaltic lava that spread over lower parts of the lunar surface during the period from about 3.8 to 3.1 billion years ago. The moon was much less active after the last of these eruptions. Meteorite impacts continued to sculpture its surface, sometimes forming spectacular craters such as Copernicus (see Fig. 17–13).

## The Earth's Neighboring Inner Planets

As noted in Chapter 5, the first four planets in our solar system (Mercury, Venus, Earth, Mars) are sometimes called **terrestrial planets** because, like the earth, they are small, dense, rocky, and rich in metals. Jupiter, Saturn, Uranus, and Neptune are large and dominated by hydrogen and helium. They are called **Jovian planets.**

### *Mercury*

Mercury has the distinction of being the smallest and swiftest of the inner planets. It revolves rapidly, making a complete journey around the sun every 88 earth days. Mercury makes a complete rotation on its axis every 59.65 days, which means that the planet rotates three times while encircling the sun twice.

*Mariner 10,* launched in 1973, provided our first close look at Mercury (Fig. 17–19). Following its launching, *Mariner 10* flew by Venus and, using the gravity of that planet to deflect its trajectory, went on to Mercury. Indeed, the spacecraft's orbit permitted visits over Mercury on three occasions. Photographs transmitted back to earth revealed a planet with a moonlike surface. Densely cratered terrains as well as

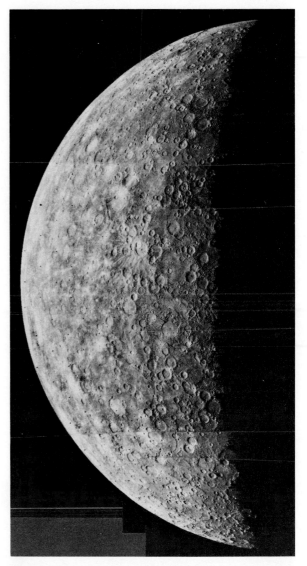

**Figure 17–19** Photomosaic of Mercury as photographed by the *Mariner* spacecraft. *(Courtesy of NASA.)*

smooth areas resembling maria were quite apparent. Some of the craters exhibited rays of light-colored impact debris, just as do craters on the moon. Long linear scarps were discerned, and these are thought to be fracture zones along which crustal adjustments occurred. Perhaps tidal forces or shrinkage of the planet as it cooled may have produced these elongate scarps.

Mercury has a weak magnetic field, a lightweight crust covered with fine dust, and an iron core

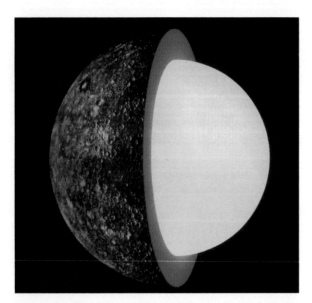

**Figure 17–20** The interior of Mercury is dominated by a metallic core about the same size as our moon. *(University of Arizona, courtesy of Robert Strom.)*

that comprises nearly 75 percent of its radius (Fig. 17–20). The existence of the iron core is inferred from Mercury's high density, its magnetic field, and other evidence suggesting that the planet experienced melting and differentiation.

## Venus

Although Venus is similar to the earth in size and shape, its surface conditions are quite different from those on earth. Indeed, our planet is a paradise compared with Venus. Maximal surface temperature on our nearest planetary neighbor is a searing 475°C. The planet is shrouded in a thick layer of clouds that extend 40 km above the surface (Fig. 17–21). Carbon dioxide is the principal gas in the atmosphere of Venus. The large amount of carbon dioxide serves as an insulating blanket in trapping solar energy by the greenhouse effect (the heating of an atmosphere by the absorption of infrared energy remitted by the planet as it receives light energy in the visible range from the sun). Surface atmospheric pressure on Venus is 90 times greater than it is on earth. The yellowish appearance of the planet as seen from the earth is the result of droplets of sulfuric acid in the planet's clouds.

Venus rotates through one complete turn every 243 earth days. Its direction of rotation is retrograde with respect to the stars, that is, opposite to the paths of revolution of the planets around the sun. It's period of revolution around the sun is 225 earth days. From its size and density, planetary scientists infer an interior rather like that of earth. Unlike the earth, however, Venus has no oceans. It is too close to the

**Figure 17–21** The clouds of Venus as photographed by *Pioneer Venus Orbiter* spacecraft. In visible light, no structure at all can be discerned. *(Courtesy of NASA.)*

**Figure 17–22**   A radar map of Venus from the *Pioneer Venus Orbiter*. Most of Venus is covered by a rolling plain, shown in green and blue. The highlands (yellow and brown contours) sit atop the plain, like continents. Aphrodite Terra is half as big as Africa. Ishtar Terra is the size of the United States, but looks relatively large in this Mercator projection. The lowlands, although resembling earth's ocean basins, cover only 16 percent of the planet. *(Experiment and data from MIT; map from U.S. Geological Survey; NASA/Ames spacecraft.)*

sun for water vapor to condense and form oceans. Without oceans in which carbon dioxide can be combined with calcium and magnesium to form carbonate rocks, there is no apparent mechanism to remove carbon dioxide from the atmosphere. The gas persists and is held in the atmosphere by gravity.

Because of its dense cloud cover, direct photography of the hard surface of Venus from orbiting spacecraft is not possible. The American *Pioneer Venus Orbiter*, however, was equipped with cloud-piercing ground-scanning radar, and with this device it was able to map the entire surface of the planet. These maps reveal that over 60 percent of Venus consists of gently rolling hills. Over most of this terrain, elevation varies only about 1000 meters between high and low points. There are, however, large highland areas on Venus that look like continents (Fig. 17–22). The largest of these, called Aphrodite Terra, is about the size of Africa and extends nearly halfway around the planet along the equator. Next in size is a region called Ishtar Terra which is as large as the United States (Fig. 17–23). Ishtar Terra contains the Maxwell Mountains, which appear to have been uplifted and deformed by compressional forces.

Radar images from the *Venera* and the more recently launched *Magellan* spacecraft have been of great value in the interpretation of the surface of Venus. Relatively large meteorites and asteroids have had no difficulty in penetrating the atmosphere of the planet, and the impact craters they have pro-

**Figure 17–23** An artist's conception of Ishtar Terra, a magnificent plateau about the size of the United States in area and higher than the Tibetan plateau. The reconstruction is based on measurements taken by the *Pioneer Venus Orbiter* spacecraft. *(Courtesy of NASA.)*

**Figure 17–24** High-resolution radar mosaic of the Lavinia region of Venus transmitted by *Magellan* spacecraft in September 1990. Three large impact craters with diameters ranging from 37 km (23 miles) to 50 km (30 miles) can be seen in a region of fractured plains. The lighter material around the craters is ejected debris. Volcanic domes with central pits can be seen in the southeast. These are 1 to 12 km (0.6 to 7.0 miles) in diameter. *(Courtesy of NASA.)*

duced (Fig. 17–24) can be counted to obtain an age of the surface of the planet. The number of craters on the plains of Venus, for example, are only 10 to 20 percent of the number in the moon's mare. This indicates that the surface is probably less than a billion years old. *Magellan* has also transmitted images of hills and valleys, as well as volcanoes. Some of the volcanoes on Venus are nearly 400 km across and 10 km high. Old lava flows can be detected on their slopes.

The first close look at the ground surface of Venus was provided in 1975 when the Russian robot spacecraft *Venera 9* and *Venera 10* succeeded in landing on the planet. In spite of the crushing atmospheric pressure and scorching temperatures (sufficient to melt tin), these spacecraft survived for about an hour. Even more successful, however, were the landings of *Venera 13* and *Venera 14* in March 1982. Like their predecessors, the two robots separated from their mother ships and drifted under parachute through the corrosive clouds to touch down on a mountainous region called Phoebe just below the Venusian equator. *Venera 13* survived 2 full hours, whereas its sister craft stopped functioning in less than an hour. The electronic eyes on both landers began working immediately and transmitted to earth remarkable photographs of a landscape strewn with rust-colored slabby rocks and rubble (Fig. 17–25). In addition to accomplishing their photographic mission, the landers were able to drill out a few cen-

**Figure 17–25** A view of the surface of Venus taken by *Venera 14*. The rocks are interpreted to be flow-layered basalts. The photograph was given by the Soviet Union to scientists at the U.S. Geological Survey in appreciation of help in choosing the landing sites for *Venera 13* and *Venera 14*.

timeters of Venusian soil and determine its composition. The analyses indicated the samples were basalt. Indeed, they had a strong resemblance to the basalts that are extruded from fissures along mid-oceanic ridges on earth. Many scientists believe that this discovery, along with the observation of great rift valleys, indicates that Venus, like the earth, is still a very dynamic planet with internal and surficial changes occurring constantly.

## Mars

### General Characteristics

The planet Mars (Fig. 17–26) has always been of special interest to humans because of speculation that some form of life, however humble and microscopic, may exist or at one time may have existed there. It is a planet that is only a little farther from the sun than is our own, it has an atmosphere that includes clouds, it has developed white polar caps, and it has seasons and a richly varied landscape. That landscape (Fig. 17–27) has been splendidly revealed to us by the *Mariner* and *Viking* missions. It includes magnificent craters, colossal volcanic peaks, deep gorges, sinuous channels, and an extensive fracture system. Evidence of wind erosion is clearly seen on photographs of the Martian surface, and there are indications of the work of ice as well. The planet has a diameter of only about 50 percent of that of the earth and has only 10 percent of the mass of the earth.

A year on Mars is 687 earth days long, and the planet's axial tilt is similar to that of the earth. Car-

bon dioxide is the principal gas in the Martian atmosphere, although small amounts of nitrogen, oxygen, and carbon monoxide are also present. Atmospheric pressure, however, is less than one percent that of the earth. The planet has virtually no greenhouse effect, and temperatures vary at the equator from about $+21°$ to $-85°C$.

**Figure 17–26** Planet Mars as photographed from the approaching *Viking* spacecraft in 1976. Red coloration is the result of dust particles containing iron oxide. *(Courtesy of NASA.)*

**Figure 17–27** Heavily eroded canyonlands on Mars. This Viking photograph shows an area about 60 km across. *(NASA/USGS, courtesy of Alfred McEwen)*

Mars has two moons. Their names are Phobos (Fig. 17–28), meaning fear, and Deimos, meaning panic. They are large chunks of rock about 21 and 15 km across, respectively.

The somewhat lower density of Mars (compared with that of other terrestrial planets) and its lack of a detectable magnetic field suggest that most of its iron may be scattered throughout the planet. This interpretation is strengthened by the color photographs transmitted from the *Viking* landers. Those photographs show the rusty red and orange hues typically developed on earth by limonite and hematite. Indeed, the sky often takes on a reddish hue, probably because of the rust-colored dust. Apparently, iron at

**Figure 17–28** Photomosaic of Phobos, the inner satellite of Mars. The photographs incorporated into this mosaic were obtained when *Viking Orbiter 1* approached to within 480 km of Phobos in 1977. Phobos is about 21 km across and 19 km from top to bottom. North is at the top. *(Courtesy of NASA.)*

the Martian surface has reacted with water vapor and oxygen to provide the color that so intrigued early planetary observers.

## Martian Landscape Features

As a result of the work accomplished by the *Mariner* and *Viking* orbiters, nearly the entire surface of Mars had been photographed by 1976. Crisp high-resolution photographs revealed craters, dunes, volcanoes, and canyons in unsurpassed clarity. On a grand scale, the planet's physiography is divisible into two rather different halves. The southern hemisphere is mostly heavily cratered rough terrain. This surface is inferred to be older than the smoother, lightly cratered surface of the northern hemisphere.

The two Viking spacecraft that reached Mars in the summer of 1976 each contained two parts, namely, a lander and an orbiter. The landers provided scientists with the first ground view of Mars. The photographs transmitted from the landers to the orbiters and thence to the earth show a sandy-looking terrain littered with cobbles and boulders (Fig. 17–29). Here and there, undisturbed bedrock was visible. Many of the rocks strewn about were clearly of volcanic origin. Some had spherical holes made by gas bubbles, whereas others were finely crystalline like many igneous rocks on earth. The devices on the lander that analyzed the soil indicated that the fine, loose material was rich in iron, magnesium, and calcium. It was very likely derived from parent rocks rich in ferromagnesian minerals.

**Figure 17–29** *Viking* lander photograph of the surface of Mars. The device on right is part of the lander that was purposely dropped for scale. It is about 20 cm long. Note the two trenches made by the lander's arm in taking samples of Martian regolith. *(Courtesy of NASA; computer color balancing by M. A. Dale-Bannister.)*

**Figure 17–30**   The gigantic Martian volcano Olympus Mons, photographed by *Viking Orbiter* in July 1976. The original black-and-white photograph has been colored in by an artist. *(Courtesy of NASA/JPL; from Abell, G. O.,* Realm of the Universe, *3rd ed. Philadelphia: Saunders College Publishing, 1983.)*

## Craters and Dunes

The density of craters in the southern half of Mars is similar to the crater density in the lunar highlands. Most of these craters were probably excavated between 4 and 4.5 billion years ago by the same torrential barrage of meteorites that pelted the lunar highlands.

Photographs clearly show row upon row of dunes on the floors of some of the larger craters. If the thinness of the Martian atmosphere is taken into account, then the size and spacing of these dunes can be used to estimate the velocity of the winds that formed them. Calculations indicate that the dune-forming winds blew at speeds of up to 320 km per hour.

Dust storms on Mars are prodigious events. As *Mariner 9* began its orbit of the planet, a dust storm was in progress, and useful photographs of the Martian surface could not be transmitted until the dust subsided. *Viking* orbiters encountered similar problems. The dust storms are not driven by excessively violent winds, however. One of the *Viking* landers recorded a gust of 120 km per hour, but average wind velocities were much lower. Dust on Mars is sometimes swirled upward into towering tornado-like columns that, on a smaller scale, would be called "dust devils." Unlike those on earth, Martian dust devils loom a spectacular 6 km above the parched terrain.

## Volcanic Features

Although most of the craters on Mars have been formed by the impact of meteorites, a smaller number are certainly volcanic. Easily the most spectacular of the volcanic craters is Olympus Mons (Fig. 17–30). This gigantic feature is more than 500 km across at its base and rises 30 km above the surrounding terrain. It is equal in volume to the total mass of all the lava extruded in the entire Hawaiian island chain. Around its summit are lava flows and the narrow rills interpreted as old lava channels or possibly collapsed lava tubes. Flood lavas are also evident, not only in the smoother northern hemisphere but also among the craters of the rugged southern half of Mars.

## Channels and Canyons

So little water exists today in the Martian atmosphere that if all of it were condensed in one place, it would probably provide only enough water to fill a community swimming pool. Yet the surface of the planet is dissected with channels and canyons that grow wider and deeper in the downslope direction. Some of these channels have braided patterns, sinuous form, and tributary branches such as those characteristically developed by streams on earth (Fig. 17–31).

These features suggest that flowing water was an important geologic agent at some earlier time in Martian history. There may once have been a considerable reserve of water frozen in the subsurface as a kind of permafrost. Permafrost is ground that remains below the freezing temperature and contains ice throughout the year. Many investigators theorize

**Figure 17–31**  A large channel system named Mangala Vallis on Mars. The system, which resembles some dry river beds on earth, flowed into the lower plains of the northern Martian hemisphere. Note the streamlined islands and intricate connections between channels. *(Courtesy of NASA.)*

that the larger channels, some of which are over 5 km wide and 500 km long, were excavated by water gushing out onto the surface as the subsurface layer melted. Large tongues of debris closely resembling mudflows may also have received their water from such a process. Many channels on Mars dwarf our own Grand Canyon in size and, in order to form, would have required torrential floods so spectacular as to be hard to visualize by earth standards.

## Ice Caps

Long before spacecraft landed on Mars, observers using telescopes noticed the planet had ice caps.

There are actually two types of ice caps on Mars. Seasonal ice caps composed of frozen carbon dioxide develop only during Martian winters by direct condensation from the atmosphere. Beneath this seasonal ice, there are also permanent ice caps at both poles. The southern permanent ice cap has a diameter of 350 km and is composed of carbon dioxide. The northern permanent cap is 1000 km in diameter and is composed of ordinary water ice. It represents a large reservoir of water relative to the very small amount of water present as vapor in the Martian atmosphere.

## Martian History

Like the earth, Mars was transformed from a protoplanet by accretion of a multitude of smaller objects. This accretion was followed by an episode of differentiation during which the once homogeneous body became partitioned into masses having different chemical compositions and physical properties. Very likely, the differentiation process involved melting, so that fluids could move from one region of the planet to another.

During the initial billion years or so of Martian history, the planet experienced heavy bombardment of meteorites and asteroids, with the result that the exposed crust became densely cratered. The craters have been modified by relatively recent wind erosion. The crust was disturbed, not only by cratering, but also by the faulting and fracturing that accompanied volume changes in the mantle.

It is likely that a dense atmosphere formed on Mars as a result of volcanic outgassing (removal of embedded gas by heating) and that the production of water and carbon dioxide far exceeded its loss into space. The planet was warmer at that time, perhaps because of a greenhouse effect. For a time, running water was an important geologic agent, as evidenced by the sinuous furrows cut into the cratered terrain. In certain regions of the planet, a small decline in surface temperatures may have caused the water contained in surface materials to freeze, thus trapping still liquid subsurface water in pockets and channels. With local melting, the subsurface water may have gushed upward and eroded large channels while simultaneously causing entire areas of surficial material to collapse in mudflows.

Eventually, the supply of atmospheric water and carbon dioxide became so depleted that the planet began to cool. Remaining carbon dioxide and water migrated to polar caps and to subsurface reservoirs, where it has remained.

Because Mars was once warmer, had more water, and possessed the elemental raw materials thought to be required for the development of organisms, scientists have long speculated on the possibility of finding primitive forms of life on the planet. For this reason, a series of experiments aboard the *Viking* lander was designed to search for signs of present or former life. The results were disappointing. Not a single microorganism or even a trace of organic molecules could be detected in the Martian soil at the landing site. However, it is still too early to rule out the possibility that continuing research and exploration of other sites may provide evidence of organisms.

# The Outer Planets

Beyond the orbit of Mars lies the ring of asteroids, and beyond the asteroids are the orbits of Jupiter, Saturn, Uranus, Neptune, and Pluto. The first four of these planets are also called the **Jovian** (for Jupiter) **planets**. They are similar to one another in having large size and relatively low densities. Pluto is not a Jovian planet. It is so far away and small that it is difficult to observe. As a result, there is considerable uncertainty about its origin.

## Jupiter

Jupiter, named for the leader of the ancient Roman gods, is the largest planet in our solar system. The diameter of this huge planet is 11 times greater than the earth's diameter, and its volume exceeds that of all the other planets combined. Its mass is 318 times that of the earth, yet it is only $\frac{1}{4}$ as dense. Jupiter spins rapidly on its axis, making one full rotation in slightly less than 10 earth hours. This rapid rotation results in the formation of the encircling colored atmospheric bands for which the planet is known (Fig. 17–32). Hydrogen, helium, and lesser quantities of ammonia and methane are the predominant gases in Jupiter's atmosphere. On the basis of experiments with these gases, some investigators believe that the reddish and orange colors in the atmosphere are derived from sulfur or phosphorus-based compounds. It is impossible to discern the surface of the

**Figure 17–32** (A) Jupiter, photographed by *Voyager 1* spacecraft from a distance of 33,000,000 km. (B) A close-up of Jupiter's famous Great Red Spot. Note the turbulence to the west (left) of the Great Red Spot. *(Courtesy of NASA.)*

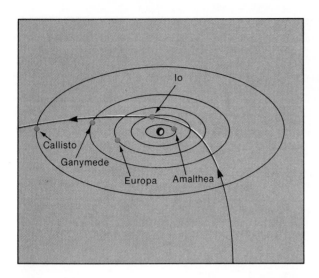

**Figure 17–33** *Voyager 1*'s flight path allowed scientists to obtain close-range photographs of five of Jupiter's 13 satellites. Each is shown at its closest point to *Voyager 1*'s outbound flight away from Jupiter. The spacecraft approached within 280,000 km of Jupiter. *(Courtesy of NASA.)*

solid interior of Jupiter with optical telescopes. The atmosphere, which is several hundred kilometers thick, passes gradually to liquid and eventually to solid matter. The interior of the giant planet may consist of highly compressed hydrogen, possibly surrounding a rocky core. Jupiter has at least 16 moons. The large planet with its many moons rather resembles a miniature solar system.

We have learned about Jupiter not only from earth-based telescopes but from close encounters by unmanned spacecraft. The most successful of the robot explorations occurred in 1979 when the two *Voyager* spacecraft swept past Jupiter and transmitted splendid images of the planet's brightly colored atmosphere, the tempestuous Great Red Spot (see Fig. 17–32), five of Jupiter's moons (Fig. 17–33), and a ring of debris less than 0.6 km thick that circles the planet about 55,000 km above the tops of the clouds. The discovery of the ring makes Jupiter the fourth planet in the solar system known to possess such a feature (Saturn, Uranus, and Neptune are the other planets known to have rings).

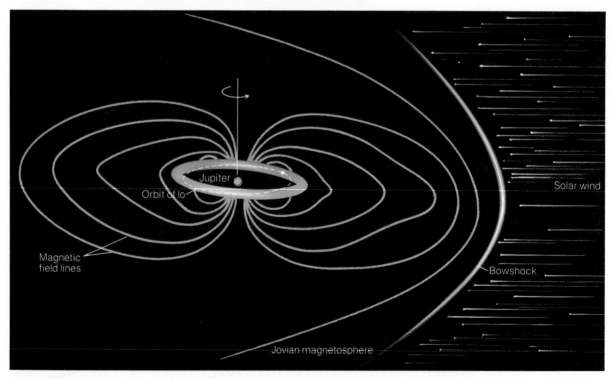

**Figure 17–34** The magnetosphere of Jupiter as revealed by the *Pioneer* and *Voyager* spacecraft explorations. The blue zone consists of hot ionized gases (plasma). *(From Abell, G. O., Morrison, D., and Wolff, S. C.* Exploration of the Universe. *Philadelphia: Saunders College Publishing, 1991.)*

Jupiter's Great Red Spot, which has twice the diameter of the earth, is an intriguing feature known to astronomers for centuries. As clearly revealed by the *Voyager* mission, this many-hued disturbance is a gargantuan atmospheric storm that extends deep into the cloud cover and turns in a complete counterclockwise revolution every six earth days.

In addition to the Great Red Spot, devices aboard the *Voyager* spacecraft radioed back the distinct images of great bolts of lightning, whose occurrence had only been suspected before the flyby. Another discovery was an auroral display far brighter than any northern lights ever seen on earth. *Voyager* also confirmed the presence of an immense magnetic field surrounding Jupiter (Fig. 17–34). First detected by the *Pioneer 10* spacecraft in 1973, the magnetic field extends outward from the planet for 16 million km and encompasses all of the larger satellites. The presence of the field indicates that Jupiter has a magnetic core resulting from fluid movements within its spinning interior.

The two *Voyager* spacecraft were not launched at the same time. *Voyager 2* reached Jupiter four months after its companion spacecraft and was thus able to fill in many of the gaps in information. *Voyager 2* also provided even better images of Jupiter's larger moons than had been transmitted by its sister ship. These larger satellites are called the Galilean moons to commemorate their discoverer. Individually, they are named Callisto, Ganymede, Europa, and Io. Each has characteristics quite different from those of its neighbors. **Callisto** (Fig. 17–35), the outermost of the Gallilean moons, is riddled with craters resulting from meteorite impact. On Callisto's icy, dirt-laden surface *Voyager* cameras detected huge, concentric rings that outline extraordinarily broad impact basins. Callisto's nearest neighbor is **Ganymede** (Fig. 17–36), the largest of Jupiter's satellites. Ganymede also shows the effects of meteorite bombardment in its dark, cratered terrain. Individual craters have bright streaks or rays that fan out from the crater rims like huge splash marks. Sinuous ridges have been detected on Ganymede's surface, as well as criss-crossing fractures suggesting that fault movements have occurred on the satellite.

**Europa** (Fig. 17–37), the brightest of the Galilean moons, has a surface crust of ice, but is otherwise a predominantly rocky satellite about the size of our own moon. It has few impact craters, and its surface is marked by huge fractures and ridges.

**Figure 17–35** Callisto, the outermost of the Galilean moons and the most heavily cratered body in our solar system. *(Courtesy of NASA.)*

**Figure 17–36** Ganymede, Jupiter's largest satellite, as photographed by *Voyager 1* from 216 million km. *(Courtesy of NASA.)*

**Figure 17–37** A view of the surface of Europa. The satellite's surface is composed of ice crossed by a complex system of cracks and low ridges. *(Courtesy of NASA.)*

These observations indicate that Europa's surface is being continuously renewed, perhaps by the development and slow migration of glacierlike masses of ice.

**Io** (Fig. 17–38) is the innermost of Jupiter's large satellites. Like Europa, it is an essentially rocky body, but it differs in having a considerable amount of active volcanism (Fig. 17–39). Eight volcanoes were erupting during the *Voyager* flybys of Jupiter. Graceful plumes of sulfur and sulfur dioxide ejected from the volcanoes extended for hundreds of kilometers into space. On cooling, the sulfur and sulfur dioxide in the plumes condense into solid particles and fall steadily onto Io's surface. The blanket of sulfurous deposits has probably buried impact craters that may have once existed on the surface of the satellite. There is evidence of former volcanism as well, for one can detect lava flows and shield volcanoes on the surface of Io. Judging from their color, Io's lava flows are probably composed of sulfur.

*Voyager 2* also passed near Amalthea, Jupiter's innermost satellite. The spacecraft's camera showed that Amalthea is a reddish-colored nonspherical satellite about 165 km long and 150 km wide. It is

**Figure 17–38** The satellite Io. The relatively smooth surface of Io and its active volcanoes suggest it may be Jupiter's youngest satellite. Large amounts of sulfur and sulfur dioxide frost are thought to be present on the surface. *(Courtesy of NASA.)*

## COMMENTARY

### Io's Volcanoes

One of the most interesting results of the Voyager flybys of Jupiter was the discovery of active volcanoes on the Galilean satellite Io (see Fig. 17–39). Before the images had been received, most planetary scientists believed Io would be a cold, dead satellite, rather like our own moon. On earth, volcanoes are ultimately powered by heat from decaying radioactive materials in the interior. Io, however, has only about one sixth of the mass of the earth, and like our moon, should have exhausted most of its original radioactivity billions of years ago. How then did the satellite acquire the heat needed to power the volcanoes seen by the Voyager spacecraft? It now appears that Io's heat is generated by alternate squeezing and unsqueezing in response to gravitational forces. We know that Io has a huge tidal bulge resulting from the gravitational attraction of Jupiter. This is to be expected, for Jupiter is hundreds of times more massive than the earth, and yet is about the same distance from Io as the earth is from its moon. If Io were Jupiter's only satellite, it would keep the same face toward Jupiter, just as our moon does while orbiting the earth, and the tidal bulge on Io would not generate heat. However, as the Galilean satellites Europa and Ganymede pass Io, they cause perturbations in Io's orbit. As a result, the huge tidal bulge rises and falls, alternately compressing and relaxing the satellite. The process generates a tremendous amount of frictional heat, and that heat powers Io's volcanoes.

Io's volcanoes are apparently of the explosive type. On earth, the major driving force for explosive volcanoes is steam. Io, however, is a waterless, rocky body. It appears that sulfurous gases emanating from molten sulfur deep beneath the crust cause the explosive activity of Io's volcanoes.

**Figure 17–39**  *Voyager 2* photograph of Io showing two active volcanoes. *(Courtesy of NASA.)*

without doubt the most irregularly shaped satellite known in our solar system.

### Saturn

Saturn's density is only about 70 percent that of water. Its mass, however, is equivalent to 95 earth masses. Measurements of Saturn's size and density suggest that the planet may have a central core of heavy elements (probably mostly iron) surrounded by an outer core of hot, compressed volatiles such as methane, ammonia, and water. These two core regions, however, constitute only a small part of the total volume of the planet, in which hydrogen and helium are overwhelmingly the predominant elements.

As was true for Jupiter, our knowledge of Saturn has been greatly improved as a result of data collected by unmanned spacecraft (Fig. 17–40). In particular, the electronic eyes of *Voyager 2* in August 1981 provided truly spectacular views of this second

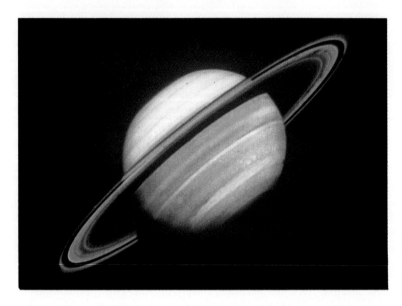

**Figure 17–40**    Saturn and its rings as photographed by *Voyager 2* from a distance of 18 million km. *(Courtesy of NASA.)*

largest planet, its rings, and its moons. The data transmitted back to earth helped planetary scientists confirm that Saturn has a magnetic field, trapped radiation belts, and an internal source of heat.

The general appearance of Saturn's atmosphere is similar to that of Jupiter. The alternating light and dark bands and swirls of gas, however, are partially veiled by a haze layer above the denser clouds. Temperatures near the cloud tops range from 86°K (−305°F) to 92°K (−294°F) with the coolest temperatures occurring near the center of the equatorial zone. The planet is well illuminated, for even its dark face receives a substantial amount of light reflected from the rings. On this dark side, *Voyager 2* detected radio emissions rather like those given off during lightning discharges.

### The Rings of Saturn

There is perhaps no more beautiful sight in the solar system than the rings of Saturn (Fig. 17–41). With his primitive telescope, Galileo was the first to recognize these rings, but he was unable to discern them clearly. Fifty years later, in 1655, Huygens was able to differentiate three concentric rings, the brightest and broadest of which was in the center. There are, however, far more than three rings around Saturn. As a result of pictures taken by *Pioneer 11,* and the more recent and truly dazzling photographs taken by

*Voyager 1* and *Voyager 2,* we now know that the once neatly defined grouping of three rings is really a complex system of thousands of rings and rings within rings. Saturn's rings are composed of billions of orbiting particles of dust and ash, as well as many larger fragments measuring tens of meters in diameter. All of these revolve around Saturn in the approximate plane of the equator. Saturn's 17 or more moons also revolve in this plane. The entire ring system has an outside diameter of about 275,000 km. Although the area of the rings is truly enormous, most of the material is concentrated within a thickness of only about 16 km. Thus, the entire system reminds one of a phonograph record with the rings corresponding to the record's many tiny grooves. Scientists are uncertain about the origin of Saturn's rings. Some suggest that the particles in the rings are remnants of an exploded satellite. Others argue that the rings are a remnant of the particle cloud from which satellites form by accretion.

### Saturn's Moons

Largely as a result of the Voyager missions, we know that Saturn ranks first among the planets in the number of its satellites (Fig. 17–42). Altogether, over 17 confirmed moons have been recognized. Some have been formally named, whereas others bear only numbers. The largest of the named satellites is **Titan.**

**Figure 17–41**  Computer-enhanced photograph of Saturn's rings as photographed by *Voyager 2*. The exaggerated color shows the reflection and transmission of different wavelengths of light by particles of different sizes in various parts of the ring system. *(Courtesy of NASA.)*

Titan is about 5120 km in diameter and has a relatively low density about twice that of water ice. The satellite's surface is obscured by thick dense haze. Titan has a nitrogen-rich atmosphere, and lakes of liquid nitrogen may exist at the poles where surface temperatures are an estimated 90°K (−300°F). The more important inner satellites of Saturn are named Mimas, Enceladus, Tethys, Dione, and Rhea (Fig. 17–43). Their densities and surface brightness suggest that they too are composed mainly of water ice. The five satellites range in size from the 390-km diameter of Mimas to the 1530-km diameter of Rhea. All are densely cratered. The outer satellites of Saturn include tiny Phoebe, potato-shaped Hyperion, and Iapetus.

## Uranus

Beyond the orbit of Saturn lies the planet Uranus (Fig. 17–44). Like Saturn and Jupiter, Uranus also has rings. They have been photographed from earth with infrared film. Uranus has a low density (1.2 g/cm$^3$) and a frigid surface temperature of about −185°C. Five large moons (and five recently detected small moons) circle Uranus. The larger moons are named Ariel, Miranda, Oberon, Titania, and

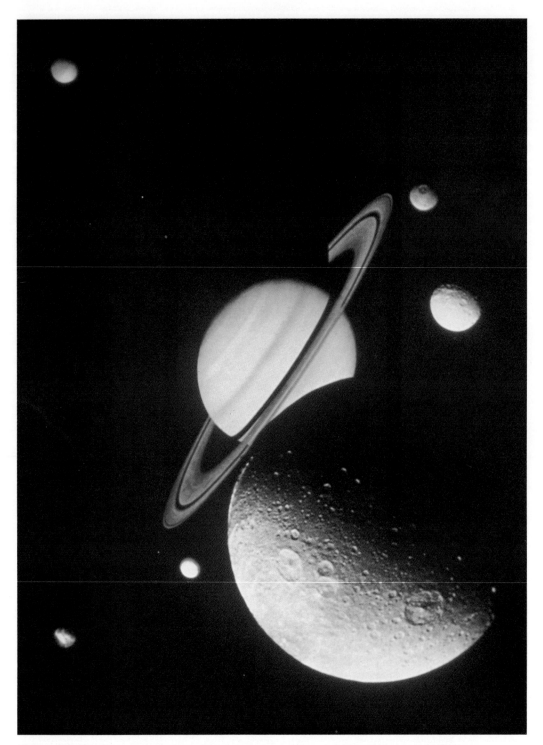

**Figure 17–42** Saturn and some of its moons are shown in this montage compiled from *Voyager 1* and *Voyager 2* photographs. The satellites shown clockwise from upper left are Titan, Iapetus, Tethys, Mimas, Enceladus, Dione, and Rhea. *(Courtesy of NASA.)*

**Figure 17–43** *Voyager 1* approached within 124,000 km (77,000 miles) of Saturn's cloud tops. Six of the satellites that were photographed are shown in their approximate positions at closest approach by the spacecraft. *(Courtesy of NASA.)*

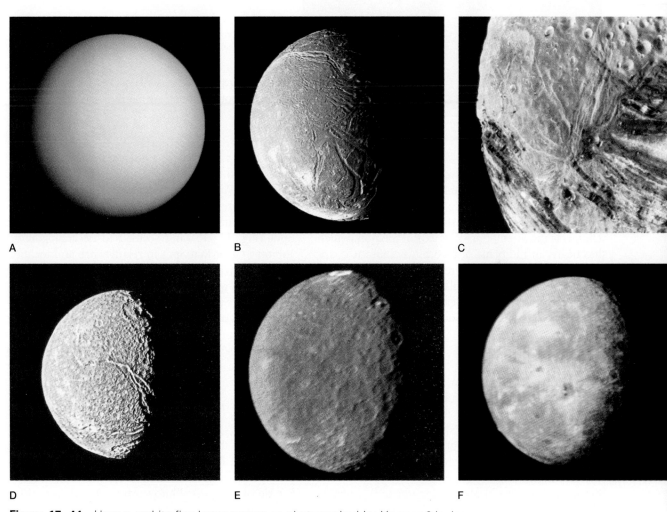

A

B

C

D

E

F

**Figure 17–44** Uranus and its five larger moons as photographed by *Voyager 2* in January 1986. (A) Uranus. (B) Mosaic of Ariel, with its many faults and valleys. (C) Miranda, whose grooved and cratered surface indicates a complex geologic history. (D) Titania, with prominent fault valleys nearly 1600 km long. (E) Umbriel, with cratering, at a distance of 557,000 km. (F) Oberon, showing craters and a large peak along the lower left edge of the photograph. *(Courtesy of NASA.)*

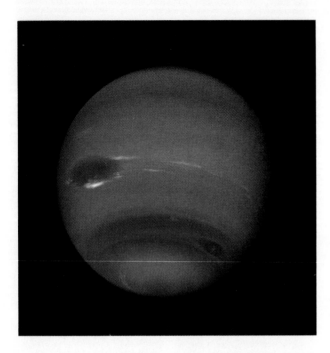

**Figure 17–45** The planet Neptune as photographed in 1989 by *Voyager. (Courtesy of NASA/JPL.)*

Umbriel. Thus far, earth-based instruments have only been able to detect hydrogen and methane on Uranus, although helium may also be present. The methane may be responsible for the planet's greenish hue. Perhaps the most distinctive characteristic of Uranus is the orientation of its axis of rotation, which lies nearly in the plane of its orbit. Because it is upside down, the planet's direction of rotation is the reverse of that of all other planets except Venus. This retrograde direction is also characteristic of the orbits of the five larger satellites around Uranus.

### Neptune

Because they are similar in size and other physical properties, Neptune and Uranus are often called the twin planets. Through the telescope, Neptune, like Uranus, appears as a greenish orb. Neptune's density of 1.7 g/cm$^3$ is just slightly greater than the density of Uranus. Neptune's atmosphere may also resemble that of Uranus. Both hydrogen and helium have been detected spectroscopically, and methane is probably also present. Two moons, named Triton and Nereid, circle the planet. Triton is one of the four largest moons in the solar system (the other three being Ganymede, Titan, and Callisto). Its surface may contain ice made of frozen methane, and it may even have a thin atmosphere of methane gas. Recently, astronomers have found evidence that Neptune may have a third moon and possibly even a ring system.

When *Voyager 2* approached Neptune in 1989, a huge, counterclockwise rotating area of atmosphere became noticeable. It was named the Great Dark Spot (Fig. 17–45) and seemed to be analogous to Jupiter's Great Red Spot in relative size and location in the southern hemisphere. Cirrus clouds composed of methane were detected around the margins of the Great Dark Spot. Other areas of atmospheric circulation on Neptune were also recognized, indicating that the planet has active weather systems, perhaps generated by internal heat. *Voyager 2* was also able to confirm the presence of two circular rings that lie very close to Neptune's equator.

### Pluto

Far out on the outer limits of our solar system lies the orbit of Pluto. It is so small and so distant that less is known about it than about the other planets. We do observe that Pluto's orbit is tilted an unusual 17° to the ecliptic plane and that its orbit crosses that of Neptune. As a result, Pluto is actually closer to the sun than Neptune during certain periods (as during the current period from 1979 to 1999). Because of the high inclination of Pluto's orbit, however, there is no danger of a collision with Neptune. Its small size and orbital peculiarities suggest that Pluto may not be a true planet but rather a former satellite of Neptune that escaped the gravitational pull of that planet and now occupies a separate orbit. Evidence for this theory, however, is not conclusive.

In 1978, scientists of the U.S. Naval Observatory found that Pluto has a moon. They named it Charon after the mythologic boatman who ferried the souls of the dead across the river Styx to Hades. Some scientists believe the existence of Pluto's moon makes the "escaped satellite" theory for Pluto's origin unlikely. Others argue that Pluto and Charon may have been ejected from the Neptunian system as a satellite pair.

# The Future of the Solar System

Anyone interested in the history of the earth is naturally led to a consideration of its future. Clearly, all that has happened, is happening, and will happen to this planet is inevitably linked to the history and destiny of the sun. The sun is a star, and like other stars, it will run its course from birth to death. By observing other stars that are at various stages in their histories, astronomers have pieced together a story of what happens when these bodies begin to deplete their store of nuclear fuel. Our star has been "burning" for about 5 billion years. In this span of time, it has consumed about half of its nuclear fuel and so has about another 5 billion years of life remaining.

What will "doomsday" for the sun and earth be like? Most astronomers believe that once the hydrogen in the sun's core has been largely converted to helium by fusion, the core will begin to contract. As the core contracts, the outer layers of the sun will expand and cool, causing the sun to puff up into an enormous sphere and to turn blood-red. Our sun will have become what astronomers call a red giant. As the great ball of rarefied gases continues to expand, it will engulf and vaporize Mercury and Venus.

What will happen to the earth can be readily imagined. The searing heat will boil away the oceans and heat the crust even to the melting point of some surface rocks. This purgatorial state of affairs may continue for a billion years or so, until the sun finally dissipates its energy. The dying sun may then begin to shed its outer layers, leaving behind a dense, hot white dwarf star about the size of the earth. Such white dwarf stars have exhausted their supply of nuclear fuels and remain incandescent only because of their residual internal heat. Eventually, even this light will fade as the sun becomes a black dwarf, so called because it will be too cool to radiate visible light. The scorched earth will continue as a barren, frigid sphere unable to free itself from the gravitational grasp of the dead sun.

# SUMMARY

There are a great many chunks of matter in interplanetary space, ranging up to tens of meters in diameter. When one of these chunks survives its passage through the earth's atmosphere, or strikes the surface of a moon or planet, it is called a meteorite. Meteorites are important to planetary geologists because they are truly extraterrestrial samples of the solar system. They can be classified as ordinary chondrites, carbonaceous chondrites, achondrites, irons, or stony-irons and are believed to be fragments resulting mostly from collisions of one or more small planetary bodies.

The moon, which has been studied by telescope for centuries, was first visited by humans in July 1969. Its general features include the relatively smooth, dark mare basins and the rugged lunar highlands. Samples of the loose regolith on the moon have been brought back to earth. They include basaltic rocks from the maria and specimens of anorthosite from the highlands. Study of the age and distribution of these rocks suggests that lunar history began about 4.6 billion years ago, when the moon had developed its surficial skin of anorthositic igneous rock. This surface was then flooded in various regions by the basalts of the maria about 1 billion years later.

The planets of our solar system can be divided into two main groups. Those resembling the earth are the inner (or terrestrial) planets. The outer planets resemble Jupiter and are termed Jovian planets. The terrestrial planets tend to be small, dense, and rocky, whereas the Jovian planets are large and have many satellites and dense, thick atmospheres.

The terrestrial planet closest to the sun is Mercury. It is the smallest of the inner planets and is very moonlike in appearance. Next in the progression outward from the sun is Venus. Venus is about the size of the earth. It has a dense carbon dioxide atmosphere, which traps heat from the sun. The surface temperature of Venus is higher than that of any other planet in the solar system. Its rocky surface is enshrouded by a cloud cover too dense to see through except by radar.

Mars is of interest because of its great variety of surface features. It has a cratered topography reminiscent of the moon's and also has features that appear to be the result of processes similar to those operating on earth. Dunes, stratified sediments, and sinuous rills suggest erosion by wind, water, and ice. Mars has a thin carbon dioxide atmosphere and polar caps composed of water ice and frozen carbon dioxide.

Among the outer planets, Jupiter is known for its tremendous size. It is larger than all other planets combined; has a massive hydrogen, helium, methane, and ammonia atmosphere; and rotates so rapidly as to form beautiful arrays of colored cloud bands. The planet has an internal heat supply. Its strong magnetic field suggests the presence of a metallic core.

Saturn is rather similar to Jupiter although smaller. Saturn's most distinctive feature is the thin, complex ring system that encircles its equator like the brim of a straw hat. At least 17 moons accompany Saturn on its revolution around the sun.

Uranus and Neptune are similar in size and general characteristics. Both have atmospheres rich in hydrogen compounds, and both appear to have small ring systems. Beyond Neptune lies Pluto. Pluto is roughly the size of the earth's moon and is thus the smallest planet in the solar system. It has a highly elliptical orbit.

The earth's environment and history are inexorably tied to the history of the sun. Our sun will continue to shine with a hot yellow light for about another 5 billion years as its hydrogen is slowly converted to helium. When the hydrogen fuel nears exhaustion and helium becomes the principal material in the solar core, the sun will expand to become a red giant and its heat will sear the earth's surface. After this phase of its history, the sun's emission of energy will gradually decline until it becomes negligible.

## QUESTIONS FOR REVIEW

1. What are the principal kinds of meteorites that have been found on the earth? In what way are they thought to have originated? What are chondrites?

2. What evidence indicates that, unlike the earth, the moon does not have an iron–nickel core?

3. How do the rocks of the lunar maria and lunar highland differ in composition and age?

4. What is the derivation of the reddish coloration of the planet Mars?

5. How does Mars differ from the earth in density, internal structure, and average range of equatorial temperatures?

6. What is the origin of the craters on the earth's moon?

7. How do lunar geologists determine the relative ages of craters on the moon?

8. What is lunar regolith? How is it thought to have originated?

9. The surface of Io is noted for its bright yellow and orange colors. What is the cause of this coloration?

10. Compare Mercury, Mars, and the moon. What do they have in common, and how do they differ?

11. Why does one side of the moon have more craters than the other? In what two ways have lunar craters been developed?

12. How do conditions on the surface of Venus differ from those on earth?

13. How do the volcanoes on Io differ from those on earth?

14. How is the absence of an ocean on Venus related to the persistence of carbon dioxide in the atmosphere of Venus?

## TERMS TO REMEMBER

| | | | |
|---|---|---|---|
| achondrite | iron meteorite | meteor | stony-iron meteorite |
| carbonaceous chondrite | Jovian planet | meteorite | stony meteorite |
| chondrite | lunar highlands | micrometeorite | terrestrial planet |
| chondrule | mare basins | regolith (lunar) | zap pit |

## SUPPLEMENTAL READINGS AND REFERENCES

Abell, G. O., Morrison, D., and Wolff, S. C., 1991. *Exploration of the Universe*. 6th ed. Philadelphia: Saunders College Publishing.

Cadogan, P., 1981. *The Moon: Our Sister Planet*. New York: Cambridge Univ. Press.

Carr, M. H., 1981. *The Surface of Mars*. New Haven: Yale University Press.

Hartmann, W. K., 1983. *Moons and Planets: An Introduction to Planetary Science*. 2nd ed. Belmont, Calif.: Wadsworth Publishing Co., Inc.

Pasachoff, J. M., 1991. *Astronomy: From the Earth to the Universe*. 4th ed. Philadelphia: Saunders College Publishing.

Skinner, B. J. (ed.), 1981. *The Solar System and Its Strange Objects: Readings from American Scientist*. Los Altos, Calif.: Wm. Kaufmann, Inc.

# Appendix A

# Classification of Living Things

Early students of biology found it convenient to divide all organisms into two great realms or kingdoms, designated the Animalia and the Plantae. However, by the late nineteenth century, many biologists suggested that perhaps a third kingdom—the Protista—should be established for certain single-celled organisms that seemed to be neither plant nor animal. Even this was considered inadequate by some biologists, for as this three-kingdom classification was gaining adherents, evidence was accumulating that clearly indicated the need for still further major groupings. Studies of unicellular organisms had revealed two quite different forms of cell structures. There were, for example, prokaryotic unicellular organisms, such as bacteria and blue-green algae, that lacked a cell nucleus and possessed other traits that set them clearly apart from eukaryotic unicellular organisms, which had a true nucleus, well-defined chromosomes, and organelles. Any classification that sought to categorize organisms according to similarity of origin and fundamental differences could not ignore the contrast between eukaryotes and prokaryotes. Finally, it was recognized that the fungi deserved special taxonomic consideration also,

for they are dependent upon a supply of organic molecules in their environment, as are animals, yet they absorb their food through cell membranes, as do plants. The fungi appear to have had an evolutionary radiation quite distinct from the other major groups.

In an effort to account for these differences and better represent the evolutionary relationships of organisms, R. H. Whittaker of Cornell University proposed a five-kingdom system of classification.[*] In this classification, plants, fungi, and animals are regarded as distinct in terms of being specialized for different modes of nutrition, photosynthesis, absorption, and ingestion. The Protista is composed of unicellular eukaryotic organisms, and the Monera include the simplest of organisms that are inferred to be similar to the primitive forms of life from which other kingdoms evolved. A modified version of Whittaker's classification, which is steadily gaining proponents, is presented here. (Geologic ranges of major fossil groups are indicated.)

---

[*]Whittaker, R. H. 1969. New concepts of kingdoms of organisms. *Science* 163:150–160.

**KINGDOM MONERA**........................ Prokaryotes (*Archean to Recent*)

PHYLUM CYANOPHYTA........................ Cyanobacteria (*Archean to Recent*)
PHYLUM MYXOBACTERIAE .................... Unicellular or filamentous gliding bacteria
PHYLUM SCHIZOPHYTA ...................... True bacteria
PHYLUM ACTINOMYCOTA..................... Certain branching, filamentous bacteria
PHYLUM SPIROCHAETAE ..................... Spirochetes

**KINGDOM PROTOCTISTA**
**(formerly Protista)**

Solitary or colonial unicellular eukaryotic organisms that do not form tissues

PHYLUM EUGLENOPHYTA ..................... Euglenoid organisms
PHYLUM XANTHOPHYTA ...................... Yellow-green algae
PHYLUM CHRYSOPHYTA....................... Golden brown algae, diatoms, and coccolithophor-ids (*Paleozoic? Triassic to Recent*)
PHYLUM PYRROPHYTA ....................... Dinoflagellates and cryptomonads (*Triassic to Recent*)
PHYLUM HYPHOCHYTRIDIOMYCOTA ........... Hypochytrids
PHYLUM PLASMODIOPHOROMYCOTA............ Plasmodiophores
PHYLUM SPOROZOA.......................... Sporozoans (parasitic protists)
PHYLUM CNIDOSPORIDIA ..................... Cnidosporidians
PHYLUM ZOOMASTIGINA...................... Animal flagellates, protozoa that have whiplike cyto-plasmic protrusions (flagellae)
PHYLUM SARCODINA ........................ Rhizopods, protozoa with pseudopodia for locomo-tion (*Cambrian to Recent*)
PHYLUM CILIOPHORA ....................... Ciliates and suctorians, movement accomplished by beating of cilia (adult suctorians are attached to objects)

**KINGDOM PLANTAE**

DIVISION* RHODOPHYTA...................... Red algae, usually marine, multicellular
DIVISION PHAEOPHYTA...................... Brown algae, multicellular often with large bodies, as in seaweeds and kelps
DIVISION CHLOROPHYTA .................... Green algae (*Proterozoic? Paleozoic to Recent*)
DIVISION CHAROPHYTA...................... Stoneworts
DIVISION BRYOPHYTA ...................... Liverworts, mosses, and hornworts (*late Paleozoic to Recent*)
DIVISION PSILOPHYTA ...................... Extinct leafless, rootless, vascular plants (*middle Pa-leozoic*)
DIVISION LYCOPODOPHYTA ................. Club mosses, with simple vascular systems and small leaves, including scale trees of Paleozoic (lycopsids) (*Silurian to Recent*)
DIVISION EQUISETOPHYTA ................... Horsetails, scouring rushes, including sphenopsids
    (ARTHROPHYTA)                        such as *Calamites* and *Annularia* of the late Paleo-zoic (*Devonian to Recent*)
DIVISION POLYPODIOPHYTA.................. The true ferns or pteropsids (*Devonian to Recent*)

---

*In botany, it is conventional to use the terms *division*, *subdivision*, and so on, in place of *phylum* and *subphylum*.

DIVISION PINOPHYTA. . . . . . . . . . . . . . . . . . . . . . The "gymnosperms" including conifers, cycads, and many evergreen plants; no true flowers (*middle Paleozoic to Recent*)

    Class Lyginopteriodopsida. . . . . . . . . . . . . . . . . The seed ferns, known from fossils of the late Paleozoic and including such forms as *Neuropteris* and *Glossopteris* (*late Paleozoic to Recent*)

    Class Bennettitopsida. . . . . . . . . . . . . . . . . . . . The extinct cycadeoids (*Triassic to Recent*)

    Class Cycadopsida . . . . . . . . . . . . . . . . . . . . . . Cycads (*Triassic to Recent*)

    Class Ginkgoopsida . . . . . . . . . . . . . . . . . . . . . Ginkgoes (*early Miocene to Recent*)

    Class Pinopsida . . . . . . . . . . . . . . . . . . . . . . . . Conifers, as well as the extinct *Cordaites* (*late Paleozoic to Recent*)

    Class Gnetopsida . . . . . . . . . . . . . . . . . . . . . . . Certain climbing shrubs and small tropical trees

DIVISION MAGNOLIOPHYTA. . . . . . . . . . . . . . . . Flowering plants or "angiosperms"; seeds enclosed in an ovary (*Cretaceous to Recent*)

    Class Magnoliopsida. . . . . . . . . . . . . . . . . . . . . Dicotyledonous plants: embryos with two cotyledons or seeds; leaves with netlike veins (*Cretaceous to Recent*)

    Class Liliopsida . . . . . . . . . . . . . . . . . . . . . . . . Monocotyledonous plants: embryos with only one seed leaf; leaves with parallel veins (*Cretaceous to Recent*)

## KINGDOM FUNGI

DIVISION MYXOMYCOPHYTA . . . . . . . . . . . . . . . . Slime molds

DIVISION EUMYCOPHYTA . . . . . . . . . . . . . . . . . . True fungi

## KINGDOM ANIMALIA

PHYLUM MESOZOA. . . . . . . . . . . . . . . . . . . . . . . Mesozoans

PHYLUM PORIFERA . . . . . . . . . . . . . . . . . . . . . . . Sponges; includes forms with calcareous spicules, siliceous spicules, and proteinaceous spicules; may also include the extinct stromatoporoids (*Cambrian to Recent*)

PHYLUM ARCHAEOCYATHA. . . . . . . . . . . . . . . . . Extinct spongelike organisms (*Cambrian*)

PHYLUM CNIDARIA (COELENTERATA). . . . . . . . . . Jellyfishes, corals; radially symmetric, aquatic, with body wall of two layers of cells, in the outer of which are stinging cells (*Proterozoic to Recent*)

    Class Hydrozoa . . . . . . . . . . . . . . . . . . . . . . . . . Hydralike animals

    Class Scyphozoa. . . . . . . . . . . . . . . . . . . . . . . . True jellyfishes

    Class Anthozoa. . . . . . . . . . . . . . . . . . . . . . . . . Corals and sea anemones

    *Subclass Zoantharia*. . . . . . . . . . . . . . . . . . . . . . Hexacorals of modern seas (*Triassic to Recent*)

    *Subclass Rugosa*. . . . . . . . . . . . . . . . . . . . . . . . Paleozoic tetracorals (*Cambrian to Permian*)

    *Subclass Tabulata* . . . . . . . . . . . . . . . . . . . . . . . Paleozoic tabulate corals (*Ordovician to Permian*)

PHYLUM CTENOPHORA . . . . . . . . . . . . . . . . . . . . Modern comb jellies or "sea walnuts." Not known as fossils

PHYLUM PLATYHELMINTHES . . . . . . . . . . . . . . . . Flatworms

PHYLUM NEMERTEA. . . . . . . . . . . . . . . . . . . . . . . Proboscis worms

PHYLUM NEMATODA . . . . . . . . . . . . . . . . . . . . . . Roundworms

PHYLUM ACANTHOCEPHALA . . . . . . . . . . . . . . . . Hook-headed worms

PHYLUM NEMATOMORPHA. . . . . . . . . . . . . . . . . . Horsehair worms

PHYLUM ROTIFERA . . . . . . . . . . . . . . . . . . . . . Small, wormlike animals with a circle of cilia on the head

PHYLUM GASTROTRICHA. . . . . . . . . . . . . . . . . . Small, wormlike animals resembling rotifers but lacking circle of cilia

PHYLUM BRYOZOA . . . . . . . . . . . . . . . . . . . . The bryozoans, sometimes considered two phyla: *Entoprocta* and *Ectoprocta* (*Ordovician to Recent*)

PHYLUM BRACHIOPODA . . . . . . . . . . . . . . . . . . Marine animals with two parts (valves) to their shell (dorsal and ventral) (*Cambrian to Recent*)

    Class Inarticulata . . . . . . . . . . . . . . . . . . . . Primitive brachiopods having phosphatic or chitinous valves, lacking hinge (*Cambrian to Recent*)

    Class Articulata . . . . . . . . . . . . . . . . . . . . Advanced calcareous brachiopods with valves that are hinged (*Cambrian to Recent*)

PHYLUM PHORONIDA. . . . . . . . . . . . . . . . . . . Wormlike marine animals that secrete and live within a leathery tube (*early Mesozoic to Recent*)

PHYLUM ANNELIDA . . . . . . . . . . . . . . . . . . . The segmented worms (*Proterozoic to Recent*)

PHYLUM ONYCOPHORA. . . . . . . . . . . . . . . . . . Rare tropical animals considered intermediate between annelids and arthropods (*Cambrian to Recent*)

PHYLUM ARTHROPODA . . . . . . . . . . . . . . . . . . Segmented animals with jointed appendages

Subphylum Trilobita . . . . . . . . . . . . . . . . . . . Trilobites, common marine arthropods of the Paleozoic Era (*Cambrian to Permian*)

Subphylum Chelicerata

    Class Xiphosura . . . . . . . . . . . . . . . . . . . . Horseshoe crabs (*Silurian to Recent*)

    Class Eurypterida. . . . . . . . . . . . . . . . . . . . Eurypterids (*Ordovician to Permian*)

    Class Pycnogonida . . . . . . . . . . . . . . . . . . . Sea spiders

    Class Arachnida . . . . . . . . . . . . . . . . . . . . Scorpions, spiders, ticks, and mites

Subphylum Crustacea . . . . . . . . . . . . . . . . . . . Lobsters, crabs, barnacles, and ostracodes (*Cambrian to Recent*)

Subphylum Labiata

    Class Chilopoda. . . . . . . . . . . . . . . . . . . . Centipedes

    Class Diplopoda. . . . . . . . . . . . . . . . . . . . Millipedes

    Class Insecta. . . . . . . . . . . . . . . . . . . . . The 24 orders of insects (*Devonian to Recent*)

PHYLUM MOLLUSCA. . . . . . . . . . . . . . . . . . . . Unsegmented, soft-bodied animals, usually with shells

    Class Monoplacophora. . . . . . . . . . . . . . . . . . Primitive forms with cap-shaped shells (*Cambrian to Recent*)

    Class Amphineura . . . . . . . . . . . . . . . . . . . Chitons, marine forms with shells composed of eight segments

    Class Scaphopoda . . . . . . . . . . . . . . . . . . . Tusk shells, curved tubular shells open at both ends

    Class Gastropoda. . . . . . . . . . . . . . . . . . . . Snails, abalones, asymmetric animals with single spiral conch or no shell (*Cambrian to Recent*)

    Class Pelecypoda (Bivalvia) . . . . . . . . . . . . . . . Shells of two valves (right and left); includes clams, mussels, oysters, and scallops (*Cambrian to Recent*)

    Class Cephalopoda. . . . . . . . . . . . . . . . . . . Marine animals with tentacles around head and well-developed eyes and nervous system

        ORDER NAUTILOIDEA . . . . . . . . . . . . . . . . . . Nautiloids; cephalopods with simple suture lines (*Cambrian to Recent*)

        ORDER AMMONOIDEA. . . . . . . . . . . . . . . . . . Ammonoids; cephalopods with complexly folded sutural lines (*Devonian to Cretaceous*)

        ORDER BELEMNOIDEA . . . . . . . . . . . . . . . . . Belemnites (*late Mississippian to early Tertiary*)

        ORDER SEPIOIDEA . . . . . . . . . . . . . . . . . . . Cuttlefishes (*Jurassic to Recent*)

saurs, ichthyosaurs, and placodonts (*Permian to Cretaceous*)

*Subclass Diapsida* . . . . . . . . . . . . . . . . . . . . . . . Reptilian group that includes the extinct dinosaurs and crocodilians, lizards, snakes, and the modern tuatara (*Permian to Recent*)

Class Aves. . . . . . . . . . . . . . . . . . . . . . . . . . . Birds; warm-blooded, feathered, and typically winged animals; primitive forms with reptilian teeth, modern forms toothless (*Jurassic to Recent*)

Class Mammalia . . . . . . . . . . . . . . . . . . . . . . . Warm-blooded animals with hair covering; females with mammary glands that secrete milk for nourishing young (*Triassic to Recent*)

*Subclass Eotheria* . . . . . . . . . . . . . . . . . . . . . . Primitive, extinct Triassic and Jurassic mammals

*Subclass Prototheria* . . . . . . . . . . . . . . . . . . . . Monotremes such as the duck-billed platypus and spiny anteater. Egg-laying mammals (*Triassic to Recent*)

*Subclass Allotheria* . . . . . . . . . . . . . . . . . . . . . Extinct early mammals with multicusped teeth; multituberculates (*Triassic to early Cenozoic*)

*Subclass Metatheria* . . . . . . . . . . . . . . . . . . . . Pouched mammals or marsupials (*Cretaceous to Recent*)

*Subclass Eutheria* . . . . . . . . . . . . . . . . . . . . . . The placental mammals; young develop within uterus of female, obtain moisture via the placenta (*Cretaceous to Recent*)

    ORDER INSECTIVORA. . . . . . . . . . . . . . . . . Primitive insect-eating mammals, including moles and shrews (*Cretaceous to Recent*)

    ORDER DERMOPTERA . . . . . . . . . . . . . . . . . The colugo

    ORDER CHIROPTERA. . . . . . . . . . . . . . . . . . Bats (*early Tertiary to Recent*)

    ORDER PRIMATES . . . . . . . . . . . . . . . . . . . Lemurs, tarsiers, monkeys, apes, and humans (*early Tertiary to Recent*)

    ORDER EDENTATA . . . . . . . . . . . . . . . . . . . Living armadillos, anteaters, and tree sloths; extinct glyptodonts and ground sloths

    ORDER RODENTIA . . . . . . . . . . . . . . . . . . . Squirrels, mice, rats, beavers, and porcupines

    ORDER LAGOMORPHA . . . . . . . . . . . . . . . . . Hares, rabbits, and pikas

    ORDER CETACEA . . . . . . . . . . . . . . . . . . . . Whales and porpoises (*early Tertiary to Recent*)

    ORDER CREODONTA. . . . . . . . . . . . . . . . . . Extinct, ancient carnivorous placentals

    ORDER CARNIVORA . . . . . . . . . . . . . . . . . . Modern carnivorous placentals, including dogs, cats, bears, hyenas, seals, sea lions, and walruses

    ORDER CONDYLARTHA . . . . . . . . . . . . . . . . Extinct ancestral hoofed placentals (ancestral ungulates) (*early Tertiary*)

    ORDER AMBLYPODA . . . . . . . . . . . . . . . . . Extinct primitive ungulates (*early Tertiary*)

    ORDER TUBULIDENTATA . . . . . . . . . . . . . . . Aardvarks

    ORDER PHOLIDOTA . . . . . . . . . . . . . . . . . . Pangolins

    ORDER PERISSODACTYLA . . . . . . . . . . . . . . . Odd-toed hoofed mammals, including living horses, rhinoceroses, and tapirs; and extinct titanotheres and chalicotheres (*early Tertiary to Recent*)

    ORDER ARTIODACTYLA . . . . . . . . . . . . . . . . Even-toed hoofed animals, including living antelopes, cattle, deer, giraffes, camels, llamas, hippos, and pigs; and extinct entelodonts and oreodonts (*early Tertiary to Recent*)

    ORDER PROBOSCIDEA . . . . . . . . . . . . . . . . . Elephants and extinct mastodons and mammoths (*early Tertiary to Recent*)

    ORDER SIRENIA . . . . . . . . . . . . . . . . . . . . Sea cows

# Appendix B

# Formation Correlation Charts For Representative Sections

The following charts may be used to ascertain the approximate equivalence of some of the better known rock formations deposited during the Phanerozoic Eon. The vertical scale does not indicate thickness but rather is an approximation of time. Time–stratigraphic terms are indicated on either side of the chart. The diagonal lines indicate formations missing by nondeposition or erosion. Fm = formation (generally including two or more lithologic types); Ss = sandstone; Sh = shale; Ls = limestone; Dol = dolomite; Qtzite = quartzite; Congl = conglomerate; Grp = group (two or more formations); Mem = member. Solid horizontal lines between the formations indicate well-established boundaries, whereas broken lines indicate formation boundaries that are not firmly established.

## CAMBRIAN

| SERIES | STAGES | Southern Appalachian | Missouri, Ozark Region | Arbuckle, Wichita Mtns., Okla. | Sawback Range, British Columbia | House Range, Utah | Nopah Range, Calif. | Northern Wales | | STAGES | SERIES |
|---|---|---|---|---|---|---|---|---|---|---|---|
| CROIXAN | Trempealeauan | Copper Ridge Dol | Eminence Fm / Potosi Dol | Butterfly Dol / Signal Mtn Fm / Royer Dol / Fort Sill Lms | Lyell Fm | Notch Peak Fm | | | Dolgelly Beds | Trempealeauan | UPPER |
| CROIXAN | Franconian | Copper Ridge Dol | Elvins Grp (Doe Run Fm / Derby Fm / Davis Fm) | Fort Sill Lms / Honey Creek Fm | Lyell Fm | Orr Fm | Nopah Fm | Lingula Flags | Festiniog Beds | Franconian | UPPER |
| CROIXAN | Dresbachian | Conasauga Fm | Bonneterre Dol / Lamotte Ss | Reagan Ss | Sullivan Fm | Weeks Fm | | | Maentwrog Beds | Dresbachian | UPPER |
| ALBERTAN | Stages not defined | Conasauga Fm | | | Eldon Dol | Marjum Fm | Cornfield Sprs. Fm | Menevian Beds | Clogan Shales | Stages not defined | MIDDLE |
| ALBERTAN | Stages not defined | Conasauga Fm | | | Stephen Fm | Wheeler Sh / Swasey Ls / Whirlwind Fm / Dome Ls / Chisolm Sh | Bonanza King Fm | | Cefn Goch Grit | Stages not defined | MIDDLE |
| ALBERTAN | Stages not defined | Conasauga Fm | | | Cathedral Dol / Ptarmigan Fm | Howell Ls / Tatow Fm | Cadiz Fm | Harlech Grits | | Stages not defined | MIDDLE |
| WAUCOBAN | Stages not defined | Rome Fm | | | Mount Whyte Fm | Pioche Fm | Wood Canyon Fm | Harlech Grits | | Stages not defined | LOWER |
| WAUCOBAN | Stages not defined | Shady Dol | | | St. Piran Ss / Fort Mtn Ss | Prospect Mtn. Qtzite | Stirling Qtzite | | | Stages not defined | LOWER |
| WAUCOBAN | Stages not defined | Weisner Qtzite | | | | | | | | Stages not defined | LOWER |

ORDOVICIAN

| SERIES | STAGES | Taconic Area of Western Vt. | Northwestern New York | Virginia | Missouri | Oklahoma | Colorado | Utah | GREAT BRITAIN SERIES |
|---|---|---|---|---|---|---|---|---|---|
| CINCINNATIAN | Gamachian | | | | | | | | ASHGILL |
| CINCINNATIAN | Richmondian | | Queenston Sh (Red beds) | Juniata Fm | Girardeau Ls / Orchard Ck Sh / Thebes Ss / Maquoketa Sh / Fernvale Ls | Polk Ck Sh | Fremont Ls | Fish Haven Dol | ASHGILL |
| CINCINNATIAN | Maysvillian | | Oswego Ss | Martinsburg Fm | | | | | |
| CINCINNATIAN | Edenian | | Lorraine Grp | Martinsburg Fm | | | | | |
| CHAMPLAINIAN (Mohawkian) | Trentonian | Snake Hill Sh / Rysedorf Congl | Utica Sh / Trenton Ls | Bays Fm | Kimmswick Ls / Decorah Grp | Big Fork Chert | Harding Ss | | CARADOC |
| CHAMPLAINIAN (Mohawkian) | Black River | Normanskill Sh | Black River Ls | Edinburg Fm | Plattin Grp | | | | CARADOC |
| CHAMPLAINIAN | Chazyan | Deepkill Sh | | Lincolnshire Ls / Whistle Ck Sh / New Market Ls | Joachim Ls / St. Peter Ss / Everton Fm | Womble Sh | | Swan Peak Qtzite | LLANDEILO / LLANVIRN |
| CANADIAN | Stages not yet established | Deepkill Sh | | Smithville Fm / Powell Fm / Cotter Dol / Knox Dol (upper beds) | Smithville Fm / Powell Fm / Cotter Dol / Jefferson City Grp / Roubidoux Fm | Blakely Ss / Mazarn Sh | Manitou Ls | Garden City Ls | ARENIG |
| CANADIAN | Stages not yet established | Shaghticoke Sh | | | Gasconade Dol | Crystal Mtn Ss | | | TREMADOC |

**A10**

## SILURIAN

| SERIES IN NORTH AMERICA | Eastern Penn. | Western N.Y. | W. Ohio to E. Kentucky | Oklahoma | Northern Ill. | Central Nevada | Southern Wales | STAGES IN EUROPE |
|---|---|---|---|---|---|---|---|---|
| CAYUGAN | Keyser Group: Manlius Ls / Rondout Ls / Decker Ls; Bossardville Ls; Poxono Island Sh; Bloomsburg red beds | CoblesKill Dol; Salina Grp | Bass Island Group | Missouri Mtn. Slate; Henryhouse Sh | | | Red Marls; Temeside beds; Downton Castle Ss | PRIDOLIAN |
| NIAGARAN | Shawangunk Congl | Albemarle Group: Guelph Dol / Lockport Dol; Decew Dol; Rochester Sh; Irondequoit Lms; Merriton Ls | Peebles Dol; Durbin Grp; Lilley Fm; Bisher Fm; Ribolt Sh; Estill Sh | St. Clair Ls; Blaylock Ss | Racine Dol; Waukesha Dol; Joliet Dol | Lone Mtn. Ls; Roberts Mtn. Fm | Upper Ludlow Sh; Aymestry Ls; Lower Ludlow Sh; Wenlock Ls; Wenlock Sh | LUDLOVIAN / WENLOCKIAN |
| MEDINIAN | | Neahga Sh; Thorold Ss; Grimsby Ss; Power Glen Fm; Whirlpool Ss | Neland Fm; Brassfield Ls; Centerville Clay | Chimneyhill Ls | Kankakee Dol; Edgewood Dol | | Llandovery Sh; Llandovery Ss | LLANDOVERIAN |

# DEVONIAN

| SERIES NA | STAGES NA | New York Catskill Mtns. | New York Western | Virginia, Md., Penn. | Western Tenn. | Nevada (Eureka Area) | Devon, England | Scotland | STAGES IN EUROPE |
|---|---|---|---|---|---|---|---|---|---|
| UPPER | BRADFORDIAN | | Oswayo Grp | | | | Pilton Beds | | FAMENNIAN |
| | | | Cattaraugus Congl | Catskill Red beds | Chattanooga Shale | Pilot Shale | Baggy Beds | Upper Old Red Ss | |
| | CASSADAGIAN | | Conneaut Grp | | | | Upcott Beds | | |
| | | | Canadaway Grp | | | | Pickwell Downs Ss | | |
| | | | Java Fm | | | | Morte Slates | | |
| | COHOC-TONIAN | Wittenburg Congl | West Falls Fm | Chemung Ss | | | | | FRASNIAN |
| | FINGER-LAKESIAN | Walton Sh | Sonyea Fm | Brallier Sh | | Devils Gate Ls | | | |
| | | Oneonta Fm | Genesee Fm | Harrell Fm | | | Ilfracombe Beds | | |
| MIDDLE | TAGHAN-ICAN | | | Burket Sh | | | | | GIVETIAN |
| | | | | Tully Fm | | | | | |
| | TIOUGH-NIOGAN | Hamilton Grp | Hamilton Grp | Mahantango Fm | | | | | |
| | CAZE-NOVIAN | | | Marcellus Sh | | | Hangman Grits | | |
| | ONESQUETH-AWAN | Onondaga Ls | Onondaga Ls | Onondago Ls | | Nevada Fm | Lynton Beds | | EIFELIAN |
| LOWER | ESPUSIAN | Schohari Fm | Bois Blanc Fm | Needmore Sh | Camden Chert | | | | EMSIAN |
| | | Carlisle Fm | | | | | | | |
| | | Esopus Sh | | | | | | | |
| | DEER-PARKIAN | Oriskany Ss | | Oriskany Ss | Harriman Fm | | | Lower Old Red Ss | SIEGENIAN |
| | | | | Shriver Chert | Flat Gap Ls | | | | |
| | | | | Licking Ck Ls | | | | | |
| | HELDERBERGIAN | Helderberg Grp | | New Scotland Ls | | | | | GEDINNIAN |
| | | | | Elbow Ridge Ss | Ross Fm | | | | |
| | | Rondout Ls | | Keyser Ls | | | | | |

# MISSISSIPPIAN

| STAGES IN N.A. | West Va. & Penn. | Alabama | Mississippi Valley (Composite sequence) | | Ohio | Arkansas Ouachita Mtns. | Montana (Central) | Lower Carb. England (Bristol) | STAGES IN EUROPE |
|---|---|---|---|---|---|---|---|---|---|
| CHESTERIAN | Mauch Chunk Sh | Pennington Sh | Elvira Group | | | Jackfork Ss | Lower Amsden Fm | Upper Cromwell Ss | VISEAN |
| | | | | Vienna Ls | | Stanley Sh | | | |
| | | | | Tar Sprs. Ss | | Hot Springs Ss | Heath Fm | | |
| | | Bangor Ls | | Glen Dean Ls | | | | | |
| | | | | Hardinsburg Ss | | | | | |
| | | | Gasper Ls | Golconda Fm | | | Otter Fm | Hotwells Ls | |
| | | | | Cypress Ss | | | | | |
| | | | | Paint Ck Fm | | | | | |
| | | | | Bethel Ss | | | | | |
| | | | | Renault Ls | | | | | |
| | | | | Aux Vases Ss | | | | | |
| MERAMECIAN | | Ste. Genevieve Ls | Ste. Genevieve Ls | | Maxville Ss | | Kibby Fm | Clifton Down Ls | |
| | | | St. Louis Ls | | | | | | |
| | | Tuscumbia Ls | Salem Ls | | | | Charles Ls | | |
| | | | Warsaw Ls | | | | | | |
| OSAGEAN | | Fort Payne Chert | Keokuk Ls | | Logan Fm | | Mission Canyon Ls | Goblin Combe Oolite | |
| | | | Burlington Ls | | | | | Clifton Down Muds | |
| | | | Fern Glen Fm | | Cuyahoga Fm | | Woodhurst Ls | Gully Oolite | |
| | Pocono Grp | | Gilmore City Ls | | | | | Black Rock Fm | |
| KINDERHOOKIAN | | | Sedalia Ls | | Sunbury Sh | | | | |
| | | | | | Berea Ss | | | | TOURNAISIAN |
| | | | Chouteau Ls | | Bedford Sh | | Paine Ls | | |
| | | | Maple Hill Ls | | | Arkansas Novaculite (Upper & Middle Members) | | Lower Limestone Sh | |
| | | Chattanooga Sh (upper) | Dev-Miss. Unassigned | Louisiana Ls | | | | | |
| | | | | Saverton Sh | | | | Shire Hampton Beds | |
| | | | | Grassy Creek Sh | | | | | |

*(Note: "Big Snowy Group" and "Madison Group" appear as vertical group labels spanning the Montana (Central) column.)*

PENNSYLVANIAN

| Stages in N.A. | Pennsylvania | West Virginia | Arkansas (Ouachita Mtns.) | Southern Illinois | Missouri | Colorado | Northeast England (Upper Carboniferous) | Stages in Europe |
|---|---|---|---|---|---|---|---|---|
| VIRGILIAN | | Monongahela Series | | | Waubaunsee Grp | | | STEPHANIAN |
| VIRGILIAN | | Conemaugh Series | | | Shawnee Grp | | | STEPHANIAN |
| VIRGILIAN | | Conemaugh Series | | | Douglas Grp | | | STEPHANIAN |
| MISSOURIAN | Conemaugh Series | Conemaugh Series | | McLeansboro Fm | Pedee Grp | | | STEPHANIAN |
| MISSOURIAN | Conemaugh Series | Conemaugh Series | | McLeansboro Fm | Lansing Grp | | | STEPHANIAN |
| MISSOURIAN | Conemaugh Series | Conemaugh Series | | McLeansboro Fm | Kansas City Grp | | | STEPHANIAN |
| MISSOURIAN | Conemaugh Series | Conemaugh Series | | McLeansboro Fm | Pleasanton Grp | | | STEPHANIAN |
| DESMOINESIAN | Allegheny Series | Allegheny Series | | | Marmaton Grp | Fountain Fm | Upper Coal Measures | WESTPHALIAN |
| DESMOINESIAN | Allegheny Series | Allegheny Series | | Carbondale Fm | Cherokee Grp | Fountain Fm | Middle Coal Measures | WESTPHALIAN |
| ATOKAN | Pottsville Series | Pottsville Series | Atoka Fm | Tradewater Fm | Riverton Fm | Fountain Fm | Lower Coal Measures | WESTPHALIAN |
| ATOKAN | Pottsville Series | Pottsville Series | Atoka Fm | Tradewater Fm | Burgner Fm | Fountain Fm | Lower Coal Measures | WESTPHALIAN |
| ATOKAN | Pottsville Series | Pottsville Series | Atoka Fm | Tradewater Fm | McLouth Fm | Fountain Fm | Lower Coal Measures | WESTPHALIAN |
| ATOKAN | Pottsville Series | Pottsville Series | Atoka Fm | Caseyville Fm | Cheltenham Fm | Glen Eyrie Sh | Lower Coal Measures | WESTPHALIAN |
| MORROWAN | Pottsville Series | Pottsville Series | Johns Valley Sh | Caseyville Fm | Hale Fm | Glen Eyrie Sh | | NAMURIAN |
| MORROWAN | | | Jackfork Ss | Caseyville Fm | | Glen Eyrie Sh | | NAMURIAN |
| MORROWAN | | | Stanley Sh | | | | Millstone Grit | NAMURIAN |
| MORROWAN | | | Hot Springs Ss | | | | Millstone Grit | NAMURIAN |

PERMIAN

| Stages in N.A. | Ohio and W. Va. | Western Texas | Arizona | Wyoming | South Africa | Germany | Russia (Ural Mtns.) | Stages in Europe |
|---|---|---|---|---|---|---|---|---|
| OCHOAN | | Dewey Lake Fm | | | | | Tartarian | TARTARIAN (in part) |
| | | Rustler Fm | | | | Zechstein Fm | | |
| | | Salado Fm | | | | | Kazanian | KAZANIAN |
| | | Castile Fm | | | Beaufort Series | | | |
| GUADALUPIAN | | Bell Canyon Fm | | | | Upper Rotliegend | Kungurian | KUNGURIAN |
| | | Cherry Canyon Fm | | Phosphoria Fm | | | | |
| | | Brushy Canyon Fm | | | | | | |
| LEONARDIAN | | Bone Spring Ls | Kaibab Ls | | Ecca Series | Lower Rotliegend | Artinskian | ARTINSKIAN |
| | | | Toroweap Fm | | | | | |
| | | | Coconino Ss | | | | | |
| | | | Hermit Sh | | | | | |
| WOLFCAMPIAN | Dunkard Grp | Hueco Ls | Supai Fm | | Dwyka Series | | Sakmarian | SAKMARIAN |
| | | | | | | | | ASSELIAN |

# TRIASSIC

| NA SERIES | STAGES | Connecticut Valley | Pennsylvania (York Co.) | Wyoming | Utah-Idaho | Colo.-S. Dak. | Arizona | Nevada | Germany (Type Section) |
|---|---|---|---|---|---|---|---|---|---|
| UPPER | Rhaetian | Portland Arkose (Newark Grp) | Gettysburg Fm (Newark Grp) | Nugget Ss | Ankareh Fm | Dolores Fm (Redbeds) | Chinle Fm | Gabbs Fm | Rhät |
| UPPER | Norian | Meriden Fm | | Popo Agie Mem | Deadman Ls | | | | Gyps-keuper |
| UPPER | Carnian | New Haven Arkose | New Oxford Fm | Crow Mtn. Ss / Alcova Ls | Higman Grit (Congl) | | Shinarump Congl | Luning Fm | Letten-kohle |
| MIDDLE | Ladinian | | | Chugwater Fm | | | | Grantsville Fm | Muschelkalk |
| MIDDLE | Anisian | | | | | | | Excelsior Fm | |
| LOWER | Sythian | | | Red Peak Mem / Dinwoody Fm | Thaynes Fm (Woodside Ss Mem / Portneuf Ls Mem / Woodside Red beds / Dinwoody Fm) | | Moenkopi Fm | Candelaria Fm | Bunter |

## JURASSIC

| SERIES | STAGES | Gulf of Mexico Region | Utah | Idaho | Wyoming | California Coast Ranges | Alberta Canada | England |
|---|---|---|---|---|---|---|---|---|
| UPPER | Portlandian | Cotton Valley Grp — Shuler Fm | | | | Knoxville Fm | Basal Ss of Kootenay Fm | Purbeck Beds |
| UPPER | Portlandian | | | | | | | Portland Beds |
| UPPER | Kimmeridgian | Bossier Fm | Morrison Fm | | Morrison Fm | Franciscan Fm | "Passage Beds" | Kimmeridge Clay |
| UPPER | Oxfordian | Buckner Fm | Summerville Fm | Stump Ss | | | | Corallian Beds |
| UPPER | Oxfordian | Smackover Fm | Curtis Fm | | | | | |
| MIDDLE | Callovian | Eagle Mills Fm | Entrada Ss | Preuss Ss | Sundance Fm | | "Green Beds" | Oxford Clay |
| MIDDLE | Callovian | | | | | | (Shale) | Kellaways Beds |
| MIDDLE | Bathonian | | Carmel Fm | Twin Creek Ls | Gypsum Springs Fm | | Gryphaea Bed | Cornbrash Beds |
| MIDDLE | Bathonian | | | | | | "Corbula Munda Bed" | Great Oolite |
| MIDDLE | Bajocian | | | | | | Lille Mem? | Inferior Oölite |
| MIDDLE | Bajocian | | | | | | Sandstones | |
| LOWER | Toarcian | | Navajo Ss | Nugget Ss | | | Sandstones | Upper Lias |
| LOWER | Toarcian | | | | | | Base Congl | Middle Lias |
| LOWER | Pliensbachian | | | | | | | |
| LOWER | Sinemurian | | | | | | | Lower Lias |
| LOWER | Hettangian | | Kayenta Fm | | | | | |

CRETACEOUS

| SERIES | STAGES | New Jersey | Alabama (Tuscaloosa) | Texas | Kansas (Wallace Co.) | Colorado | Montana | England |
|---|---|---|---|---|---|---|---|---|
| UPPER CRETACEOUS | Maastrichtian | Monmouth Grp | Prairie Bluff / Ripley Fm | Navarro Grp: Escondido Fm / Corsicana Marl | Pierre Sh | Animas Fm / McDermott Fm / Kirkland Fm / Fruitland Fm / Picture Cliffs Ss / Lewis Sh | Hell Ck Fm / Fox Hills Ss / Bear Paw Sh | Upper Chalk |
| UPPER CRETACEOUS | Campanian | Matawan Grp: Wenonah Sand / Marshalltown Fm / Englishtown Sand / Woodbury Clay / Merchantville Clay | Selma Chalk | Taylor Marl | Pierre Sh | Mesaverde Grp | Judith River Fm / Claggett Sh | Upper Chalk |
| UPPER CRETACEOUS | Santonian | Magothy Fm | | Austin Chalk | Niobrara Fm | Mancos Sh | Eagle Ss / Telegraph Ck Fm | Upper Chalk |
| UPPER CRETACEOUS | Coniacian | Magothy Fm | Eutaw Fm | Austin Chalk | Niobrara Fm | Mancos Sh | | Upper Chalk |
| UPPER CRETACEOUS | Turonian | | | Eagle Ford Shale | Carlile Sh / Greenhorn Ls | Mancos Sh | Colorado Sh | Middle Chalk |
| UPPER CRETACEOUS | Cenomanian | Raritan Fm | Tuscaloosa Fm | Woodbine Ss | Graneros Sh / Dakota Ss | Dakota Ss | Colorado Sh | Lower Chalk |
| LOWER CRETACEOUS | Albian | | | Washita Grp / Fredericksburg Grp | Kiowa Sh / Cheyenne Ss | Dakota Ss | Kootenai Fm | Upper Greensand |
| LOWER CRETACEOUS | Aptian | | | Trinity Grp | | Burro Canyon Fm | Kootenai Fm | Lower Greensand |
| LOWER CRETACEOUS | Neocomian: Barremian | | | | | | | Wealden Beds |
| LOWER CRETACEOUS | Neocomian: Hauterivian | | | | | | | Wealden Beds |
| LOWER CRETACEOUS | Neocomian: Valanginian | | | | | | | Wealden Beds |
| LOWER CRETACEOUS | Neocomian: Berriasian | | | | | | | Wealden Beds |

## CENOZOIC

| "SYSTEM" | SERIES | STAGES | Gulf Coast | Atlantic Coast | Colorado Wyoming | S. Dakota Nebraska | California (S. Great Valley) | Western Washington | Paris Basin |
|---|---|---|---|---|---|---|---|---|---|
| Quaternary | Hol. | Unnamed Holocene | Recent stream and lake deps | Pamlico Fm | Recent sediments | Recent alluvial deposits | Recent alluvial deposits | Recent alluvium | Recent alluvial deposits |
| Quaternary | Pleistocene | Wisconsinan | Prairie Fm | Talbot Fm | | Pleistocene sand & clay | Local terrace deposits | Marine & river terrace deposits | |
| Quaternary | Pleistocene | Sangamon | Montgomery Fm | Wicomico Fm | | | | | |
| Quaternary | Pleistocene | Illinoian | Bentley Fm | Sunderland Fm | | | | | |
| Quaternary | Pleistocene | Yarmouth | | | | | Tulare Fm | | |
| Quaternary | Pleistocene | Kansan | Williana Fm | Brandywine Fm | | | | | |
| Quaternary | Pleistocene | Aftonian | | | | | | | |
| Quaternary | Pleistocene | Nebraskan | | | | | | | |
| Tertiary | Pliocene | Astian | Citronelle Fm | Wacamaw Fm | | Ogallala Grp | San Joaquin Fm | | |
| Tertiary | Pliocene | Piacenzan | | | | | Etchegoin Fm | | |
| Tertiary | Pliocene | Piacenzan | | | | | Jacalitos Fm | | |
| Tertiary | Miocene | Pontian | | Yorktown Fm | | Sheep Ck Fm | Reef Ridge Sh | | |
| Tertiary | Miocene | Sarmatian | Choctawatchee Fm | | | | McClure Sh | | |
| Tertiary | Miocene | Tortonian | | St. Mary's Fm | | Marsland Fm | | | |
| Tertiary | Miocene | Helvetian | Shoal River Fm | Choptank Fm | | | Temblor Fm | Astoria Fm | Touraine Marls |
| Tertiary | Miocene | Burdigalian | Oak Grove Ss | Calvert Fm | | Arikaree Grp | | | |
| Tertiary | Miocene | Burdigalian | Chipola Fm | Hawthorn Fm | | | | | Orléanais Sands |
| Tertiary | Miocene | Aquitainian | Catahoula Ss | Trent Marl | | | | Twin Rivers Fm | |
| Tertiary | Oligocene | Chattian | | Flint River | | White River Series | | Blakely Fm | Beauce Ls |
| Tertiary | Oligocene | Rupelian | Vicksburg Ls | | Florissant Fm | | | Lincoln Fm | Stampian Ss |
| Tertiary | Oligocene | Tongrian | | | White River Grp | | | | Sannosian Ss |
| Tertiary | Eocene | Ludian | Jackson Fm | | Duchesne River Fm | | Tumey Ss | | "Marls" |
| Tertiary | Eocene | Bartonian | | Castle Hayne Marl | Uinta Fm | | Kreyenhagen Sh | Cowlitz Fm | Cresnes and Beauchamp sands |
| Tertiary | Eocene | Auversian | Yegua Fm / Cook Mtn. Fm / Sparta Ss (Claiborne Grp) | Shark River Marl | Bridger Fm | | | | |
| Tertiary | Eocene | Lutetian | Mt. Selma Fm (Claiborne Grp) | | | | Domingine Fm | | Provins Ls |
| Tertiary | Eocene | Cuisian | Carrizo Ss (Claiborne Grp) | Manasquan Marl | Green River Fm | | Yokut Ss | | |
| Tertiary | Eocene | Ypresian | Wilcox Grp | Aquia Fm | Cedar Breaks Fm | | Lodo Fm | Metachosin Volcanics | Cuise sands |
| Tertiary | Paleocene | Thanetian | Wills Pt. Fm (Midway Grp) | Clayton Ls | Fort Union Fm | Fort Union Fm | | | Plastic Clay / Bracheux sands |
| Tertiary | Paleocene | Montian | Kincaid Fm (Midway Grp) | | | | | | Meudon Ls |
| Tertiary | Paleocene | Danian | | | | | | | |

LANDFORMS OF THE UNITED STATES
by ERWIN RAISZ

Central Stable Region
1. Canadian Shield
2. Interior Plains and
   Plateaus

Appalachian and
Related belts
3a. New England
3b. Piedmont
4. Ridge and Valley
5. Ouachita

Cordilleran Belt
6. Southern Rockies
7. Northern Rockies
8. Columbia Province
9. Colorado Plateaus
10. Basin and Range
11. Sierra Nevada and
    Cascade Range
12. Pacific Coast

Coastal Plains

# Appendix D

# Periodic Table and Symbols for Chemical Elements

Scientists generally group or organize things that are similar so as to better understand them and determine how they relate to one another. An important step in the organization of elements was made in 1869 by Dimitri Mendeleev. Mendeleev showed that when elements are arranged in order of their increasing atomic weights, their physical and chemical properties tend to be repeated in cycles. The arrangement of elements was depicted on a chart called the periodic table of elements, a modern version of which is shown as Table D–1. In the periodic table, each box contains the symbol of the element and its atomic number. Except for two long sequences set apart at the bottom, the elements appear in the increasing order of their atomic numbers. The vertical columns are called *groups* and contain elements with similar properties. For example, in column VIIA, one finds fluorine, chlorine, bromine, and iodine. All of these elements are colored and highly reactive and share other similarities. Fluorine, however, is chemically the most active, chlorine somewhat less active, bromine still less active, and iodine the least active of the four. Except for hydrogen, all of the elements in group IA are soft, shiny metals and very reactive chemically. The horizontal rows on the table are called *periods* and contain sequences of elements having electron configurations that vary in characteristic patterns. It is thus apparent that the periodic table shows relationships among elements rather well and certainly better than an arbitrary listing.

Table D–2 provides the chemical names for the symbols of the elements used in the periodic table.

# Table D–1  Periodic Table of the Elements

Key:

| 26 | Atomic number (Z) |
|---|---|
| **Fe** | Element symbol |
| 55.847 | Atomic mass of naturally occurring isotopic mixture; for radioactive elements, numbers in parentheses are mass numbers of most stable isotopes |

| IA | IIA | IIIB | IVB | VB | VIB | VIIB | VIII | VIII | VIII | IB | IIB | IIIA | IVA | VA | VIA | VIIA | O |
|---|---|---|---|---|---|---|---|---|---|---|---|---|---|---|---|---|---|
| 1<br>**H**<br>1.0079 | | | | | | | | | | | | | | | | 1<br>**H**<br>1.0079 | 2<br>**He**<br>4.00260 |
| 3<br>**Li**<br>6.941 | 4<br>**Be**<br>9.01218 | | | | | | | | | | | 5<br>**B**<br>10.81 | 6<br>**C**<br>12.011 | 7<br>**N**<br>14.0067 | 8<br>**O**<br>15.9994 | 9<br>**F**<br>18.998403 | 10<br>**Ne**<br>20.179 |
| 11<br>**Na**<br>22.98977 | 12<br>**Mg**<br>24.305 | | | | | | | | | | | 13<br>**Al**<br>26.98154 | 14<br>**Si**<br>28.0855 | 15<br>**P**<br>30.97376 | 16<br>**S**<br>32.06 | 17<br>**Cl**<br>35.453 | 18<br>**Ar**<br>39.948 |
| 19<br>**K**<br>39.0983 | 20<br>**Ca**<br>40.08 | 21<br>**Sc**<br>44.9559 | 22<br>**Ti**<br>47.90 | 23<br>**V**<br>50.9415 | 24<br>**Cr**<br>51.996 | 25<br>**Mn**<br>54.9380 | 26<br>**Fe**<br>55.847 | 27<br>**Co**<br>58.9332 | 28<br>**Ni**<br>58.70 | 29<br>**Cu**<br>63.546 | 30<br>**Zn**<br>65.38 | 31<br>**Ga**<br>69.72 | 32<br>**Ge**<br>72.59 | 33<br>**As**<br>74.9216 | 34<br>**Se**<br>78.96 | 35<br>**Br**<br>79.904 | 36<br>**Kr**<br>83.80 |
| 37<br>**Rb**<br>85.4678 | 38<br>**Sr**<br>87.62 | 39<br>**Y**<br>88.9059 | 40<br>**Zr**<br>91.22 | 41<br>**Nb**<br>92.9064 | 42<br>**Mo**<br>95.94 | 43<br>**Tc**<br>(98) | 44<br>**Ru**<br>101.07 | 45<br>**Rh**<br>102.9055 | 46<br>**Pd**<br>106.4 | 47<br>**Ag**<br>107.868 | 48<br>**Cd**<br>112.41 | 49<br>**In**<br>114.82 | 50<br>**Sn**<br>118.69 | 51<br>**Sb**<br>121.75 | 52<br>**Te**<br>127.60 | 53<br>**I**<br>126.9045 | 54<br>**Xe**<br>131.30 |
| 55<br>**Cs**<br>132.9054 | 56<br>**Ba**<br>137.33 | 57<br>***La**<br>138.9055 | 72<br>**Hf**<br>178.49 | 73<br>**Ta**<br>180.9479 | 74<br>**W**<br>183.85 | 75<br>**Re**<br>186.207 | 76<br>**Os**<br>190.2 | 77<br>**Ir**<br>192.22 | 78<br>**Pt**<br>195.09 | 79<br>**Au**<br>196.9665 | 80<br>**Hg**<br>200.59 | 81<br>**Tl**<br>204.37 | 82<br>**Pb**<br>207.2 | 83<br>**Bi**<br>208.9804 | 84<br>**Po**<br>(209) | 85<br>**At**<br>(210) | 86<br>**Rn**<br>(222) |
| 87<br>**Fr**<br>(223) | 88<br>**Ra**<br>226.0254 | 89<br>†**Ac**<br>227.0278 | 104<br>**Unq**<br>(261) | 105<br>**Unp**<br>(262) | 106<br>**Unh**<br>(263) | 107<br>**Uns** | 108 | 109 | | | | | | | | | |

*Lathanide Series

| 58<br>**Ce**<br>140.12 | 59<br>**Pr**<br>140.9077 | 60<br>**Nd**<br>144.24 | 61<br>**Pm**<br>(145) | 62<br>**Sm**<br>150.4 | 63<br>**Eu**<br>151.96 | 64<br>**Gd**<br>157.25 | 65<br>**Tb**<br>158.9254 | 66<br>**Dy**<br>162.50 | 67<br>**Ho**<br>164.9304 | 68<br>**Er**<br>167.26 | 69<br>**Tm**<br>168.9342 | 70<br>**Yb**<br>173.04 | 71<br>**Lu**<br>174.967 |
|---|---|---|---|---|---|---|---|---|---|---|---|---|---|

†Actinide Series

| 90<br>**Th**<br>232.0381 | 91<br>**Pa**<br>231.0359 | 92<br>**U**<br>238.029 | 93<br>**Np**<br>237.0482 | 94<br>**Pu**<br>(244) | 95<br>**Am**<br>(243) | 96<br>**Cm**<br>(247) | 97<br>**Bk**<br>(247) | 98<br>**Cf**<br>(251) | 99<br>**Es**<br>(252) | 100<br>**Fm**<br>(257) | 101<br>**Md**<br>(258) | 102<br>**No**<br>(259) | 103<br>**Lr**<br>(260) |
|---|---|---|---|---|---|---|---|---|---|---|---|---|---|

Note: Atomic masses shown here are 1977 IUPAC values.

## Table D–2  Elements and Their Chemical Symbols

| | | | | | | | |
|---|---|---|---|---|---|---|---|
| Actinium | Ac | Erbium | Er | Mercury | Hg | Samarium | Sm |
| Aluminum | Al | Europium | Eu | Molybdenum | Mo | Scandium | Sc |
| Americium | Am | Fermium | Fm | Neodymium | Nd | Selenium | Se |
| Antimony | Sb | Fluorine | F | Neon | Ne | Silicon | Si |
| Argon | Ar | Francium | Fr | Neptunium | Np | Silver | Ag |
| Arsenic | As | Gadolinium | Gd | Nickel | Ni | Sodium | Na |
| Astatine | At | Gallium | Ga | Niobium | Nb | Strontium | Sr |
| Barium | Ba | Germanium | Ge | Nitrogen | N | Sulfur | S |
| Berkelium | Bk | Gold | Au | Nobelium | No | Tantalum | Ta |
| Beryllium | Be | Hafnium | Hf | Osmium | Os | Technetium | Tc |
| Bismuth | Bi | Helium | He | Oxygen | O | Tellurium | Te |
| Boron | B | Holmium | Ho | Palladium | Pd | Terbium | Tb |
| Bromine | Br | Hydrogen | H | Phosphorus | P | Thallium | Tl |
| Cadmium | Cd | Indium | In | Platinum | Pt | Thorium | Th |
| Calcium | Ca | Iodine | I | Plutonium | Pu | Thulium | Tm |
| Californium | Cf | Iridium | Ir | Polonium | Po | Tin | Sn |
| Carbon | C | Iron | Fe | Potassium | K | Titanium | Ti |
| Cerium | Ce | Krypton | Kr | Praseodymium | Pr | Tungsten | W |
| Cesium | Cs | Lanthanum | La | Promethium | Pm | Uranium | U |
| Chlorine | Cl | Lawrencium | Lr | Protactinium | Pa | Vanadium | V |
| Chromium | Cr | Lead | Pb | Radium | Ra | Wolfram | |
| Cobalt | Co | Lithium | Li | Radon | Rn | Xenon | Xe |
| Copper | Cu | Lutetium | Lu | Rhenium | Re | Ytterbium | Yb |
| Curium | Cm | Magnesium | Mg | Rhodium | Rh | Yttrium | Y |
| Dysprosium | Dy | Manganese | Mn | Rubidium | Rb | Zinc | Zn |
| Einsteinium | Es | Mendelevium | Md | Ruthenium | Ru | Zirconium | Zr |

# Appendix E

## Convenient Conversion Factors

| To Convert from | To | Multiply by[*] |
|---|---|---|
| Centimeters | Feet | 0.0328 ft/cm |
| | Inches | 0.394 in/cm |
| | Meters | 0.01 m/cm |
| | Microns (micrometers) | 10,000 $\mu$m/cm |
| | Miles (statute) | $6.214 \times 10^{-6}$ mi/cm |
| | Millimeters | 10 mm/cm |
| Feet | Centimeters | 30.48 cm/ft |
| | Inches | 12 in/ft |
| | Meters | 0.3048 m/ft |
| | Microns (micrometers) | 304800 $\mu$m/ft |
| | Miles (statute) | 0.000189 mi/ft |
| | Yards | 0.3333 yd/ft |
| Kilometers | Miles | 0.6214 mi/km |
| Gallons (U.S., liq.) | Cu. centimeters | 3785 cm$^3$/gal |
| | Cu. feet | 0.133 ft$^3$/gal |
| | Liters | 3.785 L/gal |
| | Quarts (U.S., liq.) | 4 qt/gal |
| Grams | Kilograms | 0.001 kg/g |
| | Micrograms | $1 \times 10^6$ $\mu$g/g |
| | Ounces | 0.03527 oz/g |
| | Pounds | 0.002205 lb/g |
| Inches | Centimeters | 2.54 cm/in |
| | Feet | 0.0833 ft/in |
| | Meters | 0.0254 m/in |
| | Yards | 0.0278 yd/in |
| Kilograms | Ounces | 35.27 oz/kg |
| | Pounds | 2.205 lb/kg |
| Meters | Centimeters | 100 cm/m |
| | Feet | 3.2808 ft/m |
| | Inches | 39.37 in/m |
| | Kilometers | 0.001 km/m |
| | Miles (statute) | 0.0006214 mi/m |
| | Millimeters | 1000 mm/m |
| | Yards | 1.0936 yd/m |
| Miles (statute) | Centimeters | 160934 cm/mi |
| | Feet | 5280 ft/mi |
| | Inches | 63360 in/mi |
| | Kilometers | 1.609 km/mi |
| | Meters | 1609 m/mi |
| | Yards | 1760 yd/mi |
| Ounces | Grams | 28.3 g/oz |
| | Pounds | 0.0625 lb/oz |
| Pounds | Grams | 453.6 g/lb |
| | Kilograms | 0.454 kg/lb |
| | Ounces | 16 oz/lb |

[*]Values are generally given to three or four significant figures.
(From Turk, J., and Turk, A. *Physical Science*. Philadelphia: W.B. Saunders Company, 1977.)

# Glossary

*For terms not included in this glossary, students may wish to consult the* Dictionary of Geologic Terms, *published in 1976 by Anchor/Doubleday under the direction of the American Geological Institute. A more comprehensive reference is the* Glossary of Geology, *edited by R. L. Bates and J. A. Jackson, 2d ed., Falls Church, Virginia, American Geological Institute.*

**Absaroka sequence**  A sequence of Permian–Pennsylvanian sediments bounded both above and below by a regional unconformity and recording an episode of marine transgression over an eroded surface, full flood level of inundation, and regression from the craton.

**Absolute geologic age**  The actual age, expressed in years, of a geologic material or event.

**Acadian orogeny**  An episode of mountain building in the northern Appalachians during the Devonian Period.

**Acanthodians**  The earliest known vertebrates (fishes) with a movable, well-developed lower jaw, or mandible; hence, the first jawed fishes.

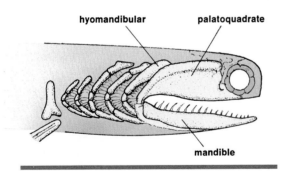

Acanthodian

**Adaptation**  A modification of an organism that better fits it for existence in its present environment or enables it to live in a somewhat different environment.

**Adaptive radiation**  The diversity that develops among species as each adapts to a different set of environmental conditions.

**Adenosine diphosphate (ADP)**  A product formed in the hydrolysis of adenosine triphosphate that is accompanied by release of energy and organic phosphate.

**Adenosine triphosphate (ATP)**  A compound that occurs in all cells and that serves as a source of energy for physiologic reactions such as muscle contraction.

**Aerobic organism**  An organism that uses oxygen in carrying out respiratory processes.

**Age**  The time represented by the time-stratigraphic unit called a stage. (Informally, may indicate any time span in geologic history, as "age of cycads.")

**Agnatha**  The jawless vertebrates, including extinct ostracoderms and living lampreys and hagfishes.

**Algae**  Any of a large group of simple plants (thallophyta) that contain chlorophyll and are capable of photosynthesis.

**Allegheny orogeny**  The late Paleozoic episodes of mountain building along the present trend of the Appalachian Mountains.

**Alluvium**  Unconsolidated, poorly sorted detrital sediments ranging from clay to gravel sizes and characteristically fluvial in origin.

**Alpha particle**  A particle, equivalent to the nucleus of a helium atom, emitted from an atomic nucleus during radioactive decay.

**Alpine orogeny**  In general, the sequence of crustal disturbances beginning in the middle Mesozoic and continuing into the Miocene that resulted in the geologic structures of the Alps.

**Amino acids**   Nitrogenous hydrocarbons that serve as the building blocks of proteins and are thus essential to all living things.

**Ammonites**   Ammonoid cephalopods having more complex sutural patterns than either ceratites or goniatites.

**Ammonoids**   An extinct group of cephalopods, with coiled, chambered conch(s) and having septa with crenulated margins.

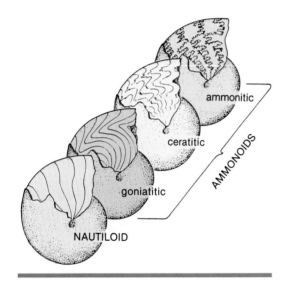

Ammonoids

**Amniotic egg**   That type of egg produced by reptiles, birds, and monotremes. In this type, the developing embryo is maintained and protected by an elaborate arrangement of shell membranes, yolk, sac, amnion, and allantois.

**Amphibians**   "Cold-blooded" vertebrates that utilize gills for respiration in the early life stages but that have air-breathing lungs as adults.

**Amphibole**   A ferromagnesium silicate mineral that occurs commonly in igneous and metamorphic rocks.

**Anaerobic organism**   An organism that does not require oxygen for respiration, but rather makes use of processes such as fermentation to obtain its energy.

**Andesite**   A volcanic rock that in chemical composition is intermediate between basalt and granite.

**Angiosperms**   An advanced group of plants having floral reproductive structures and seeds in a closed ovary. The "flowering plants."

**Anthropoidea**   The suborder of primates that includes monkeys, apes, and humans.

**Anticline**   A geologic structure in which strata are bent into an upfold or arch.

Anticline

**Antler orogeny**   A Late Devonian and Mississippian episode of mountain building involving folding and thrusting along a belt across Nevada to southwestern Alberta.

**Archean**   Division of Precambrian time beginning 3.8 billion years ago and ending 2.5 billion years ago.

**Archaeocyatha**   A group of extinct marine organisms having double, perforated, calcareous, conical-to-cylindric walls. Archaeocyathids lived during the Cambrian.

**Archosaurs**   Advanced reptiles of a group called diapsids, which includes thecodonts, "dinosaurs," pterosaurs, and crocodiles.

**Arcoids**   A group of pelecypods exemplified by species of *Arca*.

Arca

**Artiodactyl**   Hoofed mammals that typically have two or four toes on each foot.

**Asteroid**   One of numerous relatively small planetary bodies (less than 800 km in diameter) revolving around the sun in orbits lying between those of Mars and Jupiter.

**Asthenosphere**  A zone between 50 and 250 km below the surface of the earth where shock waves of earthquakes travel at much reduced speeds, perhaps because of less rigidity. The asthenosphere may be a zone where convective flow of material occurs.

**Atoll**  A ringlike island or a series of islands formed by corals and calcareous algae around a central lagoon.

**Atom**  The smallest divisible unit retaining the characteristics of a specific element.

**Atomic fission**  A nuclear process that occurs when a heavy nucleus splits into two or more lighter nuclei, simultaneously liberating a considerable amount of energy.

**Atomic fusion**  A nuclear process that occurs when two light nuclei unite to from a heavier one. In the process, a large amount of energy is released.

**Atomic mass**  A quantity essentially equivalent to the number of neutrons plus the number of protons in an atomic nucleus.

**Atomic number**  The number of protons in the nuclei of atoms of a particular element. (An element is thus a substance in which all of the atoms have the same atomic number.)

**Australopithecines**  A general term applied loosely to Pliocene and early Pleistocene primates whose skeletal characteristics place them between typically apelike individuals and those more obviously human.

**Autotroph**  An organism that uses an external source of energy to produce organic nutrients from simple inorganic chemicals.

**Basin**  A depressed area that serves as a catchment area for sediments (basin of deposition). A structural basin is an area in which strata slope inward toward a central location. Structural basins tend to experience periodic downsinking and thus receive a thicker and more complete sequence of sediments than do adjacent areas.

**Belemnites**  Members of the molluscan Class Cephalopoda, having straight internal shells.

**Benioff seismic zone**  An inclined zone along which frequent earthquake activity occurs and that marks the location of the plunging, forward edge of a lithospheric plate during subduction.

**Benthic**  A botton-dwelling organism.

**Bentonite**  A layer of clay, presumably formed by the alteration of volcanic ash.

**Beta particle**  A charged particle, essentially equivalent to an electron, emitted from an atomic nucleus during radioactive disintegration.

**Bivalvia**  A class of the Phylum Mollusca also known as the Class Pelecypoda. (The term Pelecypoda is preferred in this text so that the pelecypods are not confused with other "bivalves," such as brachiopods and ostracods.)

**Blastoids**  Sessile (attached) Paleozoic echinoderms having a stem and an attached cup or calyx composed of relatively few plates.

Blastoid

**Brachiating**  Swinging from branch to branch and tree to tree by using the limbs, as among monkeys.

**Brachiopod**  Bivalved (doubled-shelled) marine invertebrates. They were particularly common and widespread during the Paleozoic and persist in fewer numbers today.

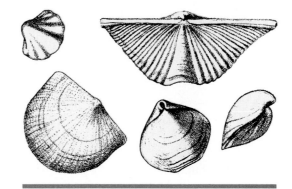

Brachiopods

**Breccia**   A clastic sedimentary rock composed largely of angular fragments of granule size or larger.

Breccia

**Breeder reactor (nuclear)**   An atomic reactor that uses uranium-238, but that creates additional fuel by producing more fissionable material than it consumes.

**Bryozoa**   A phylum of attached and incrusting colonial marine invertebrates.

Fossil Bryozoa

**Burgess Shale fauna**   A beautifully preserved fossil fauna of soft-bodied Cambrian animals discovered in 1910 by Charles Walcott in Kicking Horse Pass, Alberta.

**Caledonian orogeny**   A major early Paleozoic episode of mountain building affecting Europe, which created an orogenic belt, the Caledonides, extending from Ireland and Scotland northwestward through Scandinavia.

**Carbon-14**   A radioactive isotope of carbon with an atomic mass of 14. Carbon-14 is frequently used in determining the age of materials less than about 50,000 years old.

**Carbonate**   A general term for a chemical compound formed when carbon dioxide dissolved in water combines with oxides of calcium, magnesium, potassium, sodium, and iron. The most common carbonate minerals are calcite (which forms the carbonate rock limestone) and dolomite.

**Carbonization**   The concentration of carbon during fossilization.

**Cast (natural)**   A replica of an organic object, such as a fossil shell, formed when sediment fills a mold of that object.

**Cataract Group**   A group of formations deposited during the Early Silurian and including the Whirlpool Sandstone, Manitoulin Dolomite, Cabot Head Shale, Dyer Bay Limestone, and Wingfield Shale of southern Ontario. The Cataract Group is correlative with the Albion Group of western New York.

**Catskill delta**   A buildup of Middle and Upper Devonian clastic sediments as a broad, complex clastic wedge derived from the erosion of highland areas formed largely during the Acadian orogeny.

**Ceboidea**   The New World monkeys, characterized by prehensile tails, and including the capuchin, marmoset, and howler monkeys.

**Centrifugal force**   The apparent outward force experienced by an object moving in a circular path. Centrifugal force is a manifestation of inertia, the tendency of moving things to travel in straight lines.

**Ceratites**   One of the three larger groups of ammonoid cephalopods having sutural complexity intermediate between goniatites and ammonites.

**Ceratopsians** The quadrupedal ornithischian dinosaurs characterized by the development of prominent horns on the head.

**Cercopithecoidea** The Old World monkeys (of Asia, southern Europe, and Africa), including macaques, guenons, langurs, baboons, and mandrills.

**Cetaceans** The group of marine mammals that includes whales and porpoises.

**Chalicotheres** Extinct perissodactyls having robust claws rather than hoofs.

**Chalk** A soft, white, fine-grained variety of limestone composed largely of the calcium carbonate skeletal remains of marine microplankton.

**Chert** A dense, hard sedimentary rock or mineral composed of submicrocrystalline quartz. Unless colored by impurities, chert is white, as opposed to flint, which is dark or black.

**Chlorophyll** The catalyst that makes possible the reaction of water and carbon dioxide in green plants to produce carbohydrates. Photosynthesis is the reaction.

**Choanichthyes** That group of fishes that includes both the dipnoans (lungfishes with weak pelvic and pectoral fins) and crossopterygians (lungfishes with stout lobe-fins).

**Chondrichthyes** The broad category of fishes with cartilaginous skeletons that is exemplified by sharks, skates, and rays.

The Devonian chondrichthyan (shark)
*Cladoselache*

**Chondrites** Stony meteorites that contain rounded silicate grains or chondrules. Chondrules are believed to have formed by crystallization of liquid silicate droplets.

**Chromosome** Threadlike microscopic bodies composed of chromatin. Chromosomes appear in the nucleus of the cell at the time of cell division. They contain the genes. The number of chromosomes is normally constant for a particular species.

**Chromosphere** One of the concentric shells of the sun, lying above the photosphere and telescopically visible as a thin, brilliant red rim around the edge of the sun for a second or so at the beginning and end of a solar eclipse.

**Clastic texture** Texture that characterizes a rock made up of fragmental grains such as sand, silt, or parts of fossils. Conglomerates, sandstones, and siltstones are clastic rocks; the individual clastic grains are termed *clasts*.

**Clastic wedge** An extensive accumulation of largely clastic sediments deposited adjacent to uplifted areas. Sediments in the wedge become finer and the section becomes thinner in a direction away from the upland source areas. The Queenston and Catskill "deltas" are examples of clastic wedges.

**Coccolithophorids** Marine, planktonic, biflagellate, golden brown algae that typically secrete coverings of discoidal calcareous platelets called *coccoliths*.

Coccolithophorid

**Colorado mountains** Highlands uplifted in Pennsylvanian time in Colorado. Sometimes inappropriately termed the "ancestral Rockies."

**Conodonts** Small, toothlike fossils composed of calcium phosphate and found in rocks ranging from Cambrian to Triassic in age. Although the precise nature of the conodont-producing organism has not been determined, this uncertainty does not detract from their usefulness as guide fossils.

**Contact metamorphism** The compositional and textural changes in a rock that result from heat and pressure emanating from an adjacent igneous intrusion.

**Convergence (in evolution)** The process by which similarity of form or structure arises among different organisms as a result of their becoming adapted to similar habitats.

**Cordaites** A primitive order of treelike plants with long, bladelike leaves and clusters of naked seeds. Cordaites were in some ways intermediate in evolutionary stage between seed ferns and conifers.

Fossil cordaite leaves

**Core** Central part of the earth that lies beneath the mantle.

**Coriolis effect** The deflection of winds and water currents to the right in the Northern Hemisphere and to the left in the Southern Hemisphere as a consequence of the earth's rotation.

**Cosmic rays** Extremely high-energy particles, mostly protons, which move through the galaxy and frequently strike the earth's atmosphere.

**Craton** The long-stable region of a continent, commonly with Precambrian rocks either at the surface or only thinly covered with younger sedimentary rocks.

**Creodonta** Primitive, early, flesh-eating placental mammals.

**Crinoids** Stalked echinoderms with a calyx composed of regularly arranged plates from which radiate arms for gathering food.

**Cross-bedding (cross-stratification)** An arrangement of laminae or thin beds transverse to the planes of stratification. The inclined laminae are usually at inclinations of less than 30° and may be either straight or concave.

Cross-bedding

**Crossopterygii** That group of choanichthyan fishes ancestral to earliest amphibians and characterized by stout pectoral and pelvic fins as well as lungs.

The Devonian crossopteryian *Eusthenopteron*

**Crust (earth)** The outer part of the lithosphere; it averages about 32 km in thickness.

**Crustacea** A subphylum of the Phylum Arthropoda that includes such well-known living animals as lobsters and crayfishes.

**Curie temperature** The temperature at which a cooling mineral acquires permanent magnetic properties that record the surrounding magnetic field orientation and strength at the time of cooling. Magnetic properties of a mineral above the Curie temperature will change as the surrounding field changes. Below the Curie temperature, magnetic characteristics of the mineral will not alter.

**Cycadales** A group of seed plants that were especially common during the Mesozoic and were characterized by palmlike leaves and coarsely textured trunks marked by numerous leaf scars.

**Cyclothem** A vertical succession of sedimentary units reflecting environmental events that occurred in a constant order. Cyclotherms are particularly characteristic of the Pennsylvanian System.

**Cystoids**   Attached echinoderms with generally irregular arrangement and number of plates in the calyx and perforated by pores or slits.

**Deccan traps**   A thick sequence (3200 meters) of Upper Cretaceous basaltic lava flows that cover about 500,000 sq km of peninsular India.

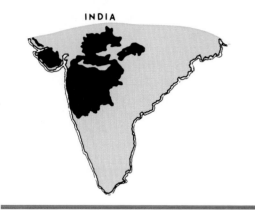

Deccan traps (outcrop map)

**Décollement**   Feature of stratified rocks in which upper formations may become "unstuck" from lower formations, deform, and slide thousands of meters over underlying beds.

**Diatoms**   Microscopic   golden   brown   algae (chrysophytes) that secrete a delicate siliceous frustule (shell).

Diatom

**Differentiation**   The process by which a planet becomes internally zoned as when heavy materials sink toward its center and light materials accumulate near the surface.

**Dinoflagellates**   Unicellular marine algae usually having two flagella and a cellulose wall.

**Diploid cells**   Cells having two sets of chromosomes that form pairs, as in somatic cells.

**Dipnoi**   An order of lungfishes with weak pectoral and pelvic fins; not considered ancestral to land vertebrates.

**Disconformity**   A variety of unconformity in which bedding planes above and below the plane of erosion or nondeposition are parallel.

**DNA**   The nucleic acid found chiefly in the nucleus of cells that functions in the transfer of genetic characteristics and in protein synthesis.

**Docodonts**   A group of small, primitive Late Jurassic mammals possibly ancestral to the living monotremes.

**Dolomite (or dolostone)**   A carbonate sedimentary rock that contains more than 50 percent of the mineral dolomite, CaMg $(CO_3)_2$.

**Dome**   An upfold in rocks having the general configuration of an inverted bowl. Strata in a dome dip outward and downward in all directions from a central area. An example is the Ozark dome.

**Dryopithecine**   In general, a group of lightly built primates that lived during the Miocene and Pliocene in mostly open savannah country and that includes *Dryopithecus,* a form considered to be in the line leading to apes.

**Dynamothermal (regional) metamorphism**   Metamorphism that has occurred over a wide region, caused by deep burial and high temperatures associated with pressures resulting from overburden and orogeny.

**Echinoderms**   The large group (Phylum Echinodermata) of marine invertebrates characterized by prominent pentamerous symmetry and skeleton frequently constructed of calcite elements and including spines. Cystoids, blastoids, crinoids, and echinoids are examples of echinoderms.

**Edentata**   An order of placental mammals that includes extinct ground sloths and gyptodonts, as well as living armadillos, tree sloths, and South American anteaters.

**Elastic limit**   The greatest amount of stress that can be applied to a body without causing permanent strain.

**Endemic population**   The native fauna of any particular region.

**Entelodonts**   A group of extinct artiodactyls bearing a superficial resemblance to giant wild boars.

Entelodont

**Eocambrian**   Name applied to the poorly fossiliferous sequence of sedimentary rocks that lie generally above Precambrian basement rocks but below readily identifiable fossiliferous Cambrian strata.

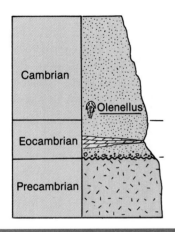

Stratigraphic position of the Eocambrian

**Eon**   A major division of the geologic time scale. All of the geologic periods from the Cambrian to the Holocene comprise the Phanerozoic Eon. The term is also sometimes used to denote a span of 1 billion years.

**Epifaunal organisms**   Organisms living *on,* as distinct from *in,* a particular body of sediment or another organism.

**Epoch**   A chronologic subdivision of a geologic period. Rocks deposited or emplaced during an epoch constitute the *series* for that epoch.

**Era**   A major division of geologic time, divisible into geologic periods.

**Eukaryote**   A type of living cell containing a true nucleus, enclosed within a nuclear membrane, and having well-defined chromosomes and cell organelles.

**Eupantotheres**   A group of Jurassic mammals with dentition similar to that of primitive representatives of later marsupial and placental mammals and thus thought to be ancestral to these groups.

**Eurypterids**   Aquatic arthropods of the Paleozoic, superficially resembling scorpions, and probably carnivorous.

**Eustatic**   Pertaining to worldwide simultaneous changes in sea level, such as might result from change in the volume of continental glaciers.

**Evaporites**   Sediments precipitated from a water solution as a result of the evaporation of that water. Evaporite minerals include anhydrite gypsum ($CaSO_4$) and halite ($NaCl$).

**Evolution**   The continuous genetic adaptation of organisms or species to the environment.

**Facies**   A particular aspect of sedimentary rocks that is a direct consequence of sedimentation in a particular depositional environment.

**Fault**   A fracture in the earth's crust along which rocks on one side have been displaced relative to rocks on the other side.

Fault in dipping strata

**Fermentation**   The partial breakdown of organic compounds by an organism in the absence of oxygen. The final product of fermentation is alcohol or lactic acid.

**Filter-feeders (organisms)**   An animal that obtains its food, which usually consists of small particles or organisms, by filtering it from the water.

**Fissility**   That property of rocks that causes them to split into thin slabs parallel to bedding. Fissility is particularly characteristic of shale.

**Flood basalts**   Regionally extensive layers of basalt that originated as low-viscosity lava pouring from fissure eruptions. The lavas of the Colombia Plateau and the Deccan Plateau are flood basalts.

**Fluvial**   Pertaining to sediments or other geologic features formed by streams.

**Flysch**   Term describing thick sequences of rapidly deposited, poorly sorted marine clastics.

**Focus (earthquake)**   The location at which rock rupture occurs so as to generate seismic waves.

**Foliation**   A textural feature especially characteristic of metamorphic rocks in which laminae develop by growth or realignment of minerals in parallel orientation.

**Foraminifers**   An order of mostly marine, unicellular protozoans that secrete tests (shells) that are usually composed of calcium carbonate.

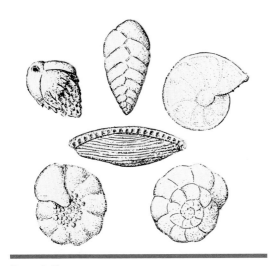

Foraminifers

**Formation**   A mappable, lithologically distinct body of rock having recognizable contacts with adjacent rock units.

**Fossil**   The remains or indications of an organism that lived in the geologic past.

**Fractional crystallization**   The separation of components of a cooling magma by sequential formation of particular mineral crystals at progressively lower temperatures.

**Fusulinids**   Primarily spindle-shaped foraminifers with calcareous, coiled tests divided into a complex of numerous chambers. Fusulinids were particularly abundant during the Pennsylvanian and Permian periods.

**Galaxy**   An aggregate of stars and planets, separated from other such aggregates by distances greater than those between member stars.

**Gamete**   Either of two cells (male or female) that must unite in sexual reproduction to initiate the development of a new individual.

**Gamma rays**   Very high-frequency electromagnetic waves.

**Garnet**   A family of aluminosilicates of iron and calcium that are particularly characteristic of metamorphic rocks.

Garnet crystal

**Gene**   The unit of heredity transmitted in the chromosome.

**Genus**   The major subdivision of a taxonomic family or subfamily of plants or animals, usually consisting of more than one species.

**Geochronology**   The study of time as applied to earth and planetary history.

**Geologic range**   The geologic time span between the first and last appearance of an organism.

**Glacier**   A large mass of ice, formed by the recrystallization of snow, that flows slowly under the influence of gravity.

**Glauconite**   A green clay mineral frequently found in marine sandstones and believed to have formed at the site of deposition.

**Glossopteris flora**   An assemblage of fossil plants found in rocks of late Paleozoic and Early Triassic age in South Africa, India, Australia, and South

America. The flora takes its name from the seed fern *Glossopteris*.

**Gondwanaland**    The great Permo-Carboniferous Southern Hemisphere continent, comprising the assembled present continents of South Africa, India, Australia, Africa–Arabia, and Antarctica.

**Goniatites**    One of the three large groups of ammonoid cephalopods with sutures forming a pattern of simple lobes and saddles and thus not as complex as either the ceratites or the ammonites.

**Gowganda Conglomerate**    An apparent tillite of the Canadian Shield. The Gowganda rests upon a surface of older rock that appears to have been polished by glacial action.

**Graptolite facies**    A Paleozoic sedimentary facies composed of dark shales and fine-grained clastics that contain the abundant remains of graptolites and that are associated with volcanic rocks.

**Graptolites**    Extinct colonial marine invertebrates considered to be protochordates. Graptolites range from the Late Cambrian to the Mississipian.

**Gravity anomaly**    The differences between the observed value of gravity at any point on the earth and the calculated theoretic value.

**Greenhouse effect**    A process in which incoming solar radiation that is absorbed and reradiated cannot escape back into space because the earth's atmosphere is not transparent to the reradiated energy, which is in the form of infrared radiation (heat).

Greenhouse effect

**Greenstones**    Low-grade metamorphic rocks containing abundant chlorite, epidote, and biotite and developed by metamorphism of basaltic extrusive igneous rocks. Great linear outcrops of greenstones are termed greenstone belts and are thought to mark the locations of former volcanic island arcs.

**Guide fossil**    Fossil with a wide geographical distribution but narrow stratigraphic range and thus useful in correlating strata and for age determination.

**Gutenberg discontinuity**    The boundary separating the mantle of the earth from the core below. The Gutenberg discontinuity lies about 2900 km below the surface.

**Gymnosperms**    An informal designation for flowerless seed plants in which seeds are not enclosed (hence "naked seeds"). Examples are conifers and cyads.

**Hadrosaurs**    The ornithischian duck-billed dinosaurs of the Cretaceous.

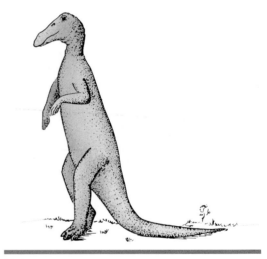

Hadrosaur

**Half-life**    The time in which one-half of an original amount of a radioactive species decays to daughter products.

**Haploid cell**    One having a single set of chromosomes, as in gametes (see *diploid*).

**Hercynian orogeny**    A major late Paleozoic orogenic episode in Europe that formed the ancient Hercynian mountains. Today, only the eroded stumps of these mountains are exposed in areas where the cover of Mesozoic and Cenozoic strata has been removed by erosion.

**Heterotroph**    An organism that depends upon an external source of organic substances for its nutrition and energy.

**Holocene**    A term sometimes used to designate the period of time since the last major episode of glaciation. The term is equivalent to Recent.

**Homologous organs**    Organs having structural and developmental similarities due to genetic relationship.

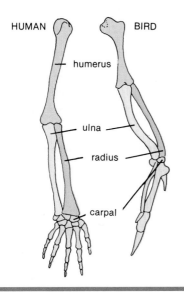

Homologous bones in forelimbs of human and bird

**Hydrosphere**    The water and water vapor present at the surface of the earth, including oceans, seas, lakes, and rivers.

**Hylobatidae**    A group of persistently arboreal, small apes that are exemplified by the gibbons and siamangs.

**Hyomandibular bone (in fishes)**    The modified upper bone of the hyoid arch, which functions as a connecting element between the jaws and the brain case in certain fishes.

**Ichthyosaurs**    Highly specialized marine reptiles of the Mesozoic, recognized by their fishlike form.

**Ictidosauria**    A group of extinct mammal-like reptiles or therapsids whose skeletal characteristics are considered to be very close to those of mammals.

**Igneous rock**    A rock formed by the cooling and solidification of magma or lava.

Igneous rocks

**Infaunal organisms**    Organisms that live and feed within bottom sediments.

**Ion**    An atom that, because of electron transfers, has excess positive or negative charges.

**Island arcs**    Chains of islands, arranged in arcuate trends on the surface of the earth. The arcs are sites of volcanic and earthquake activity and are usually bordered by deep oceanic trenches on the convex side.

**Isopachous map**    A map depicting the thickness of a sedimentary unit.

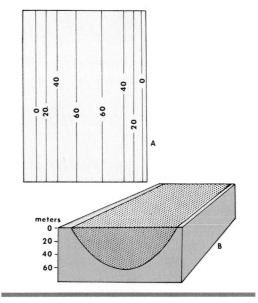

Isopach map (A) of sandstone body (B)

**Isotope**   Atoms of an element that have the same number of protons in the nucleus, the same atomic number, and the same chemical properties but different atomic masses because they have different numbers of neutrons in their nuclei.

**Karoo system**   A sequence of Permian to Lower Jurassic rocks, primarily continental formations, which outcrop in Africa and are approximately equivalent to the Gondwana system of peninsular India.

**Kaskaskia sequence**   A sequence of Devonian–Mississippian sediments, bounded above and below by regional unconformities and recording an episode of transgression followed by full flooding of a large part of the craton and by subsequent regression.

**Lacustrine**   Relating to lakes, as in lacustrine sediments (lake sediments).

**Lagomorph**   The order of small placental mammals that includes rabbits, hares, and pikas.

Lagomorph (pika)

**Laramide orogeny**   In general, those pulses of mountain building that were frequent in Late Cretaceous time and were in large part responsible for producing many of the structures of the Rocky Mountains.

**Lateral fault**   A fault in which the movement is largely horizontal and in the direction of the trend of the fault plane. Sometimes called a *strike-slip* fault.

**Laurasia**   A hypothetical supercontinent composed of what is now Europe, Asia, Greenland, and North America.

**Limestone**   A sedimentary rock consisting mainly of calcium carbonate.

**Lithofacies map**   A map that shows the areal variation in lithologic attributes of a stratigraphic unit.

Lithofacies map (A) of time-rock unit (B)

**Lithosphere**   The outer shell of the earth, lying above the asthenosphere and comprising the crust and upper mantle.

**Litoptern**   South American ungulates whose evolutionary history somewhat paralleled that of horses and camels.

**Logan's line**   A zone of thrust-faulting produced during the Taconic orogeny that extends from the west coast of Newfoundland along the trend of the St. Lawrence River to near Quebec and southward along Vermont's western border. (Named after the pioneer Canadian geologist Sir William Logan.)

**Lophophore**   An organ, located adjacent to the mouth of brachiopods and bryozoans, which bears ciliated tentacles and has as its primary function the capture of food articles.

**Low-velocity zone**   An interior zone of the earth characterized by lower seismic wave velocities than the region immediately above it.

**Lunar maria**   Low-lying, dark lunar plains filled with volcanic rocks rich in iron and magnesium.

**Lycopsids**   Leafy plants with simple, closely spaced leaves bearing sporangia on their upper surfaces. They are represented by living club mosses and vast numbers of extinct late Paleozoic "scale trees."

**Lepidodendron (Dev.-Perm.)**      2/5 x

**Sigillaria (Dev.-Trias.)**      2/5 x

Lycopsids

**Magnetic declination** The horizontal angle between "true," or geographic, north and magnetic north, as indicated by the compass needle. Declination is the result of earth's magnetic axis being inclined with respect to the earth's rotational axis.

Magnetic declination

**Magnetic inclination** The angle between the magnetic lines of force for the earth and the earth's surface; sometimes called "dip." Magnetic inclination can be demonstrated by observing a freely suspended magnetic needle. The needle will lie parallel to the earth's surface at the equator but is increasingly inclined toward the vertical as the needle is moved toward the magnetic poles.

**Mammoth** The name commonly applied to extinct elephants of the Pleistocene Epoch.

**Marsupials** Mammals of the Order Marsupialia. Female marsupials bear mammary glands, and carry their immature young in a stomach pouch.

**Mass spectrometer** An instrument that separates ions of different mass but equal charge and measures their relative quantities.

**Mastodon** The group of extinct proboscideans (elephantoids), early forms of which were characterized by long jaws, tusks in both jaws, and low-crowned teeth.

**Medusa** The free-swimming, umbrella-shaped jellyfish form of the Phylum Cnidaria.

**Meiosis** That kind of nuclear division, usually involving two successive cell divisions, that results in daughter cells having one-half the number of chromosomes that were in the original cell.

**Mélange** A body of intricately folded, faulted, and severely metamorphosed rocks, examples of which can be seen in the Franciscan rocks of California.

**Mesosphere** A zone of the earth's mantle where pressures are sufficient to impart greater strength and rigidity to the rock.

**Metamorphic rock** A rock formed from a previously existing rock by subjecting the parent rock to high temperature and pressure but without melting.

**Metamorphism** The transformation of previously existing rocks into new types by the action of heat, pressure, and chemical solutions. Metamorphism usually takes place at depth in the roots of mountain chains or adjacent to large intrusive igneous bodies.

**Metazoa** All multicellular animals whose cells become differentiated to form tissues (all animals except Protozoa).

**Meteorites** Metallic or stony bodies from interplanetary space that have passed through the earth's atmosphere and hit the earth's surface.

**Meteors** Sometimes called "shooting stars," meteors are generally small particles of solid material from interplanetary space that approach close enough to the earth to be drawn into the earth's atmosphere, where they are heated to incandescence. Most disintegrate, but a few land on the surface of the earth as meteorites.

**Micrite** A texture in carbonate rocks that, when viewed microscopically, appears as murky, fine-grained calcium carbonate. Micrite is believed to develop from fine carbonate mud or ooze.

**Milankovitch effect**   The hypothetical long-term effect on world climate caused by three known components of earth motion. The combination of these components provides a possible explanation for repeated glacial to interglacial climatic swings.

**Miliolids**   A group of foraminifers with smooth, imperforate test walls and chambers arranged in various planes around a vertical axis. Miliolids are common in shallow marine areas.

**Mineral**   A naturally occurring element or compound formed by inorganic processes that has a definite chemical composition or range of compositions, as well as distinctive properties and form that reflect its characteristic atomic structure.

**Mitosis**   The method of cell division by means of which each of the two daughter nuclei receives exactly the same complement of chromosomes as had existed in the parent nucleus.

**Mobile belt**   An elongate region of the earth's crust characterized by exceptional earthquake and volcanic activity, tectonic instability, and periodic mountain building.

**Mohorovičić discontinuity**   A plane that marks the boundary separating the crust of the earth from the underlying mantle. The "moho," as it is sometimes called, is at a depth of about 70 km below the surface of continents and 6 to 14 km below the floor of the oceans.

**Molasse**   Accumulations of primarily nonmarine, relatively light-colored, irregularly bedded conglomerates, shales, coal seams, and cross-bedded sandstones that are deposited subsequent to major orogenic events.

**Mold**   An impression, or imprint, of an organism or part of an organism in the enclosing sediment.

**Mollusk**   Any member of the invertebrate Phylum Mollusca, including pelecypods, cephalopods, gastropods, scaphopods, and chitons.

**Monoplacophorans**   Primitive marine molluscans with simple, cap-shaped shells.

**Monotremes**   The egg-laying mammals.

**Morganucodonts**   Early mammals found in Triassic beds of Europe and Asia and characterized by their small size and their retention of certain reptilian osteologic traits.

**Mosasaurs**   Large marine lizards of the Late Cretaceous.

Mosasaur

**Multituberculates**   An early group (Jurassic) of Mesozoic mammals with tooth cusps in longitudinal rows and other dental characteristics that suggest they may have been the earliest herbivorous mammals.

**Mutation**   A stable and inheritable change in a gene.

**Mytiloids**   Pelecypods having rather triangular shells that are most commonly identical but in some forms unequal. The edible mussel *Mytilus* is a representative form.

**Nappe**   A large mass of rocks that have been moved a considerable distance over underlying formations by overthrusting, recumbent folding, or both.

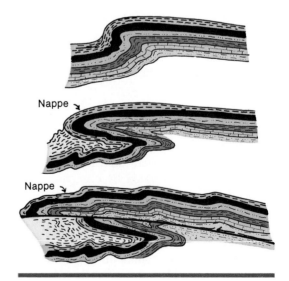

Nappe development

**Nektic**   Pertaining to swimming organisms.

**Neogene**   A subdivision of the Cenozoic that encompasses the Miocene, Pliocene, Pleistocene, and Holocene epochs.

**Neutron**   Electrically neutral (uncharged) particles of matter existing along with protons in the atomic nucleus of all elements except the mass 1 isotope of hydrogen.

**Nevadan orogeny**   In general, those pulses of mountain building, intrusion, and metamorphism that were most frequent during the Jurassic and Early Cretaceous along the western part of the Cordilleran orogenic belt.

**Newark series**   A series of Upper Triassic, nonmarine red beds (shales, sandstones, and conglomerates), lava flows, and intrusions located within downfaulted basins from Nova Scotia to South Carolina.

**New World monkeys**   Monkeys whose habitat is today confined to South America. They are thoroughly arboreal in habit and have prehensile tails by which they hang and swing from tree limbs.

**Nonconformity**   An unconformity developed between sedimentary rocks and older plutons or massive metamorphic rocks that had been exposed to erosion before the overlying sedimentary rocks were deposited.

Nonconformity

**Normal fault**   A fault in which the hanging wall appears to have moved downward relative to the footwall; normally occurring in areas of crustal tension.

**Nothosaurs**   Relatively small early Mesozoic sauropterygians that were replaced during the Jurassic by the plesiosaurs.

**Notochord**   A rod-shaped cord of cartilage cells forming the primary axial structure of the chor-

date body. In vertebrates, the notochord is present in the embryo and is later supplanted by the vertebral column.

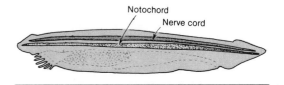

Notochord and dorsal nerve cord in the living lancelet

**Notungulates**   A group of ungulates that diversified in South America and persisted until Plio-Pleistocene time.

**Novaculites**   A term applied originally to rocks suitable for whetstones and, in America, to white chert found in Arkansas. Now applied to very tough, uniformly grained cherts composed of microcrystalline quartz.

**Nuclear fission tracks**   Submicroscopic "tunnels" in minerals produced when high-energy particles from the nucleus of uranium are forcibly ejected during spontaneous fission.

**Nucleic acid**   Any of a group of organic acids that control hereditary processes within cells and make possible the manufacture of proteins from the amino acids ingested by the cells as food.

**Nuclides**   The different weight configurations of an element caused by atoms of that element having differing numbers of neutrons. Nuclides or isotopes of an element differ in number of neutrons but not in chemical properties.

**Nummulites**   Large, coined-shaped foraminifers, especially common in Tertiary limestones.

**Offlap**   A sequence of sediments resulting from a marine regression and characterized by an upward progression from offshore marine sediments (often limestones) to shales and finally sandstones (above which will follow an unconformity).

**Oil shale**   A dark-colored shale rich in organic material that can be heated to liberate gaseous hydrocarbons.

**Old World monkeys**   Monkeys of Asia, Africa, and southern Europe that include macaques, guenons, langurs, baboons, and mandrills.

**Onlap**  A sequence of sediments resulting from a marine transgression. Normally, the sequence begins with a conglomerate or sandstone deposited over an erosional unconformity and followed upward in the vertical section by progressively more offshore sediments.

**Oölites**  Limestones composed largely of small, round or ovate calcium carbonate bodies called oöids.

Oölites

**Ophiolite suite**  An association of radiolarian cherts, pelagic muds, basaltic pillow and flow lavas, gabbros, and ultramafic rocks such as periodotite regarded as surviving masses of former oceanic crust largely destroyed in former subduction zones.

**Oreodonts**  North American artiodactyls of the middle and late Tertiary.

The oreodant *Merycoidodon*

**Orogenic belt**  Great linear tracts of deformed rocks, primarily developed near continental mar-

gins by compressional forces accompanying mountain building.

**Ornithischia**  An order of dinosaurs characterized by birdlike pelvic structures and including such herbivores as the ornithopods, stegosaurs, ankylosaurs, and ceratopsians.

**Orogeny**  The process by which great systems of mountains are formed. (Orogenesis means mountain building.)

**Ostracoderms**  Extinct jawless fishes of the early Paleozoic.

**Ostracods**  Small, bivalved, bean-shaped crustaceans.

Ostracod

**Ostreoids**  The "oyster family" of pelecypods.

The ostreoid *Exogyra* (Jur.-Cret.)

**Outcrop**  An area where specific rock units are exposed at the earth's surface or occur at the surface but are covered by surficial deposits.

**Paleoecology**  The study of the relationship of ancient organisms to their environment.

**Paleogene**  A subdivision of the Cenozoic that encompasses the Paleocene, Eocene, and Oligocene epochs.

**Paleogeography**   The geography as it existed at some time in the geologic past.

**Paleolatitudes**   The latitude that once existed across a particular region at a particular time in the geologic past.

**Paleomagnetism**   The earth's magnetic field and magnetic properties in the geologic past. Studies of paleomagnetism are helpful in determining positions of continents and magnetic poles.

**Paleontology**   The study of all ancient forms of life, their interactions, and their evolution.

**Pangea**   In Alfred Wegener's theory of continental drift, the supercontinent that included all present major continental masses.

**Panthalassa**   The great universal ocean that surrounded the supercontinent Pangea prior to its breakup.

**Paraconformity**   A rather obscure unconformity in which no erosional surface is discernible, and in which beds above and below the break are parallel.

**Partial melting**   Process by which a rock subjected to high temperature and pressure is partially melted, and the liquid fraction moved to another location. Partial melting results from the variations in melting points of different minerals in the original rock mass.

**Pectenoids**   Pelecypods exemplified by the scallops. They have generally subcircular shells and straight hinge lines.

Pecten

**Pelycosaurs**   Early mammal-like reptiles exemplified by the sail-back animals of the Permian Period.

**Period**   A subdivision of an era.

**Perissodactyl**   Progressive, hoofed mammals, characteristically having an odd number of toes on the hind feet and usually on the front feet as well.

**Permineralization**   A manner of fossilization in which voids in an organic structure (such as bone) are filled with mineral matter.

**Petrification**   The process of converting organic structures, such as bone, shell, or wood, into a stony substance, such as calcium carbonate or silica.

**Phanerozoic**   The eon of geologic time during which the earth has been populated by abundant and diverse life. The Phanerozoic Eon followed the Cryptozoic (or Precambrian) Eon and is divided into Paleozoic, Mesozoic and Cenozoic eras.

**Phosphorite**   A sediment composed largely of calcium phosphate.

**Photosphere**   A relatively thin gaseous layer on the sun that emits nearly all the light that the sun radiates into space.

**Photosynthesis**   The process of synthesizing carbohydrates from carbon dioxide and water, utilizing the radiant energy of light captured by the chlorophyll in plant cells.

**Phytoplankton**   Microscopic marine planktonic plants, most of which are various forms of algae.

**Phytosaurs**   Extinct aquatic crocodilelike thecodonts of the Triassic.

Phytosaur skull

**Pillow lava**   Type of lava that is extruded under water and in which many pillow-shaped lobes break through the chilled surface of the flow and solidify. (Resembles a pile of pillows.)

**Pinnipeds**   Marine carnivores such as seals, sea lions, and walruses.

**Placer deposit**   An accumulation of sediment rich in a valuable mineral or metal that has been concentrated because of its greater density.

**Placoderms**   Extinct primitive jawed fishes of the Paleozoic Era.

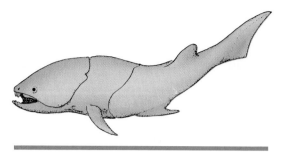

*Coccosteus,* a Devonian placoderm

**Placodonts**    Extinct walruslike marine reptiles that fed principally upon shellfish.

Skull of the placodone *Placodus.* Note blunt teeth for crushing shellfish.

**Plankton**    Minute, free-floating aquatic organisms.

Plankton (foraminifers, diatoms, and radiolaria)

**Plate tectonics**    The theory that explains the tectonic behavior of the crust of the earth in terms of several moving plates that are formed by volcanic activity at oceanic ridges and destroyed along great ocean trenches.

**Platform**    That part of a craton covered thinly by layered sedimentary rocks and characterized by relatively stable tectonic conditions.

**Plesiosaurs**    The group of extinct Mesozoic marine reptiles (sauropterygians) characterized by large, paddle-shaped limbs and broad bodies, with either very long or relatively short necks.

Plesiosaur

**Pluvial lake**    A lake formed in an earlier climate when rainfall was greater than at present.

**Polyp**    The hydralike form of some coelenterates in which the mouth and tentacles are at the top of the body.

**Porphyry**    A textural term used to describe an igneous rock in which some of the crystals, called phenocrysts, are distinctly larger than others.

Porphyry

**Precambrian**    All of geologic time and its corresponding rocks before the beginning of the Paleozoic Era.

**Primary earthquake waves**    Seismic waves that are propagated through solid rock as a train of compressions and dilations. Direction of vibration is parallel to direction of propagation.

**Principle of cross-cutting relations**    Principle that states that geologic features such as faults, veins, and dikes must be younger than the rocks or features they cut across.

**Principle of biologic succession**    The principle that states that the observed sequence of life forms has changed continuously through time, so that the total aspect of life (as recognized by fossil evidence) for a particular segment of time is distinct and different from that of life of earlier and later times.

**Principle of temporal transgression**    The principle that stipulates that sediments of advancing (transgressing) or retreating (regressing) seas are not necessarily of the same geologic age throughout their lateral extent.

**Proboscidea**    The elephants and their progenitors.

**Prokaryotes**    Organisms that lack membrane-bounded nuclei and other membrane-bounded organelles.

**Prosimii**    The less advanced primates, such as lemurs, tarsiers, and tree shrews.

**Proteinoids**    Extra-large organic molecules containing most of the 20 amino acids of proteins and produced in laboratory conditions simulating those found in nature.

**Proteins**    Giant molecules containing carbon, hydrogen, oxygen, nitrogen, and usually sulfur and phosphorus; composed of chains of amino acids and present in all living cells.

**Proton**    An elemental particle found in the nuclei of all atoms that has a positive electric charge and a mass similar to that of a neutron.

**Pyroclastics**    Fragments of volcanic debris that have usually been fragmented during eruptions.

**Pyroxene group**    A group of dark-colored iron and magnesium-rich silicate minerals.

**Queenston delta**    A clastic wedge of red beds shed westward from highlands elevated in the course of the Taconic orogeny.

**Radioactive decay**    The spontaneous emission of a particle from the atomic nucleus, thereby transforming the atom from one element to another.

**Radiolaria**    Protozoa that secrete a delicate, often beautifully filigreed skeleton of opaline silica.

**Recumbent fold**    A fold in which the axial plane is essentially horizontal; a fold that has been turned over by compressional forces so that it lies on its side.

**Red beds**    Prevailing red, usually clastic sedimentary deposits.

**Regolith**    Any solid materials, such as rock fragments or soil, lying on top of bedrock.

**Regression**    A general term signifying that a shoreline has moved toward the center of a marine basin. Regression may be caused by tectonic emergence of the land, eustatic lowering of sea level, or prograding of sediments, as in deltaic build-outs.

**Relative geologic age**    The placing of an event in a time sequence without regard to the absolute age in years.

**Replacement**    A fossilization process in which the original skeletal substance is replaced after burial by inorganically precipitated mineral matter.

**Reverse fault**    A fault formed by compression in which the hanging wall appears to move up relative to the foot wall.

**Rhynchonellids**    A group of brachiopods having pronounced beaks, accordionlike plications, and triangular outline.

**Rift valley**    A valley formed by faulting, usually involving a central fault block that moves downward in relation to adjacent blocks.

Rift valley

**Rudists**    Peculiarly specialized Mesozoic pelecypods often having one valve in the shape of a horn coral, covered by the other valve in the form of a lid.

Rudistoid pelecypods

**Rugosa**  The large group of solitary and colonial Paleozoic horn corals.

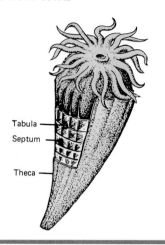

Rugosa

**Ruminant**  A herbivorous, cud-chewing ungulate.

**Salt dome**  A structural dome in sedimentary strata resulting from the upward flow of a large body of salt.

**Sauk sequence**  A sequence of upper Precambrian to Ordovician sediments bounded both above and below by a regional unconformity and recording an episode of marine transgression, followed by full flooding of a large part of the craton, and ending with a regression from the craton.

**Saurischia**  An order of dinosaurs with triradiate pelvic structures, including both the gigantic herbivorous sauropods and the carnivorous theropods.

**Scleractinid coral**  Coral belonging to the Order Scleractinia, which includes most modern and post-Paleozoic corals.

**Sea floor spreading**  The process by which new sea floor crust is produced along mid-oceanic ridges (divergence zones) and slowly conveyed away from the ridges.

Sea floor spreading

**Secondary earthquake wave**  A seismic wave in which the direction of vibration of wave energy is at right angles to the direction the wave travels.

**Sedimentary rock**  A rock that has formed as a result of the consolidation (lithification) of accumulations of sediment.

**Sessile**  The bottom-dwelling habit of aquatic animals that live continuously in one place.

**Series**  The time–rock term representing the rocks deposited or emplaced during a geologic epoch. Series are subdivisions of systems.

**Shelly facies**  In general, sedimentary deposits consisting primarily of carbonate rocks containing the abundant fossil remains of marine invertebrates.

**Silicoflagellates**  Unicellular, tiny, flagellate marine algae that secrete an internal skeleton composed of opaline silica.

Silicoflagellates

**Silicon tetrahedron**  An atomic structure in silicates consisting of a centrally located silicon atom linked to four oxygen atoms placed symmetrically around the silicon at the corners of a tetrahedron.

**Sonoma orogeny**  Middle Permian orogenic movements, the structural effects of which are most evident in western Nevada.

**Sorting**  A measure of the uniformity of the sizes of particles in a sediment or sedimentary rock.

**Spar carbonate**  As viewed microscopically, the clear, crystalline carbonate that has been deposited in a carbonate rock as a cement between clasts or has developed by recrystallization of clasts.

**Species**   A unit of taxonomic classification of organisms. In another sense, a species is a population of individuals that are similar in structural and functional characteristics and that in nature breed only with one another.

**Sphenodon (tuatara)**   Large, lizardlike reptiles that have persisted from Triassic to the present and now inhabit islands off the coast of New Zealand.

**Sphenopsids**   A group of spore-bearing plants that were particularly common during the late Paleozoic and were characterized by articulated stems with leaves borne in whorls at nodes. Only one genus, *Equisetum,* survives.

**Spiracle**   In cartilaginous fishes, a modified gill opening through which water enters the pharynx.

**Spontaneous fission**   Spontaneous fragmentation of an atom into two or more lighter atoms and nuclear particles.

**Spores**   A usually asexual reproductive body, such as occurs in bacteria, ferns, and mosses.

**Stage**   The time–rock unit equivalent to an age. A stage is a subdivision of a series.

**Stapes**   The innermost of the small bones in the middle ear cavity of mammals; also recognized in amphibians and reptiles.

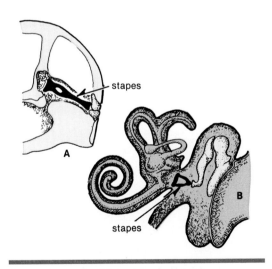

Stapes in reptile (A) and mammal (B)

**Strain**   Deformation of a rock mass in response to stress.

**Stratification**   The layering in sedimentary rocks that results from changes in texture, color, or rock type from one bed to the next.

**Stratified drift**   Deposits of glacial clastics that have been sorted and stratified by the action of meltwater.

**Stratigraphy**   The study of rock strata, with emphasis on their succession, age, correlation, form, distribution, lithology, fossil content, and all other characteristics useful in interpreting their environment of origin and geologic history.

**Stromatolites**   Distinctly laminated accumulations of calcium carbonate having rounded, branching, or frondose shape and believed to form as a result of the metabolic activity of marine algae. They are usually found in the high intertidal to low supratidal zones.

**Stromatoporoids**   An extinct group of reef-building organisms now believed to have affinities with the Porifera and noted for the large, often laminated masses constructed by the colonies.

*Stromatopora*

**Subaerial**   Formed or existing at or near a sediment surface significantly above sea level.

**Subduction zone**   An inclined planar zone, defined by high frequency of earthquakes, that is thought to locate the descending leading edge of a moving oceanic plate.

**Sublittoral zone**   The marine bottom environment that extends from low tide seaward to the edge of the continental shelf.

**Surface earthquake waves**   Seismic or earthquake waves that move only about the surface of the earth.

**Symmetrodonts**   A group of primitive Mesozoic mammals characterized by a symmetric triangular arrangement of cusps on cheek teeth.

**Syncline**   A geologic structure in which strata are bent into a downfold.

**System**   The time–rock unit representing rock deposited or emplaced during a geologic period.

**Taconic orogeny**   A major episode of orogeny that affected the Appalachian region in Ordovician time. The northern and Newfoundland Appalachians were the most severely deformed during this orogeny.

**Taxon** (pl. **taxa**)   Any unit in the taxonomic classification, such as a phylum, class, order, or family.

**Taxonomy**   The science of naming, describing, and classifying organisms.

**Tectonics**   The structural behavior of a region of the earth's crust.

**Teleosts**   The most advanced of the bony fishes, characterized by thin, rounded scales, completely bony internal skeleton, and symmetric tail. Teleosts range from Cretaceous to Recent.

Teleost fish

**Terebratulids**   A group of Silurian to Recent, mostly smooth-shelled brachiopods having a loop-shaped attachment for the lophomore. Terebratulids were most abundant during the Jurassic and Cenozoic.

**Terrane**   A three-dimensional block of crust having a distinctive assemblage of rocks (as opposed to *terrain*, implying topography, such as rolling hills or rugged mountains).

**Tethys seaway**   A great east–west trending seaway lying between Laurasia and Gondwanaland during Paleozoic and Mesozoic time and from which arose the Alpine–Himalayan Mountain ranges.

Tethys seaway

**Thecodonts**   An order of primarily Triassic reptiles considered to be the ancestral archosaurians.

The Triassic thecodont *Hesperosuchus*

**Therapsids**   An order of advanced mammal-like reptiles.

*Cynognathus,* a Permian therapsid

**Thermal plume**   A "hot spot" in the upper mantle believed to exist where a huge column of upwelling magma lies in a fixed position under the lithosphere. Thermal plumes are thought to cause volcanism in the overlying lithosphere.

**Thermoremanent magnetism** Permanent magnetization acquired by igneous rocks as they cool past the Curie point while in the earth's magnetic field.

**Theropods** The carnivorous saurischian dinosaurs.

The small Jurassic theropod *Ornitholestes*

**Thrust fault** A low-angle reverse fault, with inclination of fault plane generally less than 45°.

**Till** Unconsolidated, unsorted, unstratified glacial debris.

**Tillite** Unsorted glacial drift (till) that has been converted into solid rock.

**Time–stratigraphic unit (chronostratigraphic unit)** The rocks formed during a particular unit of geologic time. Also called time-rock unit.

**Tippecanoe sequence** A sequence of Ordovician to Lower Devonian sediments bounded above and below by regional unconformities and recording an episode of marine transgression, followed by full flooding of a large region of the craton and subsequent regression.

**Titanotheres** Large, extinct perissodactyls (odd-toed ungulates) that attained the peak of their evolutionary development during the Oligocene. *Brontotherium* is a widely known titanothere.

**Trace fossils** Tracks, trails, burrows, and other markings made in now lithified sediments by ancient animals.

**Transcontinental arch** An elongate, uplifted region extending from Arizona northeastward toward Lake Superior. During the Cambrian, the arch was emergent, as indicated by the observation that Cambrian marine sediments are located on either side of the arch but are missing above it. (Along the crest of the arch, post-Cambrian rocks rest upon Precambrian basement.)

**Transform fault** A strike-slip fault bounded at each end by an area of crustal spreading that tends to be more or less perpendicular to the trace of the fault.

**Triconodonts** A group of primitive Mesozoic mammals recognized primarily by the arrangement of three principal cheek tooth cusps in a longitudinal row.

**Trilobites** Paleozoic marine arthropods of the Class Crustacea, characterized by longitudinal and transverse division of the carapace into three

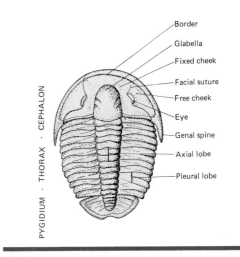

Trilobite

parts, or lobes. Trilobites were especially abundant during the early Paleozoic.

**Tuff** Volcanic ash that has become consolidated into rock.

**Turbidites** Sediment deposited from a turbidity current and characterized by graded bedding and moderate to poor sorting.

**Turbidity current**   A mass of moving water that is denser than surrounding water because of its content of suspended sediment and that flows along slopes of the sea floor as a result of that higher density.

Turbidity current

**Tylopod**   The artiodactyl group to which camels and llamas belong.

*Alticamelus,* a Pliocene tylopod

**Unconformity**   A surface separating an overlying younger rock formation from an underlying formation and representing an episode of erosion or nondeposition. Because unconformities represent a lack of continuity in deposition, they are gaps in the geologic record.

**Ungulate**   Four-legged mammals whose toes bear hoofs.

**Uniformitarianism**   A general principle that suggests that the past history of the earth can be interpreted and deciphered in terms of what is known about present natural laws.

**Vagile**   The bottom-dwelling habit of aquatic animals capable of locomotion.

**Varve**   A thin sedimentary layer or pair of layers that represent the depositional record of a single year.

Varves

**Vascular plants**   Plants, including all higher land plants, that have a system of vessels and ducts for distributing moisture and nutrients.

**Veneroids**   A group of pelecypods exemplified by the common clam, *Mercenaria.*

Veneroid Pelecypod (Venus)

**Vestigial organ**   An organ that is useless, small, or degenerate but representing a structure that was more fully developed or functional in an ancestral organism.

**Vindelician arch**   A highland area believed to have formed a barrier separating the Germanic and Alpine depositional areas during the Triassic and Jurassic.

**Williston basin**   A large structural basin extending from South Dakota and Montana northward into

Canada; well known for the petroliferous Devonian formations deposited therein.

**Zechstein sea** An arm of the Atlantic that extended across part of northern Europe during the Late Permian and in which were deposited several hundred meters of evaporites, including the well-known potassium salts of Germany.

**Zone** A bed or group of beds distinguished by a particular fossil content and frequently named after the fossil or fossils it contains. More formally known as a *biozone*.

**Zygote** The cell formed by the union of two gametes. Thus, a zygote is a fertilized egg.

# Index

## Common Rock-Forming Silicate Minerals

| Silicate Mineral | Composition | Physical Properties |
|---|---|---|
| Quartz | Silicon dioxide (silica, $SiO_2$) | Hardness of 7 (on scale of 1 to 10)*; will not cleave (fractures unevenly); specific gravity: 2.65 |
| Potassium feldspar group | Aluminosilicates of potassium | Hardness of 6.0–6.5; cleaves well in two directions; pink or white; specific gravity: 2.5–2.6 |
| Plagioclase feldspar group | Aluminosilicates of sodium and calcium | Hardness of 6.0–6.5; cleaves well in two directions; white or gray; may show striations on cleavage planes; specific gravity: 2.6–2.7 |
| Muscovite mica | Aluminosilicates of potassium with water | Hardness of 2–3; cleaves perfectly in one direction, yielding flexible thin plates; colorless; transparent in thin sheets; specific gravity: 2.8–3.0 |
| Biotite mica | Aluminosilicates of magnesium, iron, potassium, with water | Hardness of 2.5–3.0; cleaves perfectly in one direction, yielding flexible, thin plates; black to dark brown; specific gravity: 2.7–3.2 |
| Pyroxene group | Silicates of aluminum, calcium, magnesium, and iron | Hardness of 5–6; cleaves in two directions at 87° and 93°; black to dark green; specific gravity: 3.1–3.5 |
| Amphibole group | Silicates of aluminum, calcium, magnesium, and iron | Hardness of 5–6; cleaves in two directions at 56° and 124°; black to dark green; specific gravity: 3.0–3.3 |
| Olivine | Silicates of magnesium and iron | Hardness of 6.5–7.0; light green; transparent to translucent; specific gravity: 3.2–3.6 |
| Garnet group | Aluminosilicates of iron, calcium, magnesium, and manganese | Hardness of 6.5–7.5; uneven fracture; red, brown, or yellow; specific gravity: 3.5–4.3 |

(Muscovite mica, Biotite mica, Pyroxene group, Amphibole group, and Olivine are bracketed as Ferromagnesian minerals.)

*The scale of hardness used by geologists was formulated in 1822 by Frederich Mohs. Beginning with diamond as the hardest mineral, he arranged the following table:

| | | |
|---|---|---|
| 10 | Diamond | 5 Apatite |
| 9 | Corundum | 4 Fluorite |
| 8 | Topaz | 3 Calcite |
| 7 | Quartz | 2 Gypsum |
| 6 | Feldspar | 1 Talc |